From a Grain of Salt to the Ribosome

The History of Crystallography as Seen Through the Lens of the Nobel Prize

Series in Structural Biology

Series Editor: Anders Liljas *(Lund University, Sweden)*

Published

Vol. 1 Structural Aspects of Protein Synthesis, 2nd ed.
by Anders Liljas and Måns Ehrenberg

Vol. 2 Found in Translation: Collection of Original Articles on Single-Particle Reconstruction and the Structural Basis of Protein Synthesis
by Joachim Frank

Vol. 3 Selected Papers of Michael G. Rossmann with Commentaries: The Development of Structural Biology
by Michael G. Rossmann

Vol. 4 From a Grain of Salt to the Ribosome: The History of Crystallography as seen Through the Lens of the Nobel Prize
edited by Ivar Olovsson, Anders Liljas and Sven Lidin

Series in Structural Biology — Vol. 4

From a Grain of Salt to the Ribosome

The History of Crystallography as Seen Through the Lens of the Nobel Prize

Editors

Ivar Olovsson
Uppsala University, Sweden

Anders Liljas
Lund University, Sweden

Sven Lidin
Lund University, Sweden

World Scientific

NEW JERSEY • LONDON • SINGAPORE • BEIJING • SHANGHAI • HONG KONG • TAIPEI • CHENNAI

Published by

World Scientific Publishing Co. Pte. Ltd.
5 Toh Tuck Link, Singapore 596224
USA office: 27 Warren Street, Suite 401-402, Hackensack, NJ 07601
UK office: 57 Shelton Street, Covent Garden, London WC2H 9HE

Library of Congress Cataloging-in-Publication Data
From a grain of salt to the ribosome : the history of crystallography as seen through the lens of the Nobel Prize / editors Ivar Olovsson, Uppsala University, Sweden, Anders Liljas, Sven Lidin, Lund University, Sweden.
pages cm -- (Series in structural biology ; vol. 4)
Includes bibliographical references.
ISBN 978-9814623117 (hardcover : alk. paper) -- ISBN 9814623113 (hardcover : alk. paper)
1. Crystallography--History. 2. Nobel Prize winners--Biography. I. Olovsson, Ivar, 1928– editor. II. Liljas, Anders, editor. III. Lidin, Sven, editor.
QD903.F76 2014
548.09--dc23

2014035029

British Library Cataloguing-in-Publication Data
A catalogue record for this book is available from the British Library.

Printed in Singapore by Mainland Press Pte Ltd.

Introduction

A Brief History of Crystallography

Crystals have caught the imagination of man and are featured frequently not only in science but also in art and literature. Crystals have conveyed character traits such as purity and innocence but also opulence (abundance?) and endurance. We think well of crystals. Crystals surround us, from the very large gypsum crystals in Naica, Mexico, deep down in caverns, where geothermally heated water has helped shape these juggernauts of the crystal world, to the nanoscopic images of atomically resolved clusters photographed through an electron microscope. The steel and concrete, the brick and mortar that form our buildings are all crystalline, as is the very bedrock on which all rests.

In his book, *A New Era in Chemistry: Some of the More Important Developments in General Chemistry during the Last Quarter of a Century*, published in 1913, Harry Clary Jones wrote: "We do not know what is the formula of rock salt, or of ice; and we have no reliable means of finding out these simplest matters about solids. Our ignorance of solids is very nearly complete."

That was written in the nick of time. Since their discovery in 1895 by Röntgen (Nobel Prize in Physics, 1901), the nature of X-rays remained a source of debate until the iconic experiment by von Laue, Knipping and Friedrich in 1912 (Nobel Prize to Laue in Physics, 1914). In a single, decisive experiment the wave nature of X-rays was shown and the structures of crystals became accessible to analysis. The sample used in the original experiment was a specimen of $CuSO_4$.

This experiment was the key for Lawrence and Henry Bragg on how to use X-ray diffraction to determine the exact positions of atoms in crystals, and the statement of H. C. Jones was revised and reversed: There are few systems about which we know as much as we know about crystals.

Crystallography developed beyond the use of X-rays, and the prizes to Davisson and Thomson in 1937, and to Brockhouse and Shull in 1994, celebrate the use of electrons and neutrons that provide complementary information to that of X-ray diffraction, and the Chemistry Prize in 1982 to Klug showed the strength of applying diffraction techniques to electron microscopy.

The use of crystallography spread quickly in mineralogy, materials science and inorganic chemistry and provided crucial information on chemical bonding. The work by Pauling for which he was awarded the 1954 Chemistry Prize was very much based on his far-reaching crystallographic studies; the same is true for Lipscomb's Prize in 1976. Today, crystallography enables us to approach the center of the earth by means of experiments at pressures exceeding half a million atmospheres, and to the surface of Mars, where the Mars rover, using a portable diffractometer, samples the surface minerals and provides answers to past and present geology of the red planet, revealing, among other things, the likelihood of surface water once being present on the red planet. Today, crystallography has come far enough to question its very foundations and the Nobel Prize awarded to Shechtman in 2011 celebrates the discovery that periodicity is no prerequisite for crystallinity, an enigmatic and profound fact.

Perhaps the area where X-ray crystallography has had the largest impact is in structural biology. 1962 was a pivotal year when the Nobel Prize in Chemistry was awarded to Perutz and Kendrew for their work on the structure of globular proteins, and the Prize in Physiology or Medicine was awarded to Crick, Watson and Wilkins for their study of DNA. However, important achievements in structural biology were awarded much earlier. The Chemistry Prize in 1946 went to Sumner, Northrop and Stanley for their work on the purification and crystallization of enzymes and virus proteins and for showing that enzymes are proteins.

The 1962 Prize may have heralded a revolution, but initially progress was slow. X-ray sources were weak, detectors were inefficient and protein purification and crystallization was painfully slow. The first low-resolution protein structure, that of myoglobin, was published in 1958. The high-resolution structure was published in 1960, and the first structure of an enzyme, that of lysozyme, appeared in 1965. In the meantime, Hodgkin was awarded the Chemistry Prize in 1964 for work on the structures of biologically important substances such as penicillin and vitamin B_{12}. Since then things have changed and today hundreds of new structures have become available every month through the protein database. This development is also evident in the important crystallographic components in the Chemistry Prizes to Deisenhofer, Huber and Michel in 1988; Walker in 1997; MacKinnon in 2003; Kornberg in 2006; Ramakrishnan, Steitz and Yonath in 2009; and Kobilka in 2012. A family tree of crystallographers in the field of macromolecular crystallography is shown in Fig. 1.

The techniques of determining structures by crystallography are an integral part of modern physics, chemistry, geology, biology and medicine and with the advent of ever more powerful sources for X-rays, electrons and neutrons, we are now experiencing a revolution in time-dependent studies where we are able to study not only structure, but also structural evolution.

The year 2014 has been proclaimed by the United Nations as *The International Year of Crystallography*, IYCr 2014. The importance of crystallography is perhaps best illustrated by the fact that more than 20 Nobel Prizes with crystallographic profile have been awarded during the period 1901–2012. In the present book, we have summarized the achievements, which formed the basis for the awards of these Nobel prizes, together

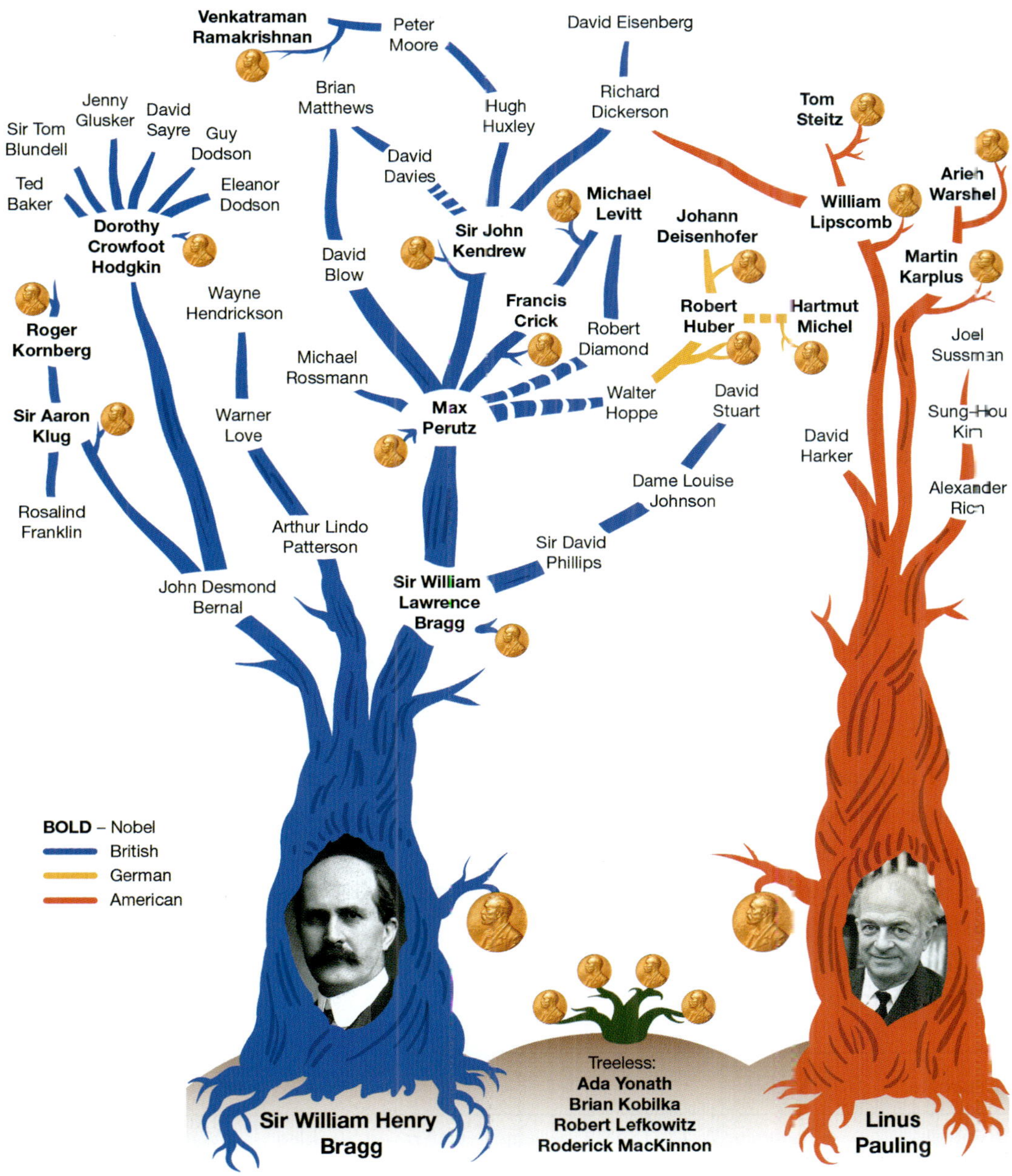

Fig. 1. A family tree of macromolecular crystallography. The tree includes both Nobel laureates and other pioneers or notable crystallographers. (The figure is reproduced with permission from M. Jaskolski, Z. Dauter and A. Wlodawer, 2014. *A brief history of macromolecular crystallography, illustrated by a family tree and its Nobel fruits. FEBS J.* 2014, 1–25. Copyright John Wiley and Sons.)

with short biographies of the laureates. One or two relevant publications have been selected for reprinting; these are marked with a star (*) in the laureates' reference list. A few references, which are common to several articles, are listed below and are not repeated in the separate reference lists.

References Relating to Many Laureates

Authier, A. (2013) *Early Days of X-ray Crystallography.* Oxford University Press.
Brown, A. (2005) *J. D. Bernal. The Sage of Science.* Oxford University Press.
Dickerson, RE. (2005) *Present at the Flood. How Structural Molecular Biology Came About.* Sinauer Associates Inc., Publishers, Sunderland, Massachusetts, U.S.A. ISBN 0-87893-168-6 (pbk).
Ewald, PP. (1962) *Fifty Years of X-ray Diffraction.* N.V.A. Oosthoek Uitgeversmaatschappi, Utrecht, The Netherlands. (A book dedicated to the International Union of Crystallography on the occasion of the commemoration meeting in München, July 1962).
Friedman, RM. (2001) *The Politics of Excellence. Behind the Nobel Prize in Science.* A W.H. Freeman Books. Times books. Henry Holt and Co, New York. ISBN 0-7167-3103-7.
Les Prix Nobel. The Nobel lectures can be accessed under www.nobelprize.org
Norrby, E. (2010) *Nobel Prizes and Life Sciences.* World Scientific, Singapore.
Norrby, E. (2013) *Nobel Prizes and Nature's Surprises.* World Scientific, Singapore.

Acknowledgments

First of all, special thanks to the Carl Trygger Foundation for Scientific Research for economic support, without which it would have been impossible to republish many of the articles in this book. We would also like to thank Mrs Maria Asp Dahlbäck, The Royal Swedish Academy of Sciences/Center for History of Science, for valuable assistance.

We would like to thank the following publishers for their permission to reproduce the articles found in this volume:

Laureate	Article	Publisher
Röntgen	Röntgen, W. C. (1898) Über eine neue Art von Strahlen (Erste Mittheilung). *Annalen der Physik und Chemie*, Band **300**: 1–11.	Wiley-VCH Verlag GmbH & Co. KGaA
	Röntgen, W. C. 1898. Über eine neue Art von Strahlen (Zweite Mittheilung). *Annalen der Physik und Chemie*, Band **300**: 12–17.	Wiley-VCH Verlag GmbH & Co. KGaA

von Laue	Friedrich, W., Knipping, P. and Laue, M. (1913) Interferenzerscheinungen bei Röntgenstrahlen. *Annalen der Physik und Chemie*, **346**: 971–988.	Wiley-VCH Verlag GmbH & Co. KGaA
	Laue, M. (1913) Eine quantitative Prüfung der Theorie für die Interferenzerscheinungen bei Röngenstrahlen. *Annalen der Physik und Chemie*, **346**: 989–1002.	Wiley-VCH Verlag GmbH & Co. KGaA
Bragg & Bragg	Bragg, W. L. (1913) The Structure of Some Crystals as Indicated by their Diffraction of X-rays. *Proc. Roy. Soc. Lond. A* **89**: 248–277.	The Royal Society, London
	Liljas, A. (2013) Background to the Nobel Prize to the Braggs. *Acta Cryst.* **A69**: 10–15.	IUCr
	Jaskolski, M., Dauter, Z. and Wlodawer, A. (2014) A Brief History of Macromolecular Crystallography, Illustrated by a Family Tree and Its Nobel Fruits. *FEBS J.* 2014, 1–25.	John Wiley and Sons
Davisson & Thompson	Davisson, C. J. (1928) The Diffraction of Electrons by a Crystal of Nickel. *Bell System Technical Journal* **7**: 90–105.	Alcatel-Lucent
	Thomson, G. P. (1931) The Diffraction of Electrons by Single Crystals. *Proc. Roy. Soc. Lond. A* **133**: 1–25.	The Royal Society, London
Sumner, Northrop & Stanley	Sumner, J. B. (1926) The Isolation and Crystallization of the Enzyme Urease: Preliminary Paper. *J. Biol. Chem.* **69**: 435–441.	American Society for Biochemistry and Molecular Biology
	Northrop, J. H. (1930) Crystalline Pepsin. I. Isolation and Tests of Purity. *J. Gen. Physiol.* **13**: 739–766.	Rockefeller University Press
	Stanley, W. (1935) Isolation of a Crystalline Protein Possessing the Properties of Tobacco-Mosaic Virus. *Science* **81**: 644–645	The American Association for the Advancement of Science
Pauling	Pauling, L. (1931) The Nature of the Chemical Bond. *J. Am. Chem. Soc.* **53**: 1367–1400.	American Chemical Society
Watson, Crick & Wilkins	Watson, J. D. and Crick, F. H. C. (1953) Molecular Structure of Nucleic Acids. *Nature* **171**: 737–738.	Nature Publishing Group
	Wilkins, M. H. F., Stokes, A. R. and Wilson, H. R. (1953) Molecular Structure of Deoxypentose Nucleic Acids. *Nature* **171**: 738–740.	Nature Publishing Group
	Franklin, R. E. and Gosling, R. G. (1953) Molecular Configuration in Sodium Thymonucleate. *Nature* **171**: 740–741.	Nature Publishing Group

Kendrew & Perutz	Perutz, M. F., Rossmann, M. G., Cullis, A. F., Muirhead, H., Will, G. and North, A. C. T. (1960) Structure of Hemoglobin. A Three-Dimensional Fourier Synthesis at 5.5-Å Resolution, Obtained by X-ray Analysis. *Nature* 185: 416–422.	Nature Publishing Group
	Kendrew, J. C., Dickerson, R. E., Strandberg, B. E., Hart, R. G., Davies, D. R., Phillips, D. C. and Shore, V. C. (1960) Structure of Myoglobin: A Three-Dimensional Fourier Synthesis at 2 Å Resolution. *Nature* **185**: 422–427.	Nature Publishing Group
Hodgkin	Brink, C., Hodgkin, D. C., Lindsey, J., Pickworth, J., Robertson, J. H. and White, J. G. (1954) X-ray Crystallographic Evidence on the Structure of Vitamin B_{12}. *Nature* **174**: 1169–1171.	Nature Publishing Group
Lipscomb	Lipscomb, W. N. (1977) The Boranes and Their Relatives. *Les Prix Nobel*, pp. 224–245.	The Nobel Foundation
Klug	DeRosier, D. J. and Klug, A. (1972) Structure of the Tubular Variants of the Head of Bacteriophage T4 (Polyheads) I. Arrangement of Subunits in Some Classes of Polyheads. *J. Mol. Biol.* **65**: 469–488.	Elsevier B. V.
Hauptman & Karle	Hauptman, H. and Karle, J. (1950) A Geometric Approach to the Crystal Structure Problem. *Acta Cryst.* **3**: 478.	IUCr
	Karle, J. and Hauptman, H. (1950) The Phases and Magnitudes of the Structure Factors. *Acta Cryst.* **3**: 181–187	IUCr
Deisenhofer, Huber & Michel	Deisenhofer, J., Epp, O., Miki, K., Huber, R. and Michel, H. (1985) Structure of the Protein Subunits in the Photosynthetic Reaction Centre of *Rhodopseudomonas viridis* at 3Å Resolution. *Nature* **318**: 618–624.	Nature Publishing Group
Shull & Brockhouse	Brockhouse, B. (1995) Slow Neutron Spectroscopy and the Grand Atlas of the Physical World. *Rev. Mod. Phys.* **67**: 735–751.	The Nobel Foundation
	Shull, C. G. (1995) Early Development of Neutron Scattering. *Rev. Mod. Phys.* **67**: 753–757.	The Nobel Foundation
Walker	Abrahams, J. P., Leslie, A. G. W., Lutter, R. and Walker, J. E. (1994) Structure at 2.8 Å Resolution of F_1-ATPase from Bovine Heart Mitochondria. *Nature* **370**: 621–628.	Nature Publishing Group
MacKinnon	Doyle, D. A., Cabral, J. M., Pfuetzner, R. A., Kuo, A., Gulbis, J. M., Cohen, S. L., Chait, B. T. and MacKinnon, R. (1998) The Structure of the Potassium Channel: Molecular Basis of K^+ Conduction and Selectivity. *Science* **280**: 69–77.	The American Association for the Advancement of Science

Kornberg	Cramer, P., Bushnell, D. A. and Kornberg, R. D. (2001) Structural Basis of Transcription: RNA Polymerase II at 2.8 Ångström Resolution. *Science* **292**: 1863–1876.	The American Association for the Advancement of Science
	Gnatt, A. L., Cramer, P., Fu, J., Bushnell, D. A. and Kornberg, R. D. (2001) Structural Basis of Transcription: RNA Polymerase II Elongation Complex at 3.3 Å Resolution. *Science* **292**: 1876–1882.	The American Association for the Advancement of Science
Ramakrishnan, Steitz & Yonath	Yonath, A. and Wittmann, H. G. (1988) Approaching the Molecular Structure of Ribosomes. *Biophys. Chem.* **29**: 17–29.	Elsevier B. V.
	Ogle, J. M., Brodersen, D. E., Clemons Jr, W. M., Tarry, M. J., Carter, A. P. and Ramakrishnan, V. (2001) Recognition of Cognate Transfer RNA by the 30S Ribosomal Subunit. *Science* **292**: 897–902.	The American Association for the Advancement of Science
	Ban, N., Nissen, P., Hansen, J., Moore, P. B. and Steitz, T. A. (2000) The Complete Atomic Structure of the Large Ribosomal Subunit at 2.4 Å Resolution. *Science* **289**: 905–920.	The American Association for the Advancement of Science
Shechtman	Shechtman, D., Blech, I., Gratias, D. and Cahn, J. W. (1984) Metallic Phase with Long-Range Orientational Order and No Translational Symmetry. *Phys. Rev. Lett.* **53**: 1951–1954.	American Physical Society
Kobilka	Rasmussen, S. G. F., DeVree, B. T., Zou, Y., Kruse, A. C., Chung, K. Y., Kobilka, T. S., Thian, F. S., Chae, P. S., Pardon, E., Calinski, D., Mathiesen, J. M., Shah, S. T. A., Lyons, J. A., Caffrey, M., Gellman, S. H., Steyaert, J., Skiniotis, G., Weis, W. I., Sunahara, R. K. and Kobilka, B. K. (2011) Crystal Structure of the β_2 Adrenergic Receptor–Gs Protein Complex. *Nature* **477**: 549–555.	Nature Publishing Group

Contents

Reprints

Biographies

Wilhelm Conrad Röntgen (P1901)

"In recognition of the extraordinary services he has rendered by the discovery of the remarkable rays subsequently named after him."

Wilhelm Conrad Röntgen
March 27 1845–February 10 1923
(From Les Prix Nobel 1901)

Wilhelm Conrad Röntgen was born in 1845 in Lennep in the Rhine province of Germany. He was the only child of the respected merchant and cloth manufacturer Friedrich Röntgen and his wife Charlotte Constanze. When he was three years old, his family moved to Apeldoorn in The Netherlands, where he went to a boarding school. In 1862, he entered a technical school in Utrecht from which he was unfairly expelled (due to being accused of producing a caricature of one of the teachers, which was in fact done by somebody else). In 1865, he entered the University of Utrecht to study physics. As he had not obtained the credentials from his technical school that were required for a regular student, he left after one year and, after very tough entry examinations, he was accepted at ETH in Zürich to study mechanical engineering. During his studies, he also worked in the laboratory of Kundt, who had a great influence on his development. In Zürich, he also met his beautiful future wife, Anna Bertha Ludwig. In 1869, he obtained his Ph.D. at the University of Zürich, and was appointed assistant to Kundt. He went with Kundt to Würtzburg the same year, and three years later, they continued to Strasbourg. After one year in Würtemberg, he returned in 1876 to Strasbourg as professor of physics, but three years later, he accepted a position as professor of physics at the University in Giessen. In 1888, he moved to the University in Würzburg, where he succeeded Kohlrausch. Among his colleagues there we note Helmholtz and Lorentz. He held the office of Rector from 1893 to 1894.

During the first two decades of his scientific career, Röntgen studied a variety of topics, including the specific heats of gases, the Faraday Effect in gases, magnetic effects associated with dielectric materials, and the compressibility of water. In Würzburg (1888–1900), Röntgen published 18 papers, the earlier of which dealt with the effects of pressure on the physical properties of solids and liquids. In the final three papers, he described the discovery of X-rays. In his will, Röntgen determined that all of his scientific notes should be destroyed and consequently very few of his laboratory notes and documents are now available.

In Würzburg, Röntgen's family lived in a seven-room apartment on the top floor of the Institute of Physics. They enjoyed the cultural scene, going to the theatre and concerts, and Röntgen leased hunting grounds in a nearby forest. Ever since his time of study in Zürich, Röntgen found recreation in the mountains, and even shortly before his death, he continued his wanderings in the mountains.

In 1900, he accepted the Chair of Physics at the University of München, where he remained for the rest of his life. He spent his last few years at his country home at Weilheim, near München. He died there on February 10 1923, after a short illness resulting from intestinal cancer. Germany's defeat in World War I also had its effect on Röntgen: The inflationary period following the war resulted in his bankruptcy.

In München, a very large institute was set up in his honor. In addition to science students, numerous medical students were also required to go to Röntgen's lecture course and first-year laboratory, and this demanded a large number of lecturers and assistants. Because of his exceptionally high demands, graduation under Röntgen was attempted only by highly devoted and serious students. Administrative work took up so much of his time that only seven papers were produced by him between 1900 and 1921. These were concerned with the physical properties of crystals, their electrical conductivity, and the influence of radiation on them.

A large number of other honors were bestowed upon Röntgen, as well as many medals. He was ordered to demonstrate X-rays to the Kaiser on January 13 1896, and was immediately awarded the Prussian Order of the Crown, Second Class. Today, the former cloth-making town of Lennep calls itself the "Röntgen town." Ever since the centenary of Röntgen receiving the Nobel Prize in 2001, the town of Remscheid arranges a "Röntgen race" in October. In addition, Asteroid 6401 is named after him. The chemical element with atomic number 111 has been named Roentgenium, with the symbol Rg.

The Discovery of X-rays

The position of physical science at the end of the nineteenth century was summarized by Prof. Manne Siegbahn (1962) as follows:

> Towards the end of the last century physical science had arrived at a picture of the natural phenomena and their interplay which was apparently very satisfactory and to a certain extent complete. Everything seemed to fit well into a mechanical conception of the universe, including electric, magnetic and optical phenomena. It was hardly expected that there would be any revolutionary innovations, when unexpectedly at the middle of the 1890's some laboratory experiments suddenly gave research an impulse the effects of which were extremely far-reaching…

One of the most remarkable achievements, Röntgen's discovery of the mysterious X-ray radiation, is summarized in the following paragraphs.

Around 1880, many physicists were involved in studies of discharges in gases. Among them were William Crookes, Joseph John Thomson, Heinrich Hertz, and Philipp Lenard. When a voltage of a few thousand volts was applied between the anode and cathode in a partially evacuated tube, they found that the cathode emanated some sort of rays, which were called cathode rays. If the wall of the tube was sufficiently thin, the cathode rays could even be observed outside the tube. However, the range of these rays was very short — only a few centimeters in air. It took a long time before the nature of these rays was clarified; it was not until 1895 that one could definitely demonstrate that these rays consisted of charged particles, electrons

When Wilhelm Conrad Röntgen was a professor in Würzburg, he also became interested in this type of research. He had previously carried out careful work in classical physics, but now decided to go in a completely different direction and take part in these experiments. He never gave a satisfactory reason for this change of direction: "There were some aspects of Hertz–Lenard's experiments with cathode rays that needed a closer investigation."

Philipp Lenard, then a professor in Kiel, was one of the leading scientists in this field, and Röntgen first wanted to reproduce Lenard's results. He wrote to Lenard in order to obtain some advice concerning his experiments and even asked him about a "reliable" tube, which he also received. This is something Lenard most probably bitterly regretted later. Röntgen purchased a few cathode tubes and, late in the fall of 1895, Röntgen had reproduced all of the experiments of Lenard and was ready for new experiments. According to what Röntgen mentioned on another occasion: "To try to detect a so far unknown type of high-frequent electromagnetic radiation..." which was theoretically predicted by Helmholtz on the basis of Maxwell's theories. In his experiments with the cathode ray equipment he had obtained from Lenard,

Röntgen chose a paper covered with barium platinum cyanide crystals as a detector, a substance that was known to give good fluorescence. To eliminate all disturbing light, he covered the tube completely with thick black paper and drew the curtains in his laboratory. He applied a high voltage current over the tube and checked carefully that no stray light emanated from the equipment. He then saw something that lit up at the border of his field of vision. Was it some ray of light that was not completely screened off? He disconnected the current — the light phenomenon disappeared. He repeated the experiment with the same result. With the help of a match, he could establish that the light phenomenon emanated from the fluorescent substance, which was located more than one meter from the discharge tube. He put a piece of black cardboard between the tube and the detector, and tried again with an additional piece of cardboard — the substance continued to fluoresce. Even if he placed a pack of cards in the way, a book with 1000 pages, or a 2.5-centimeter thick piece of wood, the light phenomenon was still there. He moved the screen closer to the tube and finally placed it up against the tube. The intensity of the light increased. Röntgen moved the sample further and further away; even at a distance of two meters, a weak light could be observed. He repeated and repeated the experiment and always obtained the same result.

Later the same evening, still working in darkness, Röntgen placed thin plates of aluminum, copper, lead, and platinum between the tube and the detector substance. Only lead and platinum absorbed the radiation completely. Subsequently, Röntgen made one of the most bizarre observations in the history of science. He held a small piece of lead in his hand in front of the shining screen. He then saw not only the expected black shadow from the lead sheet, but also the shadow of the bones in his own fingers. As he later mentioned in his lecture in Würzburg, he mistrusted his own senses: "I still thought that I was the victim of an illusion when I observed the phenomenon..." He then stretched out his hand and the dark shadow of his fingers moved slowly across the lighting screen. To register permanently the light phenomenon, he went over to some photographic film. He told nobody, not even his students or closest collaborators, what he was doing. He was quite reserved and an utterly methodical researcher, who did not leave anything to the hazard.

During the following weeks, his feverishly intense research was performed in all secrecy. He ate and slept in his laboratory as he investigated the properties of these new rays. It is clear that Röntgen performed his research with obvious risks to his health. In one notice, he write: "The retina of the eye is not sensitive to these rays; one does not see anything, even if the eye is brought near to the discharge tube." Not until the end of 1895 was he convinced that he had detected a new type of ray with strange properties. He called these rays "X-rays," as he did not know anything about their nature. Just before Christmas, on December 22, he invited his wife Bertha to his laboratory and had her hand placed in the way of the X-rays for 15 minutes in front of a film-holder. Hardly could he have suspected that this picture of Bertha's fingers would soon make him world famous. The experiments were kept strictly secret until at the end of December, when he sent his first announcement of ten printed pages to the Physical and Medical Society of Würzburg. The proof was copied and sent as a New Year's Greeting to friends and colleagues.

He announced his discovery orally in a lecture for the Physical-Medical Society in Würzburg in December 1895. Due to the New Year's Greeting that had been sent to a friend in Wien, the discovery was published on the first page of the Sunday edition of the *Newspaper Die Neue Presse* in Wien on January 5 1896. In this way, the discovery was quickly spread in the world press long before the scientific journal with Röntgen's first report came to print.

Röntgen published only three short papers on his discovery of X-rays (Röntgen, 1895, 1896, 1897). These treaties exhausted the subject so completely that, for almost a decade, nothing new could be said about it. They were written with ingenious care and Röntgen was correct in every detail.

Röntgen made several attempts to determine the nature of the X-rays. Were they of the same nature as ordinary light or a stream of particles, or corpuscles? In the *Third Communication, March 1897*, he writes:

> The experiments on the permeability (for X-rays) of plates of constant thickness cut from the same crystal in different orientations, which were mentioned in my first *Communication*, have been continued. Plates were cut from calcite, quartz, tourmaline, beryl, aragonite, apatite and barytes. Again no influence of the orientation on the transparency could be found.
>
> Ever since I began working with X-rays, I have repeatedly sought to obtain diffraction with these rays; several times, using narrow slits, I observed phenomena which looked very much like diffraction. But in each case a change of experimental conditions, undertaken for testing the correctness of the explanation, failed to confirm it, and in many cases I was able to directly show that the phenomenon had arisen in an entirely different way than by diffraction. I have not succeeded to register a single experiment from which I could gain the conviction of the existence of diffraction of X-rays with a certainty which satisfies me.

It was Max von Laue who finally succeeded in determining the nature of X-rays, almost two decades after Röntgen's discovery. The name "X-rays" has remained in the English and French language, even after the nature of the X-rays became clear.

It should be remarked that many other physicists working with cathode ray tubes had actually also observed phenomena that no doubt were caused by the then-unknown penetrating radiation, and these experiments were carried out many years before Röntgen. For example, on February 22 1890, Arthur Goodspeed at the Institute of

Physics at the University of Pennsylvania was carrying out photographic experiments using the sparks from a vacuum tube as a light source. A photographic film showed the shadow of some coins, which happened to lie on an unexposed film in the neighborhood of the discharge tube. However, he could not explain the phenomenon and did not perform further investigations. Not until after Röntgen's discovery six years later did it become obvious that the picture was in reality an X-ray image of the coins.

Why was it Röntgen and not Lenard who discovered the new radiation? He was no pioneer in studying electric discharges in rarefied gases. Lenard had precisely the equipment needed to be able to detect the new radiation, and it was only the equipment Röntgen had obtained from Lenard that was used in his experiments. The intensity of the light from the fluorescing substance in Röntgen's experiments was several orders of magnitude weaker than the fluorescence in Lenard's detector samples. But Röntgen's careful way of working probably was the determining factor, and perhaps also good luck. In very weak light, only the poles are functional, but there are only cones in the central part of the yellow spot, and these are not functional in weak light. This means that, in darkness, a weak shining spot is best discovered by indirect vision. The eye should not be directed straight towards a weak light spot! Astronomers are familiar with this phenomenon, and if they wish to observe a weak shining star, then they look indirectly at the star. As soon as the observer becomes aware of the light spot and tries to focus on it, the visual impression disappears. This may be a reason as to why Röntgen was able to detect the weak light from his fluorescing sample. His eye was directed towards the tube, which was located some distance from the fluorescing sample, and the weak light from there entered just the corner of his field of vision. Another important factor was the wavelength of the blue–green fluorescence color from the barium platinum cyanide, which corresponds to the most light-sensitive spectral region of the poles. Furthermore, Röntgen's fluorescent sample contained heavy atoms, which made his detector 100-times more sensitive to X-rays than Lenard's detector, which only contained light elements. Philipp Lenard became very upset when he did not receive the Nobel Prize in 1901 together with Röntgen. Lenard was an internationally well-recognized physicist in the gas discharge field and it was bad luck that it was not he himself who had detected the X-rays.

Only four years later, Lenard received the Nobel Prize in Physics with the brief description, "For his work on the cathode rays." But this prize did not completely suppress Lenard's indignation. It may be added that the Nobel Prize 1906 was awarded to Joseph John Thomson for the discovery of the electron. Thus, Lenard again experienced bad luck when he also missed out on the honor for this discovery by such small margins.

The excitement over the discovery of X-rays and the Nobel Prize resulted in Röntgen not having time to devote himself to work in the laboratory as before — this during his most productive time of his life. Evidently, he had planned to increase the photographic effect of the X-rays by placing the paper with barium platinum cyanide directly against the photographic plate. If he had done that, maybe he would have discovered that the film was blackened even when the X-rays were not put on. Such a general blackening is caused by another, then unknown, type of radiation. The sample Röntgen was using contained small amounts of radium, and radiation is emitted from this and its short-lived decay products. In this situation, Röntgen would probably have tested other fluorescing substances that were used in cathode ray experiments, like uranium phosphate. Then, he would perhaps have found that uranium was a source of a new type of activity. Would Röntgen have been a candidate to receive another Nobel Prize "for the discovery of the natural radioactivity"?

References to Röntgen's Discoveries

*Röntgen WC. (1895) Über eine neue Art von Strahlen. Vorläufige Mitteilung. In: *Aus den Sitzungsberichten der Würzburger Physik-medic. Gesellschaft Würzburg*. pp. 137–147.

*Röntgen WC. (1896) Eine neue Art von Strahlen. 2. Mitteilung. In: *Aus den Sitzungsberichten der Würzburger Physik.-medic. Gesellschaft Würzburg*. pp. 11–17.

Röntgen WC. (1897) Weitere Beobachtungen über die Eigenschaften der X-Strahlen. In: *Mathematische und Naturwissenschaftliche Mitteilungen aus den Sitzungsberichten der Königlich Preussischen Akademie der Wissenschaften zu Berlin. Physikalisch-Mathematische Klasse*. pp. 392–406.

These articles have been reprinted in *Annalen der Physik*, **300**: 1–37.

General References

Glasser O. (1959) *Wilhelm Conrad Röntgen*. Springer Verlag, Berlin, Göttingen, Heidelberg.

Les prix Nobel en 1901. P.A. Norstedt & Söner, Stockholm (1904).

Siegbahn M. (1962) *The Physics Prize. Nobel, the Man and his Prizes*. Elsevier publ. comp., Amsterdam.

The Nobel Prize

When the time for nominations of the Nobel Prize for 1901 having expired, 29 proposals had been submitted, 12 of them suggesting Röntgen, one for Lenard, and five for a division of the prize between the two. The Nobel Committee for Physics, in its report to the Academy of Sciences, concluded that Röntgen's discovery provided knowledge about a previously unknown phenomenon, had opened up a new field of research, and had important practical results. It was also concluded that the credit for the differentiation of the X-rays from the cathode rays belonged wholly to Röntgen. At the same time, the Committee found that the importance of Lenard's systematic

The Nobel diploma of Wilhelm Conrad Röntgen, 1901.

research into the properties of the cathode rays was comparable to Röntgen's discovery. As a consequence, it was difficult to decide which of the two should be preferred if the prize were to be awarded to one of them alone. The Committee stressed that Lenard's work paved the way for Röntgen's experiments. When this point had been made clear, the natural consequence was to propose that the prize be divided between them.

The report presented by the Nobel Committee to the Academy was as follows (in translation):

> After having examined the merits of the candidates proposed for the Nobel Prize in Physics for the year 1901, the undersigned, members of the Nobel Committee for Physics of the Academy of Sciences, have the honor to recommend that, for reasons given below, the prize in question be divided equally between
>
> Prof. W. C. Röntgen, München and Prof. Ph. Lenard, Kiel
>
> Stockholm, 9 Sept. 1901
> C. B. Hasselberg
> Knut Ångström — H. H. Hildebrand
> Rob. Thalén — Svante Arrhenius

The whole Physics class of the Academy agreed with this proposal.

No minutes were kept at the meetings of the Academy at which the Nobel Committee's proposal was discussed. But when the whole Academy was assembled on November 12 1901 to take the final decision, the admiration of Röntgen's discovery was so great that it was considered impossible to split the prize. This is one of the very few cases when the Academy has voted against a proposal from the Nobel Committee for Physics and a united Physics class.

Reference

Minutes from the meetings of the Nobel Committee for Physics on February 2, May 7, and Sept 8, Stockholm 1901. Nobel archive.

Consequences

Röntgen donated his prize money (SEK50,000) to the University in Würzburg, to be used for research. He refused to patent his results and thus made the use of X-rays freely available to everybody. He was of the opinion that his discoveries belonged to the public and should not be restricted by patents and licenses. His discovery has had far-reaching consequences in many fields. X-ray equipment quickly became an indispensable tool in medicine all over the world. The discovery of X-rays also formed the necessary foundation for the determination of crystal structures and may thus be considered as the start of modern crystallography.

Max von Laue (P1914)

"For his discovery of the diffraction of X-rays by crystals."

Max von Laue
October 9 1879–April 24 1960
(Photograph from Les Prix Nobel 1914)

Laue was born in 1879 in Pfaffendorf, now a part of Koblenz. From the start, in 1899, he studied mathematics, physics, and chemistry at the University of Strasbourg, the University of Göttingen, and the Ludwig Maximilians University of München (LMU). He went to the Friedrich-Wilhelms-University of Berlin in 1902. There, he studied under Max Planck and received his doctorate in 1903. Thereafter, Laue spent 1903–1905 at Göttingen. Laue completed his habilitation in 1906 under Arnold Sommerfeldt at LMU.

In 1906, Laue became Privatdozent in Berlin and assistant to Planck. He also met Albert Einstein for the first time; they became friends and Laue went on to contribute to the acceptance and development of Einstein's theory of relativity. Laue continued as assistant to Planck until 1909. In Berlin, he worked on the application of entropy to radiation fields and on the thermodynamic significance of the coherence of light waves. From 1909 to 1912, Laue was Privatdozent at the Institute for Theoretical Physics at LMU under Arnold Sommerfeldt. In 1912, Laue, Paul Knipping, and Walter Friedrich proved the nature of X-rays in an experimental verification of the diffraction of X-rays, for which Laue was awarded the Nobel Prize in Physics in 1914. In 1912, Laue was called to the University of Zurich as professor of physics. In 1913, his father was raised to the ranks of hereditary nobility; Laue then became "Max von Laue."

From 1914 to 1919, Laue worked as professor of theoretical physics at the University of Frankfurt. During this time, he was also engaged in vacuum tube development at the University of Würzburg for use in military telephony and wireless communications. In 1919, Laue was called to the University of Berlin as professor of theoretical physics, a position he held until his retirement in 1943. Among notables at the university in 1919 were Walther Nernst, Fritz Haber, and James Franck. During this time, Laue wrote many papers on superconductivity and also published a book on the subject.

The Kaiser-Wilhelm Gesellschaft zur Förderung der Wissenschaften was founded in 1911. Its purpose was to promote the sciences by founding and maintaining research institutes. One such institute was the Kaiser-Wilhelm Institut für Physik (KWIP) founded in Berlin-Dahlem in 1914, with Einstein as director. In 1922, Laue took over the administrative duties from Einstein. Einstein was traveling abroad when Adolf Hitler became Chancellor in January 1933, and he did not return to Germany. Laue then became acting director of the KWIP, a position he held until 1945 or 1948, except for the period 1935–1939, when Peter Debye was director. In 1943, in order to avoid casualties to personnel during the heavy bombardment of Berlin, the KWIP moved to Hechingen.

Laue was encouraged to start a military career by his father, who was an official in the German military administration, but by that time he had become more and more opposed to all kinds of military activities. Laue strongly opposed National Socialism in general and their "Deutsche Physik" in particular. The National Socialists persecuted the Jews, and the advocates of "Deutsche Physik," prominent members of whom included Philipp Lenard and Johannes Stark, considered the upcoming modern physics to be too mathematical and theoretical. They stressed the role of experimental physics and particularly they were opposed to Einstein's theory of relativity (rejected as "Jüdische Physik") and quantum mechanics, the statements of which they considered to be too unclear and unintuitive.

Laue secretly helped many scientific colleagues who were persecuted by National Socialist policies to emigrate from Germany, and he also openly opposed National Socialist policies. When Laue was chairman of the Deutsche Physikalische Gesellschaft, he gave the opening address at the 1933 physics convention. In it, he compared the persecution of Galileo and the oppression of his scientific views on the solar theory of Copernicus to the ongoing conflict and persecution regarding Einstein's theory of relativity by the proponents of "Deutsche Physik."

When the allied troops entered Germany in 1945, they were followed by a contingent of Operation Alsos. This operation was part of the Manhattan project with the purpose of investigating the German nuclear energy effort, seizing equipment, and preventing German scientists from being captured by the Soviets. On April 23 1945, French troops entered Hechingen. The scientific advisor to the operation was the physicist Samuel Goudsmit, who, adorned in a steel helmet, appeared at Laue's home. Laue and other scientists thought to be involved in nuclear research and development were taken to Huntingdon in England, and interned at Farm Hall. Operation Alsos took most of the senior German research personnel into custody, including Otto Hahn, Werner Heisenberg, and Carl Friedrich von Weizsäcker.

In the internment camp, Laue was a reminder to the other detainees that one could survive the Nazi reign without being "compromised." During his internment, Laue wrote a paper on the absorption of X-rays under interference conditions, which was later published in *Acta Crystallographica.* On October 2 1945, Laue, Otto Hahn, and Werner Heisenberg were taken to meet with Henry Hallett Dale, president of the Royal Society, and other members of the Society. Laue was also invited to attend the November 9 1945 Royal Society meeting in memory of Röntgen. However, permission for this was not granted by the military authorities.

Laue returned to Germany in early 1946 and became director of the KWIP, which had in the meantime been moved to Göttingen. In 1946, the Kaiser-Wilhelm Gesellschaft was renamed the Max-Planck Gesellschaft, and, similarly the Kaiser-Wilhelm Institut für Physik became the Max-Planck Institut für Physik. In addition to his administrative and teaching responsibilities, Laue wrote his book on superconductivity, *Theorie der Supraleitung*, and revised his books on electron diffraction, *Materiewellen und ihre Interferenzen* and the first volume of his two-volume book on relativity.

In July 1946, Laue went back to England, only four months after having been interned there, to attend an international conference on crystallography. This was a distinct honor, as he was the only German who was invited to attend. He was extended many courtesies by the British officer who escorted him there and back, and a well-known English crystallographer was his host; Laue was even allowed to wander around London of his own free will.

After the war, there was much to be done in re-establishing and organizing German scientific activities, and Laue played a key role in some of this work. For example, in 1946, he initiated the founding of the Deutsche Physikalische Gesellschaft. Around 1948, the president of the American Physical Society asked Laue to report on the status of physics in Germany; his report was published in 1949 in the *American Journal of Physics.* In April 1951, Laue became director of the Max-Planck Institut für physikalische Chemie und Elektrochemie, a position he held until 1959.

Laue married Magdalene Degen while he was a Privatdozent at LMU. They had two children. Among Laue's chief recreational activities were mountaineering, motoring in his automobile, motorcycling, sailing, and skiing. While not a mountain climber, he did enjoy hiking on the Alpine glaciers with his friends. On April 8 1960, while driving to his laboratory in Berlin, Laue's car was struck by a motorcyclist who had received his license only two days earlier. The cyclist was killed and Laue's car was overturned. He died from his injuries 16 days later, on April 24.

The Nature of X-rays

In the year 1906, Arnold Sommerfeldt became professor in München and head of the Institute for Theoretical Physics. He was selected for this position by Röntgen. Until the end of the 1800s and beginning of the 1900s, experimental physics in Germany was considered as having a higher status than theoretical physics, but at this point the field of theoretical physics had started to be introduced in several places in Germany. Sommerfeldt pioneered developments in atomic and quantum physics, and also educated a large number of students for the new era of theoretical physics.

In the fall of 1909, Laue joined Sommerfeldt's group. Being unmarried and devoted to physics, he soon became a leading member of the activities of the group. His interests covered the whole of physics. For instance, from his previous association with Planck in Berlin, he had a deep understanding of thermodynamics and optics.

After two years as student in Sommerfeldt's group, Paul Ewald was accepted as graduate student in 1910. In his thesis work, he came up with a theory on the optical properties of crystals, which seemed to be a radical departure from the traditional theory. He wanted to discuss his theory with Laue and they agreed to meet; this was probably late in January 1912. During their walk to Laue's home to have supper, Ewald began telling Laue about the problem he was working on. He explained how, in contrast to the usual theory of dispersion, he assumed that resonators were situated in a lattice array. Laue asked for the reasoning behind this assumption. Ewald replied that crystals were thought to have such an internal regularity. Laue asked: "What is the distance between the resonators?" Ewald answered that it was very small compared with the wavelength of ordinary light, perhaps 1/1000 of the wavelength, but an exact value could not be given because of the unknown nature of the "particles" of the structure theory (the existence of "atoms" was not generally accepted by physicists at this time). After supper, Ewald continued to explain his theory but found Laue to be listening in a slightly distracted way. Laue asked: "What would happen if you assumed very much shorter waves to travel through the crystal?" It was evident that there was no use in further discussions of his theory, and as he was busy finishing his thesis, he forgot about Laue's interest in the passage of very short waves through a crystal.

In his Nobel Prize lecture in Stockholm on June 3 1920, entitled *On the Discovery of X-ray Interference*, Laue mentions that his question to Ewald about the fate of short waves in a crystal was prompted by the expectation that if the wavelength of the X-rays was of the same order of magnitude as the atomic distances in a regular arrangement in a crystal, this must lead to some diffraction effect. Laue had discussed his idea with Sommerfeldt and others, but was met with strong disbelief. It was argued that the thermal motion of the atoms would inevitably destroy the regularity of the crystal grating to such an extent that no pronounced diffraction effect could be expected. Sommerfeldt was therefore strongly opposed to allowing his newly appointed experimental assistant Walter Friedrich to help Laue with any experiments on this topic. However, in discussions at the Café Lutz physics table, the opinion was that experiment is better than theory. Paul Knipping from the Röntgen institute volunteered to assist. It was an easy task to arrange the experimental setup. A crystal of copper sulfate ($CuSO_4$) was tested as it was found in the laboratory. A photographic plate was placed between the X-ray tube and the crystal on the assumption that the crystal would act as a reflection grating. The first exposure produced no effect. They concluded that it would probably be better to place the plate behind the crystal, to act as a transmission grating. The second attempt obtained a positive result. Some fuzzy spots appeared on the plate, which could be taken as a proof of a diffraction effect. This happened on April 23 1912. Laue learned about the result at the Café Lutz and hurried to the institute to convince himself about the correctness of his "optical feeling." Further experiments were made with zincblende (ZnS), rock salt (NaCl), and galena (PbS). The spots in the diffraction pictures were much more distinct in these cases and they all confirmed the result from copper sulfate. New and better X-ray tubes were developed and more photographs were

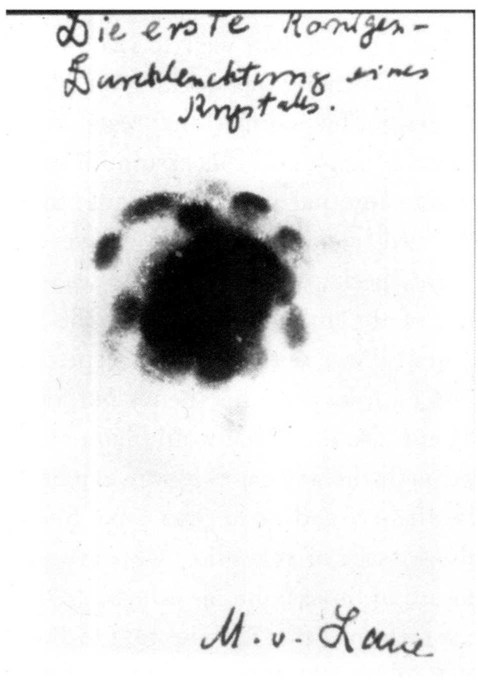

The first diffraction photograph taken in 1912 by Friedrich and Knipping on the suggestion by Max Laue. (From Ewald, P. P. Fifty years of X-ray diffraction, with permission from IUCr.)

taken. However, this work was soon interrupted by the war in 1914. The entire X-ray plant was set up as an emergency hospital and used for medical purposes.

The research by Laue, Friedrich, and Knipping was communicated to the Bavarian Academy of Sciences at the meetings of June 8 and July 6 1912 by Sommerfeldt as a Fellow of the Academy. Röntgen was originally skeptical about the interpretation of these experimental results, but after some time, he became quite convinced, and he seconded the acceptance at the meeting and stressed the importance of the work. Two papers were published in the Proceedings of the Academy: *Interferenz-Erscheinungen bei Röntgenstrahlen* by Friedrich, Knipping, and Laue and *Eine quantative Prüfung der Theorie für die Interferenz-Erscheinungen* by Laue (Friedrich *et al.*, 1912; Laue, 1912). Both papers were sent out in reprint form with a common paper cover.

It should be mentioned the many people radiated crystals with X-rays before Friedrich and Knipping. However, their observations were limited to the directly transmitted ray, which revealed nothing remarkable beyond the weakening produced by the crystal. They completely missed the less strong diffracted rays. As Laue writes in his autobiography (Ewald, 1962): "It was the theory of space lattice which provoked the idea of investigating the neighborhood of the direct ray."

It may be remarked that Sommerfeldt also concerned himself with the theory of X-rays. In his autobiography, Laue writes (Ewald, 1962, p. 291):

> [Sommerfeldt] developed a theory of the creation of X-rays at the anti-cathode of the X-ray tube, which proved very useful. It was based completely on the wave theory of these rays, therefore stood in contrast to W. H. Bragg's corpuscular theory, which was very popular in England. The basis for making a choice between the two was supplied by Walter and Pohl in Hamburg in their photographs of the diffraction of X-rays by a wedge-shaped slit, which P. P. Koch, Röntgen's chief assistant, was measuring photometrically. Sommerfeldt applied the traditional diffraction theory, with the result that he obtained a rough but very useful estimate of the mean wave-length which he published in 1912.

It may be argued that the major impact of Laue's work was the discovery that a crystal functions as a three-dimensional grating for X-rays. The resulting diffraction pattern forms the clue as to the possibility of determining the atomic arrangement in crystals.

Myths regarding the Discovery of the Diffraction of X-rays by Crystals?

The historian Paul Forman made a detailed examination of the facts behind Laue's discovery in *The Discovery of the Diffraction of X-rays by Crystals; A Critique of the Myths* (Forman, 1969). In this paper, a thorough and well-documented discussion of the situation at the time of Laue's discovery is presented. Forman argues that:

- The concept of the space lattice was taken by the physicists as a matter of course, not only in München.
- The wave nature of X-rays was not a necessary condition for conceiving of the experiment.
- It was not believed that thermal motion would destroy a possible diffraction effect.

Paul Ewald received the galley proofs of Forman's paper and replies in *The Myth of Myths: Comments on P. Forman's paper on the Discovery of the Diffraction of X-rays in Crystals* (Ewald, 1969). In summary:

> In the paper here commented upon, its author discusses the various versions of the discovery of X-ray diffraction under the aspect of the formation and embellishment of myths of origin of the clan formed by crystallographers. To that effect, he endeavors to show that most details of the traditional descriptions of motives and obstacles encountered in the planning of the Laue–Friedrich–Knipping experiment are fictitious and fabricated for the purpose of giving the clan a separate identity. This interpretation by the author is possible only because he fails to appreciate the vagueness of the physical information confronting Laue before the experiment. To make his interpretation plausible, the author repeatedly aggrandizes statements taken from the literature, especially from Laue's Nobel Lecture and from the Festschrift for the semicentennial of the discovery. These unjustified accents, needed in support of his main thesis, show that his scheme is pre-conceived, artificial, and unnecessary.

Forman emphasizes that it is the mission of the historian to provide a correct version of how scientific discoveries are actually made. Ewald concludes (ironically): "It is beyond the physicist's competence to clarify historical happenings, i.e. has to be left to historian." These two papers are also very interesting from a more general point of view, pointing out the importance of an independent, critical examination of reports on scientific discoveries. The scientist might be tempted to highlight his discovery, to appear as "the hero who has to overcome numerous and formidable hurdles before slaying the dragon" (Ewald, p. 73; cf. Forman, p. 69); see also Hildebrandt (1993).

Selected References to the Work by Laue

*Friedrich W, Knipping P, Laue M. (1913) Interferenz-Erscheinungen bei Röntgenstrahlen. *Annalen der Physik* **346:** 971–988.

*Laue M. (1913) Eine quantative Prüfung der Theorie für die Interferenz-Erscheinungen. *Annalen der Physik* **346:** 989–1002.

Laue, M. (1915) Concerning the detection of X-ray interferences. Nobel Lecture, December 12.

Other References

Ewald PP. (1962) *Fifty years of X-ray diffraction.* N.V.A. Oosthoek Uitgeversmaatschappi, Utrecht. (A book dedicated to the International Union of Crystallography on the occasion of the commemoration meeting in München, July 1962). An extensive autobiography by Laue is included (pp. 278–307). Part of the above text on *The nature of the X-rays* has been extracted from this book.

Ewald PP. (1969) The myth of myths; comments on P. Forman's paper on the discovery of the diffraction of X-rays in crystals. *Archive for History of Exact Sciences* **6:** 72–78.

Forman P. (1969) The discovery of the diffraction of X-rays by crystals; a critique of the myths. *Archive for History of Exact Sciences* **6:** 38–71.

Hildebrandt G. (1993) The discovery of the diffraction of X-rays in crystals — A historical review. *Crystal Research and Technology*, **28:** 747–766.

The Nobel Prize

In 1914, there were 44 nominations for the Nobel Prize in Physics, including Max von Laue and William Henry Bragg. Some details of the work by the Nobel Committee and the Royal Swedish Academy of Sciences are discussed in the associated paper: *The background to the Nobel Prize to the Braggs* (Liljas, 2013). Max von Laue was awarded the Nobel Prize for Physics in 1914 "for his discovery of the diffraction of X-rays by crystals." It thus took almost two decades after Röntgen's discovery to definitively establish that X-rays were of the same nature as ordinary light, only with approximately 1/1000-times shorter wavelengths.

Reference

*Liljas A. (2013) The background to the Nobel Prize to the Braggs. *Acta Cryst.* **A69:** 10–15.

The Nobel Medal

There is an interesting story regarding the fate of Laue's Nobel medal. The following is a brief summary of an article by Stephan Schwarz entitled '*The case of the bottled Nobel medals*' (Schwartz, 2008).

As Hitler came into power in Germany, Max von Laue secretly sent his medal to Niels Bohr in Copenhagen, as he was afraid it would be confiscated if it had been found in his possession. The Nobel medal of James Franck (Nobel Prize 1925) had also been transferred to Bohr. When Denmark became occupied on April 9 1940, Bohr became worried about the two medals kept in his institute.

In Hitler's Germany, it was illegal to take gold out of the country, and if it had been found that Laue had done so, he could have faced prosecution in Germany. As his name was engraved on the medal, it would have been easy to trace the owner. In addition, if the medal was discovered by the German invading forces in Copenhagen, it would also have had serious consequences for Bohr himself. Various alternatives for hiding the medal were discussed, such as burying it. But this was deemed to be too risky, as the metal could have been detected. George de Hevesy (Nobel Laureate 1943), who was in Copenhagen at this time, came up with the idea of dissolving the medals in aqua regia (a mixture of nitric acid and hydrochloric acid, known to dissolve both platinum and gold). As the invading forces were marching in the streets of Copenhagen, de Hevesy was busy dissolving the medals and placing the resulting solution on a shelf in his laboratory at the Bohr Institute. Three years later, the German army occupied the Bohr Institute and made detailed searches for valuable documents and other items. After the war, de Hevesy returned to find the solution undisturbed, and so the gold could be precipitated out of the acid. Laue's medal was recast and the Nobel Foundation in Stockholm presented him with a new medal.

Laue inadvertently evaded a third confiscatory law. Immediately after the German capitulation in Denmark on April 5 1945, the legal foundations for the confiscation of German property in Denmark were established. In principle, the bottle with the aqua regia solution at the Niels Bohr Institute should have been delivered to the Danish authorities. Perhaps the Royal Swedish Academy of Sciences might have acted, and a case might have been made for an exception, on the hardly credible assertion that the bottle contained Laue's Nobel medal.

Reference

Schwartz S. (2008) The case of the bottled Nobel medals. *Gamma* **150:** 8–17. Link to the article: http://www.gamma.nbi.dk/Galleri/gamma150/stephanschwarz.pdf

Consequences

In 1937, Bragg WH and Bragg WL wrote:

> The discovery of X-ray diffraction ranks as one of the epoch-making discoveries in the history of science. It has extended the power of observing minute structure ten thousand times beyond that given us by the microscope. The number and the range of the published papers which deal with the methods and results of the new X-ray analysis are in themselves an indication of its power. Yet its importance is more correctly assessed when it is realized that in almost any investigation dealing with the structure and properties of matter, the findings of X-ray analysis, and its recent derivative analysis by electron diffraction, are used as a foundation. In physics, chemistry, bio-chemistry, metallurgy, mineralogy,

> and other sciences new points of view have been adopted as a result of this more intimate knowledge of the ultimate structure of matter. *Current Sci.* (1937), p. 9.

Since then, X-ray, electron, and neutron diffraction have played a key role in the studies of matter, and there has been a tremendous experimental as well as theoretical development of the field. Very powerful radiation facilities are now available, such as X-ray synchrotrons, nuclear reactors, and spallation sources for neutrons. Diffraction work involves extensive calculations and the development of powerful computers and programs have been prerequisites for this development. It has become possible to study the most minute details of matter — not only the atomic, but also the electron distribution in crystals — including very complicated biological substances, such as proteins and viruses. In many cases, an obstacle to a structural determination has been the difficulty of crystallizing a particular substance, but with the development of free-electron lasers with extremely high intensity, it may become possible to study even single molecules.

William Henry Bragg and William Lawrence Bragg (P1915)

"For their services in the analysis of crystal structure by means of X-rays."

William Henry Bragg
July 2 1862–March 10 1942
(Photograph from Les Prix Nobel 1915)

William Lawrence Bragg
March 31 1890–July 1 1971

William Henry Bragg (WHB) was born in Westward, Cumbria, in 1862. He obtained his education at King William's College, Isle of Man, and subsequently at Trinity College, Cambridge. He initially studied mathematics and was Third Wrangler in the Mathematical Tripos, Part I, in June 1884. He was placed in the first class in Part II of the course in the following January. He also studied physics at the Cavendish Laboratory during part of 1885.

At the end of that year, at the age of 23, WHB was elected to be Professor of Mathematics and Physics at the University of Adelaide, South Australia. The boat trip to Australia took enough time for him to delve deeper into his physics in order to be ready for his teaching. As Jenkin (2008) vividly describes, he developed an unusual skill for oral presentation. His early years in Adelaide were spent primarily on teaching and keeping abreast of scientific developments by reading journals. During the first years of the new century, his interest in radiation grew. This included the physics of X-rays and the radiation from radioactive materials. Initially, this ended up as new lectures at his university that were illustrated with experiments that greatly impressed the audience. In 1904, his experiments during lectures developed into research. WHB was of the opinion that X-rays were corpuscles and not waves. However, he lacked real progress in his experiments.

In 1909, he became Cavendish Professor of Physics at Leeds; in 1915, he became Quain Professor of Physics at University College London; and finally, in 1925, he became Fullerian Professor of Chemistry at the Royal Institution, London.

William Lawrence Bragg (WLB) was born in Adelaide, Australia, on March 31 1890. He obtained his early education at St. Peters College and continued to Adelaide University at the age of 15. In 1908, he finished his degree in mathematics with first-class honors. In 1909, he moved to England with his family and continued his studies in Natural Sciences at Trinity College, Cambridge. His efforts in crystallography started in 1912.

In 1914, WLB was appointed Fellow and Lecturer in Natural Sciences at Trinity College. During World War I, he served as a Technical Advisor on the Sound Ranging to the Map Section in France. In 1919, he was appointed Langworthy Professor of Physics at Manchester University, and in 1938, WLB succeeded Rutherford as Cavendish Professor of Experimental Physics at the University of Cambridge. His final appointment was as Director of the Royal Institution from 1954 to 1965.

WLB had a very significant role in British science in general, not only in physics. Among his other activities, he was very supportive of biological crystallography, which he started in an extension, called The Hut, of the Cavendish laboratory. Here, Max Perutz performed much of his work together with John Kendrew. Francis Crick and Jim Watson also carried out their work on the structure of DNA in The Hut. Furthermore, Fred Sanger developed methods to determine the amino acid sequences in The Hut.

The Birth of Crystallography

WHB was interested in the nature of radiation. The mysterious X-ray radiation was a particular challenge; it was not deflected by a magnet, it was not reflected, and it did not change direction when passing from one medium to another. When WHB had settled in Leeds, in 1909, he recruited a young Norwegian, Lars Vegard, to assist him with his experiments for about a year (Schwalbe, 2014). Subsequently, Vegard continued his training in Germany. He heard a lecture by Laue about the discovery of X-ray diffraction. In the summer of 1912, when the Bragg family were on vacation at the seaside at Cloughton, WHB received a letter from Vegard reporting the experiments by Laue. This generated a great deal of interest for the Braggs.

Later in the summer, WLB performed experiments in Leeds in order to investigate whether the spots on the photographs could be due to particles traveling down avenues in the crystal lattice, rather than by diffraction of waves (Helliwell *et al.*, 2012). In the fall, WLB studied the Laue photographs and realized that the explanations given by Laue needed to be modified. Instead of five discrete wavelengths being source of the diffracted spots as Laue had thought, he concluded that they must be due to a continuous "white" radiation. Without realizing it, Laue had obtained what is now called "Laue" diffraction patterns. WLB also attended a course in optics and realized that if the atoms in a crystal are regularly arranged, different planes through the atoms could function just like mirrors and, at suitable angles, reflect the X-ray beam (North, 2014). This was supported by the observation that the reflections had different shapes depending on the distance between the crystal and the photographic film. This could be nicely explained by the mirror effect of a divergent beam. WLB presented a lecture at a meeting of the Cambridge Philosophical Society in November 1912. There, he showed how this led to the formula:

$$n\lambda = 2d \cdot sin\theta$$

which was later called Bragg's law. The lecture was subsequently published (Bragg W. L., 1912). WLB also recorded Laue photographs of crystals of sodium and potassium chloride and solved their structures. In Leeds, WHB had constructed an X-ray spectrometer, which was much more useful than the Laue photographs. The spectrometer was used during the Christmas vacation of 1912/1913 for the Braggs' joint recording of the diffraction pattern of diamond and determination of its structure (Helliwell *et al.*, 2012). The next summer (1913), WHB and WLB worked together on the structures of zincblende, fluorspar, pyrites, and some of the carbonates.

Nobel Prize

In 1914, there were 44 nominations for the Nobel Prize in Physics, including Max von Laue and WHB. In the subsequent year, both WHB and his son WLB were nominated. Some details of the work by the Nobel Committee and the Royal Academy of Sciences are discussed in the associated paper entitled *The background to the Nobel Prize to the Braggs* (Liljas, 2013).

Consequences of their Discovery

With the advent of the methods developed by the two Braggs, a whole new world had been opened up — X-ray crystallography. However impressive the significance of this method may seem, its potential could hardly have been fully anticipated. At that time, many physicists questioned the existence of atoms, but the structural determinations now proved that crystals were indeed composed of atoms. Furthermore, it was proven that sodium chloride, like many other compounds, was made up of ions and not molecules. This was hard for some chemists to accept, even a long time after this structure was determined. Even as late as 1927, in a letter to *Nature*, Professor H. E. Armstrong wrote:

> Professor W. L. Bragg asserts that in Sodium Chloride there appears to be no molecules represented by NaCl. The equality in number of sodium and chlorine atoms is arrived at by a chess-board pattern of these atoms; it is a result of geometry and not a pairing-off of the atoms. This statement is more than 'repugnant to common sense'. It is absurd to the nth degree, not chemical cricket. Chemistry is neither chess nor geometry, whatever X-ray physics might be. Such unjustified aspersion of the molecular character of our most necessary condiment must not be allowed any longer to pass unchallenged. It was time that chemists took charge of chemistry once more and protected neophytes against the worship of false gods; at least taught them to ask for something more than chess-board evidence.

From the first determination of the structures of a few simple compounds, X-ray crystallography has undergone an enormous development in the following 100 years. No other experimental method has had such an impact on our knowledge of the composition and properties of inorganic, organic, and biological substances. This has been of fundamental importance in our search for compounds with new and improved properties. In order to understand the function of biologically active substances, the determination of their structures is one of the crucial steps. Today, thousands of protein structures are solved each year in our attempts to understand their functions in living matter. By combining X-ray crystallography with large-scale computations, it has become possible to follow chemical reactions at active sites in great detail. With the advent of highly intense X-ray sources, it has also become possible to study chemical reactions experimentally in real time.

Selected References to Works by the Braggs

Bragg WH. (1913) X-rays and crystals. *Nature* **90:** 572.

Bragg WH. (1913) The reflexion of X-rays by crystals. *Nature* **91:** 477.

Bragg WH. (1913) The reflection of X-rays by crystals. (II.) *Proc. R. Soc. Lond. A* **89:** 246–248.

Bragg WH, Bragg WL. (1913) The reflection of X-rays by crystals *Proc. R. Soc. Lond. A* **88:** 428-438.

Bragg WH, Bragg WL. (1913) The structure of the diamond. *Proc. R. Soc. Lond. A* **89:** 277–291.

Bragg WH. (1914) The influence of the constituents of the crystal on the form of the spectrum in the X-ray spectrometer. *Proc. R. Soc. Lond. A* **89:** 430–438.

Bragg WH. (1914) The intensity of reflexion of X-rays by crystals. *Philos. Mag.* **27:** 881–899.

Bragg WH. (1914) The X-ray spectrometer. *Nature* **94:** 199–200.

Bragg WH, Bragg WL. (1915) *X-rays and crystal structure.* Bell and Sons, London.

Bragg WL. (1912) The diffraction of short electromagnetic waves by a crystal. *Proc. Cambridge Phil. Soc.* **17:** 43–57; *Nature* **90:** 402.

*Bragg WL. (1913) The structure of some crystals as indicated by their diffraction of X-rays. *Proc. R. Soc. Lond. A* **89:** 248–277.

Bragg WL. (1914) The analysis of crystals by the X-ray spectrometer *Proc. R. Soc. Lond. A* **89:** 468–489.

Bragg WL. (1915) The diffraction of X-rays by crystals. Nobel Lecture, September 6, 1922. http://www.nobelprize.org/nobel_prizes/physics/laureates/1915/wl-bragg-lecture.html

Other References

Helliwell JR, Blake AJ, Blunden-Ellis J, *et al.* (2012) Some historical extracts relevant to the discovery and application and the diffraction of X-rays by crystals to contribute to the centennial celebration and International Year of Crystallography. *Cryst. Rev.* **18:** 3–19.

Hunter GK. (2004) *Light is a messenger. The Life and Science of William Lawrence Bragg.* Oxford University Press, Oxford.

*Liljas A. (2013) The background to the Nobel Prize to the Braggs. *Acta Cryst.* **A69:** 10–15.

Jenkin J. (2008) *William and Lawrence Bragg, Father and Son.* Oxford University Press, Oxford.

North AT. (2014) A seaside holiday with unexpected consequences: The Braggs' visit to Cloughton in 1912. *Cryst. Rev.* **20:** 3–8.

Perutz MF. (1990) How W.L. Bragg invented X-ray analysis. A lecture given at the Royal Institution in London on 2 February 1990, in celebration of the centenary of Bragg's birth on 31 March 1890. *Acta Cryst.* **A46:** 633–643.

Schwalbe CH. (2014) Lars Vegard: Key communicator and pioneer crystallographer. *Cryst. Rev.* **20:** 9–24.

Thomas JM, Sir Phillips D (eds.). (1990) *Selections and Reflections: The Legacy of Sir Lawrence Bragg.* The Royal Institution of Great Britain, London.

Bragg centennial

Acta Crystallographica Section A
Foundations of Crystallography
ISSN 0108-7673

Received 8 June 2012
Accepted 9 July 2012

Background to the Nobel Prize to the Braggs

Anders Liljas

Biochemistry and Structural Biology, Lund University, Lund, Sweden. Correspondence e-mail: anders@liljas.net

The Nobel Committees have to follow the nominations submitted for a specific year. During the early phase of X-ray crystallography, a limited number of scientists were active. In 1914 Max von Laue and William Henry Bragg were both nominated and could have been awarded a joint Nobel Prize. However, a member of the Nobel Committee for Physics, Allvar Gullstrand, was well aware of the activities in the field and strongly recommended that only von Laue should receive the prize since a main contributor, William Laurence Bragg, was not nominated. Next year, when the First World War had started, there were few nominations, but now both Braggs, father and son, were nominated. Gullstrand was very pleased and recommended them both for the 1915 Nobel Prize in Physics. The rest of the committee agreed and this then became the decision of the Royal Academy for Sciences, Stockholm.

The will of Alfred Nobel states that the prizes shall be awarded in the fields of physics, chemistry, physiology or medicine, literature and peace. The organizations which are to award the prizes are also stated. In the case of physics and chemistry the Royal Academy of Sciences, Stockholm, was given the task of deciding on these. Here the classes of physics and chemistry elect a five-member committee, possibly with some adjunct members, who will work with nominations received from all over the world in order to recommend to the Academy a recipient for the prize. The invited nominators in a particular year are expected to reply to the relevant committee before 1 February. To be awarded a prize, a scientist has to have been nominated in that specific year. The relevant Nobel Committee performs extensive work to identify the strongest candidate or candidates (maximally three) from the nominations for that specific year. Many of the nominees are thoroughly evaluated by members of the committee or knowledgeable scientists. The relevant class of the Academy (chemistry or physics) subsequently discusses the proposal by the Nobel Committee and gives their recommendations to the Academy, where the final voting is done. The Royal Academy of Sciences in Stockholm safely stores all of the protocols of the Academy and its Nobel Committees. In the case of the oldest protocols they are all written in a clear and legible hand.

In the protocols from the Nobel Committee for Physics for 1914 one finds that 44 nominations were made for the Nobel Prize and 24 scientists were nominated (some further details are presented by Eckert, 2012). Among the nominators was Svante Arrhenius (Stockholm, a member of the Academy and Swedish Nobel Prize winner in chemistry 1903 for his electrolytic theory of dissociation) who nominated William Henry Bragg (Leeds) together with Henry Gwyn Jeffreys Moseley (Manchester and Oxford) and Charles Galton Darwin (Manchester) for their studies of the periodic system using various elements as anticathodes for generating X-rays. At the time, quite a number of scientists were engaged in exploring the nature and use of X-rays. It was well known, through the observations by Charles Glover Barkla (Nobel Prize in Physics 1917, but not nominated 1914 or 1915), that when X-rays hit some material secondary (fluorescent) X-rays emerge. These secondary X-rays are characteristic of the material being irradiated. He also could conclude that the hardness (wavelength) of this secondary radiation was related not to the atomic weight, but to the atomic number. Professor Emil Warburg (Charlottenburg) nominated Max von Laue (Munich) jointly with W. H. Bragg (Leeds). Professor Adolf von Baeyer (Munich) nominated von Laue alone. Allvar Gullstrand (Uppsala) was given the task of evaluating the work of von Laue. Gullstrand was an MD who specialized in ophthalmology and a Nobel laureate in physiology or medicine for 1911. His primary expertise was in geometrical and physiological optics. He was a member of the Academy and the Nobel Committee for Physics from 1911 to 1929 and was the chairman of the committee from 1922 to 1929 (see http://www.nobelprize.org).

The evaluation by Gullstrand (3 July 1914) has the title (my translation from Swedish): 'Evaluation concerning the discovery by von Laue of the bending of X-rays in crystal lattices and the usage by W. H. Bragg of this phenomenon for research on crystal structures'. The report discusses attempts to identify the nature of X-rays. In particular, the failure to bend X-rays by strong magnets suggested that they could not be charged particles. W. H. Bragg, who had been publishing on the topic since 1908, was of the opinion that X-rays could be particles composed of two neutralizing charges such as a positive alpha particle (then not known to be doubly charged) and a negatively charged electron. Furthermore, it was observed that by passing X-rays from one medium to another the direction of the beam was not changed. Thus, if the X-rays were a wave phenomenon, like light, the wavelengths must be very small.

Gullstrand continues on to say that this was the state of understanding of the nature of X-rays when von Laue suggested irradiating a crystal and studying the consequent effects. If X-rays had wavelengths of a similar magnitude to the lattice spacings in crystals, interference would occur. von Laue worked out a theory for the experiment that predicted what result would be expected. W. Friedrich and P. Knipping performed the simple experiment on several different types of crystals. von Laue reported the successful results together with the theory in the session documents of the Munich Academy 1912 (Friedrich *et al.*, 1912; von Laue, 1912). This epoch-making discovery of the diffraction of X-rays by crystals gave science a new tool of great significance and at the same time allowed von Laue to conclude that X-rays were a wave phenomenon. The wavelengths were evidently of the same range as the distances between atoms in the crystal. The possibility that the observations were due to X-ray fluorescence was easily disproven. However, some of the expected maxima of blackening of the photographic plates from a zincblende (sphalerite) crystal did not appear. von Laue assumed that this could be due to several cubic lattices displaced from each other in the crystal (von Laue, 1912).

Gullstrand further discusses letters to *Nature* by W. H. Bragg concerning the crystal interferences observed by von Laue. Bragg suggested that 'avenues' in the crystal could guide the X-rays in directions different from the initial one (W. H. Bragg, 1912). He also claimed that his son, William Lawrence Bragg, had confirmed his views in a presentation to the Cambridge Philosophical Society. However, in the written form of the presentation (W. L. Bragg, 1913*a*) Gullstrand finds no support of the idea of avenues but rather a new theory, where the dark spots on the photographic plate could be considered as reflections of the X-ray beam by planes in the crystal lattice. The theory of Bragg was a simple way to express the theory developed by von Laue.

In June 1913 the two Braggs submitted separate notes to the Royal Society, where the younger Bragg further describes the reflection aspect of diffraction (W. L. Bragg, 1913*b*). Planes that are richly populated by atoms will give more prominent diffraction spots. The possibility to determine the atomic arrangement from the spots and their intensities then becomes a natural consequence. W. L. Bragg in this way had developed means by which he could determine crystal structures. He then used these methods to determine the structures of crystals of alkali metal halides, and subsequently the structures of diamond and zincblende and a range of other types of crystals. At the same time, W. H. Bragg investigated the spectra produced by a range of anticathode elements (W. H. Bragg, 1913*b*).

It is remarkable that Gullstrand thoroughly discusses observations by W. L. Bragg who was not nominated for the prize in 1914. This is probably partly due to the joint nomination of von Laue and W. H. Bragg by Professor E. Warburg. This could have resulted in the Academy jointly awarding the prize to these two as an appropriate way to deal with two closely related discoveries. Gullstrand was very much against such a decision since it would miss a very central contributor, W. L. Bragg. Rather Gullstrand's conclusion was, along with the nomination by Professor von Baeyer, to recommend awarding the prize to von Laue alone. Gullstrand, with significant pleasure, also comments on the short time between the time of the discovery and the time of a possible award as well as the great value for mankind that X-ray diffraction analysis will have. Rarely would the Academy have the opportunity to find a laureate in better agreement with the wording of the will of Alfred Nobel.

The Nobel Committee for Physics (Gustaf Granquist, Allvar Gullstrand, B. Hasselberg, V. Carlheim-Gyllensköld and Svante Arrhenius) on 15 September 1914 summarized their conclusions concerning the nominations for 1914. The committee mentions the joint nomination of W. H. Bragg and von Laue, but also highlights the contributions by W. L. Bragg, even though he was not nominated in that year. The committee concluded that through these discoveries 'a new means has been generated for the studies of the inner construction of crystals, with immense consequences, no less for the chemical than for the mineralogical sciences'. Their recommendation to the Academy was that the Nobel Prize in Physics 1914 should be awarded to Max von Laue 'for his discovery of the diffraction of X-rays in crystals'. The physics class of the Academy agreed with this recommendation at their meeting on 31 October 1914.

However, 1914 became a year of great problems due to the outbreak of the First World War. This also affected scientific activities and the work of scientific academies. For this reason, the Royal Academy requested in a letter dated 16 October 1914 to the Board of the Nobel Foundation a wish to discuss whether the prizes for the year should be delayed. The Board of the Nobel Foundation, together with members of the awarding institutions, convened on 20 October and suggested that, if the awarding institutions all were of the same opinion, permission should be requested from the Swedish government to award the Nobel Prizes for 1914 together with the prizes for 1915 on 1 June 1916. Presumably they expected that the war would be over by then. It was also explicitly stated that the recipients should benefit from the additional interest on their prize money due to this delay.

The Swedish government decided that the Nobel Prizes for 1914 could be awarded in 1915, but not delayed until 1916. Therefore, on 28 October 1914 the Academy postponed the decision on the award of the prizes for 1914. The Nobel Prizes for 1914 were reserved to be awarded in 1915. This created some confusion among nominating scientists. Since no winner had been announced for 1914, Max von Laue was again nominated for the prize in 1915.

In 1915 there were only 17 nominators for the prize in physics (compared to 1914 when there were 44) and 21 scientists were nominated (in 1914 there were 24). T. W. Richards (Cambridge, MA) nominated both W. H. and W. L. Bragg. H. A. Bumstead (New Haven) nominated Max von

Bragg centennial

Laue alone or jointly with the two Braggs. Stefan Meyer (Vienna) nominated W. H. Bragg together with Max von Laue. Furthermore, Svante Arrhenius again nominated W. H. Bragg, now in both physics and chemistry, as before together with Henry Gwyn Jeffreys Moseley and Charles Galton Darwin (Arrhenius, 1915).

The Nobel Committee again gave Allvar Gullstrand the task of reviewing the work of W. H. Bragg, now together with his son W. L. Bragg. Since the Academy had not yet decided about the prize for 1914, it was Gullstrand's opinion that if von Laue was not awarded in 1914 he should get the prize for 1915.

Gullstrand's report from 25 June 1915 has the title (my translation): 'Evaluation concerning the investigations by W. H. Bragg and W. L. Bragg of crystal structures and X-ray radiation'. He reminds the readers that the possibility of rewarding von Laue together with W. H. Bragg had already been discussed a year ago. He makes it clear that this would have been a mistake, since the contributions of the younger Bragg were highly significant and he was not nominated for the prize in 1914. Gullstrand further states that in the work on crystal structures W. L. Bragg had made the most significant contributions. The father and son published their findings in a rapid flow of scientific reports and also in a book, *X-rays and Crystal Structure* (Bragg & Bragg, 1915).

Gullstrand states that when von Laue initially made his discovery and the first applications were made, two theories got a severe blow. One was the view, supported by W. H. Bragg, that X-rays were a radiation of particle nature. The second concerned the zincblende structure, where von Laue had assumed a primitive lattice. According to Barlow and Pope face-centred lattices would give the densest packing of balls. W. L. Bragg, in his first report on the bending of X-rays and his interpretation of the zincblende structure, finds it to be face centred (W. L. Bragg, 1913*a*) and thanks Professor Pope for advice. The same paper contains the outlines of methods by which the discovery by von Laue can become useful for crystallographers. Bragg the younger describes a simple formula for the principle that determines the direction of the bent X-rays and shows that it is in full agreement with the theory by von Laue. The diffraction of X-rays can be described as specular reflection by a set of parallel planes through all lattice elements. A diffracted beam is obtained if

$$2d\sin\theta = n\lambda,$$

where d is the distance between the planes, θ is the angle of incidence, n is an integer and λ is the wavelength. Through this simple approach the determination of crystal structures was made possible. In the first step the zincblende diffraction pattern by von Laue was correctly interpreted by W. L. Bragg as a face-centred atomic arrangement irradiated by a continuous spectrum. This also led W. H. Bragg to convert his X-ray spectrometer for a more accurate approach to structure determinations (W. H. Bragg, 1913*a*).

Continuing his report, Gullstrand feels he has to educate his readers in basic crystallography and space geometry to be able to describe the crystallographic analysis of the alkali metal halides. It becomes evident from the work of W. L. Bragg that the crystals are not built of molecules of these compounds, but rather of the metal and halide ions forming separate lattices in which one type of ion is surrounded by six ions of the other type, all at the same distance The determination of a number of more complex structures, including diamond and quartz, is also described.

Gullstrand also discusses the only known critical claim at the time, by William Barlow (1914). He felt that the alkali halides must be treated as molecules. Gullstrand finds his discussion to be incorrect and comments on the lack of response from Bragg as a kindness to Professor Pope, with whom Barlow was a co-worker

The report also describes the work by W. H. Bragg on the development of the X-ray spectrometer and his studies of the relationship between the X-ray wavelengths, originating from different metal targets, and the absorption coefficients of different materials (W. H. Bragg, 1914*a*). He also studied the effect on the diffraction pattern due to change in temperature, which he found to agree with the theory by Debye (W. H. Bragg, 1914*b*).

Gullstrand then tried to evaluate the individual contributions by the two Braggs. He firstly concludes that in the determination of crystal structures W. L. Bragg has been the leading figure. However, the father had developed the equipment and investigated the properties of X-ray spectra from different metals. He was probably also the one who carried out most of the measurements. The contributions therefore seemed inseparable and any other outcome than a shared prize would be impossible. With the advent of the methods developed by the two Braggs, a whole new world had been opened up and already partly explored. However impressive the significance of these methods may seem, their potential can hardly be fully anticipated. To reward the discoverers of the new methods after the discoverer of the new means, Max von Laue, has been rewarded, will create full justice. Thus the prize should be shared between W. H. Bragg and his son W. L. Bragg.

In their summary for the year (18 September 1915) the Nobel Committee for Physics discussed the analysis of X-ray spectra by Moseley and Darwin and considered their work to be more chemistry than physics Thomas Alva Edison was nominated for wireless telegraphy. The committee concluded that such a prize would honour a lifetime of achievements, which was against the rules. This was also considered to be the case for Augusto Righi (Bologna). Max Planck (Berlin) was nominated for the quantum theory, but the committee found that they needed further clarification. Walter Nernst (Göttingen) was nominated for the heat theorem and Otto Lehmann (Karlsruhe) for liquid crystals. The two latter nominations were considered not to be competitive enough. Charles Thomson Rees Wilson (Cambridge) was nominated for his method of making the paths of electrically charged particles visible by condensation of vapour. Fredrik Carl Mülertz Størmer (Christiania) was nominated for the insights into the movements of electric particles close to a magnet and for his studies of the Northern Lights. Aime Cotton (Paris) was nominated for his work on optics and dichroism, and

Louis Carl Heinrich Friedrich Paschen (Tübingen) for black-body radiation. George Ellery Hale (Carnegie Observatory) was also nominated. In addition, one of the nominations concerned a very much appreciated way of coupling railway cars.

In their conclusion, the committee went entirely along with the suggestion by Gullstrand in recommending that the Academy award Max von Laue the physics prize for 1915, in the case that he had not already obtained the prize for 1914. If von Laue was already awarded the prize for 1914 then the two Braggs should be given the prize for 1915 (Fig. 1).

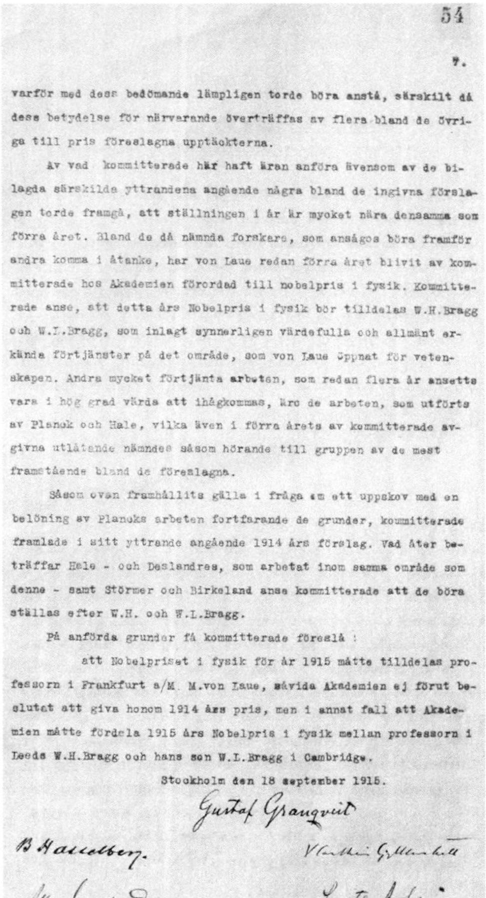

54

7.

varför med dess bedömande lämpligen torde böra anstå, särskilt då dess betydelse för närvarande överträffas av flera bland de övriga till pris föreslagna upptäckterna.

Av vad kommitterade här haft äran anföra ävensom av de bilagda särskilda yttrandena angående några bland de ingivna förslagen torde framgå, att ställningen i år är mycket nära densamma som förra året. Bland de då nämnda forskare, som ansågos böra framför andra komma i åtanke, har von Laue redan förra året blivit av kommitterade hos Akademien förordad till nobelpris i fysik. Kommitterade anse, att detta års Nobelpris i fysik bör tilldelas W.H.Bragg och W.L.Bragg, som inlagt synnerligen värdefulla och allmänt erkända förtjänster på det område, som von Laue öppnat för vetenskapen. Andra mycket förtjänta arbeten, som redan flera år ansetts vara i hög grad värda att ihågkommas, äro de arbeten, som utförts av Planck och Hale, vilka även i förra årets av kommitterade avgivna utlåtande nämndes såsom hörande till gruppen av de mest framstående bland de föreslagna.

Såsom ovan framhållits gälla i fråga om ett uppskov med en belöning av Plancks arbeten fortfarande de grunder, kommitterade framlade i sitt yttrande angående 1914 års förslag. Vad åter beträffar Hale - och Deslandres, som arbetat inom samma område som denne - samt Störmer och Birkeland anse kommitterade att de böra ställas efter W.H. och W.L.Bragg.

På anförda grunder få kommitterade föreslå :

att Nobelpriset i fysik för år 1915 måtte tilldelas professorn i Frankfurt a/M. M.von Laue, såvida Akademien ej förut beslutat att giva honom 1914 års pris, men i annat fall att Akademien måtte fördela 1915 års Nobelpris i fysik mellan professorn i Leeds W.H.Bragg och hans son W.L.Bragg i Cambridge.

Stockholm den 18 september 1915.

Figure 1
The last page of the summary by the Nobel Prize Committee for Physics dated 18 September 1915 with the signatures of its five members. They concluded by recommending that the Academy should award Max von Laue the Nobel Prize in Physics for 1915, if he had not already been awarded the prize for 1914. If von Laue had already been awarded the prize for 1914, the prize for 1915 should be shared between Professor W. H. Bragg, Leeds, and his son W. L. Bragg, Cambridge. Reproduced with the permission of the Centre for History of Science, The Royal Swedish Academy of Sciences.

At the Royal Academy meeting on 11 November 1915 (present: President Eriksson, Secretary Aurivillius and 82 additional members) it was announced that the Academy was not permitted to delay the awarding of the Nobel Prizes for 1914 and 1915 until 1916. Thus, the documents from the Nobel Committees for 1914 and 1915 should be discussed to reach decisions on the award of the prizes for both these years. It was then concluded that Professor Max von Laue, University of Frankfurt, was to obtain the prize for physics for 1914 'for his discovery of the diffraction of X-rays by crystals'.

The Academy met again the next day (12 November 1915) to discuss the Nobel Prize in Physics for 1915. In addition to the President and Secretary 78 members were present. It was decided that Professor W. H. Bragg, University of Leeds, and his son W. L. Bragg, Cambridge, would be awarded the prize jointly 'for their services in the analysis of crystal structure by the means of X-rays'. The news was rapidly communicated to the recipients by telegram.

In a responding telegram to the Royal Academy on 15 November 1915, W. H. Bragg gratefully acknowledged the information that he and his son were to receive the Nobel Prize in Physics for the year (Fig. 2).

However, the war meant that the Nobel lectures and the prize ceremonies could not be held in the regular way. A ceremony for all laureates from during the war was delayed until 1920. Both W. H. and W. L. Bragg excused themselves from being able to come. This may illustrate the strong connections between science and politics during this period (Widmalm, 1995).

Before the First World War, Germany was a leading scientific nation. With the war national tensions grew drama-

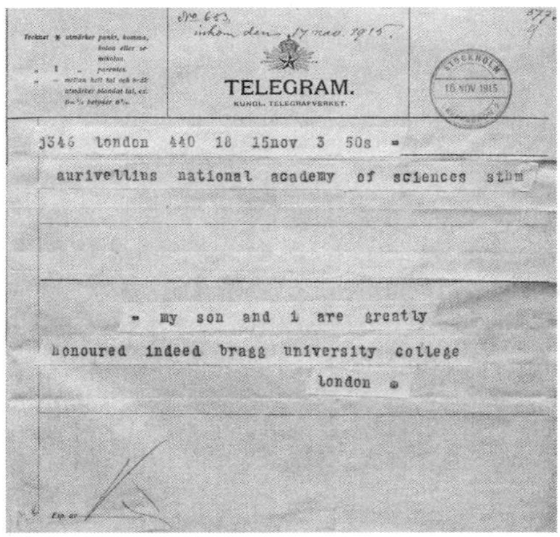

TELEGRAM.

KUNGL. TELEGRAFVERKET.

3346 london 440 18 15nov 3 50s =

aurivellius national academy of sciences sthm

= my son and i are greatly honoured indeed bragg university college london =

Figure 2
The telegram from W. H. Bragg acknowledging the message about the Nobel Prize in Physics, 1915. Reproduced with the permission of the Centre for History of Science, The Royal Swedish Academy of Sciences.

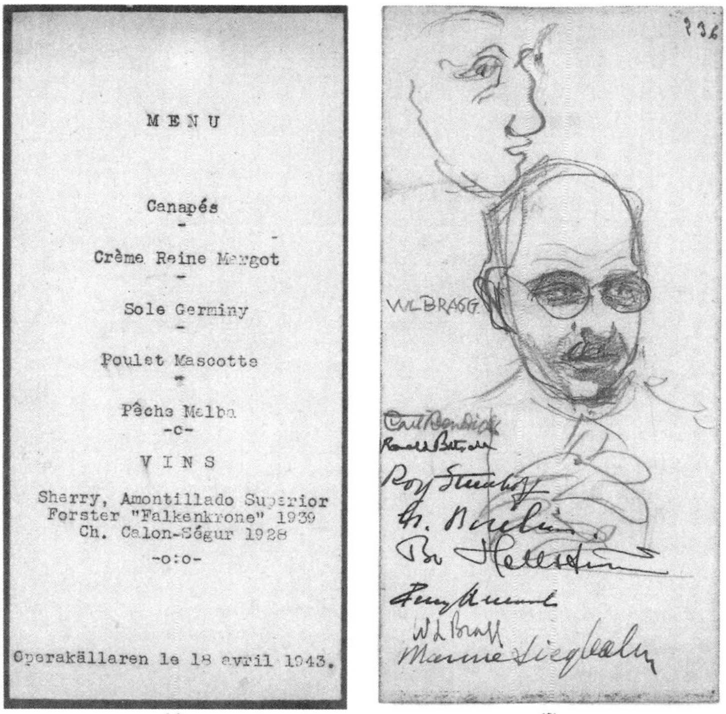

Figure 3
A drawing by Professor Carl Axel Fredrik Benedicks of W. L. Bragg on the reverse side of the menu from a dinner at the Opera cellar in Stockholm, 22 April 1943. The dinner participants signed the drawing. Reproduced with the permission of the Centre for History of Science, The Royal Swedish Academy of Sciences.

tically. Several German scientists who were awarded prizes during the war would participate in the ceremony. The absence of the Braggs could be due to nationalistic sentiments (Jenkin, 2008). In the case of W. L. Bragg it was agreed in communications with Svante Arrhenius that he would come to present his Nobel lecture in September 1922 (W. L. Bragg, 1915). W. H. Bragg never gave a Nobel lecture (Jenkin, 2008).

Sir William Lawrence Bragg again visited Sweden, this time in April and May 1943, in the midst of the Second World War (Phillips, 1990). He flew to Stockholm on 16 April. The British Council arranged the trip (British Council, annual report, 1943–1944). This organization was established to promote a wider appreciation of British culture abroad to combat German and Italian cultural propaganda. Among its activities was the organization of lecture tours abroad by established British scientists. Sir William Henry Bragg founded the Council's Science Advisory Committee, of which he was chairman from 1940 until his death in 1942.

Much of Europe was inaccessible to the British Council during the war, either because of active conflict or German occupation. Sweden was one of only four European countries (the others were Portugal, Spain and Iceland) in which the Council could operate. An office in Stockholm was opened in 1941, which offered support to anglophile societies across the country. Lecturers were sometimes sent over from London, although transport to Sweden was frequently disrupted.

During the visit to Sweden in 1943, W. L. Bragg gave 14 lectures. He visited four universities and spoke on X-ray diffraction and in addition gave lectures at several anglophile societies. His ability to interpret science for the benefit of laymen made him a most popular visiting lecturer. During the

Figure 4
A gathering with the chemistry professors at Uppsala University, 22 April 1943. The Svedberg (Nobel laureate in chemistry, 1926), Arne Fredga, Arne Tiselius (Nobel laureate in chemistry, 1948), Lawrence Bragg, Gunnar Hägg and Axel Lindh.

Figure 5
W. L. Bragg presenting the Nobel jubilee lecture in Uppsala, 1965.

visit to Stockholm a dinner was held at the Opera cellar by a group of scientists, among them Manne Siegbahn, Nobel laureate in physics in 1924. Professor Benedicks frequently used his skills at making drawings of people in his company. At this occasion he made a sketch of W. L. Bragg on the reverse side of the menu (Fig. 3). The participants, including Bragg, signed the drawing.

During the visit to Uppsala on the same trip the chemistry professors all enjoyed the visit by their famous colleague, probably one of few visitors during the war (Fig. 4).

W. L. Bragg returned to Sweden on at least two more occasions. One was for the IUCr congress in 1951 and subsequently to celebrate the 50th anniversary of his Nobel Prize, 1965. During the last occasion he gave lectures at several universities (Fig. 5). He described the developments of X-ray crystallography aimed at solving an ever-increasing complexity of crystal structures. He had the fortune to bring along a model of the first enzyme structure, hen egg-white lysozyme, that had just been determined in his laboratory by David Phillips and colleagues.

I received valuable assistance in my search for documents in the Nobel Archives by Maria Asp Dahlbäck. I also received kind assistance with the photographs by Professors Ivar Olovsson and Bror Strandberg. Joanna Loxton at The British Council, London has provided details about the wartime visit to Sweden by W. L. Bragg. Professor Steve Wilkins provided a very valuable check of the content and presentation.

References

Arrhenius, S. (1915). Nomination to the Nobel Committees of Chemistry and Physics. January 30.
Barlow, W. (1914). *Proc. R. Soc. London Ser. A*, **91**, 1–16.
Bragg, W. H. (1912). *Nature (London)*, **90**, 219.
Bragg, W. H. (1913*a*). *Nature (London)*, **90**, 572.
Bragg, W. H. (1913*b*). *Nature (London)*, **91**, 477.
Bragg, W. H. (1914*a*). *Proc. R. Soc. London Ser. A*, **89**, 430–438.
Bragg, W. H. (1914*b*). *Philos. Mag.* **27**, 881–899.
Bragg, W. H. & Bragg, W. L. (1915). *X-rays and Crystal Structure.* London: Bell and Sons.
Bragg, W. L. (1913*a*). *Proc. Cambridge Philos. Soc.* **17**, 43–57.
Bragg, W. L. (1913*b*). *Proc. R. Soc. London Ser. A*, **89**, 248–277.
Bragg, W. L. (1915). *The diffraction of X-rays by crystals.* Nobel lecture, September 6, 1922. http://www.nobelprize.org/nobel_prizes/physics/laureates/1915/wl-bragg-lecture.html.
British Council (1943–1944). *Annual Report*, pp. 35, 117.
Eckert, M. (2012). *Acta Cryst.* A**68**, 30–39.
Friedrich, W., Knipping, P. & Laue, M. (1912). *Sitzungsber. Bayerische Akad. Wiss.* **42**, 303–322.
Gullstrand, A. (1914). Special report, July 3, 1914. Nobel Archive, Stockholm.
Gullstrand, A. (1915). Special report, June 25, 1915. Nobel Archive, Stockholm.
Jenkin, J. (2008). *William and Lawrence Bragg, Father and Son.* Oxford University Press.
Laue, M. von (1912). *Sitzungsber. Bayerische Akad. Wiss.* **42**, 363–373.
Nobel Committee for Physics. Report to the Royal Academy September 15, 1914. Nobel Archive, Stockholm.
Nobel Committee for Physics. Report to the Royal Academy September 18, 1915. Nobel Archive, Stockholm.
Phillips, D. (1990). *Selections and Reflections: the Legacy of Sir Lawrence Bragg*, edited by J. M. Thomas & Sir D. Phillips, pp. 1–69. The Royal Institution of Great Britain.
Royal Academy of Sciences, Stockholm. Protocol, October 16, 1914. Nobel Archive.
Royal Academy of Sciences, Stockholm. Protocol, October 28, 1914. Nobel Archive.
Royal Academy of Sciences, Stockholm. Protocol of the Physics class. October 31, 1914. Nobel Archive.
Widmalm, S. (1995). *Minerva*, **33**, 339–360.

Clinton Joseph Davisson and George Paget Thomson (P1937)

"For their experimental discovery of the diffraction of electrons by crystals."

Clinton Joseph Davisson
October 22 1881–February 1 1958
(Photograph from Les Prix Nobel 1937)

George Paget Thomson
May 3 1892–September 10 1975
(Photograph from Les Prix Nobel 1937)

Clinton Davisson was born 1881 in Bloomington, Illinois, where he attended high school and graduated in 1902. He was then given the opportunity to attend the University of Chicago on a scholarship, but as this was only short term, he left the university in the following year to work for a local telephone company. Thanks to the influence of Prof. Millikan, of oil droplet fame and future Nobel Prize winner in Physics (1923), whom he met during his year at the University of Chicago, he was appointed as Assistant in Physics at Purdue University in the beginning of 1904. He held this position until 1908, when he received his BS from Chicago University, and then went back and forth between Purdue and Chicago for his studies when his duties as assistant at Purdue allowed.

He was awarded a one-year fellowship at Princeton in 1910–1911, during which time he completed the requirements for a Ph.D. His thesis work was completed under the supervision of Prof. O. W. Richardson, who was to receive the Nobel Prize in Physics in 1928. His first employment after obtaining his Ph.D. was at the Carnegie Institute in Pittsburg. He spent the summer of 1913 at the Cavendish laboratory working under Prof. J. J. Thomson, the 1906 Physics Laureate. After this, Davisson held an employment at the Western Electric Company, the forerunner of Bell Labs, and remained there until his retirement in 1946.

In 1911, he married Charlotte Sara Richardson, the sister of his thesis supervisor. When he died in 1958, his wife and four children survived him.

Reference

*Davisson CJ. (1928) The diffraction of electrons by a crystal of nickel. *Bell System Tech. J.* **7:** 90–105.

George Paget Thomson, the son of J. J. Thomson and Rose Elisabeth Paget, was born in 1892 in Cambridge. He went to The Perse School in Cambridge, which counts among its alumni the 1967 Chemistry Laureate Ronald Norrish and the Pink Floyd guitarist David Gilmour. At the school, he won a scholarship to Trinity College. Before World War I broke out, he had done a year of research with his father, who was by then a Nobel Laureate in Physics. After the war, he spent three more years in Cambridge as a lecturer at Corpus Christi College and as researcher in physics. He was appointed professor in Aberdeen in 1922 and at Imperial College in 1930. He returned to his alma mater in 1952, where he became Master of Corpus Christi.

In 1924, he married Kathleen Buchanan Smith. The couple had four children.

Reference

*Thomson GP. (1931) The diffraction of electrons by single crystals. *Proc. R. Soc. Lond. A* **133:** 1–25.

The Nobel Prize

The two laureates both worked with J. J. Thomson in 1913, but the work for which they were awarded the Nobel Prize was done independently in Aberdeen and in New York. The citation for the prize read "for their experimental discovery of the diffraction of electrons by crystals." These findings, like those from the experiment of Laue, served a double purpose: To provide experimental evidence of the wave-particle duality of matter; and to provide a proof of principle that electrons may be used as a probe to examine matter. The wave nature of matter had been theoretically established by de Broglie (Nobel Prize in Physics, 1923), but the discoveries awarded the 1937 physics prize consisted of the first experimental evidence for the wave nature of matter. This was a fundamental discovery that helped forge the new physics. From the point of view of structural chemistry, the other aspect of the discovery is more important. Up to this point, X-rays had reigned supreme as the only useful source for diffraction studies. The Thomson and Davisson work opened the door for electron diffraction studies, and although relatively few results had been obtained by the time the two received their Nobel Prize, the award speech emphasized the fact that the strong interaction between electrons and matter yields very different properties for electron diffraction compared with X-ray diffraction.

Consequences

The immediate consequences of the discovery are given in the award speech: Experimental evidence for the theory of de Broglie and a new structural probe being made available. However, the most important consequence in terms of structural analysis was not yet fully understood. Since electrons are charged particles, they are relatively easily manipulated by an electric field, which means that an electron beam may not only be diffracted, but also condensed using an electric lens. The logical consequence of this was the electron microscope. Ernst Ruska was awarded the 1986 physics prize for the invention of the electron microscope. This work originated before the award of the prize to Thomson and Davisson, but slowly developed from relatively humble beginnings into what it is today: A generally used structural method for imaging structures from the atomic scale and upwards, enriching all fields of science and, in particular, starting the nanoscopic revolution. Electron microscopy was also the method used by Aaron Klug to study biologically important complexes, and the method is being increasingly used today in tandem with X-ray diffraction.

James B. Sumner, John H. Northrop, and Wendell M. Stanley (C1946)

James B. Sumner

"For his discovery that enzymes can be crystallized."

James Batcheller Sumner
November 19 1887–August 12 1955
(Photograph from Les Prix Nobel 1946)

James Batcheller Sumner was born in 1887 in Canton, Massachusetts. He first attended the Eliot Grammar School for a few years and was then sent to Roxbury Latin School. At school, almost every subject except physics and chemistry bored him. However, he was interested in firearms and often went hunting. While grouse hunting at the age of 17, he was accidentally shot in the left arm by a companion; as a consequence, his arm had to be amputated just below the elbow. Having been left-handed, he then had to learn to do things with his right hand. The loss of his arm made him exert every effort to excel in all sorts of athletic sports, such as tennis, skiing, skating, billiards, and clay pigeon shooting. Sumner graduated from Harvard College in 1910 having specialized in chemistry. In 1912, he started to study biochemistry with Prof. Otto

Folin at Harvard Medical School and obtained his Ph.D. in 1914. He was then invited to be Assistant Professor of Biochemistry at Cornell Medical School.

Enzymes had long been known as very efficient catalysts, but their nature was unknown. Richard Willstätter (C1915), in particular, was of the opinion that enzymes were neither lipids, nor carbohydrates nor proteins. In 1917, Sumner decided to try to isolate and chemically characterize an enzyme, something nobody had succeeded at doing so far. His enzyme of choice was urease, an enzyme he had already been working with. His choice of species to use as a source was the jack bean, which is unusually rich in urease. After many different approaches, the best results were obtained if he dissolved the urease from jack beans with acetone and subsequently used precipitation and crystallization. Crystallization had long been used to purify organic substances and had also been shown to work for proteins. In 1926, crystals with high urease activity were obtained after a long struggle, and their activity could be increased by recrystallization. The protein nature of the enzyme was shown by digestion with the proteolytic enzymes pepsin and papain. The results were met with general disbelief. Willstätter, during a lecturing visit to Cornell in 1927, maintained that the isolated protein could be a carrier of the enzyme. However, in 1929, Sumner's success finally made him full Professor of Biochemistry at Cornell. In 1929, he worked with Prof. von Euler in Stockholm, and in 1937, he spent five months in Uppsala with Prof. The Svedberg. The success of Sumner led to efforts by many other scientists to purify and crystallize enzymes. Among the first was John Northrop (see below). Northrop's work on pepsin led to a general acceptance that enzymes are proteins. The crystal structure of urease from jack beans, which turned out to be a nickel-containing enzyme, was determined in 2010, 83 years after Sumner's work (Balasubramanian & Ponnuraj, 2010).

Selected References to Sumner's Work

*Sumner JB. (1926). The isolation and crystallization of the enzyme urease: Preliminary paper. *J. Biol. Chem.* **69:** 435–441.

Sumner JB. (1937) The story of urease. *J. Chem. Educ.* **14:** 255–259.

Other References of Interest

Maynard LA. (1958) James Batcheller Sumner. *Biographical Memoirs. Nat. Acad. Sci.* **31:** 375–396.

McPherson A. (1991) A brief history of protein crystal growth. *J. Cryst. Growth* **111:** 1–10.

Fruton JS. (2002) A history of pepsin and related enzymes. *Q. Rev. Biol.* **77:** 127–147.

Balasubramanian A, Ponnuraj K. (2010) Crystal structure of the first plant urease from jack bean: 83 years of journey from its first crystal to molecular structure. *J. Mol. Biol.* **400:** 274–283.

J. H. Northrop and W. M. Stanley

shared the second half of the prize:

"For their preparation of enzymes and virus proteins in a pure form."

John Howard Northrop
July 5 1891–May 27 1987
(Photograph from Les Prix Nobel 1946)

Wendell Meredith Stanley
August 16 1904–June 15 1971
(Photograph from Les Prix Nobel 1946)

John Howard Northrop was born 1891 in Yonkers, New York. His father was a zoologist and instructor at Columbia University and his mother was a teacher of botany at Hunter College. His father died in a lab explosion two weeks before Northrop was born. He was educated at Columbia University and obtained his Ph.D. in chemistry there in 1915. After the war, he was recruited to the Rockefeller Institute, where he stayed until he became Professor of Bacteriology at University of California, Berkeley, and later Professor of Biophysics.

Early workers such as Emil Fischer and Eduard Buchner, but not Willstätter or William Maddock Bayliss, believed that enzymes were proteins. In 1896, Cornelis Pekelharing isolated a protein from gastric juice that he considered to be the enzyme pepsin. However, his suggestion was never accepted. With the success of Sumner, Northrop decided to work along similar lines. In 1930, he managed to crystallize pepsin. He also managed to crystallize a number of other enzymes and, in some cases, their precursors: Pepsin, pepsinogen, trypsin, trypsinogen, chymotrypsin, chymotrypsinogen, carboxypeptidase, ribonuclease, and hexokinase. However, proving the purity and protein nature of the enzyme remained an important issue. The solubility curves supported the presence of a single substance and the loss of activity of the enzymes was paralleled by the loss of protein during proteolysis. Denaturation or chemical modification of the enzymes also led to a loss of activity. New methods were developed for the analysis of proteins and enzymes. The ultracentrifuge could show the homogeneity of the samples, and so could electrophoresis. Additional strong evidence for the proteins being enzymes was the spontaneous activation of inactive zymogens or by proteolytic enzymes.

Through Northrop's work, it became clear that enzymes are proteins.

Selected References to Northrop's Work

*Northrop JH. (1930) Crystalline pepsin. I. Isolation and tests of purity. *J. Gen. Physiol.* **13:** 739–751.

Northrop JH. (1961) Biochemists, biologists and William of Occam. *Annu. Rev. Biochem.* **30:** 1–12.

Other References of Interest

McPherson A. (1991) A brief history of protein crystal growth. *J. Cryst. Growth* **111:** 1–10.

Fruton JS. (2002) A history of pepsin and related enzymes. *Q. Rev. Biol.* **77:** 127–147.

Wendell Meredith Stanley was born in 1904 in Ridgeville, Indiana. He first studied at Earlham College and continued at the University of Illinois, where he obtained his Ph.D. in 1929. After some time in Germany with Heinrich Wieland, he got a position in 1931 at the Rockefeller Institute in Princeton, where Northrop was already established. Stanley remained there until he became Professor of Biochemistry at the University of California, Berkeley, in 1948.

Stanley focused on the nature of small, infectious, non-bacterial agents that caused disease: Viruses. When Stanley began his research, it was not known whether viruses were inorganic, carbohydrate, lipid, protein, or some kind of living organism. Furthermore, at this time, it was not known that nucleic acids are the substance of inheritance. Proteins and enzymes, on the other hand, were very much in fashion, due to a large extent to the work of Sumner and Northrop. Initially, Stanley chose to work on tobacco mosaic virus (TMV), a very well-known infectious agent. In 1935, he managed to isolate and crystallize material that retained the viral multiplying activity. First of all, the high infectivity of the crystalline material suggested that the crystals were infections. Furthermore, treatment with a very large number of chemicals made it clear that the virus was something other than a bacterial organism, but it might be a protein. However, Vinson and Petre had already obtained similar results in 1929 and 1931 (as discussed by Pennazio & Roggero, 2000). Stanley could not initially identify any components in the preparations other than proteins. Certainly the virus was not a living organism, but since it could multiply, this led to the question, "What is life?" In 1936, Stanley started a collaboration with Eriksson-Quensel and Svedberg in Uppsala. They found that the virus was homogenous with electrophoresis experiments, using Tiselius's equipment. By sedimentation analysis in Svedberg's ultracentrifuge, a molecular weight of approximately 17 MDa was found. The virus was thought to be the largest protein so far identified. However, in 1936, FC Bawden and NW Pirie showed that TMV, as well as several other viruses, also contained nucleic acid, but Stanley maintained his view that the infectious agent was a very large and elongated protein (see the titles of the referred papers). In 1938, he agreed that TMV also contained nucleic acid. Bawden and Pirie also questioned the crystalline nature of TMV. They observed that the highly elongated virus could form fibers or gels, which was confirmed by Bernal and Fankuchen (1937). In fact, the complete virus has never been crystallized, but its structure has been determined from oriented gels.

Stanley, like Bawden and Pirie, also purified other viruses and managed to crystallize tomato bushy stunt virus (TBSV). The shapes of the isolated viruses were investigated by electron microscopy. It was found that TMV was fibrous while TBSV was globular.

In conclusion, Stanley was one of several scientists who, at about the same time, isolated and purified viruses, but several of his conclusions were wrong, and this became clear very soon after he published his findings. He won numerous prizes and became a very prominent scientist. Stanley reluctantly referred to Bawden and Pirie. However, their responses had a sarcastic and bitter tone (Pennazio & Roggero, 2000).

Selected References to Stanley's Work

Stanley W. (1935) Isolation of a crystalline protein possessing the properties of tobacco-mosaic virus. *Science* **81:** 644–645.

Stanley W. (1936) Chemical studies on the virus of tobacco mosaic. VI. The isolation from diseased Turkish tobacco plants of a crystalline protein possessing the properties of tobacco-mosaic virus. *Phytopathology* **26:** 305–320.

Loring HS, Stanley W. (1937) Isolation of crystalline tobacco mosaic virus protein from tomato plants. *J. Biol. Chem.* **117:** 733–754.

Loring HS, Stanley W. (1938) The properties of virus proteins. *Cold Spring Harb. Symp. Quant. Biol.* **6:** 341–360.

Ross AF, Stanley WM. (1939) The sulfur and phosphorus contents of tobacco mosaic virus. *J. Am. Chem. Soc.* **61:** 535–540.

Lauffer MA, Stanley WM. (1939) The physical chemistry of tobacco mosaic virus protein. *Chem. Rev.* **24:** 303–321.

Other References of Interest

Vinson CG, Petre A. (1929) Mosaic disease of tobacco. *Bot. Gaz.* **87:** 14–38.

Vinson CG, Petre A. (1929) Mosaic disease of tobacco. II. Activity of the virus precipitate by lead acetate. *Contr. Boyce Thompson Inst.* **3:** 131–145.

Pirie NW, Bawden FC, Bernal JD, Fankuchen I. (1936) Liquid crystalline substances from virus infected plants. *Nature* **138:** 1051–1052.

Bernal JD, Fankuchen I. (1937) Structure types of protein crystals from virus-infected plants. *Nature* **139:** 923–924.

Bawden FC, Pirie NW. (1937) The isolation and some properties of liquid crystalline substances from solanecous plants infected with three strains of tobacco mosaic virus. *Proc. R. Soc. Lond. B* **123:** 274–320.

Colvig R. (1972) Wendell M. Stanley, PhD. *Cancer* **29:** 541–542.

Pennazio S, Roggero P. (2000) The discovery of the chemical nature of tobacco mosaic virus. *Riv. Biol.* **93:** 253–282.

Nominations for the Nobel Prize, Reviews and Committee Work

Sumner was nominated for the Nobel Prize eight times between 1941 and 1946 and Northrop was nominated 17 times between 1940 and 1946. In 1940, Sumner nominated Northrop for the prize. It is interesting to note that, in 1946, except for Svedberg, Tiselius, and their Committee colleague H. von Euler-Chelpin, no one nominated Sumner and Northrop. Furthermore, they made the nomination at the last possible date. Norrby briefly mentions Sumner and Northrop, but Stanley is discussed in greater detail (2010, 2013). Stanley was nominated 32 times for a Nobel Prize in Physiology or Medicine between 1937 and 1946. His work was evaluated eight times, with a mixed outcome. He was also nominated for a Nobel Prize in Chemistry 13 times between 1938 and 1946. In 1939, Svedberg nominated him together with Bawden and Pirie.

Stanley was evaluated by Svedberg three times (in 1938, 1939, and 1945) and by Tiselius in 1946. The evaluation by Tiselius in 1946 regarded Stanley together with Sumner and Northrop. In May 1946, Tiselius's conclusion was that Sumner and Northrop could be awarded the prize, but he suggested in the case of Stanley that it would be advisable to wait and see. He was of the opinion that Stanley would be better evaluated together with Bawden and Pirie. In their final recommendation to the Academy in September 1946, the Nobel Committee for Chemistry discussed how to deal with the suggestion from Tiselius to wait with the award to Stanley. Recent publications had made Tiselius seem less hesitant. Furthermore, since Bawden and Pirie were not nominated in 1946, and because of Stanley's close association with Northrop in his institute and methodological association with both Sumner and Northrop, it seemed logical that the three should be awarded jointly. Neither the evaluators nor the Committee discussed the papers that argued that the crystals of TMV that Stanley claimed to have obtained were not real crystals. Likewise, Stanley's reluctance to agree to the presence of nucleic acid in his virus preparations was not discussed. It has also been commented by several persons (Friedman, 2001; Norrby, 2010) that Svedberg and Tiselius deviated from common practice by evaluating and strongly speaking for someone that they had collaborated with, a person who furthermore was working for an institution from which they received considerable economic support.

Consequences of the Research

These studies showed for the first time that enzymes are proteins and led to the understanding that purified viruses also contain nucleic acid. These findings were of enormous consequence in biochemistry and related disciplines. In addition, when Perutz and Kendrew (C1962) had demonstrated their method of determining the structure of proteins, the enzymes that were crystallized by Northrop and Sumner were among the first structures to be determined. This book is a witness to the notion that crystallization and structure determination are of fundamental importance, as the long list of Nobel Prizes in this area suggests.

Linus Carl Pauling (C1954)

"For his research into the nature of the chemical bond and its application to the elucidation of the structure of complex substances."

Linus Carl Pauling
February 28 1901–August 19 1994
(Photograph from Les Prix Nobel 1954)

Linus Pauling was born in 1901 in Portland, Oregon. Pauling first planned to become a chemist and, at high school, he conducted many chemistry experiments, using much of the equipment and material from an abandoned steel plant. By the fall of 1916, he entered Oregon State University. During the summer, he worked part-time at a grocery store, earning $8 per week, and he later worked as an apprentice machinist at a manufacturing plant with a salary of $40 per month. Pauling excelled at his job, and saw his salary soon rise to $50 per month. In his spare time, he set up a photography laboratory with two friends and obtained business from a local photography company. He hoped that the business would earn him enough money to pay for his future college expenses. In his first semester at Oregon State University, Pauling registered for two courses in chemistry, two in mathematics, mechanical drawing, introduction to mining and use of explosives, modern English prose, gymnastics, and military drill. After his second year, he planned to take a job in Portland to help support his mother, but he was offered a position teaching quantitative analysis, a course that he had just finished taking himself.

During his last two years at the university, Pauling became aware of the work of Gilbert N. Lewis and Irving Langmuir on the electronic structure of atoms and their bonding to form molecules. He decided to focus his research on how the physical and chemical properties of substances are related to the structure of the atoms of which they are composed, so as to become one of the founders of the new science of quantum chemistry. In 1922, Pauling graduated with a degree in chemical engineering and went on to graduate school at the California Institute of Technology (Caltech) in Pasadena, California. His graduate research involved the use of X-ray diffraction in order to determine the structure of crystals. He published seven papers on the crystal structure of minerals during this time. He received his Ph.D. in physical chemistry and mathematical physics in 1925.

Chemistry and Quantum Mechanics

In 1926, Pauling was awarded a Guggenheim Fellowship to work with three experts in the new field of quantum mechanics: Arnold Sommerfeldt in München, Niels Bohr in Copenhagen, and Erwin Schrödinger in Zürich. Pauling became interested in how quantum mechanics might be applied in his chosen field of interest: The electronic structure of atoms and molecules. In Zürich, Pauling was also exposed to one of the first quantum mechanical analyses of bonding in the hydrogen molecule by Walter Heitler and Fritz London. Pauling devoted the two years of his European trip to this work and decided to make it the focus of his future research. He became one of the first scientists in the field of quantum chemistry and a pioneer in the application of quantum theory to the structure of molecules.

In 1927, Pauling became assistant professor in theoretical chemistry at Caltech. During a very productive five years, he continued with X-ray crystal studies and also performed quantum mechanical calculations on atoms and molecules. He published approximately 50 papers during this time, and created the five rules now known as Pauling's rules. By 1929, he was promoted to associate professor, and by 1930 to full professor.

In 1931, the American Chemical Society awarded Pauling the Langmuir Prize for the most significant work in pure science by a person 30 years of age or younger. The following year, Pauling published what he regarded as his most important paper, in which he first laid out the concept of the hybridization of atomic orbitals and analyzed the tetravalency of the carbon atom. He developed a close friendship with the theoretical physicist Robert Oppenheimer, who spent part of his research and teaching time away from UC Berkeley at Caltech every year. The two men planned to mount a joint attack on the nature of the chemical bond: Oppenheimer would supply the mathematics and

Pauling would interpret the results. However, this relationship with Oppenheimer came to an end due to family reasons.

In the summer of 1930, Pauling made another European trip, during which he learned about the use of electron diffraction for crystal structure studies, similar to X-ray studies. He built an electron diffraction instrument at Caltech and used it to study the molecular structure of a large number of chemical substances.

Pauling introduced the concept of electronegativity in 1932. Using the various properties of molecules, such as the energy required to break bonds and the dipole moments of molecules, he established a scale and an associated numerical value for most of the elements — the Pauling Electronegativity Scale — which is useful in predicting the nature of bonds between atoms in molecules.

The Nature of the Chemical Bond

In the late 1920s, Pauling began publishing papers on the chemical bonds, leading to his famous textbook *The Nature of the Chemical Bond*, the first edition of which appeared in 1939. This is one of the most influential chemistry books to have ever been published. In the 30 years after publication of its first edition, the book was cited more than 16,000 times. Even today, many scientific papers cite this work, more than 70 years after the first publication. It is primarily due to work in this area that he received the Nobel Prize in Chemistry in 1954.

Part of Pauling's work on the nature of the chemical bond involved the relationship between ionic and covalent bonding. Pauling showed that these are merely extremes, between which most actual cases of bonding fall. Another topic that Pauling attacked was the structure of aromatic hydrocarbons, particularly the prototype, benzene. Pauling showed that the structure of benzene was a superposition of structures. The term "resonance" was later applied to this phenomenon.

Biological Substances and Structure of DNA

In the mid-1930s, Pauling decided to enter into new areas of chemistry. Although Pauling's early interest had focused almost exclusively on inorganic molecular structures, he had occasionally thought about molecules of biological importance, in part because of Caltech's growing strength in biology. His early work included studies of the structure of hemoglobin. He demonstrated that the hemoglobin molecule changes structure when it gains or loses an oxygen atom. As a result of this observation, he decided to conduct a more thorough study of protein structures in general, and so returned to his earlier use of X-ray diffraction analysis. In 1951, based on the structures of amino acids and peptides and the planar nature of the peptide bond, Pauling, Robert Corey, and Herman Branson correctly proposed the alpha helix and the beta sheet as primary structural motifs in the secondary structures of proteins. Pauling also proposed that deoxyribonucleic acid (DNA) was a triple helix. His model contained several basic mistakes, including a proposal of neutral phosphate groups, an idea that conflicted with the acidity of DNA. He organized the phosphate groups of the DNA strands facing into the helical core, with the nitrogenous bases facing out. Three such strands then intertwined in order to make one helical DNA molecule. Pauling had forgotten the negative charges of the oxygen atoms of the phosphate groups. Located in the center of the helix, these charges would repel one another, making it impossible for the molecule to hold together.

William Lawrence Bragg was disappointed that Pauling had won the race to identify several kinds of secondary structures of proteins. Furthermore, Pauling had a serious interest in how antibodies bound to their antigens. His conclusion was that there was no special kind of chemistry involved, but simply complementarity — the classic hand and glove analogy. This was also proposed as the way enzymes recognized their substrates. Bragg's team had made a fundamental error in their models of proteins by not recognizing the planar nature of the peptide bond. When it was learned at the Cavendish Laboratory that Pauling was working on the structure of DNA, James D. Watson and Francis Crick also started to work on a molecular model of DNA. They later benefitted from unpublished data from Maurice Wilkins and Rosalind Franklin at King's College, which showed evidence for a helix and planar base stacking along the helical axis. Early in 1953, Watson and Crick proposed a correct structure for the DNA double helix. Pauling later cited several reasons in order to explain how he had been incorrect about the structure of DNA, among them misleading density data and his lack of high-quality X-ray diffraction photographs. During the time Pauling was working on this problem, Franklin was creating the world's best DNA images, which were key to Watson's and Crick's success. Pauling did not see these images before devising his mistaken DNA structure, although his assistant Robert Corey did see at least some of them while taking Pauling's place at a protein conference in England in the summer of 1952. Pauling had been prevented from attending because his passport was withheld by the State Department on suspicion that he had communist sympathies. This led to the legend that Pauling missed out on discovering the structure of DNA because of the politics of the day (this was at the start of the McCarthy period in the United States). However, politics did not play a critical role. Not only did Corey see the images at the time, but Pauling himself regained his passport within a few weeks and toured English laboratories well before writing his DNA paper. He had ample opportunity to visit Franklin's lab and see her work, but he chose not to.

The Nobel Prize and the Impact of Pauling's Work on Chemistry

Pauling's scientific output is very extensive and covers broad areas of physical chemistry. Pauling was nominated for a Nobel Prize in chemistry every year from 1940, except for 1942, 1945, and 1947, until he was finally awarded the prize in 1954. The nominations varied from crystal chemistry, quantum mechanics, chemical bonding, molecular structure, and immunology to the structure of proteins and nucleic acids. In their reports to the Academy by the members of the Nobel Committee for Chemistry, Profs. Stig Claesson (1950) and Gunnar Hägg (1954) highlighted the high quality of Pauling's work, but no specific discovery was identified that would by itself justify a Nobel Prize. However, Pauling's achievements in the fields of chemical structure and bonding, which have had such a strong influence on the development of a large part of modern chemistry, were considered to be of such importance that a Nobel Prize in chemistry was strongly recommended. Hägg particularly emphasized Pauling's efforts at improving the understanding of protein structures.

It may be added that, according to the will of Alfred Nobel, the prize for chemistry shall be awarded to "the person who shall have made the most important chemical discovery or improvement" (during the preceding year). The award of a Nobel Prize to Pauling is just one example of the pragmatic application of the will that has become necessary with modern developments in science.

References

The number of crystal structures determined by Pauling and collaborators is very large. Only a few selected references regarding some general aspects of chemical structure and bonding are listed.

Pauling L. (1926) The dynamic model of the chemical bond and its application to the structure of benzene. *J. Am. Chem. Soc.* **48:** 1132–1143.

Pauling L. (1928) Influence of relative ionic sizes on the properties of ionic compounds. *J. Am. Chem. Soc.* **50:** 1036–1045.

Pauling L. (1930) Rotational motion of molecules in crystals. *Phys. Rev.* **36:** 430–443.

Pauling L. (1931) Quantum mechanics and the chemical bond. *Phys. Rev.* **37:** 1185–1186.

*Pauling L. (1931) The nature of the chemical bond. *J. Am. Chem. Soc.* **53:** 1367–1400. (This is the first in a series of seven papers on the same subject that were published during the years 1931–1933.)

Pauling L. (1935) The structure and entropy of ice and other crystals with some randomness of atomic arrangement. *J. Am. Chem. Soc.* **57:** 2680–2684.

Pauling L. (1939). *The Nature of the Chemical Bond and the Structure of Molecules and Crystals* (first edition). Cornell University Press, Ithaca.

Pauling L, Niemann C. (1939) Structure of proteins. *J. Am. Chem. Soc.* **61:** 1860–1867.

Pauling L. (1947) Atomic radii and interatomic distances in metals. *J. Am. Chem. Soc.* **69:** 542–553.

Pauling L, Itano HA, Singer SJ, Wells IC. (1949) Sickle cell anemia, a molecular disease. *Science* **110:** 543.

Pauling L, Corey RB, Branson HR. (1951) The structure of proteins: Two hydrogen-bonded helical configurations of the polypeptide chain. *Proc. Natl. Acad. Sci. USA* **37:** 205–211.

Pauling L, Corey RB. (1951) Configurations of polypeptide chains with favored orientations around single bonds: Two new pleated sheets. *Proc. Natl. Acad. Sci. USA* **37:** 729–740.

Francis Crick, James Watson, and Maurice Wilkins (M1962)

"For their discoveries concerning the molecular structure of nucleic acids and its significance for information transfer in living material."

Francis Harry Compton Crick
June 6 1916–July 28 2004
(Photograph from Les Prix Nobel 1962)

James Dewey Watson
Born April 6 1928
(Photograph from Les Prix Nobel 1962)

Maurice Wilkins
December 15 1916–October 5 2004
(Photograph from Les Prix Nobel 1962)

The life and exceptional success of Francis Crick, James Watson and Maurice Wilkins has been reported over and over again. Here, we will only provide brief biographies and a summary of their journey to the structure of DNA with references to more extensive work.

Francis Crick was born in 1916 in Northampton, England. He studied physics at University College London and started research for his Ph.D. in 1937. His studies were interrupted due to World War II. He did not leave his military connection until 1947. In 1949, he started to study biology in Cambridge where he joined the Medical Research Council (MRC) unit led by Max Perutz. The unit was housed in an extension of the Cavendish laboratory and William Lawrence Bragg was its head. The small laboratory was commonly called "The Hut." Crick started as a graduate student a second time in 1950 and initially worked on protein crystallography with Perutz and coworkers. He obtained his Ph.D. in 1954. During this time, he was part of an effort to develop a general theory for X-ray diffraction of a helix. This led to proposals for the structures of polyglycine II and collagen, but became particularly useful when, in 1951, he developed a friendship with James Watson, who had a strong interest in DNA. Their collaboration, described in detail in various publications, led to the single-page article (published in 1953) in which they described their model of DNA. Beyond the structure of DNA, Crick was also intensely involved in research into the genetic code and its role in the translation of proteins. The format of the code and the code itself were very interesting. By 1955, he had already suggested that some type of adaptor would read the message in the mRNA and, at the same time, carry the corresponding amino acid to the site of synthesis. He suggested that these adaptors could be small RNA molecules, which led to the physical discovery of tRNA (initially called sRNA).

Selected References to the Work by Crick

*Watson JD, Crick FHC. (1953) Molecular structure of nucleic acids. *Nature* **171:** 737–738.

Watson JD, Crick FHC. (1953) Genetic implications of the structure of deoxyribonucleic acid. *Nature* **171:** 964–967.

Crick FHC, Watson JD. (1954) The complementary structure of deoxyribonucleic acid. *Proc. R. Soc.* **223:** 80–96.

Crick FHC. (1958) The biological replication of macromolecules. On protein synthesis. *Symp. Soc. Exp. Biol.* **12:** 138–163.

Crick FHC. (1966) Codon-anticodon pairing: The wobble hypothesis. *J. Mol. Biol.* **19:** 548–555.

Crick FHC. (1968) The origin of the genetic code. *J. Mol. Biol.* **38:** 367–379.

Other References of Interest

Crick FHC. (1988) *What Mad Pursuit: A Personal View of Scientific Discovery*. Basic Books, New York.

Olby RC. (2008). *Francis Crick. Hunter of Life's Secrets*. Cold Spring Harbor Laboratory Press, New York.

James Watson was born in 1928 in Chicago. He started at the University of Chicago at the age of 15 and soon focused his interests on genetics. He obtained his Bachelor's degree in 1947 and moved on to become a graduate student at Indiana University. He obtained his Ph.D. in 1950, being supervised by Salvador Luria. During this time, he was exposed to the old view that proteins were genes that were able to reproduce themselves and that nucleic acids only had a structural role in supporting proteins. At the same time, he learnt about Avery's experiments suggesting that genes were located in DNA. Parts of Watson's postdoctoral experience took place in Copenhagen together with Kalkar and Maaløe. At a conference in Italy, Watson heard Maurice Wilkins talk about X-ray diffraction data from DNA, suggesting that a structure could be obtained. In the summer of 1951, Watson met John Kendrew, who arranged a postdoctoral position for him in Cambridge. The collaboration with Crick and the use of Franklin's and Wilkins's diffraction data led to the step-by-step elucidation of the structure of DNA.

In 1956, Watson accepted a position at Harvard University. Here, his focus was the transfer of the genetic information into proteins. In a review on protein synthesis via the ribosome (published in 1964), he prophetically stated: "Unfortunately, we cannot accurately describe at the chemical level how a molecule functions unless we know first its structure." Watson has published several highly appreciated textbooks, as well as more biographical publications. His book *The Double Helix* describes the work on the structure of DNA from his point of view.

Selected References to the Work by Watson

*Watson JD, Crick FHC. (1953). Molecular structure of nucleic acids. *Nature* **171:** 737–738.

Watson JD, Crick FHC. (1953) Genetic implications of the structure of deoxyribonucleic acid. *Nature* **171:** 964–967.

Crick FHC, Watson JD. (1954) The complementary structure of deoxyribonucleic acid. *Proc. R. Soc.* **223:** 80–96.

Watson JD. (1964) The synthesis of proteins upon ribosomes. *Bull. Soc Chim. Biol.* **46:** 1399–1425.

Another Reference of Interest

Watson JD. (1968) *The Double Helix. A Personal Account of the Discovery of the Structure of DNA.* Weidenfeld and Nicolson, London.

Maurice Wilkins was born in 1916 in New Zealand. His family moved back to England when Maurice was 6 years old. He received his basic education in Birmingham and his academic training at St. John's College in Cambridge. He obtained his Ph.D. with John Randall in Birmingham in 1940, and together they worked on the development of the microwave radar. In 1943, he moved to Berkeley to work with Robert Oppenheimer on the Manhattan project for the development of nuclear weapons. His experiences made him a strong opponent of nuclear weapons. He claimed that the Nazis lost the war due to the lack of brainpower they incurred due to wartime emigration.

After the war, he continued to work with Randall, initially at St. Andrew's University in Scotland. They switched fields from solid-state physics to the new field of biophysics in general and X-ray crystallography in particular. In 1946, Randall became a full professor at King's College London. Here, Wilkins was assigned to work on the structure of DNA. Their first successful diffraction images of aligned fibers of DNA were recorded in 1950. The Randall laboratory was sometimes called the "Randall's circus." This may explain the confusion that occurred when Rosalind Franklin started her work with Randall. Wilkins thought that she was a new assistant, while she had the impression that she was to take over the X-ray laboratory.

Selected Reference to the Work by Wilkins

*Wilkins MHF, Stokes AR, Wilson HR. (1953) Molecular structure of deoxypentose nucleic acids. *Nature* **171:** 738–740.

Another Reference of Interest

Wilkins M. (2003) *The Third Man and the Double Helix: The Autobiography of Maurice Wilkins.* Oxford University Press, Oxford.

Rosalind Elsie Franklin
July 25 1920–April 16 1958
(Photograph published with the permission of the
Cold Spring Harbor Laboratory Archives)

Rosalind Franklin was born in 1920 in Notting Hill, London. She never received a Nobel Prize, but her contributions were such that if she had been alive at the time of the awarding of the prize, her name would have been seriously discussed. Much has been written about her contributions, so we will briefly summarize her main contributions in determining the structure of DNA.

Franklin was a British biophysicist and crystallographer. She was educated at Newnham College, Cambridge, and after completing her basic education, she spent a year at Cold Spring Harbor Laboratory. After a wartime period of research on the porosity of coal (for usage in gas masks, etc), she learnt to perform X-ray crystallography on amorphous substances in Paris. In 1951, she started to work at King's College London at the MRC unit directed by Randall. She and her graduate student Raymond Gosling obtained the best diffraction patterns of DNA fibers at that time and discovered that there were two forms, which they called A and B, depending on the level of hydration. Franklin focused her attention to the A-form while Wilkins worked on the B-form. The B-form was thought to be helical, but Franklin was initially uncertain as to whether the A-form was also helical. Franklin was distinctly against models of DNA where the phosphates and sugars were placed in the center. In early 1953, she came very close to correctly determining the structure of DNA. She undoubtedly made major contributions to the Watson and Crick model of DNA. She left King's College in March 1953 for a position at Birkbeck College to study virus structure. Here, she drew remarkable conclusions on the structure of tobacco mosaic virus (1955) and seduced Aaron Klug into the field of virus structure.

Franklin died at the age of 38 due to cancer, possibly resulting from X-ray exposure.

Selected References to the Work by Franklin

*Franklin RE, Gosling RG. (1953) Molecular configuration in sodium thymonucleate. *Nature* **171:** 740–741.

Franklin RE. (1955) Structure of tobacco mosaic virus. *Nature* **175:** 379–382.

Other References of Interest

Sayre A. (1975) *Rosalind Franklin and DNA.* Norton, New York.

Maddox B. (2002) *Rosalind Franklin: The Dark Lady of DNA.* Harper Collins, London.

Steps on the Way to the DNA Structure

For a considerable time, it was thought that proteins performed the role of genes and DNA and RNA had only secondary roles (Reichard, 2002). However, starting with the experiments of Avery, MacLeod, and McCarty (1944) in the early 1950s, some scientists gradually accepted that DNA could be the matter of inheritance. Thus, DNA became interesting, and a race developed between a few groups aiming to determine its structure (Dickerson, 2005; Norrby, 2013). Astbury (1947) recorded an early fiber diffraction pattern, which showed a strong meridional reflection corresponding to a repeat distance of 3.4 Å along the axis of the fiber. Better diffraction images were subsequently recorded at King's College London, where Wilkins, Franklin, and her graduate student Gosling were working. However, no method was available for determining structures from fiber data.

A general consensus developed between the competing groups that DNA fibers were of a helical nature, as suggested by their X-ray pattern. If the structure was helical, how many chains were involved and were the bases or the phosphates and riboses on the outside? Furberg (1952), having determined the crystal structure of cytidine, suggested that DNA would be arranged like the available nucleoside structures, with the bases stacked with a 3.4 Å distance and with the sugar residues in a perpendicular orientation. He suggested a single-stranded helix with the bases stacked and with sugars and phosphates on the outside of the helix. Several proposals for triple-stranded helices, some with riboses and phosphates in the center, were suggested to explain the density of the fiber. One of these originated from Pauling and Corey (1953). The triple-stranded helix overlooked the fact that bound water would also be part of the helix and contribute to its density.

With the strong meridional reflection at 3.4 Å and one turn of the helix being observed to be 34 Å, there would be ten stacked bases for each full turn of the spiral ladder. In the model suggested by Watson and Crick (1953), the two chains are anti-parallel (Fig. 1). Thus, perpendicular to the helical axis are numerous twofold symmetry axes relating to the phosphates and sugar residues, but not the sequence of bases. One key point in the Watson–Crick structure is the base pairing. Here, two elements were of fundamental importance. One was the observation by Chargaff that the ratios between the nucleotides adenosine and thymine, as well as between cytosine and guanine, are close to one. This would fit with a base paring of adenosine to thymine and cytosine to guanine (Fig. 2). Jerry Donohue provided a critical contribution in making it clear that the pairing between the bases was not due to the enol tautomers, but rather to the keto forms. An essential and verifying outcome of this structure is that it explains how replication and transcription can take place. This also convinced the scientific world that the model was correct and that DNA, rather than proteins, is the genetic material.

The Nobel Prize

Norrby (2013) has extensively covered the work behind the Nobel Prize awarded for the structure of DNA. Only a brief summary seems relevant here. Even though the first nominations for a prize for the structure of DNA did not appear until 1960, the Chemistry Committee had already commented on DNA structure in 1957. In 1962, Watson and Crick received several nominations, both for a Nobel Prize in Chemistry and in Physiology. Wilkins was nominated once for a Nobel Prize in Chemistry and twice for a Nobel Prize in Physiology. Franklin was never nominated for a Nobel Price. Norrby (2013) has carried out an extensive analysis of whether Franklin could have been included in the list of prize recipients if she had been alive and whether Wilkins should be considered to be a fully worthy laureate, not having contributed to the main insights into the structure of DNA. In October 1962, the Royal Academy, having the possibility of choosing the winner of the prize in chemistry before the Karolinska Institute, was cautious and made a comment that the structure presented by Watson and Crick (1953) was only a hypothesis. It fitted with the information available, but it could not be proven that it was the only possible interpretation. The Nobel Prize in Chemistry for 1962 was instead given to Perutz and Kendrew (see the next chapter). Therefore, the work towards elucidating the structure of DNA was awarded the Nobel Prize in Medicine.

Consequences

The Watson–Crick model (1953) was immediately accepted as a very attractive model for the structure of DNA that also suggested how the copying mechanism would occur.

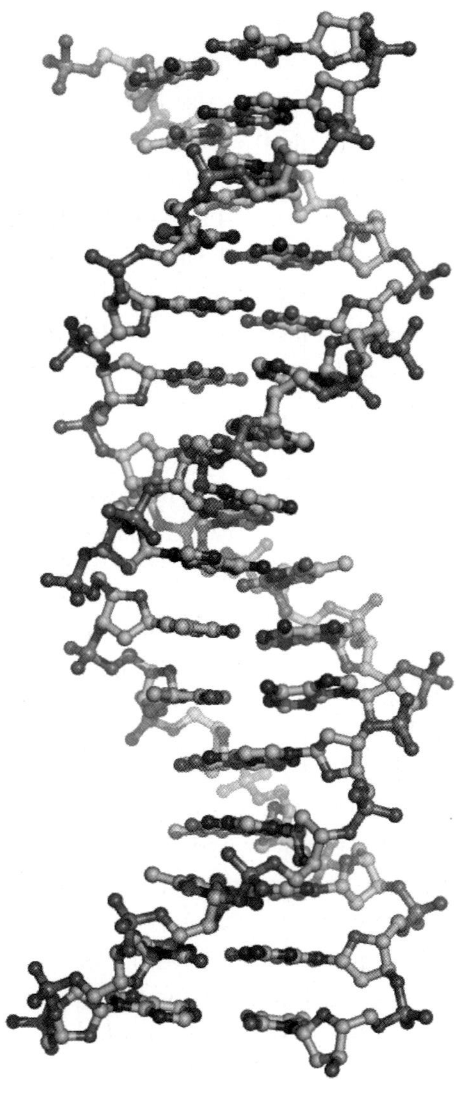

Fig. 1. The structure of B-DNA. The two strands run in opposite directions and are held together by the hydrogen bonds of the base pairs (seen as steps of a spiral staircase). (Illustration kindly provided by M. Kjeldgaard.)

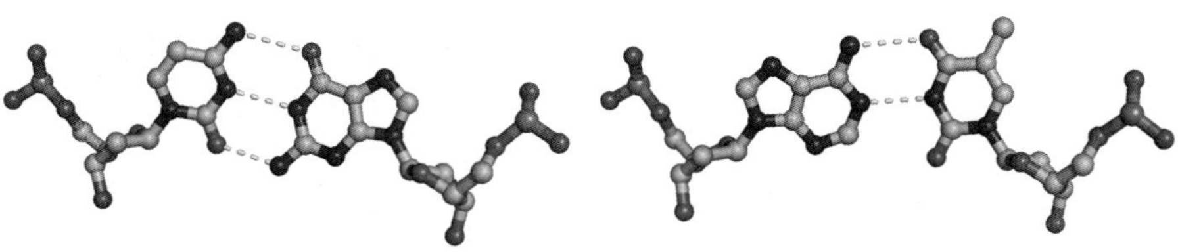

Fig. 2. The Watson–Crick base pairs. Left: C (cytosine)—G (guanine); right: A (adenine)—T (thymine). (Illustration kindly provided by M. Kjeldgaard.)

The model is probably the single biological finding with the richest consequences over the twentieth century. Determining the structure of DNA can also be said to be the beginning of structural biology, in which finding the structures of biological molecules has led to a new understanding of their function.

However, some questions remained concerning the structure of DNA. Among other things, how could the helical structure be unwound during replication or transcription? A side-by-side model avoiding this problem was proposed that made use of the same type of base pairing, but where the right-handed helix was turned into a left-handed helix after about half a full turn (Rodley *et al.*, 1976). It was not until 1981 (Drew *et al.*) that a crystal structure of a DNA dodecamer showed that the original model was correct in all of the essential details.

Relevant References

Astbury WT. (1947) X-ray studies of nucleic acids. *Symp. Soc. Exp. Biol.* **1:** 66–76.

Avery GT, MacLeod CM, McCarty M. (1944) Studies on the chemical nature of the substance inducing transformation of pneumococcal types. Induction of transformation by a desoxyribonucleic acid fraction isolated from *Pneumococcus* type III. *J. Exp. Med.* **79:** 137–158.

Chargaff E. (1950) Chemical specificity of nucleic acids and mechanism of their enzymatic degradation. *Experientia* **6:** 201–209.

Drew HR, Wing RM, Takano T, *et al.* (1981) Structure of a B-DNA dodecamer: Conformation and dynamics. *Proc. Natl. Acad. Sci. USA* **78:** 2179–2183.

Furberg S. (1952) On the structure of nucleic acids. *Acta Chem. Scand.* **6:** 634–640.

Reichard P. (2002) Osvald T. Avery and the Nobel Prize in medicine. *J. Biol. Chem.* **277:** 13355–13362.

Rodley GA, Scobie RS, Bates RHT, Lewitt RM. (1976) A possible conformation for double-stranded polynucleotides. *Proc. Natl. Acad. Sci. USA* **73:** 2959–2963.

Max Ferdinand Perutz and John Cowdery Kendrew (C1962)

"For their studies of the structures of globular proteins."

Max Ferdinand Perutz
May 19 1914–February 6 2002
(Photograph from Les Prix Nobel 1962)

John Cowdery Kendrew
March 24 1917–August 23 1997
(Photograph from Les Prix Nobel 1962)

Max Perutz was born in 1914 in Vienna, Austria, where he studied chemistry and obtained his first university degree. In her biography, Georgina Ferry (2008) manages to portray the person, his life, and his work in an accurate way. Early in his career, Perutz was very interested in the biological world. In 1936, he joined John Bernal's group at the Cavendish laboratory in Cambridge, where he started to learn crystallography (Brown, 2005). In searching for material for his research, he was recommended to begin working on hemoglobin, as crystals were available at a different institute in Cambridge. Perutz managed to obtain the first diffraction photographs of hemoglobin and could determine the cell parameters and space group. In 1937, Bernal moved to London, where he became a professor at Birkbeck College. At the same time, William Lawrence Bragg succeeded Ernest Rutherford as professor at the Cavendish laboratory. Perutz decided to stay in Cambridge. Bragg was enthusiastic about crystallographic work on biological specimens and obtained a grant for Perutz from the Rockefeller Foundation. However, Perutz work was interrupted by World War II. First, he was interned in Canada as an "enemy alien" and was later involved in wartime research. After the war, he returned to Cambridge and to the hemoglobin project.

In order to determine the structure of hemoglobin, a number of approaches were attempted in order to calculate the phase angles (Perutz, 1997). Among these were attempts to identify arrangements of helices by Patterson function calculations and the swelling and shrinkage of the crystals. Success gradually came when the possibility of isomorphous replacement was adopted. This method had previously been successfully used on organic molecules, but it was a novel approach for the phase angle determination of protein crystals. Using a single mercury compound attached to a cysteine molecule, the signs of the centrosymmetric h0l reflections could be determined. However, general reflections required two or more derivatives. After considerable efforts, this approach was successful (Green *et al.*, 1954). W. L. Bragg was deeply engaged in all of this work and was a coauthor with Perutz on several publications (Perutz, 1997)

During the 1990s, Perutz developed an interest in a different protein, huntingtin, the source of Huntington's disease (Perutz, 1995; 1997). He was studying an unusual octameric hemoglobin with a repeated sequence of Glu–Glu–His–Lys. This repeat sequence formed a zipper in the form of beta-strands. In searching for other repeated sequences that could form similar zippers, Perutz found huntingtin. This protein does not cause any disease so long as the stretch of glutamines at the amino terminal region does not exceed 35 in number. However, Huntington's disease becomes increasingly severe and starts earlier as the number of glutamines increases. Despite having entered this new field at around 80 years of age, Perutz became one of its leading scientists.

During his life, Perutz was also an active author of well-written and interesting books on science or related topics. A list of these books can be found below. One of his important duties was to be in charge of the Medical Research Council (MRC) unit, initially located in The Hut of the Cavendish laboratory, where W. L. Bragg was the leader, and subsequently in the MRC Laboratory of Molecular Biology in Cambridge. This laboratory has been a leading center of developments in molecular biology and not least in structural work. The remarkable success of the laboratory, with numerous Nobel Prizes, is, to a significant extent. due to Perutz's leadership. As one member of the laboratory expressed it, one essential factor was "the unique environment of intellectual and technological sophistication." One important factor was the vivid exchange of ideas between scientists with different competencies in the laboratory canteen, which, for a long time, was run by Perutz's wife. Many have said: "The Nobel Prizes are born in the canteen!"

Selected References to the Work by Perutz

*Perutz MF, Rossmann MG, Cullis AF *et al.* (1960) Structure of hemoglobin. A three-dimensional fourier synthesis at 5.5-Å resolution, obtained by X-ray analysis. *Nature* **185:** 416–422.

Green DW, Ingram M, Perutz MF. (1954) The structure of haemoglobin: IV. Sign determination by the isomorphous replacement method. *Proc. R. Soc. Lond. A* **225:** 287-307.
Perutz MF. (1995) Glutamine repeats as polar zippers: Their role in inherited neurodegenerative disease. *Mol. Med.* **1:** 718–721.

Books by Perutz

Perutz MF. (1962) *Proteins and Nucleic Acids: Structure and Function.* Elsevier, The Netherlands.
Perutz MF. (1989) *Is Science Necessary? Essays on Science and Scientists.* Barrie and Jenkins, London.
Perutz MF. (1990) *Mechanisms of Cooperativity and Allosteric Regulation in Proteins.* Cambridge University Press, Cambridge.
Perutz MF. (1992) *Protein Structure: New Approaches to Disease and Therapy.* W.H. Freeman, New Jersey.
Perutz MF. (1997) *Science is Not a Quiet Life: Unravelling the Atomic Mechanism of Haemoglobin.* World Scientific Publishing, New Jersey, London, Singapore.
Perutz MF. (2002) *I Wish I'd Made You Angry Earlier.* Cold Spring Harbor Laboratory Press, New York.
Perutz MF. (2009) *What a Time I Am Having: Selected Letters of Max Perutz edited by Vivien Perutz.* Cold Spring Harbor Laboratory Press, New York.

Other References of Interest

Ferry G. (2008) *Max Perutz and the Secret of Life.* Cold Spring Harbor Laboratory Press, New York.
Strandberg B, Dickerson RE, Rossmann MG. (2009) 50 years of protein structure analysis. *J. Mol. Biol.* **392:** 2–32.

John Kendrew was born in Oxford in 1917. His father was a reader in climatology at the University of Oxford and his mother was an art historian. Kendrew was educated at Clifton College in Bristol. He continued at Trinity College, Cambridge, and graduated in chemistry in 1939. During World War II, he worked on radar, but with a growing interest in biology inspired by Bernal and Pauling. In 1946, he joined Perutz at the Cavendish laboratory under the leadership of William Lawrence Bragg. Myoglobin, a protein that stores oxygen in muscle cells, was identified as a suitable project, and in 1957, Kendrew's investigations of this protein led to a structure of myoglobin at low resolution (Kendrew *et al.*, 1958). This was the first three-dimensional structure of any protein. In 1959, the structure was solved at a 2 Å resolution (Kendrew *et al.*, 1960; Strandberg, Dickerson, & Rossmann, 2009). In 1959, he also became the founder and the first editor of the *Journal of Molecular Biology*, a scientific journal that has covered much of the rapidly growing activity in molecular biology and structural studies of macromolecules from an early stage. Fluent in German, French, and Italian, he became one of the founders of the European Molecular Biology Organization (EMBO) in 1963, and in 1974, he became the founding director of the newly established European Molecular Biology Laboratory in Heidelberg. He remained there for about 20 years.

Selected References to the Work by Kendrew

Kendrew JC, Bodo G, Dintzis HM *et al.* (1958). A three-dimensional model of the myoglobin molecule obtained by X-ray analysis. *Nature* **181:** 662–666.

*Kendrew JC, Dickerson RE, Strandberg BE *et al.* (1960). Structure of myoglobin: A three-dimensional Fourier synthesis at 2Å resolution. *Nature* **185:** 422–427.

Another Reference of Interest

Strandberg B, Dickerson RE, Rossmann MG. (2009) 50 years of protein structure analysis *J. Mol. Biol.* **392:** 2–32.

The Nobel Prize

Perutz was nominated for the Nobel Prize for Chemistry in 1951 and was nominated with John Kendrew in 1961 and 1962 (Norrby, 2013; Table 7.1). In 1959–1961, Perutz was also nominated together with Kendrew and Dorothy Hodgkin. Professor Pauling nominated Robert Corey alone or together with Perutz and Kendrew in 1961 and 1962. Perutz and Kendrew were also nominated for the Nobel Prize in Physiology or Medicine from 1960 to 1962.

Professor A. Westgren reviewed the work of Kendrew and Perutz in 1959 and Prof. G. Hägg made a review of Corey, Kendrew, and Perutz in 1962. Hägg describes Corey's work in Pauling's laboratory that attempted to understand the structure of proteins from studying crystal structures of amino acids or short peptides. Among other results, they deduced the structure of the alpha-helix, which subsequently was found to be a dominant motif in the structure of myoglobin. Hägg concluded that the work by Corey, partly together with Pauling, did not merit a Nobel Prize. However, in the case of Kendrew and Perutz, Hägg was enthusiastic, in particular with regard to their work on myoglobin at a 2 Å resolution, the first protein structure in which atomic details are seen. Here, although further details became available when the resolution was improved to 1.5 Å, everything that was observed agreed with details known from other related studies. At this time, Perutz's work on hemoglobin had generated a structure at a 5.5 Å resolution. Even though fewer details were known, the similar structure of the four polypeptide chains and their similarity with the folding of myoglobin proved that the structure must have been correct. Hägg did not consider the development of the phase determination method — multiple isomorphous replacement — to be novel, but he was impressed with Perutz's efforts and successes in labeling the proteins with heavy atoms for the phase angle determination.

In the recommendation by the Nobel Committee for Chemistry to the Academy, the work by Watson and Crick was compared with that of Kendrew and Perutz. The latter had the advantage of a richer experimental set of data, but the richness of consequences of the former for a wide range of scientific problems gives them a distinctive advantage. The optimal choice would have been to reward both groups in the same year. The Committee also reiterated the statement that no new methodological developments had been made for the structure determinations by Kendrew and Perutz. This is contrary to the common belief that the award was both for the development of multiple isomorphous replacement and the structure determinations. The Committee noted with pleasure that a prize to Kendrew and Perutz would occur 50 years after the dramatic development of crystallography by Laue and W. H. and W. L. Bragg. The Committee also expected new important results along the same lines that could be proposed for awards.

Consequences

The consequences of the work by Perutz and Kendrew, as expected by the Nobel Committee for Chemistry, led to the dramatic development of a field that could be called "structural biology." At the time of the award, crystallographers were already working on a range of other proteins. It is interesting to note that many of the second series of protein structures made use of crystallizations developed by Sumner and Northrop (C1946). This type of work has increased in speed due to numerous technical developments in biochemistry, crystallization methods, and data collection using synchrotrons as well as computational methods. The crucial phase angle determination methods have been extended by two new avenues: Molecular replacement (Rossmann, 1962) and multiple wavelength anomalous dispersion (Hendrickson, 1991). Taken together, all biochemical disciplines now regard structural analysis as a highly essential element in research. The old saying of "seeing is believing" has become highly evident.

References

Hendrickson WA. (1991) Determination of macromolecular structures from anomalous diffraction of synchrotron radiation. *Science* **254:** 51–58.

Rossmann MG, Blow DM. (1962) The detection of sub-units within the crystallographic asymmetric unit. *Acta Cryst.* **15:** 24–31.

Dorothy Crowfoot Hodgkin (C1964)

"For her determinations by X-ray techniques of the structures of important biochemical substances."

Dorothy Crowfoot Hodgkin
May 12 1910–July 29 1994
(Photograph from Les Prix Nobel 1964)

Dorothy Mary Crowfoot was born in 1910 in Cairo, Egypt. Her father was working in the Egyptian Education Service. Subsequently, he moved to Sudan, where he became Director of both Education and Antiquities. After his retirement, he spent much of his time conducting archeology in Jerusalem. Dorothy's mother was constantly engaged in the various activities of her husband. Dorothy studied at the Sir John Leman School in Beccles from 1921 to 1928. At the age of 10, she was already interested in chemistry and crystals, and that interest grew during her school years. In 1928–1932, she studied at Oxford University in Somerville College, and some of her courses covered crystallography.

During these years, she met the legendary crystallographer John Bernal. This led to two happy years in Cambridge working with him as her supervisor in the determination of the crystal structures of sterols and other biologically interesting molecules. Jointly, they studied more than a hundred steroids. During this period, she conducted a pioneering study of the first protein crystal put into an X-ray beam. In 1934, one of

Bernal's friends, Glenn Millikan, made a brief visit to the laboratory of The Svedberg in Uppsala (Brown, 2005). Here, among others, he met an English scientist, John Philpot, who was learning ultracentrifuge techniques. The enzyme pepsin that he was working on happened by accident to crystallize. Millikan thought of Bernal and obtained a test tube with large pepsin crystals in order to bring it to Cambridge. These became the first protein crystals to be examined by X-rays. Northrop (1930) had initially crystallized pepsin. Two important lessons were learnt from these experiments: Protein crystals can diffract and their structures must therefore be possible to determine, but they have to be mounted in their mother liquor. In 1937, Crowfoot obtained her Ph.D. in Cambridge.

In 1934, she returned to Oxford and Somerville College where she established her laboratory and continued to work for the rest of her life. Her laboratory was in the Ruskin Science Museum and in the gothic style, but lacked fully adequate rooms. In 1936, she became a Tutor and Fellow of the College. The work she had started on sterols with Bernal was continued. In 1942, she initiated her work on penicillin, and in 1948, she begun work on vitamin B_{12}. Her work on the hormone insulin had started by 1934 (Crowfoot, 1935) and its structure was determined in 1968. For a while, one of the research students in her group was Margaret Roberts, later Margaret Thatcher, who became the Prime Minister of Great Britain.

Dorothy Crowfoot Hodgkin was a pioneer in using Patterson methods and she also used isomorphous replacement in her work on penicillin. Here, she made use of the differences between sodium, potassium, and the rubidium derivatives of benzylpenicillin, and showed that penicillin has a four-membered ring. This met with strong opposition, since a beta-lactam structure was supposed to be too unstable to be found in penicillin, which is a fairly stable molecule. The formula derived from her work was the starting point for chemical modifications leading to new antibiotics.

The work on cobalamin or vitamin B_{12} was a really challenging task. A deficiency of vitamin B_{12} leads to pernicious anemia, which can be fatal. The vitamin contains cobalt and the crystals have a beautiful red color. The cobalt atom was used as a heavy atom in the crystallographic analysis. This was the first organometallic compound to be analyzed. Cobalamin has a corrin ring that is similar to porphyrin, but two of the pyrrole rings are linked directly. The cobalt atom is bound in the center and linked to the pyrrole nitrogens, a cyano group and a 5'-deoxyadenosyl group. Bragg (1962) described the vitamin B_{12} structure as "breaking the sound barrier," with a new level of complexity.

The work on insulin also has an interesting history. Insulin was the first protein for which the amino acid sequence was determined from work done by Fred Sanger (C1958). The structure was reported at a 2.8 Å resolution in 1969. In parallel, the insulin structure was analyzed in China and there were valuable exchanges between the two groups involved.

Dorothy Hodgkin's qualities extended in many directions. She was working in many ways to ease the tensions between East and West in the Pugwash organization. Max Perutz described her as "a great chemist, a saintly, gentle and tolerant lover of people, and a devoted protagonist of peace."

Selected References to the Work by Hodgkin

Bernal JD, Crowfoot DM. (1934) X-ray photographs of crystalline pepsin. *Nature* **133:** 794–795.

*Brink C, Hodgkin DC, Lindsey J *et al.* (1954) The X-ray crystallographic evidence on the structure of vitamin B_{12}. *Nature* **174:** 1169–1171.

Crowfoot D, Bunn CW, Rogers-Low BW, Turner-Jones A. (1949) X-ray crystallographic investigation of the structure of penicillin. In: Clarke HT, Johnson JR, Robinson R (eds), *Chemistry of Penicillin.* Princeton University Press, New Jersey, pp. 310–367.

Adams MJ, Blundell TL, Dodson EJ *et al.* (1969) Structure of rhombohedral 2-zinc insulin crystals. *Nature* **224:** 491–495.

Other References of Interest

Bragg WL. (1962) The growing power of X-ray analysis. In: Ewald P (ed.), *Fifty years of X-ray Diffraction.* Oosthoek, Utrecht, pp. 120–135.

Glusker JP. (1994) Dorothy Crowfoot Hodgkin (1910–1994). *Protein Sci.* **3:** 2465–2469.

Johnson LN, Phillips D. (1994) Professor Dorothy Hodgkin, OM, FRS. *Nat. Struct. Biol.* **1:** 573–576.

Ferry G. (1998) *Dorothy Hodgkin. A life.* Granta Books, London.

Dodson GG. (2002) Dorothy Mary Crowfoot Hodgkin, OM, 12 May 1910–29 July 1994. *Biographical Memoirs of Fellows of the Royal Society* **48:** 179–219.

Consequences

The work by Dorothy Crawfoot Hodgkin has had very significant practical consequences. The correct formula of penicillin with a beta-lactam ring, as determined by Hodgkin, was vital as a starting point for the further development of new antibiotics, which were particularly needed during World War II. The structure of vitamin B_{12}, with its surprising deviations from what was expected, was also necessary for the subsequent synthesis of the vitamin.

William Nunn Lipscomb (C1976)

"For his studies on the structure of boranes illuminating problems of chemical bonding."

William Nunn Lipscomb
December 9 1919–April 14 2011
(Photograph taken from http://wlipscomb.tripod.com)

William Lipscomb was born in 1919 in Cleveland, Ohio. The family moved to Kentucky the following year. At age 11 he received a chemistry set as a birthday gift and became obsessed with science. He later recalled creating "evil smells" using hydrogen sulfide to drive his two sisters out of his room and nearly blowing up the house while concocting gunpowder for homemade fireworks. His father was a prosperous physician who lost his career and livelihood when young William's sister Helen contracted polio at age 17. No one would take their child to a doctor from a house infected by polio.

Lipscomb lived in Lexington through his undergraduate university years. Because of the family's reduced circumstances, he attended the inexpensive University of Kentucky on a clarinet scholarship, but graduated with a degree in chemistry in 1941. After his Bachelor's degree, he hitchhiked to Los Angeles to enroll in graduate school in physics at Caltech, but, stimulated by Linus Pauling, he switched to physical chemistry after a year. From then until the end of 1945, he was involved in research and development related to the war, working for the War Department's Office of Scientific

Research and Development during the day and on his doctoral research at night. After completing his Ph.D., he joined the faculty of the University of Minnesota in 1946, and moved to Harvard University in 1959.

A skilled clarinetist who often played in chamber music groups, Lipscomb was also a tennis enthusiast and practical joker. At mealtimes, he thought it funny to steal butter off other people's butter knives and was known to remove the fruit from walnuts and glue the shells back together before offering them to guests. In his later years, Lipscomb was a regular participant in the annual Ig Nobel Prize awards in Cambridge, sponsored by the *Annals of Improbable Research*. "We have a bunch of Nobel Prize winners who hand out our prizes," said Marc Abrahams, the founder of the Ig Nobel Prizes. "Bill ended up as our grand star. He played clarinet at the beginning and end of each show, and he narrated funny videos we post on our website." Many interesting episodes from his life and examples of his sense of humor may be found at wlipscomb.tripod.com.

Boron Chemistry

When Lipscomb began his studies in the 1940s, boron was considered to be a relatively uninteresting atom. Although right next to carbon on the periodic table of the elements, researchers knew of few natural compounds containing it and showed little interest in it. Lipscomb was puzzled by a lecture of his mentor, Linus Pauling, in which he proposed a theory of chemical bonding for boron that did not seem correct. Lipscomb set out to prove him wrong, but getting support for his boron research was not so easy. The Mansfield Amendment of 1969 stopped the military from funding research with no acknowledged military application. Other government agencies would not fund Lipscomb's boron research. The Office of Naval Research wrote to him suggesting that he withdraw his request for a continuation of funding, because the inevitable refusal would reflect badly on him. Lipscomb wrote back saying, no, he would not, and that their funding refusal would reflect badly not on him but on them. After he won the Nobel Prize in Chemistry for his boron work, a year or so of additional funding was made available.

Boron belongs to a family of "electron-deficient" atoms; that is, these atoms do not have enough free electrons to form the same types of bonds as carbon. However, no one could deduce how they actually bonded. Moreover, most boron-containing compounds, such as the boranes, are highly unstable, or even explosive at room temperatures. Lipscomb pioneered the use of X-ray crystallography for studying the structures of boranes at low temperatures. In 1951, he made his first key discovery. In the conventional mode of bonding, such as in hydrocarbons, each hydrogen atom shares two electrons with the carbon atom it is joined to. In boranes, however, two boron atoms are linked together by a hydrogen atom sitting between them, with the three atoms sharing two electrons. The structures, he wrote, "looked like a geometer's dream." Three boron atoms could be linked in the same fashion.

He and his colleagues went on to elucidate the chemistry of boron compounds using sophisticated quantum mechanical calculations to describe the structures of the molecules and to predict how they would react chemically. His work led to the synthesis of a variety of boron-containing compounds that have myriad uses, especially in the synthesis of complicated organic molecules that do not contain it. Boron compounds are also used for reinforcing lightweight structural materials. The chemistry of the boranes and their relatives is summarized in Lipscomb's Nobel Lecture, which was given on December 11 1976.

Biological Structures

In later years, Lipscomb worked on a variety of biochemical problems, particularly the determination of the structures of enzymes in order to reveal their modes of functioning. The first enzyme he analyzed was carboxypeptidase, one of the first enzyme structures to be determined and the biggest structure at that time (Lipscomb *et al.*, 1968). In this work, a large conformational change of a tyrosyl residue was observed in the enzyme at the binding of substrate analogues, which became the first clear example of the "induced fit" hypothesis by Daniel Koshland. This also led to a lengthy discussion of its possible role in the catalytic process (Lipscomb, 1973; Gardell *et al.*, 1985; Cho *et al.*, 2001). Among many other interesting structures that Lipscomb analyzed was a more complicated enzyme, aspartate transcarbamoyltransferase (ACTase), which is involved in the synthesis of pyrimidines. The enzyme is composed of six catalytic and six regulatory subunits. The enzyme has two main states: a T (tight) state with low activity and an R (relaxed) state with high activity. The shift between the two states depends on the binding of substrate to the catalytic subunits or nucleotides to the allosteric site of the regulatory subunits. Lipscomb's group analyzed the crystal structures of both states and were able to describe the regulation of the enzyme's activity (Honzatko *et al.*, 1983).

Selected Publications by Lipscomb

Honzatko RB, Crawford JL, Monaco HL *et al.* (1983) Crystal and molecular structures of native and CTP-liganded aspartate carbamoyltransferase from *Escherichia coli*. *J. Mol. Biol.* **160:** 219–263.

Lipscomb WN. (1963) *Boron Hydrides*. W. A. Benjamin: New York.

Lipscomb WN. (1964) Geometrical theory of boron hydrides. *Inorg. Chem.* **3:** 1683–1685.

Lipscomb WN. (1972) Bonding in boron hydrides. *Pure Applied Chem.* **29:** 493–511.

Lipscomb WN. (1973) Enzymatic activities of carboxypeptidase A's in solution and in crystals. *Proc. Natl. Acad. Sci. USA* **70:** 3797–3801.

*Lipscomb WN. (1977) The boranes and their relatives. *Les Prix Nobel*, pp. 224–245.

Lipscomb WN, Hartsuck JA, Reeke GN, Jr *et al.* (1968) The structure of carboxypeptidase A. VII. The 2.0-Angstrom resolution studies of the enzyme and of its complex with glycyltyrosine, and mechanistic deductions. *Brookhaven Symp. Biol.* **21:** 24–90.

Other References of Interest

Cho JH, Kim DH, Lee KJ *et al.* (2001) The role of Tyr248 probed by mutant bovine carboxypeptidase A: Insight into the catalytic mechanism of carboxypeptidase A. *Biochemistry* **40:** 10197–10203.

Gardell SJ, Craik CS, Hilvert D *et al.* (1985) Site-directed mutagenesis shows that tyrosine 248 of carboxypeptidase A does not play a crucial role in catalysis. *Nature* **317:** 551–555.

http://wlipscomb.tripod.com — Webpage for WN Lipscomb with very rich content.

Aaron Klug (C1982)

"For his development of crystallographic electron microscopy and his structural elucidation of biologically important nucleic acid-protein complexes."

Aaron Klug
Born August 11 1926
(Photograph from Les Prix Nobel 1982)

Aaron Klug was born in 1926 in Lithuania, but grew up in South Africa from the age of two. He received his academic training at the University of Witwatersrand, where he studied physiological chemistry (biochemistry), chemistry, physics, and mathematics. He continued his education in Cape Town and learnt crystallography from Prof. R. W. James, who had brought his knowledge from the Bragg school in Manchester. In 1949, he moved to Cambridge, where he obtained his Ph.D.

Klug's interest in biological materials grew and he obtained a postdoctoral fellowship in 1953 to work in J. D. Bernal's department at Birkbeck College (Brown, 2005). Here, he started to work with Rosalind Franklin on tobacco mosaic virus (TMV). He continued this work after Franklin's unfortunate death in 1958. In 1962, Klug moved with his collaborators to the newly built Medical Research Council laboratory for Molecular Biology, where Perutz was in charge. TMV did not crystallize, as Wendell Stanley once had thought it would, but rather formed gels, where the long axis of the virus was easily aligned, but where each particle was differently rotated. One difficulty was the uneven number of protein subunits of each turn, being at 16⅓. Attempts to crystallize the viral protein without RNA led to crystals containing two rings with 17 viral protein subunits in each ring. Gradually, the understanding grew of how the viral RNA bound and the viral protein aggregated in the helical form.

When Klug moved back to Cambridge, he started working with electron microscopy (EM), and spherical viruses also came into focus. A few X-ray data were available, but EM pictures gave a direct image and further insights. Here, Klug collaborated with Don Caspar (1960). They found that the spherical viruses have a common icosahedral arrangement of their protein shells. The general arrangement suggested that there was some underlying principle to this arrangement. This led to the formulation of the quasi-equivalence description of this structural arrangement. An icosahedron has 20 triangular surfaces, which meet at 12 fivefold symmetry axes. A virus particle with only 60 identical subunits could form a perfect icosahedron with three subunits on each triangular surface. With the different sizes and different numbers of subunits of the spherical viruses, Klug and Caspar described the way by which protein subunits could be arranged in an almost perfect icosahedral arrangement: The quasi-equivalence principle. The number of subunits that could be arranged in this way follows the formula:

$$T = h^2 + k^2 + hk$$

where h and k are real numbers and T multiplied by 60 gives the possible number of identical protein subunits of the viral shell.

The usage of EM was not as simple and direct as Klug had hoped for. Thus, he and his group developed a range of methods to extract more information from the images. A distinct problem with the negative stain was that heavy-metal salt formed a cast where both the front and back sides could be seen. For spherical viruses, the symmetry was known and models of the particles could be projected and compared with the real data. Furthermore, tilting both the specimens and the models by specific angles could validate the initial observations. In the case of helical structures, the particles were flattened, but the images contained both a front side and a back side. Through an analysis using optical diffraction, it was possible to sort out the two patterns and generate images of only one side, which makes them easier to interpret. This method can be applied more easily using computational methods. In addition to providing separate front and back sides, this method averages over numerous particles, making it very powerful. To obtain a three-dimensional structure, particles lying in different orientations on a grid could provide the information needed, but this information can also be obtained by tilting the sample. During these efforts, a new approach was developed: Crystallographic EM. Here, new approaches for determining the structures of large molecular complexes started to be developed by combining EM data with crystallographic computational analyses.

Throughout his scientific life, Klug also had a broad interest in X-ray crystallography. Thus, an early important contribution to his work was one of the first two structures of tRNA, and it came as a surprise that the structure did not display the simple cloverleaf structure derived from secondary structural analysis, but rather was folded in a complex fashion. Another research theme that was carried out in his laboratory had been determining the structure of chromatin. Yet another topic that engaged Klug's interest for a long period was the zinc-finger protein structures that most commonly bind specifically to nucleic acids.

Selected References to the Work by Klug

Klug A, Caspar DLD. (1960) The structure of small viruses. *Adv. Virus Res.* **7:** 225–325.

Klug A, DeRosier DJ. (1966) Optical filtering of electron micrographs: reconstruction of one-sided images. *Nature* **212:** 29–32.

*DeRosier DJ, Klug A. (1972) Structure of the tubular variants of the head of bacteriophage T4 (polyheads) I. Arrangement of subunits in some classes of polyheads. *J. Mol. Biol.* **68:** 469–488.

Klug A. (2010) From virus structure to chromatin: X-ray diffraction to three-dimensional electron microscopy. *Ann. Rev. Biochem.* **79:** 1–35.

Other References of Interest

Unwin PNT, Henderson R. (1975) Molecular structure determination by electron microscopy of unstained crystalline specimens. *J. Mol. Biol.* **94:** 425–440.

Henderson R, Unwin PNT. (1975) Three-dimensional model of purple membrane obtained by electron microscopy. *Nature* **257:** 28–32.

Consequences

The work initiated by Klug has had rich consequences. One early development was electron diffraction, where a monolayer crystal was used to obtain crystallographic diffraction patterns. The phase angles were then derived by taking images for subsequent digitalization and extraction of Fourier series components. Subsequent methodological developments have involved EM tomography and single-particle reconstructions of three-dimensional structures by low-dose EM under cryo conditions. Here, the three-dimensional data are recorded from grids with tens of thousands particles in different orientations.

Herbert A. Hauptman and Jerome Karle (C1985)

"For their outstanding achievements in the development of direct methods for the determination of crystal structures."

Herbert A. Hauptman
February 14 1917–October 23 2011
(Photograph from Les Prix Nobel 1985)

Jerome Karle
June 18 1918–June 6 2013
(Photograph from Les Prix Nobel 1985)

Herbert Hauptman was born in 1917 in New York City. He obtained his BA from City College of New York (1935) and his MA from Columbia University (1939). Trained as a mathematician, he took up an early partnership with his old classmate from City College, Jerome Karle, and as Hauptman himself writes:

> The collaboration with Dr. Karle proved to be fruitful because his background in physical chemistry and mine in mathematics complemented each other nicely. Not only did this combination enable us to tackle head-on the phase problem of X-ray crystallography, but this work suggested also the topic of my doctoral dissertation, *"An N-Dimensional Euclidean Algorithm"*. By 1954 I had received my Ph.D. degree and Dr. Karle and I had laid the foundations of the direct methods in X-ray crystallography.

In 1970, Hauptman joined the crystallography group in Buffalo (later the Hauptman Woodward Medical Research Institute, Inc.), where he became Research Director in 1972 and where he worked for the remainder of his active life.

Jerome Karle was born in 1918 in New York City. He received his BA from City College of New York and his MA from Harvard University in 1938. After a brief period working at the New York Health Department, Karle returned to academia and graduate school at the University of Michigan in 1940. He had completed the work for his Ph.D. in 1943, and he spent the last year before completing his dissertation working for the Manhattan Project, together with his wife and long-time collaborator, Isabella Karle, née Lugoski. He later returned to the University of Michigan, but in 1946, he and Isabella both started working at the Naval Research Laboratory in Washington, where Herbert Hauptman joined them. Together, they identified the phase problem of crystallography as one of the most important bottlenecks for structural science and set off to work on this formidable challenge.

The Phase Problem

The phase problem is often said to be the most significant problem of crystallography. In essence, this problem arises because the diffraction pattern of a crystalline material is the Fourier transform of its electron density. The Fourier transform is a complex-valued function consisting of two parts — amplitudes and phases — but a diffraction experiment only supplies direct information on the amplitudes, not on the phases. If the amplitudes and phases are known for the diffraction pattern, the structure is easily retrieved by inverting the Fourier transform. The problem is that when the phases are missing, the inversion is no longer possible, and ways must be found to retrieve the missing phases.

The first structural solutions from the Braggs and others relied on various ways of avoiding the phase problem by arriving at partial structures that could then be extended until the whole structure was known, achieving phasing from the structural model rather than directly from the diffraction data. For some particularly simple cases, the structure may be inferred from simple algebraic and geometrical methods. Once the principles had been established, the solution of the structure of rock salt, NaCl, was essentially trivial. Given the size of the unit cell and the density of NaCl, it is clear that there must be four formula units in the unit cell. The symmetry of the unit cell is Fm-3m, and in this symmetry there are only two special positions that allow for such a small number of atoms; hence, these must be the loci of Na and Cl.

Another early method for retrieving structural information from diffraction data was devised by Ludo Patterson. He showed that a Fourier summation over the unit cell of the intensities of the diffraction pattern (without phases) produces a function with maxima corresponding to the inter-atomic vectors of the structure. The intensities of the maxima scale with the product of the electron density of the two atoms that define the inter-atomic vector, and therefore the presence of a single heavy atom is a great help in determining a structure.

Direct methods are based on two assumptions that can be made for any crystal structure. These are positivity and atomicity; the electron density is real and positive everywhere and in most places is zero. These assumptions can be shown to be equivalent to a number of inequalities that say relatively little for certain, but can be translated into a set of probabilities for the relations between phases of different reflections. The realization of Hauptman and Karle that these probabilities could be used to generate a limited number of starting structures provided the first general means of *ab initio* phasing of X-ray diffraction data.

In the first 10 years or so after the initial publication by Hauptman and Karle, direct methods were predominantly used only by Hauptman and Karle. but as computing power became cheaper and computer programs for direct methods become more user-friendly, these methods evolved to become the choice method for problems where the data quality was sufficiently good.

Consequences of the Work

Direct methods completely dominate small-molecule and materials crystallography today. It may be noted that when *Acta Crystallographica A* published a paper on a much-used software for solving crystal structures based on direct methods, the impact factor of this journal rose from the normal value of around 2 to above 50 for the two years when this paper was counted in the statistics. It is still difficult to use direct methods for very large problems, but direct methods are used for the locating of heavy atoms in direct phasing of new protein structures. It is fair to say that direct methods have transformed crystallography from a specialist's method into a tool that is used across the sciences.

References

Hauptman H, Karle J. (1954) Solution of the phase problem for space group P1. *Acta Cryst.* **7:** 369–374.

Karle J, Hauptman H. (1954) Probability distributions for atomic coordinates. *Acta Cryst.* **7:** 375–376.

Hauptman H, Karle J. (1953) Locating the principal maxima of a Fourier series. *Acta Cryst.* **6:** 469–473.

Karle J, Hauptman H. (1953) Application of statistical methods to the naphthalene structure. *Acta Cryst.* **6:** 473–476.

Karle J, Hauptman H. (1953) The probability distribution of the magnitude of a structure factor. I. The centrosymmetric crystal. *Acta Cryst.* **6:** 131–135.

Hauptman H, Karle J. (1953) The probability distribution of the magnitude of a structure factor. II. The non-centrosymmetric crystal. *Acta Cryst.* **6:** 136–141.

Hauptman H, Karle J. (1952) Crystal-structure determination by means of a statistical distribution of interatomic vectors. *Acta Cryst.* **5:** 48–59.

Hauptman H, Karle J. (1951) Multiplicity of solutions of crystal-structure equations. *Acta Cryst.* **4:** 383.

Karle J, Hauptman H. (1951) A note on the solution of the structure-factor equations. *Acta Cryst.* **4:** 188–189.

*Hauptman H, Karle J. (1950) A geometric approach to the crystal structure problem. *Acta Cryst.* **3:** 478.

*Karle J, Hauptman H. (1950) The phases and magnitudes of the structure factors. *Acta Cryst.* **3:** 181–187.

Hauptman H, Karle J. (1953) *Solution of the Phase Problem. I, The Centrosymmetric Crystal.* ACA Monograph No 3.

Johann Deisenhofer, Robert Huber, and Hartmut Michel (C1988)

"For the determination of the three-dimensional structure of a photosynthetic reaction center."

Hartmut Michel
Born July 18 1948
(Photograph from Les Prix Nobel 1988)

Robert Huber
Born February 20 1937
(Photograph from Les Prix Nobel 1988)

Johann Deisenhofer
Born September 30 1943
(Photograph from Les Prix Nobel 1988)

Membrane proteins have many essential functions in communicating information across membranes and transferring material through membranes. They have been notoriously difficult to crystallize. The first membrane protein whose structure was studied at low resolution was bacteriorhodopsin. Two-dimensional crystals were analyzed by electron diffraction by Richard Henderson and Nigel Unwin (1975) at the Medical Research Council (MRC) Laboratory for Molecular Biology in Cambridge. During the 1980s, progress was made in obtaining three-dimensional crystals, and in 1985, stunning results were obtained on the first structure of a membrane protein, the photosynthetic reaction center from the purple bacterium *Rhodopseudomonas viridis*, which was determined at 3 Å resolution. Johann Deisenhofer, Robert Huber, and Hartmut Michel were central to this achievement.

Hartmut Michel was born in 1948 in Ludwigsburg, Baden-Württemberg, Germany, where his ancestors had been farmers for many generations. Michel was an active boy who enjoyed outdoor life before getting into the habit of spending most of his time reading books on a number of his favorite topics. When the time came for his university studies, his main interest was molecular biology. Therefore, he applied for admission to the University of Tübingen, where biochemistry could be studied from the first year. He started there in 1969 and obtained his degree in 1974. After that, he started his experimental research together with Dieter Oesterhelt at the Friedrich Mischer-Laboratorium in Tübingen. Friedrich Mischer was the discoverer of DNA in the previous century. In 1975, the group moved to Würzburg. Here, Michel worked on ATP synthesis in halobacteria and obtained his doctorate in 1977. While conducting this work, he made observations that convinced him that the crystallization of membrane proteins (e.g. bacteriorhodopsin) should be possible. At this time, such a feat was considered impossible. After a short while, he was successful in obtaining a new type of two-dimensional crystals of bacteriorhodopsin. This led to a visit to Richard Henderson at the MRC laboratory in Cambridge, where work on such crystals had been going on for some time. After his return to Würzburg, he managed to obtain the first three-dimensional crystals of bacteriorhodopsin in 1979. This was of definitive importance for his continued research. Instead of a postdoctoral position in California, Michel continued to work together with Oesterhelt, who had now moved to the Max-Planck-Institute in München. The possibility of a crystallographic collaboration with Robert Huber at the same place made this move even more attractive. Another visit to Cambridge and Richard Henderson led to improved crystallization methods. Due to a lack of real progress on bacteriorhodopsin, Michel started to purify and crystallize other membrane proteins, primarily those related to photosynthesis. In 1981, he managed to obtain excellent crystals of the photosynthetic reaction center of the purple bacterium *Rhodopseudomonas viridis*. This led to the beginning of the collaboration with Johann Deisenhofer and Robert Huber. Their continued work was done using a range of methods in addition to crystallography. The biochemical characterization and determination of the amino acid sequence were important for their final success. The target protein is composed of four different subunits with a range of chlorophyll molecules and chromophores bound to it.

With the significant success of solving the crystal structure of the reaction center, Michel received a range of offers. He accepted one from Frankfurt, where he became Director of the Max-Planck-Institut für Biophysik and continued his successful work on membrane proteins. One of these proteins was the mitochondrial enzyme cytochrome c oxidase. In this case, the first structure was determined in 1995. This protein catalyzes the final electron transfer step from cytochrome c to molecular oxygen. In addition to the four protons used to produce water from oxygen, another four protons are pumped across the membrane. These protons are subsequently used by ATP synthase to generate ATP from ADP and inorganic phosphate (see Walker 1997). Determining the path through which these protons are pumped by cytochrome c oxidase remains a challenging problem.

Selected References to the Work by Michel

*Deisenhofer J, Epp O, Miki K *et al.* (1985) Structure of the protein subunits in the photosynthetic reaction center of *Rhodopseudomonas viridis* at 3Å resolution. *Nature* **318:** 618–624.

Iwata S, Ostermeier C, Ludwig B, Michel H. (1995) Structure at 2.8 Å resolution of cytochrome c oxidase from *Paracoccus denitrificans*. *Nature* **376:** 660–669.

Kirchberg K, Michel H, Alexiev U. (2013) Exploring the entrance of proton pathways in cytochrome c oxidase from *Paracoccus denitrificans*: Surface charge, buffer capacity and redox-dependent polarity changes at the internal surface. *Biochim. Biophys. Acta* **1827:** 276–284.

Other References of Interest

Unwin PNT, Henderson R. (1975) Molecular structure determination by electron microscopy of unstained crystalline specimens. *J. Mol. Biol.* **94:** 425–440.

Henderson R, Unwin PNT. (1975) Three-dimensional model of purple membrane obtained by electron microscopy. *Nature* **257:** 28–32.

Robert Huber was born in 1937 in München. During World War II, everything, including going to school, was difficult and Huber could not start his education until 1947. During his early teaching, he expanded the school curriculum with any textbooks in chemistry he could acquire. He completed gymnasium in 1956 and started at the Technische Hochschule, later the Technische Universität, in München, where chemistry was the main topic. Subsequently, he joined Walter Hoppe's group in order to learn crystallography. The first protein structure he analyzed was erythrocruorin. In 1971, he accepted an invitation to become Director of the Max-Planck-Institut für Eiweiss und Lederforschung (later Biochemie) associated with the Technische Universität München. He became a professor in 1976. Huber's group has covered a wide range of structural biology, including numerous enzymes and proenzymes, immunoglobulins, and light-harvesting and electron transfer proteins. One important topic of their investigation has been the dynamics and flexibility of proteins. Many widely used computer programs have been developed by the group. One peak in the activity of the laboratory was the structural determination of the reaction center, leading to the Nobel Prize. Many additional large and complex protein aggregates have been characterized by the group in atomic detail. They have given a particular focus to the studies of proteolytic enzymes or larger aggregates, such as different proteasome structures.

Selected References to the Work by Huber

*Deisenhofer J, Epp O, Miki K *et al.* (1985) Structure of the protein subunits in the photosynthetic reaction center of *Rhodopseudomonas viridis* at 3Å resolution. *Nature* **318:** 618–624.

Bochtler M, Ditzel L, Groll M, Huber R. (1997) Crystal structure of heat shock locus V (HslV) from *Escherichia coli. Proc. Natl. Acad. Sci. USA* **94:** 6070–6074.

Lowe J, Stock D, Jap B *et al.* (1995) Crystal structure of the 20S proteasome from the Archaeon *T. acidophilum* at 3.4 A resolution. *Science* **268:** 533–539.

Groll M, Bajorek M, Kohler A *et al.* (2000) A gated channel into the proteasome core particle. *Nat. Struct. Biol.* **7:** 1062–1067.

Brandstetter H, Kim JS, Groll M, Huber R. (2001) Crystal structure of the tricorn protease reveals a protein disassembly line. *Nature* **414:** 466–470.

Groll M, Huber R. (2004) Inhibitors of the eukaryotic 20S proteasome core particle: A structural approach. *Biochim. Biophys. Acta* **1695:** 33–44.

Johann Deisenhofer was born in 1943 in the farming community of Zusamaltheim, Bayern. He showed no interest in farming and was therefore given the opportunity to continue his school years beyond the mandatory number. In 1963, he completed his "Abitur" and after military service, he was given the opportunity to continue to study physics at Technische Universität München. His reason for studying physics was an interest in astronomy, but university astronomy was different from what popular books had communicated to him. The Diplomarbeit in physics was successful and led to a publication in *Physical Review Letters* in 1971. During this time, Deisenhofer's interest in biophysics grew. In looking for a place to do a Ph.D., Robert Huber's laboratory seemed like a good place to start. His work began with the successful crystallographic refinement of the bovine pancreatic trypsin inhibitor. After completing his Ph.D. in

1974, he was offered a postdoctoral position in the laboratory and joined the work on immunoglobulin-related structures and a range of other projects. In 1982, Dieter Oesterhelt and Hartmut Michel moved to the lab in Martinsried outside München. At a seminar, they reported their success in crystallizing the bacterial photosynthetic reaction center and Deisenhofer was invited by Michel to join the project, which Huber gave his agreement to. The initial rapid success in the structural analysis required a few more years before the whole structure could be solved and refined. The structure met with strong interest from the scientific community.

Deisenhofer's success led to opportunities to start a group in a new place. In 1988, he accepted an invitation to become a professor at the Medical Center at the University of Texas. Here, the crystal structures of a wide range of biomolecules have been solved. One field that has received special interest from this group is the low-density lipoprotein and its receptor. Here, a combination of crystallography and cryo-electron microscopy has been successfully applied.

Selected References to the Work by Deisenhofer

*Deisenhofer J, Epp O, Miki K *et al.* (1985) Structure of the protein subunits in the photosynthetic reaction center of *Rhodopseudomonas viridis* at 3Å resolution. *Nature* **318:** 618–624.

Hunt JF, Weaver AJ, Landry SJ *et al.* (1996) The crystal structure of the GroES co-chaperonin at 2.8Å resolution. *Nature* **379:** 37–45.

Xia D, Yu C-A, Kim H *et al.* (1997) Crystal structure of the cytochrome bc 1 complex from bovine heart mitochondria. *Science* **277:** 60–66.

Ren G, Ludtke SJ, Chiu W *et al.* (2010) Model of human low-density lipoprotein and bound receptor based on CryoEM. *Proc. Natl. Acad. Sci. USA* **107:** 1059–1064.

Consequences

A number of observations were made by the Nobel Laureates, which not only relate to the specificities of the photosynthetic reaction center, but are also general for membrane proteins. One of the common features is that helices that are more or less perpendicular to the plane of the membrane make up the part of the protein that goes through the membrane. Furthermore, hydrophobic side chains are located on the outside of the parts of the protein that are in contact with the membrane. Aromatic groups are also frequently situated at the membrane–solvent interface. The Laureates' success in crystallization and structure determination has sparked intense activity in crystallizing membrane proteins and determining their structures. In several cases, this has led to additional Nobel Prizes: John Walker (1999), Roderick MacKinnon (2003), and Brian Kobilka (2012). The most complex membrane protein structures to be determined so far are perhaps photosystems I and II, which are more complex relatives of the bacterial photosynthetic reaction centers.

Bertram N. Brockhouse and Clifford G. Shull (P1994)

"For pioneering contributions to the development of neutron scattering techniques for studies of condensed matter."

jointly, with one half to Bertram N. Brockhouse,

"For the development of neutron spectroscopy,"

and with one half to Clifford G. Shull,

"For the development of the neutron diffraction technique."

Bertram N. Brockhouse
July 15 1918–October 13 2003
(Photograph from Les Prix Nobel 1994)

Clifford G. Shull
September 23 1915–March 31 2001
(Photograph from Les Prix Nobel 1994)

Bertram Brockhouse was born in 1918 in Lethbridge, Alberta. In 1924, the family moved to Vancouver, British Columbia, where he finished high school. Due to the financial situation during the Great Depression in 1935, the family moved again to Chicago. Part of his time there was spent working as a lab assistant in a small electronic firm. Brockhouse was interested in the technical aspects of radios and he learned to repair and build them. This and his talent for mathematics were probably what eventually led him in the direction of physics. When the electronic firm he worked for failed in the recession of 1937, the family decided to return to Vancouver in 1938. Brockhouse's anti-totalitarian and anti-communist feelings inspired him to enlist in the Royal Canadian Navy in 1939, with the intention of becoming a radio telegrapher. He spent most of his six years in the navy servicing ASDIC equipment at a shore base. In 1944, he was enrolled in a six-month course in electrical engineering at Nova Scotia Technical College and later assigned to the test facilities at the National Research Council in Ottawa. There he met Doris Miller, who later became his wife.

When World War II ended in late August 1945, Brockhouse began his studies at the University of British Columbia in Vancouver. With his earlier experience, the obvious choice of subject to study was either electrical engineering or physics, and so he chose to enroll in physics and mathematics. He completed his BA in April 1946. He had planned to continue with a Master's degree course in Vancouver, but instead he went to the Low Temperature Laboratory of the University of Toronto, as this was one of the two universities in Canada to offer Ph.D. programs at that time. He started to work on the effects of stress and temperature on ferro-magnetism and finished a Master's program. His thesis subject was a contribution to solid-state physics, which involved experiments at both low and high temperatures. He passed his Ph.D. exam in 1950.

At the end of his thesis work, he was offered work at Chalk River. One major advance in early 1951 was the realization that phonons ("vibrational motions") could be studied by inelastic neutron scattering and that such experiments might be feasible at Chalk River, where the world's highest flux beam reactor at that time was located. In the following years, he was involved in inelastic neutron scattering studies, both at Chalk River and Brookhaven National Laboratory. In the late 1950s and early 1960s, scientists from many countries visited Chalk River in order to observe and learn the new "methods for neutron spectrometry."

Brockhouse did not have very much spare time during his highly productive years at Chalk River, but he did find time to take part in a number of amateur dramatic presentations, including three Gilbert and Sullivan operettas, and a production of Shaw's *Arms and the Man*. He enjoyed music enormously, and many of his colleagues have told stories of his predilection for singing musical excerpts, whether from an opera or from a Broadway musical, while working at one of his neutron experiments.

In 1962, Brockhouse took up a position as professor of physics at McMaster University, in Hamilton, Ontario. The research program he had embarked upon 11 years earlier had been successful beyond expectations and the field was becoming well established. The laboratory at Chalk River facilitated his transfer to the university and encouraged him to continue a research program based at McMaster. He could use the "swimming pool" reactor there for training students and for preliminary work on experiments to be carried out at the Chalk River reactor with stronger neutron flux. This arrangement turned out to be quite successful. At McMaster, talented students put together all of the equipment needed for their research, such as a neutron diffractometer and a triple-axis instrument. Brockhouse completely left neutron scattering in about 1979 and retired as a professor emeritus in 1984.

Exactly 10 years after his retirement, Brockhouse awoke on a Wednesday morning to find a message on his answering machine. The message said that B. N. Brockhouse and C. G. Shull had been chosen for the 1994 Nobel Prize for Physics. Brockhouse was stunned and called his wife in to listen to the message. Because of the award, he was promptly called out of retirement for a year of lectures, awards, and travel, all for work he had completed over 40 years ago.

To honor the former professor, in 1995, McMaster dedicated its materials research school to him — The Brockhouse Institute for Materials Research. The "absent-minded professor" stories are plentiful and amusing, but the stories of his insistence on good experimental technique and of his concern that time and money be efficiently used are perhaps more to the point. Those who have had the pleasure of working with him frequently mention his intuition, his dedication to research, and his kindness and concern for his fellow workers.

References

*Brockhouse B. (1995) Slow neutron spectroscopy and the grand atlas of the physical world. *Rev. Mod. Phys.* 67, No. 4. (Lecture delivered December 8 1994, on the occasion of the presentation of the 1994 Nobel Prize in Physics.) In this paper, all of the references to the major publications by Bertram Brockhouse are listed.

Biography by John RD Copley in Physica 136B (1986) xxvii–xxxi (North-Holland, Amsterdam).

Clifford Shull was born in 1915 in Pittsburgh, Pennsylvania. He studied at the Carnegie Institute of Technology (now Carnegie Mellon University) for his Bachelor's degree and graduated with a Ph.D. from New York University in 1941. He worked for The Texas Company at Beacon, New York, during the war, followed by a position in the Clinton Laboratory (now Oak Ridge National Laboratory). He joined the Massachusetts Institute of Technology (MIT) in 1955 and retired in 1986.

During his time at The Texas Company, Shull had the opportunity to learn about diffraction processes, crystallography, and the new field of solid-state physics. Through visits and early meetings of the American Society for X-ray and Electron Diffraction, he came into contact with established crystallographers such as Warren, Buerger, Fankuchen, Zachariasen, Ewald, Harker, Gingrich, and Donnay. He became interested in the new developments in nuclear physics, and the activity at the Clinton Laboratory at Oak Ridge fascinated him. Scientists at Oak Ridge were very anxious to find important scientific uses for the information and technology that had been developed during the war at Oak Ridge and at other places associated with the wartime Manhattan Project. He moved there in 1946 along with his wife Martha-Nuel Summer and their one and a half-year-old son.

It was arranged that he would work with Ernest Wollan, who had been at the laboratory since its formation during the war period and who had just assembled a rudimentary two-axis spectrometer for obtaining neutron diffraction patterns of crystals and materials. Shull became aware that neutron diffraction could be a very important method for supplementing the results obtained with X-rays or electrons. His collaboration with Wollan continued for nearly a decade until he left Oak Ridge in 1955 for MIT. Wollan's contributions certainly would have deserved sharing the Nobel Prize together with Brockhouse and Shull in 1994, but unfortunately he died in 1984.

Shull was attracted to MIT by the prospect of working together with excellent graduate students at the research reactor that was just being finished on the campus. This reactor was among the early reactors using isotopically enriched fuel, which was being explored in that period. The group around Shull performed neutron scattering studies in many fields, including internal magnetization in crystals, the development of polarized beam technology, dynamical scattering in perfect crystals, interferometry, and fundamental properties of the neutron.

Shull was awarded the Gregori Aminoff prize from the Royal Swedish Academy of Sciences in 1993 for his "development and application of neutron diffraction methods for studies of atomic and magnetic structures of solids."

Selected Publications by Shull

*Shull CG. (1995) Early development of neutron scattering. *Rev. Mod. Phys.* **67:** 753–757. (Nobel Prize Lecture.)

Shull CG, Wollan EO. (1947) The diffraction of neutrons by crystalline powders. *Oak Ridge National Laboratory, April 28 1947.*

Davidson WL, Morton GA, Shull CG, Wollan EO. Neutron diffraction analysis of NaH and NaD. *United States Department of Energy, April 28 1947.*

Shull CG, Wollan EO, Marney MC. (1948) Neutron diffraction studies. *Oak Ridge National Laboratory, October 22 1948.*

Rundle RE, Shull CG, Wollan EO. (1951) The crystal structure of thorium and zirconium dihydrides by X-ray and neutron diffraction. *Ames Laboratory and Oak Ridge National Laboratory, April 20 1951.*

Nathans R, Riste T, Shirane G, Shull CG. (1958) Polarized Neutron studies on antiferromagnetic single crystals. *Massachusetts Institute of Technology and Brookhaven National Laboratory, November 26 1958.*

Shull CG. (1989) Low temperature and neutron physics studies. *Massachusetts Institute of Technology, July 27 1989.*

Other References

Bacon GE. ed. (1986) *Fifty Years of Neutron Diffraction.* Adam Hilger, Bristol.

Yeater ML. ed. (1962). *Neutron Physics.* Academic Press, New York.

Mason TE, Gawne TJ, Nagler SE *et al.* (2013) The early development of neutron diffraction: Science in the wings of the Manhattan Project. *Acta Cryst. A* **69:** 37–44.

Neutron Diffraction: Elastic Scattering of the Neutrons

The neutron scattering power of atoms differs radically from their X-ray scattering, which roughly follows the atomic number. The (coherent) neutron scattering power of hydrogen is, for instance, nearly of the same size as for potassium, but of the opposite phase. Deuterium scatters nearly twice as strongly as potassium, but with the same phase. Since isotopes may have entirely different scattering powers for neutrons, this opens up a new type of "heavy atom method" for structural analysis with the substitution of chemically identical atoms.

Neutron diffraction has developed to become a very powerful method for studies of matter in general as a complement to X-ray diffraction. Certain types of problems are best solved with neutrons, such as in the determination of the atomic positions of light atoms in the presence of heavier elements. This is especially valuable for compounds containing hydrogen, such as proteins. Most biological processes involve the forming and breaking of hydrogen bonds, and these processes often involve the transfer of the protons in the bonds. It is therefore important to be able to locate the proton positions. Starting in the 1980s, many of the medium-flux sources in the world have been shut down, and as high flux is essential in many cases, neutron research is now concentrated at a few world-leading high-flux sources. More recently, spallation neutron sources also offer very high flux. High flux is essential in studies of proteins, for instance, where it is mostly very difficult to obtain large crystals.

The neutron has a dipole moment, and its interaction with the magnetic field from unpaired electrons has made neutron scattering an indispensable tool in studies of magnetic structures. Since neutrons are electrically neutral, they penetrate matter more deeply than electrically charged particles of comparable kinetic energy; therefore, they are valuable probes of bulk properties.

Neutron Spectroscopy: Inelastic Scattering of the Neutrons

Shull's research involved (coherent) elastic scattering of neutrons, or neutron diffraction. By contrast, Brockhouse's work involved inelastic scattering, where the atomic and magnetic motions of the atoms are measured, which is also called neutron spectroscopy.

Their Nobel citation could be read as:

> *Clifford G. Shull has helped answer the question of where atoms "are" and Bertram N. Brockhouse, the question of what atoms "do".*

Inelastic neutron scattering measures the change in the energy of the neutron as it scatters from a sample. This can be used to study a wide variety of physical phenomena, such as the diffusional or hopping motions of atoms, the rotational modes of molecules, sound modes and molecular vibrations, magnetic and quantum excitations, or even electronic transitions.

Knowledge of the atomic structure of a material is, in many cases, sufficient for understanding its nature. However, in order to gain a deeper insight into the underlying physics of, for example, a phase change, it is necessary to also understand the atomic dynamics. The vibrational motion of atoms is — entirely or in part — responsible for a large number of the characteristic properties of a material, such as its specific heat, thermal conductivity, optical and dielectric properties, and electrical resistance. but it is also a direct way of understanding the nature of atomic bonding. The properties of smart or functional materials are often determined by a complex balance or strong coupling between competing phenomena, and therefore understanding the atomic and magnetic dynamics is essential.

Paul Boyer, John E. Walker, and Jens C. Skou (C1997)

The two first Laureates were awarded the prize:

"For their elucidation of the enzymatic mechanism underlying the synthesis of adenosine triphosphate (ATP)."

Only in the case of Walker was the prize related to crystallography. Therefore, our focus will be on him alone.

John E. Walker
Born January 7 1941
(Photograph from Les Prix Nobel 1997)

John E. Walker was born in 1941 in Halifax, Yorkshire. He was brought up in a rural setting and was active in sports. In 1960, he went to St. Catherine's College in Oxford and obtained his BA degree in 1964. The following year, he started his research education focusing on peptide antibiotics and obtained his DPhil degree in 1969. During this period, he was thrilled by the developments in structural molecular biology in Cambridge. Interesting lectures on protein structure were delivered by David Phillips, who had just become a professor in molecular biophysics at Oxford. Subsequently, Walker worked abroad for five years. At a research workshop in Cambridge, Fred Sanger, double Nobel Laureate in Chemistry (C1958, C1980), and Ieuan Harris asked him whether he would like to give the Medical Research Council (MRC) laboratory in Cambridge a chance. This was a turning point in Walker's scientific career.

With an interest in mitochondria, Walker decided to start analyzing membrane proteins using chemical methods. He considered the enzymes involved in oxidative phosphorylation from the inner membrane of mitochondria, particularly ATP synthase, to be a very interesting topic. Driven by a gradient of protons called the proton motive force, as described by Nobel Laureate Peter Mitchell (C1978), ATP is synthesized from ADP and inorganic phosphate. The proton gradient is produced by the electron transportation in mitochondria. The energy provided by NADH is used by three enzymes called complex I, complex II, and complex IV (cytochrome c oxidase; see Hartmut Michel chapter). These three enzymes work consecutively and all participate in generating the proton gradient. Poul Boyer (1993) had been studying the biochemistry and the catalytic mechanisms of the ATP synthase for a long time. An unusual aspect had emerged during this work. He concluded that parts of the membrane-bound enzyme must rotate in order to synthesize the ATP. His proposal got the name "the binding change mechanism." ATP synthase is built of two units called F_1 (the catalytic part) and F_0 (the membrane-bound part). Furthermore, the enzyme has multiple subunits and, as an initial step in Walker's work, the subunit sequences from several species were determined, as summarized in Les Prix Nobel (Walker, 1998).

The efforts required to obtain suitable crystals of the F_1 unit of ATP synthase took several years. The structure determination was more rapid. The crystallographic work had throughout been done in collaboration with Andrew Leslie at the MRC laboratory in Cambridge. The remarkable structure of the F_1 unit immediately suggested that the binding change mechanism proposed by Boyer could be correct. The structure shows six alternating α and β subunits, three of each kind, with the γ subunit forming an asymmetric stalk at the center. Yoshida and colleagues (Noji *et al.*, 1997) performed a spectacular illustration of the rotation of the γ subunit in the pseudo-symmetric $\alpha_3\beta_3$ arrangement. By fixing the $\alpha_3\beta_3$ structure on a microscope slide, binding a fluorescently labeled actin filament to the γ subunit, and inspecting the performance of the system in a microscope when ATP was added, they could demonstrate a remarkable rotation of the actin filament.

After receiving the Nobel Prize in 1997, Walker has continued the crystallographic analysis of the system together with Leslie. In particular, the F_0 part has been of interest in order to learn how the proton gradient generates the rotation of the γ subunit in the F_1 part. F_0 is, in the simplest case, composed of three subunit — a, b and c — with the composition $a_1b_2c_{8-15}$. The number of c subunits differs between species (Walker, 2013). A crystal structure of the F_1 part with a ring of 10 c subunits gave a first insight into this area of research (Stock *et al.*, 1999). Work focusing on the stator part of the structure has also been reported (Dickson *et al.*, 2006; Rees *et al.*, 2009). A more complete structure has been characterized using cryo-electron microscopy (Baker *et al.*, 2012). In this case, the ring of c subunits was composed of no more than eight protein molecules. The a subunit is located on the side of the ring in the membrane, while the b subunit, called the peripheral stalk, extends from the a subunit to the top of the F_1 part. The protons generating the rotation are channeled through part of the a subunit to an acidic residue of the c subunit. In order to escape on the other side of the membrane, they force the c-ring to rotate and thereby force the γ subunit to induce the conformational changes of the β-subunit that lead to ATP synthesis or, in the opposite direction, ATP hydrolysis (Fig. 1).

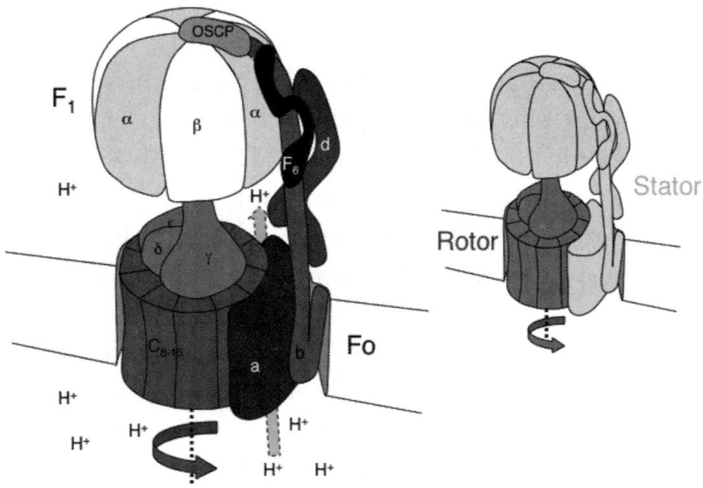

Fig. 1. A simplified view of ATP synthase from the work of Walker. Left: The individual subunits are indicated, as well as the gradient of protons that drives the rotating part of the enzyme. Right: The stator and rotor parts of ATP synthase. (Reproduced with permission from Liljas *et al.*, *Textbook of Structural Biology*. Copyright World Scientific Publishing, 2009.)

Selected References to the Work by Walker

Lutter R, Abrahams JP, van Raaij MJ *et al.* (1993) Crystallization of F1-ATPase from bovine heart mitochondria. *J. Mol. Biol.* **229:** 787–790.

*Abrahams JP, Leslie AG, Lutter R, Walker JE. (1994) Structure at 2.8Å resolution of F1-ATPase from bovine heart mitochondria. *Nature* **370:** 621–628.

Walker JE. (1997) ATP synthesis by rotatory catalysis. *Les Prix Nobel*, pp. 146–168.

Stock D, Leslie AGW, Walker JE. (1999) Molecular architecture of the rotary motor in ATP synthase. *Science* **286:** 1700–1705.

Dickson VK, Silvester JA, Fearnley IM *et al.* (2006) On the structure of the stator of the mitochondrial ATP synthase. *EMBO J.* **25:** 2911–2918.

Rees DM, Leslie AG, Walker JE. (2009) The structure of the membrane extrinsic region of bovine ATP synthase. *Proc. Natl. Acad. Sci. USA* **106:** 21597–21601.

Baker LA, Watt IN, Runswick MJ *et al.* (2012) Arrangement of subunits in intact mammalian mitochondrial ATP synthase determined by cryo-EM. *Proc. Natl. Acad. Sci. USA* **109:** 11675–11680.

Walker JE. (2013) The ATP synthase: The understood, the uncertain and the unknown. *Biochem. Soc. Trans.* **41:** 1–16.

Other References of Interest

Boyer PD. (1993) The binding change mechanism for ATP-synthase — some probabilities and possibilities. *Biochim. Biophys. Acta* **1140:** 215–250.

Noji H, Yasuda R, Yoshida M, Kinoshita Jr. K. (1997) Direct observation of the rotation of F_1-ATPase. *Nature* **386:** 299–302.

Consequences

The structural work on ATP synthase is, first of all, an important step forward in terms of understanding how certain enzymes work. Here, the kinetic energy of the rotational movement is converted into chemical energy in the form of ATP. The hypothesis by Boyer therefore meets with a strong confirmation in this research. The analogy of classical water-driven wheels, where the kinetic energy of different levels of water is used to generate work and useful products, is obvious. The consequences of the understanding of ATP synthase are primarily academic. Rotational motors have been known in biology before. The study of flagellar motors generating bacterial or sperm motion may be more accessible for structural analysis after the achievements by the group that was working with Walker.

Peter Agre and Roderick MacKinnon (C2003)

These two Laureates were awarded the Nobel Prize:

"For discoveries concerning channels in cell membranes,"

with one half going to Peter Agre:

"For the discovery of water channels,"

and with one half going to Roderick MacKinnon:

"For structural and mechanistic studies of ion channels."

Only the ion channels involved crystallographic efforts, so only MacKinnon will be discussed here.

Roderick MacKinnon
Born February 19 1956
(Photograph from Les Prix Nobel 2003)

Roderick MacKinnon was born in 1956 in Burlington, Massachusetts, in the middle of a snowstorm. He was one of seven children in the family. He was curious as a child and enjoyed to explain things to others, whether they had asked for an explanation or not. His father commented that young Roderick was a compendium of useless information.

During his school years, gymnastics also became an important activity. His university studies began at the University of Boston and continued at Brandeis University. The intellectual environment stimulated his passion for science and biochemistry in particular. The next station of his educational journey was Tufts University School of Medicine, where he obtained his MD in 1982.

A desire to return to basic science led to a postdoctoral period in which he was working on the biophysics of potassium channels. During this time, MacKinnon recovered what he had missed during his medical training. In 1989, he obtained a position as an assistant professor at Harvard. Much could be learnt about K^+ channels by various methods, but without their structure, it would be impossible to understand the chemical principles that led to their selectivity for potassium He then decided to learn crystallography in order to be able to visualize the K^+ channels. To be able to focus on this new line of research and obtain sufficient resources, in 1996, he accepted an offer of a professorship at Rockefeller University. Despite the small size of the group and the difficulty of working with and crystallizing membrane proteins, his success in crystallizing and determining the structure of K^+ channels provided insights into many questions, including their selectivity for K^+ ions.

The first medium-resolution structure of K^+ channels showed the narrow 12 Å-long selectivity filter where the carbonyl oxygens from the four-subunit signature sequence for the K^+ channels are aligned along the fourfold symmetric channel (Doyle *et al.*, 1998). At high and low K^+ content, the structures were at 2.0 Å and 2.3 Å resolution, respectively, and using a new crystal form, the structure could be determined with greater precision (Zhou *et al.*, 2001). Here, K^+ ions are seen in four sites in the selectivity filter and, in addition, one K^+ ion is seen in a hydrophobic cavity just preceding the selectivity filter. This K^+ ion is fully hydrated by eight water molecules, with the water molecules fitting nicely with the fourfold symmetry. This hydrated ion, located in the middle of the membrane, is ready to be stripped of its water molecules and enter the selectivity filter. In the selectivity filter, all four potassium sites are never occupied simultaneously; either positions 1 and 3 or 2 and 4 are occupied during the process of transfer. The distances between the K^+ ions and the carbonyl oxygens are very close to the corresponding distances of these ions to the water molecules. Smaller or bigger ions would not be accepted. In essence, the channel acts like an enzyme in stripping K^+ ions of their water molecules and transporting them to the outside of the cell, where they again become hydrated.

MacKinnon has continued his focused research on K^+ channels of different kinds. Many such channels are gated either by an electric potential or by Ca^{2+}. Detailed crystallographic structures have been generated for a range of K^+ channels.

Selected References to the Work by MacKinnon

*Doyle D, Cabral J, Pfuetzner R *et al.* (1998) The structure of the potassium channel: Molecular basis of K^+ conduction and sensitivity. *Science* **280:** 69–77.

Zhou Y, Morais-Cabral JH, Kaufman A, MacKinnon R. (2001) Chemistry of ion coordination and hydration revealed by a K^+ channel–Fab complex at 2.0Å resolution. *Nature* **414:** 43–48.

Gouaux E, MacKinnon R. (2005) Principles of selective ion transport in channels and pumps. *Science* **310:** 1461–1465.

Consequences

Humans possess approximately 80 different K^+ channels. The basic knowledge of ion transport during nerve signaling and the identification of specific ion channels in nerve cells were already established before MacKinnon's structural studies. However, how this could be achieved and how the channels could open and shut was a mystery. A series of crystal structures from the MacKinnon group has shown how K^+ ions are desolvated and transported through a channel. Furthermore, the gating of different channels could be characterized. As the sodium and potassium channels along nerve cells open and close, an action potential is created. The malfunctioning of these channels can lead to severe diseases, such as cardiac arrhythmia or non-functional release of hormones. A range of blockers can be administered in order to block K^+ channels, but some are activated by muscarinic receptors (GPCR proteins). Understanding the structures of K^+ channels presents the possibility of identifying drugs that can block or activate specific K^+ channels in order to find therapies related to a wide range of metabolic, neurological, or cardiovascular disorders.

Roger D. Kornberg (C2006)

"For his studies of the molecular basis of eukaryotic transcription."

Roger D. Kornberg
Born April 24 1947
(Photograph from Les Prix Nobel 2006)

Roger D. Kornberg was born in 1947 in St. Louis, Missouri, to parents who were biochemists. His father Arthur was awarded the Nobel Prize for Physiology or Medicine in 1959 for his discovery of DNA polymerase. Kornberg obtained a Bachelor's degree in chemistry from Harvard University in 1967 and a Ph.D. in chemical physics from Stanford in 1972. He received postdoctoral training with Aaron Klug at the Medical Research Council Laboratory for Molecular Biology in Cambridge, England. Part of his ambition was to learn protein crystallography at Cambridge. During this period, one focus of the laboratory was to understand the packaging of DNA with histones in the chromosomes. Kornberg was challenged to obtain some further insight into this question. Previous investigations suggested that there must be some type of repeating structure. Mild purification methods of histones indicated that they combined in a pairwise manner: H2A/H2B and H3/H4. Gradually, it was realized that H3 and H4 form a tetramer of the two proteins, whereas H2A and H2B form heterodimers, one of each kind. In 1975, Kornberg arrived at a picture of the nucleosome — the smallest entity of the chromosomal structure — as being composed of a compact octamer of histones around which some 200 base pairs of DNA are wrapped.

Subsequently, Kornberg took a position as an Assistant Professor at Harvard Medical School in 1976, but moved to Stanford Medical School as Professor of Structural Biology in 1978. Here, Kornberg started his work on transcription. Transcription of the genetic information contained in the DNA to RNA is a central process in all organisms. The key enzyme is called RNA polymerase and is found in all living organisms and even in some viruses. The regulation of which genes should be transcribed is a highly controlled process guided by a range of transcription factors, which are proteins that can either activate or repress the transcription of a specific gene. The study of the eukaryotic transcription machinery met with numerous problems. First of all, there are three different forms of RNA polymerase (I–III), each synthesizing specific types of RNA. The protein-coding genes are transcribed by RNA polymerase II. Furthermore, the RNA polymerases are composed of at least 12 subunits. Other complications are also due to the multitude of general factors involved in the regulation and initiation of transcription (transcription factors TFIIB, D, E, F, and H, as well as the large and complex Mediator). For Kornberg's structural work, started in the early 1980s, the size of the RNA polymerase seemed to be beyond what could be managed by crystallography. On the other hand, the electron microscopy techniques developed in Cambridge seemed to be a way forward. Kornberg decided to try to generate two-dimensional crystals for electron diffraction analysis. The approach was to attach the RNA polymerase to lipids in a lipid layer, with the expectation that the polymerase would crystallize through lateral diffusion. After many failures, this technique eventually worked, and 1991, Kornberg produced an electron microscopy structure at low resolution. Subsequently, three-dimensional crystals were obtained, and in 1999, an initial X-ray structure could be presented. A better-resolved structure of the yeast RNA polymerase II became available a year later. In a series of publications, the structure of this enzyme were described. Furthermore, the structure demonstrates how the enzyme binds to the template and non-template strands of DNA. The separation of the two strands to form the transcription bubble before the template strand can enter the active site is also described. In the synthesis of mRNA, a short stretch of hybrid RNA–DNA is formed, which is also described in detail. The selection and incorporation of the correct nucleotides in the active site, assisted by two magnesium ions, as well as the translocation of the template and product RNA–DNA hybrid has also been studied.

The five different general transcription factors are composed of a total of approximately 30 different protein molecules, which are needed for the initiation of transcription. Mediator, with approximately 20 protein molecules, assists in the binding of an activator — DNA complex to RNA polymerase. Kornberg's group has characterized the crystal structures of several of these factors in complex with RNA polymerase. However, in other cases, they have used single-particle cryo-electron microscopy. The complex picture of the synthesis of mRNAs has come a long way, but many challenges remain.

Selected References to the Work by Kornberg

Kornberg RD. (1974) Chromatin structure: A repeating unit of histones and DNA. *Science* **184:** 868–871.

Kornberg RD. (1977) Structure of chromatin. *Ann. Rev. Biochem.* **46:** 931–954.

*Cramer P, Bushnell DA, Kornberg RD. (2001) Structural basis of transcription: RNA polymerase II at 2.8 Ångström resolution. *Science* **292:** 1863–1876.

*Gnatt AL, Cramer P, Fu J *et al.* (2001) Structural basis of transcription: An RNA polymerase II elongation complex at 3.3 Å resolution. *Science* **292:** 1876–1882

Liu X, Bushnell DA, Wang D *et al.* (2010) Structure of an RNA polymerase II–TFIIB complex and the transcription initiation mechanism. *Science* **327:** 206–209.

Imasaki T, Calero G, Cai G *et al.* (2011) Architecture of the Mediator head module. *Nature* **475:** 240–243.

Murakami K, Elmlund H, Kalisman N *et al.* (2013) Architecture of an RNA polymerase II transcription pre-initiation complex. *Science* **342:** 1238724.

Consequences

The consequences of Kornberg's work on transcription are partly academic. This research provides a large step towards an insight into the mechanism of eukaryotic transcription and it has given the first molecular details of the highly complex selection of when and how to produce specific mRNAs for subsequent protein synthesis. The complex interplay between the special and general transcription factors, Mediator, and RNA polymerase is a current focus of research in this area. Kornberg's work on the nucleosome structure of chromatin has paved the way for understanding chromatin remodeling as a new level for the regulation of gene expression. In addition, a range of antibiotics, such as the family of rifamycins, can inhibit bacterial RNA polymerases. These drugs have been important in the treatment of drug-resistant tuberculosis. Crystal structures of rifamycins in complex with bacterial RNA polymerase have helped to identify their mode of inhibition.

Venkatraman Ramakrishnan, Thomas A. Steitz, and Ada Yonath (C2009)

"For studies of the structure and function of the ribosome."

Ada Yonath was the pioneer who crystallized ribosomes and ribosomal subunits, showing the first signs of success 1980. Therefore, we will describe her work first and continue with the biographies of Venkatraman Ramakrishnan and Thomas Steitz.

Ada Yonath
Born June 22 1939
(Photograph from Les Prix Nobel 2009)

Venkatraman Ramakrishnan
Born 1952
(Photograph from Les Prix Nobel 2009)

Thomas A. Steitz
Born August 23 1940
(Photograph from Les Prix Nobel 2009)

Ada Yonath was born in Jerusalem in 1939. Her parents emigrated from Poland. Despite the poverty of the family, Yonath received a good basic education in a secular school. After army service in the Medical Forces, she started at Hebrew University in Jerusalem and studied chemistry, biochemistry, and biophysics. Yonath obtained her Ph.D. from structural work on collagen with Dr. W. Traub at the Weizmann Institute in Rehovot. As a postdoc, she initially continued to work on fibrous proteins in Pittsburg and subsequently moved to the Massachusetts Institute of Technology (MIT) in Cambridge, Massachusetts, in order to work on staphylococcal nuclease with Prof. F. Albert Cotton.

In 1970, Yonath returned to the Weizmann Institute and established a crystallographic laboratory, where many different topics were studied. A collaboration with Prof. H. G. Wittmann at the Max-Planck-Institut für Molekulare Genetik in Berlin-Dahlem on the structure of ribosomes and ribosomal subunits became a real challenge. There was a general belief that ribosomes could not be crystallized. They would be too heterogeneous in their composition of ribosomal components and were probably attached to non-ribosomal proteins. To everyone's surprise, Yonath and collaborators managed to crystallize the large ribosomal subunit from *Bacillus stearothermophilus* (Yonath *et al.*, 1980). Subsequently, they were able to crystallize ribosomes from *Escherichia coli*. A Russian group (Trakhanov *et al.*, 1987), soon followed by Yonath's group (Glotz *et al.*, 1987), succeeded in crystallizing ribosomes from *Thermus thermophilus*. Later, Yonath and Wittmann (1988) managed to obtain crystals of the large ribosomal subunit from the archaebacterium *Haloarcula marismortui* that could be used to produce a structure at close to atomic resolution. However, many difficulties had to be overcome in order to solve the structure of the ribosome or its subunits.

Some of the problems were the short lifetime of the crystals in the synchrotron beam. Here, the situation was greatly improved by Yonath's freezing of the crystals with liquid nitrogen in a fruitful collaboration with Dr. H. Hope (Hope *et al.*, 1989). Another problem that was difficult to overcome was the variability in cell parameters of the crystals. However, the main problem was to obtain phase angles of the diffraction data at sufficiently high resolution. Here, Yonath took advantage of derivatives with ultra-heavy compounds, such as tetrakis(acetoxymercury)methane or undecagold (Au_{11}; Yonath & Wittmann, 1988). Of the multitude of approaches that were adopted, none seemed to lead to the goal.

Up until 1999, Yonath published many reports about improved crystals, attempts using cryo-electron microscopy to obtain low-resolution structures that could assist in phasing, as well as the binding of heavy atom clusters, all without reaching her goal (Harms *et al.*, 1999). However, later in 1999, the group managed to determine a low-resolution structure of the small ribosomal subunit (30S) from *T. thermophilus* (Tocilj *et al.*, 1999). Here, the crystal quality had been significantly improved by soaking in a polytungstate solution. A long list of heavy atom derivatives was used for the phase angle determination and the heavy atom positions were obtained through classical difference Patterson and difference electron density calculations. Next year, the structure was presented at 3.3 Å resolution, with most of the RNA and protein structures interpreted (Schlünzen *et al.*, 2000). A more thorough analysis of the structure was reported in 2001 (Pioletti *et al.*, 2001), and in that same year, the group presented the structure of the large ribosomal subunit (50S) from *Deinococcus radiodurans* at a 3.1 Å resolution (Harms *et al.*, 2001).

Subsequently, the group has identified the binding sites of many of the antibiotics that bind to the ribosome (Harms *et al.*, 2003). One of the remarkable observations made by Yonath is the twofold symmetry of the peptidyl transfer site of the 50S ribosomal subunit (Agmon *et al.*, 2003). Not only are the acceptor ends of the tRNAs binding to their A and P sites related by a twofold axis, as observed by the Steitz group, but the ribosomal RNA also shows the same symmetry. It is hard to find any other explanation for this other than a gene duplication and gene fusion far back in the early evolution of ribosomes.

Selected References to the Work by Yonath

Yonath A, Müssig J, Tesche B *et al.* (1980) Crystallization of the large ribosomal subunit from *B. stearothermophilus*. *Biochem. lnt.* **1:** 428–435.

Glotz C, Mussig J, Gewitz HS *et al.* (1987) Three-dimensional crystals of ribosomes and their subunits from eu- and archaebacteria. *Biochem. Int.* **15:** 953–960.

*Yonath A, Wittmann HG. (1988) Approaching the molecular structure of ribosomes. *Biophys. Chem.* **29:** 17–29.

Hope H, Frolow F, von Böhlen K *et al.* (1989) Cryocrystallography of ribosomal particles. *Acta Cryst. B* **45:** 190–199.

Harms J, Tocilj A, Levin I *et al.* (1999) Elucidating the medium-resolution structure of ribosomal particles: An interplay between electron cryo-microscopy and X-ray crystallography. *Structure* **7:** 931–941.

Tocilj A, Schlunzen F, Janell D *et al.* (1999) The small ribosomal subunit from *Thermus thermophilus* at 4.5Å resolution: Pattern fittings and the identification of a functional site. *Proc. Natl. Acad. Sci. USA* **96:** 14252–14257.

Schlünzen F, Tocilj A, Zariwach R *et al.* (2000) Structure of functionally activated small ribosomal subunit at 3.3Å resolution. *Cell* **102:** 615–623.

Pioletti M, Schlunzen F, Harms J *et al.* (2001) Crystal structures of complexes of the small ribosomal subunit with tetracycline, edeine and IF3. *EMBO J.* **20:** 1829–1839.

Harms J, Schluenzen F, Zarivach R *et al.* (2001) High resolution structure of the large ribosomal subunit from a mesophilic eubacterium. *Cell* **107:** 679–688.

Harms JM, Bartels H, Schlünzen F, Yonath A. (2003) Antibiotics acting on the translational machinery. *J. Cell Sci.* **116:** 1391–1393.

Agmon I, Auerbach T, Baram D *et al.* (2003) On peptide bond formation, translocation, nascent protein progression and the regulatory properties of ribosomes. *Eur. J. Biochem.* **270:** 2543–2556.

Another Reference of Interest

Trakhanov SD, Yusupov MM, Agalarov SC *et al.* (1987) Crystallization of 70 S ribosomes and 30 S ribosomal subunits from *Thermus thermophilus*. *FEBS Lett.* **220:** 319–322.

Venkatraman (Venki) Ramakrishnan was born in 1952 in Chidambaram in Tamil Nadu, India. His parents were both scientists and his father was away on a postdoctoral fellowship in Madison, Wisconsin, at the time of his birth. On several occasions, young Ramakrishnan was left in the hands of his extended Indian family while his parents were abroad for long sessions of work. The life in a scientifically oriented family with visitors from near and far made science look like an interesting way forward for Venki. His school years were spent in an English language school. The distractions of a broad range of other interests could have lowered his school grades. However, a challenging teacher in mathematics and science caught his full attention. After graduation, he started to study physics at the age of 16 at the University of Baroda in his hometown.

Having finished his undergraduate studies, he joined his parents at the University of Illinois, Urbana, instead of continuing his studies in India. He was admitted to the graduate program in physics at Ohio University. His research project did not engage his interest and he felt that he might have chosen the wrong field. By reading the journal *Scientific American*, he realized that life sciences was a field of numerous exciting discoveries. When he obtained his Ph.D. in physics in 1976, it was not easy to switch to a career in life sciences. However, he was accepted onto a graduate program in biology at the University of California, San Diego.

After two years of graduate studies, Ramakrishnan already felt sufficiently comfortable with biology not to pursue another Ph.D. in this subject area. When reading an article by Don Engelman and Peter Moore in *Scientific American* regarding their work on ribosomes, he felt this to be an exciting new direction. Moore offered him a postdoctoral position in locating the positions of the ribosomal proteins in the small ribosomal subunit by reconstituting pairs of deuterated proteins into an otherwise protonated subunit. The measurements were carried out by small-angle neutron scattering at Brookhaven National Laboratory (Capel *et al.*, 1987). The work at Yale on ribosomes made a lasting impression on Ramakrishnan. His next position, at the Neutron Reactor Center at Oak Ridge, was not the type of job Ramakrishnan had been looking for, and after a year, he started a more suitable job at Brookhaven in 1983. Here, he had the time and resources to perform his own research. By a lucky coincidence, another ribosomologist, Steve White, moved to Brookhaven somewhat later. Jointly, they cloned and expressed ribosomal proteins for crystallographic structure determination (Ramakrishnan & White, 1998). At this point, Ramakrishnan felt that he really needed to learn crystallography. He took a course in crystallography at Cold Spring Harbor and succeeded in crystallizing ribosomal proteins as well as histones.

To learn more crystallography, there was no better place than the Medical Research Council (MRC) laboratory in Cambridge, the birthplace of crystallography. Venki wrote a letter to Aaron Klug, the director at that time, and he invited Venki to spend a sabbatical year there in 1991. The excellence of the lab, the interesting projects, and the friendships he made there left a lasting impression on Venki. As a result of the sabbatical year, the two structures he had brought along were solved and subsequently published in *Nature*. Back in Brookhaven, it was obvious that the laboratory there could not compete with the one in Cambridge, but there were no positions open in Cambridge at that time. Instead, Ramakrishnan was invited to a position in Salt Lake City, Utah. Here, the laboratory focused on the structures of ribosomal proteins, some in complex with the parts of ribosomal RNA they bind to in the ribosome.

During his time in Salt Lake City, a plan to determine the structure of the whole 30S subunit took shape. Venki, together with two graduate students, started the project. At a second application to the laboratory in Cambridge, Venki received a positive response, and in 1999, he moved from Utah to Cambridge, but parts of his group remained in Salt Lake City. At that time, they already had crystals of 30S subunits as well as several low-resolution data sets. An important step in obtaining good crystals was to start from a preparation of small subunits lacking the ribosomal protein S1. For the next step, phase angle determination was crucial. A range of approaches was tried. From the experience of others, it seemed that a derivative with osmium hexamine could be a possible candidate for MAD experiments working at the LIII absorption edge of osmium. Other types of derivatives contained different tungsten clusters or tantalum bromide ($Ta_6Br_{12}^{2+}$). The Utah and Cambridge teams solved the phasing problem rapidly and a low-resolution structure was published in August 1999 (Clemons *et al.*, 1999). Since at least three other groups were working on ribosomal structures, the race to obtain a high-resolution structure was on. This required a large quantity of appropriate crystals and also obtaining beam time at suitable synchrotrons. Data were collected and analyzed in February 2000. Everything worked out well and the entire structure was solved and published (Wimberly *et al.*, 2000), along with separate studies of the binding sites of some antibiotics.

A very important contribution made by Ramakrishnan is the analysis of the decoding of mRNA by tRNAs (Ogle & Ramakrishnan, 2005). This was performed with small subunits and short pieces of mRNA and fragments of tRNA. It had been a classical problem as to how the ribosome could contribute to the accuracy of tRNA selection. A central part of the observations by Ramakrishnan was that three bases (G530, A1492, and A1493) of the small subunit ribosomal RNA (16S RNA) could change their conformation in order to stabilize the binding of cognate tRNA (Carter *et al.*, 2000; Ogle *et al.*, 2001). This is primarily done through hydrogen bonding to both riboses of a Watson–Crick base pair. The ribosome functions like a measuring device since the distance between the riboses will be different depending on the type of base pairing. Only the Watson–Crick base pairs would permit the observed hydrogen bonds. Antibiotics like paromomycin disturb this process by inducing the same conformational changes regardless of whether the codon–anticodon interaction is cognate or not.

Similarly to the studies of isolated proteins presented earlier, the isolated ribosomal subunits could not contribute to a full understanding of protein synthesis. Thus, efforts had to focus on full 70S ribosomes. Ramakrishnan and his group succeeded in this and published very interesting work on the binding to the ribosome of tRNAs, elongation factors Tu (EF-Tu) and G (EF-G), release factors RF2 and RF3, as well as the ribosome recycling factor, RRF (for a detailed discussion, see Liljas & Ehrenberg, 2013).

More recently, Ramakrishnan has also started work on eukaryotic ribosomes. Thus, structures of yeast ribosomes and yeast mitochondrial large ribosomal subunits have been reported (Amunts *et al.*, 2014). The method of structural analysis being used now is single-particle reconstructions from cryo-electron microscopy observations, where the resolution attained is comparable with crystallography.

Selected References to the Work by Ramakrishnan

Capel MS, Engelman DM, Freeborn BR *et al.* (1987) A complete mapping of the proteins in the small ribosomal subunit of *Escherichia coli*. *Science* **238:** 1403–1406.

Ramakrishnan V, White SW. (1998) Ribosomal protein structures: Insights into the architecture, machinery and evolution of the ribosome. *Trends Biochem. Sci.* **23:** 208–212.

Clemons WM Jr., May JL, Wimberly BT *et al.* (1999) Structure of a bacterial 30S ribosomal subunit at 5.5 Å resolution. *Nature* **400:** 833–840.

Wimberly BT, Brodersen DF, Clemons WM Jr. *et al.* (2000) Structure of the 30S ribosomal subunit. *Nature* **407:** 327–339.

Carter AP, Clemons WM, Brodersen DE *et al.* (2000) Functional insights from the structure of the 30S ribosomal subunit and its interactions with antibiotics. *Nature* **407:** 340–348.

*Ogle JM, Brodersen DE, Clemons WM Jr. *et al.* (2001) Recognition of cognate transfer RNA by the 30S ribosomal subunit. *Science* **292:** 897–902.

Ogle JM, Ramakrishnan V. (2005) Structural insights into translational fidelity. *Annu. Rev. Biochem.* **74:** 129–177.

Selmer M, Dunham CM, Murphy FV *et al.* (2006) Structure of the 70S ribosome complexed with mRNA and tRNA. *Science* **313:** 1935–1942.

Amunts A, Brown A, Bai X *et al.* (2014) Structure of the yeast mitochondrial large subunit. *Science* **343:** 1485–1489.

Another Reference of Interest

Liljas A, Ehrenberg M. (2013) *Structural Aspects of Protein Synthesis*, 2nd ed. World Scientific Publishing, New Jersey, London, Singapore.

Thomas Steitz was born in 1940 and grew up in Milwaukee, Wisconsin, together with two younger brothers and two younger sisters. During his school years, in addition to the academic training, he enjoyed music (saxophone and singing) and training in different manual skills, which later became very important in developing various models of proteins. After high school, he spent four years at a liberal arts school, Lawrence College. Here, an inspiring teacher developed his interest for chemistry into a professional interest. During the end of his college studies, he was invited to participate in a conference for college students at MIT. The lecturers were MIT and Harvard professors. Their presentations were truly inspiring and the contact with equally interested students was a significant bonus.

Steitz was admitted to graduate school at Harvard in 1963. A lecture by Max Perutz changed his initial direction. Seeing the myoglobin structure made it obvious that in order to understand chemistry on a molecular level, you would have to *see* the molecules of interest. Steitz approached William Lipscomb and was invited to participate in the crystallographic analysis of carboxypeptidase. Some years later, a low-resolution structure was produced. Studies of an inhibitor complex demonstrated the first case of substrate-induced conformational changes in a protein studied by crystallography. In 1967, the high-resolution structure was solved, only two years after the first enzyme structure (of hen egg white lysozyme) was solved.

For three years from 1967, Steitz worked with Richard Henderson in David Blow's group at the MRC laboratory in Cambridge. The work focused on the binding of substrates to chymotrypsin. The canteen was one of the most valuable places in the laboratory. Here, one would get to know the whole laboratory and obtain advice on how to avoid performing experiments poorly. During his years in Cambridge, Steitz developed a particular interest in the molecules involved in Crick's central dogma: DNA → RNA → protein. However, it was suggested to him that he first work on something more easily achievable. The suggested project was hexokinase, which he successfully worked on for 10 years.

Steitz was offered a job at Berkeley, but they would not offer a position to his scientist wife Joan Steitz. However, Yale offered positions to the two Steitzes, and Tom started there in 1970. His work revealed interesting conformational changes during the catalytic cycle of hexokinase and offered a clear example of an induced fit mechanism.

From the early 1980s, Steitz's work was focused on structures of proteins associated with the central dogma. The first was a gene activator protein, CAP, which gave insights into DNA bending, among other observations. Subsequently, work on DNA polymerase was initiated. The Klenow fragment of DNA polymerase I with bound DNA led to the understanding of the role of metals — two magnesium ions — in the catalytic site.

The next step in the understanding of this large family of molecules was determining the structure of glutamine-tRNA synthetase in complex with $tRNA^{Gln}$. This enzyme belonged to the class (I) of synthetases that had already been shown to have an active site domain with a Rossmann fold. A year later, the first structure of the class II synthetases was published. Here, the domain with the active site had a fold with antiparallel β-strands (Cusack *et al.*, 1990; Eriani, 1990). The next group of Steitz's crystallographic targets was the RNA polymerases. Work on T7 RNA polymerase yielded many structures in different states.

In 1995, Steitz and his coworkers decided to use the best crystals available in an effort to solve the structure of the ribosome. In this, Steitz benefitted from having a close collaborator nearby who had spent his entire scientific life studying the process of mRNA translation in ribosomes. This was Peter Moore, who was also the postdoc PI of Venkatraman Ramakrishnan. Steitz and Moore decided to work on the structure of the 50S subunit of ribosomes from *Haloarcula marismortui*. Ada Yonath's group had published a 7 Å structure with crystals of this subunit, which, however, were thin and frequently twinned.

A postdoc in the lab, Poul Nissen, found a way to solve these problems and obtained much improved crystals that could diffract beyond 2.4 Å. For obtaining the phase angles, the obvious choice was different heavy atom clusters. A single W_{18} cluster was located using classical difference Patterson methods. Its position was confirmed using molecular replacement from a 20 Å cryo-electron microscopy map of the subunit obtained from Joachim Frank. From the W_{18} cluster, the positions of other heavy atom clusters were determined and a 9 Å map was published (Ban *et al.*, 1998). Here, RNA helices could be identified, which proved that the approach worked. Subsequently, a 5 Å map was published (Ban *et al.*, 1999), and a year later, a 2.4 Å structure was published with atomic details of the whole large subunit (Ban *et al.*, 2000; Nissen *et al.*, 2000). Among other features, the map showed a tunnel through the subunit. This had been seen before via cryo-electron microscopy and identified as the tunnel through which the growing peptide reaches the cytoplasm. Another observation was that the peptidyl transfer center seemed to be without any nearby protein molecule. This suggested that the ribosome functions as a ribozyme, as had previously been suggested by several workers in the ribosome field. An obvious continuation of the 50S subunit work was to identify sites for antibiotics and to perform studies of the binding of substrate analogues in order to understand the peptidyl transfer activity (Hansen *et al.*, 2002, 2003). The highly interesting proposals from Steitz's laboratory regarding the catalytic process have been intensely discussed and refined. A chemical group that is one of the most important contributors to catalysis is the terminal 2'-OH of the P-site tRNA. This suggests that peptidyl transfer is a case of substrate-induced catalysis.

The work by Steitz on ribosomes has continued along several paths. This includes work on bacterial 70S particles to which a range of translation and ribosomal factors are bound, for instance during ribosome hibernation. The binding of ribosomal antibiotics continues to be of significant interest, and work on eukaryotic subunits is also highly interesting.

Selected References to the Work by Steitz

Brautigam CA, Steitz TA. (1998) Structural and functional insights provided by crystal structures of DNA polymerases and their substrate complexes. *Curr. Opin. Struct. Biol.* **8:** 54–63.

Cheetham GMT, Steitz TA. (1999) Structure of a transcribing T7 RNA polymerase initiation complex. *Science* **286:** 2305–2309.

Ban N, Freeborn B, Nissen P *et al.* (1998) A 9Å resolution X-ray crystallographic map of the large ribosomal subunit. *Cell* **93:** 1105–1115.

Ban N, Nissen P, Hansen J *et al.* (1999) Placement of protein and RNA structures into a 5Å resolution map of the 50S ribosomal subunit. *Nature* **400:** 841–847.

*Ban N, Nissen P, Hansen J *et al.* (2000) The complete atomic structure of the large ribosomal subunit at 2.4Å resolution. *Science* **289:** 905–920.

Nissen P, Ban N, Hansen J *et al.* (2000) The structural basis of ribosome activity in peptide bond synthesis. *Science* **289:** 920–930.

Hansen JL, Schmeing TM, Moore PB, Steitz TA. (2002) Structural insights into peptide bond formation. *Proc. Natl. Acad. Sci. USA* **99:** 11670–11675.

Hansen JL, Moore PB, Steitz TA. (2003) Structure of five antibiotics bound at the peptidyl transferase center of the large ribosomal subunit. *J. Mol. Biol.* **330:** 1061–1075.

Schmeing TM, Kitchen D, Scaringe SA *et al.* (2005) Structural insights into the roles of water and the 2' hydroxyl of the P-site tRNA in the peptidyl transferase reaction. *Mol. Cell* **20:** 437–448.

Summary and Consequences of the Work

The structure of the ribosome has been of significant interest for a long time. A number of groups have been involved and many different techniques have been employed (for an extensive account, see Liljas & Ehrenberg, 2013). The most distinct progress has been made due to cryo-electron microscopy and crystallography. Most of the work in cryo-electron microscopy has been carried out by Joachim Frank's group. The crystallographic work was initially performed in four places after the early start by Ada Yonath. In addition to her work and the work by Ramakrishnan and Steitz, the group led by Harry Noller published a 7 Å structure of the complete 70S ribosomal structure from *T. thermophilus* (Cate *et al.*, 1999) and subsequently a 5.5 Å structure (Yusupov *et al.*, 2001). Cryo-electron microscopy and crystallography are now performed in several more laboratories. Now that well-diffracting crystals of both *E. coli* and *T. thermophilus* 70S ribosomes can be produced, there are advantages to working with the complete system. For example, structure–function aspects are normally best analyzed in a complete system. Among other observations, it has now been observed that there is a *T. thermophilus* protein, L27, close to the peptidyl transfer site. This may seem to question the ribozyme nature of the ribosome — its role may rather be one of stabilization. The main role of the ribosome in the catalytic process seems to be to place the substrates close to each other and in the correct orientation. A highly critical component is the 2' ribose hydroxyl of the terminal adenine of the peptidyl tRNA in the active site, suggesting that peptide transfer is substrate assisted. The studies following these structure determinations have illustrated many of the steps involved in protein synthesis in the ribosome, including structures of eukaryotic ribosomes. Ribosomes are the targets of numerous antibiotics, which, due to the growing problem of antibiotic resistance in bacteria, have become of limited clinical use. Studies of the binding of old and the search for new antibiotics are some of the very important consequences of crystallography in ribosomes.

Other References of Interest

Liljas A, Ehrenberg M. (2013) *Structural Aspects of Protein Synthesis*, 2nd ed. World Scientific Publishing, New Jersey, London, Singapore.

Frank J, Zhu J, Penczek P *et al.* (1995) A model of protein synthesis based on cryo-electron microscopy of the *E. coli* ribosome. *Nature* **376:** 441–444.

Cate JH, Yusupov MM, Yusupova GZ *et al.* (1999) X-ray crystal structures of 70S ribosome functional complexes. *Science* **285:** 2095–2104.

Yusupov MM, Yusupova GZ, Baucom A *et al.* (2001) Crystal structure of the ribosome at 5.5Å resolution. *Science* **292:** 883–896.

Franceschi F, Duffy EM. (2006) Structure-based drug design meets the ribosome. *Biochem Pharmacol.* **71:** 1016–1025.

Dan Shechtman (C2011)

"For the discovery of quasicrystals."

Dan Shechtman
Born 24 January 1941
(Obtained through private correspondence)

In 1906, **Dan Shechtman**'s grandfather came from Russia to Israel, where he set up a printing house. He was one of the leaders of the Labor Movement. Dan Shechtman was born in Tel Aviv in 1941 and grew up in a neighboring village. As a child, he read the book *The Mysterious Island* by Jules Verne and was fascinated by the hero Cyrus Smith. Danny also wanted to become an engineer, like Cyrus, who, with knowledge in physics and mechanics, seemed to be able to create a fantastic life on the island from nothing. Shechtman went to Technion in Haifa to study mechanical engineering, which he completed in 1968. He obtained his Ph.D. in 1972 in material techniques. Soon afterwards, he moved with his family — his wife and, at that time, three daughters — to the Wright-Patterson Air Force Base in Ohio. There, he studied the microstructure and metallurgical properties of titanium aluminide, a Ti-Al alloy with high strength and thermal stability. After three years, he returned to Technion in Haifa and joined the department of materials science, where he is now Philip Tobias Professor of Materials Science and heads the Louis Edelstein Centre for Quasicrystals.

In 1981–1983, he was working at the National Bureau of Standards (presently National Institute of Standards and Technology [NIST]), where he studied rapidly solidified aluminum transition metal alloys. During this period, he discovered the icosahedral phase, which opened up the new field of quasiperiodic crystals. He currently spends about five months a year in Ames, Iowa, on a part-time appointment, and joined the Iowa State faculty in 2004. In 1992–1994, he was on sabbatical at NIST, during which time he studied the effects of the defect structure of chemical vapor-deposited diamond on its growth and properties.

Between 2001 and 2004, Shechtman served as chairperson of the Sciences Division, Israel Academy of Sciences and Humanities. Now as a member, he continues to oversee the translation of the Nobel Prize scientific posters into Hebrew, which are distributed to schools throughout the country.

Shechtman's Nobel Prize follows many other prestigious awards, including the Aminoff Prize from the Royal Swedish Academy of Sciences (2000), the Wolf Foundation Prize in Physics (1999), the Israel Prize in Physics (1998), the Weizmann Prize in Science (1993), the Rothschild Prize in Engineering (1990), and the International Award for New Materials of the American Physical Society (1987).

The Discovery of Quasicrystals

When Shechtman worked at the National Bureau of Standards in 1981–1983, he studied rapid cooling of certain Al alloys with Fe and Mn. On April 8 1982, he investigated an alloy with 14 at% Mn with electron diffraction and saw something highly remarkable and unexpected: The diffraction pattern indicated fivefold or tenfold symmetry of the sample (Fig. 1; diffraction patterns are always centrosymmetric and patterns with fivefold symmetry will exhibit diffractograms with tenfold symmetry). The occurrence of fivefold or tenfold symmetry was contrary to all of the established concepts in crystallography. The simplest explanation at hand was twinning of the sample, but after very careful investigation, Shechtman could eliminate this possibility (Fig. 2). At first, his colleagues did not believe him when he claimed to have discovered a new type of ordered phase with classically forbidden symmetry. For these reasons, it was a long time before he published the discovery. The first report was sent in 1984 to the *Journal of Applied Physics* and was returned very quickly — it was a boring metallurgical report of no interest to physicists! After discussions with some colleagues, a modified report was sent to *Physical Review Letters*. This paper, entitled *Metallic phase with long range orientational order and no translational symmetry*, by Shechtman, Blech, Gratias, and Cahn, appeared in November 1984.

A frantic burst of activity began almost immediately in the United States and Europe, particularly in France. Many so-called quasicrystals were prepared by rapid solidification of different metal alloys. A quasicrystal group was formed at the National

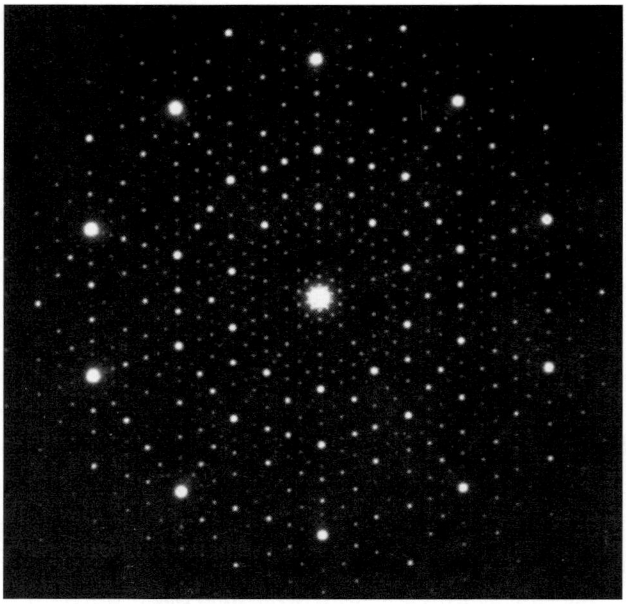

Fig. 1. The electron diffraction pattern with forbidden tenfold symmetry discovered by Shechtman in 1982. Note that the golden ratio τ appears in the relative distances between successive "Bragg" spots in the diffractogram.

Fig. 2. From Shechtman's notebook dated April 8 1982, showing the observation of tenfold symmetry.

Bureau of Standards, and in the first year alone, 300 preprints were received. The field expanded very quickly and a large amount of quasicrystalline intermetallic compounds were prepared, many without much difficulty. We now know of more than 100 binary and a much greater number of ternary combinations of metals that form quasiperiodic crystals. Almost all quasicrystalline materials known today are in intermetallic phases; one reason for this is that the absence of directional forces in metals allows a great flexibility in the relative arrangement of the atoms. The first quasicrystals were as small as one micrometer and had to be studied by electron diffraction, but a few years later, single-crystal X-ray diffraction could also be applied in order to obtain precise parameters. Today, we can produce large quasicrystals that are several centimeters in length.

Disbelief in the Discovery

When, in April 1982, Shechtman told his colleagues at the NIST about his discovery, he met with a variety of reactions, some positive and some negative. His host, John Cahn, said: "Danny, this material is telling us something and I challenge you to find what it is." The head of his laboratory, on the other hand, was afraid that what Shechtman was talking about might harm the reputation of his laboratory and asked Shechtman to leave. Shechtman left the lab and found another group that adopted the scientific orphan.

Shechtman's discovery was quickly accepted by many in the scientific community in general as soon as the first report came into print in 1984, but Linus Pauling was not among them. Over many years, Pauling and Shechtman discussed the interpretation of the diffractograms. Pauling persistently maintained that the crystals must be twins. He was never convinced regarding Shechtman's interpretation, in spite of Shechtman's efforts. Shechtman has said that at a science conference in front of an audience of hundreds where Shechtman was also present, Pauling claimed: "Danny Shechtman is talking nonsense; there are no quasi-crystals, just quasi-scientists." With Pauling's solid reputation, a lot of people also came to doubt that anything new and revolutionary had been detected.

At a later time, Pauling asked Shechtman if he would like to publish a paper together with him on twinning. Shechtman replied that he would be delighted to write the joint paper, but coming to a mutual agreement that quasiperiodic crystals exist was his precondition. Pauling replied that it was too early for that. No joint paper was ever published.

From the above, one will get the impression that Pauling more or less took it for granted that quasicrystals are impossible. However, an examination of his research notebooks (available at Oregon State University) reveals that he was actually seriously working on the problem and critically examined data from various samples. The following note from July 13 1984 is one example (Fig. 3):

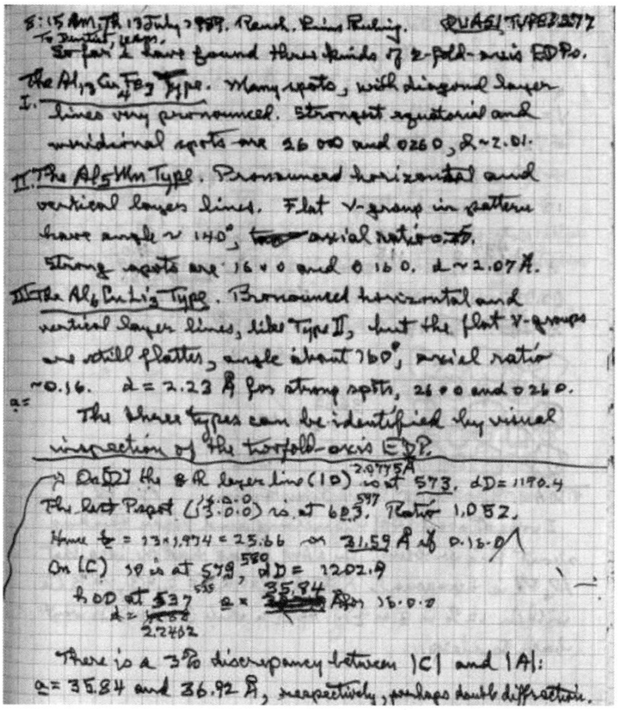

Fig. 3. Pauling's research notebook from July 13 1984.

Pauling and Shechtman had a lively correspondence for several years. It reflected cordial but tense scientific competition. After a year of silence, Pauling wrote to Shechtman on October 6 1987. By now, his tone had shifted. In addition to thanking Shechtman for the many glossy X-ray photographs and diffraction data calculations he had provided him with over the years, Pauling also thanked Shechtman for his discovery of quasicrystals. He wrote:

> This discovery has resulted in a great contribution to crystallography and metallurgy, in that it has stimulated hundreds of investigators to study alloys and has led to much additional knowledge about intermetallic compounds… Your discovery has also made me happier… For over two years I have worked on this problem, and have enjoyed myself while doing it. I estimate that I have spent nearly 1,000 hours just thinking on this whole question, and more than 1,000 hours making calculations, and writing papers.

This letter is the last archived piece of correspondence between Pauling and Shechtman.

How should the above letter be interpreted? Pauling seems to have defended his multiple-twinning theory until his death in 1994, despite growing evidence for the fact that quasicrystals do exist. Maybe he simply wanted to thank Shechtman for a discovery "which stimulated worldwide research in this area as well as stimulating his own work." Pauling's dedication to his own ideas and mind to question the results of colleagues illustrate the intellectual drive and dynamic spirit that made Pauling such a brilliant scientist.

For an interesting summary of the Pauling–Shechtman debate, go to: http://paulingblog.wordpress.com/tag/quasicrystals (cf. also Hargittai & Hargittai, 2000).

The Structure of Quasicrystals

In a conventional crystal, the atoms are located according to a certain basic pattern, the contents of the unit cell, which is identical throughout the crystal by translation in three dimensions and the pattern is thus strictly periodic. Quasicrystals are non-periodic objects (without translation symmetry) and a quasicrystal has no underlying average structure. Whereas the atomic arrangement in a classical crystal can be described by the contents in one single unit cell, a complete description of the structure of a quasicrystal means that the position of every single atom in the whole crystal has to be specified. In spite of many years of great efforts, the determination of the atomic arrangement in quasicrystals remains a significant challenge.

The remarkable discovery by Shechtman was to find diffraction patterns with classically forbidden symmetry. How is it possible that a diffraction pattern with discrete spots can be obtained from a quasicrystal? Crystallographers used to believe that a diffraction pattern with discrete spots only occurs on scattering from a regularly repeated arrangement — a lattice. However, Alan MacKay had already demonstrated in 1982 that a two-dimensional Penrose arrangement (Fig. 4) gives rise to a sharp optical diffraction pattern. This can be mathematically explained by the properties of nearly periodic (quasiperiodic) functions: A Penrose pattern ("tiling") may be considered to be a picture of such a nearly periodic function. The Fourier transform of a quasiperiodic function is a series of Dirac δ-maxima in the same way as for an ideal crystalline arrangement. In a diffraction experiment, quasicrystals will, in a similar way, give rise to discrete reflections, but these cannot be indexed in a three-dimensional lattice. For several years, it seemed as if the symmetries that were possible for aperiodic tilings were extremely limited, but recent discoveries have revealed aperiodic tilings with local symmetries that are eightfold, tenfold, twelvefold, and even sevenfold (cf. Senechal, 1996).

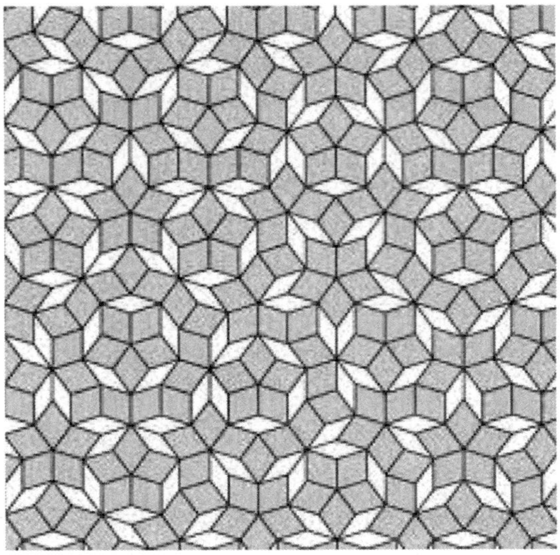

Fig. 4. The quasiperiodic pattern discovered by Penrose in 1974.

The Golden Ratio

It is interesting to note that the distance between successive "Bragg" spots in Shechtman's diffractogram are in the golden ratio $\tau = \left(1+\sqrt{5}\right)/2 \sim 1.6180\ldots$ Several other features of the diffractogram are also related to the golden ratio. This is due to the fivefold symmetry of the underlying structure. The golden ratio occurs in all diffractograms of icosahedral quasicrystals. In the quasicrystals with eightfold and twelvefold symmetry, however, the golden ratio does not appear. It is fascinating to note that the golden ratio τ appears in so many cases in Penrose tilings:

- The areas of the two rhombs are in the golden ratio τ.
- You need approximately τ as many "fat" rhombs as "thin" rhombs in order to form a tiling. For an infinite tiling, the relative number needed is exactly τ.

Recall also the Fibonacci numbers F_n: 1, 1, 2, 3, 5, 8, 13, 21, 34... defined by $F_{n+1} = F_n + F_{n-1}$. The ratio F_n/F_{n-1} approaches τ as n goes to infinity.

Applications

Quasicrystals have unique physical and chemical properties and many proposals for practical applications have been advanced. They are hard and corrosion resistant, and coatings made from quasicrystals are therefore one important area of application. Quasicrystals remain hard and tough at high temperatures and their increased ductility reduces the risk of cracking in substrates with different thermal expansions. Using quasicrystals for coatings on the wings, bodies, and turbine blades in aeroplanes is an interesting area of potential application. The low thermal conductivity of the quasicrystalline layer makes it possible to run turbine blades at higher temperatures without overheating the underlying construction material. Quasicrystals are able to reinforce other materials, for instance in the very tough and strong "supersteel" that has become a worldwide product and is used by surgeons in eye operations. In this instance, quasicrystals appear as a secondary phase in the steel and prevent dislocations from propagating through the steel and thus weakening it. Lightweight materials with high fractural strength have also been produced by embedding fine-grained quasicrystals in an Al matrix, and such alloys are commercially available. Another potential application of quasicrystals is based on their capacity to store hydrogen atoms. A quasicrystalline Ti–Zr–Ni alloy can absorb almost two hydrogen atoms per host metal atom. Absorption and desorption is fast and the alloy can be produced from low-cost materials. Several other applications will no doubt appear in the future.

Impact of the Discovery

Translational symmetry is a characteristic property of a normal crystalline arrangement and it is easy to prove that fivefold, sevenfold, and higher symmetries cannot occur in ideal crystals. The discovery of quasicrystals in 1982 of an ordered material with classically forbidden symmetry was something of a crystallographic bomb. It meant no less than a paradigm shift in our old concepts of what constitutes a crystalline material and a new definition of a crystal had to be introduced. The International Union of Crystallography formed a commission on Aperiodic Crystals and proposed the following working definition in 1992: "A crystal is any solid with an essentially discrete diffraction diagram." The focus has thus been shifted from the features of the structure itself to those of its diffraction image.

A quasiperiodic arrangement is evidently a *sufficient* condition for "Bragg" spots. However, a more general question is: What is the *necessary* condition for obtaining a discrete diffraction diagram? To our knowledge, this has not yet been solved mathematically. Tiling and quasicrystal research are fields that are currently experiencing significant development. A lot of exciting and difficult problems remain to be solved!

The discovery of quasicrystals has had significant impacts on many fields. The theory of tiling has been very important for the development of models of the structures of quasicrystals. This has attracted great interest among mathematicians, and in recent years, scientists in seemingly unrelated fields of research have also discovered that many problems translate into tiling problems. There is a wide variety of difficult and mathematically interesting tiling problems to be solved, and this field of mathematics is growing very rapidly. The physical properties of quasicrystals are often quite different from ordinary crystalline substances, and many of these are still not well understood.

The very interesting electronic properties of quasicrystals have, for instance, led to intensive research in theoretical and experimental solid-state physics.

References

The bibliography of quasicrystals and tiling is very extensive and, for more complete listings, the monographs below should be consulted.

Cahn JW, Gratias D, Shechtman D. (1986) Pauling's model not universally accepted. *Nature* **319:** 102–103.

Grünbaum B, Shephard G. (1987) *Tilings and Patterns.* W.H. Freeman & Co, New York.

Hargittai I, Hargittai M. (2000) *In Our Own Image: Personal Symmetry in Discovery.* Kluwer/Plenum, New York.

Mackay A. (1982) Crystallography and the Penrose pattern. *Physica A* **114:** 609–613.

Schattschneider D. (1981) In: Klarner D (ed.), *The Mathematical Gardner*, Prindle, Weber & Schmidt, Boston, pp. 140–166.

Senechal M. (1996) *Quasicrystals and Geometry.* Cambridge University Press, Cambridge.

*Shechtman D, Blech I, Gratias D, Cahn JW. (1984) Metallic phase with long-range orientational order and no translational symmetry. *Phys. Rev. Lett.* **53:** 1951–1954.

Steinhardt PJ, Jeong H-C. (1996) A simpler approach to Penrose tiling with implications for quasicrystal formation. *Nature* **382:** 431–433.

Steinhardt PJ, Jeong H-C, Saitoh K *et al.* (1998) Experimental verification of the quasi-unit-cell model of quasicrystal structure. *Nature* **396:** 55–57.

Brian K. Kobilka and Robert J. Lefkowitz (C2012)

were awarded the Nobel Prize

"For studies of G-protein-coupled receptors."

Since only Kobilka employed crystallography, his research alone will be presented here.

Brian Kobilka
Born May 30 1955
(Photograph from Les Prix Nobel 2012)

Brian Kobilka was born in 1955 in Little Falls, Minnesota. Both his father and grandfather were bakers and Brian and his sister used to help in their bakery. After having finished high school in Little Falls, where his teachers encouraged him to focus on science, he studied biology and biochemistry at the University of Minnesota in Duluth and obtained his Bachelor's degree there. He continued his studies at Yale School of Medicine, where he obtained his MD. In 1984, he began his postdoctoral studies of the β_2-adrenergic receptor (β_2AR) at Duke University, together with Robert Lefkowitz. In 1989, he moved to Stanford after being offered a position there.

Ligand-activated receptors have been known for a long time. One group of such receptors in eukaryotes is called G-protein-coupled receptors (GPCRs). They transmit signals from the outside of the cell to the inside, where the signal activates different types of responses. As is evident from the name, G-proteins are involved in this process. The G-proteins are then located inside the cell and are activated by binding to guanosinetriphosphate (GTP). At some point in this process, GTP is hydrolyzed to guanosinediphoshate (GDP) and inorganic phosphate and so become inactivated. There are different classes of G-proteins. Those that interact with GPCRs are trimeric, where the α-subunit binds the GTP, whereas the β- and γ-subunits dissociate from the activated α-subunit. The GPCRs are located in the membrane and composed of seven trans-membrane helices (7TM). When the GPCR receives a signal from outside of the cell, such as the binding of a hormone, it changes its conformation. This leads to the binding of a G-protein that is specific for this GPCR. The G-protein is thereby induced to exchange its GDP for a GTP molecule. The activated α-subunit then targets and interacts with its effector in order to induce the cellular activity that the initial signal was meant to produce. The ternary complex of a GPCR with an attached signaling molecule on the outside and a G-protein on the inside is evidently of particular interest.

In humans, there are in the order of 800 different GPCRs, which corresponds to approximately 4% of all protein-coding genes. A classical GPCR molecule is rhodopsin, which is involved in our visual system. A different and large group of receptors are olfactory (smell) and pheromone (chemicals that affect behavior) receptors. Other GPCRs bind neurotransmitters such as serotonin, dopamine, or γ-aminobutyric acid. Furthermore, the heart rate, blood pressure, and digestive processes are controlled by GPCRs. It is obvious that GPCRs are of great pharmacological interest. Almost half of all prescribed drugs are directed to GPCRs directly or indirectly. Problems such as high blood pressure, ulcers, allergies, migraines, and bipolar disorders are treated with GPCR-related medicines. The structures of GPCRs are therefore very important for attempts to develop new drugs against well-known and new disorders.

For these reasons, the structural determination of GPCRs is a hot field with many competing groups. The first structure of a 7TM protein was bacteriorhodopsin, which was initially investigated via electron crystallography from a crystalline monolayer. The first structure of a mammalian GPCR was that of bovine rhodopsin (Palczewski *et al.*, 2000). In 2007, Kobilka and his group characterized the first human GPCR, β_2AR (Rasmussen *et al.*, 2007), followed by a series of others. The different kinds of 7TM or GPCR molecules differ primarily in the structure of the internal and external loops that connect the seven helices. Such differences have important effects, since the three loops on the outside of the cell need to identify different signal substances, whereas the three loops on the inside recognize their specific G-proteins. Once these basic structures were available, the interplay between the signal, the receptor, and the G-protein became a central issue. Solving this problem turned out to be a very difficult task.

The quantity of specific GPCRs in cells is extremely low. After successful purification and determination of the amino acid sequence of some peptides, work could be aimed at cloning and over-expression. One of Kobilka's first tasks during his time at Duke with

Lefkowitz was to clone β_2AR (Dixon *et al.*, 1986). One surprising observation was the presence of 7TM regions that were previously identified for light-sensitive molecules (Lefkowitz, 2013). In order to obtain an understanding of the essential components of GPCRs, Kobilka continued with comparative site-directed mutagenesis studies of related receptors. Determining their structures was the obvious route forwards.

The crystallization attempts started around 1998 and the first crystals were obtained in 2004. However, the crystals were very small and diffracted only to approximately 20 Å. The molecular flexibility of the protein, particularly involving the intracellular loop between the trans-membrane helices TM5 and TM6, could have been a reason for this difficulty in producing high-quality crystals. Two approaches were tried successfully. One was to bind the antigen-binding fragment of the antibody Fab from monoclonal antibodies to this region, and another was to replace the loop with lysozyme from bacteriophage T4. Both led to well-resolved structures (Cherezov *et al.*, 2007; Rosenbaum *et al.*, 2007). Kobilka's group, as well as other groups, used the same fusion peptide for several structural determinations. However, studying the ligand-activated β_2AR was difficult, probably due to the flexibility in the binding area of the G-protein. Here, monomeric antibodies from llamas (popularly called "nanobodies") came to the rescue. However, the main research focus was on the ternary complex of a GPCR, a G-protein, and a ligand (also called an agonist). Here, a range of stabilizing tricks was needed: An agonist with ultra-high affinity, a new detergent for crystallization, a nanobody with a specific stabilizing binding site on the trimeric G-protein, and a fusion with T4 lysozyme to a different site. The structure of the ternary complex revealed the active state of the GPCR with distinctive conformational changes for the binding of the G-protein (Rasmussen *et al.*, 2011). It was discovered that the G-protein also undergoes a large conformational change by which it loses its affinity for GDP.

Selected References to the Work by Kobilka

Cherezov V, Rosenbaum DM, Hanson MA *et al.* (2007) High-resolution crystal structure of an engineered human beta2-adrenergic G protein-coupled receptor. *Science* **318:** 1258–1265.

Dixon RAF, Kobilka BK, Strader DJ *et al.* (1986) Cloning of the gene and cDNA for mammalian β-adrenergic receptor and homology with rhodopsin. *Nature* **321:** 75–79.

Rasmussen SG, Choi H, Rosenbaum DM *et al.* (2007) Crystal structure of the human β_2 adrenergic G-protein coupled receptor. *Nature* **450:** 383–387.

*Rasmussen SGF, DeVree BT, Zou Y *et al.* (2011) Crystal structure of the β_2 adrenergic receptor–Gs protein complex. *Nature* **477:** 549–555.

Rosenbaum DM, Cherezov V, Hanson MA *et al.* (2007) GPCR engineering yields high-resolution structural insights into beta2-adrenergic receptor function. *Science* **318:** 1266–1273.

Another Reference of Interest

Palczewski K, Kumasaka T, Hori T *et al.* (2000) Crystal structure of rhodopsin: A G protein-coupled receptor. *Science* **289:** 739–745.

Conclusions

The work by Kobilka has provided the first insight into the complex interactions that a GPCR undergoes. This will naturally be followed by many more related studies, making use of similar techniques for crystallization. This will provide the necessary knowledge for the design of new and improved drugs against a multitude of diseases and medical problems.

Reprints

1898. ANNALEN № 1.
DER
PHYSIK UND CHEMIE.
NEUE FOLGE. BAND 64.

1. *Ueber eine neue Art von Strahlen;*[1]
von W. C. Röntgen.

(Erste Mittheilung.)

Aus den Sitzungsber. der Würzburger Physik.-Medic. Gesellsch. Jahrg. 1895.

1. Lässt man durch eine Hittorf'sche Vacuumröhre, oder einen genügend evacuirten Lenard'schen, Crookes'schen oder ähnlichen Apparat die Entladungen eines grösseren Ruhmkorff gehen und bedeckt die Röhre mit einem ziemlich eng anliegenden Mantel aus dünnem, schwarzem Carton, so sieht man in dem vollständig verdunkelten Zimmer einen in die Nähe des Apparates gebrachten, mit Baryumplatincyanür angestrichenen Papierschirm bei jeder Entladung hell aufleuchten, fluoresciren, gleichgültig ob die angestrichene oder die andere Seite des Schirmes dem Entladungsapparat zugewendet ist. Die Fluorescenz ist noch in 2 m Entfernung vom Apparat bemerkbar.

Man überzeugt sich leicht, dass die Ursache der Fluorescenz vom Entladungsapparat und von keiner anderen Stelle der Leitung ausgeht.

2. Das an dieser Erscheinung zunächst Auffallende ist, dass durch die schwarze Cartonhülse, welche keine sichtbaren oder ultravioletten Strahlen des Sonnen- oder des electrischen Bogenlichtes durchlässt, ein Agens hindurchgeht, das im Stande ist, lebhafte Fluorescenz zu erzeugen, und man wird deshalb

1) Nachdem durch Entgegenkommen von Seiten des Verlegers der Sitzungsber. der Würzburger Physik.-Medic. Gesellsch. der unveränderte Abdruck meiner beiden ersten Mittheilungen über X-Strahlen in den Annalen möglich geworden war, haben sich Redacteur und Verleger der Annalen zu meiner Freude bereit erklärt, die beiden älteren Arbeiten zusammen mit der dritten Mittheilung aufzunehmen, wofür ich den genannten Herren zu Dank verpflichtet bin.

wohl zuerst untersuchen, ob auch andere Körper diese Eigenschaft besitzen.

Man findet bald, dass alle Körper für dasselbe durchlässig sind, aber in sehr verschiedenem Grade. Einige Beispiele führe ich an. Papier ist sehr durchlässig[1]): hinter einem eingebundenen Buch von ca. 1000 Seiten sah ich den Fluorescenzschirm noch deutlich leuchten; die Druckerschwärze bietet kein merkliches Hinderniss. Ebenso zeigte sich Fluorescenz hinter einem doppelten Whistspiel; eine einzelne Karte zwischen Apparat und Schirm gehalten macht sich dem Auge fast gar nicht bemerkbar. — Auch ein einfaches Blatt Stanniol ist kaum wahrzunehmen; erst nachdem mehrere Lagen übereinander gelegt sind, sieht man ihren Schatten deutlich auf dem Schirm. — Dicke Holzblöcke sind noch durchlässig; 2 bis 3 cm dicke Bretter aus Tannenholz absorbiren nur sehr wenig. — Eine ca. 15 mm dicke Aluminiumschicht schwächte die Wirkung recht beträchtlich, war aber nicht im Stande, die Fluorescenz ganz zum Verschwinden zu bringen. — Mehrere Centimeter dicke Hartgummischeiben lassen noch Strahlen[2]) hindurch. — Glasplatten gleicher Dicke verhalten sich verschieden, je nachdem sie bleihaltig sind (Flintglas) oder nicht; erstere sind viel weniger durchlässig als letztere. — Hält man die Hand zwischen den Entladungsapparat und den Schirm, so sieht man die dunkleren Schatten der Handknochen in dem nur wenig dunklen Schattenbild der Hand. — Wasser, Schwefelkohlenstoff und verschiedene andere Flüssigkeiten erweisen sich in Glimmergefässen untersucht als sehr durchlässig. — Dass Wasserstoff wesentlich durchlässiger wäre als Luft, habe ich nicht finden können. — Hinter Platten aus Kupfer, bez. Silber, Blei, Gold, Platin ist die Fluorescenz noch deutlich zu erkennen, doch nur dann, wenn die Plattendicke nicht zu bedeutend ist. Platin von 0,2 mm Dicke ist noch durchlässig; die Silber- und Kupferplatten können schon stärker sein. Blei

1) Mit „Durchlässigkeit" eines Körpers bezeichne ich das Verhältniss der Helligkeit eines dicht hinter dem Körper gehaltenen Fluorescenzschirmes zu derjenigen Helligkeit des Schirmes, welcher dieser unter denselben Verhältnissen, aber ohne Zwischenschaltung des Körpers zeigt.

2) Der Kürze halber möchte ich den Ausdruck „Strahlen" und zwar zur Unterscheidung von anderen den Namen „*X*-Strahlen" gebrauchen. Vgl. p. 9.

in 1,5 mm Dicke ist so gut wie undurchlässig und wurde deshalb häufig wegen dieser Eigenschaft verwendet. — Ein Holzstab mit quadratischem Querschnitt (20 × 20 mm), dessen eine Seite mit Bleifarbe weiss angestrichen ist, verhält sich verschieden, je nachdem er zwischen Apparat und Schirm gehalten wird; fast vollständig wirkungslos, wenn die *X*-Strahlen parallel der angestrichenen Seite durchgehen, entwirft der Stab einen dunklen Schatten, wenn die Strahlen die Anstrichfarbe durchsetzen müssen. — In eine ähnliche Reihe, wie die Metalle, lassen sich ihre Salze, fest oder in Lösung, in Bezug auf ihre Durchlässigkeit ordnen.

3. Die angeführten Versuchsergebnisse und andere führen zu der Folgerung, dass die Durchlässigkeit der verschiedenen Substanzen, gleiche Schichtendicke vorausgesetzt, wesentlich bedingt ist durch ihre Dichte: keine andere Eigenschaft macht sich wenigstens in so hohem Grade bemerkbar als diese.

Dass aber die Dichte doch nicht ganz allein maassgebend ist, das beweisen folgende Versuche. Ich untersuchte auf ihre Durchlässigkeit nahezu gleich dicke Platten aus Glas, Aluminium, Kalkspath und Quarz; die Dichte dieser Substanzen stellte sich als ungefähr gleich heraus, und doch zeigte sich ganz evident, dass der Kalkspath beträchtlich weniger durchlässig ist als die übrigen Körper, die sich untereinander ziemlich gleich verhielten. Eine besonders starke Fluorescenz des Kalkspathes (vgl. unten p. 4) namentlich im Vergleich zum Glas habe ich nicht bemerkt.

4. Mit zunehmender Dicke werden alle Körper weniger durchlässig. Um vielleicht eine Beziehung zwischen Durchlässigkeit und Schichtendicke finden zu können, habe ich photographische Aufnahmen (vgl. unten p. 4) gemacht, bei denen die photographische Platte zum Theil bedeckt war mit Stanniolschichten von stufenweise zunehmender Blätterzahl; eine photometrische Messung soll vorgenommen werden, wenn ich im Besitz eines geeigneten Photometers bin.

5. Aus Platin, Blei, Zink und Aluminium wurden durch Auswalzen Bleche von einer solchen Dicke hergestellt, dass alle nahezu gleich durchlässig erschienen. Die folgende Tabelle enthält die gemessene Dicke in Millimetern, die relative Dicke bezogen auf die des Platinbleches und die Dichte.

	Dicke	relat. Dicke	Dichte
Pt	0,018 mm	1	21,5
Pb	0,05 „	3	11,3
Zn	0,10 „	6	7,1
Al	3,5 „	200	2,6

Aus diesen Werthen ist zu entnehmen, dass keineswegs gleiche Durchlässigkeit verschiedener Metalle vorhanden ist, wenn das Product aus Dicke und Dichte gleich ist. Die Durchlässigkeit nimmt in viel stärkerem Maasse zu, als jenes Product abnimmt.

6. Die Fluorescenz des Baryumplatincyanürs ist nicht die einzige erkennbare Wirkung der *X*-Strahlen. Zunächst ist zu erwähnen, dass auch andere Körper fluoresciren; so z. B. die als Phosphore bekannten Calciumverbindungen, dann Uranglas, gewöhnliches Glas, Kalkspath, Steinsalz etc.

Von besonderer Bedeutung in mancher Hinsicht ist die Thatsache, dass photographische Trockenplatten sich als empfindlich für die *X*-Strahlen erwiesen haben. Man ist im Stande, manche Erscheinung zu fixiren, wodurch Täuschungen leichter ausgeschlossen werden; und ich habe, wo es irgend anging, jede wichtigere Beobachtung, die ich mit dem Auge am Fluorescenzschirm machte, durch eine photographische Aufnahme controllirt.

Dabei kommt die Eigenschaft der Strahlen, fast ungehindert durch dünnere Holz-, Papier- und Stanniolschichten hindurchgehen zu können, sehr zu Statten; man kann die Aufnahmen mit der in der Cassette, oder in einer Papierumhüllung eingeschlossenen photographischen Platte im beleuchteten Zimmer machen. Andererseits hat diese Eigenschaft auch zur Folge, dass man unentwickelte Platten nicht bloss durch die gebräuchliche Hülle aus Pappendeckel und Papier geschützt längere Zeit in der Nähe des Entladungsapparates liegen lassen darf.

Fraglich erscheint es noch, ob die chemische Wirkung auf die Silbersalze der photographischen Platte direct von den *X*-Strahlen ausgeübt wird. Möglich ist es, dass diese Wirkung herrührt von dem Fluorescenzlicht, das, wie oben angegeben, in der Glasplatte, oder vielleicht in der Gelatineschicht erzeugt wird. „Films" können übrigens ebenso gut wie Glasplatten verwendet werden.

Dass die *X*-Strahlen auch eine Wärmewirkung auszuüben im Stande sind, habe ich noch nicht experimentell nachgewiesen; doch darf man wohl diese Eigenschaft als vorhanden annehmen, nachdem durch die Fluorescenzerscheinungen die Fähigkeit der *X*-Strahlen, verwandelt zu werden, nachgewiesen ist, und es sicher ist, dass nicht alle auffallenden *X*-Strahlen den Körper als solche wieder verlassen.

Die Retina des Auges ist für unsere Strahlen unempfindlich; das dicht an den Entladungsapparat herangebrachte Auge bemerkt nichts, wiewohl nach den gemachten Erfahrungen die im Auge enthaltenen Medien für die Strahlen durchlässig genug sein müssen.

7. Nachdem ich die Durchlässigkeit verschiedener Körper von relativ grosser Dicke erkannt hatte, beeilte ich mich, zu erfahren, wie sich die *X*-Strahlen beim Durchgang durch ein Prisma verhalten, ob sie darin abgelenkt werden oder nicht. Versuche mit Wasser und Schwefelkohlenstoff in Glimmerprismen von ca. 30° brechendem Winkel haben gar keine Ablenkung erkennen lassen, weder am Fluorescenzschirm, noch an der photographischen Platte. Zum Vergleich wurde unter denselben Verhältnissen die Ablenkung von Lichtstrahlen beobachtet; die abgelenkten Bilder lagen auf der Platte um ca. 10 mm bez. ca. 20 mm von dem nicht abgelenkten entfernt. — Mit einem Hartgummi- und einem Aluminiumprisma von ebenfalls ca. 30° brechendem Winkel habe ich auf der photographischen Platte Bilder bekommen, an denen man vielleicht eine Ablenkung erkennen kann. Doch ist die Sache sehr unsicher, und die Ablenkung ist, wenn überhaupt vorhanden, jedenfalls so klein, dass der Brechungsexponent der *X*-Strahlen in den genannten Substanzen höchstens 1,05 sein könnte. Mit dem Fluorescenzschirm habe ich auch in diesem Fall keine Ablenkung beobachten können.

Versuche mit Prismen aus dichteren Metallen lieferten bis jetzt wegen der geringen Durchlässigkeit und der infolge dessen geringen Intensität der durchgelassenen Strahlen kein sicheres Resultat.

In Anbetracht dieser Sachlage einerseits und andererseits der Wichtigkeit der Frage, ob die *X*-Strahlen beim Uebergang von einem Medium zum anderen gebrochen werden können oder nicht, ist es sehr erfreulich, dass diese Frage noch in anderer Weise untersucht werden kann, als mit Hülfe von Prismen. Fein pulverisirte Körper lassen in genügender Schichtendicke das auffallende Licht nur wenig und zerstreut hindurch infolge von Brechung und Reflexion: erweisen sich nun die Pulver für die *X*-Strahlen gleich durchlässig, wie die cohärente Substanz — gleiche Massen vorausgesetzt — so ist damit nachgewiesen, dass sowohl eine Brechung als auch eine regelmässige Reflexion nicht in merklichem Betrage vorhanden ist. Die Versuche wurden mit fein pulverisirtem Steinsalz, mit feinem, auf electrolytischem Wege gewonnenem Silberpulver und dem zu chemischen Untersuchungen vielfach verwandten Zinkstaub angestellt; es ergab sich in allen Fällen kein Unterschied in der Durchlässigkeit der Pulver und der cohärenten Substanz, sowohl bei der Beobachtung am Fluorescenzschirm, als auch auf der photographischen Platte.

Dass man mit Linsen die *X*-Strahlen nicht concentriren kann, ist nach dem Mitgetheilten selbstverständlich; eine grosse Hartgummilinse und eine Glaslinse erwiesen sich in der That als wirkungslos. Das Schattenbild eines runden Stabes ist in der Mitte dunkler als am Rande; dasjenige einer Röhre, die mit einer Substanz gefüllt ist, die durchlässiger ist als das Material der Röhre, ist in der Mitte heller als am Rande.

8. Die Frage nach der Reflexion der *X*-Strahlen ist durch die Versuche des vorigen Paragraphen als in dem Sinne erledigt zu betrachten, dass eine merkliche regelmässige Zurückwerfung der Strahlen an keiner der untersuchten Substanzen stattfindet. Andere Versuche, die ich hier übergehen will, führen zu demselben Resultat.

Indessen ist eine Beobachtung zu erwähnen, die auf den ersten Blick das Gegentheil zu ergeben scheint. Ich exponirte eine durch schwarzes Papier gegen Lichtstrahlen geschützte photographische Platte, mit der Glasseite dem Entladungsapparat zugewendet, den *X*-Strahlen; die empfindliche Schicht war bis auf einen frei bleibenden Theil mit blanken Platten aus Platin, Blei, Zink und Aluminium in sternförmiger Anordnung bedeckt. Auf dem entwickelten Negativ ist deutlich zu erkennen, dass die Schwärzung unter dem Platin, dem Blei und besonders unter dem Zink stärker ist als an den anderen Stellen; das Aluminium hatte gar keine Wirkung ausgeübt. Es scheint somit, dass die drei genannten Metalle die Strahlen reflectiren; indessen wären noch andere Ursachen für die stärkere Schwärzung denkbar, und um sicher zu gehen, legte ich bei einem zweiten Versuch zwischen die empfindliche Schicht und die Metallplatten ein Stück dünnes Blattaluminium, welches für ultraviolette Strahlen undurchlässig, dagegen für die X-Strahlen sehr durchlässig ist. Da auch jetzt wieder im wesentlichen dasselbe Resultat erhalten wurde, so ist eine Reflexion von X-Strahlen an den genannten Metallen nachgewiesen.

Hält man diese Thatsache zusammen mit der Beobachtung, dass Pulver ebenso durchlässig sind, wie cohärente Körper, dass weiter Körper mit rauher Oberfläche sich beim Durchgang der X-Strahlen, wie auch bei dem zuletzt beschriebenen Versuch ganz gleich wie polirte Körper verhalten, so kommt man zu der Anschauung, dass zwar eine regelmässige Reflexion, wie gesagt, nicht stattfindet, dass aber die Körper sich den X-Strahlen gegenüber ähnlich verhalten, wie die trüben Medien dem Licht gegenüber.

Da ich auch keine Brechung beim Uebergang von einem Medium zum anderen nachweisen konnte, so hat es den Anschein, als ob die X-Strahlen sich mit gleicher Geschwindigkeit in allen Körpern bewegen, und zwar in einem Medium, das überall vorhanden ist, und in welchem die Körpertheilchen eingebettet sind. Die letzteren bilden für die Ausbreitung der X-Strahlen ein Hinderniss und zwar im allgemeinen ein desto grösseres, je dichter der betreffende Körper ist.

9. Demnach wäre es möglich, dass auch die Anordnung der Theilchen im Körper auf die Durchlässigkeit desselben einen Einfluss ausübte, dass z. B. ein Stück Kalkspath bei gleicher Dicke verschieden durchlässig wäre, wenn dasselbe in der Richtung der Axe oder senkrecht dazu durchstrahlt wird. Versuche mit Kalkspath und Quarz haben aber ein negatives Resultat ergeben.

10. Bekanntlich ist Lenard bei seinen schönen Ver-

suchen über die von einem dünnen Aluminiumblättchen hindurchgelassenen Hittorf'schen Kathodenstrahlen zu dem Resultat gekommen, dass diese Strahlen Vorgänge im Aether sind, und dass sie in allen Körpern diffus verlaufen. Von unseren Strahlen haben wir ähnliches aussagen können.

In seiner letzten Arbeit hat Lenard das Absorptionsvermögen verschiedener Körper für die Kathodenstrahlen bestimmt und dasselbe u. a. für Luft von Atmosphärendruck zu 4,10, 3,40, 3,10 auf 1 cm bezogen gefunden, je nach der Verdünnung des im Entladungsapparat enthaltenen Gases. Nach der aus der Funkenstrecke geschätzten Entladungsspannung zu urtheilen, habe ich es bei meinen Versuchen meistens mit ungefähr gleichgrossen und nur selten mit geringeren und grösseren Verdünnungen zu thun gehabt. Es gelang mir mit dem L. Weber'schen Photometer — ein besseres besitze ich nicht — in atmosphärischer Luft die Intensitäten des Fluorescenzlichtes meines Schirmes in zwei Abständen — ca. 100 bez. 200 mm — vom Entladungsapparat miteinander zu vergleichen, und ich fand aus drei recht gut miteinander übereinstimmenden Versuchen, dass dieselben sich umgekehrt wie die Quadrate der bez. Entfernungen des Schirmes vom Entladungsapparat verhalten. Demnach hält die Luft von den hindurchgehenden X-Strahlen einen viel kleineren Bruchtheil zurück als von den Kathodenstrahlen. Dieses Resultat ist auch ganz in Uebereinstimmung mit der oben erwähnten Beobachtung, dass das Fluorescenzlicht noch in 2 m Distanz vom Entladungsapparat wahrzunehmen ist.

Aehnlich wie Luft verhalten sich im allgemeinen die anderen Körper: sie sind für die X-Strahlen durchlässiger als für die Kathodenstrahlen.

11. Eine weitere sehr bemerkenswerthe Verschiedenheit in dem Verhalten der Kathodenstrahlen und der X-Strahlen liegt in der Thatsache, dass es mir trotz vieler Bemühungen nicht gelungen ist, auch in sehr kräftigen magnetischen Feldern eine Ablenkung der X-Strahlen durch den Magnet zu erhalten.

Die Ablenkbarkeit durch den Magnet gilt aber bis jetzt als ein charakteristisches Merkmal der Kathodenstrahlen; wohl ward von Hertz und Lenard beobachtet, dass es verschiedene Arten von Kathodenstrahlen giebt, die sich durch ihre „Phosphorescenzerzeugung, Absorbirbarkeit und Ablenkbarkeit durch den Magnet voneinander unterscheiden", aber eine beträchtliche Ablenkung wurde doch in allen von ihnen untersuchten Fällen wahrgenommen, und ich glaube nicht, dass man dieses Characteristicum ohne zwingenden Grund aufgeben wird.

12. Nach besonders zu diesem Zweck angestellten Versuchen ist es sicher, dass die Stelle der Wand des Entladungsapparates, die am stärksten fluorescirt, als Hauptausgangspunkt der nach allen Richtungen sich ausbreitenden *X*-Strahlen zu betrachten ist. Die *X*-Strahlen gehen somit von der Stelle aus, wo nach den Angaben verschiedener Forscher die Kathodenstrahlen die Glaswand treffen. Lenkt man die Kathodenstrahlen innerhalb des Entladungsapparates durch einen Magnet ab, so sieht man, dass auch die *X*-Strahlen von einer anderen Stelle, d. h. wieder von dem Endpunkte der Kathodenstrahlen ausgehen.

Auch aus diesem Grund können die *X*-Strahlen, die nicht ablenkbar sind, nicht einfach unverändert von der Glaswand hindurchgelassene bez. reflectirte Kathodenstrahlen sein. Die grössere Dichte des Glases ausserhalb des Entladungsgefässes kann ja nach Lenard für die grosse Verschiedenheit der Ablenkbarkeit nicht verantwortlich gemacht werden.

Ich komme deshalb zu dem Resultat, dass die *X*-Strahlen nicht identisch sind mit den Kathodenstrahlen, dass sie aber von den Kathodenstrahlen in der Glaswand des Entladungsapparates erzeugt werden.

13. Diese Erzeugung findet nicht nur in Glas statt, sondern, wie ich an einem mit 2 mm starkem Aluminiumblech abgeschlossenen Apparat beobachten konnte, auch in diesem Metall. Andere Substanzen sollen später untersucht werden.

14. Die Berechtigung, für das von der Wand des Entladungsapparates ausgehende Agens den Namen „Strahlen" zu verwenden, leite ich zum Theil von der ganz regelmässigen Schattenbildung her, die sich zeigt, wenn man zwischen den Apparat und den fluorescirenden Schirm (oder die photographische Platte) mehr oder weniger durchlässige Körper bringt.

Viele derartige Schattenbilder, deren Erzeugung mitunter einen ganz besonderen Reiz bietet, habe ich beobachtet und theilweise auch photographisch aufgenommen; so besitze ich

z. B. Photographien von den Schatten der Profile einer Thüre, welche die Zimmer trennt, in welchen einerseits der Entladungsapparat, andererseits die photographische Platte aufgestellt waren; von den Schatten der Handknochen; von dem Schatten eines auf einer Holzspule versteckt aufgewickelten Drahtes; eines in einem Kästchen eingeschlossenen Gewichtsatzes; einer Bussole, bei welcher die Magnetnadel ganz von Metall eingeschlossen ist; eines Metallstückes, dessen Inhomogenität durch die X-Strahlen bemerkbar wird; etc.

Für die geradlinige Ausbreitung der X-Strahlen beweisend ist weiter eine Lochphotographie, die ich von dem mit schwarzem Papier eingehüllten Entladungsapparat habe machen können; das Bild ist schwach, aber unverkennbar richtig.

15. Nach Interferenzerscheinungen der X-Strahlen habe ich viel gesucht, aber leider, vielleicht nur infolge der geringen Intensität derselben, ohne Erfolg.

16. Versuche, um zu constatiren, ob electrostatische Kräfte in irgend einer Weise die X-Strahlen beeinflussen können, sind zwar angefangen, aber noch nicht abgeschlossen.

17. Legt man sich die Frage vor, was denn die X-Strahlen — die keine Kathodenstrahlen sein können — eigentlich sind, so wird man vielleicht im ersten Augenblick, verleitet durch ihre lebhaften Fluorescenz- und chemischen Wirkungen, an ultraviolettes Licht denken. Indessen stösst man doch sofort auf schwerwiegende Bedenken. Wenn nämlich die X-Strahlen ultraviolettes Licht sein sollten, so müsste dieses Licht die Eigenschaft haben:

a) dass es beim Uebergang aus Luft in Wasser, Schwefelkohlenstoff, Aluminium, Steinsalz, Glas, Zink etc. keine merkliche Brechung erleiden kann;

b) dass es von den genannten Körpern nicht merklich regelmässig reflectirt werden kann;

c) dass es somit durch die sonst gebräuchlichen Mittel nicht polarisirt werden kann;

d) dass die Absorption desselben von keiner anderen Eigenschaft der Körper so beeinflusst wird als von ihrer Dichte.

Das heisst, man müsste annehmen, dass sich diese ultravioletten Strahlen ganz anders verhalten, als die bisher bekannten ultrarothen, sichtbaren und ultravioletten Strahlen.

Dazu habe ich mich nicht entschliessen können und nach einer anderen Erklärung gesucht.

Eine Art von Verwandtschaft zwischen den neuen Strahlen und den Lichtstrahlen scheint zu bestehen, wenigstens deutet die Schattenbildung, die Fluorescenz und die chemische Wirkung, welche bei beiden Strahlenarten vorkommen, darauf hin. Nun weiss man schon seit langer Zeit, dass ausser den transversalen Lichtschwingungen auch longitudinale Schwingungen im Aether vorkommen können und nach Ansicht verschiedener Physiker vorkommen müssen. Freilich ist ihre Existenz bis jetzt noch nicht evident nachgewiesen, und sind deshalb ihre Eigenschaften noch nicht experimentell untersucht.

Sollten nun die neuen Strahlen nicht longitudinalen Schwingungen im Aether zuzuschreiben sein?

Ich muss bekennen, dass ich mich im Laufe der Untersuchung immer mehr mit diesem Gedanken vertraut gemacht habe und gestatte mir dann auch diese Vermuthung hier auszusprechen, wiewohl ich mir sehr wohl bewusst bin, dass die gegebene Erklärung einer weiteren Begründung noch bedarf.

Würzburg, Physik. Institut d. Univ., Dec. 1895.

2. *Ueber eine neue Art von Strahlen;* *von W. C. Röntgen.*

(Zweite Mittheilung.)

Aus den Sitzungsber. der Würzburger Physik.-Medic. Gesellschaft. Jahrg. 1895.

Da meine Arbeit auf mehrere Wochen unterbrochen werden muss, gestatte ich mir im Folgenden einige neue Ergebnisse schon jetzt mitzutheilen.

18. Zur Zeit meiner ersten Publication war mir bekannt, dass die *X*-Strahlen im Stande sind, electrische Körper zu entladen, und ich vermuthe, dass es auch die *X*-Strahlen und nicht die von dem Aluminiumfenster seines Apparates unverändert durchgelassenen Kathodenstrahlen gewesen sind, welche die von Lenard beschriebene Wirkung auf entfernte electrische Körper ausgeübt haben. Mit der Veröffentlichung meiner Versuche habe ich aber gewartet, bis ich in der Lage war, einwandsfreie Resultate mitzutheilen.

Solche lassen sich wohl nur dann erhalten, wenn man die Beobachtungen in einem Raum anstellt, der nicht nur vollständig gegen die von der Vacuumröhre, den Zuleitungsdrähten, dem Inductionsapparat etc. ausgehenden electrostatischen Kräfte geschützt ist, sondern der auch gegen Luft abgeschlossen ist, welche aus der Nähe des Entladungsapparates kommt.

Ich liess mir zu diesem Zweck aus zusammengelötheten Zinkblechen einen Kasten anfertigen, der gross genug ist, um mich und die nöthigen Apparate aufzunehmen, und der bis auf eine durch eine Zinkthüre verschliessbare Oeffnung überall luftdicht verschlossen ist. Die der Thüre gegenüber liegende Wand ist zu einem grossen Theil mit Blei belegt; an einer dem ausserhalb des Kastens aufgestellten Entladungsapparat nahe gelegenen Stelle wurde die Zinkwand mit der darüber gelegten Bleiplatte in einer Weite von 4 cm ausgeschnitten, und die Oeffnung ist mit einem dünnen Aluminiumblech wieder luftdicht verschlossen. Durch dieses Fenster können die *X*-Strahlen in den Beobachtungskasten eindringen.

Ich habe nun folgendes wahrgenommen:

a) In der Luft aufgestellte, positiv oder negativ electrisch geladene Körper werden, wenn sie mit *X*-Strahlen bestrahlt werden, entladen und zwar desto rascher, je intensiver die Strahlen sind. Die Intensität der Strahlen wurde nach ihrer Wirkung auf einen Fluorescenzschirm oder auf eine photographische Platte beurtheilt.

Es ist im allgemeinen gleichgültig, ob die electrischen Körper Leiter oder Isolatoren sind. Bis jetzt habe ich auch keinen specifischen Unterschied in dem Verhalten der verschiedenen Körper bezüglich der Geschwindigkeit der Entladung gefunden; ebensowenig in dem Verhalten von positiver und negativer Electricität. Doch ist es nicht ausgeschlossen, dass geringe Unterschiede bestehen.

b) Ist ein electrisirter Leiter nicht von Luft, sondern von einem festen Isolator, z. B. Paraffin, umgeben, so bewirkt die Bestrahlung dasselbe, wie das Bestreichen der isolirenden Hülle mit einer zur Erde abgeleiteten Flamme.

c) Ist diese isolirende Hülle von einem eng anliegenden, zur Erde abgeleiteten Leiter umschlossen, welcher wie der Isolator für *X*-Strahlen durchlässig sein soll, so übt die Bestrahlung auf den inneren, electrisirten Leiter keine mit meinen Hülfsmitteln nachweisbare Wirkung aus.

d) Die unter a, b, c mitgetheilten Beobachtungen deuten darauf hin, dass die von den *X*-Strahlen bestrahlte Luft die Eigenschaft erhalten hat, electrische Körper, mit denen sie in Berührung kommt, zu entladen.

e) Wenn sich die Sache wirklich so verhält, und wenn ausserdem die Luft diese Eigenschaft noch einige Zeit behält, nachdem sie den *X*-Strahlen ausgesetzt war, so muss es möglich sein, electrische Körper, welche selbst nicht von den *X*-Strahlen getroffen werden, dadurch zu entladen, dass man ihnen bestrahlte Luft zuführt.

In verschiedener Weise kann man sich davon überzeugen, dass diese Folgerung in der That zutrifft. Eine, wenn auch nicht die einfachste, Versuchsanordnung möchte ich mittheilen.

Ich benutzte eine 3 cm weite, 45 cm lange Messingröhre; in einigen Centimenter Entfernung von dem einen Ende ist ein

Theil der Röhrenwand weggeschnitten und durch ein dünnes Aluminiumblech ersetzt; am anderen Ende ist unter luftdichtem Abschluss eine an einer Metallstange befestigte Messingkugel isolirt in die Röhre eingeführt. Zwischen der Kugel und dem verschlossenen Ende der Röhre ist ein Seitenröhrchen angelöthet, das mit einer Saugvorrichtung in Verbindung gesetzt werden kann; wenn gesaugt wird, so wird die Messingkugel umspült von Luft, die auf ihrem Wege durch die Röhre an dem Aluminiumfenster vorüber gegangen ist. Die Entfernung vom Fenster bis zur Kugel beträgt über 20 cm.

Diese Röhre stellte ich im Zinkkasten so auf, dass die *X*-Strahlen durch das Aluminiumfenster der Röhre, senkrecht zur Axe derselben eintreten konnten, die isolirte Kugel lag dann ausserhalb des Bereiches dieser Strahlen, im Schatten. Die Röhre und der Zinkkasten waren leitend miteinander, die Kugel mit einem Hankel'schen Electroskop verbunden.

Es zeigte sich nun, dass eine der Kugel mitgetheilte Ladung (positive oder negative) von den *X*-Strahlen nicht beeinflusst wurde, solange die Luft in der Röhre in Ruhe blieb, dass die Ladung aber sofort beträchtlich abnahm, wenn durch kräftiges Saugen bestrahlte Luft der Kugel zugeführt wurde. Erhielt die Kugel durch Verbindung mit Accumulatoren ein constantes Potential, und wurde fortwährend bestrahlte Luft durch die Röhre gesaugt, so entstand ein electrischer Strom, wie wenn die Kugel mit der Röhrenwand durch einen schlechten Leiter verbunden gewesen wäre.

f) Es fragt sich, in welcher Weise die Luft die ihr von den *X*-Strahlen mitgetheilte Eigenschaft wieder verlieren kann. Ob sie sie von selbst, d. h. ohne mit anderen Körpern in Berührung zu kommen, mit der Zeit verliert, ist noch unentschieden. Sicher dagegen ist es, dass eine kurz dauernde Berührung mit einem Körper von grosser Oberfläche, der nicht electrisch zu sein braucht, die Luft unwirksam machen kann. Schiebt man z. B. einen genügend dicken Pfropf aus Watte in die Röhre so weit ein, dass die bestrahlte Luft die Watte durchstreichen muss, bevor sie zu der electrischen Kugel gelangt, so bleibt die Ladung der Kugel auch beim Saugen unverändert.

Sitzt der Pfropf an einer Stelle, die vor dem Aluminiumfenster liegt, so erhält man dasselbe Resultat wie ohne Watte; ein Beweis, dass nicht etwa Staubtheilchen die Ursache der beobachteten Entladung sind.

Drahtgitter wirken ähnlich wie Watte; doch muss das Gitter sehr eng sein, und viele Lagen müssen übereinander gelegt werden, wenn die durchgestrichene, bestrahlte Luft unwirksam sein soll. Sind diese Gitter nicht, wie bisher angenommen, zur Erde abgeleitet, sondern mit einer Electricitätsquelle von constantem Potential verbunden, so habe ich immer das beobachtet, was ich erwartet hatte; doch sind diese Versuche noch nicht abgeschlossen.

g) Befinden sich die electrischen Körper statt in Luft in trockenem Wasserstoff, so werden sie ebenfalls durch die X-Strahlen entladen. Die Entladung in Wasserstoff schien mir etwas langsamer zu verlaufen, doch ist diese Angabe noch unsicher wegen der Schwierigkeit, bei aufeinander folgenden Versuchen gleiche Intensität der X-Strahlen zu erhalten.

Die Art und Weise der Füllung der Apparate mit Wasserstoff dürfte die Möglichkeit ausschliessen, dass die anfänglich auf der Oberfläche der Körper vorhandene verdichtete Luftschicht bei der Entladung eine wesentliche Rolle gespielt hätte.

h) In stark evacuirten Räumen findet die Entladung eines direct von den X-Strahlen getroffenen Körpers viel langsamer — in einem Fall z. B. ca. 70 mal langsamer — statt, als in denselben Gefässen, welche mit Luft oder Wasserstoff von Atmosphärendruck gefüllt sind.

i) Versuche über das Verhalten einer Mischung von Chlor und Wasserstoff unter dem Einfluss der X-Strahlen sind in Angriff genommen.

j) Schliesslich möchte ich noch erwähnen, dass die Resultate von Untersuchungen über die entladende Wirkung der X-Strahlen, bei welchen der Einfluss des umgebenden Gases unberücksichtigt blieb, vielfach mit Vorsicht aufzunehmen sind.

19. In manchen Fällen ist es vortheilhaft, zwischen den die X-Strahlen liefernden Entladungsapparat und den Ruhmkorff einen Tesla'schen Apparat (Condensator und Transformator) einzuschalten. Diese Anordnung hat folgende Vorzüge: erstens werden die Entladungsapparate weniger leicht

durchschlagen und weniger warm; zweitens hält sich das Vacuum, wenigstens bei meinen selbstangefertigten Apparaten, längere Zeit, und drittens liefern manche Apparate intensivere *X*-Strahlen. Bei Apparaten, die zu wenig oder zu stark evacuirt waren, um mit dem Ruhmkorff allein gut zu functioniren, leistete die Anwendung des Tesla'schen Transformators gute Dienste.

Es liegt die Frage nahe — und ich gestatte mir deshalb sie zu erwähnen, ohne zu ihrer Beantwortung vorläufig etwas beitragen zu können — ob auch durch eine continuirliche Entladung mit constant bleibendem Entladungspotential *X*-Strahlen erzeugt werden können; oder ob nicht vielmehr Schwankungen dieses Potentials zum Entstehen derselben durchaus erforderlich sind.

20. In § 13 meiner ersten Veröffentlichung ist mitgetheilt, dass die *X*-Strahlen nicht bloss in Glas, sondern auch in Aluminium entstehen können. Bei der Fortsetzung der Untersuchung nach dieser Richtung hin hat sich kein fester Körper ergeben, welcher nicht im Stande wäre, unter dem Einfluss der Kathodenstrahlen *X*-Strahlen zu erzeugen. Es ist mir auch kein Grund bekannt geworden, weshalb sich flüssige und gasförmige Körper nicht ebenso verhalten würden.

Quantitative Unterschiede in dem Verhalten der verschiedenen Körper haben sich dagegen ergeben. Lässt man z. B. die Kathodenstrahlen auf eine Platte fallen, deren eine Hälfte aus einem 0,3 mm dicken Platinblech, deren andere Hälfte aus einem 1 mm dicken Aluminiumblech besteht, so beobachtet man an dem mit der Lochcamera aufgenommenen photographischen Bild dieser Doppelplatte, dass das Platinblech auf der von den Kathodenstrahlen getroffenen (Vorder-) Seite viel mehr *X*-Strahlen aussendet, als das Aluminiumblech auf der gleichen Seite. Von der Hinterseite dagegen gehen vom Platin so gut wie gar keine, vom Aluminium aber relativ viel *X*-Strahlen aus. Letztere Strahlen sind in den vorderen Schichten des Aluminiums erzeugt und durch die Platte hindurch gegangen.

Man kann sich von dieser Beobachtung leicht eine Erklärung verschaffen, doch dürfte es sich empfehlen, vorher noch weitere Eigenschaften der *X*-Strahlen zu erfahren.

Zu erwähnen ist aber, dass der gefundenen Thatsache auch eine praktische Bedeutung zukommt. Zur Erzeugung von möglichst intensiven *X*-Strahlen eignet sich nach meinen bisherigen Erfahrungen Platin am besten. Ich gebrauche seit einigen Wochen mit gutem Erfolg einen Entladungsapparat, bei dem ein Hohlspiegel aus Aluminium als Kathode, ein unter 45° gegen die Spiegelaxe geneigtes, im Krümmungscentrum aufgestelltes Platinblech als Anode fungirt.

21. Die *X*-Strahlen gehen bei diesem Apparat von der Anode aus. Wie ich aus Versuchen mit verschieden geformten Apparaten schliessen muss, ist es mit Rücksicht auf die Intensität der *X*-Strahlen gleichgültig, ob die Stelle, wo diese Strahlen erzeugt werden, die Anode ist oder nicht.

Speciell zu den Versuchen mit den Wechselströmen des Tesla'schen Transformators wird ein Entladungsapparat angefertigt, bei dem beide Electroden Aluminiumhohlspiegel sind, deren Axen miteinander einen rechten Winkel bilden; im gemeinschaftlichen Krümmungscentrum ist eine die Kathodenstrahlen auffangende Platinplatte angebracht. Ueber die Brauchbarkeit dieses Apparates soll später berichtet werden.

Würzburg, Physikal. Institut d. Universität, 9. März 1896.

3. *Interferenzerscheinungen bei Röntgenstrahlen; von W. Friedrich, P. Knipping und M. Laue.*[1])

(Hierzu Taf. I—IV, Figg. 1—10.)

Theoretischer Teil von M. Laue.

Barklas Untersuchungen in den letzten Jahren haben gezeigt, daß die Röntgenstrahlen in der Materie eine Zerstreuung erfahren, ganz entsprechend der Zerstreuung des Lichtes in trüben Medien, daß sie aber noch daneben im allgemeinen die Atome des Körpers zur Aussendung einer spektral homogenen Eigenstrahlung (Fluoreszenzstrahlung) anregen, welche ausschließlich für den Körper charakteristisch ist.

Andererseits ist schon seit 1850 durch Bravais in die Kristallographie die Theorie eingeführt, daß die Atome in den Kristallen nach Raumgittern angeordnet sind. Wenn die Röntgenstrahlen wirklich in elektromagnetischen Wellen bestehen, so war zu vermuten, daß die Raumgitterstruktur bei einer Anregung der Atome zu freien oder erzwungenen Schwingungen zu Interferenzerscheinungen Anlaß gibt; und zwar zu Interferenzerscheinungen derselben Natur, wie die in der Optik bekannten Gitterspektren. Die Konstanten dieser Gitter lassen sich aus dem Molekulargewicht der kristallisierten Verbindung, ihrer Dichte und der Zahl der Moleküle pro Grammolekül, sowie den kristallographischen Daten leicht berechnen. Man findet für sie stets die Größenordnung 10^{-8} cm, während die Wellenlänge der Röntgenstrahlen nach den Beugungsversuchen von Walter und Pohl [2]) und nach den Arbeiten von Sommerfeld und Koch [3]) von der Größenordnung 10^{-9} cm sind. Eine erhebliche Komplikation freilich bedeutet es, daß bei den

1) Wenig veränderter Abdruck aus den Münchener Sitzungsber. 1912. p. 303.

2) B. Walter u. R. Pohl, Ann. d. Phys. **25.** p. 715. 1908; **29.** p. 331. 1908.

3) R. Sommerfeld, Ann. d. Phys. **38.** p. 473. 1912; P. P. Koch, Ann. d. Phys. **38.** p. 507. 1912.

Raumgittern eine dreifache Periodizität vorliegt, während man bei den optischen Gittern nur in einer Richtung, höchstens (bei den Kreuzgittern) in zwei Richtungen periodische Wiederholungen hat.

Die Herren Friedrich und Knipping haben auf meine Anregung diese Vermutung experimentell geprüft. Über die Versuche und ihr Ergebnis berichten sie selbst im zweiten Teil der Veröffentlichung.

Die Theorie und ihr qualitativer Vergleich mit der Erfahrung.

Wir wollen den oben angedeuteten Gedanken mathematisch zu fassen suchen. Den Ort der Mittelpunkte der Atome bestimmen wir durch die rechtwinkligen Koordinaten x, y, z, deren Achsenkreuz seinen Ursprung im Mittelpunkt eines beliebigen Atoms im durchstrahlten Teil des Kristalles hat. Das Raumgitter gehört dem allgemeinsten, d. h. triklinen Typus an; die Kanten seiner Elementarparallelepipede mögen also beliebige Längen haben und beliebige Winkel miteinander einschließen. Durch spezielle Wahl dieser Längen und Winkel kann man immer zu Raumgittern anderer Typen übergehen. Stellen wir diese Kanten nach Länge und Richtung durch die Vektoren $\mathfrak{a}_1$, $\mathfrak{a}_2$, $\mathfrak{a}_3$ dar, so liegen die Mittelpunkte eines Atoms an einem Ort

$$(1)\qquad \begin{cases} x = m\,\mathfrak{a}_{1x} + n\,\mathfrak{a}_{2x} + p\,\mathfrak{a}_{3x} \\ y = m\,\mathfrak{a}_{1y} + n\,\mathfrak{a}_{2y} + p\,\mathfrak{a}_{3y} \\ z = m\,\mathfrak{a}_{1z} + n\,\mathfrak{a}_{2z} + p\,\mathfrak{a}_{3z}, \end{cases}$$

wobei m, n, p positive oder negative, das Atom numerierende ganze Zahlen (einschließlich der Null) sind.

Für die Schwingung eines einzelnen Atoms wollen wir zunächst einmal die Annhme machen, daß sie rein sinusförmig verläuft. Dies kann hier natürlich ebensowenig wie in der Optik streng richtig sein. Aber ebenso wie in der Optik kann man spektral inhomogene Strahlung durch Fouriesche Zerlegung aus Sinusschwingungen zurückführen. Die von einem Atom ausgehende Welle können wir dann in großer Entfernung vom Atom darstellen durch den Ausdruck

$$(2)\qquad \Psi \frac{e^{-ikr}}{r},$$

wo r den Betrag des Radiumvektors vom Atom zum Aufpunkt bedeutet, Ψ eine Funktion seiner Richtung und $k = 2\pi/\lambda$ ist, wo λ die Wellenlänge der später interferierenden Röntgenstrahlen darstellt. Wäre, wie man es in der Optik gewohnt ist, das Atom klein gegen die Wellenlänge, so wäre Ψ eine Konstante.[1]) Hier aber muß man mit der Möglichkeit rechnen, und die Versuchsergebnisse legen die Vermutung nahe, daß infolge der mit der Wellenlänge vergleichbaren Abmessung des Atoms Richtungsunterschiede auftreten. Berücksichtigt man ferner noch, daß die anregende Schwingung in (als eben vorausgesetzten) Wellen mit Lichtgeschwindigkeit fortschreitet, so sieht man, daß man noch den Faktor $e^{-ik(x\alpha_0 + y\beta_0 + z\gamma_0)}$ hinzufügen muß, wenn α_0, β_0, γ_0 die Richtungskosinus der einfallenden primären Röntgenstrahlen sind. Über Einzelheiten des Anregungsvorganges brauchen wir dabei keine weitere Voraussetzung zu machen, als daß er bei allen Atomen gleich erfolgt. Insbesondere ist es für den weiteren Verlauf des Vorganges vollkommen dasselbe, ob die Schwingungen im Atom durch Schwingungen gleicher Frequenz in der primären Strahlung erzwungen, oder ob es durch die primäre Strahlung einmal angeregte, sonst aber freie Eigenschwingungen sind. Stets finden wir als Superpositionseffekt aller Elementarwellen

$$\sum \Psi \frac{e^{-ik(r + x\alpha_0 + y\beta_0 + z\gamma_0)}}{r}. \tag{3}$$

Wir berechnen diese Summe nur für Aufpunkte, deren Abstand sehr groß ist gegen die Abmessungen des durchleuchteten Kristallstücks, und benutzen die auch sonst in der Gittertheorie übliche Näherung, d. h. wir setzen für das r im

---

1) Vgl. H. v. Helmholtz, Abh. 1, p. 331. Gehen von beliebig vielen, in einem endlichen Bereich liegenden Strahlungswellen Kugelwellen aus, so ist die Erregung in einem Punkte für welchen der Radiusvektor vom Mittelpunkte des Bereiches den gegen dessen Abmessungen großen Betrag r und die Richtungskosinus α, β, γ hat:

$$\sum_n \frac{e^{-ikr(r_n - \vartheta_n)}}{r_n} = \frac{e^{-ikr}}{r} \cdot \sum_n e^{ik(x_n \cdot \alpha + y_n \cdot \beta + z_n \cdot \gamma) - i\vartheta_n}.$$

Diese Summe ist im allgemeinen eine Funktion von α und β. Sind aber alle x_n, y_n, z_n klein gegen λ, so wird sie konstant.

Nenner den Betrag R des Radiumvektors vom Nullpunkt des Achsenkreuzes zum Aufpunkt und geben der Richtungsfunktion Ψ den Wert, welcher dessen Richtung (α, β, γ) entspricht. Für das r im Exponenten setzen wir aber den Näherungswert:

$$r = R - (x\alpha + y\beta + z\gamma).$$

Unter Berücksichtigung von (1) geht somit die Summe (3) über in:

$$(4) \quad \begin{cases} \Psi(\alpha\beta)\dfrac{e^{-ikR}}{R}\sum e^{ik[x(\alpha-\alpha_0)+y(\beta-\beta_0)+z(\gamma-\gamma_0)]} \\ = \Psi(\alpha\beta)\dfrac{e^{-ikR}}{R}\sum^m\sum^n\sum^p e^{i(mA+nB+pC)}, \end{cases}$$

wo zur Abkürzung

$$(6) \quad \begin{cases} A = k\,[\mathfrak{a}_{1x}(\alpha-\alpha_0) + \mathfrak{a}_{1y}(\beta-\beta_0) + \mathfrak{a}_{1z}(\gamma-\gamma_0)] \\ B = k\,[\mathfrak{a}_{2x}(\alpha-\alpha_0) + \mathfrak{a}_{2y}(\beta-\beta_0) + \mathfrak{a}_{2z}(\gamma-\gamma_0)] \\ C = k\,[\mathfrak{a}_{3x}(\alpha-\alpha_0) + \mathfrak{a}_{3y}(\beta-\beta_0) + \mathfrak{a}_{3z}(\gamma-\gamma_0)] \end{cases}$$

gesetzt ist. Denken wir uns den durchstrahlten Teil von Ebenen begrenzt, welche zu den Seiten eines Elementarparallelepipedes parallel sind, so ist die Summation nach m von einer Zahl $-M$ bis zu $+M$, nach n von $-N$ bis $+N$, nach p von $-P$ bis $+P$ auszuführen. Die Lage der Intensitätsmaxima ist von derartigen Annahmen nicht abhängig.[1] Die Intensität der Schwingung (4) wird dann:

$$(6) \quad \frac{|\Psi(\alpha,\beta)|^2}{R^2}\,\frac{\sin^2 MA}{\sin^2\frac{1}{2}A}\,\frac{\sin^2 MB}{\sin^2\frac{1}{2}B}\,\frac{\sin^2 PC}{\sin^2\frac{1}{2}C}.$$

Jeder dieser Sinusquotienten erreicht sein Maximum bekanntlich, wenn sein Nenner verschwindet. Die Bedingungen für die Maxima lauten also:

$$(7) \quad \begin{cases} A = 2h_1\pi \\ \text{d. h. } \mathfrak{a}_{1x}\alpha + \mathfrak{a}_{1y}\beta + \mathfrak{a}_{1z}\gamma = h_1\lambda + \mathfrak{a}_{1x}\alpha_0 + \mathfrak{a}_{1y}\beta_0 + \mathfrak{a}_{1z}\gamma_0 \\ B = 2h_2\pi \\ \text{d. h. } \mathfrak{a}_{2x}\alpha + \mathfrak{a}_{2y}\beta + \mathfrak{a}_{2z}\gamma = h_2\lambda + \mathfrak{a}_{2x}\alpha_0 + \mathfrak{a}_{2y}\beta_0 + \mathfrak{a}_{2z}\gamma_0 \\ C = 2h_3\pi \\ \text{d. h. } \mathfrak{a}_{3x}\alpha + \mathfrak{a}_{3y}\beta + \mathfrak{a}_{3z}\gamma = h_3\lambda + \mathfrak{a}_{3x}\alpha_0 + \mathfrak{a}_{3y}\beta_0 + \mathfrak{a}_{3z}\gamma_0. \end{cases}$$

1) Damit stimmt die Beobachtung überein. Vgl. p. 985.

Die links stehenden Summen sind gleich der Länge je einer Kante multipliziert mit dem Kosinus des Winkels zwischen ihr und der Richtung α, β, γ. Jede der Gleichungen (7) stellt somit eine Schar von Kreiskegeln dar, deren Achse mit der Richtung der Kanten $\mathfrak{a}_1$, $\mathfrak{a}_2$, $\mathfrak{a}_3$ zusammenfällt.[1]) Nun wird es freilich nur in Ausnahmefällen vorkommen, daß eine Richtung allen drei Bedingungen gleichzeitig genügt. Darin liegt die Komplikation gegenüber dem einfachen oder dem Kreuzgitter. Dennoch ist ein sichtbares Intensitätsmaximum zu erwarten, wenn die Schnittlinie zweier Kegel der ersten beiden Scharen einem Kegel, der dritten Schar nahe liegt. Wir wollen diese Verhältnisse etwas näher betrachten für den in Fig. 5 Taf. II abgebildeten Fall, in welchem ein regulärer Kristall in Richtung einer der Kanten $\mathfrak{a}_1$, $\mathfrak{a}_2$, $\mathfrak{a}_3$ durchstrahlt wird.

In diesem Falle erhalten die drei Kanten die gleiche Länge a und stehen senkrecht aufeinander, so daß wir in ihre Richtungen die Koordinatenachsen legen können. Da dann

$$\mathfrak{a}_{1y} = \mathfrak{a}_{1z} = \mathfrak{a}_{2x} = \mathfrak{a}_{2z} = \mathfrak{a}_{3x} = \mathfrak{a}_{3y} = 0, \quad \mathfrak{a}_{1x} = \mathfrak{a}_{2y} = \mathfrak{a}_{3z} = a$$

ferner

$$\alpha_0 = 0, \quad \beta_0 = 0, \quad \gamma_0 = 1$$

wird, so gehen die Gleichungen (7) über in

$$\alpha = h_1 \frac{\lambda}{a}, \quad \beta = h_2 \frac{\lambda}{a}, \quad 1 - \gamma = h_3 \frac{\lambda}{a}. \tag{8}$$

Auf einer zum einfallenden Strahl senkrechten photographischen Platte sind die Kurven $\alpha = \text{const}$ und $\beta = \text{const}$. Hyperbeln, deren Mittelpunkt im Durchstoßpunkt des primären Strahles liegt und deren Achsen aufeinander senkrecht stehen. Wären nur die beiden ersten Bedingungen in (8) zu erfüllen, so sähe man das bekannte Kreuzgitterspektrum, bei dem an jedem Schnittpunkte zweier Hyperbeln ein Intensitätsmaximum liegt. So aber wählen die Kreise $\gamma = \text{const}$, deren Mittelpunkt ebenfalls im Durchstoßpunkt des primären Strahles liegt, aus diesen Kreuzgitterspektren diejenigen aus, welche einem von ihnen hinreichend nahe liegen, d. h. man wird auf der

1) Man macht sich wie in der Gittertheorie leicht an der elementaren geometrischen Konstruktion des Gangunterschiedes zwischen zwei parallelen, von benachbarten Teilchen ausgehenden Strahlen klar, daß dieser Gangunterschied auf den genannten Kegeln konstant ist.

Platte diese Kreise nicht ganz, sondern nur in einzelnen Punkten vertreten sehen. Dies ist aber in der Tat der Eindruck, welchen die Figur macht. Die Kreise

$$1 - \gamma = h_3 \frac{\lambda}{\alpha},$$

welche die Periodizität in der Strahlrichtung ihre Entstehung verdanken, haben in der Optik ein schon seit Newton bekanntes Analogon in den sogenannten Queteletschen Ringen;[1]) diese treten bei Bestäubung einer ebenen, auf der Rückseite stark reflektierenden Glasplatte im Fraunhoferschen Beugungsbild auf. Es interferieren nämlich die von einem und demselben Staubteilchen aus dem einfallenden und aus dem bespiegelten Licht abgebeugten Wellen miteinander. Bei senkrechter Inzidenz liegen die Maxima auf den durch Gleichung (9) angegebenen Kreisen, sind aber viel flacher als in unserem Falle, weil nur zwei Wellen interferieren. Ließen sich die Staubteilchen zum regelmäßigen Kreuzgitter anordnen, so ließe sich die Analogie zu den obigen Interferenzen an Kristallen zweifellos noch weiter treiben.

Es muß beachtet werden, daß bei gegebenem Raumgitter die Einteilung in Elementarparallelepipede keine eindeutige ist, sondern auf unendlich viele Arten vorgenommen werden kann. Beim regulären Raumgitter z. B. kann man statt in Würfel in solchen Parallelepipede zerlegen, deren eine Seite die Diagonale einer Würfelfläche ist und somit die Richtung einer zweizähligen Symmetrieachse hat. Die Intensitätsmaxima müssen sich nach den obigen auch zu durchbrochenen Kegelschnitten um solche Achsen zusammerfassen lassen, wie überhaupt zu jederartigen Einteilung eine Art, die Maxima zusammenzufassen, gehört. In der Tat bestätigt die Fig. 12, Taf. IV, daß die Flecken sich zu Kreisen um eine zweizählige Achse anordnen, wenn man in Richtung dieser den regulären Kristall durchstrahlt und die photographische Platte dazu senkrecht stellt. Auch in den komplizierteren Figuren hat man stets den Eindruck, daß sich die Intensitätsmaxima zu durchbrochenen Kegelschnitten verbinden lassen. So viel bisher

1) Vgl. z. B. A. Winkelmann, Handbuch der Physik. Leipzig 1906. p. 1083.

zu sehen ist, läßt sich die Theorie unter der Annahme mehrerer Wellenlängen von $0{,}038 \cdot a$ bis $0{,}015 \cdot a$ gut der vierzählig symmetrischen Aufnahme (Fig. 5, Taf. II) anpassen. Da a für Zinksulfid gleich $3{,}38 \cdot 10^{-8}$ cm ist, würden sich daraus die Wellenlängen in dem Intervall von $1{,}3 \cdot 10^{-9}$ bis $5{,}2 \cdot 10^{-9}$ cm ergeben.[1])

Trotz dieser Übereinstimmung darf nicht verschwiegen werden, daß unsere Theorie auf jeden Fall weitgehender Verbesserung bedarf. Die Wärmebewegung bei den Molekülen verrückt diese nämlich schon bei Zimmertemperatur um einen erheblichen Bruchteil der Gitterkonstanten und infolgedessen um ein vielfaches der Wellenlänge, ein Umstand, der durchaus der Berücksichtigung bedarf. Deshalb wäre es auch voreilig, wenn man aus dem Ausdruck (6) Schlüsse auf die Schärfe der Interferenzmaxima schließen wollte.

Daß von allen bisher untersuchten Kristallen mit Ausnahme des Diamants die Intensitätsmaxima auf einen spitzen Winkel gegen die Richtung des primären Strahles beschränkt sind, statt nach allen Richtungen auszugehen, wie man zunächst nach Gleichung (6) erwarten könnte, wird man wohl auf die Richtungsfunktion Ψ zurückführen müssen; doch ist es auch denkbar, daß die Wärmebewegung zur Erklärung herangezogen werden muß.

Allgemeine Folgerungen.

Diskutieren wir zum Schluß noch ohne allen Bezug auf die Formel die Frage, inwiefern diese Versuche für die Wellennatur der Röntgenstrahlen sprechen. Daß die vom Kristall ausgehende Strahlung Wellencharakter trägt, ist wohl durch die Schärfe der Intensitätsmaxima bewiesen, welche als Interferenzphänomen leicht auf Grund korpuskularer Vorstellungen aber wohl kaum verständlich ist; daneben auch durch ihre große Durchdringungsfähigkeit, welche von allen bekannten Korpuskularstrahlen nur die schnellsten β-Strahlen erreichen. Dennoch könnte man aber vielleicht die Wellennatur der primären bezweifeln. Denken wir nun einmal, die Atome der Kristalle würden im Fall von Fig. 5, Taf. III, durch eine Korpuskularstrahlung angeregt. (Die von manchen Forschern angenommene Lichtquantenstruktur der Röntgenstrahlen kann hier unter der Bezeichnung korpuskulare

1) Vgl. Zusatz 3 zur folgenden Arbeit von M. Laue.

Strahung mit einbegriffen werden.) Zu kohärenten Schwingungen könnten dabei nur solche Reihen von Atomen kommen, welche von demselben Korpuskel getroffen werden, d. h. Reihen, welche zur z-Richtung parallel sind. Atome, welche in der x- oder y-Richtung einen gewissen Abstand voneinander haben, würden von verschiedenen Korpuskeln angeregt; eine bestimmte Phasendifferenz zwischen ihren Schwingungen könnte nicht auftreten. Infolgedessen würde von dem Intensitätsausdruck (6) nur ein Sinusquotient übrig bleiben; wir erhielten nur eine Bedingung für ein Intensitätsmaximum und diese wäre, wie schon aus Symmetriegründen einleuchtet, auf Kreisen um den Durchstoßpunkt des primären Strahls erfüllt. Die Durchbrechung dieser Kreise, die doch tatsächlich auftritt, wäre danach unverständlich. Zudem sind aber die primären und die vom Kristall ausgehenden Strahlen allem Anschein nach so gleichartig, daß man von der Wellennatur der letzteren wohl ziemlich sicher auf die Wellennatur der ersteren schließen kann. Ein Unterschied bleibt freilich bestehen: die vom Kristall ausgehende Strahlung hat sicher eine erhebliche spektrale Homogenität, d. h. eine gewisse Periodizität. Die primäre Strahlung hingegen wird man nach Sommerfeld wohl soweit sie „Bremsstrahlung“ ist, als aus durchaus unperiodischen Impulswellen bestehend annehmen müssen; die im folgenden mitgeteilten Versuche sind jedenfalls mit dieser Annahme verträglich. Unentschieden bleibe vorläufig, ob die periodische Strahlung erst im Kristall durch Fluoreszenz entsteht, oder ob sie neben den Impulsen schon in der primären Strahlung vorhanden ist und durch den Kristall nur ausgesondert wird.[1] Doch scheint Hoffnung zu bestehen, daß weitere Versuche bald darüber Aufschluß geben werden.

Experimenteller Teil von W. Friedrich und P. Knipping.

Zur experimentellen Prüfung der im vorhergehenden Teile beschriebenen Vermutung wurde nach einigen Vorversuchen mit einem provisorischen Apparat folgende definitive Versuchsanordnung verwandt, die in Fig. 1 schematisch abgebildet ist.

1) Vgl. p. 988.

Von den von der Antikathode A einer Röntgenröhre ausgehenden Röntgenstrahlen wird ein schmales Bündel von ca. 1 mm Durchmesser durch die Blenden B_1 bis B_4 ausgeblendet. Dieses Bündel durchsetzt den Kristall Kr, der in einem Goniometer G aufgestellt ist. Um den Kristall sind in verschiedenen Richtungen und Abständen photographische Platten P angebracht,

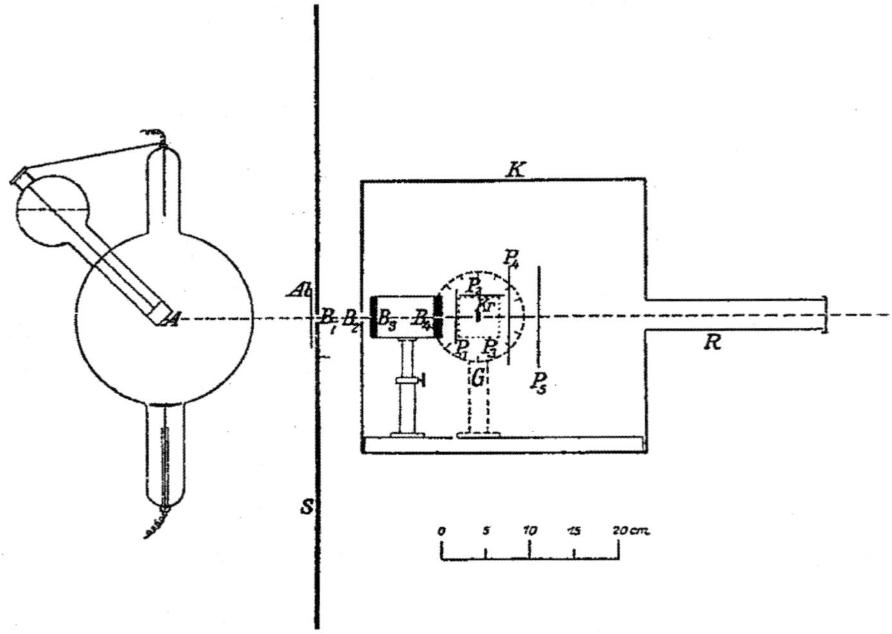

Abstand	Antikathode–Kristall	350 mm
„	Kristall–P_1 bzw. P_2 bzw. P_3 . .	25 „
„	Kristall–P_4	35 „
„	Kristall–P_5	70 „

Fig. 1.

auf denen sich die Intensitätsverteilung der vom Kristall ausgehenden Sekundärstrahlen registriert. Gegen nicht gewollte Strahlen ist die Anordnung durch einen großen Bleischirm S sowie durch den Bleikasten K in genügender Weise geschützt. Die Größe der wichtigen Teile der Versuchsanordnung lassen sich aus der Fig. 1 entnehmen, die im Maßstab 1:10 gezeichnet ist.

Die Einstellung der gesamten Versuchsanordnung wurde auf optischem Wege bewerkstelligt. Ein Kathetometer, dessen Fernrohr mit einem Fadenkreuz versehen war, hatten wir ein für allemal unverrückbar aufgestellt. Der Reihe nach wurden Brennfleck und Antikathode, Blenden und Goniometerachse in die optische Achse des Fernrohrs gebracht. Auf diese Weise konnten wir die recht umständliche Einstellung mit Röntgenstrahlen durch die viel bequemere optische ersetzen. Kontrollversuche mit Röntgenstrahlen selbst erwiesen auch, daß diese Justierung auf optischem Wege völlig ausreichend war. Die Blenden B_1 bis B_2 schirmten hauptsächlich die Sekundärstrahlen der Röhrenwände ab, während die eigentliche Begrenzung für das auf den Kristall auffallende Röntgenstrahlenbündel die Blende B_4 bildete. Diese hat gewöhnlich 0,75 mm Durchmesser, war in eine 10 mm dicke Bleischeibe gebohrt und konnte vermittelst dreier Stellschrauben (nicht gezeichnet) so justiert werden, daß die Achse des Loches genau mit der Fernrohrachse bzw. Achse des Strahlenbündels zusammenfiel. Hierdurch wurde erreicht, daß ein Strahlenbündel von kreisrundem Querschnitt auf den Kristall fiel; ferner wurde hierdurch daß Maß der von der Wand jener Blende B_4 ausgehenden Sekundärstrahlen auf ein Minimum reduziert, was nötig war, um unliebsame Schwärzungen der Platte möglichst zu unterdrücken. Zur Kontrollierung dieser Einstellung während der Aufnahme photographierten wir das Primärbündel, nachdem es den Schutzkasten durch ein langes Ansatzrohr R verlassen hatte, zu welchem Zwecke eine eigene Kassette am Ende dieses Rohres angebracht war (nicht mitgezeichnet). Das Rohr R diente sonst dazu, die Sekundärstrahlen, die bei dem Auftreffen der Primärstrahlen auf die hintere Kastenwand entstehen würden, möglichst zu vermeiden.

Nach diesen Justierungen, deren Güte wir vor jeder Aufnahme kontrollierten, wurde die Goniometerachse auf dem üblichen Wege senkrecht zum Strahlengang gerichtet. In gleicher Weise waren die verschiedenen Plattenhalter so justiert, daß die Primärstrahlen die aufgestellten Films bzw. Platten senkrecht durchsetzten bzw. parallel zu ihnen verliefen. Wenn der Apparat soweit orientiert war, wurde der zu bestrahlende Kristall, der mit einer Spur Klebwachs am Goniometertisch befestigt war, eingestellt und zwar wieder mit Hilfe des schon genannten Fernrohres in der bekannten Weise unter Benutzung eines „Signales". Diese — wie sich später zeigen wird —

sehr wesentliche Justierung konnte von uns bis auf eine Minute genau vorgenommen werden. Als Aufnahmematerial benutzten wir, nachdem wir einige andere Sorten als nicht so geeignet befunden hatten, Schleußner-Röntgenfilms, die mit Rodinal (1 : 15) entwickelt die besten Resultate lieferten.

Die Überschlagsrechnungen nach den bisherigen Erfahrungen über Sekundärstrahlen zeigten, daß recht beträchtliche Expositionszeiten notwendig waren. Die Belichtungszeiten bewegten sich bei 2—10 Milliampère Belastung (je nach der 6 — 12 Wehnelt betragenden Härte der Röhre)[1]) zwischen 1 — 20 Stunden. Als Röntgenröhren kamen teils Intensivröhren von Gundelach, teils Rapidröhren mit Wasserkühlung von Müller zur Verwendung, die von einem 50 cm Klingelfußinduktor betrieben wurden. Als Unterbrecher gelangten teils ein Wehnelt-, teils ein mechanischer Unterbrecher zur Verwendung. Vor die Röntgenröhre wurden passende Ventilröhren zur Vermeidung von Schließungslicht geschaltet. Es wurde mit Pausen exponiert, um die Röhre nicht allzusehr zu erhitzen. Für die bisher ausgeführten Versuche erwies sich eine Härte von 8—10 Wehnelt als geeiget.

Versuche mit dem provisorischen Apparat.[2])

Da wir anfangs glaubten, es mit einer Fluoreszenzstrahlung zu tun zu haben, mußte ein Kristall verwendet werden, der Metall von beträchtlichem Atomgewicht als Bestandteil enthielt, um möglichst intensive und zugleich homogene Sekundärstrahlen zu erhalten, die für die Versuche am geeignetsten zu sein schienen. Nach Barkla kamen in erster Linie die Metalle vom Atomgewicht 50—100 in Betracht. Da wir vorderhand keinen guten Kristall, der derartige Metalle enthielt, zur Verfügung hatten, benutzten wir zu den Vorversuchen einen leidlich ausgebildeten Kupfervitriolkristall. Dieser wurde ziemlich willkürlich in den Apparat eingesetzt, und zwar fielen die

1) Zwei Härtemesser waren uns in dankenswerter Weise von der Firma Reiniger, Gebbert & Schall zur Verfügung gestellt.

2) Dieser war im Prinzip wie der in Fig. 1 dargestellte, nur einfacher gehalten und ohne genaue Einstellungsmöglichkeit. Die Blendöffnungen betrugen hierbei 3 mm.

Röntgenstrahlen ungefähr senkrecht auf eine Pinakoidfläche 3. Art (110) auf. Im Abstand von 40 mm von ihm waren zwei photographische Platten — entsprechend P_2 und P_4 in Fig. 1 — aufgestellt. Nach der Exposition war die obere Platte schwach, oder gleichmäßig geschwärzt, P_4 wies außer dem Durchstoßungspunkt der Primärstrahlen eine Reihe von anscheinend geordneten Flecken auf (vgl. Taf. I, Fig. 1).

Um sicher zu sein, daß diese Flecken durch die Kristallstruktur des Kupfervitriols bedingt waren, wurden Kupfervitriolkristalle grob pulverisiert, in ein kleines Papierschächtelchen eingeschlossen und der vorhergehende Versuch bei sonst gleichen Bedingungen wiederholt. Die Platte P_2 zeigt keine Änderung, dagegen waren die großen geordneten Flecken auf P_4 verschwunden; statt dessen erschien der Durchstoßungsfleck mit einem Saum von vielen kleinen, unregelmäßig geordneten Pünktchen umgeben (vgl. Taf. I, Fig. 2). Die Vermutung, daß bei ganz fein pulverisiertem Material der Rest der Flecken verschwinden würde, fand seine Bestätigung durch den nächsten Versuch. Ein weiterer Versuch mit herausgenommenem Kristall bei sonst identischen Bedingungen zeigte, wie wohl zu erwarten war, daß die Platten bis auf den Durchstoßungspunkt der Primärstrahlen ungeschwärzt blieben.

Es folgten noch zwei Aufnahmen mit dem zuerst genannten Kupfervitriolkristall. Bei der einen wurde bei gleicher Orientierung der Kristall parallel zu sich verschoben, so daß eine andere Stelle des Kristalles durchstrahlt wurde; bei der zweiten der Kristall um einige Grade gegen die Röhre geneigt. Jene Aufnahme ist identisch mit Fig. 1, Taf. I, die Erscheinung ist unabhängig vom Ort der Durchstrahlung. Die andere Aufnahme zeigt, daß die Lage der sekundären Flecken von der Orientierung des Kristalles gegen die Einfallrichtung der Primärstrahlen abhängig ist.

Da die eben beschriebenen Versuche erwiesen hatten, daß sich die Vermutung Laues glänzend bestätigte, ließen wir den vorhin beschriebenen besseren Apparat bauen, stellten aber inzwischen weitere Versuche mit Zinkblende, Steinsalz und Bleiglanz (Spaltstücke) an. Diese ergaben ganz ähnliche Resultate wie bei Kupfervitriol. Wir werden jedoch von einer Beschreibung derselben Abstand nehmen, weil, wie sich zeigen wird, eine genaue Orientierung der Kristalle notwendig war, um einen tieferen Einblick in die Erscheinungen zu gewinnen.

Versuche mit genauer Orientierung des Kristalles zur Richtung der Primärstrahlen.

Mit dem definitiven Apparat wurde zunächst die auf Fig. 1, Taf. I abgebildete Aufnahme wiederholt, und zwar wurden sämtliche in Fig. 1 angegebenen Platten aufgestellt.[1]) Die Orientierung des Kupfervitrioles war möglichst dieselbe wie vorher. P_2 und P_3 zeigten gleichmäßige Schwärzung; P_1 wies bloß den Durchstoßungspunkt der Primärstrahlen auf. Auf P_4 und P_5 war eine ähnliche Figur wie bei Aufnahme 1 zu sehen; indessen waren jetzt entsprechend der kleinen Blendenöffnung die sekundären Flecken zusammengeschrumpft (vgl. Taf. I, Fig. 3 und 4). Bemerkenswert ist, daß sich die Abstände (Kristall — P_4) zu (Kristall — P_5) verhalten wie die Größe der Figuren auf P_4 bzw. P_5, wodurch erwiesen wurde, daß die Strahlung gradlinig vom Kristall ausgeht. Ferner ist zu beachten, daß die Größe der einzelnen Interferenzflecke trotz des größeren Abstandes der Platte P_5 vom Kristall dieselbe geblieben ist. Dies dürfte wohl ein Anzeichen dafür sein, daß die je einen Einzelfleck hervorrufenden Strahlen als paralleles Bündel aus dem Kristall heraustreten.

Es ist zu erwarten, daß die Erscheinung bei Kristallen des regulären Systems übersichtlicher und leichter verständlich wird wie bei dem triklinen Kupfervitriol, weil man wohl annehmen darf, daß das zugehörige Raumgitter von der größtmöglichen Einfachheit ist. Als geeignet erschien uns die reguläre Zinkblende, mit der wir, wie schon erwähnt, Vorversuche gemacht und beträchtliche Intensität der Sekundärstrahlen erhalten hatten. Wir ließen von Steeg & Reuter in Homburg eine planparallele Platte aus einem guten Kristall parallel zu einer Würfelfläche (100) (senkrecht zu einer kristallographischen Hauptachse) schleifen von 10×10 mm Größe und 0,5 mm Dicke. Diese Platte wurde in der oben beschriebenen Weise genau orientiert, so daß die Primärstrahlen den Kristall senkrecht zur Würfelfläche durchsetzten. Das Resultat eines solchen Versuches zeigt uns Fig. 5, Taf. II. Die Lage

1) Blendenöffnung B_4 bei diesem Versuch 1,5 mm.

der Flecke ist völlig symmetrisch geworden in bezug auf den Durchstoßungspunkt. Man kann in die Figur zwei Paare zueinander senkrecht stehende Symmetrieebenen einzeichnen. Nimmt man irgend einen der Flecken aus der Figur heraus und liegt dieser nicht auf einer der Symmetrieebenen, so kann man ihn durch Spiegelung und Drehung um 90° mit noch sieben zugehörigen Punkten zur Deckung bringen. Fällt ein Fleck mit einer Symmetrieebene zusammen, so kann man diesen naturgemäß nur mit noch drei entsprechenden zur Deckung bringen. Dies entspricht der holoedrischen Symmetrie des regulären Systems, trotzdem die Zinkblende in eine hemiedrische Klasse gehört. Diese Tatsache, daß eine völlige Vierzähligkeit auf der Platte vorhanden ist, ist wohl einer der schönsten Beweise für den gitterartigen Aufbau der Kristalle, und ein Beweis dafür, daß keine andere Eigenschaft als allein das Raumgitter hier in Betracht kommt. Denn die einfachen Raumgitter zeigen *stets* eine höhere Symmetrie als der Kristall, für dessen Darstellung sie ein Element bilden; Translationen von Raumgittern verschiedener Natur gegeneinander, wie sie zur Erklärung der Hemiedrien angenommen werden müssen, sind nach den Versuchen[1]) und unserer Theorie ohne Einfluß.

Die Anforderung der Symmetrie fand auch weiterhin ausnahmslos Bestätigung durch die Versuche. Einmal ließen wir die Primärstrahlen senkrecht auf eine Oktaederfläche (111), dann auf eine Rhombendodekaederfläche (110) auffallen. Fig. 7, Taf. II, gibt eine Dreizähligkeit wieder entsprechend der dreizähligen Achse, in der der Kristall bestrahlt wurde; Fig. 12, Taf. IV analog Zweizähligkeit, entsprechend der zweizähligen Achse.[2])

---

1) Neuere Versuche an Pyrit und Hauerit haben dies nicht bestätigt. Die betreffenden Aufnahmen (nach der vierzähligen Achse) zeigen vielmehr hemiedrische Symmetrie. W. Friedrich.

2) Die Verwaschenheit bzw. Unterteiltheit der Flecken hat ihre Ursache in der Mangelhaftigkeit des untersuchten Kristallstückes. Die Spaltflächen bestanden deutlich aus Flächenstücken, die um kleine Winkel gegeneinander geneigt waren. Das reflektierte Bild eines Gegenstandes war sehr verzerrt. Ob diese Unregelmäßigkeiten schon in dem Kristall, aus dem das untersuchte Stück herausgespalten war, enthalten waren, oder erst das regelmäßige Gefüge durch den Stoß des Spaltens zerstört wurde, wollen wir dahingestellt sein lassen. Jedenfalls tritt die Unterteilung der Einzelflecken fast stets bei Spaltstücken auf, selten an solchen Präparaten, die aus einem großen Kristall herausgesägt oder geschnitten sind.

Drehen wir den Kristall um den primären Strahl, so muß man aus dem Vorhergehenden schließen, daß sich das Bild auf der Platte mitdreht. In der Tat bestätigen dies die Versuche.

Es war noch zu zeigen, eine wie genaue Orientierung des Kristalles erforderlich ist, um bei Wiederholung eines Versuches identische Bilder zu erhalten. Wir ließen den primären Strahl um 3° gegen eine vierzählige Achse geneigt, doch so, daß die Ebene durch Strahl und Achse eine Symmetrieebene war, durch den Kristall gehen. Das Resultat dieses Versuches gibt Fig. 11, Taf. IV. Wie wir sehen, ist die Vierzähligkeit verschwunden, wenngleich man noch die Mehrzahl der Flecken in Fig. 5, Taf. II, wiedererkennen kann. Die eine Symmetrieebene zeigt sich auch hier. Ebenso ist Fig. 8, Taf. III, durch Neigung des Kristalles um 3° aus Fig. 7, Taf. III, entstanden.

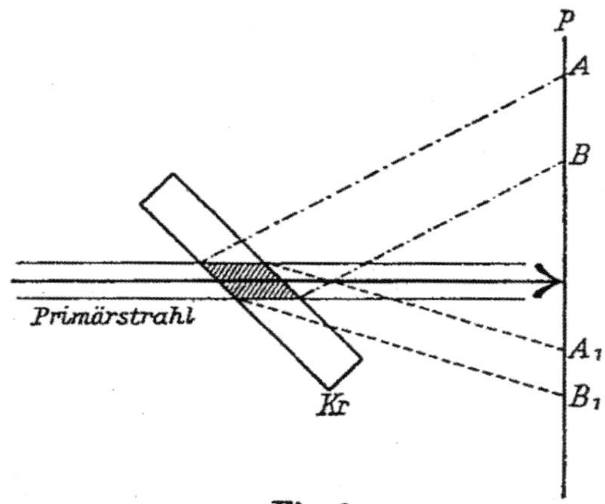

Fig. 2.

Drehten wir die senkrecht zur vierzähligen Achse geschliffene Platte so, daß der primäre Strahl sie parallel zu einer dreizähligen Achse durchsetzte, so ergab sich eine Figur die sich von Fig. 7, Taf. III, nicht in der Lage der Flecke, sondern allein durch deren Größe unterschied. Auch nach der Symmetrie gleichberechtigte Flecken zeigen jetzt Größenunterschiede. Es beweist dies, daß die Begrenzung des durchstrahlten Kristallstückes ohne Einfluß auf die Lage der Intensitätsmaxima ist, während die Unterschiede in der Fleckengröße den Unterschieden der verschieden gerichteten Projektionen des durchsetzten Metallstückes auf die photographische Platte entsprechen (vgl. Fig. 2). Das schraffierte Gebiet stellt den durchleuchteten Teil des Kristalles *Kr* dar. AB und $A_1 B_1$ sind die Projektionen dieses Teiles auf die Platte P und geben die Größe der Flecken an. Drehten wir dieselbe Kristallplatte so, daß der primäre Strahl sie jetzt in der Richtung einer zweizähligen Symmetrieachse durchsetzte, so entstand eine Figur, die sich nur durch die ungleiche Größe entsprechender Flecke von der Aufnahme Fig. 12, Taf. IV unterschied, die wir mit einem zur zweizähligen Achse senkrechten Spaltstück gemacht hatten.

Außer den eben beschriebenen Versuchen haben wir weitere mit Zinkblende, Kupferkristall, Steinsalz und einer Diamantplatte angestellt. (Die Diamantplatte ist ein höchst dankenswertes Geschenk der Firma Ginsberg in Hanau an das physikalische Institut der Universität.) Wir sehen jedoch von der Mitteilung der Photogramme ab, weil die Versuche noch nicht abgeschlossen sind, hoffen aber später darüber berichten zu können. Wir wollen indessen noch auf zwei Versuchsergebnisse aus dieser Serie hinweisen, die uns von Bedeutung erscheinen.

Erstens haben wir beim Steinsalz gefunden, daß die Intensität der Flecke von der Dicke der durchstrahlten Schicht abhängt; denn bei gleicher Expositionszeit, gleicher Härte der Röhre und sonst gleichen Bedingungen war die Schwärzung bei 15 mm durchstrahlter Schicht des Kristalles wesentlich größer als bei einer 10mal dünneren Schicht. Wir wollen diese Versuche mit anderem Material weiter fortführen.

Zweitens wollen wir einem Versuch mit dem Diamanten ein paar Zeilen widmen. Kohlenstoff zeigt nach Barkla keinerlei bisher nachgewiesene Eigenstrahlung. Es war also zu erwarten, daß selbst bei erheblichen Expositionen die Platten außer der durch die zerstreute Strahlung hervorgerufenen allgemeinen Schwärzung keine Flecken wie bei den anderen Kristallen zeigten, wenn diese auf Eigenstrahlung des Kristalles beruhten. Entgegen dieser Vermutung wissen nicht nur die hinteren Platten P_4 und P_5 ähnliche Erscheinungen wie bei den früheren Aufnahmen auf, sondern sogar die Platten P_1, P_2 und P_3 ließen deutliche Flecken erkennen.[1]) Ob diese auf-

1) Auch bei tagelanger Exposition haben wir bei anderen Kristallen niemals auf den Platten P_1, P_2 und P_3 etwas anderes als allgemeine Schwärzung erhalten.

fallende Tatsache zusammenhängt mit dem kleinen Atomvolumen im Vergleich zu dem der übrigen Kristalle oder mit dem anomalen Verhalten des Diamanten gegenüber den Wärmeschwingungen, welches sich an der spezifischen Wärme zeigt, wollen wir vorläufig dahingestellt sein lassen.[1])

Neben diesen bis jetzt beschriebenen Untersuchungen, die den Zusammenhang der Lage der Sekundärflecken und der Gitterstruktur erforschen sollen, haben wir den Charakter der die Flecken hervorbringenden Strahlen untersucht. Daß wir es in der Tat mit Röntgenstrahlen zu tun haben, zeigt der Umstand, daß die Strahlen durch beträchtliche Schichten von Metallen hindurchgehen, z. B. durch die Plattenhalter aus Stahlblech. Vorläufige Härtemessungen wurden mit dem Zinkblendekristall, der senkrecht zur vierzähligen Achse geschliffen war, auf folgende Weise angestellt. Da wir das Schwärzungsgesetz der Röntgenstrahlen nicht kennen, ist die Intensität der Strahlen proportional der durch sie hervorgerufenen Schwärzung gesetzt und diese mit einem Hartmannschen Mikrophotometer ausgemessen. Die Kristallsymmetrie bedingt, daß je acht einem Kreise zugeordnete Flecke dieselbe Schwärzung haben, was in der Tat der Fall ist. Wir brachten nun vor der Platte P_5 ein Aluminiumblech von 3 mm Dicke, welches zwei gegenüberliegende Viertel der Platte bedeckte, die anderen beiden freiließ. Das Primärbündel ging, um Sekundärstrahlen des Aluminiumblechs zu vermeiden, durch eine zentrale Öffnung hindurch. Die Schwärzungsmessungen, die in folgender Tabelle zusammengestellt sind, zeigen, daß der Absorptionskoeffizient der Strahlen ungefähr 3,84 cm^{-1} für Aluminium ist. In Kolonne 1 stehen die Durchmesser d der Kreise, auf dem die zueinander gehörenden Flecken liegen. In der zweiten die Schwärzung der Flecken ohne Aluminiumblech, während in der dritten Kolonne die Schwärzung nach dem Durchgang der Strahlen durch das Aluminiumblech verzeichnet ist. In der vorletzten Spalte sind die Werte der Schleierschwärzung der Platten eingetragen, während in die letzte die nach dem bekannten Absorptionsgesetz berechneten Absorptionskoeffizienten k für Al stehen. Das Absorptionsgesetz ist als ex-

1) Vgl. theor. Teil.

ponentiell angenommen. Barkla findet für die Durchdringungsfähigkeit der Eigenstrahlung des Zinks einen wesentlich kleineren (für k also einen wesentlich größeren) Wert. Eine nach derselben Methode ausgeführte Härtemessung beim Diamanten ergab eine Härte von ungefähr derselben Größe. Die Härtemessungen nach der photographischen Methode sind jedoch nicht genau genug, um sichere Schlüsse hierüber ziehen zu lassen, da der gefundene Härteunterschied außerhalb der Meßgenauigkeit liegt. Mit einer elektrischen Methode, die wohl von den Zufälligkeiten der photographischen Methode, besonders den Plattenfehlern, verschieden harte Entwickelung usw. frei ist, werden diese Versuche wieder aufgenommen und es wird dann darüber berichtet werden.

d	ohne Al	mit Al	Schleier	k
50 mm	1,78	0,88	0,45	3,53 cm^{-1}
57,5 „	2,02	0,79	0,45	4,71 „
74 „	0,69	0,53	0,45	3,23 „
86 „	0,84	0,55	0,45	3,89 „

Für die Beurteilung der Frage nach dem Ursprung der Strahlen ist die folgende Tatsache von Wichtigkeit: Während der Expositionszeit schwankte die Härte der Primärstrahlen besonders bei älteren Röhren, die ziemlich aufgebraucht waren und daher oft regeneriert werden mußten, in beträchtlicher Weise (von 6—12 Wehnelt). Dennoch blieben, wie die Photogramme zeigten, die sekundären Flecken scharf begrenzt und unverrückt.

Zum Schluß wollen wir besonders Hrn. Prof. A. Sommerfeld, in dessen Institut diese Arbeit ausgeführt ist, für die bereitwillige Beschaffung der Apparate, sowie sein stetiges Interesse und seine fördernden Ratschläge unseren herzlichsten Dank aussprechen. Wir möchten bei dieser Gelegenheit auch nicht versäumen, Hrn. Geheimrat Röntgen und Hrn. Geheimrat v. Groth für Überlassung von wertvollen Kristallen und Apparaten, sowie ihr reges Interesse ebenfalls verbindlich zu danken.

München, Inst. für theor. Physik der Universität.

(Eingegangen 15. März 1913.)

4. *Eine quantitative Prüfung der Theorie für die Interferenzerscheinungen bei Röntgenstrahlen; von M. Laue.*[1])

(Hierzu Taf. II, Figg. 5 u. 6.)

Vor kurzem veröffentlichten W. Friedrich, P. Knipping und der Verfasser[2]) an dieser Stelle Beobachtungen über Interferenzerscheinungen, die beim Durchgang von Röntgenstrahlen durch Kristalle entstehen, sowie eine Theorie dazu, welche sich qualitativ der Erfahrung anschloß. Jetzt soll diese Theorie einer ersten quantitativen Prüfung unterzogen werden. Mag es sein, daß es dabei nicht gelingt, zur vollen Wahrheit durchzudringen, daß besonders die Werte für die Wellenlängen der Röntgenstrahlen später ersetzt werden müssen durch andere, welche zu ihnen in einfachen rationalen Verhältnissen stehen. Das eine geht wohl aus der weitgehenden Übereinstimmung mit der Erfahrung hervor: daß die Theorie auf dem rechten Wege ist. Zugleich mag das Folgende als Beispiel für das Verfahren nützlich sein, nach welchem Photogramme, wie die dort veröffentlichten, auszuwerten sind. Wir wählen dazu die Fig. 5 auf Taf. II, bei welcher das regulär kristallisierte Zinksulfid parallel einer vierzähligen Symmetrieachse durchstrahlt wird. Die Gitterkonstante a ergibt sich aus der Überlegung, daß jedem Elementarwürfel dieses Gitters ein Zink- und ein Schwefelatom entspricht. Bezeichnet also

$$N = 6{,}20 \cdot 10^{23}$$

die Zahl der Molekeln im Mol[3]),

$$m = 65{,}4 + 32 = 97{,}4$$

1) Wenig veränderter Abdruck aus den Münchner Ber. 1912. p. 363.

2) W. Friedrich, P. Knipping u. M. Laue, Münchener Ber. 1912. p. 303 und die vorhergehende Arbeit. Die dortige Numerierung der Gleichungen und Figuren wird hier fortgesetzt.

3) M. Planck, Wärmestrahlung. Leipzig 1913. p. 166.

das Molekulargewicht und

$$\delta = 4{,}06$$

die Dichte[1]), so muß

(10) $$a^3 N = \frac{m}{\delta}$$

sein, weil beide Seiten der Gleichung das Volumen eines Mols darstellen. Daraus folgt:

(11) $$a = 3{,}38 \cdot 10^{-8}\,\text{cm}\,.$$

Wie schon erwähnt (p. 975) legt man zweckmäßig die Koordinatenachsen in die Richtungen der Würfelkanten. Die drei Sinusquotienten des Intensitätsausdruckes (6) haben dann nach (8) ihre Maxima auf den Kegelflächen.

(12) $$\alpha = h_1 \frac{\lambda}{a}, \quad \beta = h_2 \frac{\lambda}{a}, \quad 1 - \gamma = h_3 \frac{\lambda}{a}\,.$$

Die Schnitte der beiden ersteren Scharen mit der senkrecht zur z-Richtung stehenden photographischen Platte sind Hyperbeln, während die Gleichung $\gamma = \text{const}$ konzentrische Kreise um den Durchstoßpunkt des primären Strahles ergibt; die Radien dieser Kreise sind proportional zur Tangente des Achsenwinkels

$$\varphi = \arccos \gamma$$

der zugehörigen Kreiskegel.

Um das Photogramm theoretisch nachzukonstruieren, ersetzt man zweckmäßig die wirklichen Koordinaten

$$x = \alpha \sqrt{x^2 + y^2 + z^2}, \quad y = \beta \sqrt{x^2 + y^2 + z^2}$$

eines Punktes auf der Platte durch α und β als rechtwinklige Koordinaten. Dann bekommen die Kreise

$$\gamma = \sqrt{1 - (\alpha^2 + \beta^2)} = \text{const}$$

Radien gleich

$$\sqrt{\alpha^2 + \beta^2} = \sin \varphi\,.$$

In radialer Richtung ist also die Zeichnung gegen das Photogramm verzerrt. Erhalten dagegen bleiben alle Winkel, deren Spitze im Mittelpunkt liegt und damit die Winkelabstände der zu berechnenden Flecken, weil

$$\alpha : \beta = x : y$$

1) Landolt-Börnstein, Tabellen. Berlin 1894. p. 62.

ist. Vor allem aber werden die Kurven $\alpha = \text{const}$ und $\beta = \text{const}$ jetzt zu den Koordinatenachsen parallele äquidistante Gerade; benutzt man bei der Zeichnung λ/a als Längeneinheit, so findet man diese auf Koordinatenpapier unmittelbar vor. Jedem Schnittpunkt solcher Geraden entspricht ein Schnittpunkt von zwei Hyperbeln im Photogramm, d. h. ein mögliches Kreuzgitterspektrum. Es fragt sich nun, welche von diesen Punkten einem Kreise

$$1 - \gamma = h_3 \frac{\lambda}{a}$$

naheliegen. Wie nahe sie daran liegen müssen, um wirklich sichtbar zu werden, entzieht sich vorläufig unserer Beurteilung. Wir rechnen im folgenden, als ob sie genau auf ihnen liegen müßten, d. h., als ob die Gleichungen (8) alle genau erfüllt wären. Jedem Interferenzpunkt sind dann drei ganze Zahlen $h_1\ h_2\ h_2$, seine Ordnungszahlen, zugeordnet. Diese für spektralhomogene Strahlen sicherlich nicht zutreffende Annahme hat auf die zu errechnende Lage der Interferenzpunkte keinen merklichen Einfluß. Auffälliger wird ihr Einfluß auf die zu berechnenden Werte der Wellenlänge, insofern sich diese auch dann ein wenig unterscheiden werden, wenn sie sich auf dieselbe Strahlung beziehen. So lassen sich die unten gefundenen, einige Prozente betragenden Differenzen dieser Art erklären. Wie weit dabei die tatsächliche Unbestimmtheit der Wellenlänge infolge der spektralen Inhomogenität eine Rolle spielt, bleibt eine offene Frage.

Haben wir nun für ein Intensitätsmaximum die drei Ordnungszahlen h gefunden, so bestimmt sich das Verhältnis λ/a wegen der Gleichungen (12) und der Identität

$$\alpha^2 + \beta^2 + \gamma^2 = 1$$

zu

$$\frac{\lambda}{a} = \frac{2\,h_3}{h_1^2 + h_2^2 + h_3^2}\,. \tag{13}$$

Die Richtungskosinus α, β, γ, d. h. die Lage eines Interferenzpunktes, bestimmt sich wegen (8) eindeutig aus den Ordnungszahlen h nach den Formeln:

$$\alpha = \frac{2\,h_1\,h_3}{h_1^2+h_2^2+h_3^2}, \quad \beta = \frac{2\,h_2\,h_3}{h_1^2+h_2^2+h_3^2}, \quad \gamma = \frac{h_1^2+h_2^2-h_3^2}{h_1^2+h_2^2+h_3^2}. \tag{13a}$$

Die Koordinaten x, y eines Interferenzpunktes im Photogramm hängen natürlich außerdem noch vom Abstand z der Platte vom Kristall ab; für sie gelten die Formeln:

(13b) $$x = \frac{\alpha}{\gamma} z = \frac{2 h_1 h_3}{h_1^2 + h_2^2 - h_3^2} \cdot z, \quad y = \frac{\beta}{\gamma} z = \frac{2 h_2 h_3}{h_1^2 + h_2^2 - h_3^2} z.$$

Da die Zahlen h ganze Zahlen sind, so sind sie und damit der nach (13) bestimmte Wert λ/α von kleinen Beobachtungsfehlern unabhängig.

In dem genannten Photogramm fassen wir zunächst den zweiten Ring von innen ins Auge, der einen sehr stark ausgeprägten, der Symmetrie wegen achtfachen Punkt trägt. Sein Radius, bestimmt aus dem Abstand einander gegenüberligender Punkte, ist $r = 1{,}25$ cm. Da der Abstand der Platte vom Kristall $z = 3{,}56$ cm[1]) betrug, so folgt:

$$\operatorname{tg} \varphi = \frac{r}{z} = 0{,}350, \quad \varphi = 19^0 20',$$

$$\cos \varphi = \gamma = 0{,}9436,$$

$$1 - \gamma = 0{,}0564.$$

Ist der fragliche Ring von der ersten Größenordnung, so muß nach der dritten der Gleichungen (8)

(14) $$\frac{\lambda}{a} = 0{,}0564$$

sein. Dann aber müssen wir wegen der Ganzzahligkeit der h nach (13) schließen:

$$\sum h^2 = h_1^2 + h_2^2 + h_3^2 = 36.$$

Auf dem Koordinatenpapier (Fig. 3) findet man nun leicht die Punkte

$$h_1 = 3, \quad h_2 = 5,$$

$$h_1 = 5, \quad h_2 = 3,$$

1) Dieser Abstand z war experimentell nicht so genau bestimmt, wie er hier angegeben ist, schon weil der photographische Film nie ganz eben ist. Der Wert 3,56 cm ist so gewählt, daß die Radien der Ringe in der Berechnung möglichst genau mit dem am Photogramm gemessenen stimmen. Somit enthält die Messung eines einzelnen Radius noch keine Prüfung der Theorie, sondern nur die Messung der Quotienten aus den Radien verschiedener Ringe.

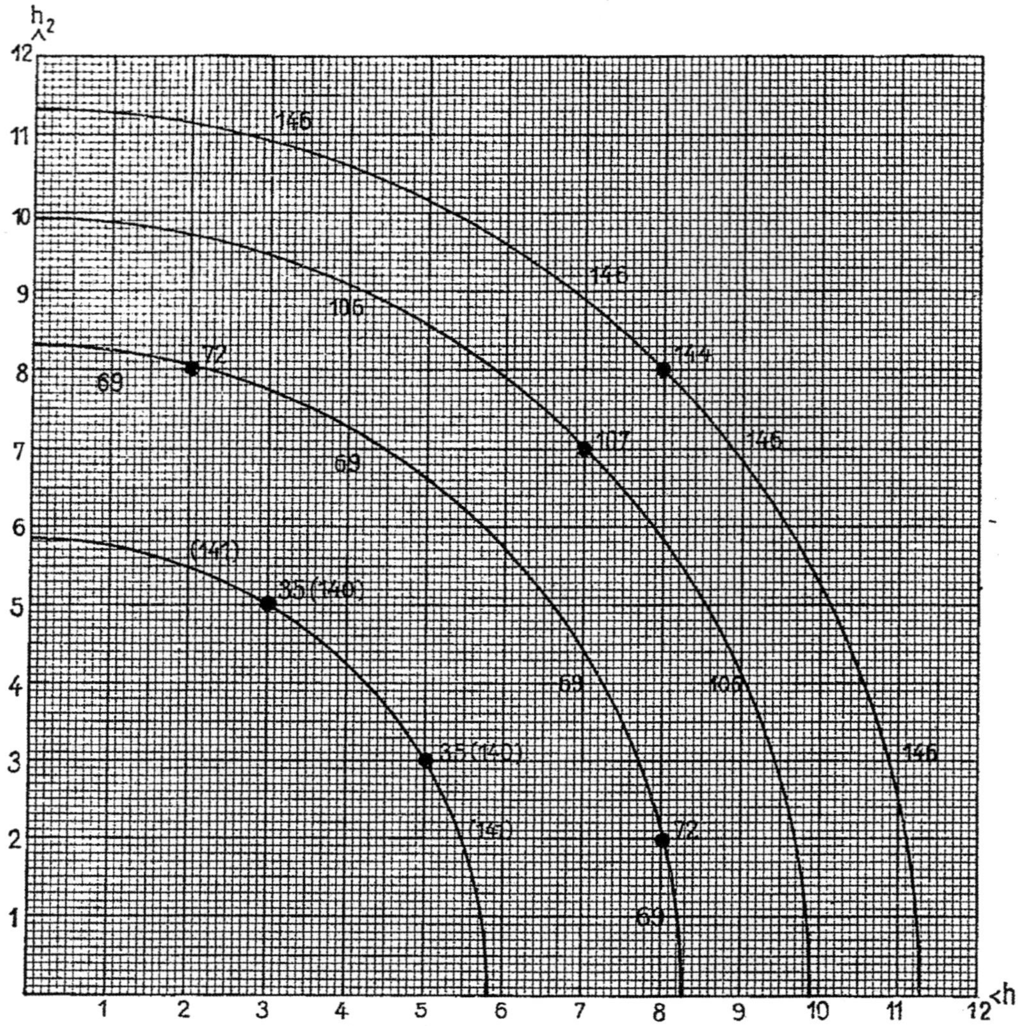

Fig. 3.

für welche, wenn $h_3 = 1$, $\sum h^2 = 35$ ist. Messen wir ihre Winkelabstände gegeneinander und gegen die Achsen, so finden wir völlige Übereinstimmung mit der Lage jener acht Punkte im Photogramm. Somit werden wir dieser Wellenlänge den achtfachen Punkt

h_1	h_2	h_3	$\sum h^2$	$\frac{\lambda}{a}$ nach (13)
3	5	1	35	$\frac{2}{35} = 0{,}0571$

zuschreiben. Die Differenz zwischen diesem Werte von λ/a und dem in (14) gegebenen liegt völlig innerhalb der Fehlergrenze.

Könnte nun dieser Ring von der zweiten Ordnung sein? Wohl kaum; denn sonst bekämen die betrachteten Punkte die Ordnungszahlen

h_1	h_2	h_3	$\sum h^2$	$\frac{\lambda}{a}$ nach (13)
6	10	2	140	

und die Figur zeigt, daß die Punkte

4	11	2	141	

dem durch sie führenden Kreis so nahe liegen, daß noch ein weiterer achtfacher Punkt zu erwarten wäre, der tatsächlich auf dem Photogramm nicht vorhanden ist.

Fragen wir weiter, ob es für die oben berechnete Wellenlänge einen Ring zweiter Ordnung gibt. Dann muß nach (13) $\sum h^2 = 70$ sein. In Fig. 3 findet man nun den achtfachen Punkt.

2	8	2	72	$\frac{4}{72} = 0{,}0555$.

Der Radius dieses Ringes im Photogramm berechnet sich nach dem Schema:

$$1 - \gamma = 2\frac{\lambda}{a} = 0{,}1111\,, \quad \gamma = 0{,}8889\,, \quad \varphi = 27^{0}\,16'\,,$$

$$\operatorname{tg}\varphi = 0{,}5154\,, \quad r = z\operatorname{tg}\varphi = 3{,}56\cdot 0{,}5154 = 1{,}84\,\mathrm{cm}\,.$$

In der Tat finden sich in dem Photogramm acht Punkte, deren Abstand vom Mittelpunkt diesen Wert hat, und welche gegen die Achsen und gegeneinander dieselben Winkelabstände haben, wie der angegebene Punkt und die sieben anderen, welche man durch Vertauschung von h_1 und h_2, sowie durch Umkehrung der Vorzeichen dieser Zahlen daraus ableitet. Dagegen fehlen die Punkte

4	8	2	69	$\frac{4}{69} = 0{,}0580$,
1	7	2		

welche nach der Zeichnung dem Kreise auch noch recht nahe liegen.

Für einen Ring dritter Ordnung ($r = 2{,}26$ cm) findet sich nach Fig. 3 ein vierfacher Punkt

h_1	h_2	h_3	$\sum h^2$	$\frac{\lambda}{a}$ nach (13)
7	7	3	107	$\frac{7}{107} = 0{,}0561$,

der im Photogramm tatsächlich zu sehen ist, während merkwürdigerweise der Punkt

4	9	3	106	$\frac{6}{106} = 0{,}0567$,

fehlt. Eine Erklärung dafür läßt sich zurzeit nicht geben.

Auf dem Ring vierter Ordnung ($r = 2{,}88$ cm) findet sich nach Fig. 3 ein auch im Photogramm vertretener vierfacher Punkt

8	8	4	144	$\frac{8}{144} = 0{,}0555$,

während die Punkte

11	3	4	146	$\frac{8}{146} = 0{,}0548$
7	9	4		

fehlen. Der Ring fünfter Ordnung liefert nach der Zeichnung keinen Punkt. Er würde im Photogramm schon vollkommen außerhalb des Bereiches der Interferenzpunkte liegen.

Übersieht man diese Zusammenstellung, so erscheint ziemlich sicher, daß λ/a zwischen 0,0555 und 0,0571 also etwa bei 0,0563 liegt, was nach (11) auf den Wert

$$\lambda = 1{,}90 \cdot 10^{-9}\ \text{cm}$$

führt. Unaufgeklärt bleibt nur das Fehlen des einen Punktes im Ring dritter Ordnung. Daß für λ/a nicht überall derselbe Wert herauskommt, liegt daran, daß die drei Beziehungen (12) nicht alle streng erfüllt sein können, wie in Gleichung (13) angenommen war. Ob die spektrale Breite der Strahlung so groß ist, daß alle Wellenlängen zwischen den angegebenen Grenzen vorkommen, muß noch dahingestellt bleiben. Dieser Wellenlänge entspricht zweifellos die stärkste von allen vom Zinksulfid ausgehenden Strahlungen.

Ganz analog findet man für den innersten schwachen Ring der vierzählig symmetrischen Aufnahme ($r = 1{,}01$ cm) die Punkte:

h_1	h_2	h_3	$\sum h^2$	$\frac{\lambda}{a}$ nach (13)
1	7	1	51	$\frac{2}{51} = 0{,}0392$,
5	5	1		

von denen der erste achtfach, der zweite vierfach vertreten ist. Dazu gehört als Ring zweiter Ordnung der stark ausgebildete dritte Ring von innen ($r = 1{,}46$ cm) mit dem achtfachen Punkt

2	10	2	108	$\frac{4}{108} = 0{,}0370$.

Weitere Ringe fehlen, trotzdem auf ihnen nach der Konstruktion noch weitere Punkte liegen könnten. Es stimmt dies damit überein, daß nach außen die Intensitäten der Interferenzpunkte schnell abnehmen.

Das Verhältnis λ / a dürfte somit für diese Gruppe von Interferenzpunkten zwischen 0,0392 und 0,0370 liegen, und zwar, da der innere Ring soviel schwächer als der äußere ist, näher an dem letzteren Wert. Nehmen wir $\lambda/a = 0{,}0377$ an, so wird nach (11)

$$\lambda = 1{,}27 \cdot 10^{-9}\,\mathrm{cm}.$$

Sehr auffällig sind in dem Photogramm die vier Gruppen von je drei engbenachbarten Flecken, welche im Abstand $r = 2{,}67$ cm vom Mittelpunkt auf den Koordinatenachsen liegen. Diese lassen sich ungezwungen durch einen Ring dritter Ordnung mit der Wellenlänge

$$\lambda = 0{,}0663\,a = 2{,}24 \cdot 10^{-9}\,\mathrm{cm}$$

erklären, auf welchem nach der Zeichnung die Punkte

0	9	3	90	$\frac{6}{90} = 0{,}0667$
1	9	3	91	$\frac{6}{91} = 0{,}0659$

liegen, von denen der erste ein vierfacher, der zweite ein achtfacher ist. Dieselbe Wellenlänge gehört zu einem Ring vierter Ordnung ($r = 3{,}31$ cm) mit den achtfachen Punkten

2	10	4	120	$\frac{8}{120} = 0{,}0667$,

während der auch noch in Betracht kommende Punkt

h_1	h_2	h_3	$\sum h^2$	$\frac{\lambda}{a}$ nach (13)
1	10	4	117	$\frac{8}{117} = 0,0683$

fehlt. Daß hier kein Ring zweiter Ordnung auftritt, erklärt die Zeichnung damit, daß der einzige dafür in Betracht kommende Punkt

$$3 \mid 7 \mid 2 \mid 62 \mid \frac{4}{62} = 0,0645$$

schon zu wenig zu dieser Wellenlänge paßt. Unerklärt bleibt dagegen, weswegen kein Ring erster Ordnung auftritt; der Punkt

$$2 \mid 5 \mid 1 \mid 30 \mid \frac{2}{30} = 0,0667$$

würde trefflich zu den oben angegebenen stimmen.

Ferner finden wir noch einen Ring erster Ordnung ($r = 1,78$ cm) für die nur schwach vertretene Wellenlänge

$$\lambda = 0,1051\, a = 3,55 \cdot 10^{-9}\,\text{cm}$$

mit dem vierfachen Punkt

$$3 \mid 3 \mid 1 \mid 19 \mid \frac{2}{19} = 0,1051\,.$$

Ein Ring zweiter Ordnung tritt nicht auf, obwohl der Punkt

$$3 \mid 5 \mid 2 \mid 38 \mid \frac{4}{38} = 0,1051$$

gut zu den angegebenen paßte. Es erklärt sich dies wohl aus dem allgemeinen Abklingen der Intensität nach außen.

Die längste der bisher berechneten Wellenlänge ist aber

$$\lambda = 0,1428\, a = 4,83 \cdot 10^{-9}\,\text{cm}\,,$$

welche mit dem achtfachen Punkt

$$2 \mid 3 \mid 1 \mid 14 \mid \frac{2}{14} = 0,1428$$

einen Ring erster Ordnung ($r = 2,12$ cm) bildet. Der zugehörige Ring zweiter Ordnung würde schon außerhalb des Gesichtsfeldes liegen.

In Fig. 6, Taf. II ist, um den Vergleich mit der Erfahrung zu erleichtern, das Photogramm 5 auf Taf. II unmittelbar aus den hier angegebenen Zahlen h_1, h_2, h_3 nach den Gleichungen (13b) und unter der Annahme $z = 3{,}56$ cm nachkonstruiert worden. Der Leser mag sich selber durch Abgreifen mit dem Zirkel von der Übereinstimmung der Figur mit dem Photogramm überzeugen. Fortgelassen sind in der Zeichnung nur die beiden Punkte, deren Fehlen im Photogramm unerklärt blieb, und einige Punkte in den äußeren Teilen des Photogrammes, welche bisher noch nicht durchgerechnet sind. Abgesehen von den beiden mehrfach erwähnten Ausnahmen, erklärt also die Theorie, weshalb gerade die beobachteten Interferenzpunkte und nur diese auftreten, sie leistet dies unter Annahme von niedrigen Ordnungszahlen ($\leqq 10$) und von Wellenlängen, welche von der erwarteten Größenordnung 10^{-9} cm sind.

An den obigen Angaben sind meiner Überzeugung nach sicher die Verhältnisse zwischen den Ordnungszahlen h_1, h_2, h_3, von denen nach (13b) die Lage der Interferenzpunkte allein abhängt. Nicht ebenso sicher sind diese Zahlen selbst und damit auch nicht die angegebenen Werte der Wellenlängen. Immerhin sei bemerkt, daß ich mich für die angegebenen Werte erst nach Prüfung mancher anderer Möglichkeiten entschieden habe. Verdächtig ist z. B., daß die angegebenen Wellenlängen sich verhalten wie

$$1{,}27 : 1{,}90 : 2{,}24 : 3{,}55 : 4{,}83 = 4 : 6 : 7 : 11 : 15 .$$

Wollte man alle errechneten Interferenzpunkte auf die Wellenlänge $^1/_4\ 1{,}27 \cdot 10 = 3{,}20 \cdot 10^{-10}$ cm zurückführen, so müßten, soweit ich sehe, noch viele andere Interferenzpunkte vorhanden sein, die tatsächlich nicht auftreten. Wie dem aber auch sein mag, in der Übereinstimmung der Fig. 6 mit dem Photogramm 5 scheint mir eine weitgehende Bestätigung der Interferenztheorie dieser Erscheinungen zu liegen.

Zusätze (März 1913).

1. In den obigen Auseinandersetzungen wird stets nur von einem einfachen Raumgitter gesprochen. Bekanntlich aber nötigt die Tatsache, daß hemiedrische Kristalle vor-

kommen, zu der Annahme, daß die Struktur im allgemeinen aus mehreren kongruenten, aber gegeneinander verschobenen Raumgittern besteht; selbst bei holoedrisch regulären Kristallen kommen Raumgitter vor, bei denen der Mittelpunkt jedes Elementarwürfels oder auch die Mittelpunkte der Würfelseiten mit Atomen besetzt sind.[1]) In diesen Fällen treten außer den Atomen, deren Orte durch die Gleichungen (1) bestimmt sind, andere, von den ersteren im allgemeinen verschiedene, auf, die in den Punkten

$$\begin{aligned} x' &= (m+\delta')\mathfrak{a}_{1x} + (n+\varepsilon')\mathfrak{a}_{2x} + (p+\zeta')\mathfrak{a}_{3x}, \\ y' &= (m+\delta')\mathfrak{a}_{1y} + (n+\varepsilon')\mathfrak{a}_{2y} + (p+\zeta')\mathfrak{a}_{3y}, \\ z' &= (m+\delta')\mathfrak{a}_{1z} + (n+\varepsilon')\mathfrak{a}_{2z} + (p+\zeta')\mathfrak{a}_{3z} \end{aligned}$$

liegen. Unter Umständen kommt noch eine dritte und vierte Art usw. hinzu, an den Orten:

$$\begin{aligned} x'' &= (m+\delta'')\mathfrak{a}_{1x} + (n+\varepsilon')\mathfrak{a}_{2y} + (p+\zeta'')\mathfrak{a}_{3z} \text{ usw.}, \\ x''' &= (m+\delta''')\mathfrak{a}_{1y} + (n+\varepsilon''')\mathfrak{a}_{2x} + (p+\zeta''')\mathfrak{a}_{3x}. \end{aligned}$$

usw. Die δ, ε, ζ sind dabei echte Brüche, deren Werte für das Punktsystem charakteristisch sind. Die Vektoren

$$(15)\quad \left\{ \begin{aligned} \mathfrak{r}' &= \delta'\,\mathfrak{a}_1 + \varepsilon'\,\mathfrak{a}_2 + \zeta'\,\mathfrak{a}_3, \\ \mathfrak{r}'' &= \delta''\,\mathfrak{a}_1 + \varepsilon''\,\mathfrak{a}_2 + \zeta''\,\mathfrak{a}_3, \\ \mathfrak{r}''' &= \delta'''\,\mathfrak{a}_1 + \varepsilon'''\,\mathfrak{a}_2 + \zeta'''\,\mathfrak{a}_3 \end{aligned} \right.$$

usw. führen von einem ungestrichenen Atom zu den benachbarten ein-, zwei- und dreigestrichenen; ihre Komponenten nach den Koordinatenachsen sind deshalb gleich $x'-x$, $y'-y$, $z'-z$ usw. Da sich die durch die Striche unterschiedenen Atome i. A. verschieden verhalten, müssen wir ihnen auch verschiedene Funktionen Ψ, Ψ' usw. zuschreiben.

In diesem Fall tritt an die Stelle der Summe (4):

$$(16)\left\{ \begin{aligned} &\frac{e^{-ikR}}{R}\{\Psi(\alpha,\beta)\sum e^{ik[x(\alpha-\alpha_0)-y(\beta+\beta_0)-z(\gamma+\gamma_0)]} \\ &\qquad + \Psi'(\alpha,\beta)\sum e^{ik[x'(\alpha-\alpha_0)+y'(\beta-\beta_0)+z'(\gamma-\gamma_0)]} + \ldots\} \\ &= \frac{e^{-ikR}}{R}\{\Psi(\alpha,\beta) + \Psi'(\alpha,\beta)e^{iR} + \Psi''(\alpha,\beta)e^{iR'} + \ldots\} \\ &\qquad \sum^m \sum^n \sum^p e^{i(mA+nB+pC)}, \end{aligned} \right.$$

---

1) P. Groth, Physikalische Kristallographie, Leipzig 1905. p. 272 u. f.

wo A, B, C die in (5) angegebenen Werte haben, während analog

$$(17)\quad \begin{cases} R' = k\,(\mathfrak{r}_x'(\alpha-\alpha_0) + r_y'(\beta-\beta_0) + r_z'(\gamma-\gamma_0))\,, \\ R'' = k\,(\mathfrak{r}_x''(\alpha-\alpha_0) + r_y''(\beta-\beta_0) + r_z''(\gamma-\gamma_0)) \end{cases}$$

ist. Man sieht, daß der einzige Unterschied gegen früher in der Ersetzung der Funktion Ψ durch die Klammer besteht. Es werden eben eine ganze Schar benachbarter Atome als *ein* Gitterelement wirken. Obwohl so die Grundzüge der Theorie erhalten beiben, kann doch diese Abänderung gelegentlich für die beobachtete Interferenzerscheinung wichtig sein; denn der Klammerausdruck kann an Stellen, an denen die Summe

$$\sum^m \sum^n \sum^p e^{i(mA+nB+pC)}$$

ihrem absoluten Wert nach ein Maximum hat, Null sein, so daß durch das Zusammenwirken der Atome eines Gitterelementes Interferenzpunkte, die sonst auftreten sollten, ausgelöscht werden. Die neuerdings von Hrn. Friedrich sowie vom Verfasser und Hrn. Tank beobachteten Folgen[1]) hemiedrischer Strukturen sind auf diese Art zu erklären.

2. Die sehr verwunderliche und noch unerklärte Tatsache, daß die spektral hochgradig inhomogenen Impulse der einfallenden Röntgenstrahlung im Kristall Schwingungen von recht genau definierter Wellenlänge hervorrufen (denn nur durch solche läßt sich die Schärfe der beobachteten Interferenzpunkte m. E. einfach erklären), hat gesprächsweise gelegentlich die Deutung gefunden, daß die Auswahl dieser Sinusschwingungen aus allen in den Impulsen vorhandenen eine Funktion des Raumgitters sei. Demgegenüber wollen wir zeigen, daß die (bei Vorhandensein der erforderlichen Schwingungen) nach (12) möglichen Interferenzmaxima überall dicht liegen, so daß die photographische Platte vollständig geschwärzt werden müßte, wenn nicht die in den Funktionen $\Psi(\alpha, \beta)$ ausgesprochenen Eigenschaften der Gitter*elemente* eine bestimmte endliche Zahl aus den unendlich vielen Wellenlängen der ursprünglichen Strahlung aussonderten. Daß wir uns dabei auf den in den Gleichungen (8) betrachteten und in Fig. 5 dargestellten besonderen Fall beschränken, ist für das Ergebnis unwesentlich.

---

1) Vgl. die Anm. auf p. 984 und eine der folgenden Arbeiten.

Führen wir in der Ebene der photographischen Platte (z = const) Polarkoordinaten ϱ, φ ein

$$x = \varrho \cos \varphi \,, \quad y = \varrho \sin \varphi \,,$$

so wird nach (13b)

$$\operatorname{tg} \varphi = \frac{h_2}{h_1}, \quad \varrho = 2z \frac{\sqrt{u}}{|1-u|}, \quad \text{wo} \quad u = \frac{h_3^2}{h_1^2 + h_2^2}.$$

Es sind also in praxi alle Werte von φ möglich, weil man mit zwei ganzen Zahlen h_1 und h_2 stets das Verhältnis $h_2 : h_1$ jedem vorgeschriebenen Werte beliebig annähern kann. Ist dies Verhältnis festgesetzt, so kann man noch immer durch geeignete Wahl von h_1 und h_3 mit u bis auf einen beliebig kleinen Fehler jeden Wert erreichen, so daß dasselbe auch für den Radiusvektor ϱ gilt.[1])

3. Die in der Arbeit „Eine quantitative Prüfung" usw. gegebene Durchrechnung der vierzählig-symmetrischen Aufnahme 5 soll hier noch ergänzt werden. Es fehlten damals noch drei achtfache Punkte, die auf dem Photogramm selbst deutlich, in der Reproduktion aber wohl nur schwer sichtbar sind. Die beiden ersten sind auch auf dem Original nur an je vier Stellen vertreten, die anderen Stellen liegen außerhalb des zur Aufnahme dienenden Films; deshalb ist auch die Bestimmung ihrer Koordinaten und die Übereinstimmung zwischen deren (unter der Annahme z = 3,56 cm) nach (13b) berechneten und beobachteten Werten geringer als sonst. Immerhin sprechen die Werte der Wellenlängen, die wir nach (13) aus ihren Ordnungszahlen finden, für die Richtigkeit unserer Angaben, weil sie zu den früher berechneten in rationalen Verhältnissen stehen. Es sind dies die Punkte:

h_1	h_2	h_3	x berechn.	y berechn.	x gemess.	y gemess.	$\frac{\lambda}{a}$ nach (13)
7	3	3	3,05	1,30	2,96	1,28	0,0895
12	9	8	4,36	3,27	4,47	3,25	0,0563
3	1	1	2,37	0,79	2,36	0,79	0,182

Man sieht, daß hier die Zahl der Wellenlängen noch um zwei vermehrt wird; denn nur der Wert 0,0563 ist schon

1) Für freundlichen Rat hierbei bin in Hrn. Kollegen P. Bernays zu Dank verpflichtet.

früher vorgekommen. Nun war aber schon früher (p. 998) darauf hingewiesen, daß sich alle berechneten Wellenlängen als ganze Vielfache eines einzigen Wertes darstellen. Wir stellen in der folgenden Tabelle alle Werte von λ/a und die entsprechenden Vielfachen von $4{,}75 \cdot 10^{-3}$ zusammen, eines Wertes, der der Hälfte des früher angegebenen ($\lambda = 3{,}20 \cdot 10^{-10}$) entspricht.

0,0377	0,0563	0,0663	0,0895	0,1051	0,1428	0,182
8	12	14	19	22	30	38
0,0380	0,0570	0,0665	0,0903	0,1045	0,1426	0,1805

In der ersten Zeile steht hier der berechnete Wert von λ/a, in der zweiten die ganze Zahl, mit der $4{,}75 \cdot 10^{-3}$ zu multiplizieren ist, um den in der dritten Zeile stehenden Wert zu erhalten. Wir wollen schon hier mitteilen, daß auch die in anderem Zusammenhang zu veröffentlichende Durchrechnung des dreizählig-symmetrischen Photogramms 7 für eine erhebliche Zahl von Interferenzpunkten auf Wellenlängen führt, die ganze Vielfache desselben Wertes sind.

Auch die Berechnung der Gitterkonstanten a läßt sich jetzt verbessern; nach Hrn. Tutton[1]) sind in der Zinkblende 32 ineinander gestellte Raumgitter von der Art des betrachteten vorhanden. Die Hälfte besteht aus Zink-, die Hälfte aus Schwefelatomen. Infolgedessen kommen nicht je ein Atom dieser beiden Arten auf einen Elementarkubus, sondern je 16; Gleichung (10) ist infolgedessen abzuändern in

$$\frac{N}{16} a^3 = \frac{m}{\delta}, \tag{10a}$$

sodaß

$$a = 8{,}53 \cdot 10^{-8} \text{ cm} \tag{11a}$$

wird. Aus den angegebenen Werten von λ/a folgt dann

$$\lambda = 3{,}24 \cdot 10^{-9};\quad 4{,}86 \cdot 10^{-9};\quad 5{,}66 \cdot 10^{-9};\quad 7{,}70 \cdot 10^{-9};$$
$$8{,}90 \cdot 10^{-9};\quad 1{,}214 \cdot 10^{-8};\quad 1{,}540 \cdot 10^{-8}.$$

1) A. E. H. Tutton, Nature 90. p. 306. 1912.

(Eingegangen 15. März 1913.)

The Structure of Some Crystals as Indicated by their Diffraction of X-rays.

By W. L. Bragg, B.A.

(Communicated by Prof. W. H. Bragg, F.R.S. Received June 21,—Read June 26, 1913.)

[Plate 10.]

A new method of investigating the structure of a crystal has been afforded by the work of Laue* and his collaborators on the diffraction of X-rays by crystals. The phenomena which they were the first to investigate, and which have since been observed by many others, lend themselves readily to the explanation proposed by Laue, who supposed that electromagnetic waves of very short wave-lengths were diffracted by a set of small obstacles arranged on a regular point system in space. In analysing the interference pattern obtained with a zincblende crystal, Laue, in his original memoir, came to the conclusion that the primary radiation possessed a spectrum consisting of narrow bands, in fact, that it was composed of a series of six or seven approximately homogeneous wave trains.

In a recent paper† I tried to show that the need for assuming this complexity was avoided by the adoption of a point system for the cubic crystal of zincblende which differed from the system considered by Laue. I supposed the diffracting centres to be arranged in a simple cubic space lattice, the element of the pattern being a cube with a point at each corner, and one at the centre of each cube face. A simpler conception of the radiation then became possible. It might be looked on as continuous over a wide range of wave-lengths, or as a series of independent pulses, and there was no longer any need to assume the existence of lines or narrow bands in its spectrum.

* W. Friedrich, P. Knipping, and M. Laue, 'Münch. Ber.,' June, 1912.
† 'Camb. Phil. Soc. Proc.,' November, 1912.

It is the object of this paper to extend the analysis used in the case of the zincblende to some other crystals, particularly those of the simple alkaline halides.

In treating the diffraction of waves by a space point system such as a crystal, that case is the most simple in which the diffraction is caused by a series of points arranged in a space lattice, of one of the 14 Bravais types. Here every point is identical with every other point of the arrangement, and it is always possible to find an element of the pattern consisting of a parallelepiped with a point at each corner: there will then be as many parallelepipeds as atoms in any space. The points can be referred to three axes parallel to the edges of the parallelepiped, and if one of the points is taken as the origin the co-ordinates of the others may be written

$$x = \pm pa, \qquad y = \pm qb, \qquad z = \pm rc,$$

where p, q, r are any integers, and a, b, c are equal to the sides of the parallelepiped, and therefore proportional to the axial ratios of the space lattice. When the axes are chosen in this way the co-ordinates of the points and the equations of the planes passing through given selections of points are expressed in the simplest manner possible.

Let a series of pulses fall on this space lattice, the direction of propagation having a given relation to the axes of the system. As any one pulse passes over each point, a diffracted wavelet spreads from it, and it will be shown that the wavelets from all the points due to one incident pulse will combine in certain directions to form trains of waves, which give rise to the patterns of spots appearing in Laue's diffraction patterns. This may be done in the following way:—

If the axial ratios of the space lattice be denoted by a, b, c, any plane which makes intercepts pa, qb, rc, on the axis (p, q, r being any integers) is parallel to a whole set of planes on which the points of the system may be considered as arranged. It is such planes as these which form the faces of a crystal. When a pulse falls on a set of points on a plane the wavelets from the points combine to build up a wave front which will appear to be regularly reflected from the plane. As there are a series of planes regularly spaced one behind the other, a single pulse falling on them gives rise to a reflected wave train.

When therefore a narrow pencil of X-rays falls on a section of a crystal, part of its energy is transmitted undeviated, but there is also a part which is reflected on the crystal planes existing potentially in the body of the crystal. It is the series of narrow beams arising by reflection in this way which gives rise to the pattern of spots appearing in the photograph.

There can be assigned to each set of parallel planes integral indices (h, k, l), as when naming the faces of a crystal, and a spot can be classified as being reflected in the (h, k, l) set of planes. Here the integers h, k, l, are reciprocal to the intercepts which a parallel plane makes on the axis of reference. They are Millerian indices, the equation of the plane being

$$h\frac{x}{a}+k\frac{y}{b}+l\frac{z}{c} = \text{an integer}.$$

They may be considered as the parameters of each spot of the pattern, and are exactly equivalent to the parameters (h_1, h_2, h_3) employed by Laue in his original treatment of the subject. Laue defines his parameters by saying that the diffracted wavelet from a point at the origin of co-ordinates lags h_1, h_2, h_3, wave-lengths in the directions under consideration behind the wavelets proceeding from its neighbours along the x, y, z axes respectively. Thus the wavelets from all atoms in the plane

$$h_1\frac{x}{a}+h_2\frac{y}{b}+h_3\frac{z}{c} = 0$$

are in phase with that from the origin, and, in general, the wavelets from all points in the plane

$$h_1\frac{x}{a}+h_2\frac{y}{b}+h_3\frac{z}{c} = \text{an integer}$$

are in the same phase. We are led back to the same conclusion as before, that the direction of the diffracted wave front is one in which the primary beam is reflected by planes whose Millerian indices are (h_1, h_2, h_3). But, as will appear presently, it is important to bear in mind the fact that crystal structure alone fixes the exact position of the interference maxima, quite independently of the existence of homogeneous components of definite wave-length in the incident rays, and therefore a method of treatment has been adopted in which all reference to the wave-length has been avoided.

It is possible for a spot to appear in a position corresponding to reflection in any set of planes having integral indices (h, k, l). In an actual pattern obtained by allowing the diffracted X-rays to fall on a photographic plate, since there is not an infinite number of spots, only a selection of the planes can be operative. The spots forming the pattern are of very different intensities, and one can never say that spots are entirely absent but merely that in certain cases the rays are too weak to make an impression. It would seem, however, that by classification of the planes which reflect the principal spots of the pattern a clue can be got to the true point system arrangement of the diffracting centres. The point system which affords the most simple interpretation of the pattern is that which ought to be taken as representing the crystal structure.

When a photograph has been taken with a crystal an analysis is necessary in order to assign to each spot of the pattern obtained the correct parameters h, k, l, the Millerian indices of the face from which the spot is reflected. Having decided on the axes of reference of the crystal, the following method will be found to apply whatever their inclinations to each other and axial ratios may be.

The more important planes of the system are those densely packed with points, and from this fact it follows that these planes contain rows along which the points are closely packed. These are the important "point-rows" of the crystal, and each of these point-rows will have a set of important planes parallel to it. The three axes themselves are the most obvious examples of these point-rows. The "zone axis" of planes belonging to a common zone is also such a direction. There is a convenient relation between the points which are reflected in the planes belonging to any one zone. The reflected beam always lies on a circular cone with apex at the crystal section, the zone axis as axis, and the direction of the primary beam as one generator. This cone cuts the photographic plate in an ellipse passing through the central point of the pattern, and all spots reflected in planes of the zone lie on it. The arrangement of the spots on ellipses is very obvious in an interference pattern (see fig. 11), and the ellipses can immediately be drawn. A little calculation shows to which zone axis each ellipse must be assigned, and, by marking a given spot as lying on the intersection of two ellipses, the calculation of the indices h, k, l, of that spot is made possible. For each zone of planes we have the relation $hU+kV+lW = 0$ where (U, V, W) is the zone symbol; that is to say, the set of direction ratios of the zone axis.

When representing diagrammatically an interference pattern it is very inconvenient to draw the ellipses at the intersections of which the spots of the pattern lie. It is simpler to employ an extension of the usual stereographic projection of crystallography. Reference to fig. 1 will make this clear. Let the section of crystal be situated at C, the centre of the sphere represented in the figure by the circle ABP, and let the direction of the incident rays be from P to C. The rays which traverse the crystal undeviated fall on the photographic plate AD at A. Let CZ represent the direction of a zone axis. The beams reflected in planes of this zone lie on a circular cone with vertex at C, of which CZ is the axis, and CA, CB are two generators.

This cone cuts the sphere in a circle of which AB is a diameter, and by the well known property of the stereographic projection, the projection of this circle on the plane AD from the pole P is also a circle. The centre of this circle is at Z, since AZ = ZS′, Z being the point where the zone axis cuts the photographic plate.

Let the pattern of spots be supposed to be made by the diffracted beams on the sphere ABP, and this pattern projected on the plane AD from the pole P. Spots corresponding to reflection in planes of a zone now lie on a circle, having its centre at the point where the zone axis cuts the plate AD. The spot at S made by the reflected beam CB becomes a spot at S′ of the

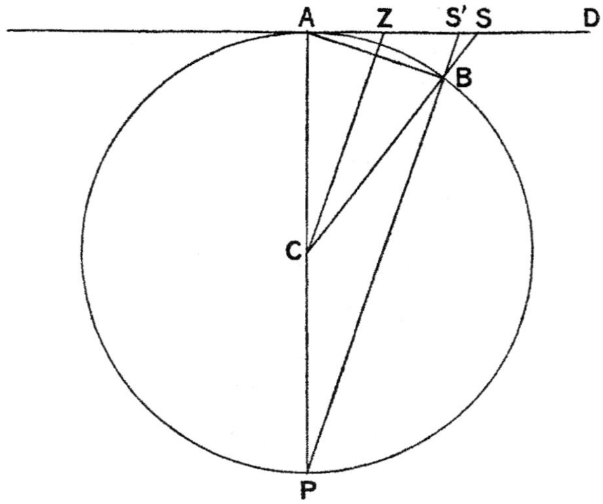

Fig. 1.

transformed pattern. The distortion of the pattern of spots by the transformation is very small except in the regions distant from the centre, and circles are much easier to draw than ellipses. This has been done in all the diagrams given in this paper. In constructing a diagram, a point is chosen as that in which the incident rays meet the photographic plate. This is A in the figure. Then the points such as Z, where the principal zone axes meet the plate, are found by a calculation, which is easy when the crystal is placed symmetrically to the pencil of X-rays. A circle is drawn with centre Z and radius ZA. This being done for each zone axis, the intersections of the circles give the stereographic projections of the reflected spots.

Let us take as an example of the application of this analysis the very simple case of potassium chloride. The diffraction pattern obtained when the X-rays fall normally on a plate cut parallel to the cube face (001) is reproduced in fig. 2, Plate 10, and its stereographic projection in fig. 3, which shows also the indices of the reflecting planes corresponding to the several points of the photograph.

Potassium chloride is cubic, and the indices given to the planes are those obtained when the edges of the cube are selected as axes of reference.

The circles correspond to zone axes for which

$$\left.\begin{array}{ll} U = 0, & V = \pm 1 \\ U = \pm 1 & V = 0 \end{array}\right\} W = 0, 1, 2, 3, 4, 5,$$

and their intersections give the spots reflected in all planes of the form $(h\ k\ 1)$ where h and k have any of the values

$$0,\ \pm 1,\ \pm 2,\ \pm 3,\ \pm 4,\ \pm 5.$$

In the diagram of the KCl pattern the spots are represented by dots, the magnitude of each dot indicating the strength of the corresponding spot in the

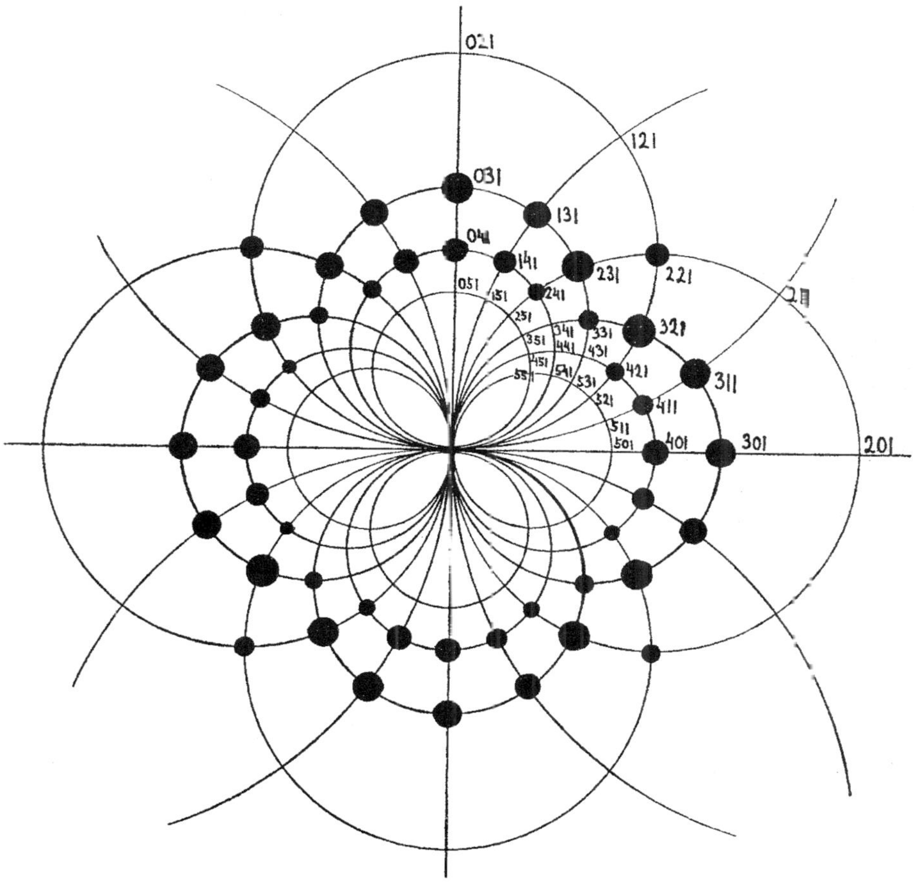

Fig. 3.—Potassium chloride.

photograph. It will be seen how complete the pattern is, and how within a certain range each intersection is represented by a spot in the photograph.

This may be put in another way. Below is a table of values of h and k, l being equal to 1. In the square corresponding to given values of h and k

is placed a dot denoting the magnitude of the corresponding spot, the square being left empty if no spot is to be seen.

Table I.

KCl.

h → / h ↓	0	1	2	3	4	5
0				✳	✶	
1				✳	✶	
2			✳	✳	+	
3	✳	✳	✳	+		
4	✶	✶	+			
5						

This table contains a list of all the spots appearing in the photograph. It suggests that the diffraction is due to a simple cubic space lattice, for when the elementary parallelepiped is a cube and its edges are taken as axes the indices of the planes naturally take these simple forms.

For, let a series of planes be taken, for which two indices are constant and one varies, such as the series 031, 131, 231, 331, 431, etc. These planes will be arranged in this order for all their properties, such as reticular density, distance apart, and so forth, if, and only if, the indices are referred to axes parallel to the sides of the elementary parallelepiped. If this condition is satisfied, then all the properties of the planes vary in an orderly manner as one passes along the series. The power of reflecting the X-rays is a particular property of the planes, which is seen from the photograph to vary continuously for any such series as the one given above, therefore the diffracting system is a space lattice with axes at right angles, and a cube for its elementary parallelepiped.

It may be objected here, that it is conceivable that a complex structure, and a radiation consisting of homogeneous components, just happen in the case of potassium chloride to give a deceptively simple pattern. That this is not the case can be seen by displacing the crystal from its symmetrical position, and obtaining the distorted interference pattern. It is still as straightforward as before, though of course no longer symmetrical. Corresponding spots in the two patterns are now made by different wave-lengths, and it is obvious that there can be nothing of the nature of homogeneous components of the incident radiation. The contrast of this pattern with those characteristic of potassium bromide and iodide, of rock salt, and of zincblende, which will shortly be given, will tend to make this more clear.

All the most intense spots of the photograph are at about the same distance from the centre of the pattern. A circle can be drawn such that the spots which lie near to it are very intense, while those further away are weaker and those at great distances are too faint to appear. All ellipse intersections lying near this circle are represented by a spot. Those planes reflect most strongly for which the glancing angle lies between 12° and 20° and it is in seeking to explain this fact that it is necessary to consider the question of the "wave-length" of the diffracted rays. This we must now do.

When a pulse falls on a series of planes regularly spaced one behind the other, it is reflected as a wave train, and the waves making any one spot must be considered to be of the lengths given by

$$n\lambda = 2d\sin\theta,$$

where θ = glancing angle, d = distance between successive planes, and n is an integer 1, 2, 3, etc. An alternative way of regarding the phenomenon is to consider the incident radiations as compounded of homogeneous wave trains of all wave-lengths, with a characteristic distribution of energy in the spectrum, when the parallel planes must be considered as reflecting only that part of the spectrum for which the relation $n\lambda = 2d\sin\theta$ holds good. In the case of the cubic space lattice considered here, this relation reduces to

$$n\lambda = \frac{2a}{l}\sin^2\theta,$$

where a is the distance between neighbouring points along the cubic axes, for a simple calculation shows that $d = a\sin\theta/l$. Since l is equal to unity for all the spots of the potassium chloride photograph, spots which have the same θ, and lie at the same distance from the centre of the pattern, correspond to the same wave-length. The ring of intense spots in the pattern indicates strong reflection of a certain part of the spectrum of the incident radiation. Either conditions are especially good* for reflecting this wave-length, or the incident rays have a large amount of their energy in this part of the spectrum.

It is interesting to compare the simple pattern of potassium chloride with those of potassium iodide, potassium bromide, fluorspar, and zincblende.

The stereographic projection of KBr (100) is given in fig. 4, that of KI being very similar. Both of these are like the patterns of zincblende and fluorspar, and are in marked contrast to that of KCl.

* The most important condition in this connection is probably the thickness of the crystal section used. As will be shown later, the use of a thin crystal section favours the reflection of the longer wave-lengths, the corresponding spots coming out strongly, while the use of a thick section favours the reflection of the shorter.

It is evident that some factor has now entered which destroys the simplicity of the arrangement of spots characteristic of potassium chloride. Spots no longer appear at every intersection of the ellipses within a certain region, as

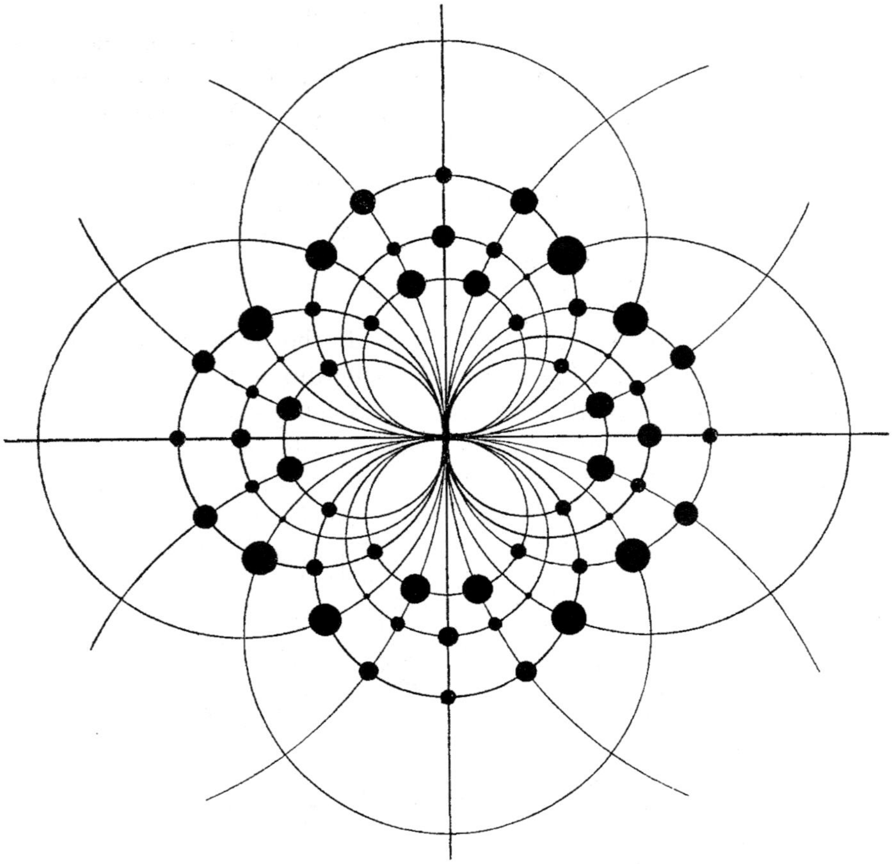

Fig. 4.—Potassium bromide.

in fig. 3. A tabulation of the intensities of the spots in the zincblende and potassium bromide patterns will make this clear. (See Tables II and III.)

There are also in the case of zincblende several spots for which a value 3 must be assigned to l, if h and k are to be integers, but not many. It was a consideration of Table II which led in a former paper* to the conclusion that the diffracting centres of the zincblende crystal are arranged on the face-centred cubic space lattice.

An examination of the planes in which reflection takes place shows that there is a differentiation between those whose indices are wholly odd and

* 'Camb. Phil. Soc. Proc.,' November, 1912.

Table II.

Zn.S.

k ↓ / h →	0	1	2	3	4	5	6	7
0				+				
1			•	+	✱	✳		+
2		•	•	✳	+			
3	+	+	✳	+	+	✳		+
4		✱	+	+				
5		✳		✳		+		
6								
7		+		+				

Table III.

K.Br.

k ↓ / h →	0	1	2	3	4	5
0				+	✱	
1				✱	+	✳
2				✳		
3	+	✱	✳	+		✱
4	✱	+				
5		✳		✱		
6						
7						

those which have one or more even indices. If those planes alone are considered which have odd indices, as in the following table, the scheme is as complete as it was for the spots of the KCl crystal.

Table IV.

Zn.S. (odd indices only)

k ↓ / h →	1	3	5	7	9
1		+	✳	+	
3	+	✱	✳	•	
5	✳	✳	+		
7	+	•			
9					

Odd planes, ZnS.

Table V.

Zn.S. (even indices) only

k ↓ / h →	0	1	2	3	4	5
0				+		
1			•		✱	
2		•	+	✳	•	
3	+		✳		•	
4		✱	•	•		
5						

Even planes, ZnS.

The same is true for those with an even index* except that the intense spots are all further removed from the centre of the picture.

This difference can be explained without our being forced to assume that the diffracting system is anything more than a simple system of identical points. It is sufficient to suppose that the point system has points at the centres of the cube faces as well as at the cube corners. Let a cubic point system of the first kind be taken which has points at cube corners alone, and let points be introduced at the cube face centres in order to turn it into a

* The spot 041 forms an exception, indeed evidence is not lacking that the assumption of diffraction by the face-centred space lattice does not completely account for the pattern. The reason for this will appear when the parts played by the atoms of two kinds in the diffraction are discussed.

point system of the so-called "third kind." (The "second kind" is the centred cube.) The spacing of the planes which have odd indices (h, k, l) is not altered by the introduction of the new points, for they all lie on the original planes and only increase their point density. On the other hand, in the case of planes having an even index, some of the new points lie halfway between the original planes, the distance between the successive planes of this type must now be halved, and so must therefore the wave-lengths of the reflected beams.

The interpretation of the zincblende pattern is now simple. We have seen that the planes with odd indices only are a complete set. There are fewer spots corresponding to reflection in planes with an even index, for these planes are, relatively to the former, less closely packed, and of a more complex nature. Moreover, they are, at the same time, less widely separated, and therefore the intense spots with even indices are further from the centre; the angle of incidence must be increased in order to make these planes reflect that region of the spectrum which gives intense spots. By assuming the third cubic space lattice instead of the first, all the intense spots of the pattern again correspond to the same wave-length region.

The difference between the diffraction by the two cubic space lattices may be put in a much clearer way on analysis of the patterns of threefold symmetry, obtained when the incident rays fall normally on a plate cut perpendicular to a trigonal axis. If the points of the space lattices are considered from this aspect, they are special cases of the trigonal rhombohedral space lattice, one of the Bravais types.

The axes to which the spots ought to be referred are not the same for the two cubic lattices. When the points are at cube corners alone, the axes along which they are nearest neighbours are the cube edges, and the cube itself is the elementary parallelepiped. When there are also points at the centres of the cube faces, three semi-diagonals of cube faces meeting in a corner form three edges of the elementary parallelepiped. The angles between the axes are right angles in the first case, in the second they are 60°.

It will now be clear that when the stereographic diagram is constructed, giving the positions of the spots reflected from the simple planes of a trigonal rhombohedral space lattice, one such diagram will represent the patterns of all rhombohedral lattices, the alteration of the angle between the axes, *i.e.*, the rhombohedral angle, only causing an alteration of the scale of the diagram. The radius of the sphere used in the projection is, of course, supposed to be always the same.

Given the points where the three axes of the lattice meet the diagram, the corresponding points for the other zone axes can immediately be found and the whole diagram drawn. Such a diagram is shown in fig. 5, X, Y, Z, being these axial points, Now, our study of the patterns of fourfold symmetry given by potassium chloride and zincblende has shown that the axes of the

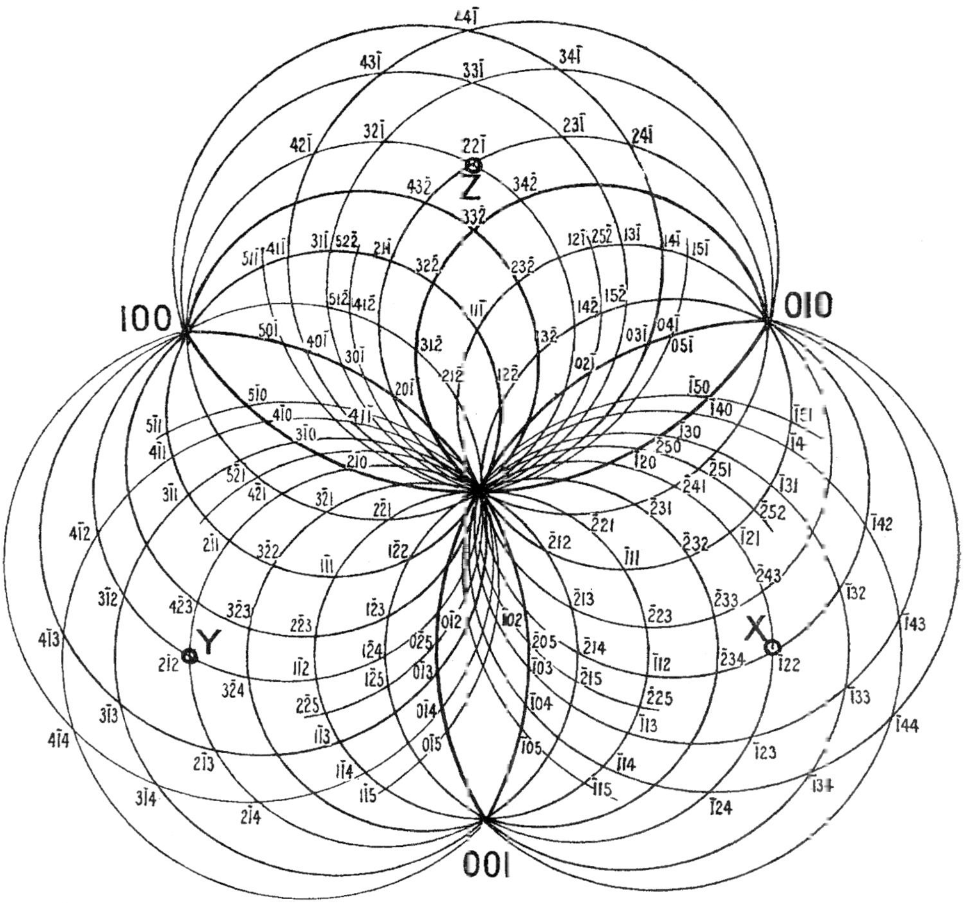

Fig. 5.

lattice, which is from this aspect a special case of the rhombohedral lattice, make angles of 90° with each other in the first case, and of 60° in the second. Therefore it ought to be possible to refer the threefold patterns of potassium chloride and zincblende to the same trigonal lattice diagram, the scale being different, however, in the two cases. This conclusion is exactly confirmed by experiment. The pattern given by zincblende was published by Laue in his original memoir. The stereographic projections of the zincblende and potassium chloride patterns are given in figs. 6 and 7. The object of this comparison is to show that the spots of the KCl pattern fall naturally on a diagram twice the size of that on which the spots of the ZnS pattern are arranged, the points where the axes making 90° with each other cut the plate, being just twice as far apart as for axes making 60° with each other.

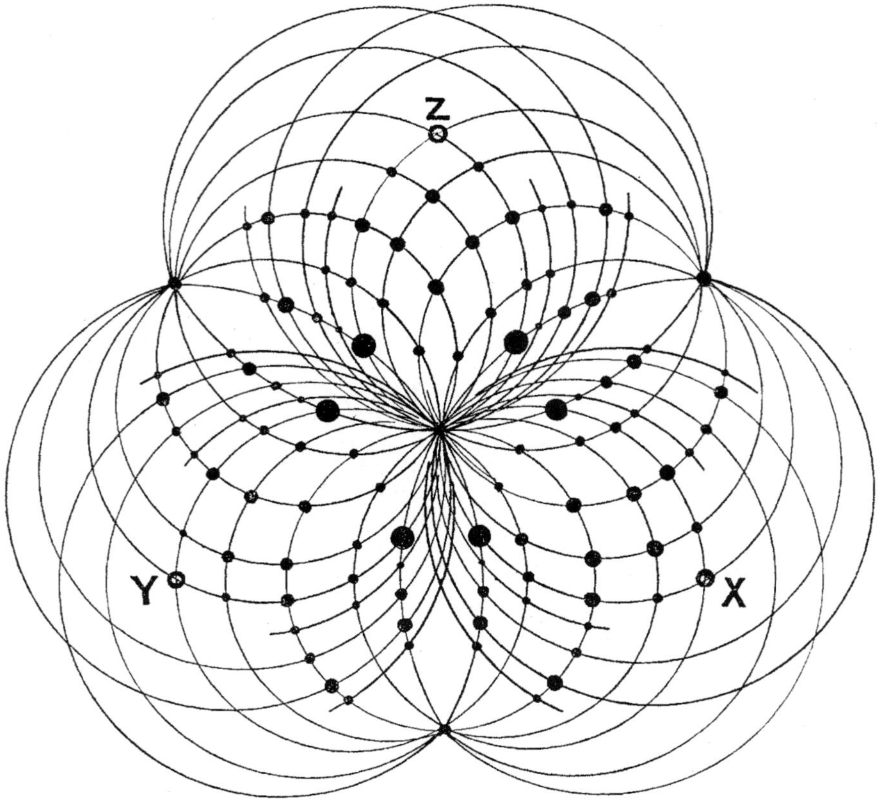

FIG. 6.—Zincblende.

The following table will illustrate the awkwardness of referring either pattern to the wrong diagram.

If the spots of the zincblende pattern are referred to the cubic axes, their indices become very much more complicated, and the pattern is no longer complete. The converse is true for the potassium chloride pattern, the indices in this case being simpler when referred to cubic axes. In Table VI a list of indices of the spots of each pattern is set down referred to (A) the cube edges, (B) the cube face diagonals, *i.e.*, three edges of the regular tetrahedron, as axes.

It will be seen that in the case of zincblende it is the B series, in the case of the potassium chloride the A series, which is simple, and which gives a complete series of indices over a certain range. I do not think there can be any doubt which space lattice is the right one in either case.

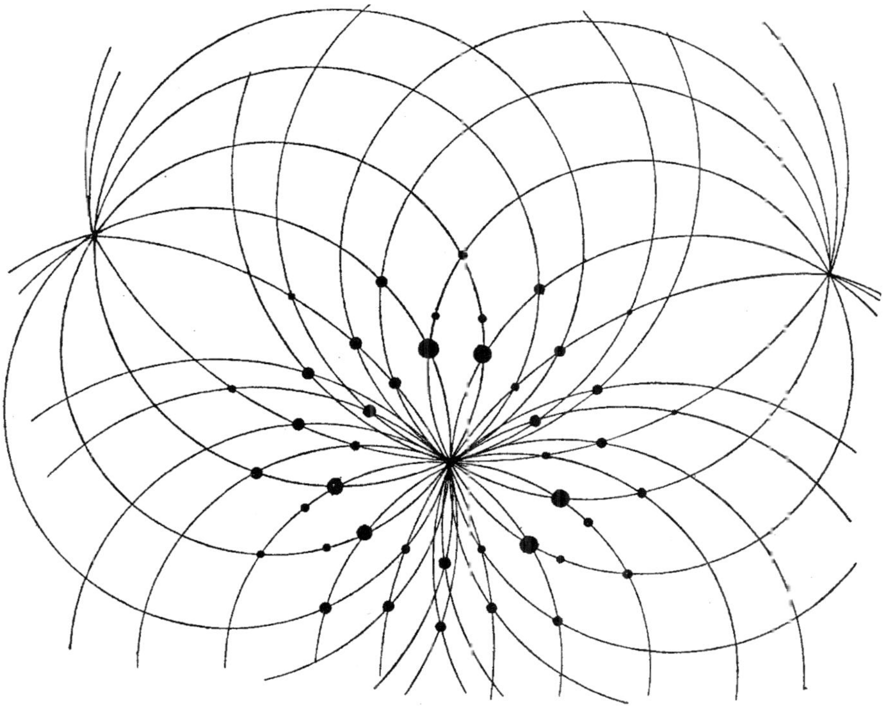

FIG. 7.—Potassium chloride.

Table VI.

Zincblende.				Potassium chloride.*	
A.	B.	A.	B.	A.	B.
$\bar{1}11$	010	$\bar{1}11$	100	$3\bar{1}\bar{1}$	$11\bar{1}$
$\bar{3}32$	$0\bar{5}1$	$\bar{3}75$	$51\bar{1}$	$20\bar{1}$	$21\bar{1}$
$\bar{5}53$	$0\bar{4}1$	$\bar{1}32$	$41\bar{1}$	$31\bar{2}$	$41\bar{1}$
$\bar{2}21$	$0\bar{3}1$	$\bar{1}53$	$31\bar{1}$	$11\bar{1}$	100
$\bar{3}31$	$0\bar{2}1$	$\bar{1}95$	$52\bar{2}$	$30\bar{2}$	$31\bar{2}$
$\bar{1}10$	$0\bar{1}1$	$0\bar{2}1$	$21\bar{1}$	$21\bar{2}$	$30\bar{1}$
		$1\bar{7}3$	$32\bar{2}$	$4\bar{1}\bar{1}$	$33\bar{2}$
		$1\bar{3}1$	$11\bar{1}$	$32\bar{3}$	$50\bar{1}$
		$3\bar{5}1$	$12\bar{2}$		

* The indices of the A series for the KCl pattern may not, at first sight, appear simpler than those for the B series. The B series, however, is curiously incomplete. For instance, $21\bar{1}$ and $41\bar{1}$ occur but no $31\bar{1}$; $50\bar{1}$ and $30\bar{1}$ but no $40\bar{1}$; $33\bar{2}$ and $31\bar{2}$ but no $32\bar{2}$. In fact, there is a tendency for the sum of the indices to be even, and such a selection of the simple planes as this always implies that the wrong axes have been taken.

In the B list for ZnS, indices such as $(52\bar{2})$ are intermediate between $31\bar{1}$ and $21\bar{1}$.

In the case of rock-salt, it is not possible to regard the pattern as completely characteristic of the one point system or the other. Photographs obtained using a thin section of crystal, about 1 mm. thick, cut parallel to a cube face, show a pattern very similar to that given by potassium chloride, though in no case are they so simple as the pattern given by the latter crystal; for a considerable amount of difference between the "odd"

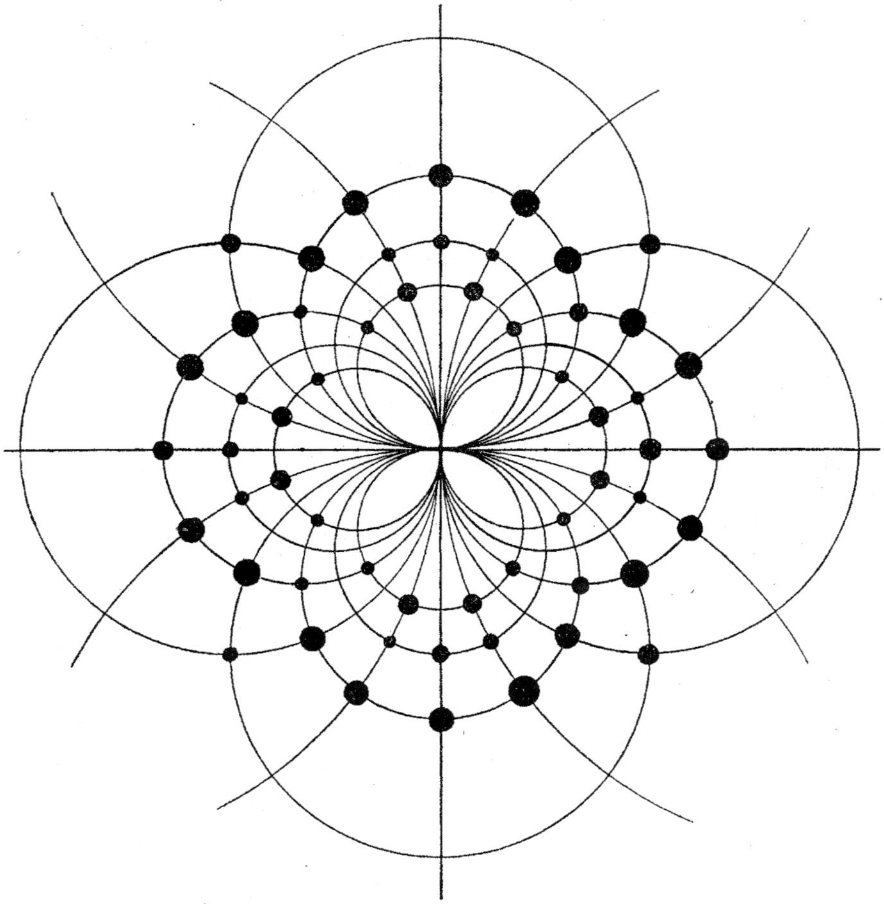

FIG. 9.—Rock-salt, 2·5 mm. thick.

and "even" planes is also evident. A photograph taken with a section 6 mm. in thickness is more like one taken with potassium bromide or iodide, the difference between odd and even planes being more marked. In fig. 8, Plate 10, is reproduced a photograph taken with a section of rock-salt, 2·5 mm. thick, cut parallel to a face (100), and in fig. 9 its stereographic diagram. In the tables on the next page are set down the intensities of the spots when different thicknesses of crystal are used.

Table VII.

$h \longrightarrow$ 1 mm.

	0	1	2	3	4	5
0				✳	+	
1			•	✳	﹢	+
2		•	✳	✶		
3	✳	✳	✶	✶		+
4	+	﹢				
5		+		+		

NaCl, 1 mm.

Table VIII.

$h \longrightarrow$ 2·5 mm.

$k \downarrow$

	0	1	2	3	4	5
0				+	+	
1				✳	﹢	✶
2			•	✳		
3	+	✳	✳	﹢		﹢
4	+	﹢				
5		✶		﹢		

NaCl, 2·5 mm.

Table IX.

$h \longrightarrow$ 6 mm.

$k \downarrow$

	0	1	2	3	4	5
0					+	
1				+	﹢	✶
2				✳		
3		+	✳	﹢		+
4	+	﹢				
5		✶		+		

NaCl, 6 mm.

I think these tables, especially the first one, suggest that the rock-salt diffracting system is in some way intermediate between those of potassium chloride and potassium bromide. The change of intensity of the spots with the thickness of the crystal is interesting. It may be laid down as a general rule that increasing the crystal thickness increases the intensity of the inner spots as compared with that of the other. This is to be expected, for the inner spots represent the shorter and more penetrating radiations.

If in passing through a thickness dx of the crystal a proportion αdx of the radiation of a certain wave-length is reflected, and if the absorption coefficient of the radiation in the crystal is k, then the radiation reflected in a slip of crystal of thickness t will be proportional to αte^{-kt} to a rough approximation. This is a maximum when $t = 1/k$, and since α is very small, this means that to get a spot strongly marked a section of crystal should be used of such thickness that it absorbs a fraction $1/e$ of the incident radiation of that wave-length. This would explain qualitatively the variation with thickness of crystal of the intensity distribution over the spots which is actually observed.

On comparing the evidence as to the nature of the diffracting systems in these crystals of sodium chloride, and of potassium chloride, bromide, and iodide, it would seem that a very simple explanation of their curious difference may be arrived at when it is considered that in each case diffraction is caused by two different atoms, and that the relative efficiencies of the two vary from crystal to crystal. Any explanation of these differences would be an extremely improbable one which did not assume a similar structure for the whole group of alkaline halides, for these crystals resemble each other very closely in their properties. Yet it has been seen that the space lattice of diffracting points is the simple cubic one in KCl; it is the face-centred cubic lattice in KBr and KI, and that in the case of NaCl the diffracting point system is in some way intermediate between the two space lattices.

Let us consider on what atomic properties the relative efficiencies might depend. It has been firmly established that the absorption of X-rays depends on the relative proportions of the various elements contained in the absorber. It is a purely additive property of the weight of each element per cubic centimetre of the absorber, and does not depend on the manner in which the elements are combined. Also the absorption of homogeneous X-rays increases steadily with the atomic weight of the absorber, except for a sudden discontinuity consisting of a large drop in the absorption coefficient when the atomic weight of the absorber passes through that of the element of which the homogeneous X-rays are the characteristic radiation. There are, however, no discontinuities in the absorption coefficient corresponding to the changes in chemical properties of the elements in the periodic table as one passes to higher atomic weights. It is reasonable, therefore, to assume provisionally that the weight of the atom in the main defines its effectiveness as a diffracting centre, and that two atoms of equal weight are equally effective. In the case of potassium chloride the atoms of potassium and chlorine, of atomic weight 39 and 35·5 respectively, are sufficiently close in atomic weight to act as identical diffracting centres. For rock salt this is no longer true; the atomic weight of sodium and chlorine differ considerably (35·5 to 23), and complications are introduced into the simple pattern characteristic of potassium chloride. In potassium bromide and iodide one atom preponderates so greatly over the other in atomic weight that the diffracting system consists practically of atoms of one kind only, and the pattern can again be assigned to a simple space lattice, but one which is of a different nature to that of potassium chloride. Yet the atoms of alkaline metal and halogen have precisely the same arrangement in all these cases.

Let us distinguish between two kinds of diffracting points by calling them black and white. Then the points must be arranged in such a way that—

1. There are equal numbers of black and white.

2. The arrangement of points black and white taken all together is that of the first cubic space lattice.

3. The arrangement of blacks alone or of whites alone is that of the third cubic space lattice.

An arrangement which gives this result is shown in fig. 10.

In this diagram we may associate black centres with the alkaline metal, and white with the halogen, or *vice versâ*. The space lattice formed by the whites is the same as that formed by the blacks, being in each case the face-centred cube. If black and white centres become identical, as in potassium chloride, the diffracting lattice becomes the simple cubic one.

The evidence for this arrangement in the alkaline halides seems very strong, but this does not by any means complete the solution of their

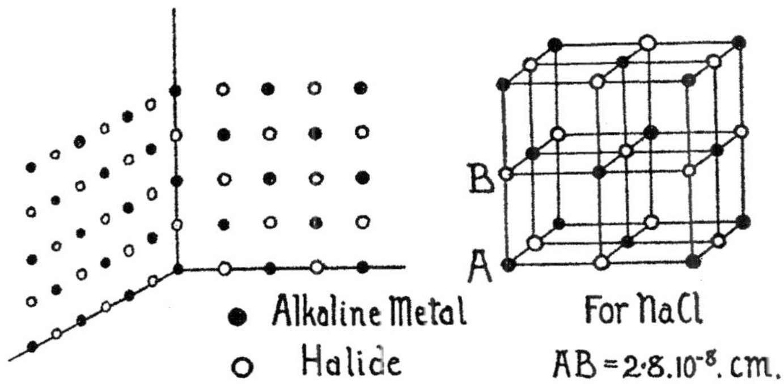

FIG. 10.

structure. It yet remains an open question whether one atom alone is to be associated with each point of the system, so that, for example, the black and white centres actually represent sodium and chlorine atoms in rock salt, or whether the crystal structure is of a more complex nature. It was with the object of discovering the complexity of the diffracting unit of these and other crystals that the series of crystal pattern analyses, which will now be described, were made; their results would seem to indicate clearly the association with each diffracting centre of a single atom.

It was shown above that the stereographic diagram, giving the position of spots reflected in the planes of a trigonal rhombohedral lattice, is the same except for scale for lattices of any rhombohedral angle. The completeness with which the spots of the zincblende pattern whose projection is given in fig. 6 represent the intersections of the circles of the diagram for the trigonal lattice would suggest that in all cases in which diffracting centres were arranged on a trigonal lattice, the pattern obtained would be very much the same. At all events, this would be so when the rhombohedral angle is not very different from 60°. That this is actually the case is well shown by the stereographic projections in figs. 12 and 13. They are the projections of photographs taken with sections of fluorspar and calcite, cut perpendicular to a trigonal axis, the fluorspar photograph being given in fig. 11, Plate 10. It is clear that in this case also, as in that of zincblende, every simple plane of the lattice reflects a spot when the angle of incidence lies within a certain range. Corresponding spots in the three diagrams vary very much in intensity, but this is to be expected. The point which it is desired to make clear is the certainty with which the correct axes of the lattice may be found from a study of the pattern. In the case of fluorspar, as in that of zincblende, the rhombohedral angle of these axes is 60°. In the case of calcite it is slightly greater, a calculation of this angle and a comparison of the orientations of crystal and pattern making it clear that the axes of the lattice are three diagonals of rhombohedron faces meeting in an obtuse corner of the calcite rhombohedron, not the edges of the rhombohedron,

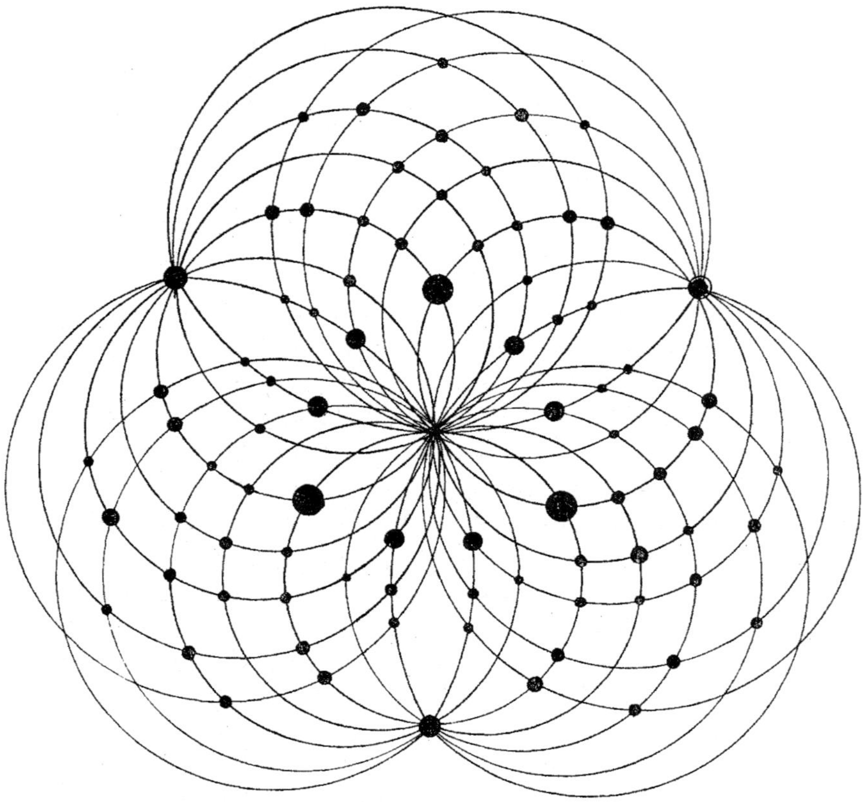

Fig. 12.—Fluorspar.

which are generally taken as the axes to which the faces of a calcite crystal are referred.

In these three cases, zincblende (ZnS), fluorspar (CaF_2) and calcite ($CaCO_3$), the diffracting centres are thus arranged on one space lattice. But, since a space lattice is an arrangement in which each point is related to its neighbours in exactly the same way as every other point, it would be impossible to arrange complex molecules in a space lattice unless only one point in each molecule is effective. It is difficult to avoid the conclusion that the molecule acts as a single point because it contains in each case one atom of much greater atomic weight than the others. A comparison with a case in which this is not so is afforded by the rock-salt pattern of threefold symmetry given in fig. 14 and Plate 10, fig. 15; it will be seen how awkwardly it fits the diagram. There is a certain want of symmetry in the figures, which must not be confused with the want of fit. Its pattern is, in fact, intermediate between that characteristic of axes of rhombohedral angle

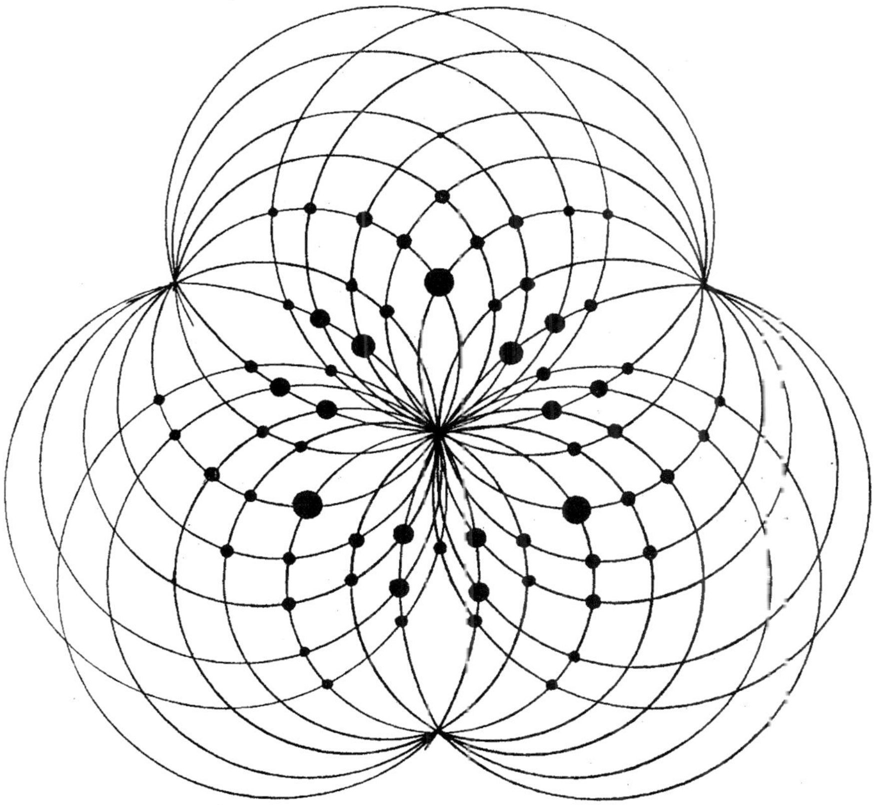

FIG. 13.—Calcite.

at 60° and that of axes at 90°, just as was the case with its pattern of fourfold symmetry.

Having found the nature of the simple diffracting lattice in these three cases and that of the alkaline halides, the final step in the argument which would assign a single heavy atom to each diffracting centre is made below by a comparison of the scale of their lattices, that is, a comparison of the length of the sides of the elementary parallelepiped for different crystals. Since the diffraction is caused by the heavy atoms, and there is only one in each molecule, this conclusion, if correct, will mean an association of one molecule with each parallelepiped of the lattice.

In a paper to the Royal Society, read in April, 1913,* a method was described of analysing the radiations from an X-ray bulb by reflecting them from the face of a crystal, and measuring the ionisation produced by the reflected beam.

The apparatus devised for this purpose resembled a spectrometer; the

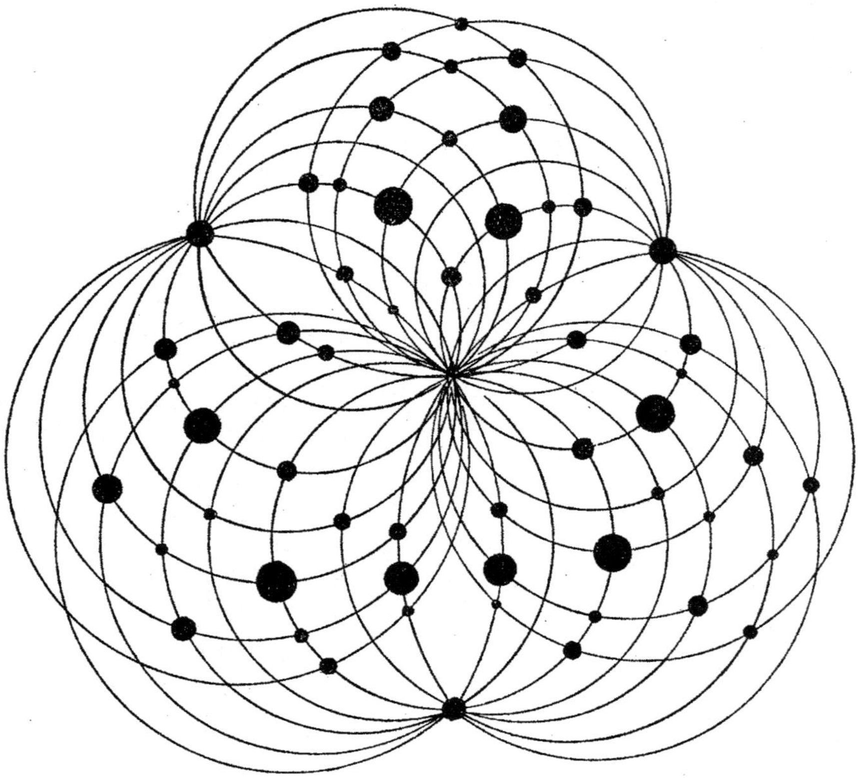

Fig. 14.—Rock-salt.

crystal was set on a revolving table at the centre of the instrument, and irradiated by a narrow beam of X-rays passing through a collimator slit. The reflected beam was received, and its ionisation measured, in a chamber mounted like the spectrometer telescope. By means of this instrument it was possible to measure the strength of the reflected beam when an approximately constant beam of rays fell on the face of the crystal at varying angles of incidence. The results of the measurement made with a crystal face may be summed up in a curve, in which the strength of the

* W. H. Bragg and W. L. Bragg, 'Roy. Soc. Proc.,' July 1, 1913.

reflected beam is plotted against the glancing angle at which the rays fell on the crystal. With most crystals, such a curve shows that at all angles there is a general reflection of the rays, the reflection being much stronger at the more glancing angles. Superimposed on this, as it were, there is at certain very sharply defined angles a special reflection of very great intensity as compared with the general reflection. This special reflection shows itself as very marked peaks on the curve, the amount of ionisation in the chamber becoming perhaps 20 times as great as the chamber is swept through the region of a peak, and then falling away to its normal value again, the whole taking place within settings a degree apart.

The results which we obtained with various crystals pointed to the existence of at least three components of the X-ray beam of definite wave-length, reflected from a crystal face when the condition

$$n\lambda = 2d \sin \theta$$

is satisfied for the component and crystal in question. Here λ is the wave-length, θ the glancing angle, and d the distance between successive planes of the crystal structure parallel to the reflecting face. The number n represents the "order" of the reflection, for several peaks, which can be recognised as belonging to the same homogeneous radiation by the identity of their absorption coefficients, appear at a series of angles whose sines are in the ratio 1 : 2 : 3.

It will possibly be objected here, that in the previous discussion of the interference patterns, the spectrum of the incident radiation has been taken to be continuous, the simplicity of the potassium chloride photograph supporting the view; while here direct evidence of the existence of homogeneous components of the radiation has been obtained. This difficulty disappears when the wave-lengths to be assigned to the spots of the interference pattern, and to the peaks of the curve, are compared. For instance, in the potassium chloride pattern of fourfold symmetry (fig. 2) the spot formed by the longest waves is reflected in the planes (221), and if a is the side of the elementary cube of the lattice of this crystal, the wave-length to be associated with the spot, from the formula

$$\lambda = 2d \sin \theta = 2a \sin^2 \theta$$

is equal to $a \times 0{\cdot}22$. On the other hand, the peak corresponding to the homogeneous radiation of shortest wave-length is reflected from the face (100) of the same crystal at an angle of about 8·4°, and corresponds to a wave-length $a \times 0{\cdot}29$ nearly. The spots of the pattern thus correspond to a region of the spectrum well inside the peaks on the reflection curve, and, in general, this is true for all the patterns. The reflection curve, indeed, bears out well the assumptions as to the continuous nature of the spectrum of the radiation giving the spots, for at all angles less than 10° a strong reflection takes place which only falls away slowly as the glancing angle is increased. The very fact that in most photographs the incident rays are parallel to an axis of symmetry ensures that no important planes occur making such an angle with the primary rays as to reflect the homogeneous components.

It is possible to recognise the same three peaks, which will be referred to as A, B, C, in the curves drawn for reflection in the faces of almost all crystals which have been tried. They are always of approximately the same relative height, they have each a characteristic absorption coefficient, and they are spaced in the same way on the curves. There appears to be no doubt that three lines* exist in the spectrum of the incident radiation, which give rise to the peaks. This being so, one has a means of finding with some accuracy the ratio of the values of d, the distance between successive planes of the structure, for different crystals and different faces of the same crystal.

The angles at which these peaks are reflected from the various faces of a crystal, and from faces of different crystals, thus afford a great insight into the crystal structure; and, in fact, they supply just that information concerning the structure which the interference patterns do not, for by their means the dimensions of the lattices of different crystals can be accurately compared. The interference patterns only supply information concerning the nature of the lattices.

An analysis of the results obtained when different faces of the same crystal are used to reflect the X-rays in the spectrometer will show how this comparison is carried out.

In the curves for reflection from the three primary planes of the rock-salt crystal (100), (110), (111), (two of which are shown in fig. 16), the peaks occur for each face, but at different angles. From the equation

$$\lambda = 2d \sin \theta,$$

θ being the angle for the peak of the first order in each case, we have, for these three faces

$$d_{(100)} : d_{(110)} : d_{(111)} :: \frac{1}{\sin \theta_{(100)}} : \frac{1}{\sin \theta_{(110)}} : \frac{1}{\sin \theta_{(111)}},$$

where $d_{(100)}$ is the distance between planes parallel to the face (100), $\theta_{(100)}$ the angle at which the most prominent peak is reflected from the same face.

* By using narrow slits in the reflection apparatus, evidence of the existence of more than three lines has been obtained, the B and C peaks being really double. See fig. 16.

The angles for these three faces are 11·4°, 16·0°, 9·8°, respectively. This gives the result

$$d_{(100)} : d_{(110)} : d_{(111)} :: 1 : 0{\cdot}718 : 1{\cdot}16,$$

whereas

$$1 : \frac{1}{\sqrt{2}} : \frac{2}{\sqrt{3}} :: 1 : 0{\cdot}707 : 1{\cdot}15,$$

these being the theoretical relations for the face-centred cubic space lattice. This does not comprise all the information which a study of these three curves yields, they will be analysed more carefully below. For the present, it is sufficient to indicate what strong reasons there are for assuming that

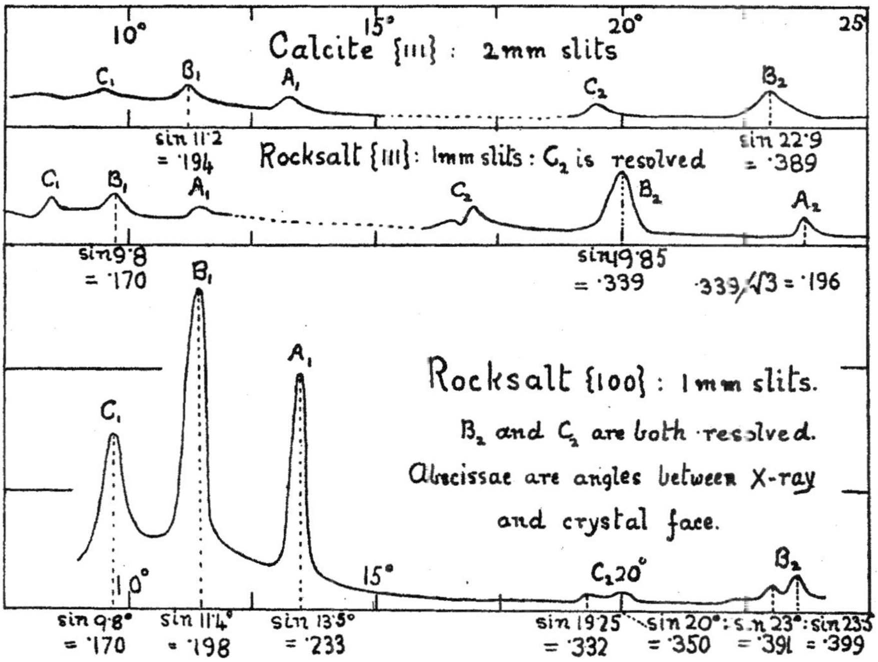

FIG. 16.

the distances d for the various faces can be accurately compared, by finding the angles at which these peaks occur.

It has been seen that the patterns given by potassium chloride, zincblende, fluorspar, and calcite can be ascribed to diffraction by points of a space lattice. It is now desired to compare the dimensions of the lattices of these crystals; since the absolute value of the wave-length λ of the radiation which forms the peaks is as yet unknown, the dimensions of each lattice will be expressed in terms of λ as unit.

Since now both the form and the dimensions (in terms of λ) of the elementary parallelepiped are known, for the space lattice arrangement of diffracting centres of these crystals, it is possible to calculate the volume of this parallelepiped; the volume is that associated with each point of the lattice, it is the inverse of the number of points per unit volume. A multiplication of this volume by the density of the crystal gives the mass associated with each diffracting centre, and it is to be expected that the comparison of these masses (for different crystals) will give some idea as to whether the centre consists of atoms, molecules, or groups of molecules. The results of these calculations for the various crystals are set forth in the table below:—

In this table—

θ = glancing angle of B peak, 1st order.

d = distance between planes parallel to the face investigated.

V = volume of elementary parallelepiped, calculated from this value of d and a knowledge of the nature of the lattice.

ρ = density of crystal.

M = molecular weight of substance.

Table X.

Crystal.	Lattice.	Face.	θ.	$\frac{d}{\lambda}$.	$\frac{V}{\lambda^3}$.	ρ.	M.	$\frac{V\rho}{\lambda^3 M}$.
Sylvine, KCl	Simple cubic...........	(100) (111)	10·2° 18·0	2·86 1·62	23·4 22·2	1·97	74·5	0·605
Rock-salt, NaCl...	Face-centred cubic ...	(100) (110) (111)	11·4 16·0 9·8	2·53 1·82 2·95	32·5 33·9 33·5	2·15	58·5	1·22
Zincblende, ZnS...	" " ...	(110)	16·5	1·76	30·8	4·06	97·0	1·28
Fluorspar, CaF_2...	" " ...	(100) (111)	11·7 10·3	2·46 2·79	29·8 28·3	3·18	78·0	1·18
Calcite, $CaCO_3$ ...	Rhombohedral	(100) (111)	10·5 11·2	2·74 2·60	44·8	2·71	100·0	1·22
Iron pyrites, FeS_2	Face-centred cubic ?*	(100)	12·1	2·39	27·3	5·03	120·0	1·15

* The interference pattern of this crystal has not yet been obtained.

The last column gives the value of $V\rho/\lambda^3 M$, the mass associated with each diffracting centre divided by the molecular weight of the substance. This quantity is, therefore, proportional to the number of molecules associated with each centre. For each of these crystals, with the exception of potassium chloride, this quantity is the same within the errors* of experiment, showing

* It must be remembered that in calculating this quantity, any percentage error in the value for sin θ is trebled, since sin θ is raised to the third power.

that in all these crystals the number of molecules associated with each diffracting centre is the same. Taking into consideration the very different constitution of these crystals, this fact seems to point to the association of one molecule, and one alone, with each diffracting centre; and since in zincblende, fluorspar, and calcite the molecule contains only one heavy atom, the conclusion is arrived at that the space lattice which the diffracting pattern indicates is that formed by the individual zinc or calcium atoms of these crystals.

Potassium chloride forms an apparent exception to this rule, for it has a value for $V\rho/\lambda^3M$ half that given by the other crystals. The reason for this is clear when it is remembered that in potassium chloride there are two atoms of very nearly equal atomic weights. Each molecule provides two diffracting centres, these being arranged on the simple cubic space lattice. The mass associated with each centre is not that of a molecule, but half of this quantity, and again it is single atoms, but now of two kinds, which form the points of the diffracting space lattice.

It is clear that the argument given here cannot pretend to be a complete proof of this important point. It is conceivable, for example, that in all these crystals it just happens that the molecules are grouped together in fours, and that these groups form the diffracting centres. It is easy to picture such an arrangement for the alkaline halides, in fact this is the arrangement given to all such binary compounds by the theory of closest packing by Pope and Barlow. Their arrangement would explain satisfactorily the patterns and peak relations of rock-salt, zincblende, and potassium bromide and iodide, for the black and white centres of the diagram given in fig. 8 are represented by tetrahedra composed of four spheres corresponding to atoms of either nature. This would also involve, however, the grouping in fours of the calcium atoms in calcite, and considerable difficulty is experienced in picturing an arrangement which does this. A similar difficulty arises in the case of fluorspar. Potassium chloride is also hard to account for on this arrangement, if it is granted that in this substance potassium and chlorine act almost identically on the X-rays, for the atoms are in the closest packed arrangement of the face-centred cube, while the diffraction pattern is characteristic of the simple cubic lattice. Many more comparisons of crystals are necessary to confirm this point, in the meantime it will be assumed that the simple structure correctly represents the truth, and that the diffracting centres are single atoms.

It has been seen how the comparison of the angles of reflection of a peak from various faces of the same crystal gives information concerning the space lattice structure of the crystal. Further information can be got by studying the dimensions of these peaks. For instance, the curves for two of the three primary planes (100), (110), (111) of rock-salt are given in fig. 16* and a reference to these curves will show the very marked difference which there is between the curves for the face (100) and that for the face (111). The (100) curve shows very marked first-order peaks, much smaller second-order peaks, and the merest indication of the peaks of the third order. The (111) curve on the other hand shows the second-order peaks very much stronger than those of the first order.

This difference of the curves corresponds to a difference in the nature of the planes parallel to these faces of the crystal. In the arrangement of black and white points given in fig. 8, it will be seen that the successive planes parallel to (100) contain equal numbers of black and white points; the same is true for the planes (110), which also give a strong first-order reflection. The planes parallel to (111), on the other hand, contain alternately all blacks and all whites. The black points alone form a face-centred lattice, for which the (111) planes are further apart than the (100) planes in the ratio $2/\sqrt{3}$. Thus the small first-order peaks reflected from the (111) face of rock-salt correspond to a periodicity of black planes alone, parallel to (111), the planes containing the heavy chlorine atoms. The presence midway between these planes of the planes containing sodium atoms does not completely destroy this reflection of the first order, but it goes a long way towards doing so, while of course the reflection of the second order is reinforced and gives a large second-order peak. This explains the abnormal relative magnitudes of the (111) peaks of different orders as compared with those reflected from the faces (100) and (110).

In accordance with this, it is found that if the sodium is replaced by potassium, the first-order peak reflected from the (111) face becomes too small to be detected, the (111) curve for sylvine appears to have a peak of the first order, at an angle corresponding to planes $1/\sqrt{3}$ as far apart as the planes parallel to (100). In fact, the peaks are where they should be for the simple cubic space lattice.

This argument may be summed up as follows: The arrangement of the heavy atoms of these crystals (potassium chloride with its two equal atoms being an exceptional case) is that of the space lattice which is the skeleton of the crystal, one molecule containing one heavy atom being associated with each point of the lattice. The first order peaks of the reflection curves are in the positions which theory would give for this space lattice, but the relative magnitudes of the peaks of the first and second orders on any curve are influenced by the positions of the lighter atoms in the crystal structure. If

* For the experimental evidence in support of this part of my argument I am indebted to my father.

these lighter atoms are so disposed as to lie on the planes of the heavy atom space lattice parallel to the face investigated, then the reflection curve may be said to be of the normal type, it will have large peaks of the first order and small ones of the second. If the lighter atoms are arranged on planes situated halfway between the planes of the lattice, the first-order peaks will be diminished and those of the second order reinforced.

The curves for the faces (100) and (111) of fluorspar show this effect in a very marked manner (fig. 17). This crystal has as its skeleton the face-centred cubic lattice, the points of the lattice being represented by the calcium atoms. The fluorine atoms are so disposed that the reflection from

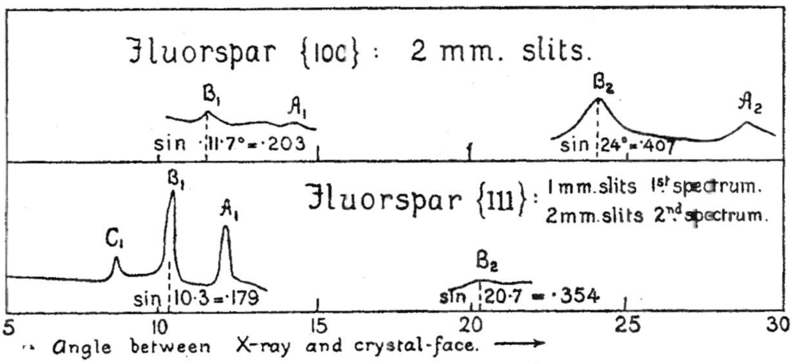

Fig. 17.

the (111) planes is now of the normal type, in strong contrast to the curves for rock-salt. On the other hand, the first-order reflection from the face (100) has almost disappeared. The fluorine atoms must be arranged so as to lie on or near the (111) planes of the fundamental lattice, not on the planes (100) as are the sodium atoms of rock-salt. The calcite curves given in fig. 16 show that for this crystal it is the (100) planes which give normal reflection, the curve for the face (111) being very like that for the face (111) of rock-salt.

It is hoped that an examination of the reflection from various faces of all these crystals may lead to the discovery of the exact positions of the lighter atoms in the crystal structure; as yet the experimental results are very incomplete. The results obtained so far seem to fix with some certainty the arrangement of the heavy atoms of these simple crystals, and in the case of the alkaline halides it is hoped that the positions assigned to atoms of both kinds are at any rate close approximations to the truth. A slight symmetrical distortion of the arrangement, which would reduce the crystal symmetry, would not affect any of the results which have been obtained here.

The analysis of crystal structures given here was initially undertaken with the object of discovering the absolute wave-length in centimetres of the homogeneous radiations issuing from the X-ray bulb. The positions of the peaks on the curves gave the wave-length of the corresponding radiations in terms of the dimensions of the crystal space lattice. As long as the complexity of the unit associated with each point of the lattice is unknown, the absolute wave-length cannot be calculated. If the arrangement here assigned to the alkaline halides is right, the dimensions of the lattice can be given in centimetres, for the mass associated with each centre of the lattice can be calculated from the known mass in grammes of the hydrogen atom.

For rock-salt, mass of 1 molecule of NaCl = $58{\cdot}5 \times 1{\cdot}64 \times 10^{-24}$; therefore

$$58{\cdot}5 \times 1{\cdot}64 \times 10^{-24} = V\rho = 33{\cdot}3 \times 2{\cdot}15 \times \lambda^3,$$

$$\lambda^3 = 1{\cdot}34 \times 10^{-24}, \qquad \lambda = 1{\cdot}10 \times 10^{-8} \text{ cm.}$$

From the value for λ, and that for d/λ given in Table X, the dimensions of the lattice for any crystal in this table can be calculated.

Summary.

For a number of simple crystals the interference patterns can be ascribed to diffraction of a "white" radiation by a set of points on a space lattice. Each of these points is a single atom; if one atom in the molecule is at least twice as heavy as any of the others, it is the lattice formed by these atoms alone which the diffraction pattern reveals. Two atoms of nearly the same atomic weight are nearly equivalent as diffracting centres. The lighter atoms of the molecule are not grouped closely round the heavy atom forming the diffracting space lattice, but occupy intermediate positions. For instance, in sodium chloride the sodium atom has six neighbouring chlorine atoms equally close with which it might pair off to form a molecule of NaCl. The reflection curves and interference patterns given by the alkaline halides agree in assigning the same structure to these salts, the atoms being arranged on a simple cubic space lattice in such a way that rows parallel to the cubic axes contain alternate atoms of either kind. The association of a single heavy atom with each point of the space lattice is indicated by the fact that the mass of each point is proportional to the molecular weight of the substance when each molecule contains one heavy atom. This relation is got from the reflection curves of different crystals. A knowledge of the mass of a hydrogen atom makes it possible to calculate the actual dimensions of a crystal lattice, and so to find the wave-length in centimetres of the homogeneous components of the X-ray beam, this being the object for which these analyses of crystal structure were undertaken.

W. L. Bragg. *Roy. Soc. Proc., A, vol.* 89, *Plate* 10.

Fig. 2.—Potassium chloride.

Fig. 8.—Rock-salt, 2·5 mm. thick.

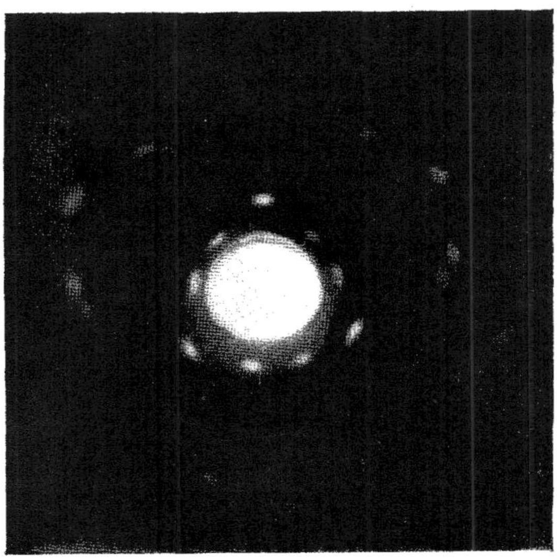

Fig. 11.—Fluorspar.

Fig. 15.—Rock-salt.

In conclusion I should like to express my indebtedness to Prof. Pope for his sympathetic interest and generous assistance. Dr. Hutchinson, with the greatest kindness, has overcome the only experimental difficulties connected with this subject by supplying the necessary crystal sections; but for his help it would have been impossible to obtain the very large number of photographs used in the investigation. These photographs were obtained at the Cavendish Laboratory, and I wish to thank Prof. Sir J. J. Thomson for his kind interest in the experiments. The measurements with the X-ray spectroscope were, as already stated, made by my father in the laboratory of the University of Leeds.

The Diffraction of Electrons by a Crystal of Nickel

By C. J. DAVISSON

This article is taken from the manuscript prepared by the author for his address at the joint meeting of Section B of the American Association for the Advancement of Science and the American Physical Society on December 28, 1927, at Nashville, Tennessee. An account of this work giving fuller experimental details is given by Davisson and Germer in the December, 1927, issue of the *Physical Review*.

These experiments are fundamental to some of the newer theories in physics. Until they were performed, it could be said that all experimental facts about the electron could be explained by regarding it as a particle of negative electricity. It now appears that in some way a "wave-length" is connected with the electron's behavior. The work thus shows an interesting contrast with the discovery of A. H. Compton that a ray of light (a light pulse) suffers a change of wave-length upon impact with an electron, the change of wave-length corresponding exactly to the momentum gained by the electron. Until Compton's work, all the known facts about light could be explained by thinking of light as a wave motion. The Compton effect seems to prove the existence of particles of light.

Physics is thus faced with a double duality. Compton showed that light is in some sense *both* a wave motion and a stream of particles. Davisson and Germer have now shown that a beam of electrons is in some sense *both* a stream of particles and a wave motion.

At the same time, theoretical advances have been made which seem to pave the way for an understanding of this curious situation. A general account of these new developments was given by K. K. Darrow in his series "Contemporary Advances in Physics" in the *Bell System Technical Journal* for October, 1927. Some remarks on the relation of the Davisson and Germer experiments to the new mechanics were given in this article, p. 692 *et seq.*—EDITOR.

THE experiments which I have been asked to describe are the most recent of an investigation of the scattering of electrons by metals on which we have been engaged in the Bell Telephone Laboratories for the last seven or eight years.

The investigation had its inception in a simple but significant observation. We observed some time in the year 1919 that when a beam of electrons is directed against a metal target, electrons having the same speed as those in the incident beam stream out in all directions from the bombarded area. It seemed to us at the time that these could be no other than particular electrons from the incident beam that had suffered large deflections in simple elastic encounters with single atoms of the target. The mechanism of scattering, as we pictured it, was similar to that of alpha ray scattering. There was a certain probability that an incident electron would be caught in the field of an atom, turned through a large angle, and sent on its way without loss of energy. If this were the nature of electron scattering it would be possible, we thought, to deduce from a statistical study of the deflections some information in regard to the field of the

Reprinted with permission from Bell System Technical Journal, Vol. 7, pp. 90–105.

deflecting atom. It was with these ideas in mind that the investigation was begun. What we were attempting, it will be seen, were atomic explorations similar to those of Sir Ernest Rutherford and his collaborators but explorations in which the probe should be an electron instead of an alpha particle. I shall not stop to recount the earlier experiments of this investigation, but shall pass at once to the most recent ones—those in which Dr. Germer and I have studied the scattering of electrons by a single crystal of nickel.

The unusual interest that attaches to these experiments is due to their revealing the phenomenon of electron scattering in a new and, I may say, fashionable rôle. Electron scattering is not, it would seem, the mildly interesting matter of flying particles and central fields that we supposed, but is instead a much more interesting phenomenon in which electrons exhibit the properties of waves. The experiments reveal that the way in which electrons are scattered by a crystal is very similar to the way in which x-rays are scattered by a crystal. The analogy is not so much with the alpha ray experiments of Sir Ernest Rutherford, as with the x-ray diffraction experiments of Professor von Laue.

My task of describing these experiments is much simplified by the fact that the experiments of Professor von Laue are so well known and so thoroughly comprehended. I remind you very briefly that in the original Laue experiment a beam of x-rays was directed against a crystal of zincblende, that about the transmitted beam was found an array of regularly disposed subsidiary beams proceeding outward from the irradiated portion of the crystal, and that these subsidiary beams could be interpreted completely and precisely in terms of the then already popular wave theory of x-radiation. They could indeed be explained as diffraction beams that resulted from the superposition of secondary wave trains expanding from the regularly arranged atoms of the crystal lattice.

There are two features of the Laue experiment which we shall need particularly to remember. The first is that diffraction beams issue not only from the far side of the crystal along with the transmitted beam, but also from the near or incidence side of the crystal—these latter being disposed in a regular array about the incident beam. The second is that each diffraction beam is characterized by a particular wave-length, and that a given beam appears in the diffraction pattern if the incident beam contains radiation of its characteristic wave-length, or of some submultiple value of this wave-length, but not otherwise. If the incident beam is monochromatic, no diffraction beams appear at all unless the wave-length of the incident beam happens to coincide with a wave-length of one or more of the diffraction beams. In that case the favored beams appear but no others.

With this picture of x-ray scattering in mind one sees at once the significance of the main results of the present experiments. A homogeneous beam of electrons is directed against a crystal of nickel, and at certain critical speeds of bombardment full speed scattered electrons issue from the incidence side of the crystal in sharply defined beams —a few beams at each of the critical speeds—the totality of such beams making up a regularly disposed array similar to the array of Laue beams that would issue from the same side of the same crystal if the incident beam were a beam of x-rays.

The electron beams are not identical in disposition with the Laue beams, and yet it is possible to treat them as diffraction beams, and from their position and from the geometry and scale of the crystal to calculate "wave-lengths" of the incident beam—just as we might do if we were dealing with x-rays or with any other wave radiation. When this is done we arrive at a definite and simple relation between the speed of the electron beam and its apparent wave-length—the wave-length is inversely proportional to the speed.

Surprising as it is to find a beam of electrons exhibiting thus the properties of a beam of waves, the phenomenon is less surprising today than it would have been a few years ago. We have been prepared, to a certain extent, by recent developments in the theory of mechanics for surprises of just this sort—for the discovery of circumstances in which particles exhibit the properties of waves. We have witnessed, during the last three years, the inception and development of the idea that all mechanical phenomena are in some sense wave phenomena—that the rigorous solution of every problem in mechanics must concern itself with the propagation and interference of waves. The wave nature of mechanical phenomena is not ordinarily apparent, we are told, because the length of the waves involved is ordinarily small compared to the dimensions of the system. It is only in such small scale phenomena as the intimate reactions between atoms and electrons that the wave-lengths are comparable with the dimensions of the system. Here only are we to expect notable departures from classical mechanics, and here only are we to find evidence of a more comprehensive wave mechanics.*

The success of this new theory has been confined, up to the present time, to explanations of certain of the data of spectroscopy. In this field the theory has appealed very strongly to all of us because of the

* It was predicted by W. Elsasser in 1925 (Naturwiss., *13*, 711 (1925)) that evidence for the wave mechanics would be found in the interaction between a beam of electrons and a crystal.

elegance of its methods and because of its remarkable facility in accounting for various of the inhibitions with which the radiating atom is afflicted. We have been prepared by these successes to view with not too great surprise—or alarm—evidence for the wave nature of phenomena involving freely moving electrons. And any reluctance we may feel in treating electron scattering as a wave phenomenon is apt to be dispelled when we find that the value calculated for the wave-length of the equivalent radiation is in acceptable agreement with that which L. de Broglie assigned to the waves which he associated with a freely moving particle—that is to say, the value h/mv (Planck's constant divided by the momentum of the particle).

In this account of the experiments I will describe the general method of the measurements and the general character of the results rather than attempt to go into these matters in detail.

Nickel forms crystals of the face centered cubic type. In Fig. 1 (*a*) the crystal which we had at our disposal is represented by a block of unit cubes of this type.

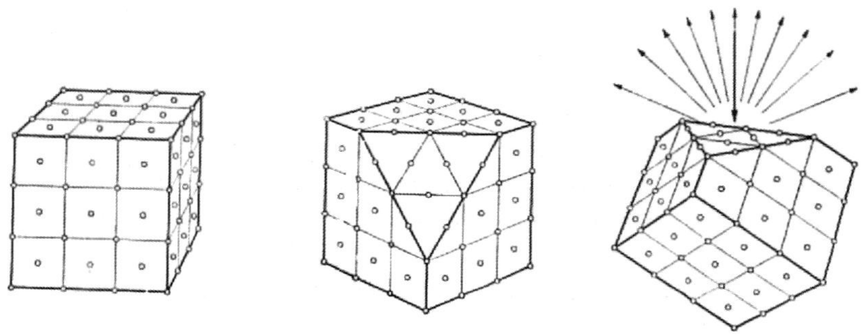

a *b* *c*

Fig. 1—Diagrams of nickel lattice, of cut lattice, and of lattice with incident and scattered beams

Our first step in preparing the crystal for bombardment was to cut through this structure at right angles to one of the cube diagonals. The appearance of the crystal after the cut was made, and the corner of the cube removed, is indicated in Fig. 1 (*b*). It is this newly formed triangular surface that was exposed to electron bombardment. The bombardment was at normal incidence as indicated in Fig. 1 (*c*). We are to think of electrons raining down normally upon this triangular surface, and of some of these emerging from the crystal without loss of energy, and proceeding from it in various directions.

What is measured is the current density of these full speed scattered electrons as a function of direction and of bombarding potential. The way in which the measurements are made is illustrated in Fig. 2. The electrons proceeding in a given direction from the crystal enter the inner box of a double Faraday collector and a galvanometer of high sensitivity is used to measure the current to which they give rise. An appropriate retarding potential between the parts of the collector excludes from the inner box all but full speed electrons.

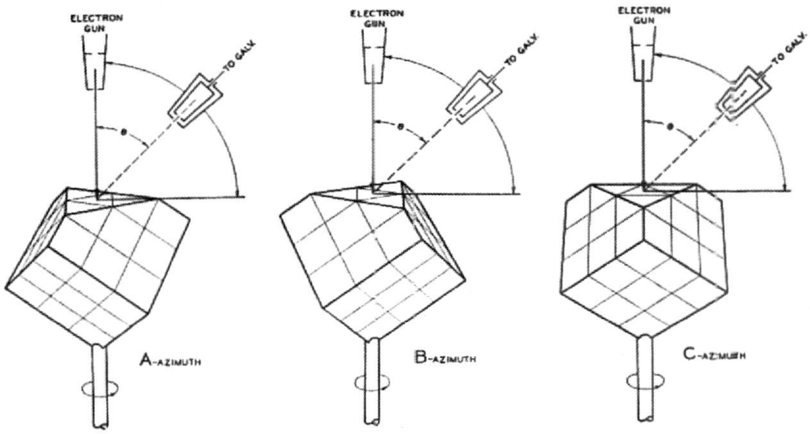

Fig. 2—Showing the three principal azimuths

The collector may be moved over an arc of a circle in the plane of the drawing as indicated, and the crystal may be rotated about an axis which coincides with the axis of the incident beam of electrons. Thus the collector may be set for measuring the intensity of scattering in any direction relative to the crystal—by turning the crystal to the desired azimuth, and moving the collector to the desired colatitude. The whole solid angle in front of the crystal may be thus explored with the exception of the region within twenty degrees of the incident beam.

Certain of the azimuths related most simply to the crystal structure we shall refer to as "principal azimuths." Thus there are the three azimuths that include the apexes of the triangle. If we find the intensity of scattering depending on colatitude in a certain way in one of these azimuths, we expect, of course, to find it depending upon colatitude in the same way in each of the other two. We shall call these the "A-azimuths." On the left in Fig. 2 the crystal has been turned to bring one of the A-azimuths into the plane of rotation of the collector.

Another triad of principal azimuths consists of the three which include the mid-points of the sides of the triangle. These we shall call the "B-azimuths." The next most important family of azimuths comprises those which are parallel to the sides of the triangle; of these there are six, the "C-azimuths."

If we turn the crystal to any arbitrarily chosen azimuth, set the bombarding potential at any arbitrarily chosen value, and measure the intensity of scattering as a function of colatitude, what we find ordinarily is the type of relation represented by the curve on the left in Fig. 3.

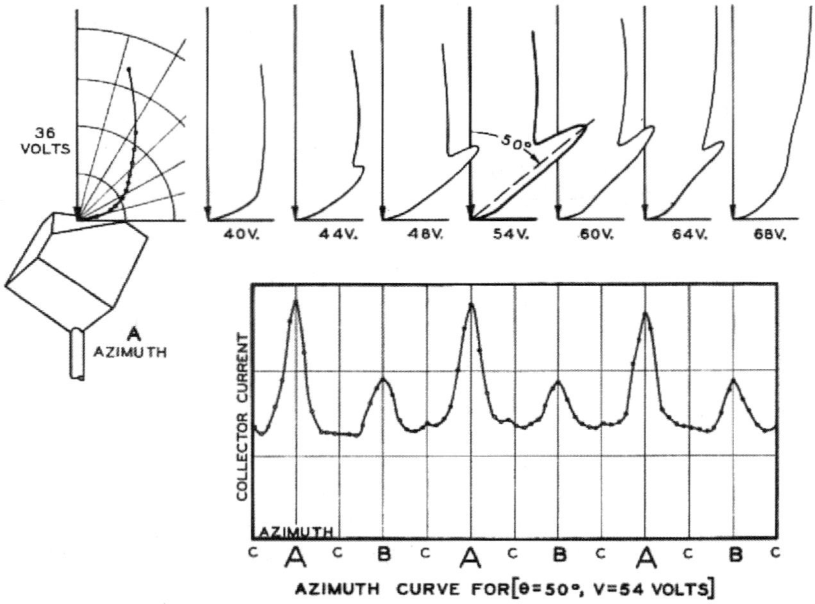

Fig. 3—Curves showing development of diffraction beam in the A-azimuth . . . and variation of intensity with the azimuth at colat. 50° for which beam is strongest in the A-azimuth

This curve is actually one found for scattering in the A-azimuth when the bombarding potential is 36 volts. It is typical, however, of the curves that are obtained when no diffraction beam is showing. The intensity of scattering in a given direction is indicated by the length of the vector from the point of bombardment to the curve. The intensity is zero in the plane of the crystal surface, and increases regularly as the colatitude angle is decreased. This type of scattering forms a background upon which the diffraction beams are superposed.

The occurrence of a diffraction beam is illustrated in the series of curves to the right in Fig. 3. When the bombarding potential is increased from 36 to 40 volts, the curve is characterized by a slight hump at colatitude 60 degrees. With further increase in bombarding potential this hump moves upward, and at the same time develops into a strong spur. The spur reaches its maximum development at 54 volts in colatitude 50 degrees, then decreases in intensity, and finally vanishes from the curve at about 70 volts in colatitude 40 degrees.

We next make an exploration in azimuth through this spur at its maximum; we adjust the bombarding potential to 54 volts, set the collector in colatitude 50 degrees, and make measurements of the intensity of scattering as the crystal is rotated. The results of this exploration are exhibited by the curve at the bottom of Fig. 3, in which current to the collector is plotted against azimuth. We find that the spur is sharp in azimuth as well as in latitude and that it is one of a set of three spurs as required by the symmetry of the crystal.

We observe also that there are small spurs showing in the B-azimuths. We turn the crystal to bring the B-azimuth under observation, and again make explorations in latitude for various speeds of bombardment. We find that the spur in the B-azimuth is similar to the "54 volt" spur in the A-azimuth, but that it attains its maximum development at a higher voltage and at a higher angle. Curves exhibiting its growth and decay are shown in Fig. 4. Maximum

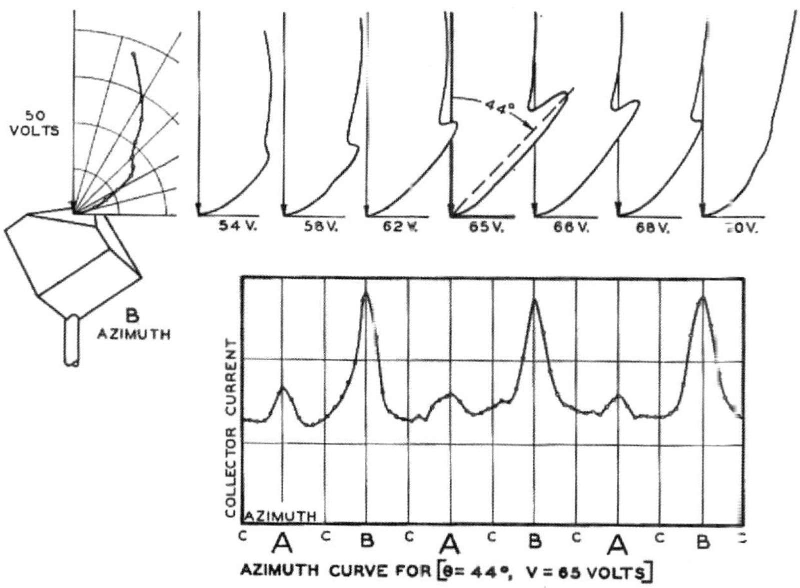

Fig. 4—Similar for the B-azimuth

development is attained at 65 volts in colatitude 44 degrees. At the bottom of the figure we show the intensity-azimuth curve through this spur at its maximum. The small maxima in the A-azimuths represent the remnants of the "54-volt" spurs.

We have thus a set of spurs at colatitude 50 degrees in the A-azimuths when the bombarding potential is 54 volts and a set of 44 degrees in the B-azimuths when the bombarding potential is 65 volts. These spurs are due to beams of full speed scattered electrons which are comparable in sharpness and definition with the beam of incident electrons. This is inferred from the widths of the spurs and the resolving power of the apparatus.

It is hardly necessary to point out that these sharply defined beams of scattered electrons are similar in their behavior to x-ray diffraction beams. If the incident beam were a beam of monochromatic x-rays of adjustable wave-length instead of a homogeneous beam of electrons of adjustable speed, quite similar effects could be produced. If the wave-length of the x-ray beam were varied, critical values would be found at which intense diffraction beams would issue from the crystal in its A-azimuths and others at which such beams would issue in the B-azimuths. The x-ray diffraction beams would indeed be more sharply defined in wave-length than the electron beams defined in voltage. No diffraction beam would be observed until the wave-length of the incident x-rays were very close indeed to its critical value, and the beam would disappear again when the wave-length had passed only very slightly beyond the critical value. This "wave-length sharpness" or "wave-length resolving power" is dependent, however, upon the number and disposition of the atoms involved in the diffraction. If the crystal were only a few atom layers in thickness, or if the x-rays were extinguished on penetrating through only a few atom layers of the crystal, then the x-ray diffraction beams would be much less sharply defined in wave-length; they would behave more like the electron beams. We may say then that the electron beams exhibit the general behavior of diffraction beams resulting from the scattering of a beam of very soft wave radiation—radiation that is very rapidly extinguished in the crystal.

Let us try now to forget that what we are measuring in these experiments is a current of discrete electrons arriving one by one at our collector. Let us imagine that what we are dealing with is indeed a monochromatic wave radiation, and that our Faraday box and galvanometer are instruments suitable for measuring the intensity of this radiation. We are to think of the incident electron beam as a beam of monochromatic waves, and of the "54-volt beam" in the A-azimuth and the "65-volt beam" in the B-azimuth as diffraction beams that owe their intensities, in the usual way, to constructive interference among elements of the incident beam scattered by the atoms of the crystal. With this picture in mind we try next to calculate wave-lengths of this electron radiation from the data of these beams and from the geometry and scale of the crystal.

To begin with, we shall need to look more closely into our crystal. The atoms in the triangular face of the crystal may be regarded as arranged in lines or files at right angles to the plane of the A- and B-azimuths. If a beam of radiation were scattered by this single layer of atoms, these lines of atoms would function as the lines of an ordinary line grating. In particular, if the beam met the plane of atoms at normal incidence, diffraction beams would appear in the A- and B-azimuths, and the wave-lengths and inclinations of these beams would be related to one another and to the grating constant d by the well-known formula, $n\lambda = d \sin \theta$, as illustrated at the top of the figure.

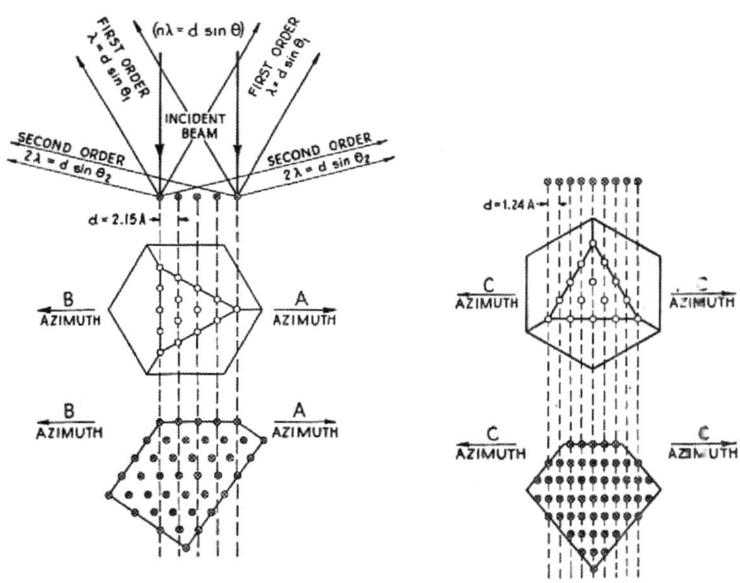

Fig. 5—Showing $n\lambda = d \sin \theta$ relation in the A-, B- and C-azimuths

In the actual experiments the diffracting system is not quite so simple. It comprises not a single layer of atoms, but many layers; it is equivalent not to a single line grating, but to many line gratings piled one above the other, as shown graphically at the bottom of the figure. What diffraction beams will issue from this pile of similar and similarly oriented plane gratings?

The answer to this question is twofold. In respect of position all the beams which appear will coincide with beams which would issue from a single grating. We get no additional beams by adding extra layers to the lattice. In respect of intensity, however, the results are greatly changed. A given beam may be accentuated or it may be diminished, both absolutely and relatively to the other beams; it may in fact be blotted out completely, or reduced to such an extent that it can no longer be perceived. These are effects of interference among the similar beams proceeding from the various plane gratings that make up the pile. Later we shall consider under what conditions these component beams combine to produce a resultant beam of maximum intensity; for the present, however, I wish only to stress the fact that whenever and wherever a space lattice beam appears its wave-length and colatitude angle θ will be related to the constant d of the plane grating through the ordinary plane grating formula. We therefore apply this formula to the 54- and 65-volt beams that have been described. The grating constant d has the value 2.15 Å., the 54-volt beam occurs at $\theta = 50°$ so that $n\lambda$ for this beam should have the value $2.15 \times \sin 50°$, or 1.65 Å. For the 65-volt beam we obtain for $n\lambda$ the value 1.50 Å.

We now compare these wave-lengths with the wave-lengths associated with freely moving electrons of these speeds in the theory of wave mechanics. Translated into bombarding potentials, de Broglie's relation

$$\lambda = h/mv \text{ becomes } \lambda = \sqrt{\frac{150}{V}}\,\text{Å.},$$

where V represents the bombarding potential in volts. The length of the phase wave of a "54-volt electron" is $(150/54)^{1/2} = 1.67$ Å., and for a 65-volt electron 1.52 Å. The 54- and 65-volt electron beams do very well indeed as first order phase wave diffraction beams.

It may be mentioned that beams occur at different voltages in the A- and B-azimuths because the plane gratings that make up the crystal are not piled one immediately above the other. There is a lateral shift from one grating to the next amounting to one third of the grating constant. Because of this shift the phase relation among the elementary beams emerging in the A-azimuth is not the same as that among those emerging in the B-azimuth—and coincidence of phase among these beams occurs at different voltages, or at different wave-lengths, in the two azimuths.

We next make similar calculations for a beam occurring in the C-azimuth. One such beam attains its maximum development in colatitude 56° when the bombarding potential is 143 volts. For diffraction into the C-azimuth we must regard the atoms in the surface layer as arranged in lines normal to the plane of this azimuth as illustrated in Fig. 5. The grating constant is 1.24 Å., and the similar gratings that make up the whole crystal are piled up without lateral shift. For this reason the C-azimuth is six-fold instead of only three-fold. For a beam occurring in this azimuth in colatitude 56°, $n\lambda$ should be equal to 1.24 × sin 56° or 1.03 Å. The value of h/mv for electrons that have been accelerated from rest through 143 volts is $(150/143)^{1/2}$ or 1.025 Å. Again the beam does very well as a first order diffraction beam.

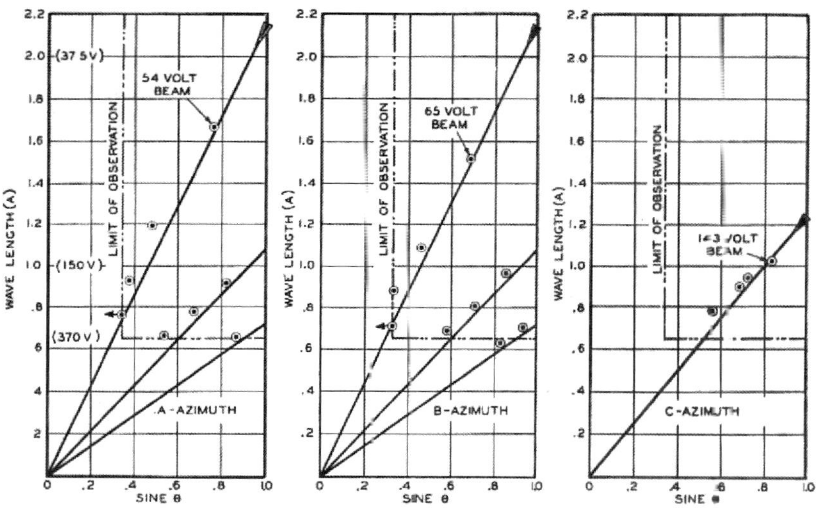

Fig. 6—Plot of λ against sin θ for various beams

The total number of such beams which we have observed in all azimuths in explorations up to 370 volts is twenty-four—nine in the A-azimuth, ten in the B-azimuth, and five in the C-azimuth. It would be possible to calculate an observed wave-length for each of these beams from $n\lambda = d \sin \theta$, and to compare this in each case with the theoretical wave-length calculated from $\lambda = h/mv$, just as we have done already for three of the beams. We have chosen, however, to display the results graphically rather than numerically.

The data for the twenty-four beams are exhibited in diagrams in Fig. 6, in which wave-length λ is plotted against the sine of the colatitude, θ. There is a separate diagram for each azimuth and in each the straight lines passing through the origin represent the plane grating formula $n\lambda = d \sin \theta$ in its different orders. Each of the twenty-four beams is represented by a point or by a wedge-shaped symbol in one of these diagrams. The quantities coordinated in each case are the wave-length of the incident beam as calculated from $\lambda = h/mv = (150/V)^{1/2}$ and the sine of the colatitude angle of the diffraction beam as observed. There are no points to the left of the line $\theta = 20°$ as this represents the lower limit of our colatitude range of observation, and none below the line $\lambda = 0.637$ Å. as this corresponds to the upper limit of our voltage range, 370 volts. The bombarding potentials corresponding to various wave-lengths are shown by figures enclosed in brackets.

When the data are exhibited in this fashion the question as to whether or not the observed wave-length of a beam agrees with its theoretical wave-length is answered by whether or not the point representing the beam falls on one or another of the lines representing the plane grating formula. If there were perfect agreement in all cases, each of the points would lie on some one of these lines.

It will be seen that the points all lie close to the lines, though not as a rule exactly on them. It is of course very important to decide whether the departures of the points from the lines are or are not too great to be attributed to uncertainties of measurements. It is our belief that they are in fact due to experimental error in the determination of the colatitude angles. If we accept the theoretical values of the wave-lengths as correct, and calculate the values of θ which we should have observed, we find that in no case do they deviate by more than 4 degrees from the values of θ actually set down. Corrections of this magnitude do not seem excessive when it is considered that we are making measurements with what amounts to a rather crude spectrometer, that the arm of the spectrometer is but 11 mm. in length, that the opening in the collector is 5 degrees in width, and that the spectrometer itself is sealed into a glass bulb. We therefore assume that in every case the value of the wave-length assigned by de Broglie is the correct one.

I now direct your attention to a particular group of these beams—the group comprising the beam of greatest wave-length in each of the three azimuths, which are represented in the figure by wedge-shaped symbols. The interpretation of these three is quite simple.

The radiation to which our electron beam is equivalent is extremely soft as already noted. Its intensity suffers a considerable decrement when the beam passes normally through only a single layer of atoms. This characteristic is inferred from the low resolving power of the crystal, and is consistent with what we know of the penetrating power of low speed electrons. When the beam passes through a layer of atoms at other than normal incidence the decrement in its intensity is greater still—and in the limit as the angle of incidence approaches grazing to the atom layer the intensity of the transmitted beam will approach zero. Thus we may expect that when a diffraction beam leaves the crystal at near grazing emergence the contributions to the resultant beam which come from the second and lower layers of atoms will be much less important than when the beam emerges from the crystal at a higher angle. Near grazing the radiation proceeding from the second and lower layers will be heavily absorbed in its passage through the overlying layers. Within a limited angular range near grazing the diffraction beam will be made up almost entirely of radiation scattered by the uppermost layer of atoms. The diffracting system becomes essentially a single plane grating and what we should observe is ordinary plane grating diffraction.

The first order diffraction beam from a line grating appears at grazing emergence when the wave-length of the incident radiation is equal to the grating constant. The grating constant for diffraction into the A- and B-azimuths is 2.15 Å. and grazing beams should appear in both azimuths when the wave-length of the incident electron beam has this value. The bombarding potential corresponding to wave-length 2.15 Å. is 32.5 volts, and at just 32.5 volts diffraction beams appear at grazing in both these azimuths. As the bombarding potential is increased the beams move up from the surface to satisfy the relation $\lambda = d \sin \theta$. Ten or fifteen degrees above the surface radiation from the second and lower layers escapes in sufficient amounts to reduce the intensity of the resultant beam through interference, and at a somewhat higher angle the beam disappears.

An exactly similar beam is found at grazing in the C-azimuth. The grating constant here is 1.24 Å. and the bombarding potential corresponding to wave-length 1.24 Å. is 97.5 volts. The beam appears at grazing at just this voltage. These three beams occurring and behaving exactly as required by the theory constitute the strongest evidence we have in favor of the wave interpretation of electron scattering.

We have been less successful in trying to account for the occurrences of the remaining 21 sets of beams. We do not know why they occur where they do. The most we have been able to do is to relate their occurrences with those of the Laue beams that would issue from the same crystal if the incident beam were a beam of x-rays.

In Fig. 7 we indicate by crossed circles in a (λ, sin θ) diagram the x-ray diffraction beams that would be observed in the B-azimuth. We show also again the electron beams as actually observed. It is obvious that the law of occurrence of electron beams is not the same as the law of occurrence of Laue beams, and yet we see that the occurrences of the two sets of beams have certain features in common. The dots representing electron beams occur along the plane grating lines at about the same intervals as the crossed circles representing the Laue beams. Other points of similarity are found with further study of the data and one is led finally to the conviction that each electron beam is the analogue of a particular Laue beam. The electron beam represented by a given dot appears to be the analogue of the Laue beam of the same order represented by the crossed circle occurring next above it in the diagram. This association of beams is indicated in the figure.

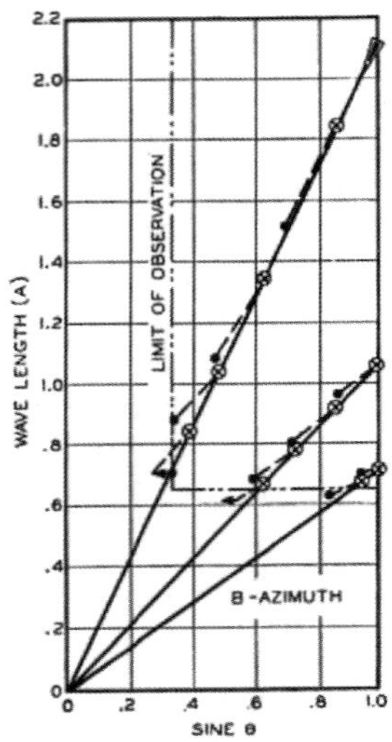

Fig. 7—λ sin θ diagram for B-azimuth

The occurrences of the Laue beams are determined in part by the separation between the atomic plane gratings that make up the crystal. If the separation between adjacent planes were increased the crossed circles representing the Laue beams would be moved upward along the plane grating lines; if the separation were decreased the crossed circles would be moved downward. Merely as a mode of description, then, we may say that a given electron beam has the wave-length and position that its Laue beam analogue would have if the separation between planes were decreased by a certain factor.

We have calculated this spacing factor for each of the 21 beams and the values found are plotted in the upper part of Fig. 8 against the voltages of the beams. The points form a very bad curve. They do indicate, however, that the factor increases with the speed of the electron, and there is the suggestion that it approaches unity as a limiting value. There is the suggestion, that is, that at high voltages the law of occurrence of electron beams is the same as the law of occurrence of Laue beams.

It has been pointed out by Eckart that if the index of refraction of the crystal for the electron radiation is other than unity diffraction beams will occur as if the separation between atom planes were other than normal. We have computed the indices of refraction that would

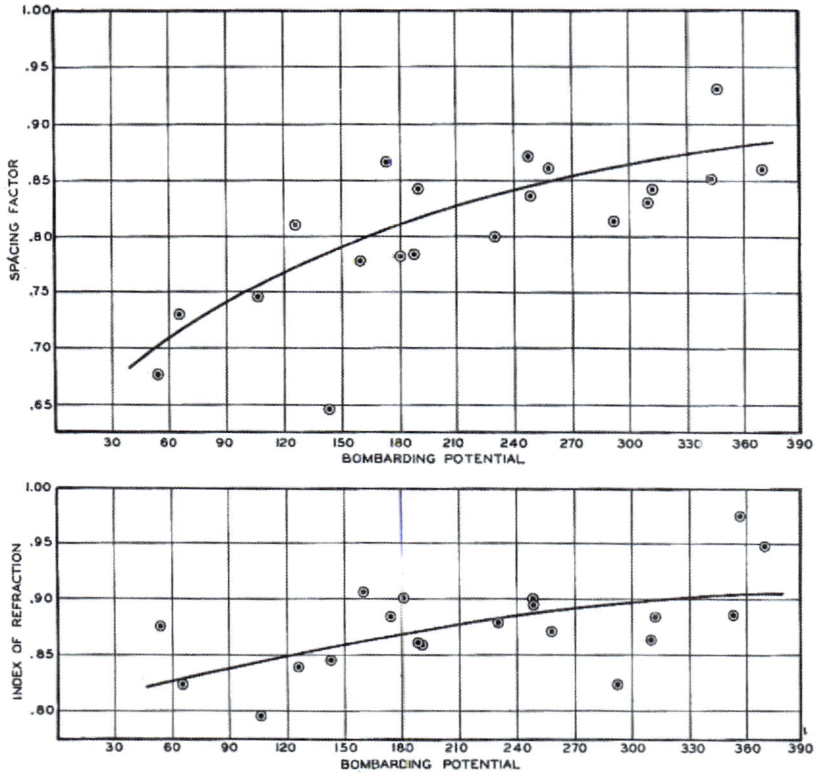

Fig. 8—Plot of values of spacing factor and associated values of refractive index for twenty-one beams

give rise to the observed occurrence of beams and these are plotted in the lower part of the diagram against bombarding potential. Again the points fall very irregularly. While it cannot be said that there is at present a satisfactory explanation of the peculiar occurrence of the space lattice electron diffraction beams, it should be clearly understood that this deficiency in no way affects either the wave-length measurements of these beams or the agreement of these wave-lengths with the values of h/mv.

The electron diffraction beams which I have described are the only ones observed when the surface of the crystal is free from gas. When the surface is not free from gas still other beams appear. These beams are due to the scattering of electrons by the adsorbed gas and therefore we shall not consider them at this time.

In closing I should like to say a few words about the conceptual difficulty in which these experiments involve us. When Laue and his collaborators investigated the scattering of x-rays by crystals the results of their observations were accepted at once as establishing the wave theory of x-rays. It was a very simple matter for W. H. Bragg and others to give up the corpuscular theory because of the hypothetical nature of the x-ray corpuscle. It was only necessary to recognize that Laue's results were contrary to hypothesis and the corpuscle disappeared.

If the electron were not the well-authenticated particle we know it to be, it is possible that the experiment I have described would cause it to vanish in like manner. We do not, however, anticipate any such event. The electron as a particle is too well established to be discredited by a few experiments with a nickel crystal. The most we are apt to allow is that there are circumstances in which it is more convenient to regard electrons as waves than as particles. We will allow perhaps that electrons have a dual nature—when they produce tracks in a C. T. R. Wilson cloud experiment they are particles, but when they are scattered by a crystal they are waves.

A quite similar situation exists, of course, in the case of x-rays. It has been evident for some years that the adherents of the corpuscular theory of x-rays were too enthusiastic in their recantations. X-rays also exhibit a dual nature—when they give rise to diffraction patterns they are waves, but when they exhibit the Compton effect or cause the emission of electrons from atoms they are particles—quanta or photons.

This state of affairs is one that should appeal to us as intolerable. There must, it would seem, be comprehensive modes of description applicable to all electron and x-ray phenomena, but what these are we do not yet know. We do not know whether we shall eventually believe with de Broglie and Schroedinger that electrons and x-rays are waves that sometimes masquerade as particles, with Duane that electrons and x-rays are particles that sometimes masquerade as waves, or whether eventually we shall believe with Born that we are dealing in both cases with actual particles and phantom waves.

I believe, however, that for the present and for a long time to come we shall, in describing experiments, worry but little about ultimate realities and logical consistency. We will describe each phenomenon in whatever terms we find most convenient.

PROCEEDINGS OF
THE ROYAL SOCIETY.

Section A.—Mathematical and Physical Sciences.

The Diffraction of Electrons by Single Crystals.

By G. P. Thomson, M.A., F.R.S., Professor of Physics in the University of London.

(Received May 26, 1931.)

[Plates 1, 2.]

The Diffraction of Electrons by Single Crystals.

1. The diffraction of electron waves has been observed both with single crystals and with polycrystalline aggregates. So far most of the work with single crystals has been done with slow electrons of a few hundred volts energy. While the results are in general agreement with wave mechanics, they differ considerably from the predictions of the simple theory, even when the latter is extended to include a consideration of the inner potential of the crystal. Part of the discrepancy is caused by undue simplification of the theory, and better results have been got with the more elaborate theory of Morse,* but there are a number of cases in which beams corresponding to half order spectra and reflection by "forbidden" crystal planes have been recorded,† and for these the present theory is quite unable to account.

Experiments with fast electrons on polycrystalline aggregates have shown results in full agreement with the simple theory, and it is hoped that electron diffraction will become a valuable method for the study of surface films.

* 'Phys. Rev.,' vol. 35, p. 1310 (1930).

† v. Laue and Rupp, 'Ann. Physik,' vol. 4, p. 1097 (1930); Farnsworth, 'Phys. Rev.,' vol. 34, p. 679 (1929), and vol. 35, p. 1131 (1930).

Clearly, however, this cannot be done satisfactorily as long as there is any doubt of the ability of the existing theory to explain the diffraction by known structures. It is therefore of importance to see whether any abnormalities exist for single crystals with fast electrons. So far the only work in this connection is that of Kikuchi and others on mica, that of Niskikawa and Kikuchi on calcite, and some work by Emslie at Aberdeen on galena and calcite.* Kikuchi's work shows complete agreement with theory, but has unusual features on account of the large spacing of mica across the cleavage face. The work on calcite is complicated by charging up of the crystal, and galena, as the present author has found,† changes under the influence of the rays, so that it is difficult to draw precise conclusions. No metals have been investigated. The present work has, therefore, been undertaken. It consists of a detailed investigation of the diffraction of electrons of from 30,000 to 50,000 volts by single crystals of rocksalt, copper and silver.

Method.

2. The apparatus used was described in the 'Proceedings of the Royal Society.'‡ The principle is shown diagrammatically in fig. 1. A narrow

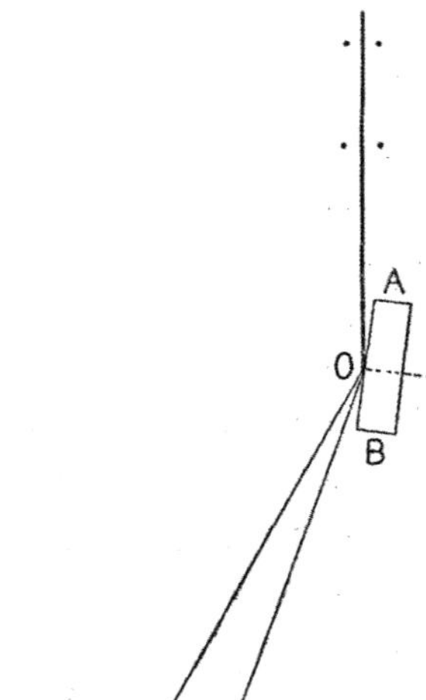

FIG. 1.

* Kikuchi, 'Proc. Imp. Acad. Jap.,' vol. 4, pp. 271, 275, 354, 471 (1928) and 'Jap. J. Phys.,' vol. 5, p. 83 (1928); Nishikawa and Kikuchi, 'Nature,' vol. 122, p. 726 (1928); 'Proc. Imp. Acad. Jap.,' vol. 4, p. 475 (1928); Emslie, 'Nature,' p. 977 (1929).

† "Wave Mechanics of Free Electrons," McGraw Hill, p. 103.

‡ Thomson and Frazer, 'Proc. Roy. Soc.,' A, vol. 128, p. 641 (1930).

beam of cathode rays is incident on the face AB of the crystal. Diffracted beams strike a willemite screen at C and produce visible fluorescence. The crystal can be rotated about an axis perpendicular to the plane of the diagram (change of incidence), and about ON the normal to the free surface (change of azimuth). It can also be moved bodily in a direction nearly parallel to ON. These movements can be made *in vacuo* with the electron beam in action. To obtain a permanent record a photographic plate (Paget process) is lowered in front of C and the rays record themselves directly by their impact on it. The exposures were of the order of 15 seconds.

Experiments on Copper and Silver.

3. These have face-centred cubic lattices. The cube side $a = 3\cdot61$ Å. for copper and $4\cdot079$ for silver. I am indebted to Sir Harold Carpenter for the single crystal of copper. This was a rod about 10 cm. long and $0\cdot6$ cm. diameter. In order to get a surface corresponding to a plane of simple indices it was necessary to find the directions of the axes and then cut the crystal. The first was done by the method of optical reflection as described by Bridgman.* The crystal was then cut with a fine saw, roughly to the required face, a piece mounted in a brass holder with plaster of Paris, and the face etched. The etched face was set approximately perpendicular to the axis of a lathe, the crystal being held in the toolholder so that it could be rotated through a small angle. A beam of light was reflected along the axis of the lathe by a small mirror on the axis, and the reflection from the etched surface observed by an eye near the lathe axis. When the etched surface was slightly turned spots of light flashed out on it due to reflection from the crystal facets, and the crystal was clamped in position when the flashes were brightest. This makes the facets perpendicular to the lathe axis. It was then ground by a wheel rotated in the lathe to give a surface perpendicular to the axis of the latter.† It is sometimes difficult to pick out the reflection from the facets against the background of diffuse reflection from the rest of the crystal surface, and the later specimens were prepared by using the electron diffraction itself to determine the direction of the axes. In this method, after the crystal had been cut approximately to the right face, it was etched and tested in the apparatus. From the patterns obtained with two settings differing by 90° in azimuth, the direction of the required face could be calculated. The crystal was then

* 'Proc. Amer. Acad. Sci.,' vol. 60, p. 303 (1924).

† Farnsworth, 'Phys. Rev.,' vol. 34, p. 677 (1929).

mounted in the jig shown in fig. 2 (the screws AB being set to give the calculated angle) and ground on a piece of plate glass. It was afterwards etched to remove the broken surface. The etching was sometimes done with ammonium

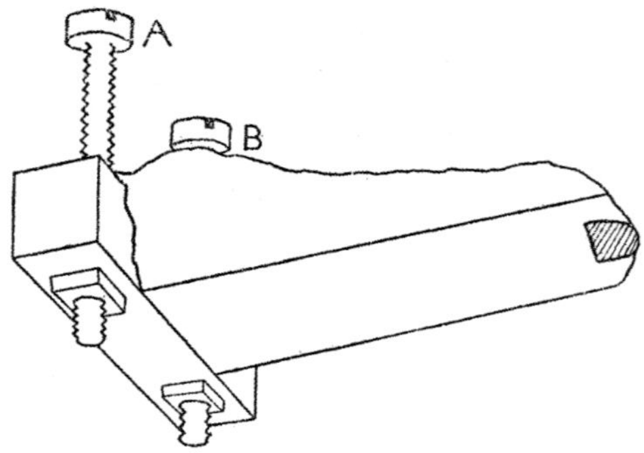

FIG. 2.

persulphate, sometimes with 50 per cent. nitric acid. The silver crystal was made accidentally by slowly cooling a bead of molten silver. It was cut by the method last described.

4. Both silver and copper gave in general a pattern of diffracted spots. Several orders of the same reflection were often visible at once (see Plate 1). The photographs were taken without rotation of the crystal, and so are equivalent to Laue patterns *with monochromatic rays.* One would, therefore, expect no spots to appear except for certain special settings of the crystal and then only one or two. Actually a pattern of many spots appears over a considerable range of angles. A similar effect was observed by Kikuchi with very thin sheets of mica. The explanation depends on a principle explained in "Wave Mechanics of the Free Electron," p. 72. Quite recently* Bragg and Kirchner have suggested that this is also the explanation of Kikuchi's results. In the present case it is necessary to make an assumption as to the nature of the surface. This must apparently consist of a number of lumps or ridges through which the electrons pass. Consider, for simplicity, a cubical lump with Bragg planes parallel to one pair of faces and rays incident on these at the glancing angle given by $2d \sin \theta = n\lambda$ (fig. 3). There will, of course, be reflection. The centre atoms of each line in the figure form a two-dimensional grating whose spacing in the plane of the paper is d. Now for a plane grating with angle of incidence ϕ, the directions ψ of the diffracted beams are

* 'Nature,' vol. 127, p. 738 (1931).

given by $d(\sin\psi + \sin\phi) = n\lambda$, where ψ, ϕ are positive on the same side of the normal. Hence

$$2d \sin\tfrac{1}{2}(\psi + \phi) \,.\, \cos\tfrac{1}{2}(\psi - \phi) = n\lambda.$$

If ϕ, ψ are small this is approximately $\chi d = n\lambda$, where χ is the deviation of the ray and is of the order 2°–12°. Thus as far as the centre atoms are concerned a ray will appear at the same deviation within a wide range of angle of incidence. If the incident and diffracted rays do not make equal angles with the Bragg plane, the wavelets from the different atoms in each separate plane will not be exactly in phase, but at small angles and for moderate sized pieces of crystal the discrepancy is not serious.

In fact, the path difference for atoms at the end of a row of length l is

$$l(\cos\phi - \cos\psi) \simeq \tfrac{1}{2}l(\psi^2 - \phi^2).$$

The resultant for the row does not vanish until the path difference between the wavelets from the end atoms is λ. This occurs when $\psi - \phi = \pm 2d/nl$, so that the width of the diffracted beam from each row of atoms is $4d/nl$. If l is not too large this may be wide enough to include values of n corresponding to several orders, all of which will then appear for the same λ and ϕ. The same expression gives the angle through which the crystal can be rotated without destroying the spot. Notice that on this view "forbidden" planes remain forbidden, and can give rise to no reflected spots. The value of l required to give patterns extending over the range of angles observed, is of the order of 2×10^{-6} cm. for 40,000 volt electrons.

It remains to describe and explain the actual arrangement of the spots in the various patterns. Specimens of copper crystals were examined with (100), (111), (331) and (311) faces and one of silver with a (100) face, besides the plates taken to determine the angle of the final grinding in which the exposed surface of the crystal had no simple indices. In all, some 50 plates were measured. The majority show regular patterns having the following characteristics. The exceptions will be dealt with later (p. 12), but (1) applies to them also.

(1) The radial distance of each spot from the origin (undeflected beam) satisfies approximately the relation

$$\chi = \lambda\sqrt{j^2 + k^2 + l^2}\,/a,$$

where a is the cube face, (jkh) are possible Miller indices, and λ is the electronic wave-length calculated from the potential of the rays by de Broglie's formula $\lambda = h/mv$.

(2) The spots in the plane of incidence correspond to reflections from the free face of the crystal. After this had been verified in one or two cases, it was used as a method of determining the orientation of the axes with respect to the crystal for the purpose of making the final cutting of the face.

(3) The spots form a portion of a two-dimensional lattice pattern of which the origin is a member.

(4) The angles between the lines joining any two spots to the origin is equal to the angle between the planes to which they correspond.

(5) The distance between any two spots is equal to the deviation due to reflection from a Bragg plane whose indices are the difference of those assigned to the two spots.

(6) The indices correspond to planes of a single zone.

(7) The patterns which showed the greatest numbers of spots for (100) and (111) faces were of the forms illustrated in figs. 4–8.

(8) If experiments are made with the crystal turned through various angles about the normal to its face, the sequence of these patterns is that to be expected from the change in azimuth.

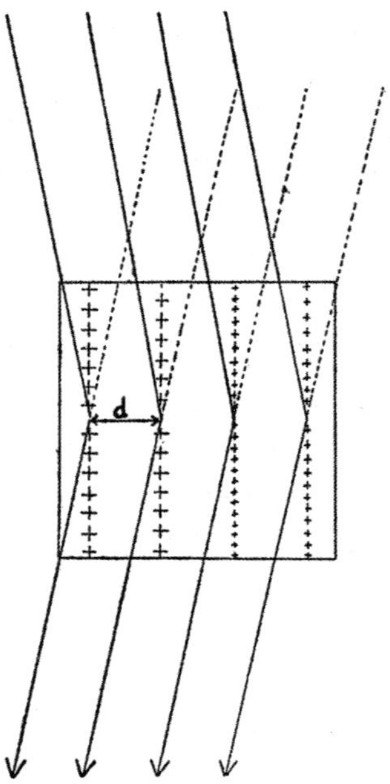

FIG. 3.

(9) The appearance on the screen changed when the angle of incidence of the beam to the free surface of the crystal was altered keeping the azimuth constant. The spots changed in intensity, some disappearing and others appearing, but they did not move, and the nature of the pattern did not change.

Some of the above are geometrically deducible from others. Thus it can be shown that if (1) and (3) are exactly true, the intervals being precisely equal at all places of the lattice pattern, (4) and (5) will inevitably follow. The truth of (7) carries with it that of (1), (3), (4), (5) and (6) and when the pattern is one of these familiar simple forms, it can easily be checked by measurements of a few of the spacings, and the indices assigned. In other cases the best method of assigning indices to the spots is a matter of experience. Thus (1) means little for the distant spots, as values of the indices can always be found to give the right radius within the error of measurements. On the other hand when the indices have been assigned, the measurements of radial distance give a useful check. Strictly speaking, any two spots determine the zone axis and so the whole pattern, but if two spots are observed it must be possible to assign indices to satisfy (1) and (4) or (5). When the pattern is determined two co-ordinates of any kind will determine if any given spot belongs to it. One of these may be the fact that it lies on a line with two other assigned spots.

5. The explanation of these laws and the assignment of indices to the spots is most shortly expressed by the use of Ewald's reciprocal lattice. This is a collection of points such that the vector to each from a common origin coincides in direction with the normal to a net plane, and is inversely proportional to the spacing between adjacent net planes parallel to it. For a face-centred cubic crystal, it is a collection of points whose rectangular cartesians co-ordinates are (jkl) where j, k, l are all even or all odd and the point corresponds to reflection from the plane of the same Miller indices. The axes in this case are the same as those of the crystal.

If now we have an incident beam parallel to a zone axis, the points of the reciprocal lattice corresponding to the zone planes will all lie on a plane OXY normal to the beam. The lines in which a given set of parallel net planes cut OXY form a two-dimensional grating of the kind considered above, for it can easily be seen that they contain the projections of the lattice points on OXY. The lattice points projected on OXY form in fact a two-dimensional array which can be referred to the lines OX, OY in which any two (non-parallel) planes of the zone cut OXY. The deviation due to such a grating, regarding the lattice points as diffracting centres, is inversely as the spacing of the lines, *i.e.*, of the net planes, and is $\chi = n\lambda/d = \lambda\sqrt{j^2+k^2+l^2}/a$, where n is included in (jkl) which are therefore not usually prime to each other. Let L be the distance of the plate from the crystal, and let all distances on the plate be measured in terms of the unit $\mathrm{L}\lambda/a$. Then, since the plate is nearly normal to the rays and all angles are small, the distance of the spot from the origin O′ (position of the undeviated beam) is $\sqrt{(j^2+k^2+l^2)}$.

For each set of indices (jkl) belonging to the zone there is one spot P on the plate, and the line O′P is normal to the planes (hjk). Corresponding to the indices (jkl) there is a point Q of the reciprocal lattice which lies on OXY, if (jkl) belongs to the zone. Further $\mathrm{OQ} = \sqrt{(j^2+k^2+l^2)}$ and is normal to the planes (hjk). Thus the pattern formed on the plate by the spots P is identical with that formed on OXY by the points Q or the *complete pattern of diffraction spots, due to the two-dimensional array, formed on the photographic plate is identical with the section of the reciprocal lattice, by a plane normal to the incident beam.* We have seen in the previous paragraph that if the lumps of the crystal are not too thick the actual pattern will be identical with that due to the two-dimensional array.

The general character of the patterns is readily seen from the consideration that if any two planes $(j_1k_1l_1)$ represented by Q_1 and $(j_2k_2l_2)$, represented by Q_2, are both members of the zone then $(aj_1 + bj_2, ak_1 + bk_2, al_1 + bl_2)$ is also a member if a and b are integers. Thus any point of the two-dimensional lattice whose unit translations are OQ_1, OQ_2 will appear as a spot. Notice that for the face-centred cubic structure of all the substances dealt with in this paper, (jkl) are all even or all odd.

These considerations enable us to determine the indices of any spot in the pattern when those of any other two spots are known, assuming, of course, that the whole pattern is due to a single crystal. The numbers found can be checked by applying the following result. The line joining any two points on the reciprocal lattice is parallel and equal to the radius vector whose indices are the difference of the indices of the first two. This is easily seen for a cubic lattice since the co-ordinates of the reciprocal lattice points are the indices of the plane. It is generally true. It follows that if indices can be assigned to any two spots, the line joining them gives the direction of the normal to a plane of the zone and is inversely proportional to the corresponding spacing.

6. In most cases the most convenient way of measuring up the pattern of spots is to use a special holder for the plate which allows the point of a needle to be brought into coincidence with the spots in succession, the needle being moved by means of two perpendicular slides with scales on each. This is equivalent to measuring rectangular cartesians co-ordinates, and the calculations can be made as follows.

Let α, β, γ be the direction cosines of the zone axis to which the incident beam is mostly nearly parallel. Then the spots are given by the points of the reciprocal lattice through which $\alpha x + \beta y + \gamma z = 0$ passes, *i.e.*, we have to find the solutions in integers of this equation. The points will be referred to rectangular axes $O\xi$, $O\eta$ in the plane $\alpha x + \beta y + \gamma z = 0$, $O\xi$ being in the direction of the normal (pqr) to the free surface of the crystal. As explained above, the crystal was cut so as to make this a net plane. The glancing angle of the rays on the crystal surface was always small.

Then the co-ordinates ξ, η, corresponding to a point (jkl) of the reciprocal lattice, are seen to be

$$\xi = pj + qk + rl$$

$$\eta = (q\gamma - r\beta)j + (r\alpha - p\gamma)k + (p\beta - q\alpha)l = \sqrt{j^2 + k^2 + l^2 - \xi^2}$$

by applying the usual formulæ for change of axes.

Clearly if (ξ_1, η_1) correspond to $(j_1k_1l_1)$ and (ξ_2, η_2) to $(j_2k_2l_2)$ the whole pattern is a lattice built up of $o \rightarrow (\xi_1, \eta_1)$ and $o \rightarrow (\xi_2, \eta_2)$ as unit translations. In particular cases the solutions are generally best found by inspection of the equation $\alpha j + \beta k + \gamma l = 0$, remembering that since the incidence is small $\alpha p + \beta q + \gamma r \simeq 0$.

Case 1. $p = q = r = 1/\sqrt{3}$, crystal cut to a (111) plane.

$$\sqrt{3}\xi = j + k + l$$

$$\sqrt{3}\eta = j(\gamma - \beta) + k(\alpha - \gamma) + l(\beta - \alpha)$$

$$= \sqrt{3(j^2 + k^2 + l^2) - (j + k + l)^2} = \sqrt{\Sigma(j - k)^2}$$

Simple solutions of $\alpha j + \beta k + \gamma l = 0$

$$\alpha + \beta + \gamma = 0$$

$$\alpha^2 + \beta^2 + \gamma^2 = 1$$

are $\alpha = 0$, $\beta = \mp 1/\sqrt{2}$, $\gamma = \pm 1/\sqrt{2}$, which with permutations of signs and axes occur six times in all. Also we have $\alpha = \pm 2/\sqrt{6}$, $\beta = \gamma = \mp 1/\sqrt{6}$ again ocurring six times.*

The first gives $k = l$ and $\sqrt{3}\eta = \pm \sqrt{2}(j - k)$. We can take as unit translations $o \rightarrow (\sqrt{3}, 0)$ and $o \rightarrow (1/\sqrt{3}, \sqrt{8/3})$. The second gives $2j - k - l = 0$, $\eta = \pm (k - l)/\sqrt{2}$ with unit translations $o \rightarrow (\sqrt{3}, 0)$ and $o \rightarrow (0, 2\sqrt{2})$. In both cases $\sqrt{3}\xi = j + k + l$.

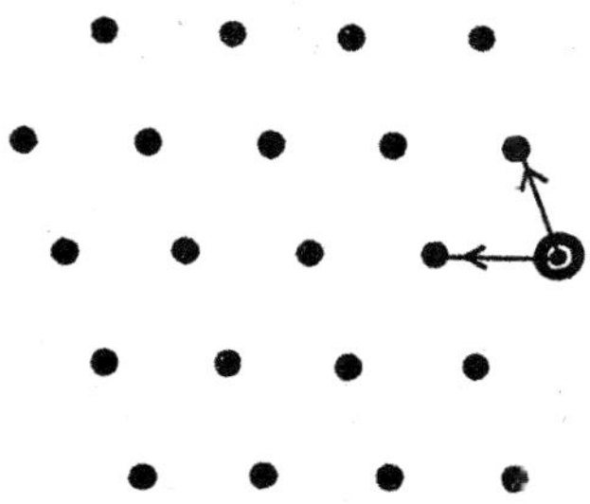

FIG. 4.

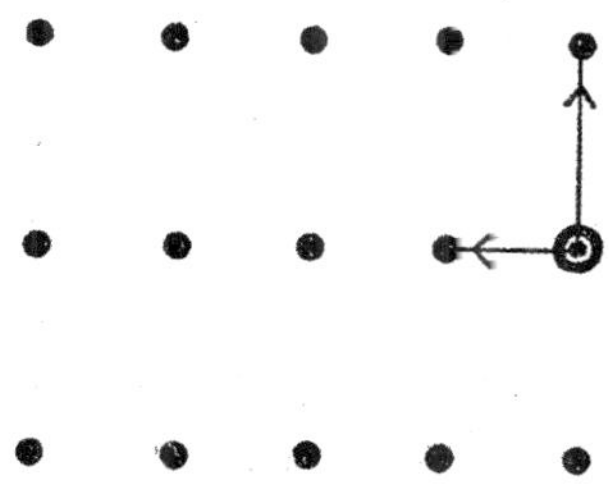

FIG. 5.

Case 2. $p = 1$, $q = r = 0$ crystal cut to a (100) face. Here $\alpha = 0$, $\xi = j$, $\eta = \beta l - \gamma k = \sqrt{k^2 + l^2}$ where kl satisfy $\beta k + \gamma l = 0$ and j is any integer. Simple cases are :—

(*a*) $\beta = 0$, $\gamma = \pm 1$, incident ray nearly parallel to a cube axis, giving $l = 0$, $\eta = \pm k$. In this case j, k must be even. Unit translations are $o \rightarrow (2, 0)$ and $o \rightarrow (0, 2)$.

* There is a difference between the orientation of the whole crystal corresponding to the plus and minus signs in the latter case, but since we are now virtually only dealing with the diffraction by a plane grating it is immaterial from which side the rays come.

(*b*) $\beta = \pm \gamma = 1/\sqrt{2}$, $k = \mp l$, giving $\eta = \sqrt{2}k$. Unit translations are $o \rightarrow (2, 0)$ and $o \rightarrow (1, \sqrt{2})$.

(*c*) $\beta = 1/\sqrt{10}$, $\gamma = \pm 3/\sqrt{10}$, $k + 3l = 0$, $\eta = \sqrt{10}l$. Unit translations $o \rightarrow (2, 0)$ and $o \rightarrow (1, \sqrt{10})$.

Each of the above occurs for four different crystal settings, as is obvious from the fourfold symmetry of the (100) axis.

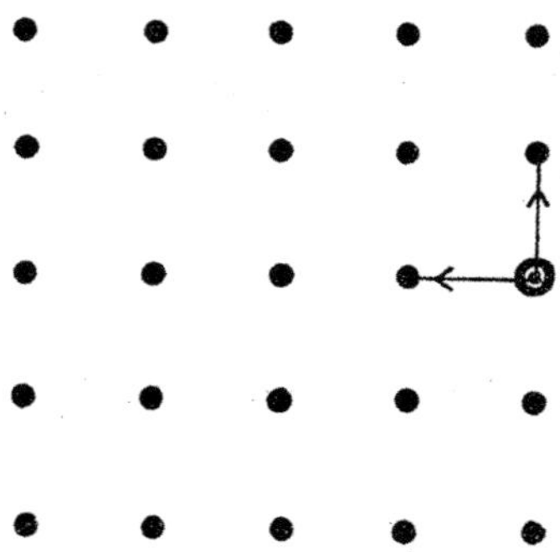

Fig. 6.

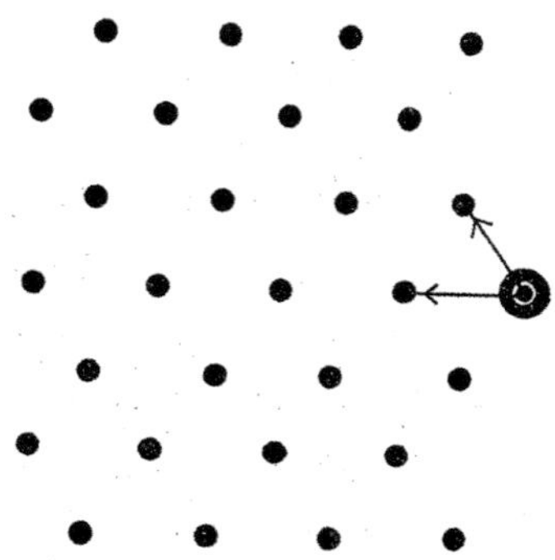

Fig. 7.

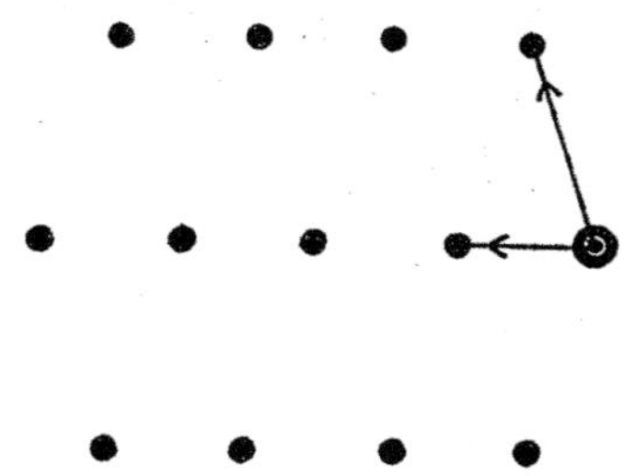

Fig. 8.

The majority of the 50 plates measured were of a regular type such as those shown in Plate 1, figs. 1–5.

While it is easy, after a little practice, to assign the correct indices to each spot of such a pattern by mere inspection, the plates were always carefully measured up, as one object of the research was to detect divergencies from theory. The degree of accuracy possible in these measurements varies greatly with the size of the spots. In some plates the spots were only about 1 mm. in diameter. These could be measured to an error of about $\pm 0{\cdot}1$ and the greatest error was in the determination of the zero which was uncertain to $\pm 0{\cdot}2$ mm. On other plates the spots were drawn out into short lines pointing to the origin, indicating that the electrons were not sufficiently homogeneous in wave-length. In some cases the spots were over-exposed and had spread owing to halation, in others they were rather fuzzy. It is probable that the last cases were due to insufficient resolving power of the crystal, the small blocks (fig. 3) not containing enough atoms in a cross section normal to the electron beam to give good resolutions. Spots about 2 mm. across are common. If 1 mm. of this width is due to the size of the crystal the width of the beam effectively used would be of the order of $0 \cdot 5 \times 10^{-6}$, which is of the same order as that found above for l. Thus blocks of crystal of this order of size would account for the observed effects. The measurements of the co-ordinates ξ, η agreed with the theory within the limits of error of measurements in the majority of cases. In some cases the values of ξ corresponding to successive orders of spots on the central line were not exactly in the ratio of the simple integers. This is almost certainly due to the error in fixing the origin (position of the undeflected beam) for it did not occur consistently.

The only consistent divergence from theory was found in the specimen of copper cut to a (111) face for which the η co-ordinates were systematically 2 per cent. larger than would correspond to the ξ values. The specimen in question was the first one prepared, and probably suffered more than the others in the process of cutting and grinding. It seems likely that the lattice had been slightly distorted by this process. After the surface had been etched several times it gave the normal ratios.

The angles between the lines of spots could be measured to about $\frac{1}{2}°$ in the best cases. The results were in agreement with the theory to this extent, except in the cases of the (111) specimen, where the discrepancy between ξ and η carried with it an error of the order of 1° as a necessary consequence.

When the crystal is rotated in azimuth the patterns repeat. This is easiest seen on the willemite screen where, for example, the square pattern of the (100) face of copper repeated at 90° intervals as nearly as could be measured (2°–3°). The more complicated patterns also recurred, thus, for example, for the (100) face of silver the following sequence was found from the photographs (which show up the more complicated patterns better than the screen) :—

Azimuth of crystal	Zero	42° / 44°	73°	87°	102°	133°
(α, β, γ) for incident ray	(001)	(011)	(031)	(010)	(03$\bar{1}$)	(01$\bar{1}$)
Calculated angle (to nearest degree)	Zero	45°	72°	90°	108°	135°

In some cases when the setting was not exact, the upper part of the photograph showed a pattern corresponding to one simple azimuth, and the lower part that of another near it.

No attempt was made to check the de Broglie law of wave-length over a wide range of voltages, which only varied between 30,000 and 60,000 volts, but within this range the values of the unit cube found from the pattern size were consistent. The mean of 32 plates gave " a " for copper as $3 \cdot 59 \times 10^{-8}$ against the X-ray value of $3 \cdot 61 \times 10^{-8}$, while for silver 10 plates gave $4 \cdot 05 \times 10^{-8}$ instead of $4 \cdot 079 \times 10^{-8}$. The relativity correction has been included in these figures. About half the plates (19) differed from the mean by $0 \cdot 03$ or less.

Exceptional Patterns.

Exceptions to the rules stated on p. 5 were of four kinds :—

(1) Isolated faint spots ; these were probably due to small pieces of broken crystal orientated at random. They were most common on the first specimen prepared.

(2) Cases in which portions of two regular patterns were superposed, generally in different parts of the field. This occurred when the setting was not quite right for either pattern ; in such a case the angle might be more nearly right for one pattern in part of the field and for another in the rest.

(3) Various spots not forming any simple pattern, but always having some regularity.

(4) A pattern, of which the origin did not form part, often not quite correct in angles and spacing, usually faint and superposed on a normal pattern, occasionally existing alone.

In all cases the radial distances of the spots gave possible indices without the use of "forbidden" combinations or half orders. The third case occurred most often when the crystal was cut to a face other than (111) or (100), *e.g.*, (331). In such a case the possible true patterns have their spots scattered sparsely, but, on the other hand, there are a number of points of the reciprocal lattice which lie *near* a plane passing through the face normal. All these cases were carefully analysed in the hope of finding some abnormality which the theory could not explain, but none was found. The general method was to assign indices to the spots, using their radial distances and their relative positions. In most cases several lay on a line, often equally spaced or nearly so ; the difference of indices between adjacent spots is then constant. Also the angles between the lines joining two pairs of spots is, at least nearly, that between the planes denoted by the difference of indices of the members of a pair. After assigning indices the test is to see if a plane can be drawn through the reciprocal lattice such that the points corresponding to the spots all lie within a small distance of it. The ideal method would be to use least squares. It is near enough in practice to take the plane through two spots near the centre of the group.

To take a particular case, illustrated in Plate 1, fig. 6. The face was approximately (331), and there is a faint spot 1a on the centre line which agrees with this. The nearest bright spot 1 corresponds to $\sqrt{(j^2+k^2+l^2)} = \sqrt{11}$. The other five gave distances in proportion to $\sqrt{40 \cdot 8}$, $\sqrt{55}$, $\sqrt{83 \cdot 5}$, $\sqrt{116}$, $\sqrt{146}$. These numbers are too large to be sure of to a few units, so we take the differences (1–3) and (1–2) which give $\sqrt{28}$ and $\sqrt{11}$. The nearest possible indices to $40 \cdot 8$ are $6^2+2^2+0^2$, to 28, $5^2+1^2+1^2$ or $3^2+3^2+3^2$, and for 11 we have $3^2+1^2+1^2$. We assign (311) to spot 1 so we must permute the indices or change signs. A possible arrangement is to take (1–3) as (151) and (1–2) as $(3\bar{1}1)$. The angle between these is $93\frac{1}{2}°$, the measured value is 93° which checks. The indices of spots 2, 3 are (602) or $\sqrt{40}$ and (462) or $\sqrt{56}$ in reasonable agreement with measurement. The distance (3–5) is again $\sqrt{11}$ and (5–6) = (1–2). We assign accordingly (884) or $\sqrt{144}$ to 6 and (593) or $\sqrt{115}$ to 5 in reasonable agreement with the measured radial distances. The angle between (5–3) and (3–4) should then be that between (131) and $(3\bar{1}1)$, *i.e.*, $\cos^{-1}(1/11) = 85°$, the measured value is 85°. Now let us see if these points are possible as one pattern, *i.e.*, if they lie near a plane. The plane through (462) and (753) has direction cosines proportional to $(4, 1, \overline{11})$. The angular distance between this plane and the radii vectors to the points of the reciprocal lattice, corresponding to the indices we have taken for the other spots, is never more than about 3°, which is well within the tolerance observed in the regular patterns.

Several cases were observed of a pattern resembling the main pattern of a regular plane but displaced so that no simple line passed through the origin and generally slightly distorted. When superposed, as it sometimes is, on a similar true pattern, it looks at first sight like a halving of the regular pattern. This, however, is not so. The spots all obey the law of radial distances and correspond to possible indices, but the angles are wrong and the corresponding points of the reciprocal lattice do not lie on a plane. As before, however, a plane can be drawn through two which *nearly* contains the others. The projection on this plane causes a distortion of the angles between the lines of spots. The divergence of the lattice points from the plane is sometimes greater than that given in the above example, reaching 11° in the worst case, but is usually only 4° or 5°. In such cases this pseudo-pattern contains all the points of the reciprocal lattice nearest to the assumed plane within the field covered by the main pattern. There is, of course, no essential difference between this case and case 3, except for the frequent presence of the regular pattern in addition, but the pseudo-patterns associated with regular planes are themselves unusually regular. This is to be expected if they are caused by a plane near a very simple one. In those cases when a true pattern and a pseudo-pattern are superposed the presence of the latter is presumably due to a small portion of the crystal having a somewhat different orientation from the rest. In one case it seems likely that a minute piece was slightly curved.

Nature of the Surface.

Microscopic examination of the etched (100) face of copper showed that the most prominent features were a number of square-sided pyramidal pits. In other cases ridges appeared. Our explanation requires the surface to consist of a series of lumps, through which the electrons can pass, of the order of 10^{-6} in linear dimensions. If the surface has etched into pits, and if on a sub-microscopic scale the pits are very close together, the intermediate portions can be regarded as projections and will serve the purpose of our explanation. It seemed of interest to try if the surface after etching could be made smooth, especially as it should be possible* to measure the inner potential of the metal if the electrons passed out from the same surface by which they entered, but this hope was not realised.

Electro-deposited Surfaces.

With this object copper was deposited electrolytically on the etched surface of single crystals of copper with (100) and (111) faces, and one with a face cut at random. The bath contained 10 gm. copper sulphate and 10 gm. sulphuric acid in 50 c.c. of water. The current density was low, about 4 milliamps. per square centimetre for 1 hour, but as the surface was irregular in shape this could only be estimated roughly. Experiments with X-rays† have shown that in some cases copper when deposited from solution continues the crystal structure of the original surface. The results showed that :—

(1) The spacing of part at least of the added substance is larger than that of normal copper but the structure is roughly similar.

* "Wave Mechanics of Free Electrons," p. 95.

† Wood, 'Proc. Phys. Soc.,' vol. 43, p. 138 (1931).

(2) The deposited substance generally, but not always, follows the orientation of the original crystal.

(3) Certain spots occurred which would correspond to " forbidden " planes of copper.

The first is shown best in the photograph of Plate 2, fig. 1. Here are two sets of points corresponding to similar structures but of different scale. One is the same as the original copper, which perhaps shows through the added layer in some places, the other is a structure of larger spacing.

The spacing of the latter structure was about $4 \cdot 5 \times 10^{-8}$. Both sets of points showed a discrepancy between the ξ and η values, the latter being disproportionately large, especially in the case of the new structure. It is difficult to say if much importance is to be attached to this. The spots are unusually diffuse, they are not very exactly arranged, and in some cases appear double. Probably the portions of crystal causing them are small, giving poor resolving power, and not quite exactly aligned with the underlying structure or with each other. The structure causing the new spots, for which the effect is most marked, may not be strictly cubic but stretched along the normal to the surface.

The forbidden spots may be seen faintly on the plate at half the distance up and down of the main spots. They correspond to planes (120), (231) and (342), but since the structure from its large spacing is certainly not normal copper their appearance is no contradiction of the theory. As it seemed possible that the new structure was due to hydrogen absorbed by the copper, I tried omitting the sulphuric acid from the electrolyte. The same spacing still occurred, however, though the spots were less good and the surface had a purple tinge. Heating *in vacuo* to a very dull red turned the pattern to rings, some of which corresponded to Cu_2O. It is possible that the deposited copper had reacted chemically with the electrolyte while it was being washed and dried. If so, the normal structure may be due to part of the deposit which had remained unchanged, and not to the original crystal showing through.

When the (100) surface was used the patterns were less extensive, but in one or two cases definitely of the (111) form with the large spacing. The spots on the ξ axis, which were always present, were in the same positions as before, corresponding to the (111) plane of the new structure. In addition one or two strong spots corresponding to the original (100) surface could be seen, and one photograph had a (100) pattern of normal spacing. One photograph also, difficult to interpret, suggested a structure of the new spacing, but arranged in conformity with the underlying (100) plane. One may conclude that the deposited copper followed the original structure in some parts of the specimen but not in others, part, at least, of the deposited substance being of the large spacing.

When the copper was deposited on a surface cut at random, the patterns showed spots due to the underlying structure unchanged in position and spacing. In addition there are short arcs due to the deposited substance. Some of these arcs lie roughly on the same radius as a spot, and have a radial distance proportional to that of the spot. They show that at least, in some places, the new structure has followed the orientation of the old but with a bigger spacing. Other arcs show no obvious relation to the underlying structure. They may be, and in some cases almost certainly are, due to planes forbidden in the case of ordinary copper, while in other cases they are due to crystals which do not follow the direction of the axes of the original specimen. The spacing from these plates is about $4 \cdot 4 \times 10^{-8}$. Silver deposited on a (100) surface followed this surface with the normal spacing.

Lines and Rings.

In contrast to rocksalt, the metal crystals usually show no sign of the Kikuchi lines. After repeated etching and grinding, however, a few appear, but they are never so strong or so sharply defined as those of rocksalt (see below). They are usually short and generally pass near to, or through, a spot. They are not sufficiently sharp for accurate measurement, but when they pass through a spot are roughly perpendicular to the line joining it to the origin. It seems likely that the double scattering required for the production of the lines does not have room to occur in the small lumps which usually constitute the metal surface. As the etching continues the small pits may be expected to coalesce, leaving the surface in larger lumps which may have been protected by some minute particles of insoluble impurity, *e.g.*, of grinding material. It is noticeable that the photographs which do show lines have generally only a few strong spots, which, however, are still part of a regular pattern. These photographs approximate in appearance to those given by rocksalt.

Rings.

Some of the photographs with silver show faint rings, apparently due to traces of the metal in a disorientated state. One or two of the copper photographs have faint rings not due to the metal, and probably caused by a compound formed on the surface by air or an impurity in the etching liquid.

Low Voltage.

A few photographs were taken at about 7000 volts with an Evershed and Vignolles generator. In one case there was a faint spot.

Experiments with Rocksalt.

Some preliminary experiments were made on rocksalt in 1929.* They showed nothing abnormal and were not continued at the time. Later, a paper by v. Laue and Rupp† described the frequent occurrence of reflections by "forbidden" planes and of half order spectra, from rocksalt and a number of other ionic crystals. Although certain unexplained effects had been observed before,‡ these new divergencies from the theory were much more startling, and the experiments were therefore continued in the summer§ and autumn of last year to see if similar effects could be found with high-speed electrons.

The difficulty with experiments of this kind on an insulator is that the specimen is liable to charge up electrostatically. This happens with rocksalt to a marked extent when the incidence of the rays on the crystal is made very small. At larger angles it is not apparent, but may still be taking place to some extent. I tried several devices to diminish the charging. Bringing a red hot spiral near the surface of the crystal was effective, but also destroyed the diffracted spots, which reappeared faintly on cooling. Bombardment by electrons from a hot filament made the charging worse. The charging sometimes seemed to grow worse with time, and it seemed possible that the crystal became a better insulator as it lost moisture in the vacuum. However, making the crystal into a thin plate covering a cell containing water was no very definite improvement. A beam of ultra-violet light focussed on the crystal with a quartz lens had no effect. It seemed that the natural blue salt was rather better than the white, and the great majority of the experiments were done with it, but there is a good deal of difference between different specimens, and as I had no numerical measure of the effect, do not attach much weight to the difference between blue and white.

In all cases a cleaved surface (100) was used. There was a considerable fatigue effect in the production of diffraction spots. Specimens cleaved more than 2 or 3 days, and kept in a desiccator in the interval, were inferior to a fresh surface. When in use, the fatigue from any small portion of the surface is very rapid. If the crystal is set so as to give a bright spot on the willemite

* "Wave Mechanics of Free Electrons," p. 161.

† von Laue and Rupp, 'Ann. Physik,' vol. 4, p. 1047 (1930).

‡ Rupp, 'Ann. Physik,' vol. 3, p. 497 (1929).

§ 'Nature,' vol. 126, p. 55 (1930).

screen, the spot will become faint in a few minutes. By turning the crystal in azimuth the spot can be brightened again, probably because another part of the surface is used. In this way, observations can be continued for some hours. There appears to be a slight recovery if the crystal is left overnight in the vacuum, but it is by no means complete. Care was therefore taken to withdraw the crystal from the beam when not actually in use. One specimen could be made to give four or five good photographs.

Diffraction Patterns.

The structure of rocksalt is similar to that of copper and silver, but the patterns observed were markedly different in appearance. Rocksalt never gave a wide pattern containing a number of spots, only a few being visible at once. Alteration of the angle of incidence which had little effect on the metals, changed the intensity of the rocksalt spots markedly and produced a quite visible shift in their position. Further, the photographs with rocksalt show strong Kikuchi lines (see Plate 2). The difference is due to the difference in surface. The rocksalt cleavage is really smooth, except possibly for steps, and the humps we have had to postulate for copper and silver are absent. We get, therefore, conditions like those in X-ray work. Each spot occurs, in theory, for only one particular setting of the crystal plane from which the (monochromatic) beam is reflected to form the spot. The shift of the spot as the crystal is moved is due to the limited resolving power of the few surface layers of atoms to which alone the electrons penetrate.

Centre Line Spots.

The brightest spots seen on the willemite screen are generally those in the plane of incidence of the rays. These "centre line spots," due to reflection from the cleavage face (100), can be followed for several orders by altering the angle of incidence of the face. They are best investigated by visual measurement on the fluorescent screen. If x is the displacement in centimetres and P the voltage of the rays, the following values were found for different specimens of blue rocksalt :—

Date.	$x\sqrt{P} \times 10^{-1}$.					
June 13	45·3	65·4	85·0	101	118	—
,, 24–25	46·4	66·1	85·0	104·5	122·5	137
,, 26	47·6	62·6	83·7	104	121	140
December 3–4	—	65·8	84·1	104	123	146
Mean	46·4	65·0	84·6	103	121	141
n'	5·24	7·33	9·55	11·7	13·7	15·9
Φ *volts*	10·1	12·1	10·7	8·5	9·4	—

Except in the last column each number is the mean of a large number of readings. In this table n' is the number found by dividing the spacing calculated from the value of $x\sqrt{P}$ into $5 \cdot 63 \times 10^{-8}$, the (100) spacing of NaCl. If the simple theory held, the values should be even integers. It will be seen that they tend to approach this law as n' increases. The considerable difference of the three lower orders is well beyond the experimental error. It may be due to a charging of the crystal, but it would require a positive charge. This is not absolutely impossible, since the secondary electrons leaving the crystal might exceed in number those striking it; yet since when charging is strong it is of negative sign, this explanation seems unlikely. Also one would hardly expect the results with different crystals to be as constant as they are, since the conditions varied considerably from time to time. It seems more likely that this is an effect of inner potential, and the potentials required to account for the effect are calculated in the last line from the formula

$$\Phi = \frac{150n^2}{4a^2}\left[1 - \left(\frac{n'}{n}\right)^2\right],$$

where a is the cube side and n is the integral order corresponding to n'. The values of Φ show tolerable agreement when it is remembered that they involve the difference of two quantities, which is small for the higher orders.

Depth of Penetration of the Rays.

Unlike the spots given by the metals, those due to rocksalt show an appreciable change in position as the angle of incidence is altered by turning the crystal. The cleavage surface probably contains a large number of atoms in regular order, and the wavelets from these can only reinforce if the angle of reflection is very exactly equal to that of incidence. There will only be one angle of deviation which exactly satisfies Bragg's law

$$2d \sin \theta = n\lambda \text{ for a given } \lambda.$$

But if either λ is variable or some latitude is allowed in Bragg's law there will be a possibility of shift. The first will give a range

$$\pm\, \delta\theta = \pm\, \theta \,.\, \delta\lambda/\lambda,$$

where $\pm\, \delta\lambda$ is the range of wave-length. The second case will arise if, as is probable, the number of layers affecting the electrons is limited by the penetrating power of the latter. If we denote by N the number of layers which would give a range $\pm\ \delta'\theta$, supposing each layer equally effective, then by the usual argument of resolving power

$$\mathrm{N} \,.\, 2d \sin(\theta + \delta'\theta) = (n \,.\, \mathrm{N} + 1)\lambda$$

giving $\mathrm{N}d = \lambda/2\delta'\theta$, taking $\cos\theta$ as unity.

We shall call Nd the effective depth of penetration.

By measurements on the willemite screen I found the distance over which the centre of the spot could be moved and still remain just visible. To find $\delta'\theta$ we must subtract $\delta\theta$ from this. If the main beam is deflected magnetically, since the deflection varies as the wave-length, the spot will spread into a short spectrum whose length, for a given average λ and given deflection, is equivalent to $\delta\theta$. Since the homogeneity of the rays varies with the potential of discharge, this experiment must be made at the same potential as the measurement of shift, but allowance can easily be made if the deflection is not the same. It was found that at the highest voltages the apparent width of the deflected beam was actually less than that of the undeflected, probably because it was much fainter, as the magnet interfered with the passage of the rays through the diaphragms defining the beam. Since the spots were fainter still, this width was taken as representing the geometrical size of the beam and was subtracted from the widths found at the lower voltages to give $\delta\theta$.

The value deduced for the effective depth of penetration is $1 \cdot 2 \times 10^{-7}$ cm. for rays of 35,000 volts. The corrected shifts due to finite resolving power were of the order of 2 mm. The probable error is large, perhaps as much as 25 per cent.; no definite variation with voltage was found. It will, of course, be understood that we are dealing with a very oblique grazing angle (*circa* 4°); the normal penetration would be much larger. Since the faster the rays the smaller the angle, the effective depth would not be expected to vary rapidly with the speed of the rays.

Photographs.

About 10 successful plates were taken showing spots and Kikuchi lines and carefully measured to look for abnormalities such as "forbidden" planes. The theory of the occurrence of spots in such a case as this is as follows. Let the principal axes of the crystal be taken as axes of co-ordinates, OZ normal to the cleavage face. Let the incident ray OC meet the plate in C (fig. 9) and (p, q, o) be the indices of the zone axis nearest to the projection of OC on OXY. Then if (jkl) is a plane giving rise to a spot it must pass through (pqo) since the angles are small, *i.e.*, $pj + qk = 0$. Take C as the origin of co-ordinates $\xi\eta$, the axis of ξ being parallel to OZ. Let the zone axis (pqo) meet the plate in A, the plane OXY meet it in AB and the plane (jkl) in AD. Draw CN parallel to AD. By the law of reflection, if a spot occurs due to (jkl) it must be at P $(\xi\eta)$, on CN produced such that CN = NP, but it will only occur if Bragg's law is obeyed. The first condition shows that P lies on a circle through C with centre A.

If, as usual, $\xi\eta$ are measured in terms of the unit $a/L\lambda$, then as in case 2, p. 9, we have $\xi = 1$, $\eta = \sqrt{(j^2 + k^2)}$ but (jkl) are integral and $pj + qk = 0$,

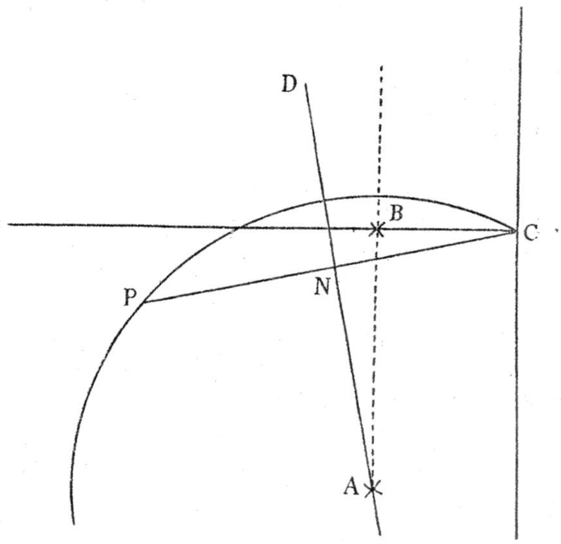

FIG. 9.

hence ξ and $\eta q/\sqrt{(p^2 + q^2)}$ must be integral, and we shall expect a spot at each point of the circle for which this is the case.

In practice the distortion of the crystal allows a little latitude in the fulfilment of the condition. The photographs which show bright spots satisfy these two rules ; an example is shown in Plate 2, fig. 4. The co-ordinates of the spots in the above units are (9·90, *o*) ; (12·1, 3·99) ; (12·15, 5·98). In the case of spots for which ξ is small, the co-ordinates on most of the plates are less than even integers, showing the refractive index effect in fair agreement with the willemite measurements, but as the setting was not necessarily that to make the spots brightest, the shift considered in the last paragraph prevents them being used to find Φ.

Kikuchi Lines.

The plates all show well-marked lines of the kind studied by Kikuchi for mica, usually showing both white and black lines, though the latter are more prominent. The distance between the white and black lines is equal on Kikuchi's theory to the deflection of a spot reflected from the corresponding planes. This is confirmed here, and the indices so assigned have been checked by measurement of the angles between the lines which should be those between the corresponding planes. The agreement is good to a degree or two, which is about the accuracy of measurement, the lines being often fuzzy. They should theoretically be affected by refraction if this exists. In fact if $\mu^2 - 1 = \Phi/P$ is small, each point (ξ, η) of the line is moved towards the axis (if Φ is positive) to a position (ξ', η) where $\xi'^2 = \xi^2 - \Phi L./P$. The effect gets small if the line is far out, as most of the clear ones are.

An effect noticed by Kikuchi* for mica is shown in Plate 2 (fig. 2). Here the incident beam is in a principal plane, and corresponding to this plane we have a broad strip of blackening of width corresponding to the distance between black and white lines, which themselves do not appear. Other black lines near it are also accompanied by similar strips, the white lines being faint or invisible. This effect has not so far been explained.

It shall be noticed that, as Kikuchi's theory shows, and as is observed in the experiment, there may be several black and white lines parallel to a plane and corresponding to different orders. The distance between a pair of black lines is *half* that between the spots of two adjacent orders due to the same plane. At first sight these look rather like half-order spectra or the "forbidden" odd orders of planes such as (100). The vertical lines on Plate 1, fig. 3, are due to this plane.†

Discussion.

Two points in the above work need further remarks : the complete absence of forbidden reflections and half orders, and the large positive potential of rocksalt.

The clearest and most certain cases of half-order reflections have been found by Rupp‡ for the (111) face of nickel and the (110) face of iron and molybdenum. These reflections occur only if the metal has been exposed to hydrogen, and can be destroyed again by heating *in vacuo*. They are undoubtedly to be explained as a pentration of the metal lattice by the hydrogen atoms in such a

* Kikuchi, 'Phys. Z.,' vol. 31, p. 783 (1930).

† *Note added in proof.*—It is worth notice that when two Kikuchi lines meet at a small angle one curves into the other, which weakens before the point of tangency. The effect is shown in the centre of fig. 3, Plate 2. It seems likely that the explanation requires a complete treatment of the diffraction problem on the lines suggested by Morse.

‡ Rupp, 'Ann. Physik,' vol. 5, p. 453 (1930).

manner as to make alternate planes different, without appreciably altering the spacing. This explanation may also account for various cases of faint reflections by forbidden planes which have been reported, if they are not to be assigned to experimental error. The case of copper is more complicated. We have seen that no such effect occurs at high voltages. Rupp* finds half-order spectra from the (100) face which apparently depend on hydrogen, but even when these are removed the "forbidden" third order spectrum persists. Farnsworth (*loc. cit.*), in a very careful and elaborate research, in which a number of azimuths are investigated, finds a large number of beams which cannot be accounted for on the simple theory, even allowing a variable Φ (7 to 30 volts).† In fact, in spite of the sharpness of the beams, his results bear only a qualitative resemblance to the requirements of the wave theory. Some of the extra beams can be assigned to half-order spectra, which in this case have a rather different significance from those in Rupp's work, since the incidence was always normal and not at an angle equal to that of reflection. In his second paper‡ he finds these beams diminished by very intense heating. Some other beams are apparently connected with the presence of hydrogen. I think that the experiments of p. 14 have a bearing on this work. They show that a structure widely different in spacing from the original copper can yet follow it in orientation. Thus beams will appear in the same azimuths as those due to copper, and will show the same fourfold symmetry round the (100) axis. Copper is easily attacked chemically, and moderate heating, such as Farnsworth used in his original work, may, if the vacuum is not quite perfect, cover the surface with a layer of a compound. Heating strong enough to vaporise the copper would probably remove the layer if the supply of gas was limited. This may also account for Rupp's result. In a later paper Boas and Rupp§ find a large number of forbidden planes and half-order spectra for polycrystalline tungsten, using the method of a fixed collector and variable λ. Here a compound is unlikely ; the higher orders might be accounted for by a different assignment of indices or choice of Φ, but the occurrence of the first order reflections from the (100) and (111) planes in two experiments is a definite challenge to the existing theory.

* *Loc. cit.*

† I wish to take this opportunity of withdrawing the criticism of Farnsworth's method of calculating Φ expressed on p. 101 of "Wave Mechanics of a Free Electron." As he has pointed out to me, the calculation is correct if the observed *voltages* are used to find Φ. If the observed *angles* are used the values found are markedly different.

‡ Farnsworth, 'Phys. Rev.,' vol. 35, p. 1131 (1930).

§ Boas and Rupp, 'Ann. Physik,' vol. 7, p. 983 (1930).

In the case of ionic crystals Rupp* reports a number of cases. The complexity of the photographs in Plate 2 shows the great difficulty of interpreting correctly experiments in which electrical observations are made, confined to a single plane of incidence and reflection, such as would correspond to exploring one of these photographs along a single line. It is not even certain in such experiments whether the actual surface is that of the original crystal or polycrystalline, as seems to have been the case for the cleaved and lightly polished surfaces of NaCl, KCl and KBr which gave reflections corresponding to planes other than the cleavage plane. In view of the theoretical importance of the point, it is very desirable that the diffraction by some ionic crystal should be measured over the whole field to see if there is really a fundamental difference between the behaviour of slow and fast electrons.

Rupp† finds —3·5 volts for the inner potential of natural rocksalt, and —2·8 for the same coloured by electron bombardment. He failed to get diffraction for the blue variety. The electrons were of the order of 100 volts energy in all cases. The large value, about 10 volts, found in the present work may be compared with 10·6 volts found by Yamaguti‡ for mica and 22 volts found by Emslie for calcite, who both used high-speed electrons. According to v. Laue's theory§ a cubic crystal should have a definite inner potential and this should be positive for NaCl. It ought to be possible to distinguish between a true inner potential, which involves a discontinuity at the surface, and a charge on the crystal as a whole, which would exert an electric intensity throughout the chamber enclosing it. The latter should produce much the same effect on all diffracted beams, while the former influences the lower orders of spectra much the most. We have seen that the different orders give values of Φ in fair agreement; this is good evidence that Φ has a real existence.

I wish to express my sincere thanks to Sir Harold Carpenter, F.R.S., for making the copper crystals for me, and to Mr. J. D. McKay and Mr. T. Riches for their most valuable assistance in working the apparatus.

Summary.

(1) Experiments are described on the diffraction of electrons of 30–50 kilovolts by single crystals of copper and silver. They show a widely extended pattern of diffraction spots.

* *Loc. cit.*, p. 21, also 'Z. Physik,' vol. 61, p. 587 (1930).

† *Loc. cit.*

‡ Yamaguti, 'Proc. Phys. Math. Soc., Japan,' vol. 12, p. 203 (1930); also quoted by Kikuchi, 'Phys. Z.,' vol. 31, p. 781 (1930).

§ v. Laue, 'Sitzungsberichten der Preuss. Acad. der Wiss.' (1930, II).

Thomson. *Proc. Roy. Soc., A., vol. 133, Pl. 1.*

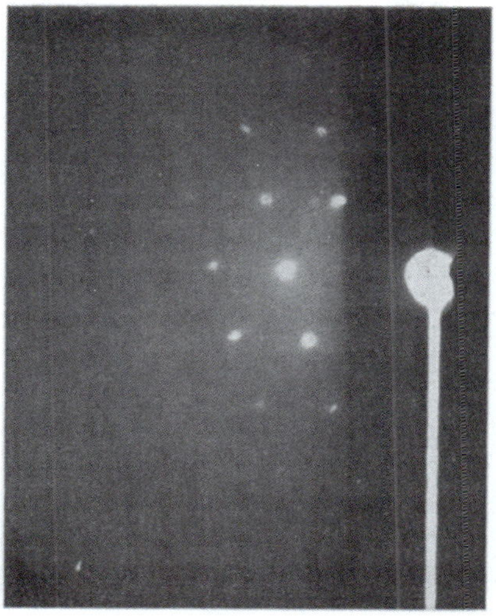

Fig. 1.—Copper (111).

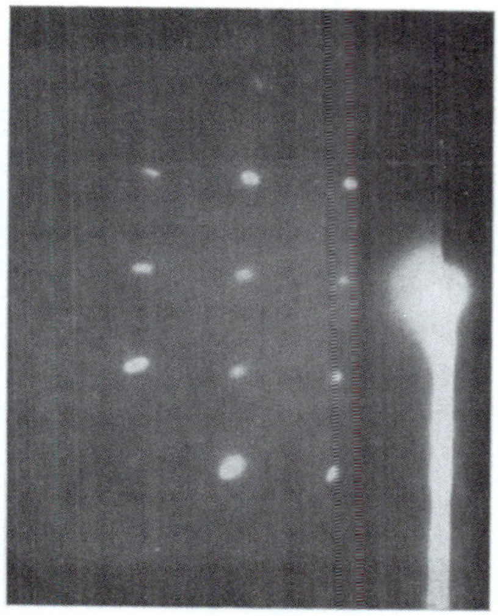

Fig. 2.—Copper (100).

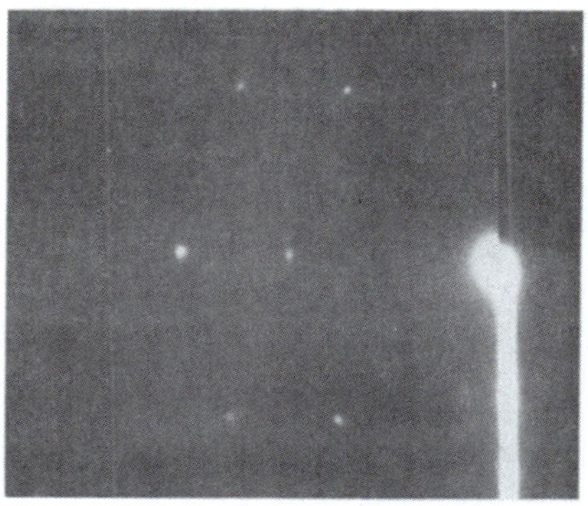

Fig. 3.—Copper (100).

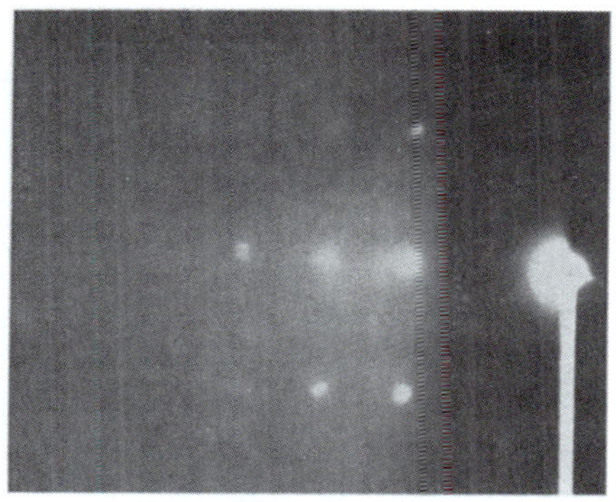

Fig. 4.—Copper (111).

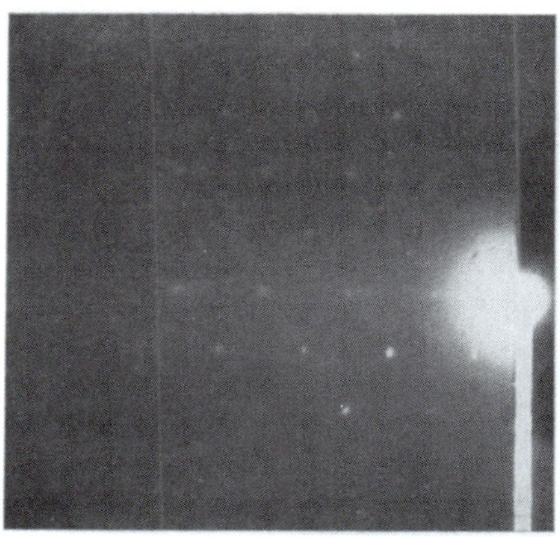

Fig. 5.—Silver (100).

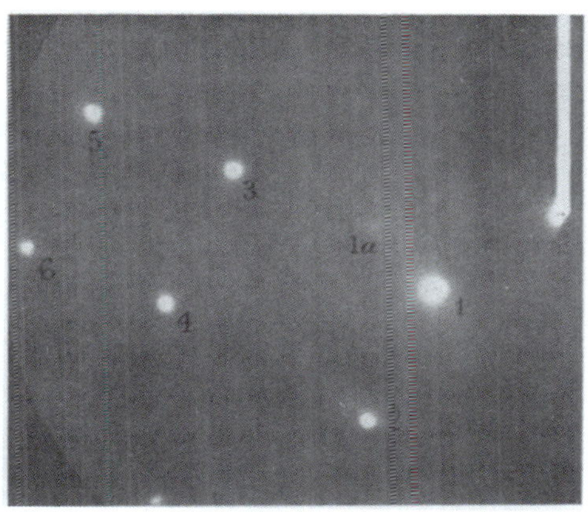

Fig. 6.—Copper (331).

Note.—The vertical white streaks were produced by withdrawing the crystal and gradually raising the plate. They help to define the origin of co-ordinates.

[*Facing p.* 24.)

Thomson. *Proc. Roy. Soc., A., vol.* 133, *Pl.* 2.

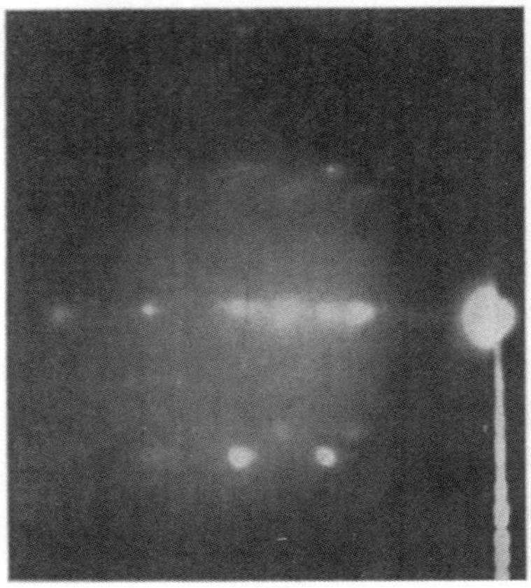

Fig. 1.—Electro-deposited copper.

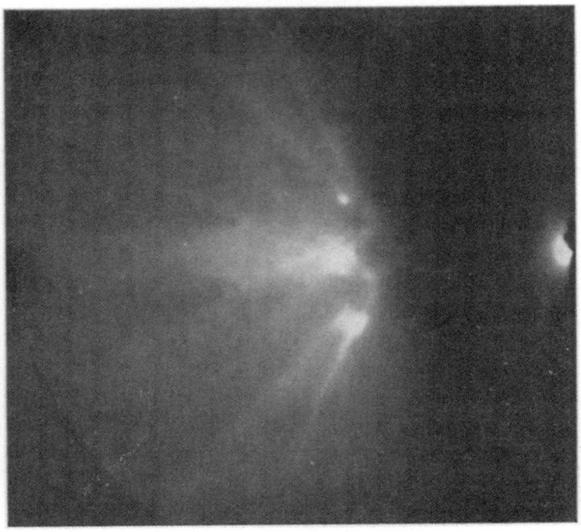

Fig. 2.—Rock salt.

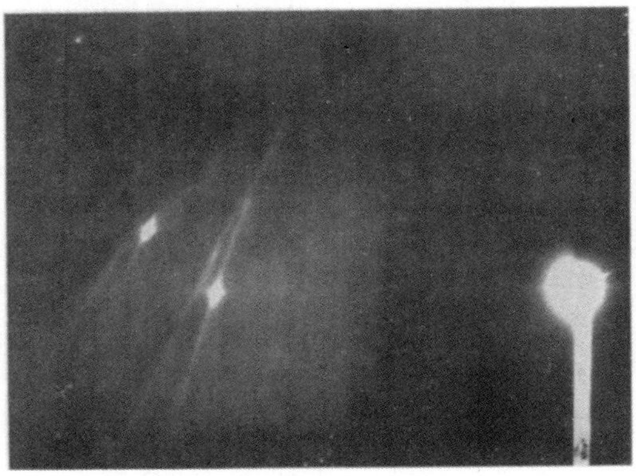

Fig. 3.—Rock salt.

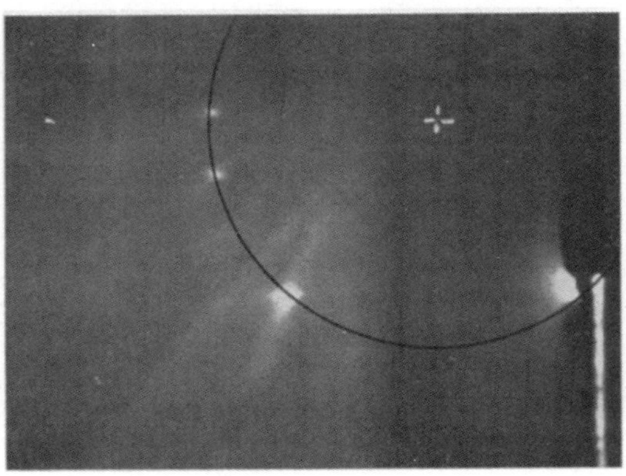

Fig. 4.—Rock salt.
(The cross shows the centre of a circle drawn through the spots.)

(2) The usual wave theory can be extended to account for these patterns by supposing that the etched surface of the crystal is composed of a number of lumps of dimensions of the order of 10^{-6} cm. These may in some cases be the portions left between adjacent etching pits.

(3) The nature of the surface prevents any effect of inner potential being detected. A detailed agreement exists between theory and experiment, even in the case of the more irregular " exceptional " patterns, the only discrepancy being probably due to a 2 per cent. distortion of the surface lattice in one specimen.

(4) No sign was found of reflection by " forbidden " crystal planes or of half-order spectra.

(5) Some experiments on copper electrolytically deposited on a single crystal of the same showed : (*a*) a strong tendency for the deposit to follow the underlying copper in orientation ; (*b*) a marked difference of spacing (4·45 instead of 3·61) in some cases and the appearance of spots not characteristic of the ordinary face-centred structure of copper.

(6) Diffraction spectra are produced from a cleavage face of rocksalt, and also well-marked Kikuchi lines.

(7) The spectra of different orders are not evenly spaced. This seems to indicate an inner potential of about +10 volts for rocksalt.

(8) The depth of penetration of the rays, under the conditions of oblique incidence used, is about 10^{-7} cm.

(9) With the exception of (7) the position of the spots and Kikuchi lines is in excellent agreement with the simple theory. In particular, there are no " forbidden " reflections or half-order spectra.

(10) In some cases the Kikuchi lines are represented by blackened strips. The cause of this is not explained.

(11) Explanations are suggested for some of the anomalies observed by other workers.

THE ISOLATION AND CRYSTALLIZATION OF THE ENZYME UREASE.

PRELIMINARY PAPER.

By JAMES B. SUMNER.

(*From the Department of Physiology and Biochemistry, Cornell University Medical College, Ithaca.*)

(Received for publication, June 2, 1926.)

After work both by myself and in collaboration with Dr. V. A. Graham and Dr. C. V. Noback that extends over a period of a little less than 9 years, I discovered on the 29th of April a means of obtaining from the jack bean a new protein which crystallizes beautifully and whose solutions possess to an extraordinary degree the ability to decompose urea into ammonium carbonate. The protein crystals, which are shown in Fig. 1, have been examined through the kindness of Dr. A. C. Gill, who reports them to be sharply crystallized, colorless octahedra, belonging by this definition to the isometric system. They show no double refraction and are from 4 to 5μ in diameter.

While the most active solutions of urease prepared in this laboratory by Sumner, Graham, and Noback[1] and by Sumner and Graham[2] possessed an activity of about 30,000 units per gm. of protein present, the octahedra, after washing away the mother liquor, have an activity of 100,000 units per gm. of dry material. In other words, 1 gm. of the material will produce 100,000 mg. of ammonia nitrogen from a urea-phosphate solution in 5 minutes at 20°C. At this temperature the material requires 1.4 seconds to decompose its own weight of urea.

The crystals, when freshly formed, dissolve fairly rapidly in distilled water, giving a water-clear solution after centrifuging from the slight amount of insoluble matter that is present. The solution coagulates upon heating and gives strongly the biuret,

[1] Sumner, J. B., Graham, V. A., and Noback, C. V., *Proc. Soc. Exp. Biol. and Med.*, 1924, xxi, 551.

xanthoproteic, Millon, Hopkins and Cole, ninhydrin, and unoxidized sulfur tests. The phenol reagent of Folin and Denis gives a strong color, while the uric acid reagent gives none. The material can be entirely precipitated by saturating with ammonium sulfate. The Molisch test is negative and Bial's test is negative also. The absence of pentose carbohydrate, as shown by Bial's

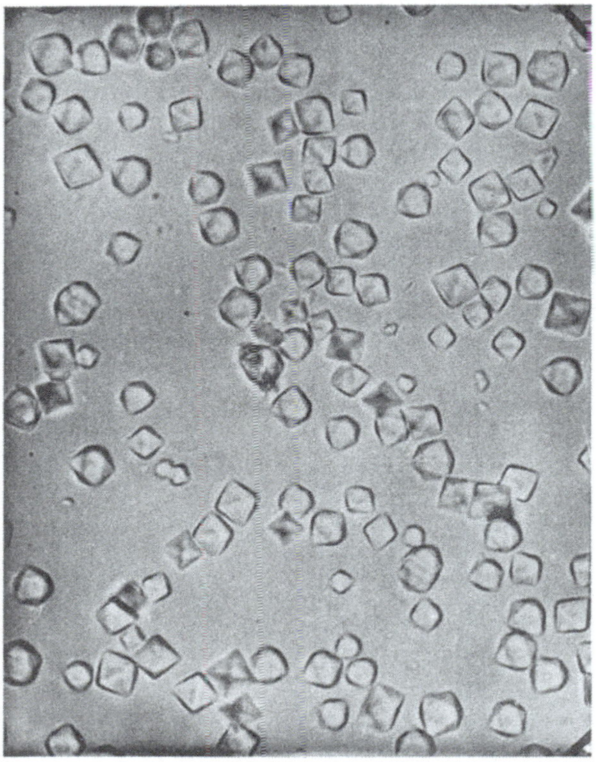

Fig. 1. Photomicrograph of urease crystals magnified 728 diameters.

test is especially pleasing as we have experienced a great deal of trouble in the past in freeing jack bean proteins from this substance.

The octahedral crystals, when freshly prepared, are very soluble in dilute alkali or dilute ammonia, and are either dissolved or coagulated by dilute mineral and organic acids, depending upon the concentration of acid. Even so weakly acid a substance as primary potassium phosphate is capable of causing an irreversible coagulation. Although the crystals dissolve in distilled water I am inclined to regard the material as globulin inasmuch as a precipitate is formed when carbon dioxide is passed into its solution and this precipitate immediately redissolves upon the addition of a drop of neutral phosphate solution.

Owing to the fact that I have not had large enough amounts of the material to work with I am unable to give accurate figures for its nitrogen content at the present time, but this can be stated to be not far from 17 per cent, as shown by micro-Kjeldahl determinations made on several preparations. The content of ash is certainly low, so low that a considerable amount of material will have to be used to obtain this figure. Determination of the enzyme activity of the crystals has been somewhat interfered with, owing to the fact that dilute solutions of the crystals produce less ammonia from urea than one would calculate from results obtained from more concentrated solutions. If this effect is real, rather than apparent, it may be due to the instability of the enzyme at great dilutions. When in concentrated solution the activity is not lost very rapidly, provided the material is kept in the ice chest.

When old the crystals are entirely insoluble in distilled water, salt solutions, and dilute ammonia. In this condition the enzyme activity is almost nil. I have made several attempts to purify the fresh crystals by a second crystallization but have never succeeded in obtaining more than traces of crystals and these have been insoluble in water and inactive.

It may be worth noting that practically all of our previous ideas concerning the nature of urease appear to be confirmed by the discovery of the octahedral crystals and by study of their properties. I undertook the task of isolating urease in the fall of 1917 with the idea that it might be found to be a crystallizable globulin, in which case the proof of its isolation would be greatly simplified. Other reasons for choosing urease were that the quantitative estimation of urease is both rapid and accurate, that urease can be reasonably expected to be an individual enzyme, rather than a mixture of enzymes, and that the jack bean appears to contain a very large amount of urease, if it is permissible to draw a parallelism between the urease content of the jack bean and the amounts of other enzymes found in other plant and animal materials.

In previous work in collaboration with Graham and Noback[1,2] and in unpublished work of my own it has been found that urease is very completely precipitated, together with the jack bean globulins, by cooling its 35 per cent alcoholic solution to −5 to −10°C., provided the reaction is sufficiently acid. We have found that urease can be precipitated by neutral lead acetate and neutralized cadmium chloride and that most of the urease can be re-extracted by decomposing the precipitate with potassium oxalate; that urease can be precipitated by tannic acid without very much inactivation and that urease can be rendered insoluble, with loss of a part of its activity, by the action of dilute alcohol or very dilute acid.

Although the literature contains numerous references to a co-enzyme of urease, I believe that no specific coenzyme exists. My evidence rests upon the fact that the loss of activity that occurs when the octahedral crystals are separated from a jack bean extract is almost exactly equal to the activity obtained when these crystals are washed with dilute acetone and then dissolved in water. If anything could separate an enzyme from its coenzyme crystallization might be expected to do so. The proteins in impure urease solutions doubtless exert a protective action as buffers and both proteins and polysaccharides may exert protective colloidal action.

I present below a list of reasons why I believe the octahedral crystals to be identical with the enzyme urease.

1. The fact that the crystals can be seen by the microscope to be practically uncontaminated by any other material.

2. The great activity of solutions of the crystals.

3. The fact that solvents which do not dissolve the crystals extract little or no urease and that to obtain solutions of urease one must dissolve the crystals.

4. The fact that the other crystallizable jack bean globulins, concanavalin A and B, carry with them very little urease when they are formed from solutions that are comparatively rich in urease.

[2] Sumner, J. B., and Graham, V. A., *Proc. Soc. Exp. Biol. and Med.*, 1925, xxii, 504.

5. The unique crystalline habit of the octahedra and their ready denaturation by acid.

6. The fact that the crystals are purely protein in so far as can be determined by chemical tests, combined with evidence from previous work to the effect that urease behaves like a protein in its reactions towards heavy metals, alkaloid reagents, alcohol, and acids.

7. The fact that the crystals are nearly free from ash and the fact that we have previously prepared solutions of urease that contained neither iron, manganese, nor phosphorus.

The method which I have used to obtain the crystals is extremely simple. It consists in extracting finely powdered, fat-free jack bean meal with 31.6 per cent acetone and allowing the material to filter by gravity in an ice chest. After standing overnight the filtrate is centrifuged and the precipitate of crystalline urease is stirred with cold 31.6 per cent acetone and centrifuged again. The crystals can be now dissolved in distilled water and centrifuged free from insoluble and inactive matter that has passed through the filter during the filtration. Of the urease extracted from the meal as much as 47 per cent may be present in the crystals. If one uses coarsely ground jack bean meal that has not been freed from fat the crystals are still obtained, but in traces only. I have carried out the process described above about fifteen times since first discovering the crystals and have always had success. The method is described in the experimental part of this paper in detail. It is probable that not all of the precautions that I have taken are necessary, but I think it best to give a description of the method exactly as it has been used by me.

Preparation of Urease Crystals.

Distill commercial acetone from a mixture of fused calcium chloride and a little soda-lime to remove water and acids. Place 158 cc. of the distillate in a 500 cc. graduate. Dilute with distilled water to the 500 cc. mark, mix, and cool to 22°C. Dilute exactly to mark and pour upon 100 gm. of "Arlco" jack bean meal which has been placed in a beaker. Stir for 3 or 4 minutes to break up the lumps. Now pour the material upon a 28 cm. Schleicher and Schüll filter, number 595. It is best to pour the mixture on the filter where there are four thicknesses and to add more only when the paper has become moistened higher up, as in preparing the Folin-Wu blood filtrate, but even if this is done the filtrate may come through not quite clear at first and in this case the first 50 cc. should be refiltered. When all, or nearly all of the material is on the filter place in an ice chest at 2–2.5°C. and allow to remain overnight. The next morning centrifuge off the crystals that are present in the filtrate, using cold 50 cc. centrifuge tubes and cold holders for the tubes. (I have centrifuged for 7 minutes at a moderate speed and then cooled the tubes and holders in the ice chest for 10 minutes before centrifuging more of the filtrate.)

When all of the crystals have been centrifuged off drain the tubes against clean filter papers to remove the last drops of the mother liquor. The crystals can now be stirred up with 5 to 10 cc. of ice-cold 31.6 per cent acetone and centrifuged again and the tubes again drained against filter papers. The crystals are now dissolved in from 15 to 40 cc. of distilled water at room temperature by stirring, and the solution is centrifuged free from insoluble matter. For examination under the microscope it is best not to separate the crystals from all of the original mother liquor as they are insoluble in this.

The Arlco jack bean meal referred to above is an extremely finely ground, fat-free meal obtained from The Arlington Chemical Company, Yonkers, N. Y.

Estimation of Urease Action.

Make a solution containing 3 per cent of urea, 5.4 per cent of K_2HPO_4, and 4.25 per cent of KH_2PO_4; preserve with toluene and place in a bath of 20.0°C. Pipette 1 cc. of the diluted enzyme solution into a test-tube that contains no adsorbed mercury and allow to remain in bath to reach its temperature. Now rapidly add to the enzyme 1 cc. of the urea-phosphate solution and mix. Allow the action to proceed for 5 minutes. At the end of this interval blow into the solution 1 cc. of normal hydrochloric acid to inactivate the enzyme and mix immediately. Wash the material into a 100 cc. volumetric flask, dilute, Nesslerize, dilute to mark, mix, and compare with a standard containing 1 mg. of ammonium nitrogen that has been prepared at the same time. If the color of the unknown is much lighter or darker than that of the standard the test must be repeated, using a more concentrated, or a more dilute enzyme solution, as the case may be.

This procedure can be used only with solutions of urease that are relatively pure. For example, urease prepared by extracting jack bean meal with 30 per cent alcohol gives results by this method that are about 3 per cent too low on account of the interfering action of the protein that is present. The protein can be removed by precipitation with potassiomercuric iodide containing hydrochloric acid and an aliquot of the filtrate can be Nesslerized in cases of this sort.[3]

Determination of Dry Weight of Urease in Solutions.

As the addition of absolute alcohol to solutions of the crystals did not precipitate all of the protein, which was probably on account of the lack of electrolytes, I was obliged to pipette the solution into a weighed glass dish and evaporate to dryness in a boiling water oven. The temperature of the oven was about 94°C. The dish and contents were then dried to constant weight in a desiccator over sulfuric acid.

Determination of Total Nitrogen.

As I have not yet prepared the material on a large scale the macro-Kjeldahl method was not used. The material was digested in a large Kjeldahl flask with a small amount of sulfuric acid and copper sulfate. After digestion the material was washed into a large glass tube and the ammonia aerated off and Nesslerized. It was found impossible to digest the material in a glass tube, such as is ordinarily used, as it foamed very badly.

CONCLUSION.

A new crystallizable globulin has been isolated from the jack bean, *Canavalia ensiformis*. From the reasons given elsewhere in this paper I am compelled to believe that this globulin is identical with the enzyme urease.

I wish to thank Professor A. C. Gill for examining the crystals and Professor S. H. Gage and Professor B. F. Kingsbury for photographing them.

[3] Sumner, J. B., *J. Biol. Chem.*, 1919, xxxviii, 59.

CRYSTALLINE PEPSIN

I. Isolation and Tests of Purity

By JOHN H. NORTHROP

(From the Laboratories of The Rockefeller Institute for Medical Research, Princeton, N. J.)

(Accepted for publication, May 9, 1930)

I

INTRODUCTION

Enzymes are in many respects connecting links between living and inanimate matter since their action is analogous to inorganic catalysts, although the enzymes themselves are found only in living organisms. As catalysts they increase the rate of one or more specific reactions and so act as directive agents for the reactions occurring in the organism. This directive property is undoubtedly essential for the existence of living cells. As a consequence of these properties the study of enzymes has been of interest to both chemists and biologists and has resulted in a great increase in the knowledge of their mode of action. The results of attempts to isolate the enzymes in pure form, however, have been singularly unsuccessful. There seems to be no convincing evidence that any enzyme has been obtained in the pure state; and only one, the urease described by Sumner (1), has been previously obtained in crystalline form. A number of methods have been found which allow the activity of an enzyme preparation to be increased almost indefinitely; at the same time, however, the preparation becomes more unstable and eventually the activity becomes lost.

In practically all the work the assumption has been made either explicitly or otherwise that the activity was a measure of the purity of the preparation and that any increase in activity was due to an increase in purity. This is not necessarily true. If the enzymes are analogous to inorganic catalysts then it is quite possible that the activity depends on the physical arrangement of the molecules or atoms (2). Evidence for this relation between the physical state and the activity was found by Fodor (3) in the case of the proteolytic enzymes of yeast, and by Kuhn and Wasserman (4) in the case of hemin. It is possible, on the other hand, that enzymes in general are of the type of hemoglobin (which might be considered an enzyme), and that they consist of an active group combined with an inert group. It might be possible under certain conditions to attach many more active groups to the inert group and so increase the activity above that of the original compound. Either of the above ideas would account for the well-known fact that crude preparations are much more stable than purified material and that the rate of inactivation of enzyme solutions practically always shows evidence of a mixture of stable and unstable forms (5, 6).

There is some reason to think, therefore, that enzymes exist in a more stable form for either physical or chemical reasons, and in view of the uniformly negative results which have been obtained in attempting to isolate the most active preparations it seemed advisable in attempting the isolation of pepsin to study the more stable as well as the most active fraction.

II

PRELIMINARY EXPERIMENTS

A number of methods have been proposed for the purification of pepsin, such as precipitation with safranin (7), etc., fractionation by various adsorbents, and precipitation by dialysis from acid solution (Pekelharing (8)). These and a number of other methods were tried and more or less active preparations obtained. The results with Pekelharing's method seemed the most encouraging, however, since the loss of activity was less and there was some indication that a constant activity was reached. This result has been reported by Pekelharing and also by Fenger, Andrew and Ralston (9) using a similar method. It was found, however, that the dialysis could be dispensed with and the process made more rapid and efficient by solution with alkali and subsequent precipitation with acid, after a preliminary precipitation with half saturated $MgSO_4$ or $(NH_4)_2SO_4$. The amorphous material so obtained contains about half the activity present in the original material and is 3 to 6 times as active as measured by the liquefaction of gelatin and about 5 times as active as measured by the digestion of casein or by the rennet action on milk.[1] Repetition of this procedure gave products of increasing activity as measured by the liquefaction of gelatin, and apparently this activity could be increased indefinitely. Several samples were obtained which were 100 times as active as the original preparation. They were also more unstable, so that each succeeding precipitation was accompanied by a larger and larger percentage loss until finally no more active material remained. This has been the fate of all previous attempts to isolate the most active fraction of a number of enzymes. When the activity of the various fractions was determined by the rate of hydrolysis of casein or by the rennet action on milk, however, it was found that the activity increased until it reached about 5 times that of the crude preparation

---

[1] *Determination of Activity.*—The rate of digestion of proteins by pepsin may be followed by determining the increase in carboxyl groups or amino nitrogen, or by the decrease in protein nitrogen or by the changes in viscosity. The increase in carboxyl groups probably represents the best measure of the progress of the reaction and has been used as the basis of the activity units used in this paper.

The most widely used unit of enzyme activity in general is that suggested by Euler (16) and is equal to the velocity constant of the reaction divided by the grams of enzyme. Theoretically this is undoubtedly correct since the activity of a catalyst can be expressed only as a velocity constant. Many enzymes do not give a velocity constant independent of the substrate concentration but instead the velocity constant decreases as the substrate concentration increases. This is especially true of pepsin and in addition the reaction does not follow any simple reaction rate. The end point of the reaction, which must be known in order to calculate the velocity constant, is also very difficult to determine. It is not practical therefore to use the velocity constant as a measure of activity of pepsin.

The activity of pepsin may conveniently be defined as the milliequivalents of carboxyl groups liberated per mole or gram of enzyme per minute at 35.5°C., optimum pH, and 5 per cent substrate concentration. Theoretically it would be better to use that substrate concentration at which the rate of reaction is a maximum, but experimentally this is difficult, owing to the high viscosity of such concentrated solutions. The enzyme concentration used should be in the range in which the activity is proportional to the enzyme concentration. In order to determine the activity in this way the rate of hydrolysis of casein, gelatin edestin, and denatured egg albumin at pH 2.0 to 2.5, was determined by means of the increase in formol titration (19).

The following abbreviations are used in this paper:

PU	=	proteolytic units = milliequivalents carboxyl groups per minute.
$[PU]_{gm.}$	=	" " per gram.
$[PU]_{gm.}^{cas.F}$	=	" " " " as determined by cas. F. method, etc.

and then remained constant instead of increasing as did the gelatin liquefying power. This was the result reported by Pekelharing and also by Fenger, Andrew and Ralston. This material appeared to be protein, as previous workers had found, and was reasonably stable. Efforts were therefore made to isolate this protein in crystalline form.

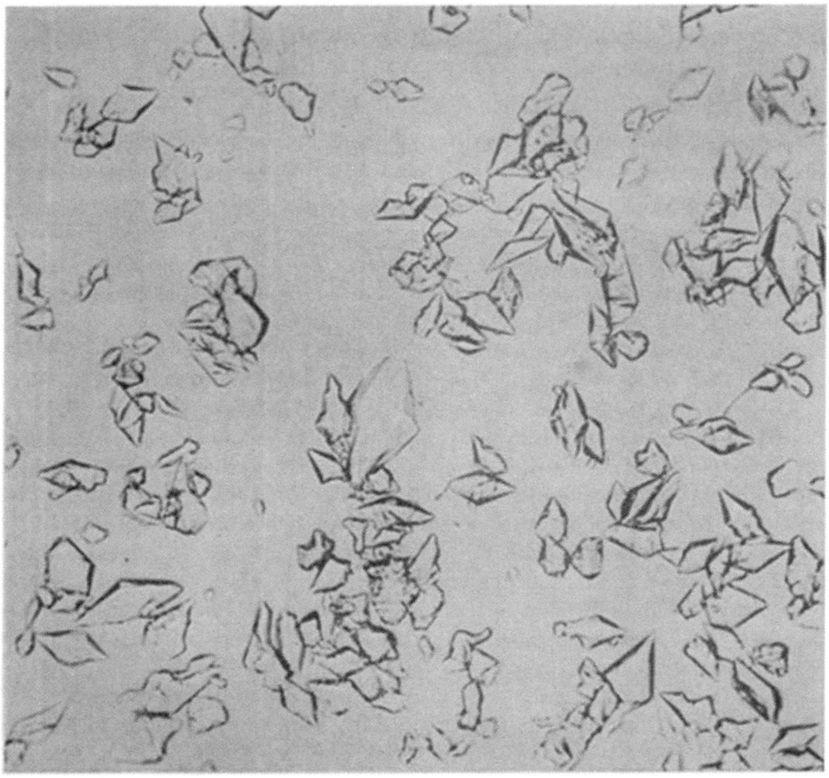

FIG. 1. Crystalline pepsin

III

Isolation of the Crystalline Enzyme

It was noticed that the precipitate which formed in the dialyzing sac when the procedure of Pekelharing was followed appeared in more or less granular form and filtered rather easily, as though it were on the verge of crystallization. This precipitate dissolved on warming the suspension and it was eventually found that it could be induced to crystallize by warming to 45°C., filtering, and allowing the filtrate to cool slowly. The crystals so obtained were regular hexahedra and showed a tendency to grow in clusters, especially when appearing from more acid solutions. They are remarkably similar to the urease crystals pictured by Sumner and differ only in that they have a hexagonal base while the urease has an octagonal base. On one occasion a few crystals with truncated pyramids were obtained. They had the same activity and optical activity as the usual form. The crystals showed positive double refraction and were optically active in solution. They possessed proteolytic activity, when dissolved, equivalent to 5 times that of the U. S. P. 1 to 10,000 pepsin as measured by hydrolysis of casein, and 2.5 times as measured by the liquefaction of gelatin.

Improved Method for the Preparation of the Crystals.—The isolation of the crystals in bulk by the above method was difficult owing to the dialysis. It was found that this could be avoided and the purification carried out as outlined above for the amorphous preparations except that the acid precipitate was dissolved at 45° C. in concentrated solution. On inoculation, this solution set to a solid paste of crystals. The following is an outline of the method as finally developed.

IV

General Properties and Analysis of the Crystals

The material prepared in this way has the general properties of a protein. It is coagulated by heat, precipitated by saturation of the solution with $MgSO_4$ or $(NH_4)_2SO_4$ and gives a strongly positive xanthoproteic test. The Molisch test is negative. The crude material contains a large amount of yellowish pigment which is removed with difficulty. It may be largely removed by reprecipitation with $MgSO_4$ and becomes less as the material is recrystallized. All the preparations however, give a slightly yellowish solution when dissolved although the dry crystals are pure white after several crystallizations. There is no relation between the activity and the color. Analysis of the material gave the results shown in Table II.

The crystals are difficult to dissolve after drying and are best kept under saturated $MgSO_4$ at 5°C. They are instantly inactivated by alkali in solution and lose activity slowly in acid solutions. The

TABLE I

Preparation of Crystalline Pepsin

Procedure	Activity per gm. dry weight Method			
	Gel. V.	Cas. S.	Rennet	E. A.
500 gm. Parke, Davis pepsin U. S. P. 1:10,000 dissolved in 500 ml. H_2O and 500 ml. 1 Normal H_2SO_4 added. 1,000 ml. saturated $MgSO_4$ added with stirring. Solution filtered through fluted paper (S. and S. No. 1450½) and then with suction. Filtrate discarded.	0.075	2.5	2.5	4.5
Precipitate 1. Wash twice with equal volume ⅔ saturated $MgSO_4$, filter with suction. Filtrate discarded.				
Precipitate 2. Stir with water to thick paste and M/2 NaOH run in until complete solution. (Great care must be taken to avoid local excess of NaOH. pH never more than 5.0.) M/2 H_2SO_4 added with stirring until heavy precipitate forms, (pH about 3.0), 3 to 6 hrs. at 8°C., filter with suction. Filtrate discarded.	.15	7.5	7.5	7.0
Precipitate 3. Stir with H_2O to thick paste at 45°C., M/2 NaOH added carefully until precipitate dissolves, (filter if cloudy and discard precipitate). Beaker containing filtrate placed in a vessel containing about 4 liters of H_2O at 45°C., inoculated and allowed to cool slowly, cooling should require 3 to 4 hrs. and heavy crystalline precipitate should form at about 30 to 35°C. Solution kept at 20°C. for 24 hours. Thick crystalline paste, filter with suction.	.15	10.0	10.0	8.0
Precipitate 4. Wash with small amount of cold H_2O and then with ½ saturated $MgSO_4$ and store under saturated $MgSO_4$ at 5°C.	.17	14.0	[14.0]	9.0
Filtrate. M/2 H_2SO_4 added to pH 3.0, amorphous precipitate filtered off and treat as Precipitate 3.				

Recrystallization

Method 1. Crystalline paste filtered with suction on large funnel so as to form a thin layer of crystals and washed 3 times with cold M/500 HCl. Filter cake stirred to a paste with ½ its weight of water, the suspension warmed to 45°C. and M/2 NaOH run in slowly with constant stirring until the precipitate dissolves (pH$<$5.0). M/2 H_2SO_4 is then run in until the solution is faintly turbid, a few crystals added and the solution allowed to cool slowly as before. A heavy crop of crystals should separate in about 24 hrs. The suspension is then warmed to 45°C. again and more H_2SO_4 added until the pH of the suspension is about 3.0. It is then allowed to cool slowly again, and filtered after 24 hrs. The crystals may be washed with M/500 HCl until free of SO_4.

Ammonium sulfate may be used in place of $MgSO_4$. Sodium acetate may be used in place of sodium hydroxide.

Method 2. Crystals dissolved with NaOH and treated as described for Precipitate 2.

inactivated material is digested by the remaining active material and a large amount of tyrosin crystallizes out. This process also occurs slowly in the ice box so that the crystals on standing become mixed with nonprotein material that is not precipitated by salt nor by heat. The crystals can be freed from this soluble material by thorough washing with H_2O or by recrystallization. When freshly prepared in this way 98 to 99 per cent of the nitrogen is precipitated from solution by heating rapidly to boiling at pH 3 with sulfuric acid and Na_2SO_4, by saturation with $MgSO_4$ or $(NH_4)_2SO_4$, by the addition of alkali

TABLE II

Analysis. Dried at 60° in vacuo for 24 Hrs.

Method	Dumas	Kjeldahl	Van Slyke							Precipitation with $BaCl_2$
	Total N	Total N	Amino N	C	H	Cl	S	P	Ash	$SO_4^{=}$
Per cent dry weight	15.5 15.3	15.15 15.30	0.80	52.3 52.6	6.66 6.64 6.70	0.23 .20	0.88 .82	0.078	0.40 .55	0

TABLE III

Activity and Composition of Various Preparations of Pepsin, Crystallized Once

Preparation	1	2	3–5	5–10	10–12	12–20	22
$[PU]^{gel. V.}_{gm.}$	0.17	0.16	0.18	0.154	0.16	0.14	0.17
Per cent N	15.2	15.1	15.1	15.3	15.2	15.0	15.1
$[\alpha]_D^{22°}$ pH 4.5	−70	−72	−74	−73	−74	−7[illegible]	−70

and subsequent neutralization, or by heating with 10 per cent trichloracetic acid.

Constant Activity of Various Preparations.—About 2 kg of the crystals have been prepared from six different lots of the commercial preparation during the course of this work and have all had the same percentage of nitrogen and the same activity within the experimental error. A summary of these properties of the various preparations is given in Table III. The preparations in some cases were combined at the final crystallization.

The relative activity with various substrates and various methods is shown in Table IV. The figures are the average of 6 to 10 determinations with different enzyme preparations. All crystalline preparations tested yielded the same result with these methods, but amorphous preparations frequently gave much higher activity as measured by gelatin hydrolysis and lower activity as measured by egg albumin,

TABLE IV

Activity of Various Enzyme Preparations

Enzyme	Substrate	Method	Activity. Milliequivalents per gm. per min. = $[PU]_{gm.}$	Milliequivalents per millimole or equivalents per mole per min. = $[PU]_{mole}$
Crystalline pepsin	Casein	Formol	14 ± 0.5	500
		Solution	[14]	
		Rennet action	[14]	
	Gelatin	Formol	.17 ± 1	6
		Viscosity	[.17]	
	Edestin	Formol	28 ± .8	1,000
	Egg albumin	Formol	9 ± .5	320
U.S.P. pepsin 1:10,000	Casein	Formol	2.5	
		Solution	[2.5]	
		Rennet action	[2.5]	
	Gelatin	Formol	.075	
		Viscosity	[.075]	
	Egg albumin	Formol	4.5	

while the activity as measured by casein hydrolysis remained constant. On recrystallization of these abnormal amorphous precipitates the activity always returned to the characteristic value for the crystalline material, shown in the table.

The activity compared to other enzymes is less than that reported by Sumner for urease, on a weight basis, and much less than for several other enzymes, but comparison should be made on the basis of activity per mole rather than per gram of enzyme.

V

Evidence Concerning the Purity of the Crystalline Material

Constant Activity and Composition on Repeated Crystallization.—The work reported above shows that a crystalline protein having proteolytic activity may be obtained in quantity from commercial preparations of pepsin. The question of interest is whether this proteolytic activity is a property of the protein molecule or of another molecular species associated with the protein. The usual criterion for the purity of a substance is constant composition and properties after repeated recrystallization. Unfortunately in the case of proteins the

TABLE V

Summary of Recrystallized Pepsin

Crystallization No. and color	Quantity of crystals	$[PU]^{gel. V.}_{gm.}$	$[PU]^{cas. S.}_{gm.}$	N	P	$[\alpha]^{22°}_{D}$ pH 5.0	Formol per gm.
	gm.			*per cent*	*per cent*		*ml. 0.1 NaOH*
1. Dark brown........	50	0.18	14.0	14.85	0.078	−70	14.0
2. Brown.............	22	.17	14.5	15.0	.080	−70	12.5
3. Yellowish...........	15	.15	15.0	15.14	.075	−68	12.5
4. Slightly yellow fine powder...........	9	.15	14.5	15.16	.076	−67.6	11.0
5. Nearly white........	5	.16	14.8	15.13	.077	−68	11.0
6. " "	3	.14	14.5		.073	−72	11.0
7. " "	.5	.17	15.0	15.15	.080	−72	12.0
Original preparation....		.075	2.5				

melting point, which is the most sensitive test of purity, cannot be used since proteins decompose before melting. Of the other properties the proteolytic activity, optical rotation, and percentage of nitrogen and of phosphorus were considered the most significant and these properties were determined on preparations obtained from a series of seven successive crystallizations. A summary of the results of this experiment is shown in Table V. There is no perceptible drift in any of the properties determined and therefore no indication that the material can ever be separated into fractions by crystallization under these conditions. This shows that the composition of the crystals is independent of the concentration and quantity of the solution from which they are formed.

Several systems could give this result (10). (1) A mixture of two or more substances present in amounts which are proportional to their solubilities. (2) A solid solution of such composition as to have a minimum solubility. This would correspond to a constant boiling point mixture of miscible liquids. (3) A solid solution of two or more substances having nearly the same solubility. (4) A pure substance.

The various possibilities noted above all depend on some definite relation between the composition of the mixture and the solubilities of the hypothetical components. If this relation were changed, the composition would be changed (except in the case of allotropic modifications). The possibility of such a relation would be greatly reduced if the recrystallization were carried out in a number of different solvents in which the solubility was different. It is quite possible that a mixture of constant composition might be obtained from one solvent just as constant boiling mixtures are frequently found, but it would be expected that the composition of the mixture would vary with different solvents just as the composition of constant boiling mixtures varies with the pressure. If it could be shown therefore that the material retained its constant proteolytic power and composition after recrystallization from a number of different solvents, it would be excellent proof that it was a pure substance. Unfortunately, however, it is not possible to crystallize the material from a series of solvents, but the same result may be obtained by studying the solubility in several solvents.

Results of the Solubility Determinations.—The values for the solubility obtained with different amounts of precipitate in various solvents are shown graphically[2] in Figs. 2 to 6. Since the solubility is found to be independent of the amount of precipitate this procedure does not affect the shape of the curve. The soluble nitrogen per milliliter of solution, the proteolytic activity per milliliter of solution as determined by the digestion of casein or gelatin ($[PU]_{ml.}$) and the optical rotation of the solution in a 1 dm. tube are plotted against the total nitrogen content per milliliter of the suspension. The values corresponding to solutions obtained from the undersaturated side are represented by solid dots and those corresponding to solutions obtained from the supersaturated side

[2] In order to save space the scale representing the total amount of nitrogen per milliliter of suspension has been changed to a logarithmic one in some of the curves after the amount has become large as compared to the amount in solution.

are represented by circles. The solid lines drawn for the curves representing the pepsin per milliliter and the optical activity were calculated on the assumption that the ratio of activity or optical activity to nitrogen was the same in the solution as in the original crystals. That

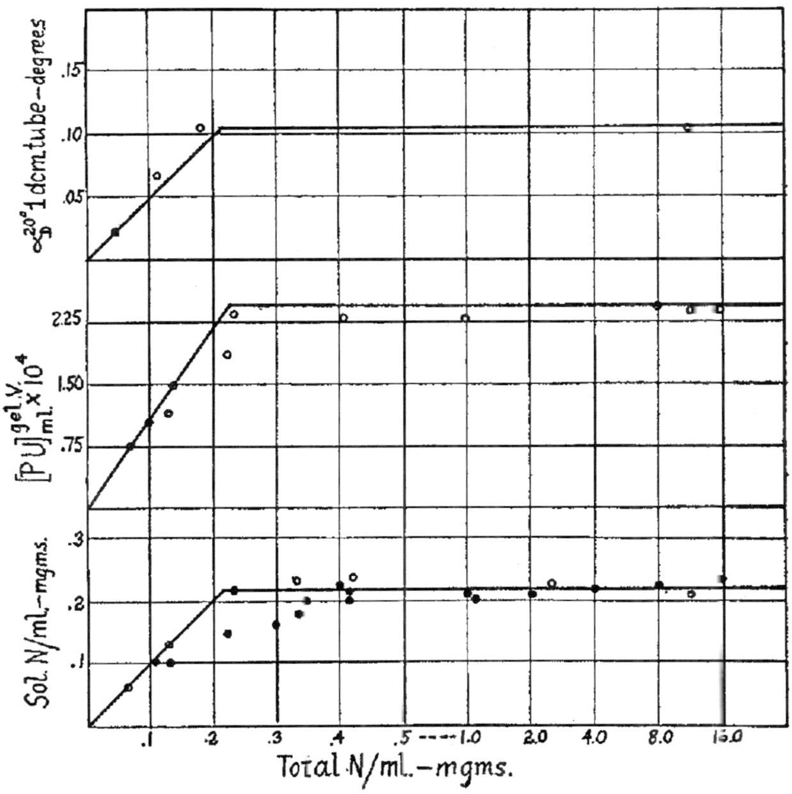

FIG. 2. Solubility of crystalline pepsin in 0.05 acetate and 0.5 saturated magnesium sulfate at 8°C.

is, 1.15 proteolytic units per gram of nitrogen as measured by the liquefaction of gelatin, 92 proteolytic units per gram of nitrogen as measured by the digestion of casein and a specific rotation per gram of nitrogen of −460° as measured at 22 to 24°C. with the sodium D line.

The results show: (1) that the solubility is independent of the quantity of solid present, (2) that the value for the solubility is an equilibrium value since it is the same whether obtained from supersaturated or undersaturated solutions, and (3) the specific proteolytic activity and the specific optical activity are the same for the material present in the solution as for the original material. The material behaves in regard to these determinations as a pure substance and there is no evidence of

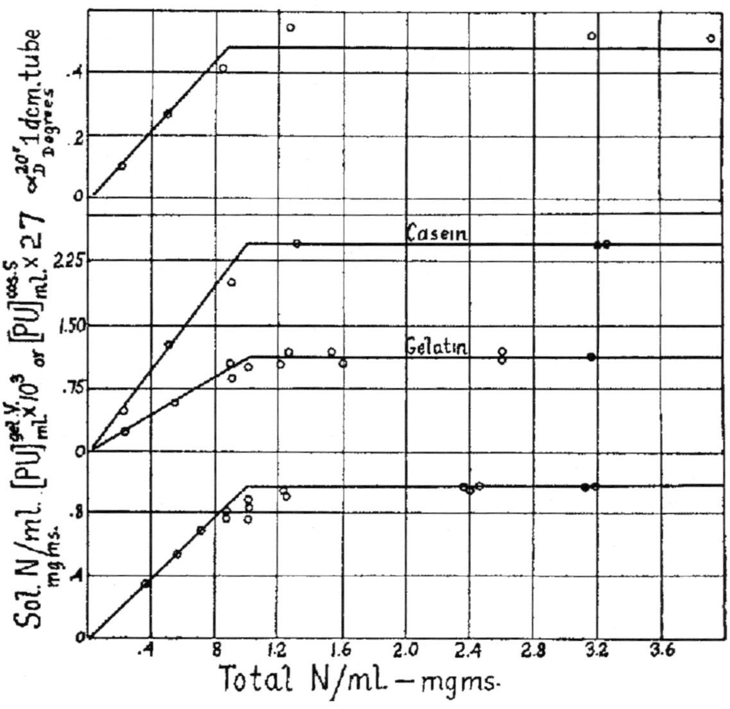

Fig. 3. Solubility of amorphous pepsin in 0.05 acetate and 0.5 saturated magnesium sulfate at 20°C.

mixture. They show conclusively that the activity cannot be ascribed to the presence of a minute amount of a very highly active material associated with the protein, unless it be further assumed either that this active material has about the same solubility as the protein or that it forms solid solutions with the protein. If the active material were soluble and *mixed* with the protein the activity per milliliter of solution would continue to increase as more and more solid material was added. While if it were present as a relatively insoluble material the activity per milliliter of the supernatant solution would become constant before the solubility of the protein had become constant. The figures show, however, that the proteolytic activity of the liquid becomes constant at the same point at which the concentration of the protein in the solution becomes constant. The only mixture which would behave in this

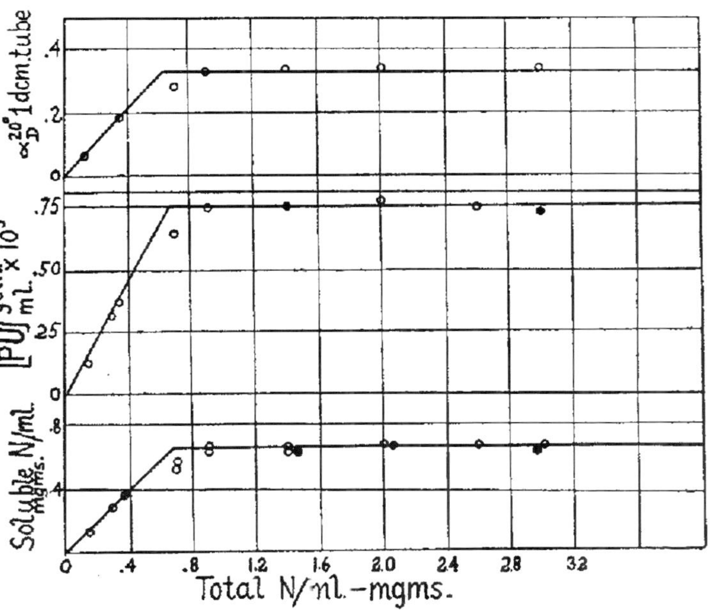

FIG. 4. Solubility of amorphous pepsin in 0.01 acetate and 0.5 saturated magnesium sulfate at 20°C.

way would be one in which the relative amounts of active material and protein were in almost exact proportion to the relative solubilities of the two substances. This might conceivably be the case in one solvent but the possibility that this relation would hold in several solvents appears so small as to be negligible.

There remains only the possibility that the active material forms a solid solution with the protein and is either present in small amount or has about the same solubility as the protein. If it were present in small amount and dissolved in the protein, its presence would not affect the total solubility curve. The ratio of the active material to the protein in the precipitate at the point where the precipitate first

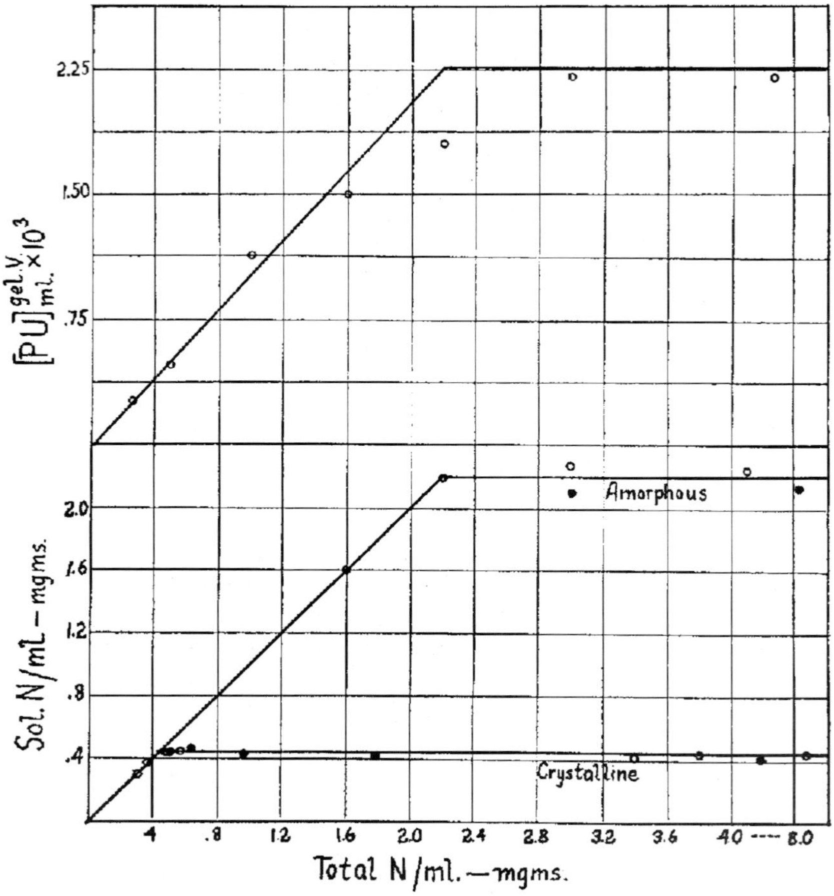

FIG. 5. Solubility of crystalline and amorphous pepsin in 0.055 acetate and 0.444 saturated magnesium sulfate at 20°C.

appears would differ, however, from the ratio in the original material. The ratio in the solution in that part of the curve where there is a large amount of solid present would also differ from the ratio in the original material. The figures show that this latter condition does not occur. The activity per gram of nitrogen in the solution in equilibrium with a

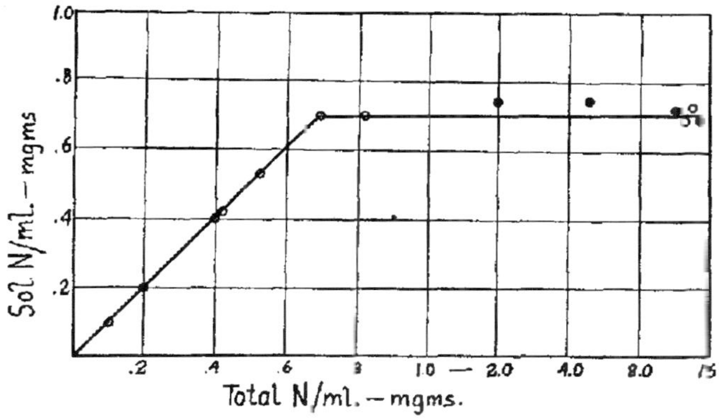

FIG. 6. Solubility of crystalline pepsin in 0.522 M sodium sulfate and 2.5 × 10^{-5} N sulfuric acid at 20°C.

TABLE VI

$$Ratio\ \frac{Activity}{Nitrogen} = [PU]^{\text{gel. V.}}_{\text{gm. N}}$$

Experiment	Pepsin	Solvent	In precipitate when nearly all soluble	In solution when large excess of solid present	In original crystals
26	Crystals	0.05 acetate pH 4.65, 0.50 sat. $MgSO_4$	0.75 1.10	1.00 1.09	1.11 ± 0.05
23	"	" " " " " " "	1.2	1.11	
22	Amorphous	" " " " " " "	1.20	1.16	
21	"	0.01 " " " " " "	1.09	1.02 1.13	
20	"	" " " " " " "	1.11	1.21	

large excess of precipitate is the same within the experimental error as the ratio in the original material. The activity-nitrogen ratio of the first precipitate to appear is also the same in the experiments with amorphous and crystalline material as is shown in Table VI. If old preparations which have not been freed from the decomposition products are used, the activity-nitrogen ratio at this point will be low. An indication of this is shown in the first figure in the table and in some of the figures.

The material is therefore either a pure substance or a solid solution the components of which have nearly the same solubility in all the solvents tried.

A summary of the solubilities in the various solvents is given in Table VII.

TABLE VII

Solubility of Pepsin

Preparation	Solvent	Temperature	Solubility N/ml.
			mg.
Crystals...........	0.522 M Na_2SO_4, 2.5 × 10^{-5} N H_2SO_4	20°	0.70
"	0.444 M Na_2SO_4, 0.0556 acetate buffer pH 4.65	20°	.43
Amorphous.........	" " " " " " "	20°	2.2
Crystals...........	0.05 acetate pH 4.65, 0.50 sat. $MgSO_4$	8°	.22
Amorphous.........	" " " " " " "	20°	1.00
"	0.01 " " " " " "	20°	.65

Inactivation.—It was pointed out by Pekelharing that the temperature coefficient for the inactivation of enzymes and especially of pepsin was extremely high and agreed with that for the denaturization of proteins. No other reaction known has such an enormous temperature coefficient and the agreement of the two values suggests that the enzymes are proteins. Pekelharing also showed that loss of activity in strong acid solution was accompanied by the appearance of denatured, insoluble protein. It can be shown that inactivation either by heat or by alkali is quantitatively proportional to the denaturization of the protein.

Inactivation by Alkali.—It is well known that pepsin is very sensitive to alkali. Goulding, Wasteneys and Borsook (13) showed that this inactivation could be separated into two reactions, one of which was instantaneous; and further that if the amount of this inactivation were plotted against the pH a titration curve with a pK of 6.85 was obtained. These experiments were repeated and the amount of inactivation compared with the amount of denatured protein found.

Experimental.—A series of 1 per cent solutions of the pepsin was made and adjusted to about pH 5. Increasing amounts of alkali were then added to these solutions so as to give a series of solutions varying in pH from about 5.5 to 9. An amount of acid equivalent to the amount of alkali in the tube, was added immediately after the addition of the alkali and the activity of the resulting solution determined after neutralization. 5 ml. portions were added to 5 ml. of a solution of half-saturated sodium sulfate and M/500 sulfuric acid. The unchanged pepsin is soluble under these conditions (to about 1 per cent) but the denatured protein is insoluble. The denatured protein therefore precipitates and the unchanged protein nitrogen is determined from the filtrate. In this way the percentage inactivation and percentage denaturization may be determined. The denatured protein is rapidly digested by the active enzyme so these solutions must be precipitated and filtered as soon as possible.

Control experiments with known mixtures of denatured and active pepsin show that very little activity is carried down with this precipitate if the activity is determined by hydrolysis of casein. If the activity is determined by hydrolysis of gelatin, however, the precipitate of denatured protein may be quite active; and this result is evidently due to the abnormal activity observed throughout in relation to gelatin hydrolysis. This complication may be avoided by determining the activity of the solution before precipitation.

The results of such an experiment are shown in Fig. 7. The inactivation of the pepsin is quantitatively parallel with the denaturization of the protein since the percentage of activity left is the same as the

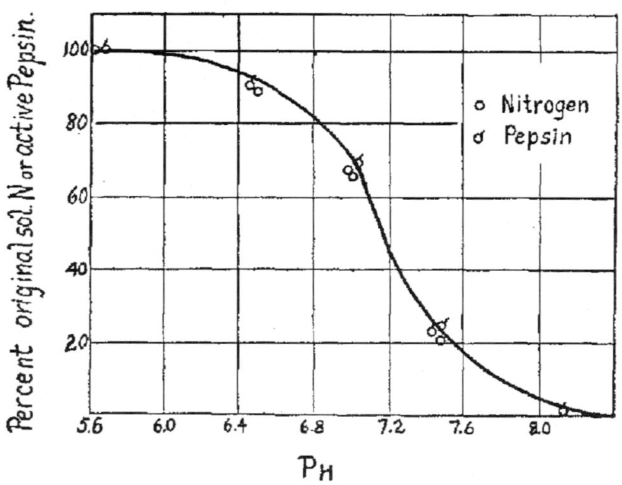

FIG. 7. Percentage inactivation and percentage denaturization of pepsin at various pH at 20°C.

percentage of soluble nitrogen remaining. The curve is very similar to that of Wasteneys and Borsook but is a little more to the alkaline side.

Inactivation by Heat.—Pekelharing noted that when pepsin solutions were allowed to stand in a range of acidity where the pepsin is active, the protein was digested and at the same time the activity of the solution decreased. He concluded from this that the pepsin digested itself. It can easily be shown, as was mentioned above, that the denatured protein is rapidly digested and it seems probable that the effect noticed by Pekelharing was the result of denaturization of the protein followed by hydrolysis. It is then unnecessary to suppose that the enzyme digests itself. Pepsin solutions when heated slowly or when maintained at a temperature at which the inactivation proceeds slowly, do not contain denatured, insoluble protein since the latter is hydrolyzed as rapidly as it is formed. Under these conditions inactivation of pepsin is accompanied by the appearance of soluble nitrogen instead of insoluble; and it is difficult to make a sharp separation of the original pepsin from the products of hydrolysis. The result of an experiment which shows the digestion of denatured pepsin by active pepsin is given in Table VIII in which mixtures of various amounts of active and denatured pepsin were made up at pH 3 and left at 35°C. for 24 hours.

TABLE VIII

Digestion of Inactivated Pepsin pH 3.0, 35°C.

Pepsin dissolved in HCl, pH 3.0 = Solution A
50 ml. A boiled, cooled = Solution D
A and D mixed in various proportions,
2 ml. sample of mixture taken and
2 ml. M/1 Na_2SO_4 added, filtered,
N/ml. determined in filtrate.

A	20	10	5	0
D	0	10	15	20
Mg. soluble N in filtrate after:				
0 hrs.	.30	.11	.07	0
20 "	.30	.30	.28	.01

The denatured pepsin is precipitated by M/1 Na_2SO_4 at pH 3, while the active pepsin is soluble. The table shows that the inactive and insoluble pepsin is dissolved by the active pepsin.

It may be shown, however, by means of solubility experiments, that the loss of inactivation in the solution is exactly paralleled by the disappearance of the original protein. If a saturated solution of pepsin in M/300 hydrochloric acid is placed at 65°C. the activity is rapidly lost so that only about half the original activity remains after 15 minutes. No precipitate occurs but the formol titration increases. If the solution is now cooled and stirred again with a large quantity of crystalline pepsin more of the crystals must go into solution to replace the material which has been changed. If the inactivation of the enzyme is accompanied by some chemical change in the protein then the amount of material which will dissolve must be proportional to the percentage loss in activity. That is, if half the activity is lost the increase in solubility must be equal to half the original solubility (assuming the solubility to be unaffected by the presence of the inactivated enzyme). On the other hand, the activity per milliliter of the solution after being stirred with the solid should return to its original value. The result of an experiment designed to test this assumption is shown in Table IX. The solution was heated for 15 minutes at 65°C. and in that time the activity per milliliter had been reduced to 50 per cent of its original value, whereas the formol titration

TABLE IX

Inactivated Pepsin and Solubility. 65°C.

	$[PU]^{gel. V.}_{ml.} \times 10^3$	N/ml.	Formol/ml.
		mg.	*ml. 0.01 NaOH*
M/300 HCl, stirred 1 hr. with pepsin, 20°C. Centrifuge	0.315	0.30	0.22
Supernatant heated at 65°C. for 15 min.	.160	.30	.33
Pepsin crystals added and stir 1 hr. Centrifuge			
Observed	.325	.45	.45
Calculated	.315	.455	.44

had increased 50 per cent and the total N per milliliter was unchanged since no precipitate appeared. The solution was then cooled and stirred at 20°C. with more crystals. The suspension was then filtered and the supernatant analyzed. The activity per milliliter is now the same as that of the original solution while the nitrogen per milliliter is 50 per cent higher than the original value. The formol titration has been increased by 50 per cent of its original value also. The results show that the loss of activity is exactly paralleled by some chemical change in the protein molecule.

The Rate of Inactivation at 65°C.—It has frequently been found that the rate of inactivation of enzymes does not follow the course of a simple monomolecular reaction but becomes slower as the reaction proceeds. It was found by the writer (6) that this peculiarity was connected with the purity of the enzyme solution and that the rate of inactivation of more highly purified solutions agreed with the monomolecular reaction rate, while those solutions containing inhibiting substances did not. On this basis the rate of inactivation of the crystalline material would be expected to be monomolecular. The result of an experiment in which the rate of inactivation of a 0.06 per cent solution of pepsin in 0.001 M hydrochloric acid (pH 3) at 65°C. is shown in

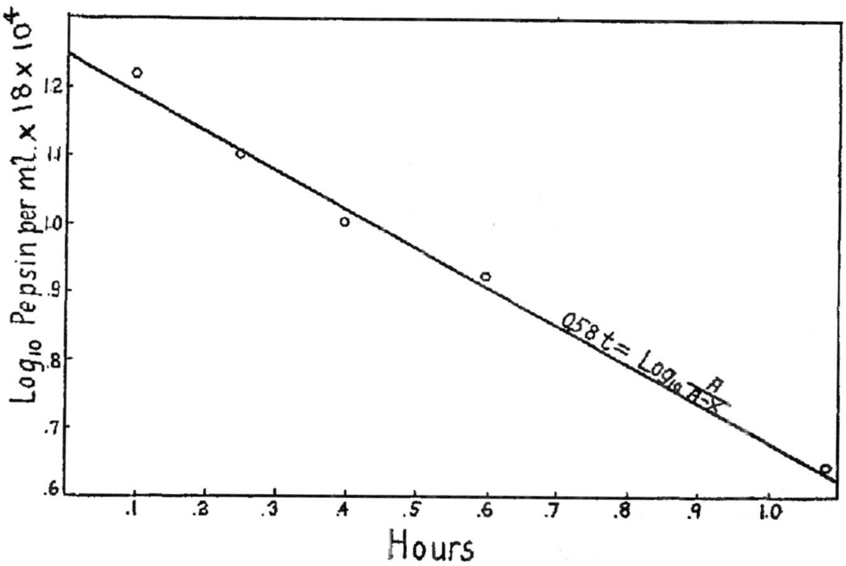

FIG. 8. Rate of inactivation of 0.06 per cent pepsin in 0.001 M hydrochloric acid at 65°C.

Fig. 8. The reaction follows the theoretical monomolecular course quite closely.

Identity of the Diffusion Coefficient of the Protein and of the Active Material.—The preceding experiments have shown that the activity cannot be separated from the protein molecule either by solubility measurements or inactivation experiments. It might be assumed, however, that the crystalline substance was analogous to a double salt and that the active material and the protein were in equilibrium with each other but were separate in solution. If this were the case it would be expected that the rate of diffusion of the two would be different. The diffusion coefficients were therefore determined by the method previously described (14). The experiments were carried out at 8°C. and the amount of material diffusing was determined by analyzing for nitrogen and also by activity determinations. The results of this experiment are shown in Table X. The diffusion coefficient was found to be the same when determined by either nitrogen or by activity. The radius (and presumably the weight) of the molecule containing the nitrogen and of the active molecule must therefore be the same, so that there is no indication that the protein molecule is different from the molecule responsible for the activity even in solution. The diffusion coefficient is found to be 0.048 ± 0.0005 cm.[2] per day which, according

TABLE X

Diffusion Coefficient, Radius of Molecule and Molecular Weight

8°C.	1 per cent solution in M/1 acetate buffer pH 4.6, viscosity of buffer = 0.017.		
D	from activity determination	=	.047 ± .0005 cm.²/day
	from N determination	=	.0468 ± .0005 cm.²/day
		=	5.44×10^{-7} cm.²/sec.
r	= 2.22×10^{-7} cm.		
	= 2.22 mμ		
M	= 37,000 ± 1,000		

to Einstein's formula, gives a molecular weight of 36,000. This figure agrees very well with the molecular weight of 37,000 as determined by osmotic pressure measurements. The agreement between these two figures shows that the molecules must be nearly all the same size since the molecular weight gives a mean value while the diffusion experiment as carried out gives a minimum value. The agreement also shows that the molecule is not hydrated to any great extent, since the value used for the specific gravity in calculating the molecular weight from the diffusion coefficient is the specific gravity of dry protein; and the value for the molecular weight so obtained agrees with that found by osmotic pressure.[3]

[3] Preliminary experiments (15) gave larger diffusion coefficients corresponding to a molecular weight of about 10,000, but this result was found to be due to an error in calibrating the diffusion cell.

Antipepsin.—If the enzyme is a protein an antibody should be formed on injection into an animal. The production of such antienzymes has been reported in the literature but the results are uncertain, as Euler has pointed out (16), because of lack of information in regard to the activity of the preparation used. In order to see if any antibody could be produced and if its production were associated with the activity of the preparation, two rabbits were immunized, one with an active preparation of the enzyme and one with a preparation of the same concentration which had been inactivated by alkali. 5 ml. injections of a 1 per cent solution of the active or inactive material were made intra-

TABLE XI

Antiserum

1 ml. 1 per cent pepsin solution in M/5 4.6 acetate buffer, diluted with buffer, + 0.15 ml. serum.

Pepsin dilution		1	1	1/2	1/4	1/8	1/16	1/32	1/64	1/128	1/256	1/512	1/1,000	1/2,000	0
Pepsin	Serum	0 serum	.15 ml. serum												
Active	Active	−	++	++	++	++	++	+	+	+	+	+	+	+	−
"	Denatured	−	++	++	++	++	+	−	−	−	−	−	−	−	−
"	Normal	−	−	−	−	−	−	−	−						
Denatured	Active	−	++	++	−	+	+	−	−	−	−	−	−	−	−
"	Denatured	−	++	++	++	+	−	−	−	−	−	−	−	−	−
"	Normal	−	−	−	−	−	−	−	−	−	−	−	−	−	−

peritoneally at intervals of about 1 week. Four injections were made; the rabbits were bled 2 weeks after the final injection and the serum obtained in the usual way. These sera gave a precipitin reaction with the pepsin solution, as shown in Table XI. Normal serum gave no precipitate with either the active or the inactive pepsin preparation. The serum obtained from the rabbit injected with active pepsin gave a positive precipitin reaction in a dilution of about 1/2,000 with the active pepsin solution and to about 1/16 with the inactive solution. The serum from the rabbit immunized with denatured pepsin gave a precipitin test with the denatured preparation in a dilution of about 1/8 per cent solution and with the active preparation to about the same extent. Both sera gave precipitin reactions with a 1 per cent solution of pepsin inactivated by boiling in a dilution of about 1/16. As might be expected, therefore, the sera were not strictly specific but gave cross precipitations with both active and denatured pepsin. The inhibiting effect of the sera on the activity was tested by mixing increasing amounts of the antiserum with a small amount of active pepsin and determining the rate of digestion of casein or gelatin with this mixture. The result of this experiment is shown in Table XII.

TABLE XII

Effect of Antiserum on Digestion of Casein or Gelatin by Pepsin

Serum diluted as noted + M/5 4.6 acetate buffer. Pepsin in 4.6 acetate buffer. 1 ml. pepsin solution + 1 ml. serum dilution.

Gelatin

Serum	Final dilution of serum					
	1/4	1/8	1/16	1/32	1/64	0
	$[PU]^{gel. V.}_{ml.}$ after 3 hrs. $\times 10^{3}$					
Normal	0.075	0.10	0.11	0.12	0.11	0.12
Antidenatured	.060	.10	.09	.10	.11	.12
Antiactive	.045	.075	.10	.10	.11	.12

Casein

Serum	0	Normal	Anti-denatured	Antiactive
Soluble N in filtrate, mg.	2.2	2.0	1.87	1.87
$\simeq [PU]^{cas. S.}_{ml.}$	.015	.008	.007	.007

Both the serum prepared by the injection of the active pepsin and that prepared by the injection of the denatured pepsin inhibit the action of the enzyme to about the same extent when tested by the solution of casein (cas. S. method). The normal serum inactivates somewhat less although there is no very great difference. It is difficult to compare the action of the various sera quantitatively from this experiment. The effect may be seen more clearly by determining the rate of hydrolysis of gelatin (gel. V. method) with a small amount of pepsin to which has been added increasing amounts of the various sera. The result of such an experiment is also shown in Table XII. The active serum inhibits quite strongly in a dilution of 1:4, the denatured serum less so and the normal serum still less. This inhibiting effect is evidently very weak compared with that of many antitoxins, although the precipitin reaction is about what would be expected. This may be due to the fact that the pepsin injected must be almost instantly inactivated since, as was shown above, the pepsin is instantly more than half inactivated at a pH of 7. It is probable therefore that the active pepsin is in the circulation of the animal for only a very short time.

The writer is indebted to Dr. F. S. Jones and Dr. A. P. Krueger for advice and assistance in connection with these experiments.

VI

Conclusions as to the Purity of the Preparation

The preceding experiments have shown that no evidence for the existence of a mixture of active and inactive material in the crystals could be obtained by recrystallization, solubility determinations in a series of solvents, inactivation by either heat or alkali, or by the rate of diffusion. It is reasonable to conclude therefore that the material is either a pure substance or a solid solution of two very closely related substances. If it is a solid solution of two or more substances it must be further assumed that these substances have about the same degree of solubility in the various solvents used, as well as the same diffusion coefficient and rate of inactivation or denaturization by heat. It must also be assumed that both substances are changed by alkali at the same rate and to the same extent. This could hardly be true with the possible exception of two closely related proteins. It is conceivable that two proteins might be indistinguishable by any of the tests applied in this work. But in this case it would follow that the enzyme itself was a protein and this, after all, is the main point. It does not necessarily follow even if the material represents the pure enzyme that it is the most active preparation that can be obtained nor that it is the only compound which has proteolytic activity. There is some evidence that the activity of the preparation may depend on its physical state as is known to be the case with the catalytic activity of colloidal metals. It is possible, on the other hand, that hemoglobin is the type structure for the enzymes and that they consist of an active group combined with a protein as suggested by Pekelharing. The active group may be too unstable to exist alone, but it is quite conceivable that a series of compounds may exist containing varying numbers of active groups combined with the protein, and that the activity of the compound would depend on the number of these active groups. This hypothetical complex would not differ much from that assumed by Willstätter (18) and his coworkers, except that it supposes a definite chemical compound with the protective group in place of an adsorption complex. It is of course possible that both types of complex may be formed under suitable conditions. The reactivation of enzymes as reported in the literature also suggests their protein nature since the conditions for this reactivation are similar to those found by Anson and Mirsky (20) to be suitable for the formation of native from denatured protein. The fact that the crystalline urease prepared by Sumner is also a protein and that the temperature coefficient for the rate of inactivation of enzymes in general is that characteristic for the denaturization of proteins, suggests that the protein fraction in the purification of enzymes be given special attention even though it may not be the most active fraction.

The crystalline pepsin resembles the amorphous preparations obtained previously by Pekelharing, Ringer, Fenger, Andrew and Ralston and other workers. It is probable that the preparations obtained by these workers were nearly pure pepsin. Both Pekelharing and Ringer (17) obtained preparations free from phosphorus so that there may be several proteolytically active forms.

Pekelharing showed that the same protein material could be obtained from gastric juice as from autolyzed gastric mucosa so that it is probable that the crystalline material could also be readily prepared from gastric juice. It seems fair to conclude therefore that the crystalline protein described in this paper is identical with the enzyme pepsin as secreted by the animal.

VII

Proteolysis Is a Homogeneous Reaction

Sørensen has shown that a solution of egg albumin is a one-phase system and the solubility measurements described in this paper show that a solution of pepsin consists also of only one phase. The reaction between dissolved egg albumin and dissolved pepsin therefore is a case of homogeneous, rather than heterogeneous catalysis.

SUMMARY

A method is described for the preparation of a crystalline protein from commercial pepsin preparations which has powerful peptic activity. The composition, optical activity, and proteolytic activity of this protein remain constant through seven successive crystallizations. No evidence for the presence of a mixture or of a solid solution is found in a study of the solubility of the protein in a series of different salt solutions, nor from the diffusion coefficient or from the rate of inactivation. These results indicate that the material is a pure substance or possibly a solid solution of two or more substances having nearly the same solubility in all the various solvents studied. It seems reasonable to conclude from these experiments that the possibility of a mixture must be limited to a mixture of proteins, so that the conclusion seems justified that pepsin itself is a protein.

BIBLIOGRAPHY

1. Sumner, J. B., *J. Biol. Chem.*, 1926, **69**, 435.
2. *Cf.* Taylor, H. S., Treatise on physical chemistry, Van Nostrand, New York, 1925, **2**, 956.
3. Fodor, A., Grundlagen der dispersoidchemie Steinkopff, Dresden, 1925.
4. Kuhn, R., and Wasserman, A., *Ber.*, 1928, **61 B**, 1550.
5. Vernon, H. M., *J. Physiol.*, 1904, **30**, 330.
6. Northrop, J. H., *J. Gen. Physiol.*, 1922, **4**, 261.
7. Marston, H. R., *Biochem. J.*, 1923, **17**, 851.
8. Pekelharing, C. A., *Z. Physiol. Chem.*, 1896, **22**, 233.
9. Fenger, F., Andrew, R. H., and Ralston, A. W., *J. Biol. Chem.*, 1928, **80**, 187.
10. Northrop, J. H., *J. Gen. Physiol.*, 1930, **13**, 781.
11. Anson, M. L., Unpublished results.
12. *Cf.* Taylor, H. S., Colloid Symposium Monograph, 1926, **4**, 19.
13. Goulding, A. M., Wasteneys, H., and Borsook, H., *J. Gen. Physiol.*, 1926–27, **10**, 451.

14. Northrop, J. H., and Anson, M. L., *J. Gen. Physiol.*, 1929, **12,** 543.
15. Northrop, J. H., *Science*, 1929, **69,** 580.
16. Euler, H., Chemie der enzyme, Bergmann, München, 3rd edition, 1925, Part 1, 390.
17. Ringer, W. A., *Z. physik. Chem.*, 1915, **95,** 195.
18. Willstätter, R., *Naturwissenschaften*, 1927, **15,** 585.
19. Northrop, J. H., *J. Gen. Physiol.*, 1926, **9,** 767.
20. Anson, M. L., and Mirsky, A. E., *J. Gen. Physiol.*, 1929, **13,** 121.

ISOLATION OF A CRYSTALLINE PROTEIN POSSESSING THE PROPERTIES OF TOBACCO-MOSAIC VIRUS

A CRYSTALLINE material, which has the properties of tobacco-mosaic virus, has been isolated from the juice of Turkish tobacco plants infected with this virus. The crystalline material contains 20 per cent. nitrogen and 1 per cent. ash, and a solution containing 1 milligram per cubic centimeter gives a positive test with Millon's biuret, xanthoproteic, glyoxylic acid and Folin's tyrosine reagents. The Molisch and Fehlings tests are negative, even with concentrated solutions. The material is precipitated by 0.4 saturated ammonium sulfate, by saturated magnesium sulfate, or by safranine, ethyl alcohol, acetone, trichloracetic acid, tannic acid, phosphotungstic acid and lead acetate. The crystalline protein is practically insoluble in water and is soluble in dilute acid, alkali or salt solutions. Solutions containing from 0.1 per cent. to 2 per cent. of the protein are opalescent. They are fairly clear between pH 6 and 11 and between pH 1 and 4, and take on a dense whitish appearance between pH 4 and 6.

The infectivity, chemical composition and optical rotation of the crystalline protein were unchanged after 10 successive crystallizations. In a fractional crystallization experiment the activity of the first small portion of crystals to come out of solution was the same as the activity of the mother liquor. When solutions are made more alkaline than about pH 11.8 the opalescence disappears and they become clear. Such solutions are devoid of activity and it was shown by solubility tests that the protein had been denatured. The material is also denatured and its activity lost when solutions are made more acid than about pH 1. It is completely coagulated and the activity lost on heating to 94° C. Preliminary experiments, in which the amorphous form of the protein was partially digested with pepsin, or partially coagulated by heat, indicate that the loss in activity is about proportional to the loss of native protein. The molecular weight of the protein, as determined by two preliminary experiments on osmotic pressure and diffusion, is of the order of a few millions. That the molecule is quite large is also indicated by the fact that the protein is held back by collodion filters through which proteins such as egg albumin readily pass. Collodion filters which fail to allow the protein to pass also fail to allow the active agent to pass. The material readily passes a Berkefeld "W" filter.

The crystals are over 100 times more active than the suspension made by grinding up diseased Turkish tobacco leaves, and about 1,000 times more active than the twice-frozen juice from diseased plants. One cubic centimeter of a 1 to 1,000,000,000 dilution of the crystals has usually proved infectious. The disease produced by this, as well as more concentrated solutions, has proved to be typical tobacco mosaic. Activity measurements were made by comparing the number of lesions produced on one half of the leaves of plants of Early Golden Cluster bean, *Nicotiana glutinosa* L., or *N. langsdorffii* Schrank after inoculation with dilutions of a solution of the crystals, with the number of lesions produced on the other halves of the same leaves after inoculation with dilutions of a virus preparation used for comparison.

The sera of animals injected with tobacco-mosaic virus give a precipitate when mixed with a solution of the crystals diluted as high as 1 part in 100,000. The sera of animals injected with juice from healthy tobacco plants give no precipitate when mixed with a solution of the crystals. Injection of solutions of the crystals into animals causes the production of a precipitin that is active for solutions of the crystals and juice of plants containing tobacco-mosaic virus but that is inactive for juice of normal plants.

The material herein described is quite different from the active crystalline material mentioned by Vinson and Petre[1] and by Barton-Wright and McBain,[2] which consisted, as Caldwell[3] has demonstrated, largely of inorganic matter having no connection with the activity. These preparations were less active than ordinary juice from diseased plants, and the activity they possessed diminished on further crystallizations.

The crystalline protein described in this paper was prepared from the juice of Turkish tobacco plants infected with tobacco-mosaic virus. The juice was brought to 0.4 saturation with ammonium sulfate and the precipitated globulin fraction thus obtained was removed by filtration. The dark brown globulin portion was repeatedly fractionated with ammonium sulfate and then most of the remaining color was removed by precipitation with a small amount of lead subacetate at pH 8.7. An inactive protein fraction was removed from the light yellow colored filtrate by adjusting to pH 4.5 and adding 2 per cent. by weight of standard celite. The celite was removed, suspended in

[1] C. G. Vinson and A. W. Petre, *Contrib. Boyce Thompson Inst.*, 3: 131, 1931.

[2] E. Barton-Wright and A. McBain, *Nature*, 132: 1003, 1933.

[3] J. Caldwell, *Nature*, 133: 177, 1934.

JUNE 28, 1935 SCIENCE 645

water at pH 8, and the suspension filtered. The active protein was found in the colorless filtrate. This procedure was repeated twice in order to remove completely the inactive protein. Crystallization was accomplished by adding slowly, with stirring, a solution containing 1 cubic centimeter of glacial acetic acid in 20 cubic centimeters of 0.5 saturated ammonium sulfate to a solution of the protein containing sufficient ammonium sulfate to cause a faint turbidity. Small needles about 0.03 millimeters long appeared immediately and crystallization was completed in an hour. Crystallization may also be caused by the addition of a little saturated ammonium or magnesium sulfate to a solution of the protein in 0.001 N acid. Several attempts to obtain crystals by dialyzing solutions of the protein gave only amorphous material. To date a little more than 10 grams of the active crystalline protein have been obtained.

Although it is difficult, if not impossible, to obtain conclusive positive proof of the purity of a protein, there is strong evidence that the crystalline protein herein described is either pure or is a solid solution of proteins. As yet no evidence for the existence of a mixture of active and inactive material in the crystals has been obtained. Tobacco-mosaic virus is regarded as an autocatalytic protein which, for the present, may be assumed to require the presence of living cells for multiplication.

W. M. STANLEY

THE ROCKEFELLER INSTITUTE FOR
MEDICAL RESEARCH,
PRINCETON, N. J.

[CONTRIBUTION FROM GATES CHEMICAL LABORATORY, CALIFORNIA INSTITUTE OF TECHNOLOGY, No. 280]

THE NATURE OF THE CHEMICAL BOND. APPLICATION OF RESULTS OBTAINED FROM THE QUANTUM MECHANICS AND FROM A THEORY OF PARAMAGNETIC SUSCEPTIBILITY TO THE STRUCTURE OF MOLECULES

BY LINUS PAULING

RECEIVED FEBRUARY 17, 1931 PUBLISHED APRIL 6, 1931

During the last four years the problem of the nature of the chemical bond has been attacked by theoretical physicists, especially Heitler and London, by the application of the quantum mechanics. This work has led to an approximate theoretical calculation of the energy of formation and of other properties of very simple molecules, such as H_2, and has also provided a formal justification of the rules set up in 1916 by G. N. Lewis for his electron-pair bond. In the following paper it will be shown that many more results of chemical significance can be obtained from the quantum mechanical equations, permitting the formulation of an extensive and powerful set of rules for the electron-pair bond supplementing those of Lewis. These rules provide information regarding the relative strengths of bonds formed by different atoms, the angles between bonds, free rotation or lack of free rotation about bond axes, the relation between the quantum numbers of bonding electrons and the number and spatial arrangement of the bonds, etc. A complete theory of the magnetic moments of molecules and complex ions is also developed, and it is shown that for many compounds involving elements of the transition groups this theory together with the rules for electron-pair bonds leads to a unique assignment of electron structures as well as a definite determination of the type of bonds involved.[1]

I. The Electron-Pair Bond

The Interaction of Simple Atoms.—The discussion of the wave equation for the hydrogen molecule by Heitler and London,[2] Sugiura,[3] and Wang[4] showed that two normal hydrogen atoms can interact in either of two ways, one of which gives rise to repulsion with no molecule formation, the other

[1] A preliminary announcement of some of these results was made three years ago [Linus Pauling, *Proc. Nat. Acad. Sci.*, **14**, 359 (1928)]. Two of the results (90° bond angles for *p* eigenfunctions, and the existence, but not the stability, of tetrahedral eigenfunctions) have been independently discovered by Professor J. C. Slater and announced at meetings of the National Academy of Sciences (Washington, April, 1930) and the American Physical Society (Cleveland, December, 1930).

[2] W. Heitler and F. London, *Z. Physik*, **44**, 455 (1927).

[3] Y. Sugiura, *ibid.*, **45**, 484 (1927).

[4] S. C. Wang, *Phys. Rev.*, **31**, 579 (1928).

to attraction and the formation of a stable molecule. These two modes of interaction result from the identity of the two electrons. The characteristic resonance phenomenon of the quantum mechanics, which produces the stable bond in the hydrogen molecule, always occurs with two electrons, for even though the nuclei to which they are attached are different, the energy of the unperturbed system with one electron on one nucleus and the other on the other nucleus is the same as with the electrons interchanged. Hence we may expect to find electron-pair bonds turning up often.

But the interaction of atoms with more than one electron does not always lead to molecule formation. A normal helium atom and a normal hydrogen atom interact in only one way,[5] giving repulsion only, and two normal helium atoms repel each other except at large distances, where there is very weak attraction.[5,6] Two lithium atoms, on the other hand, can interact in two ways,[7] giving a repulsive potential and an attractive potential, the latter corresponding to formation of a stable molecule. In these cases it is seen that only when each of the two atoms initially possesses an unpaired electron is a stable molecule formed. The general conclusion that an electron-pair bond is formed by the interaction of an unpaired electron on each of two atoms has been obtained formally by Heitler[8] and London,[9] with the use of certain assumptions regarding the signs of integrals occurring in the theory. The energy of the bond is largely the resonance or interchange energy of two electrons. This energy depends mainly on electrostatic forces between electrons and nuclei, and is not due to magnetic interactions, although the electron spins determine whether attractive or repulsive potentials, or both, will occur.

Properties of the Electron-Pair Bond.—From the foregoing discussion we infer the following properties of the electron-pair bond.

1. *The electron-pair bond is formed through the interaction of an unpaired electron on each of two atoms.*

2. *The spins of the electrons are opposed when the bond is formed, so that they cannot contribute to the paramagnetic susceptibility of the substance.*

3. *Two electrons which form a shared pair cannot take part in forming additional pairs.*

In addition we postulate the following three rules, which are justified by the qualitative consideration of the factors influencing bond energies. An outline of the derivation of the rules from the wave equation is given below.

[5] G. Gentile, *Z. Physik.*, **63**, 795 (1930).

[6] J. C. Slater, *Phys. Rev.*, **32**, 349 (1927).

[7] M. Delbrück, *Ann. Physik*, **5**, 36 (1930).

[8] W. Heitler, *Z. Physik*, **46**, 47 (1927); **47**, 835 (1928); *Physik. Z.*, **31**, 185 (1930), etc.

[9] F. London, *Z. Physik*, **46**, 455 (1928); **50**, 24 (1928); "Sommerfeld Festschrift," p. 104; etc.

4. *The main resonance terms for a single electron-pair bond are those involving only one eigenfunction from each atom.*

5. *Of two eigenfunctions with the same dependence on r, the one with the larger value in the bond direction will give rise to the stronger bond, and for a given eigenfunction the bond will tend to be formed in the direction with the largest value of the eigenfunction.*

6. *Of two eigenfunctions with the same dependence on θ and φ, the one with the smaller mean value of r, that is, the one corresponding to the lower energy level for the atom, will give rise to the stronger bond.*

Here the eigenfunctions referred to are those for an electron in an atom, and r, θ and φ are polar coördinates of the electron, the nucleus being at the origin of the coördinate system.

It is not proposed to develop a complete proof of the above rules at this place, for even the formal justification of the electron-pair bond in the simplest cases (diatomic molecule, say) requires a formidable array of symbols and equations. The following sketch outlines the construction of an inclusive proof.

It can be shown[10] that if Ψ is an arbitrary function of the independent variables in a wave equation

$$(H - W)\psi = 0$$

then the integral

$$E = \int \Psi^* H \Psi d\tau$$

called the variation integral, is always larger than W_0, the lowest energy level for the system. A function Ψ containing several parameters provides the best approximation to the eigenfunction ψ_0 for the normal state of the system when the variational integral is minimized with respect to these parameters. Now let us consider two atoms A and B connected by an electron-pair bond, and for simplicity let all the other electrons in the system be paired, the pairs being either lone pairs or pairs shared between A or B and other atoms. Let us assume that there are available for bond formation by atom A several single-electron eigenfunctions of approximately the same energy, and that the change in energy of penetration into the core is negligible compared with bond energy. Then we may take as single-electron eigenfunctions

$$\psi_{Ai} = \Sigma_k a_{ik} \psi^0_{Ak}$$

in which the a_{ik}'s are numerical coefficients and the ψ^0_{Ak}'s are an arbitrary set of single-electron eigenfunctions, such as those obtained on separating the wave equation in polar coördinates. From the ψ_{Ai}'s there is built up a group composed of atom A and the atoms to which it is bonded except atom B, such that all electrons are paired except one, corresponding to the eigenfunction ψ_{Ai}, say. From atom B a similar group with one unpaired electron is built. The interaction energy of these two groups can then be calculated with the aid of the variational equation through the substitution of an eigenfunction for the molecule built of those for the two groups in such a way that it has the correct symmetry character. The construction of this eingenfunction and evaluation of the integral would be very laborious; it will be noticed, however, that this problem is formally similar to Born's treatment[11] of the interaction of two atoms in S states, based on Slater's treatment of atomic eigenfunctions, and the value of E is found to be

$$E = W_A + W_B + J_E + J_X - \Sigma_Y J_Y - 2\Sigma_Z J_Z$$

[10] A clear discussion is given by C. Eckart, *Phys. Rev.*, **36**, 878 (1930).

[11] M. Born, *Z. Physik*, **64**, 729 (1930).

Here W_A and W_B are the energies of the separate groups, and J_E represents the Coulomb interaction of A and B, neglecting resonance. The resonance term J_X corresponds to a permutation of the two AB bond electrons; J_Y corresponds to a permutation of the AB bond electron on B with a paired electron with similarly directed spin on A, or *vice versa;* and J_Z corresponds to a permutation of a paired electron on A with one on B. (For explicit expressions for these see Born.[11]) The resonance integrals J_X, J_Y and J_Z have been found to have negative signs in the case of simple molecules for which calculations have been made, and it is probable that these signs obtain in most cases. The resonance integrals depend qualitatively on what may be called the *overlapping* of the single-electron eigenfunctions involved; if ψ_A and ψ_B are two single-electron eigenfunctions, the product $\psi_A(1)\ \psi_B(2)\ \psi_A^*(2)\ \psi_B^*(1)$ occurs in the resonance integral corresponding to the permutation involving electrons 1 and 2, and the value of the integral increases as the magnitude of this product in the region between the two nuclei increases.

Now we vary the coefficients a_{ik} in such a way as to minimize E. W_A and W_B are not affected by this variation, and J_E is not changed in case that there is one electron for every eigenfunction in a subgroup on A, and is changed relatively slightly otherwise. The resonance integrals are, however, strongly affected by changing the coefficients. The positive sign preceding J_X requires that the two bond eigenfunctions ψ_A and ψ_B show the maximum overlapping in the region between the two nuclei, while the negative sign preceding J_Y requires the minimum overlapping between ψ_A and the eigenfunctions of B other than ψ_B, and between ψ_B and the eigenfunctions of A other than ψ_A. Hence the correct zeroth-order eigenfunctions for the atom A are such that one, the AB bond eigenfunction ψ_A, extends largely in the direction of atom B, while the other A eigenfunctions avoid overlapping with ψ_B. As a consequence the integral J_X is of large magnitude, while the integrals J_Y, because of the small overlapping of the eigenfunctions involved, are small.

An extension of this argument shows that the phenomenon of *concentration of the bond eigenfunctions* further increases the magnitude of J_X and decreases J_Y. The non-orthogonality of the bond eigenfunctions as well as certain second-order perturbations leads to a shrinkage of the region in which the bond eigenfunctions have appreciable values. This is strikingly shown by a comparison of H_2^+ and H; the volume within which the electron probability function $\psi\psi^*$ for H_2^+ is greater than one-tenth of its maximum value is found from Burrau's calculations to be 0.67 Å.^{3}, *less than 10% of its value 8.6 Å.^{3} for a hydrogen atom.*[12] This concentration of the bond eigenfunctions greatly increases their interaction with one another, and decreases their interaction with other eigenfunctions, a fact expressed in Rule 4. For double or triple bonds interactions among all four or six eigenfunctions must be considered.

s and p Eigenfunctions. Compounds of Normal Atoms.—As a rule s and p eigenfunctions with the same total quantum number in an atom do not differ very much in their mean values of r (the s levels lie lower because of greater penetration of inner shells), so that Rule 6 would not lead us to expect them to differ in bond-forming power. But their dependence on θ and φ is widely different. Putting

$$\begin{aligned} \Psi_{n0}(r,\theta,\varphi) &= R_{n0}(r)\cdot s(\theta,\varphi) && \text{for } s \text{ eigenfunctions} \\ \Psi_{n1}(r,\theta,\varphi) &= R_{n1}(r)\cdot \left.\begin{matrix} p_x(\theta,\varphi) \\ p_y(\theta,\varphi) \\ p_z(\theta,\varphi) \end{matrix}\right\} && \text{for } p \text{ eigenfunctions} \end{aligned} \qquad (1)$$

[12] Compare Fig. 6 with Fig. 7, which is drawn to half the scale of Fig. 6, of Linus Pauling, *Chem. Rev.*, **5**, 173 (1928).

the parts s, p_x, p_y, p_z of the eigenfunctions depending on θ and φ, normalized to 4π, are

$$\left.\begin{aligned} s &= 1 \\ p_x &= \sqrt{3}\sin\theta\cos\varphi \\ p_y &= \sqrt{3}\sin\theta\sin\varphi \\ p_z &= \sqrt{3}\cos\theta \end{aligned}\right\} \quad (2)$$

Absolute values of s and p_x are represented in the xz plane in Figs. 1 and 2. s is spherically symmetrical, with the value 1 in all directions. $|p_x|$ consists of two spheres as shown (the x axis is an infinite symmetry axis), with the maximum value $\sqrt{3}$ along the x axis. $|p_y|$ and $|p_z|$ are similar, with maximum values of $\sqrt{3}$ along the y and z axis, respectively. From Rule 5 we conclude that *p electrons will form stronger bonds than s electrons,* and that *the bonds formed by p electrons in an atom tend to be oriented at right angles to one another.*

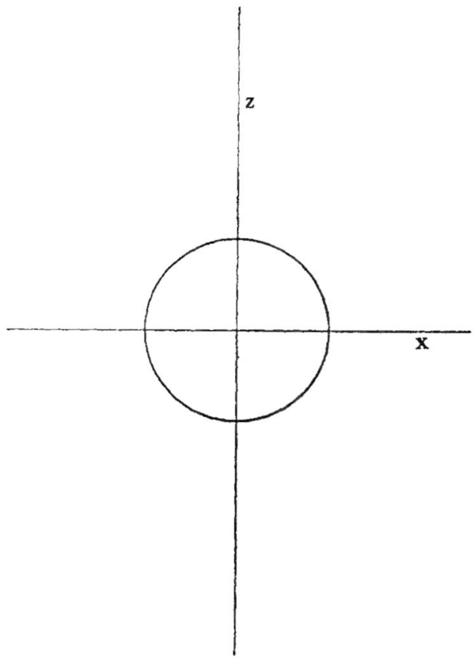

Fig. 1.—Polar graph of 1 in the xz plane, representing an s eigenfunction.

The second conclusion explains several interesting facts. Normal oxygen, in the state $2s^22p^4\ ^3P$, contains two unpaired p electrons. When an atom of oxygen combines with two of hydrogen, a water molecule will result in which the angle formed by the three atoms is 90°, or somewhat larger because of interaction of the two hydrogen atoms. It has been long known from their large electric moment that water molecules have a kinked rather than a collinear arrangement of their atoms, and attempts have been made to explain this with rather unsatisfactory calculations based on an ionic structure with strong polarization of the oxygen anion in the field of the protons. The above simple explanation results directly from the reasonable assumption of an electron-pair bond structure and the properties of tesseral harmonics.

It can be predicted that H_2O_2, with the structure $\begin{matrix} :\ddot{O}:\ddot{O}: \\ \ddot{H}\ \ \ddot{H} \end{matrix}$ involving bonds of p electrons, also consists of kinked rather than collinear molecules.

Nitrogen, with the normal state $2s^22p^3\ ^4S$, contains three unpaired p

electrons, which can form bonds at about 90° from one another with three hydrogen atoms. The ammonia molecule, with the resulting pyramidal structure, also has a large electric moment.

The crystal skutterudite, $Co_4^{3+}(As_4^{4-})_3$, contains As_4^{4-} groups with a square configuration, corresponding to the structure $\left[\begin{matrix} :\ddot{A}s:\ddot{A}s: \\ :\ddot{A}s:\ddot{A}s: \end{matrix}\right]^{4-}$. This complex has bond angles of exactly 90°.

In the above discussion it has been assumed that the type of quantization has not been changed, and that *s* and *p* eigenfunctions retain their identity. This is probably true for H_2O and H_2O_2, and perhaps for NH_3 and As_4^{4-} also. A discussion of the effect of change of quantization on bond angles is given in a later section.

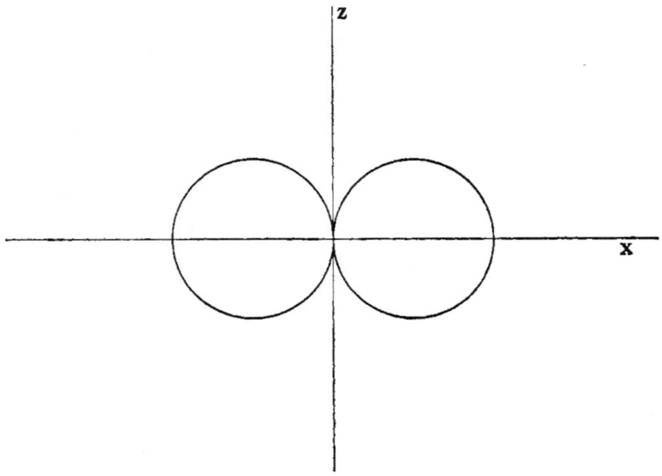

Fig. 2.—Polar graph of $|\sqrt{3}\sin\theta|$ in the *xz* plane, representing the p_z eigenfunction.

Transition from Electron-Pair to Ionic Bonds. The Hydrogen Bond.— In case that the symmetry character of an electron-pair structure and an ionic structure for a molecule are the same, it may be difficult to decide between the two, for the structure may lie anywhere between these extremes. The zeroth-order eigenfunction for the two bond electrons for a molecule MX (HF, say, or NaCl) with a single electron-pair bond would be

$$\Psi_{MX} = \frac{\psi_M(1)\,\psi_X(2) + \psi_M(2)\,\psi_X(1)}{\sqrt{2 + 2S^2}} \qquad (3)$$

in which $S = \int \psi_M(1)\,\psi_X^*(1)\,d\tau_1$. The eigenfunction for a pure ionic state would be

$$\Psi_{M^+X^-} = \psi_X(1)\,\psi_X(2) \qquad (4)$$

In certain cases one of these might approximate the correct eigenfunction closely. In other cases, however, it would be necessary to consider combinations of the two, namely

$$\Psi_+ = a\,\Psi_{MX} + \sqrt{1-a^2}\,\Psi_{M^+X^-}$$

and

$$\Psi_- = \sqrt{1-a^2}\,\Psi_{MX} - a\,\Psi_{M^+X^-} \quad (5)$$

For a given molecule and a given internuclear separation a would have a definite value, such as to make the energy level for Ψ_+ lie as low as possible. If a happens to be nearly 1 for the equilibrium state of the molecule, it would be convenient to say that the bond is an electron-pair bond; if a is nearly zero, it could be called an ionic bond. This definition is somewhat unsatisfactory in that it does not depend on easily observable quantities. For example, a compound which is ionic by the above definition might dissociate adiabatically into neutral atoms, the value of a changing from nearly zero to unity as the nuclei separate, and it would do this in case the electron affinity of X were less than the ionization potential of M. HF is an example of such a compound. There is evidence, given below, that the normal molecule approximates an ionic compound; yet it would dissociate adiabatically into neutral F and H.[13]

But direct evidence regarding the value of a can sometimes be obtained. The hydrogen bond, discovered by Huggins and by Latimer and Rodebush, has been usually considered as produced by a hydrogen atom with two electron-pair bonds, as in $\left[:\ddot{\underset{..}{F}}:H:\ddot{\underset{..}{F}}:\right]^-$. It was later pointed out[1] that this is not compatible with the quantum mechanical rules, for hydrogen can have only one unpaired 1*s* electron, and outer orbits are so much less stable that strong bonds would not be formed. With an ionic structure, however, we would expect H^+F^- to polymerize and to add on to F^-, to give H_6F_6 and $[F^-H^+F^-]^-$; moreover, the observed coördination number 2 is just that predicted[14] from the radius ratio 0. Hence the observation that hydrogen bonds are formed with fluorine supports an ionic structure for HF. Hydrogen bonds are not formed with chlorine, bromine, and iodine, so that the bonds in HCl, HBr, and HI are to be considered as approaching the electron-pair type.

Hydrogen bonds are formed to some extent by oxygen ($(H_2O)_x$, ice, etc.) and perhaps also in some cases by nitrogen. The electrostatic structure for the hydrogen bond explains the observation that only these atoms of high electron affinity form such bonds, a fact for which no explanation was given by the older conception. It is of interest that there is considerable

[13] There would, however, be a certain probability, dependent on the nature of the eigenfunctions, that actual non-adiabatic dissociation would give ions rather than atoms, and this might be nearly unity, in case the two potential curves come very close to one another at some point. See I. v. Neumann and E. Wigner, *Physik. Z.*, **30**, 467 (1929).

[14] Linus Pauling, This Journal, **51**, 1010 (1929).

evidence from crystal structure data for $[OHO]^{=}$ groups. In many crystals containing H and O, including topaz,[15] $Al_2SiO_4(F,OH)_2$; diaspore,[16] $AlHO_2$; goethite,[16] $FeHO_2$; chondrodite,[17] $Mg_5Si_2O_8(F,OH)_2$; etc., the sum of the strengths of the electrostatic bonds from all cations (except hydrogen) to an anion is either 2 or 1, indicating, according to the electrostatic valence rule,[14] the presence of $O^{=}$ and of F^{-} or $(OH)^{-}$, respectively. But in some crystals, including[18] KH_2PO_4; staurolite,[19] $H_2FeAl_4Si_2O_{12}$; and lepidocrocite,[16] $FeHO_2$, the sum of bond strengths is 2 or $^3/_2$, the latter value occurring twice for each H; the electrostatic valence rule in these cases supports the assumption of $[O^{=}H^{+}O^{=}]^{=}$ groups, the hydrogen ion contributing a bond of strength $^1/_2$ to each of two oxygen ions.

In other cases, discussed below, the lowest electron-pair-bond structure and the lowest ionic-bond structure do not have the same multiplicity, so that (when the interaction of electron spin and orbital motion is neglected) these two states cannot be combined, and a knowledge of the multiplicity of the normal state of the molecule or complex ion permits a definite statement as to the bond type to be made.

Change in Quantization of Bond Eigenfunctions.—A normal carbon atom, in the state $2s^22p^2\ ^3P$, contains only two unpaired electrons, and can hence form no more than two single bonds or one double bond (as in CO, formed from a normal carbon atom and a normal oxygen atom). But only about 1.6 v. e. of energy is needed to excite a carbon atom to the state $2s2p^3\ ^5S$, with four unpaired electrons, and in this state the atom can form four bonds. We might then describe the formation of a substituted methane CRR′R″R‴ in the following way. The radicals R, R′, and R″, each with an unpaired electron, form electron-pair bonds with the three p electrons of the carbon atom, the bond directions making angles of 90° with one another. The fourth radical R‴ then forms a weaker bond with the s electron, probably at an angle of 125° with each of the other bonds. This would give an unsymmetrical structure, with non-equivalent bonds, and considerable discussion has been given by various authors to the difference in the carbon bonds due to s and p electrons. Actually the foregoing treatment is fallacious, for the phenomenon of change in quantization of the bond eigenfunctions, first discussed in the note referred to before,[1] leads simply and directly to the conclusion that *the four bonds formed by a carbon atom are equivalent and are directed toward tetrahedron corners.*

The importance of s, p, d, and f eigenfunctions for single atoms and ions

[15] Linus Pauling, *Proc. Nat. Acad. Sci.*, **14**, 603 (1928); N. A. Alston and J. West, *Z. Krist.*, **69**, 149 (1928).

[16] Unpublished investigation in this Laboratory.

[17] W. L. Bragg and J. West, *Proc. Roy. Soc.* (London), **A114**, 450 (1927); W. H. Taylor and J. West, *ibid.*, **A117**, 517 (1928).

[18] J. West, *Z. Krist.*, **74**, 306 (1930).

[19] St. Naray-Szabo, *ibid.*, **71**, 103 (1929).

results from the fact that the interaction of one electron with the nucleus and other electrons can be represented approximately by a non-Coulombian central field, so that the wave equation can be separated in polar coördinates r, θ, and φ, giving rise to eigenfunctions involving tesseral harmonics such as those in Equation 1. The deeper penetration of s electrons within inner shells causes them to be more tightly bound than p electrons with the same total quantum number. If an atom approaches a given atom, forming a bond with it, the interaction between the two can be considered as a perturbation, and the first step in applying the perturbation theory for a degenerate system consists in finding the correct zeroth-order eigenfunctions for the perturbation, one of which is the eigenfunction which will lead to the largest negative perturbation energy. This will be the one with the largest values along the bond direction. The correct zeroth-order eigenfunctions must be certain normalized and mutually orthogonal linear aggregates of the original eigenfunctions. If the perturbation is small, the s eigenfunction cannot be changed, and the only combinations which can be made with the p eigenfunctions are equivalent merely to a rotation of axes. But in case the energy of interaction of the two atoms is greater than the difference in energy of an s electron and a p electron (or, if there are originally two s electrons present, as in a normal carbon atom of twice this difference), hydrogen-like s and p eigenfunctions must be grouped together to form the original degenerate state, and the interaction of the two atoms together with the deviation of the atomic field from a Coulombian one must be considered as the perturbation, with the former predominating. The correct zeroth-order bond eigenfunctions will then be those orthogonal and normalized linear aggregates of both the s and p eigenfunctions which would give the strongest bonds according to Rule 5.

A rough criterion as to whether the quantization is changed from that in polar coördinates to a type giving stronger bond eigenfunctions is thus that the possible bond energy be greater than the s–p (or, if d eigenfunctions are also involved, s–d or p–d) separation.[20]

This criterion is satisfied for quadrivalent carbon. The energy difference of the states[21] $2s^22p^2\ {}^3P$ and $2s2p^3\ {}^3P$ of carbon is 9.3 v. e., and a similar value of about 200,000 cal. per mole is found for other atoms in the first row of the periodic system. The energy of a single bond is of the order of 100,000 cal. per mole. Hence a carbon atom forming four bonds would certainly have changed quantization, and even when the bond energy must be divided between two atoms, as in a diamond crystal, the criterion is sufficiently well satisfied. The same results hold for quadrivalent

[20] This criterion was expressed in Ref. 1.

[21] States with the same multiplicity should be compared, for increase in multiplicity decreases the term value, the difference between $2s^22p^2\ {}^3P$ and $2s2p^3\ {}^5S$ being only about 1.6 v. e., as mentioned above.

nitrogen, a nitrogen *ion* in the state N^+ $2s2p^3$ ^5S forming four bonds, as in $(NH_4)^+$, $N(CH_3)_4^+$, etc. But for bivalent oxygen there is available only about 200,000 cal. per mole bond energy, and the s–p separation for two s electrons corresponds to about 400,000 cal. per mole, so that it is very probable that the oxygen bond eigenfunctions in H_2O, for example, are p eigenfunctions, as assumed in a previous section. Trivalent nitrogen is a border-line case; the bond energy of about 300,000 cal. per mole is sufficiently close to the s–p energy of 400,000 cal. per mole to permit the eigenfunctions to be changed somewhat, but not to the extent that they are in quadrivalent carbon and nitrogen.

It may be pointed out that the s–p separation for atoms in the same column of the periodic table is nearly constant, about 200,000 cal. per mole for one s electron. The bond energy decreases somewhat with increasing atomic number. Thus the energies of a bond in the compounds H_2O, H_2S, H_2Se, and H_2Te, calculated from thermochemical and band spectral data, are 110,000, 90,000, 73,000, and 60,000 cal. per mole, respectively. Hence we conclude that if quantization in polar coördinates is not broken for a light atom on formation of a compound, it will not be broken for heavier atoms in the same column of the periodic system. The molecules H_2S, H_2Se, and H_2Te must accordingly also have a non-linear structure, with bond angles of 90° or slightly greater.

Let us now determine the zeroth-order eigenfunctions which will form the strongest bonds for the case when the s–p quantization is broken. The dependence on r of s and p hydrogen-like eigenfunctions is not greatly different,[22] and it seems probable that the effect of the non-Coulombian field would decrease the difference for actual atoms. We may accordingly assume that $R_{n0}(r)$ and $R_{n1}(r)$ are effectively the same as far as bond formation is concerned, so that the problem of determining the bond eigenfunctions reduces to a discussion of the θ, φ eigenfunctions of Equation 1. Arbitrary sets of θ, φ eigenfunctions formed from s, p_x, p_y, and p_z are given by the expressions

$$\left.\begin{aligned} \psi_1 &= a_1s + b_1p_x + c_1p_y + d_1p_z \\ \psi_2 &= a_2s + b_2p_x + c_2p_y + d_2p_z \\ \psi_3 &= a_3s + b_3p_x + c_3p_y + d_3p_z \\ \psi_4 &= a_4s + b_4p_x + c_4p_y + d_4p_z \end{aligned}\right\} \qquad (6)$$

in which the coefficients a_1, etc., are restricted only by the orthogonality and normalization requirements

$$\int \psi_i^2 d\tau = 1 \quad \text{or } a_i^2 + b_i^2 + c_i^2 + d_i^2 = 1 \qquad i = 1,2,3,4 \qquad (7a)$$

and

$$\int \psi_i\psi_k d\tau = 0 \quad \text{or } a_ia_k + b_ib_k + c_ic_k + d_id_k = 0 \quad i,k = 1,2,3,4 \quad i \neq k \qquad (7b)$$

[22] See the curves given by Linus Pauling, *Proc. Roy. Soc.* (London), **A114**, 181 (1927), or A. Sommerfeld, "Wellenmechanischer Ergänzungsband," p. 88.

From Rule 5 the best bond eigenfunction will be that which has the largest value in the bond direction. This direction can be chosen arbitrarily for a single bond. Taking it along the x axis, it is found that the best single bond eigenfunction is[23]

$$\psi_1 = \frac{1}{2}s + \frac{\sqrt{3}}{2}p_x \tag{8a}$$

with a maximum value of 2, considerably larger than that 1.732 for a p eigenfunction. A graph of this function in the xz plane is shown in Fig. 3.

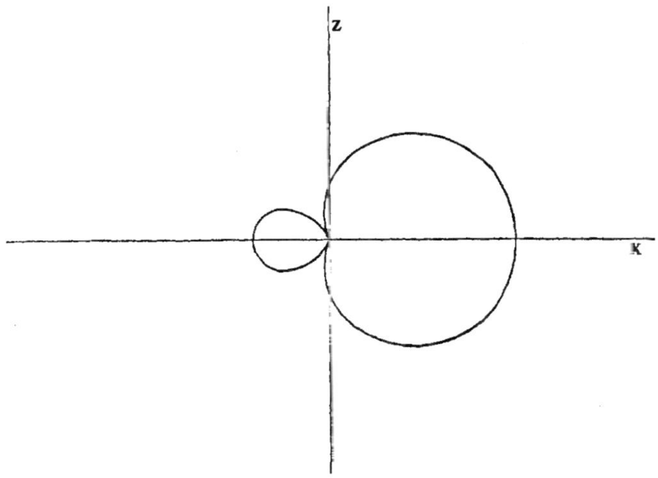

Fig. 3.—Polar graph of $\left| {}^1/_2 + {}^3/_2 \sin\theta \right|$ in the xz plane, representing a tetrahedral eigenfunction, the best bond eigenfunction which can be formed from s and p eigenfunctions.

A second bond can be introduced in the xz plane. The best eigenfunction for this bond is found to be

$$\psi_2 = \frac{1}{2}s - \frac{1}{2\sqrt{3}}p_x + \frac{\sqrt{2}}{\sqrt{3}}p_z \tag{8b}$$

[23] It is easily shown with the use of the method of undetermined multipliers that the eigenfunction with the maximum value in the direction defined by the polar angles θ_0, φ_0 has as coefficients of the initial eigenfunctions quantities proportional to $\psi_k(\theta_0,\varphi_0)$, and that the maximum value is itself equal to $\left\{\Sigma_k[\psi_k(\theta_0,\varphi_0)]^2\right\}^{1/2}$. For let $\psi(\theta,\varphi) = \Sigma_{k=1}^{n} a_k\psi_k(\theta,\varphi)$, with $\Sigma a_k^2 = 1$. We want $\psi(\theta_0,\varphi_0) = \Sigma a_k\psi_k(\theta_0,\varphi_0)$ to be a maximum with respect to variation in the a_k 's. Consider the expression

$$\Lambda = \psi(\theta_0,\varphi_0) - \frac{\lambda}{2}\left\{\Sigma a_k^2 - 1\right\} = \Sigma\left\{a_k\psi_k(\theta_0,\varphi_0) - \frac{\lambda}{2}a_k^2\right\} + \frac{\lambda}{2}$$

in which λ is an undetermined multiplier. Then we put

$$\frac{\partial\Lambda}{\partial a_k} = \psi_k(\theta_0,\varphi_0) - \lambda a_k = 0 \text{ or } a_k = \frac{\psi_k(\theta_0,\varphi_0)}{\lambda},\ k = 1,2\ldots n$$

in which λ has such a value that $\Sigma a_k^2 = 1$; *i. e.*, $\lambda = \left\{\Sigma[\psi_k(\theta_0,\varphi_0)]^2\right\}^{1/2}$. $\psi(\theta_0,\varphi_0)$ is itself then equal to $\Sigma[\psi_k(\theta_0,\varphi_0)]^2/\lambda$ or $\left\{\Sigma[\psi_k(\theta_0,\varphi_0)]^2\right\}^{1/2}$.

This eigenfunction is equivalent to and orthogonal to ψ_1, and has its maximum value of 2 at $\theta = 19°28'$, $\varphi = 180°$, that is, at an angle of 109°28′ with the first bond, *which is just the angle between the lines drawn from the center to two corners of a regular tetrahedron.* The third and fourth best bond eigenfunctions

$$\psi_3 = \frac{1}{2}s - \frac{1}{2\sqrt{3}}p_z - \frac{1}{\sqrt{6}}p_x + \frac{1}{\sqrt{2}}p_y \tag{8c}$$

and

$$\psi_4 = \frac{1}{2}s - \frac{1}{2\sqrt{3}}p_z - \frac{1}{\sqrt{6}}p_x - \frac{1}{\sqrt{2}}p_y \tag{8d}$$

are also equivalent to the others, and have their maximum values of 2 along the lines toward the other two corners of a regular tetrahedron.

An equivalent set of four tetrahedral eigenfunctions is[24]

$$\left.\begin{aligned} \psi_{111} &= \frac{1}{2}(s + p_x + p_y + p_z) \\ \psi_{1\bar{1}\bar{1}} &= \frac{1}{2}(s + p_x - p_y - p_z) \\ \psi_{\bar{1}1\bar{1}} &= \frac{1}{2}(s - p_x + p_y - p_z) \\ \psi_{\bar{1}\bar{1}1} &= \frac{1}{2}(s - p_x - p_y + p_z) \end{aligned}\right\} \tag{9}$$

These differ from the others only by a rotation of the atom as a whole.

The Tetrahedral Carbon Atom.—We have thus derived the result that *an atom in which only s and p eigenfunctions contribute to bond formation and in which the quantization in polar coördinates is broken can form one, two, three, or four equivalent bonds, which are directed toward the corners of a regular tetrahedron* (Fig. 4). This calculation provides the quantum mechanical justification of the chemist's tetrahedral carbon atom, present in diamond and all aliphatic carbon compounds, and for the tetrahedral quadrivalent nitrogen atom, the tetrahedral phosphorus atom, as in phosphonium compounds, the tetrahedral boron atom in B_2H_6 (involving single-electron bonds), and many other such atoms.

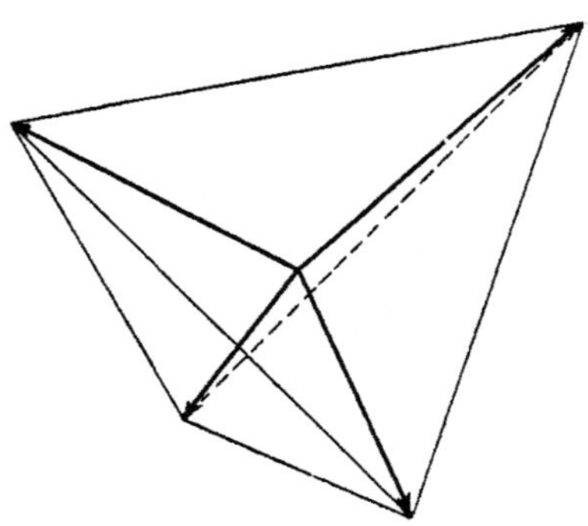

Fig. 4.—Diagram showing relative orientation in space of the directions of the maxima of four tetrahedral eigenfunctions.

Free or Restricted Rotation.—Each of these tetrahedral bond eigen-

[24] It should be borne in mind that the bond eigenfunctions actually are obtained from the expressions given in this paper by substituting for s the complete eigenfunction $\Psi_{n00}(r,\theta,\varphi)$, etc. It is not necessary that the r part of the eigenfunctions be identical; the assumption made in the above treatment is that they do not affect the evaluation of the coefficients in the bond eigenfunctions.

functions is cylindrically symmetrical about its bond direction. Hence the bond energy is independent of orientation about this direction, so that there will be *free rotation about a single bond*, except in so far as rotation is hindered by steric effects, arising from interactions of the substituent atoms or groups.

A double bond behaves differently, however. Let us introduce two substituents in the octants xyz and $\bar{x}\bar{y}z$ of an atom, a carbon atom, say, using the bond eigenfunctions ψ_{111} and $\psi_{\bar{1}\bar{1}1}$. The two eigenfunctions $\psi_{1\bar{1}\bar{1}}$ and $\psi_{\bar{1}1\bar{1}}$ are then left to form a double bond with another such group. Now $\psi_{1\bar{1}\bar{1}}$ and $\psi_{\bar{1}1\bar{1}}$ (or any two eigenfunctions formed from them) are not cylindrically symmetrical about the z axis or any direction, nor are the two eigenfunctions on the other group. Hence the energy of the double bond will depend on the relative orientation of the two tetrahedral carbon atoms, and will be a maximum when the two sets of eigenfunctions show the maximum overlapping. This will occur when the two tetrahedral atoms share an edge (Fig. 5). Thus we derive the result, found long ago by chemists, that there are two stable states for a simple compound involving a double bond, a *cis* and a *trans* state, differing in orientation by 180°. *There is no free rotation about a double bond.*[25]

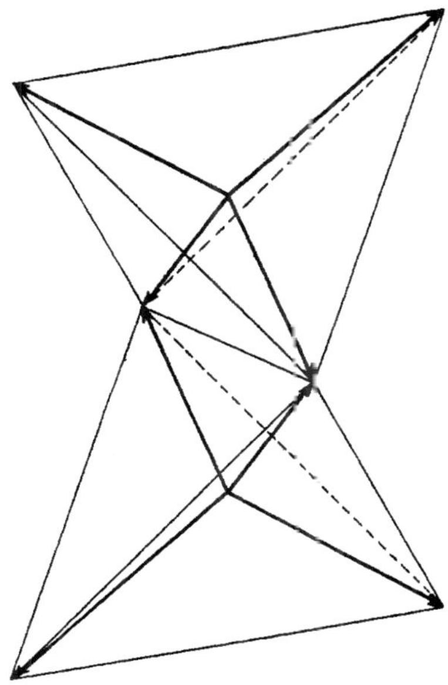

Fig. 5.—Directions of maxima of tetrahedral eigenfunctions in two atoms connected by a double bond.

The three eigenfunctions which would take part in the formation of a triple bond can be made symmetrical about the bond direction, for an atom of the type considered above, with only four eigenfunctions in the outer shell; but since the group attached by the fourth valence lies on the axis of the triple bond, there is no way of verifying the resulting free rotation about the triple bond.

The Angles between Bonds.—The above calculation of tetrahedral angles between bonds when the quantization is changed sets an upper limit on bond angles in doubtful cases, when the criterion is only approximately satisfied. For we can now state that the bond angles in H_2O and NH_3

[25] A discussion of rotation about a double bond on the basis of the quantum mechanics has been published by E. Hückel, *Z. Physik*, **60**, 423 (1930), which is, I feel, neither so straightforward nor so convincing as the above treatment, inasmuch as neither the phenomenon of concentration of the bond eigenfunctions nor that of change in quantization is taken into account.

should lie between 90 and 109°28′, closer to 90° for the first and to 109°28′ for the second compound. The same limits should apply to other atoms with an outer 8-shell (counting both shared and unshared electron pairs). Direct evidence on this point is provided by crystal structure data for non-ionic crystals, given in Table I. Every one of the angles given in this table depends on one or more parameters, which have been determined experimentally from observed intensities of x-ray reflections. The probable error in most cases is less than 5°, and in many is only about ±1°. It will be observed that quadrivalent carbon and nitrogen and trivalent nitrogen form bonds at tetrahedral angles, whereas heavier atoms forming only two or three bonds prefer smaller bond angles. The series As, Sb, Bi is particularly interesting. We expect, from an argument given earlier,

TABLE I

ANGLES BETWEEN BONDS, FROM CRYSTAL STRUCTURE DATA[a]

Compound	Atom	Number of bonds	Angles between bonds
$C_6N_4H_{12}$	C	2 C—N, 2 C—H	112° between C—N bonds
$C_6N_4H_{12}$	N	3 N—C	108°
$(NH_2)_2CO$[b]	C	2 single C—N 1 double C=O	115° between single bonds
As	As	3	97°
Sb	Sb	3	96°
Bi	Bi	3	94°
Se	Se	2	105°
Te	Te	2	102°
FeS_2[d], MnS_2, CoS_2, NiS_2	S^{++}	1 S—S 3 S—M	103° between S—S and S—M bonds 115° between two M—S bonds
MoS_2[e]	S^+	3 S—Mo	82°
$Co_4(As_4)_3$	As^-	2 As—As	90°
$CaSi_2$	Si	3 Si—Si	103° between Si—Si bonds
HgI_2	I^+	2 Hg—I	103°
GeI_4	Ge	4 Ge—I	109.5°
SnI_4	Sn	4 Sn—I	109.5°
As_4O_6	As	3 As—O	109.5°
	O	2 O—As	109.5°
Sb_4O_6	Sb	3 Sb—O	109.5°
	O	2 O—Sb	109.5°
$NaClO_3$[c]	Cl^{++}	3 Cl—O	109.5°
$KClO_3$	Cl^{++}	3 Cl—O	109.5°
$KBrO_3$	Br^{++}	3 Br—O	109.5°

[a] Data for which no reference is given are from the *Strukturbericht* of P. P. Ewald and C. Hermann. [b] R. W. G. Wyckoff, *Z. Krist.*, **75**, 529 (1930). [c] W. H. Zachariasen, *ibid.*, **71**, 501, 517 (1929). [d] The very small paramagnetic susceptibility of pyrite requires the presence of electron-pair bonds, eliminating an ionic structure $Fe^{++}S_2^{=}$. Angles are calculated for FeS_2, for which the parameters have been most accurately determined. [e] The parameter value (correct value u = 0.371) and interatomic distances for molybdenite are incorrectly given in the *Strukturbericht*.

that the bond eigenfunctions will deviate less and less from pure p eigenfunctions in this order, and this evidences itself in a closer approach of the bond angle to 90° in the series. Geometrical effects sometimes affect the bond angles, as in As_4O_6 and Sb_4O_6, where a decrease in the oxygen bond angle would necessarily be accompanied by an increase in that for the other atom, and in molybdenite and pyrite.

Many compounds with tetrahedral structures (diamond, sphalerite, wurzite, carborundum, etc.) are known, in which the four bonds have tetrahedral angles. Tetrahedral atoms in such crystals include C (diamond, SiC), Si, Ge, Sn, Cl^{3+} (in CuCl), Br^{3+}, I^{3+}, O^{++} (in Cu_2O and ZnO), S^{++}, Se^{++}, Te^{++}, N^{+} (in AlN), P^{+}, As^{+}, Sb^{+}, Bi^{+}, $Cu^{\equiv}$, $Zn^{=}$, $Cd^{=}$, $Hg^{=}$, Al^{-}, Ga^{-} and In^{-}.

The Valence of Atoms.—In the last paragraph and in Table I the atoms are represented with electrical charges which are not those usually seen. These charges are obtained by the application of Rule 1, according to which an electron-pair bond is formed by one electron from each of the two atoms (even though as the atoms separate the type of bonding may change in such a way that both electrons go over to one atom.) Accordingly in determining the state of ionization of the atoms in a molecule or crystal containing electron-pair bonds each shared electron-pair is to be split between the two atoms. In this way every atom is assigned an electrovalence obtained by the above procedure and a covalence equal to the number of its shared electron-pair bonds.

It is of interest to note that a quantity closely related to the "valence" of the old valence theory is obtained for an atom by taking the algebraic sum of the electrovalence and of the covalence, the latter being given the positive sign for metals and the negative sign for non-metals. For example, oxygen in OH^- is O^- with a covalence of 1, in H_2O it is O with a covalence of 2, in H_3O^+ it is O^+ with a covalence of 3, and in crystalline ZnO it is O^{++} with a covalence of 4; in each case the above rule gives -2 for its valence.

Trigonal Quantization.—We have seen that an atom with *s-p* quantization unchanged will form three equivalent bonds at 90° to one another. If quantization is changed, the three strongest bonds will lie at tetrahedral angles. But increase in the bond angle beyond the tetrahedral angle is not accompanied by a very pronounced decrease in bond strength. Thus three equivalent bond eigenfunctions in a plane, with maxima 120° apart, can be formed

$$\left.\begin{aligned} \psi_1 &= \frac{1}{\sqrt{3}}\,s + \sqrt{\frac{2}{3}}\,p_x \\ \psi_2 &= \frac{1}{\sqrt{3}}\,s - \frac{1}{\sqrt{6}}\,p_x + \frac{1}{\sqrt{2}}\,p_y \\ \psi_3 &= \frac{1}{\sqrt{3}}\,s - \frac{1}{\sqrt{6}}\,p_x - \frac{1}{\sqrt{2}}\,p_y \end{aligned}\right\} \qquad (10)$$

and these have a strength of 1.991, only a little less than that 2.000 of tetrahedral bonds (Fig. 6). As a result, we may anticipate that in some cases the bond angles will be larger than 109°28′. The carbonate ion in calcite and the nitrate ion in sodium nitrate are assigned a plane configuration from the results of x-ray investigations. In these ions the oxygen atoms are only 2.25 Å. from one another, so that their characteristic repulsive forces must be large, resisting decrease in the bond angle (the smallest distance observed between oxygen ions in ionic crystals is 2.5 Å.). But repulsion of the oxygen atoms would not be very effective in increasing the bond angle in the neighborhood of 120°, so that we might expect equilibrium to be achieved at a somewhat smaller angle, such as 118°. This would give $CO_3^{=}$ and NO_3^{-} a

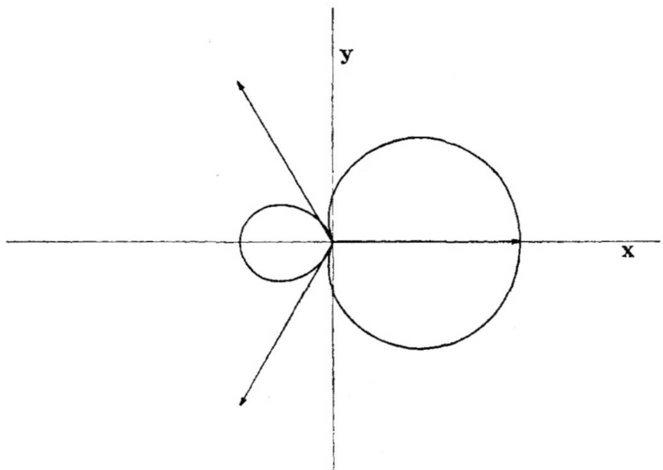

Fig. 6.—Polar graph of $\left|\frac{1}{\sqrt{3}} + \sqrt{2}\cos\varphi\right|$ in the xy plane, representing a trigonal eigenfunction. The maximum directions of the other two equivalent eigenfunctions are also shown.

pyramidal structure, like that of NH_3. There would be two configurations possible for a given orientation of the O_3 plane, one in which the carbon (or nitrogen) atom was a short distance above this plane (taken as horizontal) and one with it below the plane. If there is appreciable interaction between these two, as there will be in case the pyramid is flat, the symmetric and antisymmetric combinations of the two will be the correct eigenfunctions, corresponding to the rapid inversion of the pyramid, with a frequency of the order of magnitude of the vibrational frequency of the complex ion along its symmetry axis. This inversion would introduce an effective symmetry plane normal to the three-fold axis, so that a pyramidal structure with rapid inversion is compatible with the x-ray observations.[26]

[26] Simulation of symmetry by molecules or complex ions in crystals has been discussed by Linus Pauling, *Phys. Rev.*, **36**, 430 (1930).

Thus the x-ray data do not decide between this structure and a truly plane structure. Evidence from another source is at hand, however. A plane $CO_3^=$ or NO_3^- ion should show three characteristic fundamental vibrational frequencies. These have been observed as reflection maxima in the infra-red region. But two of the maxima, at 7μ and 14μ, are double,[27] and this doubling, which is not explicable with a plane configuration, is just that required by a pyramidal structure, the separation of the components giving the frequency of inversion of the pyramid.[28]

In graphite each carbon atom is bound to three others in the same plane; and here the assumption of inversion of a puckered layer is improbable, because of the number of atoms involved. A probable structure is one in which each carbon atom forms two single bonds and one double bond with other atoms. These three bonds should lie in a plane, with angles 109°28′ and 125°16′, which are not far from 120°. Two single bonds and a double bond should be nearly as stable as four single bonds (in diamond), and the stability would be increased by the resonance terms arising from the shift of the double bond from one atom to another. But this problem and the closely related problem of the structure of aromatic nuclei demand a detailed discussion, perhaps along the lines indicated, before they can be considered to be solved.

The Structures of Simple Molecules.—The foregoing considerations throw some light on the structure of very simple molecules in the normal and lower excited states, but they do not permit such a complete and accurate discussion of these questions as for more complicated molecules, because of the difficulty of taking into consideration the effect of several unshared and sometimes unpaired electrons. Often the bond energy is not great enough to destroy s–p quantization, and the interaction between a bond and unshared electrons is more important than between a bond and other shared electrons because of the absence of the effect of concentration of the eigenfunctions.

Let us consider an atom forming a bond with another atom in the direction of the z axis. Then p_z and s form two eigenfunctions designated σ, p_x and p_y two designated π (one with a resultant moment of $+1$ along the z axis, one with -1). If s–p quantization is not broken, the strongest bond will be formed by p_z, and weaker ones by π. If s–p quantization is broken, new eigenfunctions σ_b and σ_o will be formed from s and p_z. In this case the strongest bond is formed by the σ_b eigenfunction, which extends out toward the other atom, weaker ones are formed by π_+ and π_-, and an extremely weak one, if any, by σ_o. We can also predict the stability of

[27] C. Schaefer, F. Matossi and F. Dane, *Z. Physik*, **45**, 493 (1927).

[28] The normal states of these ions are similar to certain excited states of ammonia, which also show doubling. The frequency of inversion of the normal ammonia molecule is negligibly small.

unshared electrons; σ_o, involving s with its greater penetration of the atom core, will be more stable than π.

As examples we may discuss CO, CN, N_2 and NO. CO might be composed of normal or excited atoms, or even of ions. A neutral oxygen atom can form only two bonds. Hence a normal carbon atom, ^3P, which can also form two bonds, is at no disadvantage. We can write the following reaction, using symbols similar to those of Lennard-Jones[29] and Dunkel,[30] whose treatments of the electronic structure of simple molecules have several points of similarity with ours

$$\mathrm{C}\ 2s^2 2p2p\ ^3P + \mathrm{O}\ 2s^2 2p^2 2p2p\ ^3P \longrightarrow \mathrm{CO}\ (2\sigma_o^2)(2\sigma_o^2 2\pi^2)\{2\sigma_b 2\pi + 2\sigma_b 2\pi\}\,^1\Sigma$$

$$:\underset{\cdot}{\mathrm{C}}\cdot + \cdot\underset{\cdot}{\ddot{\mathrm{O}}}: \longrightarrow :\mathrm{C}::\ddot{\mathrm{O}}: \qquad \text{Normal state}$$

Here symbols in parentheses represent unshared electrons attached to C and O, respectively, and those in braces represent shared electrons. An excited carbon atom ^5S lies about 1.6 v. e. above the normal state, but can still form only a double bond with oxygen, so that the resultant molecule should be excited. We write

$$\mathrm{C}^*\ 2s2p2p2p\ ^5S + \mathrm{O}\ 2s^2 2p^2 2p2p\ ^3P \longrightarrow \mathrm{CO}^*\ (2\sigma_o 2\pi)(2\sigma_o^2 2\pi^2)\{2\sigma_b 2\pi + 2\sigma_b 2\pi\}\,^3\Pi \text{ or } ^1\Pi$$

The resultant states are necessarily Π, for σ_b and one π are used for the bond, leaving on C σ_o and π. These two electrons may or may not pair with one another, giving $^1\Pi$ and $^3\Pi$, respectively. Of these $^3\Pi$ should be the more stable, for the two electrons are attached essentially to one atom, and the rules for atomic spectra should be valid. This is substantiated; the observed excited states $^3\Pi$ and $^1\Pi$ lie at 5.98 and 7.99 v. e., respectively. Another way of considering these three states is the following: to go from $:\mathrm{C}::\ddot{\mathrm{O}}:$ to $\cdot\underset{\cdot}{\mathrm{C}}::\ddot{\mathrm{O}}:$ we lift an electron from the more deeply penetrating σ_o orbit to π; about 6–8 v. e. is needed for this, and the resultant state is either $^3\Pi$ or $^1\Pi$. This viewpoint does not necessitate the discussion of products of dissociation.

CN is closely similar. The normal nitrogen atom, $2s^2 2p2p2p\ ^4S$, can form three bonds, and more cannot be formed by an excited neutral atom (with five L electrons), so that there is no reason to expect excitation. But a normal carbon atom can form only a double bond, and an excited carbon atom, only 1.6 v. e. higher, can form a triple bond, which contributes about 3 v. e. more than a double bond to the bond energy. Hence we write

$$\mathrm{C}^*\ 2s2p2p2p\ ^5S + \mathrm{N}\ 2s^2 2p2p2p\ ^4S \longrightarrow \mathrm{CN}\ (2\sigma_o)(2\sigma_o^2)\{2\sigma_b 2\pi 2\pi + 2\sigma_b 2\pi 2\pi\}\,^2\Sigma$$

$$\cdot\underset{\cdot}{\dot{\mathrm{C}}}\cdot + \cdot\underset{\cdot}{\dot{\mathrm{N}}}: \longrightarrow \cdot\mathrm{C}:::\mathrm{N}: \qquad \text{Normal state}$$

The first excited state of the molecule, $:\mathrm{C}::\underset{\cdot}{\mathrm{N}}:$, is built from normal atoms, and has the term symbol $^2\Pi$. It lies 1.78 v. e. above the normal state.

[29] J. E. Lennard-Jones, *Trans. Faraday Soc.*, **25**, 668 (1929).

[30] M. Dunkel, *Z. physik. Chem.*, **B7**, 81 (1930).

Two normal nitrogen atoms form a normal molecule with a triple bond.

$$2\text{N}\ 2s^2 2p2p2p\ {}^4S \longrightarrow \text{N}_2(2\sigma_s^2)(2\sigma_o^2)\{2\sigma_b 2\pi 2\pi + 2\sigma_b 2\pi 2\pi\}\ {}^1\Sigma$$

$$:\dot{\text{N}}\cdot + \cdot\dot{\text{N}}: \longrightarrow :\text{N}:::\text{N}: \quad \text{Normal state}$$

All other states lie much higher.

A normal oxygen atom and a normal nitrogen atom form a normal NO molecule with a double bond.

$$\text{N}\ 2s^2 2p2p2p\ {}^4S + \text{O}\ 2s^2 2p^2 2p2p\ {}^3P \longrightarrow \text{NO}(2\sigma_o^2 2\pi)(2\sigma_o^2 2\pi^2)\{2\sigma_b 2\pi + 2\sigma_b 2\pi\}\ {}^2\Pi$$

$$:\dot{\text{N}}\cdot + \cdot\ddot{\text{O}}: \longrightarrow :\text{N}::\text{O}: \quad \text{Normal state}$$

This treatment sometimes fails for symmetrical molecules. Thus :Ö::Ö: $^1\Sigma$ would be predicted for the normal state of O_2, whereas the observed normal state, $^3\Sigma$, lies 1.62 v. e. below this. It seems probable that the additional degeneracy arising from the identity of the two atoms gives rise to a new type of bond, the *three-electron bond*, and that in normal O_2 there are one single bond and two three-electron bonds, :O:O:, $^3\Sigma$; a definite decision regarding this question must await a detailed quantum-mechanical treatment. Evidence regarding the oxygen–oxygen single bond is provided by O_4, with the square structure

:Ö:Ö:
:Ö:Ö:

The 90° bond angles are expected, since quantization in s and p eigenfunctions is not changed. The equality in energy of O_4 and $2O_2$ leads to an energy of 58,000 cal. per mole per single bond in O_4; the difference between this value and that for a carbon–carbon single bond (100,000 cal.) shows the greater bond-forming power of tetrahedral eigenfunctions over p eigenfunctions. Ozone, which very probably has the symmetrical arrangement

:O:
:O : O:

has 60° bond angles, and this distortion from the most favorable bond angle of 90° shows up in the bond energy, for the heat of formation of −34,000 cal. per mole leads to 47,000 cal. per mole per single bond, a decrease of 11,000 cal. over the favored O_4 bonds.

For some polyatomic molecules predictions can be made regarding the atomic arrangement from a knowledge of the electronic structure or *vice versa*. Thus ·C:::N: $^2\Sigma$ can form a bond through the unpaired σ_o electron of carbon, and this bond will extend along the CN axis. Hence the molecules H:C:::N:, :N:::C:C:::N: and :C̈l:C:::N: should be linear. This is verified by band spectral data.[31] The isocyanides, RNC, such as H_3CNC, may be given either a triple or a double bond structure: R:N:::C: or R:N̤::C:. The first of these is built of the ions N^+ ^5S and C^- ^4S, which may be an argument in favor of the second structure, built of normal

[31] Private communication from Professor Richard M. Badger of this Laboratory, who has kindly provided me with much information concerning the results of band spectroscopy.

atoms.[32] A decision between the two alternatives could be made by determining the atomic arrangement of an isocyanide, for the triple bond gives a linear molecule, bond angle 180°, and the double bond a kinked molecule, bond angle between 90 and 109°28′.

The molecules and complex ions containing three kernels and sixteen L electrons form an interesting group. Of these CO_2, formed from excited carbon ^5S and normal oxygen atoms, would have the structure $:\ddot{O}::C::\ddot{O}:$. The two double bonds make the molecule linear, which is verified by both crystal structure and band spectral data. Crystal structure data also show N_2O to be linear, although it is not known whether or not the molecule has oxygen in the middle or at one end, as first suggested by Langmuir[33] and supported by the kernel-repulsion rule.[34] The known linear arrangement eliminates structures built of neutral atoms, $:\underset{\cdot}{\dot{N}}:O:\underset{\cdot}{\dot{N}}:$ and $:\underset{\cdot}{\dot{N}}::\underset{\cdot\cdot}{N}:\ddot{O}:$, for these have bond angles between 90 and 125°. The structures $:\underset{\cdot\cdot}{N}::N::\underset{\cdot\cdot}{O}:$ and $:\underset{\cdot\cdot}{N}::O::\underset{\cdot\cdot}{N}:$, built from N N^+ O^- and N^- O^{++} N^-, respectively, would both be linear, and so compatible with the known arrangement. An *a priori* decision between them is difficult, although previously advanced arguments favor the unsymmetrical structure. Band spectra should soon decide the question.

The trinitride, cyanate, and isocyanate ions, the first two of which are known[35] to be linear, no doubt have identical electronic structures.

$$:\underset{\cdot}{\ddot{N}}\cdot^{-}\ {}^3P + \cdot\underset{\cdot}{\dot{N}}\cdot^{+}\ {}^5S + \cdot\underset{\cdot}{\ddot{N}}:^{-} \longrightarrow :\ddot{N}::N::\ddot{N}:^{-}$$

or

$$N^- + N^+ + N^- \longrightarrow N_3^- \text{ Trinitride ion}$$
$$N^- + C + O \longrightarrow NCO^- \text{ Cyanate ion}$$

The fulminate ion, CNO^-, probably has a structure intermediate between $:\ddot{C}::N::\ddot{O}:^-$ and $:C:::N:\underset{\cdot\cdot}{\ddot{O}}:^-$; for since these two bond types have the same bond angles and term symbols ($^1\Sigma$), they can form intermediate structures lying anywhere between the two extremes. Which extreme is the more closely approached could be determined from a study of the bond angles in un-ionized fulminate molecules, such as AgCNO or ONCHgCNO, for the first structure would lead to an angle of 125° between the CNO axis and the metal–carbon bond, the second to an angle of 180°.

Bonds Involving *d*-Eigenfunctions.—When *d* eigenfunctions as well as *s* and *p* can take part in bond formation, the number and variety of bonds which can be formed are increased. Thus with an *s*, a *p* and a *d* subgroup as many as nine bonds can be formed by an atom. It is found from a

[32] Thus W. Heitler and G. Rumer, *Nachr. Ges. Wiss. Göttingen, Math. physik. Klasse*, **7**, 277 (1930), in a paper on the quantum mechanics of polyatomic molecules, discuss only the second structure.

[33] I. Langmuir, This Journal, **41**, 1543 (1919).

[34] Linus Pauling and S. B. Hendricks, *ibid.*, **48**, 641 (1926).

[35] S. B. Hendricks and Linus Pauling, This Journal, **47**, 2904 (1925).

consideration of the eigenfunctions that all cannot be equivalent, but six equivalent bonds extending toward the corners of either a regular octahedron or a trigonal pyramid, four extending toward the corners of a tetrahedron or a square, etc., can be formed; and the strength and mutual orientation of the bonds are determined by the number of d eigenfunctions involved in their formation.

There are five d eigenfunctions in a subgroup with $l = 2$ and with given n. They are

$$\left.\begin{aligned} d_z &= \sqrt{5/4}\,(3\cos^2\theta - 1) \\ d_{y+z} &= \sqrt{15}\,\sin\theta\cos\theta\cos\varphi \\ d_{x+z} &= \sqrt{15}\,\sin\theta\cos\theta\sin\varphi \\ d_{x+y} &= \sqrt{15/4}\,\sin^2\theta\sin 2\varphi \\ d_x &= \sqrt{15/4}\,\sin^2\theta\cos 2\varphi \end{aligned}\right\} \qquad (11)$$

or any set of five orthogonal functions formed by linear combination of these. These functions are not well suited to bond formation. d_{y+z}, d_{x+z} and d_{x+y}, which are similarly related to the x, y and z axes, respectively, have the form shown in Fig. 7. Each eigenfunction has maxima in

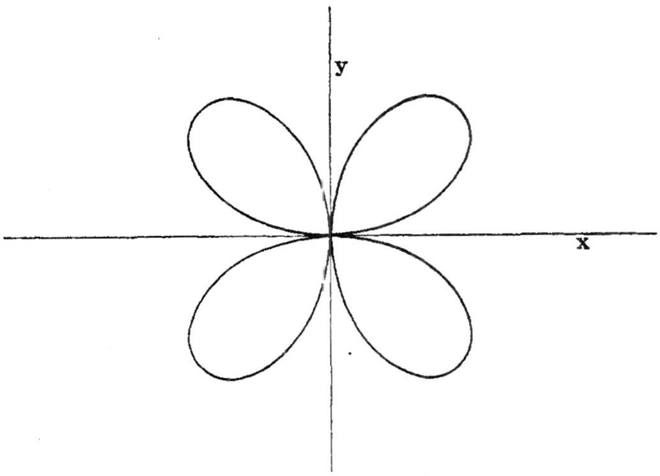

Fig. 7.—Polar graph of $\left|\frac{\sqrt{15}}{2}\sin^2\theta\sin 2\varphi\right|$ in the xy plane, representing the $d_x +$ eigenfunction.

four directions. d_x is similar in shape, differing from d_{x+y} only in a rotation of 45° about the z axis. d_z, shown in Fig. 8, has two maxima along the z axis, and a girdle about its waist.

Assuming as before that the dependence on r of the s, p and d eigenfunctions under discussion is not greatly different, the best bond eigenfunctions can be determined by the application of the treatment already applied to s and p alone, with the following results.

The best bond eigenfunction which can be obtained from s, p and d is

$$\frac{1}{3}s + \frac{1}{\sqrt{3}}p_z + \frac{\sqrt{5}}{3}d_z$$

and has a strength of 3. The best two equivalent bond eigenfunctions involving one d eigenfunction

$$\frac{1}{2\sqrt{3}}s + \frac{1}{\sqrt{2}}p_z + \frac{\sqrt{5}}{2\sqrt{3}}d_z \quad \text{and}$$

$$\frac{1}{2\sqrt{3}}s - \frac{1}{\sqrt{2}}p_z + \frac{\sqrt{5}}{2\sqrt{3}}d_z$$

are oppositely directed and have a strength of 2.96.

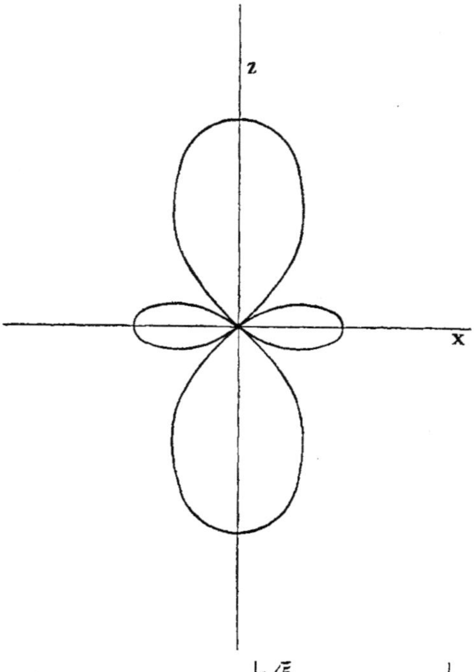

Fig. 8.—Polar graph of $\left|\frac{\sqrt{5}}{2}(3\cos^2\theta - 1)\right|$ in the xz plane, representing the d_z eigenfunction.

The atoms of the transition elements, for which d eigenfunctions need to be considered, are of such a size as usually to have a coördination number of 4 or 6, so that four or six equivalent bond eigenfunctions are here of especial interest. If there is available only one d eigenfunction to be combined with an s and three p eigenfunctions, then no more than five bond eigenfunctions can be formed. One may have the maximum strength 3, in which case the others are weak; or two may be strong and three weak; but *with a single d eigenfunction no more than four strong bonds can be formed, and these lie in a plane.* The fifth bond is necessarily weak. The four equivalent bond eigenfunctions formed from s, p and one d eigenfunction are

$$\left.\begin{aligned}
\psi_1 &= \frac{1}{2}s + \frac{1}{2}d_x + \frac{1}{\sqrt{2}}p_x \\
\psi_2 &= \frac{1}{2}s + \frac{1}{2}d_x - \frac{1}{\sqrt{2}}p_x \\
\psi_3 &= \frac{1}{2}s - \frac{1}{2}d_x + \frac{1}{\sqrt{2}}p_y \\
\psi_4 &= \frac{1}{2}s - \frac{1}{2}d_x - \frac{1}{\sqrt{2}}p_y
\end{aligned}\right\} \qquad (12)$$

One of these is shown in Fig. 9. These all have their maxima in the xy plane, directed toward the corners of a square. The strength of these bond eigenfunctions, 2.694, is much greater than that of the four tetrahedral eigenfunctions formed from s and p alone (2.00). But if three d

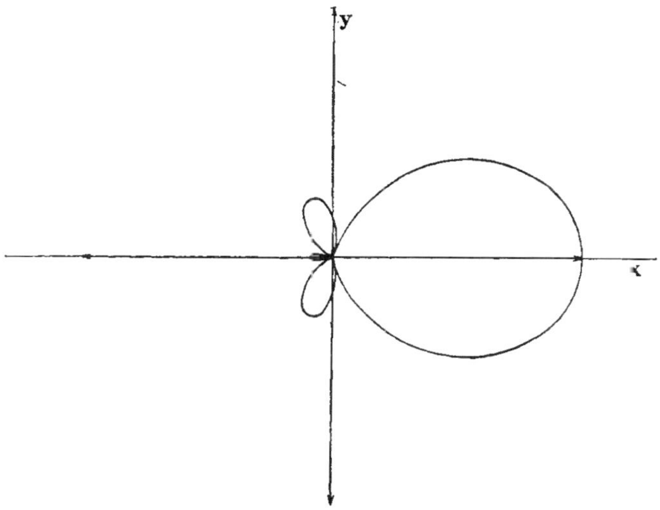

Fig. 9.—Polar graph of $\left| {}^1/_2 + \sqrt{{}^3/_2} \cos \varphi + \frac{\sqrt{15}}{4} \cos 2\varphi \right|$ in the xy plane, representing one of the four equivalent dsp^2 bond eigenfunctions. The directions of the maxima of the four are represented by arrows.

eigenfunctions are available, stronger bonds directed toward tetrahedron corners can be formed. The equivalent tetrahedral bond eigenfunctions

$$\psi_{111} = \frac{1}{2}s + \frac{\sqrt{3}}{4\sqrt{2}}(p_x + p_y + p_z) + \frac{\sqrt{5}}{4\sqrt{2}}(d_{y+z} + d_{x+z} + d_{x+y})$$

$$\psi_{1\bar{1}\bar{1}} = \frac{1}{2}s + \frac{\sqrt{3}}{4\sqrt{2}}(p_x - p_y - p_z) + \frac{\sqrt{5}}{4\sqrt{2}}(d_{y+z} - d_{x+z} - d_{x+y})$$

$$\psi_{\bar{1}1\bar{1}} = \frac{1}{2}s + \frac{\sqrt{3}}{4\sqrt{2}}(-p_x + p_y - p_z) + \frac{\sqrt{5}}{4\sqrt{2}}(-d_{y+z} + d_{x+z} - d_{x+y})$$

$$\psi_{\bar{1}\bar{1}1} = \frac{1}{2}s + \frac{\sqrt{3}}{4\sqrt{2}}(-p_x - p_y + p_z) + \frac{\sqrt{5}}{4\sqrt{2}}(-d_{y+z} - d_{x+z} + d_{x+y})$$

have a strength of 2.950, nearly equal to the maximum 3. These leave only two pure d eigenfunctions behind, however, the others being part d and part p. Thus we conclude that if there are three d eigenfunctions available, a transition group element forming four electron-pair bonds will direct them toward tetrahedron corners. Examples of such bonds are provided by $CrO_4^{=}$, $MoO_4^{=}$, etc. Only when one d eigenfunction alone is available will the four bonds lie in a plane. In compounds of bivalent nickel, palladium, and platinum, such as $K_2Ni(CN)_4$, $K_2Pd(CN)_4$, K_2PdCl_4, K_2PtCl_4, etc., there are eight unshared *d* electrons on each metal atom, which occupy four of the five *d* eigenfunctions. Hence the four added atoms or groups lie in a plane at the corners of a square about the metal atom. Such a configuration was assigned to palladous and platinous compounds by Werner because of the existence of apparent *cis* and *trans* compounds, and has been completely substantiated by the x-ray investigation of the chloropalladites and chloroplatinites.[36] The square configuration has not before been attributed to $K_2Ni(CN)_4$; it is supported by the observed isomorphism of the monoclinic crystals $K_2Pd(CN)_4 \cdot H_2O$ and $K_2Ni(CN)_4 \cdot H_2O$, and it will be shown in a following section that it is compatible with the magnetic data.

The non-existence of compounds K_3PtCl_5, etc., is explained by the weak bond-forming power (1.732) of the remaining eigenfunction p_z.

Now if two *d* eigenfunctions are available, six equivalent eigenfunctions

$$\left.\begin{aligned}
\psi_1 &= \frac{1}{\sqrt{6}} s + \frac{1}{\sqrt{2}} p_z + \frac{1}{\sqrt{3}} d_z \\
\psi_2 &= \frac{1}{\sqrt{6}} s - \frac{1}{\sqrt{2}} p_z + \frac{1}{\sqrt{3}} d_z \\
\psi_3 &= \frac{1}{\sqrt{6}} s + \frac{1}{\sqrt{12}} d_z + \frac{1}{2} d_x + \frac{1}{\sqrt{2}} p_x \\
\psi_4 &= \frac{1}{\sqrt{6}} s + \frac{1}{\sqrt{12}} d_z + \frac{1}{2} d_x - \frac{1}{\sqrt{2}} p_x \\
\psi_5 &= \frac{1}{\sqrt{6}} s + \frac{1}{\sqrt{12}} d_z - \frac{1}{2} d_x + \frac{1}{\sqrt{2}} p_y \\
\psi_6 &= \frac{1}{\sqrt{6}} s + \frac{1}{\sqrt{12}} d_z - \frac{1}{2} d_x - \frac{1}{\sqrt{2}} p_y
\end{aligned}\right\} \tag{13}$$

can be formed. These form strong bonds, of strength 2.923, directed toward the corners of a regular octahedron; and no stronger octahedral bonds can be formed even though more *d* eigenfunctions be available (Figs. 10 and 11). Hence we expect transition group atoms with six or less unshared electrons to form six electron-pair bonds. Examples of such compounds are numerous: $PtCl_6^{=}$, $Fe(CN)_6^{=}$, etc., although the definite assignment of an electron-pair bond structure rather than an ionic structure (as in $FeF_6^{\equiv}$, formed of Fe^{+++} and 6 F^-) can be made only after the discussion of paramagnetic susceptibility.

I have not succeeded in determining whether or not these octahedral eigenfunctions are the strongest six equivalent bond eigenfunctions which can be formed when more than two *d*'s are available. The known structure of molybdenite, MoS_2, suggests that six bonds directed toward the corners of a trigonal prism are stable; but only a small increase in bond strength can possibly be obtained (from 2.923 to not over 3), and the mutual re-

[36] R. G. Dickinson, THIS JOURNAL, **44**, 2404 (1922).

pulsion of the six atoms or groups will in most cases overcome this, if it does exist, and leave the octahedral configuration the stable one.

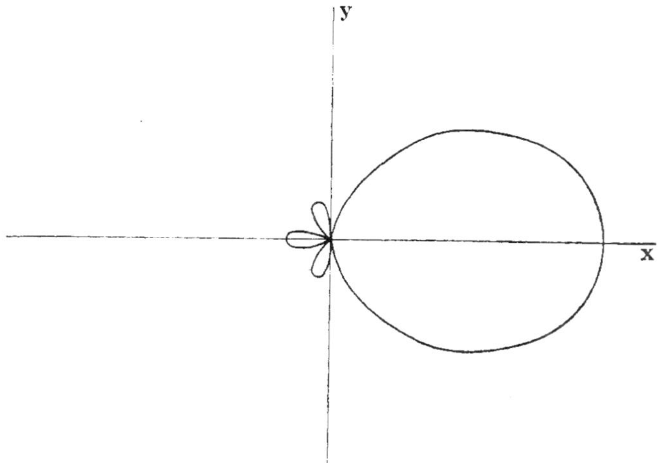

Fig. 10.—Polar graph of $|\psi_8|$ of Equation 13, in the xy plane, representing one of the six equivalent d^2sp^3 bond eigenfunctions (octahedral eigenfunctions).

II. The Magnetic Moments of Molecules and Complex Ions

The theory of the paramagnetic susceptibility of substances has been developed gradually over a long period of years through the efforts of a number of investigators. The theoretical calculation of the magnetic moments of complex molecules and ions has in particular attracted much attention recently, and both theoretical and empirical considerations have been used in developing rules applicable in various cases. The work reported in this paper provides little more than the justification and unification of previously developed rules. This finishing touch is, however, of much significance for the problem of the nature of the chemical bond; for it, in conjunction with the quantum mechanical discussion of the previous sections, permits definite conclusions to be drawn regarding type of bond in many molecules and complex ions from a knowledge of their magnetic moments, and conversely provides the basis for the definite prediction of magnetic moments from a knowledge of the type of bonds and the atomic arrangement.

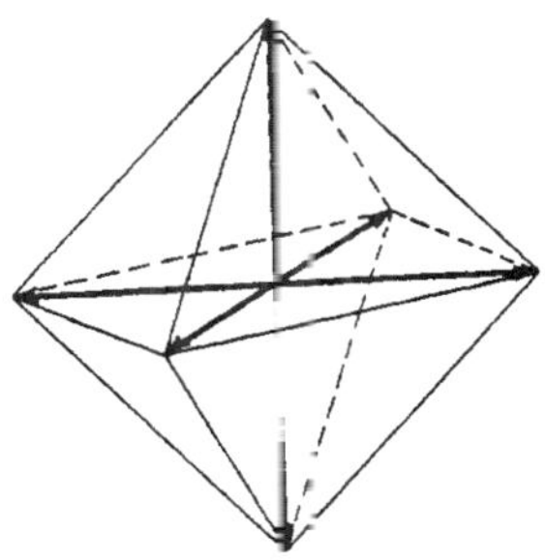

Fig. 11.—Diagram showing relative orientation in space of the directions of the maxima of the octahedral eigenfunctions.

The calculation of the magnetic moments of the rare-earth ions by

Hund[37] in 1926 and of oxygen and nitric oxide by Van Vleck[38] in 1928 were triumphs of the theory of spectra. The magnetic moment of an atom or monatomic ion with Russell–Saunders coupling of the quantum vectors is

$$\mu_J = g\sqrt{J(J+1)}$$

in which g, the Landé splitting factor, is given by

$$g = 1 + \frac{J(J+1) + S(S+1) - L(L+1)}{2J(J+1)}$$

Here L, S, and J are the quantum numbers corresponding to the total orbital angular momentum of the electrons, the total spin angular momentum, and the resultant of these two. Hund predicted values of L, S, and J for the normal states of the rare-earth ions from spectroscopic rules, and calculated μ-values for them which are in generally excellent agreement with the experimental data for both aqueous solutions and solid salts.[39] In case that the interaction between L and S is small, so that the multiplet separation corresponding to various values of J is small compared with kT, Van Vleck's formula[38]

$$\mu_{LS} = \sqrt{4S(S+1) + J(J+1)}$$

is to be used.

But similar calculations for the iron-group ions show marked disagreement with experiment, and many attempts were made to explain the discrepancies. The explanation is simple: *in many condensed systems the perturbing effect of the atoms or molecules surrounding a magnetic atom destroys the contribution of the orbital momentum to the magnetic moment, which is produced entirely by the spin moments of unpaired electrons.*[40]

This conclusion is easily deduced from the consideration of the nature of eigenfunctions giving rise to magnetic moments. In an atom containing unpaired p electrons, say, a component of orbital magnetic moment of $\pm(h/2\pi)\cdot(e/2mc)$ is obtained when an unpaired electron is in a state given by the eigenfunction $p_x \pm i\, p_y$. Now if the perturbing influence of surrounding atoms or molecules is such as to make the perturbation energy for the eigenfunction p_x or p_y or any combination of them other than $p_x \pm i\, p_y$ greater than the field energy, this will be the correct zeroth order eigenfunction, and the atom will show no orbital magnetic moment. In an atom with Russell–Saunders coupling the interaction energy of L and S takes the place of the field energy, so that the criterion to be satisfied in order that the magnetic moment due to L be destroyed is that the perturbation energy due to surrounding atoms and ions be greater than the multiplet separation, which for the iron-group ions is of the order of magnitude of 1 v. e.[41]

[37] F. Hund, *Z. Physik*, **33**, 345 (1925).

[38] J. H. Van Vleck, *Phys. Rev.*, **31**, 587 (1928).

[39] The few discrepancies have been accounted for by S. Freed [This Journal, **52**, 2702 (1930)] and J. H. Van Vleck and A. Frank [*Phys. Rev.*, **34**, 1494 (1929), and a paper delivered at the Cleveland meeting of the American Physical Society, December 31, 1930].

[40] This assumption was first made by E. C. Stoner, *Phil. Mag.*, **8**, 250 (1929), in order to account for the observed moments of iron-group ions.

[41] Essentially the same conclusion has been announced by J. H. Van Vleck at the Cleveland meeting of the American Physical Society, December 31, 1930.

If the perturbation function shows cubic symmetry, and in certain other special cases, the first-order perturbation energy is not effective in destroying the orbital magnetic moment, for the eigenfunction $p_x \pm i\, p_y$ leads to the same first-order perturbation terms as p_x or p_y or any other combinations of them. In such cases the higher order perturbation energies are to be compared with the multiplet separation in the above criterion.

In linear molecules only the component of orbital momentum normal to the figure axis is destroyed, that along the figure axis being retained. In non-linear molecules with strong interatomic interactions the concept of orbital angular momentum loses its significance.

The rare-earth ions owe their magnetic moments to an incompleted $4f$ subshell, which lies within an outer shell of $5s$ and $5p$ electrons, and is thus protected from strong perturbations by surrounding atoms. As a consequence the orbital magnetic moment is not destroyed, and the ion is not affected by its environment. But in the iron-group ions and other transition-group ions the incompleted subshell is the outermost one. Hence it is not surprising that the solvent molecules or the surrounding atoms or ions in a complex ion or a crystal interact sufficiently strongly with these atoms or ions to destroy, in whole, or in part, the orbital magnetic moment, leaving the spin moment, with perhaps a small contribution from the orbital moment in border-line cases. We can state with certainty that the formation of electron-pair bonds will destroy the orbital moment.

This greatly simplifies the theory of the magnetic moments of molecules and complex ions. *The magnetic moment of a molecule or complex ion is determined entirely by the number of unpaired electrons, being equal to*

$$\mu_S = 2\sqrt{S(S+1)}$$

in which S is one-half that number. The factor 2 is the g-factor for electron spin.

As a matter of fact, Sommerfeld[42] in 1924, a year before Hund's treatment of the rare-earth ions, noticed that the observed magnetic moments of K^+ and Ca^{++}, Ca^+ (spectroscopic), Ca (spectroscopic), Cr^{3+}, Cr^{++}, Mn^{++}, Fe^{++}, Co^{++}, Ni^{++}, Cu^{++} and Cu^+ are approximately reproduced by the above equation with $S = 0, {}^1/_2, 1, {}^3/_2, 2, {}^5/_2, 2, {}^3/_2, 1, /_2$ and 0, respectively. But with the development of spectral theory he apparently gave up this simple formula because of lack of a theoretical derivation of it, and it remained for Bose[43] in 1927 to state explicitly the assumption that only S contributes to the moment in these cases, without, however, explaining why L gives no contribution, and for Stoner[40] in 1929 to supply the explanation. The comparison of calculated and observed values is given in Table I. It may be pointed out that S increases to a maximum value of ${}^5/_2$ when the $3d$ subgroup is half filled; Pauli's principle requires that succeeding electrons decrease the spin, so that μ_S is symmetrical about

[42] A. Sommerfeld, "Atombau," 4th ed., p. 639.

[43] D. M. Bose, *Z. Physik*, **43**, 864 (1927).

this point. The agreement with experiment, while much better than for μ_J, is not perfect; ions with more than five 3*d* electrons are found to have moments larger than μ_S, while V^{3+} deviates in the other direction. Bose suggested that perhaps *S* could in some cases exceed the maximum value allowed by Pauli's principle, but the obviously correct explanation is that the perturbing effect of surrounding atoms is not sufficient completely to destroy the *L* moment. Hence the observed moment should lie between μ_S and μ_J, which it does in every case.

Since the interaction is not strong enough to destroy the *L* moment, we conclude that in aqueous solution and in some crystalline salts the atoms[44] Fe^{II}, Co^{III}, Co^{II}, Ni^{II} and Cu^{II} do not form strong electron-pair bonds with H_2O, Cl, or certain other atoms, the bonds instead being ion-dipole or ionic bonds.

The formation of a stable coördination compound involving the four tetrahedral sp^3 eigenfunctions might decrease the *L* contribution appreciably. It was indeed pointed out by Bose that in the compounds listed in the last column of Table II the observed moments approach more closely the theoretical values μ_S.

The Magnetic Moments of Complexes with Electron-Pair Bonds.—The peculiar magnetic behavior of some complex ions has attracted much attention. $[Fe(CN)_6]^{3-}$ and $[Fe(CN)_6]^{4-}$, for example, have $\mu = 2.0$ and 0.00, respectively, instead of the values 5.9 and 4.9 for Fe^{3+} and Fe^{++}. Welo and Baudisch[45] and later Sidgwick and Bose expressed essentially the following rule: the magnetic moment of a complex is the same as that of the atom with the same number of electrons as the central atom of the complex, counting two for each electron-pair bond. Fe^{++} has 24 electrons; adding 12 for the six bonds gives 36, the electron number of krypton, so that the diamagnetism of the ferrocyanide ion is explained. This rule is satisfactory in many cases, but there are also many exceptions. Thus $[Ni(CN)_4]^{=}$ is diamagnetic, although the above rule would make it as paramagnetic as $[Ni(NH_3)_4]^{++}$.

The whole question is clarified when considered in relation to the foregoing quantum mechanical treatment of the electron-pair bond. For the iron-group elements the following rules follow directly from that treatment and from the rules of line spectroscopy.

1. *Bond eigenfunctions for iron-group atoms are formed from the nine eigenfunctions $3d^5$, $4s$ and $4p^3$, as described in preceding sections. One bond eigenfunction is needed for each electron-pair bond.*

2. *The remaining (unshared) electrons are to be introduced into the 3d eigenfunctions not involved in bond formation.*

[44] The symbol Fe^{II} is used for bivalent iron, etc., when the type of bond is undetermined.

[45] L. A. Welo and O. Baudisch, *Nature*, **116**, 606 (1925).

TABLE II

MAGNETIC MOMENTS OF IRON-GROUP IONS[a]

Ion	Normal state	μ_J	μ_S	Obs. moment in aqueous soln.	Solid salts, probable coördination number 6		Solid salts, coördination number	
K^+, Ca^{++}, Sc^{3+}, Ti^{4+}	$^1S_0$	0.00	0.00	0.00				
V^{4+}	$^2D_{3/2}$	1.55	1.73	1.7				
V^{3+}	$^3F_2$	1.63	2.83	2.4				
V^{++}, Cr^{3+}	$^4F_{3/2}$	0.78	3.88	3.8–3.9	$Cr_2O_3 \cdot 7H_2O$	3.85		
					$CrCl_3$	3.81		
Cr^{++}, Mn^{3+}	$^5D_0$	0.00	4.90	4.8–4.9				
Mn^{++}, Fe^{3+}	$^6S_{5/2}$	5.91	5.91	5.8	$MnCl_2$	5.75		
					$MnSO_4$	5.87		
					$MnSO_4 \cdot 4H_2O$	5.87		
					$Fe_2(SO_4)_3$	5.86		
					$(NH_4)_2Fe_2(SO_4)_4$	5.86		
Fe^{++}, Co^{3+}	$^5D_4$	6.76	4.90	5.3	$FeCl_2$	5.23	$Fe(N_2H_4)_2Cl_2$	4.87
					$FeCl_2 \cdot 4H_2O$	5.25		
					$FeSO_4$	5.26		
					$FeSO_4 \cdot 7H_2O$	5.25		
					$(NH_4)_2Fe(SO_4)_2 \cdot 6H_2O$	5.25		
Co^{++}	$^4F_{9/2}$	6.68	3.88	5.0–5.2	$CoCl_2$	5.04	$Co(N_2H_4)_2SO_3 \cdot H_2O$	4.31
					$CoSO_4$	5.04–5.25	$Co(N_2H_4)_2(CH_3COO)_2$	4.56
					$CoSO_4.7H_2O$	5.06	$Co(N_2H_4)_2Cl_2$	4.93
					$(NH_4)_2Co(SO_4)_2 \cdot 6H_2O$	5.00		
Ni^{++}	$^3F_4$	5.64	2.83	3.2	$NiCl_2$	3.24–3.42	$Ni(N_2H_4)_2SO_3$	3.20
					$NiSO_4$	3.42	$Ni(N_2H_4)_2(NO_2)_2$	2.80
							$Ni(NH_3)_4SO_4$	2.63
							$Ni(C_2H_4(NH_2)_2)_2(SCN)_2H_2O$	2.63
Cu^{++}	$^2D_{5/2}$	3.56	1.73	1.9–2.0	$CuCl_2$	2.02	$Cu(NH_3)_4(NO_3)_2$	1.82
					$CuSO_4$	2.01	$Cu(NH_3)_4SO_4 \cdot H_2O$	1.81
Cu^+, Zn^{++}	$^1S_0$	0.00	0.00	0.00				

[a] Observed magnetic moments, other than those in the last column, are from "International Critical Tables."

3. *The normal state is the state with the maximum resultant spin S allowed by Pauli's principle.*

These rules apply also to the palladium and platinum groups, the eigenfunctions involved being $4d^5 5s5p^3$ and $5d^5 6s6p^3$, respectively.

There are several important types of molecules and complexes to be given separate discussion.

If the bonds are ionic or ion-dipole bonds, the magnetic moments are those of the isolated central ions, given in the first column of moments in Table III. If the complex involves electron-pair bonds formed from sp^3 alone, such as four tetrahedral sp^3 bonds, the magnetic moments are the same, for the five d eigenfunctions are still available for the remaining electrons. The hydrazine and ammonia complexes mentioned above come in this class.

If four strong bonds involving a d eigenfunction are formed (giving a square configuration), only four d eigenfunctions are available for the additional electrons. The magnetic moments are then those given in the second column of the table. Examples of such compounds are $K_2Ni(CN)_4$, $K_2Pd(CN)_4 \cdot H_2O$, K_2PdCl_4, K_2PtCl_4, $K_2Pt(C_2O_4)_2 \cdot 2H_2O$ and $Pt(NH_3)_4SO_4$. With eight unshared d electrons, these should all be diamagnetic. This has been experimentally verified for the first and the last three compounds; data for the others are not available. The square configuration has been experimentally verified for the chloropalladites and chloroplatinites, as mentioned before. It can be predicted that in the $[Pt(C_2O_4)_2 \cdot 2H_2O]^{=}$ complex the two oxalate groups lie in a plane, each attached to the platinum atom by two electron-pair bonds of the type dsp^2. The two water molecules, if attached to the complex, are held by ion-dipole bonds.

In complexes in which the central atom forms a coördinated octahedron of six atoms or groups, the bonds may be any of several types. If they are all ionic or ion-dipole bonds, the moments are those in the first column. If four electron-pair bonds are formed, these must be dsp^2 and lie in a plane (sp^3 gives tetrahedral bonds); the $[Pt(C_2O_4)_2 \cdot 2H_2O]^{=}$ ion is of this type, assuming that the water molecules are part of the complex. The moments are then those of the second column. If six electron-pair bonds are formed, only three d eigenfunctions are left for the additional electrons, giving the magnetic moments of the third column. It is seen that in atoms with three or fewer unshared electrons magnetic data provide no information as to bond type with coördination number six, but that in other cases a definite statement can be made as to the type of bond when magnetic data are available. The observed magnetic moments are collected in Table IV. From them we deduce that trivalent and bivalent manganese, chromium, iron, and cobalt form six strong electron-pair bonds with cyanide groups, and in some cases with other groups, including NH_3, Cl and NO_2.[46] Tri-

[46] An electron-pair bond with a water molecule may perhaps be formed when induced by other strong bond-forming groups in the complex.

TABLE III

PREDICTED MAGNETIC MOMENTS OF COMPLEXES CONTAINING TRANSITION ELEMENTS

			For ion or $4sp^3$ bonds	For 4 dsp^2 bonds	For 6 d^2sp^3 bonds	For 8 d^4sp^3 bonds
K^{I} Ca^{II} Sc^{III} Ti^{IV}, etc.	Rb^{I} Sr^{II} Y^{III} Zr^{IV} Nb^{V} Mo^{VI}	Cs^{I} Ba^{II}—Hf^{IV} Ta^{V} W^{VI}	0.00	0.00	0.00	0.00
V^{IV}	Nb^{IV} Mo^{V}	W^{V}	1.73	1.73	1.73	1.73
V^{III} Cr^{IV}	Mo^{IV} Ru^{VI}	W^{IV} Os^{VI}	2.83	2.83	2.83	0.00
V^{II} Cr^{III} Mn^{IV}	Mo^{III}		3.88	3.88	3.88	
Cr^{II} Mn^{III} Fe^{IV}	Mo^{II} Ru^{IV}	Os^{IV}	4.90	4.90	2.83	
Mn^{II} Fe^{III} Co^{IV}	Ru^{III}	Os^{III} Ir^{IV}	5.91	3.88	1.73	
Fe^{II} Co^{III}	Ru^{II} Rh^{III} Pd^{IV}	Ir^{III} Pt^{IV}	4.90	2.83	0.00	
Co^{II} Ni^{III}	Rh^{II}		3.88	1.73		
Ni^{II}	Rh^{I} Pd^{II} Ag^{III}	Pt^{II} Au^{III}	2.83	0.00		
Cu^{II}			1.73			
Cu^{I} Zn^{II} Ca^{III} Ge^{IV}, etc.	Ag^{I} Cd^{II} In^{III} Sn^{IV} Sb^{V} Te^{VI}	Au^{I} Hg^{II} Tl^{III} Pb^{IV} Bi^{V} Po^{VI}	0.00			

valent iron apparently does not form electron-pair bonds with fluorine (in $[FeF_5 \cdot H_2O]^-$); although investigation of $(NH_4)_3FeF_6$ is to be desired in order to be sure of this conclusion. Ir^{III} and Pt^{IV} form six electron-pair bonds with Cl, NO_2 or NH_3.

TABLE IV

OBSERVED MAGNETIC MOMENTS OF COMPLEXES CONTAINING TRANSITION ELEMENTS[a]

	μ		μ
$K_3[Mn(CN)_6]$	3.01	$[Co(NH_3)_6]Cl_3$	0.00
$K_4[Cr(CN)_6]$	3.3	$[Co(NH_3)_5Cl]Cl_2$	.00
$K_3[Fe(CN)_6]$	2.0	$[Co(NH_3)_4Cl_2]Cl$	.00
$K_4[Mn(CN)_6]$	2.0	$[Co(NH_3)_3(NO_2)_3]$	.00
$K_4[Fe(CN)_6] \cdot 3H_2O$	0.00	$[Co(NH_3)_5H_2O]_2(C_2O_4)_3$	.00
$Na_3[Fe(CN)_5NH_3]$	.00		
$K_3[Co(CN)_6]$	.00	$K_2Ni(CN)_4$	0.00
$(NH_4)_2[FeF_5 \cdot H_2O]$	5.97	$K_2Ni(CN)_4 \cdot H_2O$	.00
$K_4[Mo(CN)_8]$	0.00	K_2PtCl_4	.00
$K_4[W(CN)_8] \cdot 2H_2O$	.00	$K_2Pt(C_2O_4)_2 \cdot 2H_2O$	.00
$Na_3[IrCl_2(NO_2)_4]$	.00	$Pt(NH_3)_4SO_4$	.00
$[Ir(NH_3)_5NO_2]Cl_2$	.00		
$[Ir(NH_3)_4(NO_2)_2]Cl$	.00	$Na_2[Fe(CN)_5NO] \cdot 2H_2O$	.00
$[Ir(NH_3)_3(NO_2)_3]$	.00	$[Ru(NH_3)_4 \cdot NO \cdot H_2O]Cl_3$	.00
$K_2[PtCl_6]$	.00	$[Ru(NH_3)_4 \cdot NO \cdot Cl]Br_2$	.00
$[Pt(NH_3)_6]Cl_4$	.00	$[Co(NH_3)_5NO]Cl_2$	2.81
$[Pt(NH_3)_5Cl]Cl_3$	.00		
$[Pt(NH_3)_4Cl_2]Cl_2$	.00	$Ni(CO)_4$	0.00
$[Pt(NH_3)_3Cl_3]Cl$	.00	$Fe(CO)_5$	.00
$[Pt(NH_3)_2Cl_4]$	.00	$Cr(CO)_5$	.00

[a] Values quoted are from "International Critical Tables" or from W. Biltz, *Z. anorg. Chem.*, **170**, 161 (1928), and D. M. Bose, *Z. Physik*, **65**, 677 (1930). I am indebted to Mr. P. D. Brass for collecting from the literature some of the data in this table.

The moments of complexes containing NO offer a puzzling problem. The diamagnetism of compounds of iron and ruthenium suggests that Fe^{IV} and Ru^{IV} form a double bond with NO, making seven bonds in all, which woud lead to $\mu = 0$. But this structure cannot be applied to $[Co(NH_3)_5\text{-}NO]Cl_2$, which has a moment corresponding to a triplet state. Further study of such complexes is needed.

The observed diamagnetism of the ions $[Mo(CN)_8]^{4-}$ and $[W(CN)_8]^{4-}$ shows that the central atom forms eight electron-pair bonds, involving the eigenfunctions d^4sp^3 (fourth column of Table III).

The metal carbonyls $Ni(CO)_4$, $Fe(CO)_5$, and $Cr(CO)_6$ are observed to be diamagnetic. This follows from the theoretical discussion if it is assumed that an electron-pair bond is formed with each carbonyl; for the nine eigenfunctions available ($3d^54s4p^3$) are completely filled by the n bonds and $2(9-n)$ additional electrons attached to the metal atom ($n = 4, 5, 6$). The theory also explains the observed composition of these unusual substances; for the formulas $M(CO)_n$, with $n = 4$, 5, and 6, respectively, follow at once from the assumption that CO molecules add on as long as bond eigenfunctions are available. Since a single unshared electron can occupy an eigenfunction, this assumption leads to the formula $Co(CO)_4$, which is known to be correct. This substance should have $\mu = 1.73$. The compounds $Mn(CO)_5$ and $V(CO)_6$ should also exist, and have $\mu = 1.73$. $Co(CO)_4$ and $Mn(CO)_5$ should form un-ionized diamagnetic cyanides, $[Co(CO)_4CN]$ and $[Mn(CO)_5CN]$, while $V(CO)_6$ would not form a stable cyanide, since steric effects would prevent the cyanide group from forming an electron-pair bond with the vanadium atom, and ionic cyanides are formed only by strong metals. It is interesting to note the effect of the four strong bond eigenfunctions and one weak one formed from dsp^3; whereas nickel forms no lower carbonyl than $Ni(CO)_4$, iron forms $Fe(CO)_4$ and $Fe_2(CO)_9$ in addition to $Fe(CO)_5$.

The palladium and platinum metals also form carbonyl compounds. Of the expected compounds $Pd(CO)_4$, $Pt(CO)_4$, $Ru(CO)_5$, $Os(CO)_5$, $Mo(CO)_6$, and $W(CO)_6$ only $Mo(CO)_6$ has been prepared, although some unsaturated ruthenium carbonyls have been prepared. The compounds $Pd(CO)_2Cl_2$, $Pt(CO)_2Cl_2$, $K[PtCOCl_3]$, etc., show the stability of the four dsp^2 bonds. It would be interesting to determine whether or not each CO is bonded to two metal atoms in compounds such as $[Pt(CO)Cl_2]_2$, whose structure is predicted to be

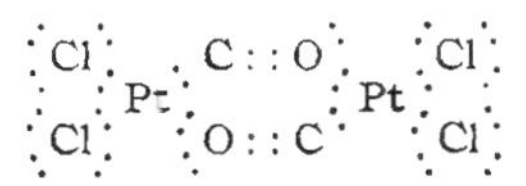

with the whole molecule in one plane. The compounds $2PdCl_2{\cdot}3CO$ and $2PtCl_2{\cdot}3CO$ probably have the structure

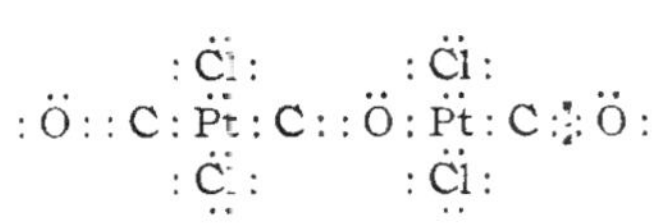

or one of the structures isomeric with this.

This by no means exhaustive discussion may serve to indicate the value of the information provided by magnetic data relative to the nature of the chemical bond. The quantum-mechanical rules for electron-pair bonds are essential to the treatment. Much further information is provided when these methods of attack are combined with crystal structure data, a topic which has been almost completely neglected in this paper. It has been found that the rules for electron-pair bonds permit the formulation of a set of structural principles for non-ionic inorganic crystals similar to that for complex ionic crystals; the statement of these principles and applications illustrating their use will be the subject of an article to be published in the *Zeitschrift für Kristallographie*.

Summary

With the aid of the quantum mechanics there is formulated a set of rules regarding electron-pair bonds, dealing particularly with the strength of bonds in relation to the nature of the single-electron eigenfunctions involved. It is shown that one single-electron eigenfunction on each of two atoms determines essentially the nature of the electron-pair bond formed between them; this effect is accentuated by the phenomenon of concentration of the bond eigenfunctions.

The type of bond formed by an atom is dependent on the ratio of bond energy to energy of penetration of the core (*s–p* separation). When this ratio is small, the bond eigenfunctions are *p* eigenfunctions, giving rise to bonds at right angles to one another; but when it is large, new eigenfunctions especially adapted to bond formation can be constructed. From *s* and *p* eigenfunctions the best bond eigenfunctions which can be made are four equivalent tetrahedral eigenfunctions, giving bonds directed toward the corners of a regular tetrahedron. These account for the chemist's tetrahedral atom, and lead directly to free rotation about a single bond but not about a double bond and to other tetrahedral properties. A single *d* eigenfunction with *s* and *p* gives rise to four strong bonds lying in a plane and directed toward the corners of a square. These are formed by bivalent nickel, palladium, and platinum. Two *d* eigenfunctions with *s* and *p* give six octahedral eigenfunctions, occurring in many complexes formed by transition-group elements.

It is then shown that (excepting the rare-earth ions) the magnetic moment of a non-linear molecule or complex ion is determined by the number of unpaired electrons, being equal to $\mu_S = 2\sqrt{S(S+1)}$, in which S is half that number. This makes it possible to determine from magnetic data which eigenfunctions are involved in bond formation, and so to decide between electron-pair bonds and ionic or ion-dipole bonds for various complexes. It is found that the transition-group elements almost without exception form electron-pair bonds with CN, ionic bonds with F, and ion-dipole bonds with H_2O; with other groups the bond type varies.

Examples of deductions regarding atomic arrangement, bond angles and other properties of molecules and complex ions from magnetic data, with the aid of calculations involving bond eigenfunctions, are given.

PASADENA, CALIFORNIA

No. 4356 April 25, 1953 NATURE 737

MOLECULAR STRUCTURE OF NUCLEIC ACIDS

A Structure for Deoxyribose Nucleic Acid

We wish to suggest a structure for the salt of deoxyribose nucleic acid (D.N.A.). This structure has novel features which are of considerable biological interest.

A structure for nucleic acid has already been proposed by Pauling and Corey[1]. They kindly made their manuscript available to us in advance of publication. Their model consists of three inter-twined chains, with the phosphates near the fibre axis, and the bases on the outside. In our opinion, this structure is unsatisfactory for two reasons: (1) We believe that the material which gives the X-ray diagrams is the salt, not the free acid. Without the acidic hydrogen atoms it is not clear what forces would hold the structure together, especially as the negatively charged phosphates near the axis will repel each other. (2) Some of the van der Waals distances appear to be too small.

Another three-chain structure has also been suggested by Fraser (in the press). In his model the phosphates are on the outside and the bases on the inside, linked together by hydrogen bonds. This structure as described is rather ill-defined, and for this reason we shall not comment on it.

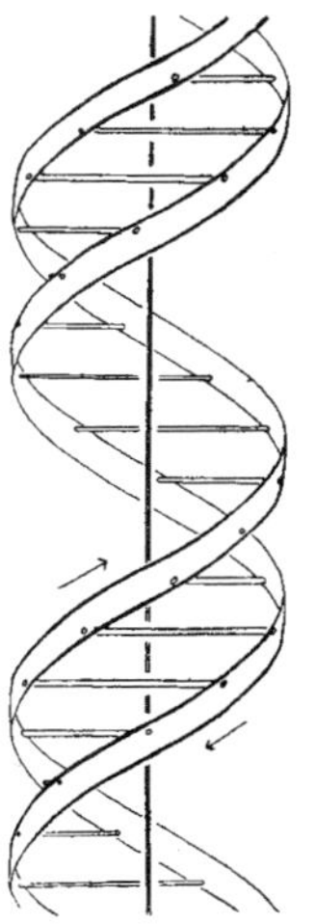

This figure is purely diagrammatic. The two ribbons symbolize the two phosphate—sugar chains, and the horizontal rods the pairs of bases holding the chains together. The vertical line marks the fibre axis

We wish to put forward a radically different structure for the salt of deoxyribose nucleic acid. This structure has two helical chains each coiled round the same axis (see diagram). We have made the usual chemical assumptions, namely, that each chain consists of phosphate diester groups joining β-D-deoxyribofuranose residues with 3′,5′ linkages. The two chains (but not their bases) are related by a dyad perpendicular to the fibre axis. Both chains follow right-handed helices, but owing to the dyad the sequences of the atoms in the two chains run in opposite directions. Each chain loosely resembles Furberg's[2] model No. 1; that is, the bases are on the inside of the helix and the phosphates on the outside. The configuration of the sugar and the atoms near it is close to Furberg's 'standard configuration', the sugar being roughly perpendicular to the attached base. There is a residue on each chain every 3·4 A. in the z-direction. We have assumed an angle of 36° between adjacent residues in the same chain, so that the structure repeats after 10 residues on each chain, that is, after 34 A. The distance of a phosphorus atom from the fibre axis is 10 A. As the phosphates are on the outside, cations have easy access to them.

The structure is an open one, and its water content is rather high. At lower water contents we would expect the bases to tilt so that the structure could become more compact.

The novel feature of the structure is the manner in which the two chains are held together by the purine and pyrimidine bases. The planes of the bases are perpendicular to the fibre axis. They are joined together in pairs, a single base from one chain being hydrogen-bonded to a single base from the other chain, so that the two lie side by side with identical z-co-ordinates. One of the pair must be a purine and the other a pyrimidine for bonding to occur. The hydrogen bonds are made as follows: purine position 1 to pyrimidine position 1; purine position 6 to pyrimidine position 6.

If it is assumed that the bases only occur in the structure in the most plausible tautomeric forms (that is, with the keto rather than the enol configurations) it is found that only specific pairs of bases can bond together. These pairs are: adenine (purine) with thymine (pyrimidine), and guanine (purine) with cytosine (pyrimidine).

In other words, if an adenine forms one member of a pair, on either chain, then on these assumptions the other member must be thymine; similarly for guanine and cytosine. The sequence of bases on a single chain does not appear to be restricted in any way. However, if only specific pairs of bases can be formed, it follows that if the sequence of bases on one chain is given, then the sequence on the other chain is automatically determined.

It has been found experimentally[3,4] that the ratio of the amounts of adenine to thymine, and the ratio of guanine to cytosine, are always very close to unity for deoxyribose nucleic acid.

It is probably impossible to build this structure with a ribose sugar in place of the deoxyribose, as the extra oxygen atom would make too close a van der Waals contact.

The previously published X-ray data[5,6] on deoxyribose nucleic acid are insufficient for a rigorous test of our structure. So far as we can tell, it is roughly compatible with the experimental data, but it must be regarded as unproved until it has been checked against more exact results. Some of these are given in the following communications. We were not aware of the details of the results presented there when we devised our structure, which rests mainly though not entirely on published experimental data and stereochemical arguments.

It has not escaped our notice that the specific pairing we have postulated immediately suggests a possible copying mechanism for the genetic material.

Full details of the structure, including the conditions assumed in building it, together with a set of co-ordinates for the atoms, will be published elsewhere.

We are much indebted to Dr. Jerry Donohue for constant advice and criticism, especially on inter-atomic distances. We have also been stimulated by a knowledge of the general nature of the unpublished experimental results and ideas of Dr. M. H. F. Wilkins, Dr. R. E. Franklin and their co-workers at

King's College, London. One of us (J. D. W.) has been aided by a fellowship from the National Foundation for Infantile Paralysis.

J. D. WATSON
F. H. C. CRICK

Medical Research Council Unit for the
Study of the Molecular Structure of
Biological Systems,
Cavendish Laboratory, Cambridge.
April 2.

[1] Pauling, L., and Corey, R. B., *Nature*, **171**, 346 (1953); *Proc. U.S. Nat. Acad. Sci.*, **39**, 84 (1953).
[2] Furberg, S., *Acta Chem. Scand.*, **6**, 634 (1952).
[3] Chargaff, E., for references see Zamenhof, S., Brawerman, G., and Chargaff, E., *Biochim. et Biophys. Acta*, **9**, 402 (1952).
[4] Wyatt, G. R., *J. Gen. Physiol.*, **36**, 201 (1952).
[5] Astbury, W. T., Symp. Soc. Exp. Biol. 1, Nucleic Acid, 66 (Camb. Univ. Press, 1947).
[6] Wilkins, M. H. F., and Randall, J. T., *Biochim. et Biophys. Acta*, **10**, 192 (1953).

738 NATURE April 25, 1953 VOL. 171

Molecular Structure of Deoxypentose Nucleic Acids

WHILE the biological properties of deoxypentose nucleic acid suggest a molecular structure containing great complexity, X-ray diffraction studies described here (cf. Astbury[1]) show the basic molecular configuration has great simplicity. The purpose of this communication is to describe, in a preliminary way, some of the experimental evidence for the polynucleotide chain configuration being helical, and existing in this form when in the natural state. A fuller account of the work will be published shortly.

The structure of deoxypentose nucleic acid is the same in all species (although the nitrogen base ratios alter considerably) in nucleoprotein, extracted or in cells, and in purified nucleate. The same linear group of polynucleotide chains may pack together parallel in different ways to give crystalline[1-3], semi-crystalline or paracrystalline material. In all cases the X-ray diffraction photograph consists of two regions, one determined largely by the regular spacing of nucleotides along the chain, and the other by the longer spacings of the chain configuration. The sequence of different nitrogen bases along the chain is not made visible.

Oriented paracrystalline deoxypentose nucleic acid ('structure *B*' in the following communication by Franklin and Gosling) gives a fibre diagram as shown in Fig. 1 (cf. ref. 4). Astbury suggested that the strong 3·4-A. reflexion corresponded to the inter-nucleotide repeat along the fibre axis. The ~ 34 A. layer lines, however, are not due to a repeat of a polynucleotide composition, but to the chain configuration repeat, which causes strong diffraction as the nucleotide chains have higher density than the interstitial water. The absence of reflexions on or near the meridian immediately suggests a helical structure with axis parallel to fibre length.

Diffraction by Helices

It may be shown[5] (also Stokes, unpublished) that the intensity distribution in the diffraction pattern of a series of points equally spaced along a helix is given by the squares of Bessel functions. A uniform continuous helix gives a series of layer lines of spacing corresponding to the helix pitch, the intensity distribution along the *n*th layer line being proportional to the square of J_n, the *n*th order Bessel function. A straight line may be drawn approximately through

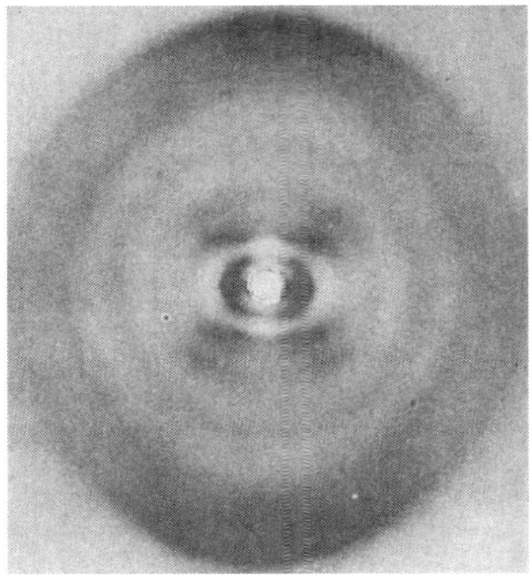

Fig. 1. Fibre diagram of deoxypentose nucleic acid from *B. coli*. Fibre axis vertical

the innermost maxima of each Bessel function and the origin. The angle this line makes with the equator is roughly equal to the angle between an element of the helix and the helix axis. If a unit repeats *n* times along the helix there will be a meridional reflexion (J_0^2) on the *n*th layer line. The helical configuration produces side-bands on this fundamental frequency, the effect[5] being to reproduce the intensity distribution about the origin around the new origin, on the *n*th layer line, corresponding to *C* in Fig. 2.

We will now briefly analyse in physical terms some of the effects of the shape and size of the repeat unit or nucleotide on the diffraction pattern. First, if the nucleotide consists of a unit having circular symmetry about an axis parallel to the helix axis, the whole diffraction pattern is modified by the form factor of the nucleotide. Second, if the nucleotide consists of a series of points on a radius at right-angles to the helix axis, the phases of radiation scattered by the helices of different diameter passing through each point are the same. Summation of the corresponding Bessel functions gives reinforcement for the inner-

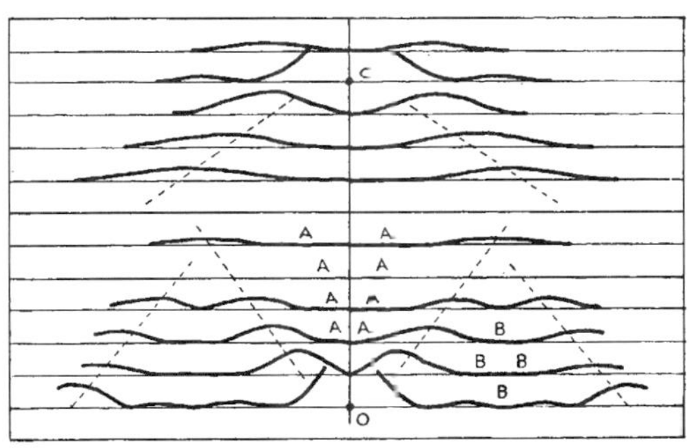

Fig. 2. Diffraction pattern of system of helices corresponding to structure of deoxypentose nucleic acid. The squares of Bessel functions are plotted about 0 on the equator and on the first, second, third and fifth layer lines for half of the nucleotide mass at 20 A. diameter and remainder distributed along a radius, the mass at a given radius being proportional to the radius. About *C* on the tenth layer line similar functions are plotted for an outer diameter of 12 A.

No. 4356 April 25, 1953 NATURE 739

most maxima and, in general, owing to phase difference, cancellation of all other maxima. Such a system of helices (corresponding to a spiral staircase with the core removed) diffracts mainly over a limited angular range, behaving, in fact, like a periodic arrangement of flat plates inclined at a fixed angle to the axis. Third, if the nucleotide is extended as an arc of a circle in a plane at right-angles to the helix axis, and with centre at the axis, the intensity of the system of Bessel function layer-line streaks emanating from the origin is modified owing to the phase differences of radiation from the helices drawn through each point on the nucleotide. The form factor is that of the series of points in which the helices intersect a plane drawn through the helix axis. This part of the diffraction pattern is then repeated as a whole with origin at C (Fig. 2). Hence this aspect of nucleotide shape affects the central and peripheral regions of each layer line differently.

Interpretation of the X-Ray Photograph

It must first be decided whether the structure consists of essentially one helix giving an intensity distribution along the layer lines corresponding to $J_1, J_2, J_3 \ldots$, or two similar co-axial helices of twice the above size and relatively displaced along the axis a distance equal to half the pitch giving $J_2, J_4, J_6 \ldots$, or three helices, etc. Examination of the width of the layer-line streaks suggests the intensities correspond more closely to J_1^2, J_2^2, J_3^2 than to $J_2^2, J_4^2, J_6^2 \ldots$ Hence the dominant helix has a pitch of $\sim$ 34 A., and, from the angle of the helix, its diameter is found to be $\sim$ 20 A. The strong equatorial reflexion at $\sim$ 17 A. suggests that the helices have a maximum diameter of $\sim$ 20 A. and are hexagonally packed with little interpenetration. Apart from the width of the Bessel function streaks, the possibility of the helices having twice the above dimensions is also made unlikely by the absence of an equatorial reflexion at $\sim$ 34 A. To obtain a reasonable number of nucleotides per unit volume in the fibre, two or three intertwined coaxial helices are required, there being ten nucleotides on one turn of each helix.

The absence of reflexions on or near the meridian (an empty region AAA on Fig. 2) is a direct consequence of the helical structure. On the photograph there is also a relatively empty region on and near the equator, corresponding to region BBB on Fig. 2. As discussed above, this absence of secondary Bessel function maxima can be produced by a radial distribution of the nucleotide shape. To make the layer-line streaks sufficiently narrow, it is necessary to place a large fraction of the nucleotide mass at $\sim$ 20 A. diameter. In Fig. 2 the squares of Bessel functions are plotted for half the mass at 20 A. diameter, and the rest distributed along a radius, the mass at a given radius being proportional to the radius.

On the zero layer line there appears to be a marked J_{10}^2, and on the first, second and third layer lines, $J_9^2 + J_{11}^2, J_8^2 + J_{12}^2$, etc., respectively. This means that, in projection on a plane at right-angles to the fibre axis, the outer part of the nucleotide is relatively concentrated, giving rise to high-density regions spaced *c.* 6 A. apart around the circumference of a circle of 20 A. diameter. On the fifth layer line two J_5 functions overlap and produce a strong reflexion. On the sixth, seventh and eighth layer lines the maxima correspond to a helix of diameter $\sim$ 12 A. Apparently it is only the central region of the helix structure which is well divided by the 3·4-A. spacing, the outer parts of the nucleotide overlapping to form a continuous helix. This suggests the presence of nitrogen bases arranged like a pile of pennies[1] in the central regions of the helical system.

There is a marked absence of reflexions on layer lines beyond the tenth. Disorientation in the specimen will cause more extension along the layer lines of the Bessel function streaks on the eleventh, twelfth and thirteenth layer lines than on the ninth, eighth and seventh. For this reason the reflexions on the higher-order layer lines will be less readily visible. The form factor of the nucleotide is also probably causing diminution of intensity in this region. Tilting of the nitrogen bases could have such an effect.

Reflexions on the equator are rather inadequate for determination of the radial distribution of density in the helical system. There are, however, indications that a high-density shell, as suggested above, occurs at diameter $\sim$ 20 A.

The material is apparently not completely para-crystalline, as sharp spots appear in the central region of the second layer line, indicating a partial degree of order of the helical units relative to one another in the direction of the helix axis. Photographs similar to Fig. 1 have been obtained from sodium nucleate from calf and pig thymus, wheat germ, herring sperm, human tissue and T_2 bacteriophage. The most marked correspondence with Fig. 2 is shown by the exceptional photograph obtained by our colleagues, R. E. Franklin and R. G. Gosling, from calf thymus deoxypentose nucleate (see following communication).

It must be stressed that some of the above discussion is not without ambiguity, but in general there appears to be reasonable agreement between the experimental data and the kind of model described by Watson and Crick (see also preceding communication).

It is interesting to note that if there are ten phosphate groups arranged on each helix of diameter 20 A. and pitch 34 A., the phosphate ester backbone chain is in an almost fully extended state. Hence, when sodium nucleate fibres are stretched[3], the helix is evidently extended in length like a spiral spring in tension.

Structure *in vivo*

The biological significance of a two-chain nucleic acid unit has been noted (see preceding communication). The evidence that the helical structure discussed above does, in fact, exist in intact biological systems is briefly as follows:

Sperm heads. It may be shown that the intensity of the X-ray spectra from crystalline sperm heads is determined by the helical form-function in Fig. 2. Centrifuged trout semen give the same pattern as the dried and rehydrated or washed sperm heads used previously[6]. The sperm head fibre diagram is also given by extracted or synthetic[1] nucleoprotamine or extracted calf thymus nucleohistone.

Bacteriophage. Centrifuged wet pellets of T_2 phage photographed with X-rays while sealed in a cell with mica windows give a diffraction pattern containing the main features of paracrystalline sodium nucleate as distinct from that of crystalline nucleoprotein. This confirms current ideas of phage structure.

Transforming principle (in collaboration with H. Ephrussi-Taylor). Active deoxypentose nucleate allowed to dry at $\sim$ 60 per cent humidity has the same crystalline structure as certain samples[3] of sodium thymonucleate.

We wish to thank Prof. J. T. Randall for encouragement; Profs. E. Chargaff, R. Signer, J. A. V. Butler and Drs. J. D. Watson, J. D. Smith, L. Hamilton, J. C. White and G. R. Wyatt for supplying material without which this work would have been impossible; also Drs. J. D. Watson and Mr. F. H. C. Crick for stimulation, and our colleagues R. E. Franklin, R. G. Gosling, G. L. Brown and W. E. Seeds for discussion. One of us (H. R. W.) wishes to acknowledge the award of a University of Wales Fellowship.

M. H. F. WILKINS

Medical Research Council Biophysics
Research Unit,

A. R. STOKES
H. R. WILSON

Wheatstone Physics Laboratory,
King's College, London.

April 2.

[1] Astbury, W. T., Symp. Soc. Exp. Biol., 1, Nucleic Acid (Cambridge Univ. Press, 1947).

[2] Riley, D. P., and Oster, G., *Biochim. et Biophys. Acta*, **7**, 526 (1951).

[3] Wilkins, M. H. F., Gosling, R. G., and Seeds, W. E., *Nature*, **167**, 759 (1951).

[4] Astbury, W. T., and Bell, F. O., Cold Spring Harb. Symp. Quant. Biol., **6**, 109 (1938).

[5] Cochran, W., Crick, F. H. C., and Vand, V., *Acta Cryst.*, **5**, 581 (1952).

[6] Wilkins, M. H. F., and Randall, J. T., *Biochim. et Biophys. Acta*, **10**, 192 (1953).

740 NATURE April 25, 1953 VOL. 171

Molecular Configuration in Sodium Thymonucleate

SODIUM thymonucleate fibres give two distinct types of X-ray diagram. The first corresponds to a crystalline form, structure A, obtained at about 75 per cent relative humidity; a study of this is described in detail elsewhere[1]. At higher humidities a different structure, structure B, showing a lower degree of order, appears and persists over a wide range of ambient humidity. The change from A to B is reversible. The water content of structure B fibres which undergo this reversible change may vary from 40–50 per cent to several hundred per cent of the dry weight. Moreover, some fibres never show structure A, and in these structure B can be obtained with an even lower water content.

The X-ray diagram of structure B (see photograph) shows in striking manner the features characteristic of helical structures, first worked out in this laboratory by Stokes (unpublished) and by Crick, Cochran and Vand[2]. Stokes and Wilkins were the first to propose such structures for nucleic acid as a result of direct studies of nucleic acid fibres, although a helical structure had been previously suggested by Furberg (thesis, London, 1949) on the basis of X-ray studies of nucleosides and nucleotides.

While the X-ray evidence cannot, at present, be taken as direct proof that the structure is helical, other considerations discussed below make the existence of a helical structure highly probable.

Structure B is derived from the crystalline structure A when the sodium thymonucleate fibres take up quantities of water in excess of about 40 per cent of their weight. The change is accompanied by an increase of about 30 per cent in the length of the fibre, and by a substantial re-arrangement of the molecule. It therefore seems reasonable to suppose that in structure B the structural units of sodium thymonucleate (molecules on groups of molecules) are relatively free from the influence of neighbouring

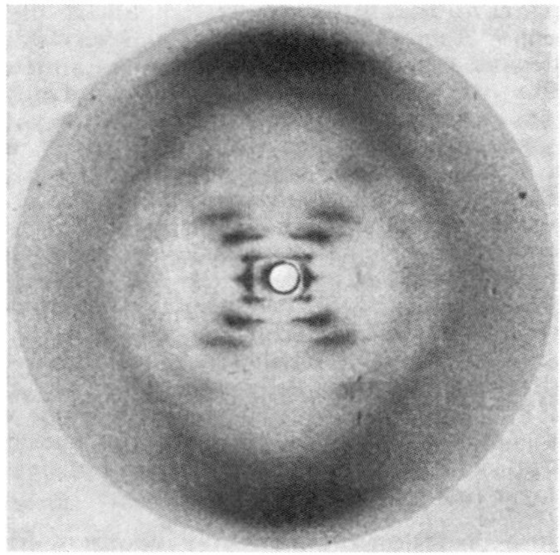

Sodium deoxyribose nucleate from calf thymus. Structure B

molecules, each unit being shielded by a sheath of water. Each unit is then free to take up its least-energy configuration independently of its neighbours and, in view of the nature of the long-chain molecules involved, it is highly likely that the general form will be helical[3]. If we adopt the hypothesis of a helical structure, it is immediately possible, from the X-ray diagram of structure B, to make certain deductions as to the nature and dimensions of the helix.

The innermost maxima on the first, second, third and fifth layer lines lie approximately on straight lines radiating from the origin. For a smooth single-strand helix the structure factor on the nth layer line is given by:

$$F_n = J_n(2\pi rR) \exp i\, n(\psi + \tfrac{1}{2}\pi),$$

where $J_n(u)$ is the nth-order Bessel function of u, r is the radius of the helix, and R and ψ are the radial and azimuthal co-ordinates in reciprocal space[2]; this expression leads to an approximately linear array of intensity maxima of the type observed, corresponding to the first maxima in the functions J_1, J_2, J_3, etc.

If, instead of a smooth helix, we consider a series of residues equally spaced along the helix, the transform in the general case treated by Crick, Cochran and Vand is more complicated. But if there is a whole number, m, of residues per turn, the form of the transform is as for a smooth helix with the addition, only, of the same pattern repeated with its origin at heights mc^*, $2mc^*$ . . . etc. (c is the fibre-axis period).

In the present case the fibre-axis period is 34 A. and the very strong reflexion at 3·4 A. lies on the tenth layer line. Moreover, lines of maxima radiating from the 3·4-A. reflexion as from the origin are visible on the fifth and lower layer lines, having a J_5 maximum coincident with that of the origin series on the fifth layer line. (The strong outer streaks which apparently radiate from the 3·4-A. maximum are not, however, so easily explained.) This suggests strongly that there are exactly 10 residues per turn of the helix. If this is so, then from a measurement of R_n the position of the first maximum on the nth layer line (for n 5$\nless$), the radius of the helix, can be obtained. In the present instance, measurements of R_1, R_2, R_3 and R_5 all lead to values of r of about 10 A.

No. 4356 April 25, 1953 NATURE 741

Since this linear array of maxima is one of the strongest features of the X-ray diagram, we must conclude that a crystallographically important part of the molecule lies on a helix of this diameter. This can only be the phosphate groups or phosphorus atoms.

If ten phosphorus atoms lie on one turn of a helix of radius 10 A., the distance between neighbouring phosphorus atoms in a molecule is 7·1 A. This corresponds to the P . . . P distance in a fully extended molecule, and therefore provides a further indication that the phosphates lie on the outside of the structural unit.

Thus, our conclusions differ from those of Pauling and Corey[4], who proposed for the nucleic acids a helical structure in which the phosphate groups form a dense core.

We must now consider briefly the equatorial reflexions. For a single helix the series of equatorial maxima should correspond to the maxima in $J_0(2\pi rR)$. The maxima on our photograph do not, however, fit this function for the value of r deduced above. There is a very strong reflexion at about 24 A. and then only a faint sharp reflexion at 9·0 A. and two diffuse bands around 5·5 A. and 4·0 A. This lack of agreement is, however, to be expected, for we know that the helix so far considered can only be the most important member of a series of coaxial helices of different radii; the non-phosphate parts of the molecule will lie on inner co-axial helices, and it can be shown that, whereas these will not appreciably influence the innermost maxima on the layer lines, they may have the effect of destroying or shifting both the equatorial maxima and the outer maxima on other layer lines.

Thus, if the structure is helical, we find that the phosphate groups or phosphorus atoms lie on a helix of diameter about 20 A., and the sugar and base groups must accordingly be turned inwards towards the helical axis.

Considerations of density show, however, that a cylindrical repeat unit of height 34 A. and diameter 20 A. must contain many more than ten nucleotides.

Since structure *B* often exists in fibres with low water content, it seems that the density of the helical unit cannot differ greatly from that of dry sodium thymonucleate, 1·63 gm./cm.[3] [1,5], the water in fibres of high water-content being situated outside the structural unit. On this basis we find that a cylinder of radius 10 A. and height 34 A. would contain thirty-two nucleotides. However, there might possibly be some slight inter-penetration of the cylindrical units in the dry state making their effective radius rather less. It is therefore difficult to decide, on the basis of density measurements alone, whether one repeating unit contains ten nucleotides on each of two or on each of three co-axial molecules. (If the effective radius were 8 A. the cylinder would contain twenty nucleotides.) Two other arguments, however, make it highly probable that there are only two co-axial molecules.

First, a study of the Patterson function of structure *A*, using superposition methods, has indicated[6] that there are only two chains passing through a primitive unit cell in this structure. Since the $A \rightleftharpoons B$ transformation is readily reversible, it seems very unlikely that the molecules would be grouped in threes in structure *B*. Secondly, from measurements on the X-ray diagram of structure *B* it can readily be shown that, whether the number of chains per unit is two or three, the chains are not equally spaced along the fibre axis. For example, three equally spaced chains would mean that the nth layer line depended on J_{3n}, and would lead to a helix of diameter about 60 A. This is many times larger than the primitive unit cell in structure *A*, and absurdly large in relation to the dimensions of nucleotides. Three unequally spaced chains, on the other hand, would be crystallographically non-equivalent, and this, again, seems unlikely. It therefore seems probable that there are only two co-axial molecules and that these are unequally spaced along the fibre axis.

Thus, while we do not attempt to offer a complete interpretation of the fibre-diagram of structure *B* we may state the following conclusions. The structure is probably helical. The phosphate groups lie on the outside of the structural unit, on a helix of diameter about 20 A. The structural unit probably consists of two co-axial molecules which are not equally spaced along the fibre axis, their mutual displacement being such as to account for the variation of observed intensities of the innermost maxima on the layer lines; if one molecule is displaced from the other by about three-eighths of the fibre-axis period, this would account for the absence of the fourth layer line maxima and the weakness of the sixth. Thus our general ideas are not inconsistent with the model proposed by Watson and Crick in the preceding communication.

The conclusion that the phosphate groups lie on the outside of the structural unit has been reached previously by quite other reasoning[1]. Two principal lines of argument were invoked. The first derives from the work of Gulland and his collaborators[7], who showed that even in aqueous solution the —CO and —NH_2 groups of the bases are inaccessible and cannot be titrated, whereas the phosphate groups are fully accessible. The second is based on our own observations[1] on the way in which the structural units in structures *A* and *B* are progressively separated by an excess of water, the process being a continuous one which leads to the formation first of a gel and ultimately to a solution. The hygroscopic part of the molecule may be presumed to lie in the phosphate groups ($(C_2H_5O)_2PO_2Na$ and $(C_3H_7O)_2PO_2Na$ are highly hygroscopic[8]), and the simplest explanation of the above process is that these groups lie on the outside of the structural units. Moreover, the ready availability of the phosphate groups for interaction with proteins can most easily be explained in this way.

We are grateful to Prof. J. T. Randall for his interest and to Drs. F. H. C. Crick, A. R. Stokes and M. H. F. Wilkins for discussion. One of us (R. E. F.) acknowledges the award of a Turner and Newall Fellowship.

ROSALIND E. FRANKLIN*
R. G. GOSLING

Wheatstone Physics Laboratory,
King's College, London.
April 2.

* Now at Birkbeck College Research Laboratories, 21 Torrington Square, London, W.C.1.

[1] Franklin, R. E., and Gosling, R. G. (in the press).
[2] Cochran, W., Crick, F. H. C., and Vand, V., *Acta Cryst.*, **5**, 501 (1952).
[3] Pauling, L., Corey, R. B., and Branson, H. R., *Proc. U.S. Nat. Acad. Sci.*, **37**, 205 (1951).
[4] Pauling, L., and Corey, R. B., *Proc. U.S. Nat. Acad. Sci.*, **39**, 84 (1953).
[5] Astbury, W. T., Cold Spring Harbor Symp. on Quant. Biol., **12**, 56 (1947).
[6] Franklin, R. E., and Gosling, R. G. (to be published).
[7] Gulland, J. M., and Jordan, D. O., Cold Spring Harbor Symp. on Quant. Biol., **12**, 5 (1947).
[8] Drushel, W. A., and Felty, A. R., *Chem. Zent.*, **89**, 1016 (1918).

416 NATURE February 13, 1960 VOL. 185

STRUCTURE OF HÆMOGLOBIN

A THREE-DIMENSIONAL FOURIER SYNTHESIS AT 5·5-Å. RESOLUTION, OBTAINED BY X-RAY ANALYSIS

By Dr. M. F. PERUTZ, F.R.S., Dr. M. G. ROSSMANN, ANN F. CULLIS, HILARY MUIRHEAD and Dr. GEORG WILL

Medical Research Council Unit for Molecular Biology, Cavendish Laboratory, University of Cambridge

AND

Dr. A. C. T. NORTH

Medical Research Council External Staff, Davy Faraday Research Laboratory, Royal Institution, London, W.1

VERTEBRATE hæmoglobin is a protein of molecular weight 67,000. Four of its 10,000 atoms are iron atoms which are combined with protoporphyrin to form four hæm groups. The remaining atoms are in four polypeptide chains of roughly equal size, which are identical in pairs[1-3]. Their amino-acid sequence is still largely unknown.

We have used horse oxy- or met-hæmoglobin because it crystallizes in a form especially suited for X-ray analysis, and employed the method of isomorphous replacement with heavy atoms to determine the phase angles of the diffracted rays[4-7]. The Fourier synthesis which we have calculated shows that hæmoglobin consists of four sub-units in a tetrahedral array and that each sub-unit closely resembles Kendrew's model of sperm whale myoglobin[6]. The four hæm groups lie in separate pockets on the surface of the molecule.

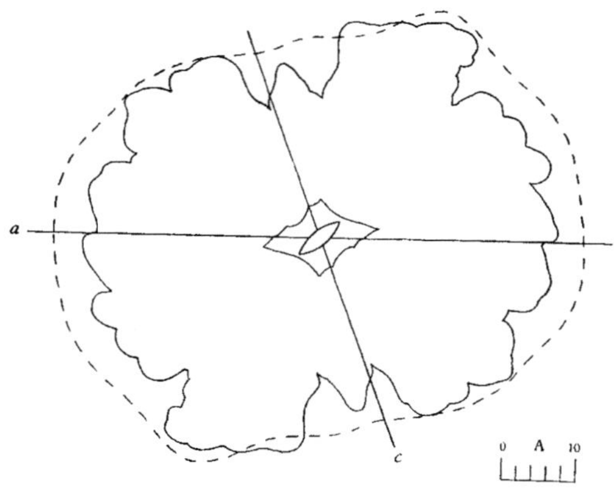

Fig. 2. External shape of the molecule. Full lines indicate the boundary derived from the contour at 0·54 electron/Å.^{3}. The broken line shows the boundary derived by Bragg and Perutz from two-dimensional data (ref. 13). Note the hole in the middle

Method of Analysis

Horse oxyhæmoglobin, crystallized from 1·9 *M* ammonium sulphate solution at *p*H 7, has the space group *C*2 with two molecules in the unit cell which lie on dyad axes[8]. In order to determine the phase angles of the 1,200 reflexions contained in the limiting sphere of 5·5 Å.$^{-1}$, six different isomorphous heavy-atom compounds were used (ref. 9 and unpublished work). Intensities were measured photographically and by counter spectrometer (Arndt, U. W., and Phillips, D. C., unpublished work). The relative positions and shapes of the heavy-atom replacement groups were found by correlation functions based on Patterson methods[10] and refined by least squares procedures. For each reflexion the structure amplitudes of all seven compounds were combined in an Argand diagram[11], and the probability of the phase angle having a value α was calculated for $\alpha = 0, 5, 10, \ldots 355°$. The centroid of the probability distribution, plotted around a circle, was then chosen as the best vector *F* in the Fourier synthesis[12]. The results were finally plotted on 32 contour maps showing the distribution of electron density in sections spaced 2 Å. apart normal to *b* (Fig. 1). The absolute configuration of the molecule was determined from anomalous dispersion[4].

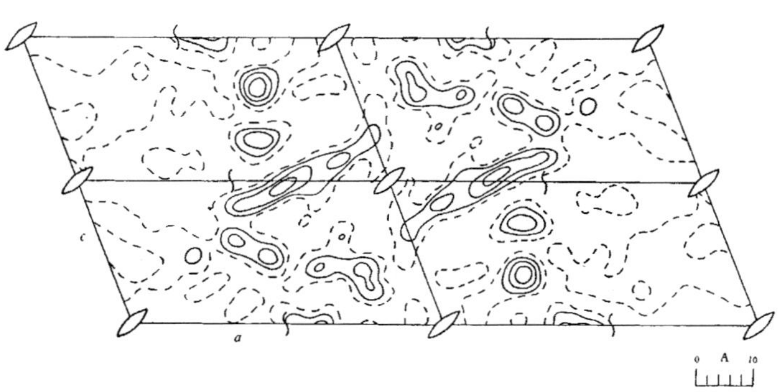

Fig. 1. Section at $y = 1/32b$. This cuts through the middle of the molecule on which the diagram is centred. 'Flat' areas indicating liquid appear on the left and right. Contours are drawn at intervals of 0·14 electron/Å.^{3}. The broken line marks 0·4 electron/Å.^{3}. Contours at lower levels are omitted

External Shape of the Molecule

More than half the volume of the crystals is taken up by liquid of crystallization, which mainly fills the spaces between the molecules and shows up on the contour maps in the form of flat, featureless regions (Fig. 1). The outlines of the two molecules in the unit cell can be traced by following the boundaries between these regions and the continuous electron-dense regions described below. In Fig. 2 the outline of one molecule, traced from the periphery of the 0·54

No. 4711 February 13, 1960 NATURE 417

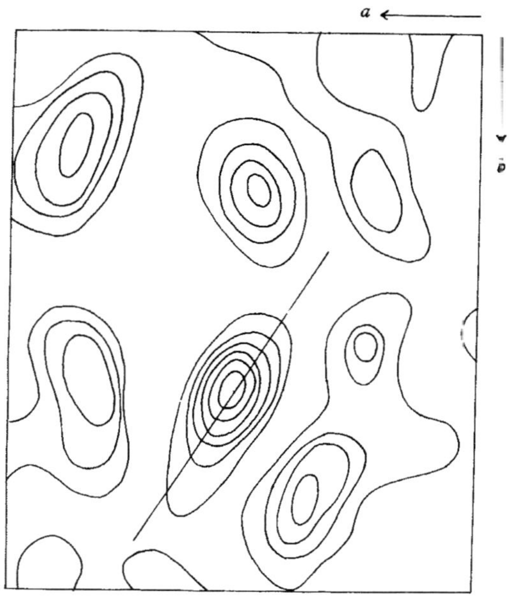

Hæm 1 and 2

Fig. 3. Section at $z = 1/4\ c$ showing one of the two hæm groups in the asymmetric unit. The straight line indicates the hæm orientation derived from electron spin resonance (ref. 14). The lowest contour shown is at 0·4 electron/Å.^{3}

electron/Å.^{3} contour, is seen in projection on the *b*-plane. To a first approximation it can be regarded as a spheroid with a length of 64 Å., a width of 55 Å. and a height of 50 Å. normal to the plane of the paper. Except for a slight shortening along *a*, this shape agrees with the earlier picture obtained from two-dimensional data, even including the dimple in the centre of the molecule (see Fig. 2 of ref. 12).

Positions of the Hæm and Sulphydryl Groups

Four peaks stand out from the rest, clearly representing the iron atoms with their surrounding porphyrin rings. A flattening of the peaks indicates the approximate orientation of the rings (Fig. 3). Fortunately, their exact orientation was already known from electron spin resonance[14], and it only remained to assign the correct one of the four alternative orientations of the hæms to each of the electron density peaks. The results are shown in Fig. 4. The iron atoms lie at the corners of an irregular tetrahedron with distances of 33·4 and 36·0 Å. between symmetrically related pairs. The closest approach between symmetrically unrelated iron atoms is 25·2 Å. (Table 1).

Table 1

	Fe_1	Fe_3	S_1
x	−6·6 Å.	12·3 Å.	5 Å.
y	7·3 Å.	−10·7 Å.	10 Å.
z	13·1 Å.	18·2 Å.	16 Å.
$Fe_1 - Fe_2$ = 33·4 Å.			
$Fe_3 - Fe_4$ = 36·0 Å.			
$Fe_1 - Fe_3$ = 25·2 Å.			
$Fe_1 - S_1$	13 Å.		
$Fe_3 - S_1$	21 Å.		

Horse hæmoglobin contains four cysteine residues, but only two sulphydryls combine with mercury in the native protein[5]. From the positions of the mercury atoms we inferred that each of the two sulphydryl groups is about 13 Å. away from one iron atom and 21 Å. away from another. (Fig. 4 and Table 1). The significance of this situation is discussed below.

Configuration of the Polypeptide Chains

The most prominent feature of the Fourier synthesis consists of more or less cylindrical clouds of high density, like the vapour trails of an aeroplane; they are curved to form intricate three-dimensional figures. Sections through various parts of these appear in Fig. 1. To build a model of the figures, we rolled out sheets of a thermo-setting plastic to the thickness of our sections on a scale of 2 Å. = 1 cm., and cut out the shape of each region on the contour maps where the density exceeds 0·54 electron/Å.^{3} (this corresponds to the first full contour line in Fig. 1. From this section, for example, 14 shapes would be cut). The shapes were then assembled in accordance with their positions and heights in the different sections, and the hæm groups were attached in the appropriate orientation. The model was then baked to set it permanently. For comparison, a Fourier synthesis of sperm whale myoglobin was calculated at a resolution of 5·5 Å., using the new X-ray data of Kendrew *et al.*[15], and a model of the electron density distribution was constructed by the method just described.

From the hæmoglobin Fourier synthesis there emerged four separate units which are identical in pairs. Fig. 5 shows one member of each pair together with the model of myoglobin on the left. In each unit the cloud of high density describes a complicated figure which, in the white unit (middle), can be traced from end to end by following the superimposed line. Except for two small gaps where the density sinks slightly below 0·54 electron/Å.^{3}, the black unit (right) closely resembles the white one. There are several gaps interrupting the myoglobin model where, probably due to increased thermal motion, the electron density falls below the contour-level chosen for cutting the model. However, we know from the recent work of Kendrew and his collaborators[15] that

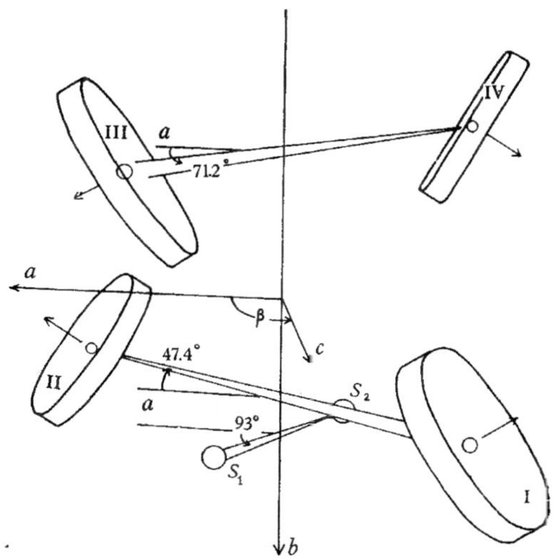

Fig. 4. Arrangement of hæm groups in hæmoglobin. Arrows indicate the reactive side of each group. The *c*-axis comes out of the paper towards the observer. This picture and all subsequent ones show the absolute configuration

418 NATURE February 13, 1960 VOL. 185

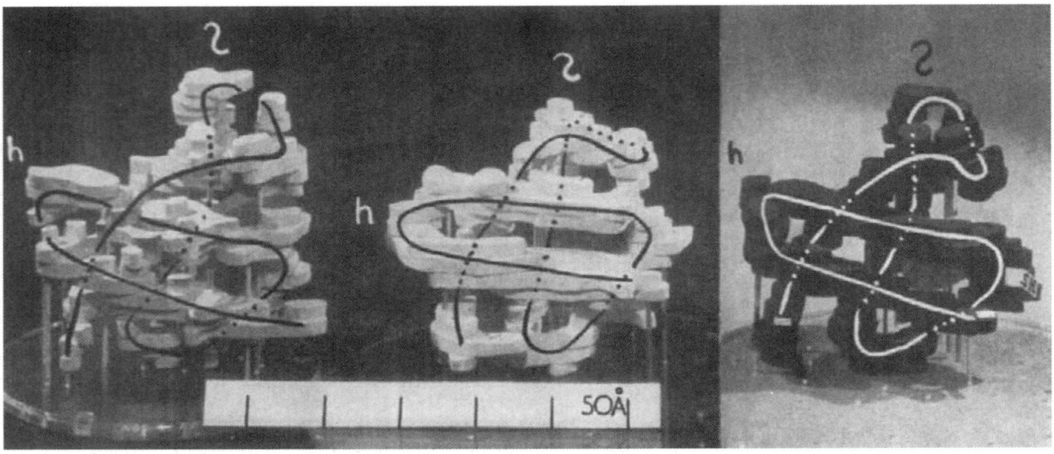

Fig. 5. Two different polypeptide chains in the asymmetric unit of hæmoglobin compared with myoglobin (left). The hæm groups are at the back of the chains

the gaps are bridged by a continuous polypeptide chain, and it is evident from Fig. 5 that apart from the gaps, the model has a configuration closely similar to the hæmoglobin units.

Clearly, the four tortuous clouds of high electron density in hæmoglobin represent the four polypeptide chains. The black and the white chains have similar, but not identical configurations. In the black chain the *S*-shaped bend at the top is more pronounced, the hæm group is lower, and the bend *h* sharper than in the white chain. These, however, are details. The most important result is their resemblance to each other and to sperm whale myoglobin.

Arrangement of the Four Sub-units

The first step in the assembly of the molecule is the matching of each chain by its symmetrically related partner (Fig. 6). It will be noted that there is comparatively little contact between the members of each pair, suggesting rather tenuous linkages. In the next step the white pair is inverted and

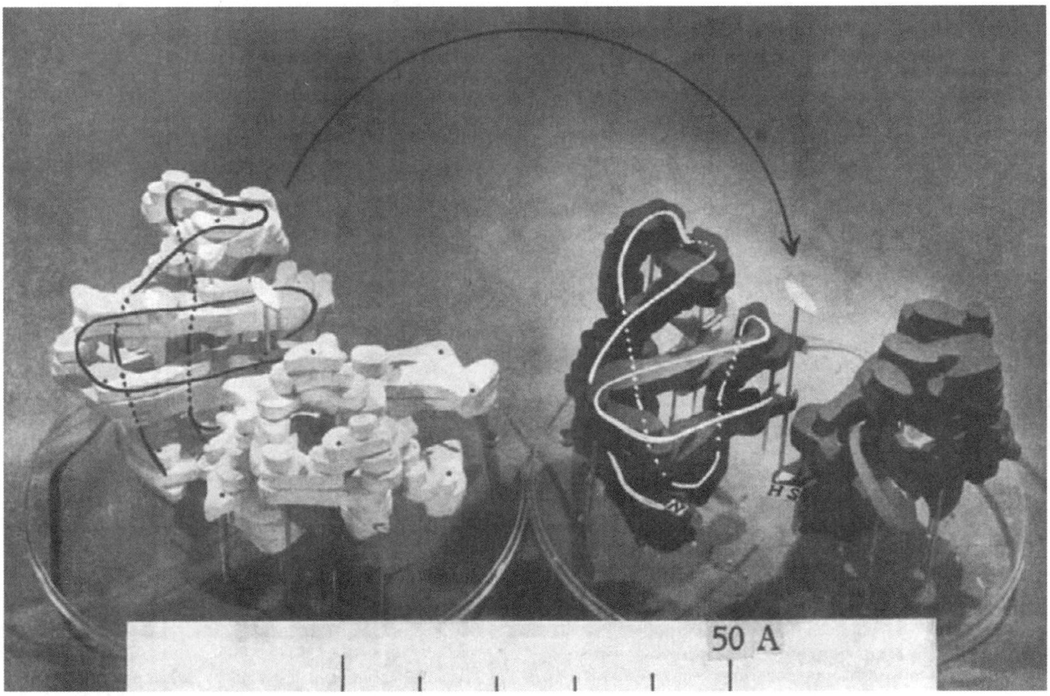

Fig. 6. Two pairs of chains symmetrically related by the dyad axis. The arrow shows how one pair is placed over the other to assemble the complete molecule

No. 4711 February 13, 1960 NATURE 419

placed over the black pair as indicated by the arrow. Fig. 7 shows one white chain placed over the pair of black ones and Fig. 8 shows the molecule completely assembled. The resulting arrangement is tetrahedral and has almost, but not quite, the orthorhombic point group symmetry 222. It contains two 'pseudo dyads' which lie approximately at right angles to each other and to the true dyad, one emerging from the centre of Fig. 8 and the other from the centre of Fig. 9. This means that, to a first approximation, each sub-unit can be generated by a rotation of 180° from any of its neighbours. Figs. 7, 8 and 9 also show that the surface contours of the white chains exactly fit those of the black, so that there is a large area of contact between them. This structural complementarity is one of the most striking features of the molecule. Fig. 10 is a view down the true dyad axis and reveals a hole going right through the centre of the molecule, as was to be expected from the Fourier projection[13]. However, the van der Waals radii of the chains are much bigger than appears in the model, and little room may, in fact, be left for water or electrolytes to pass through. Fig. 10 also reveals a dimple at the top where the white chains meet. This is matched by a similar, but larger, hollow at the bottom where the black chains meet.

Fig. 7. Partially assembled molecule showing two black chains and one white

The hæm groups are seen to lie in separate pockets on the surface of the molecule (Fig. 8). Each pocket is formed by the folds in one of the polypeptide chains, which appears to make contact with the hæm group at four different points at least. The iron atoms in the neighbouring pockets formed by the black and the white chains are 25 Å. apart.

Information from the Fourier Synthesis of Myoglobin at 2-Å. Resolution (ref. 15)

Thanks to the similarity with myoglobin, the interpretation can be carried further than would have been possible on the basis of our results alone. Kendrew *et al.* have found that the straight stretches of rod indicated in Fig. 5 are α-helices and that the N-terminal end of the chain is at the bottom left. The ends of the two hæmoglobin chains have been labelled *N* and *C* accordingly, as seen in Figs. 7 and 10. The myoglobin Fourier synthesis also reveals the hæm-linked amino-acid side-chain, probably histidine, on one side of the (ferric) iron atom and a small peak, probably representing a water molecule, on the other side. If this information is transferred to hæmoglobin, the reactive side of the hæm group is as indicated by the arrows in Fig. 4 and by the labels O_2 in Figs. 7 and 8*a*.

Figs. 7 and 8*b* show the reactive sulphydryl group (which is absent in myoglobin) to be attached to the portion of the black chain carrying the hæm-linked histidine. It is seen that the histidine and cysteine side-chains point in roughly opposite directions, one towards and the other away from the hæm group. The sulphydryl group may possibly be in contact with the loop of the white chain which lies below the hæm group on the left. This is the situation of one pair of sulphydryl groups in the molecule. The second pair is probably attached to the white chains, but it is unreactive in the native protein and its position is still unknown.

Reliability of Results

The method of isomorphous replacement makes no assumptions about the structure of the protein, and its results suffer from none of the ambiguities which bedevil the interpretation of vector maps. Ideally, the parent protein, in combination with two different isomorphous heavy-atom compounds, should give accurate phase angles and electron density maps which are free from all except series termination errors. In practice, errors and uncertainties arise from several sources and must be minimized by using more than two heavy-atom compounds. In our case it was possible to estimate the accuracy of the phase angle for each reflexion from the measure of agreement between the angles indicated by the six different compounds. From the standard error in the vector *F*, averaged over all reflexions, the standard error in the final electron density was calculated as 0·12 electron/Å.^{3}, which amounts to 0·85 of the interval between successive contours in Figs. 1 and 3 (0·14 electron/Å.^{3}), or 0·15 of the difference in density between 'peaks' and 'valleys' (0·7 electron/Å.^{3}) (ref. 12). This calculated error is borne out by the observed fluctuations in the liquid regions between the molecules and by the difference in height between the two iron peaks (0·13 electron/Å.^{3}).

420 NATURE February 13, 1960 VOL. 185

An error of at least five times this magnitude would be needed to turn a peak into a valley and thus to simulate the backbone of a polypeptide chain in a region actually occupied by side-chains.

Normally, the accuracy of a structure is checked by comparison of calculated and observed intensities. In hæmoglobin this is impossible so long as the light atoms are not resolved, but the positions of the iron atoms at least can be checked by their anomalous dispersion effect. Depending on the phase relationship between the structure factor of the four iron atoms by themselves and that of the entire molecule, the total intensity of any given reflexion $I\ (hkl)$ may be larger or smaller than $I\ (\bar{h}\bar{k}\bar{l})$. We selected 50 reflexions for which the effect of anomalous dispersion was expected to be largest and measured the intensity of each reflexion in the four symmetrically related quadrants on the counter spectrometer. Statistically significant differences between $I(hkl)+I(\bar{h}k\bar{l})$ and $I(\bar{h}\bar{k}\bar{l})+I(h\bar{k}l)$ were found in 36 reflexions. In 34 of them the signs of the differences agreed with prediction, indicating that the phase angles of the reflexions as well as the positions of the iron atoms are correct.

As a further check on the positions of the iron atoms, an attempt was made to label them with heavy atoms. This was done by allowing crystals of methæmoglobin, which are isomorphous with oxyhæmoglobin, to react with *p*-iodophenyl–hydroxylamine, which attaches itself to the iron atoms. A difference Fourier projection on the *b*-plane showed four prominent peaks which lie well within the calculated distances between the iodine and the iron atoms.

Finally, to avoid any possible bias in the interpretation of the Fourier synthesis, we have tried to construct objective models containing only those features actually found in the electron density maps. Due to series termination errors and inaccuracies in many of the phase angles, these maps must certainly contain errors of detail. Large errors, on the other hand, are unlikely because, quite apart from the checks just described, the results really prove themselves. No combination of errors could have led to the appearance of four distinct chains of roughly equal length, in agreement with chemical evidence, to the similarity between the two pairs of chains which are not related by crystal symmetry, and to the resemblance between hæmoglobin and myoglobin.

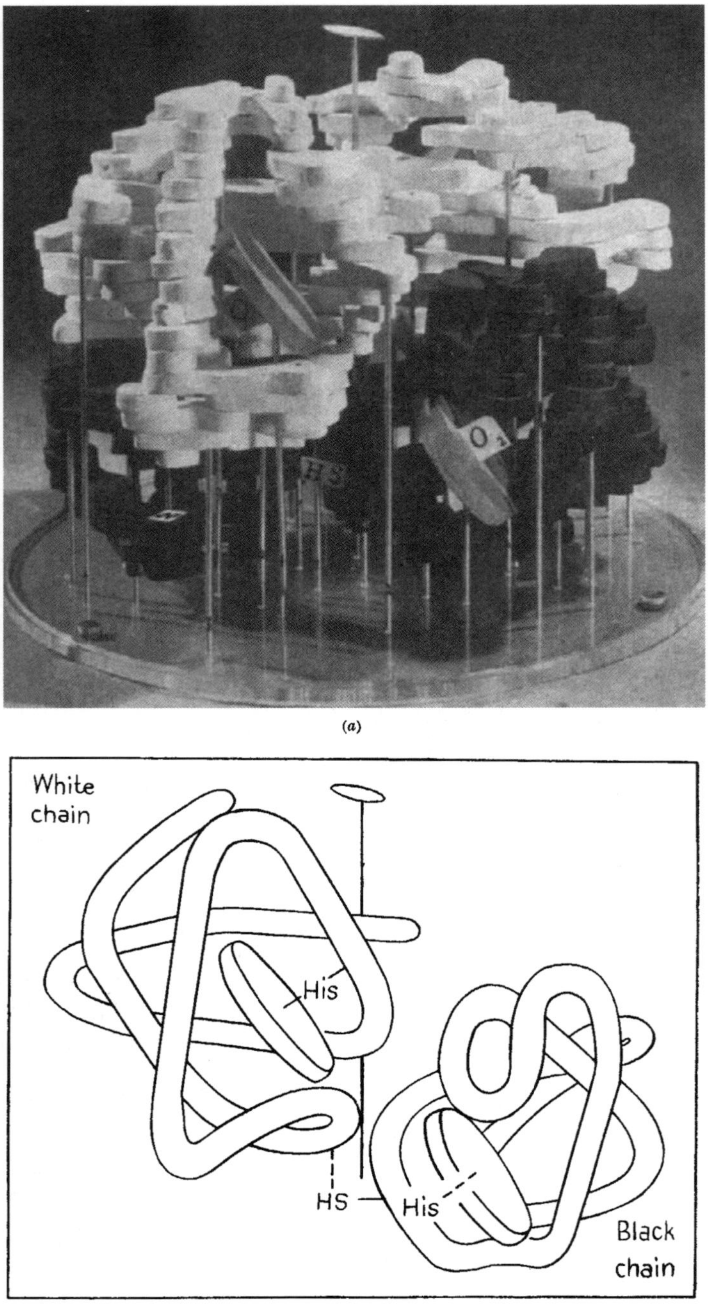

(*a*)

(*b*)

Fig. 8. (*a*) Hæmoglobin model viewed normal to *a*. The hæm groups are indicated by grey disks. (*b*) Chain configuration in the two sub-units facing the observer. The other two chains are produced by the operation of the dyad axis

Discussion

The polypeptide chain-fold which Kendrew and his collaborators first discovered in sperm whale myoglobin has since been found also in seal myoglobin[16].

No. 4711 February 13, 1960 NATURE 421

Fig. 9. Hæmoglobin model viewed normal to *c*

Its appearance in horse hæmoglobin suggests that all hæmoglobins and myoglobins of vertebrates follow the same pattern. How does this arise ? It is scarcely conceivable that a three-dimensional template forces the chain to take up this fold. More probably the chain, once it is synthesized and provided with a hæm group around which it can coil, takes up this configuration spontaneously, as the only one which satisfies the stereochemical requirements of its amino-acid sequence. This suggests the occurrence of similar sequences throughout this group of proteins, despite their marked differences in amino-acid content. This seems all the more likely, since their structural similarity suggests that they have developed from a common genetic precursor.

Little can be said as yet about the relation between structure and function. The hæm groups are much too far apart for the combination with oxygen of any one of them to affect the oxygen affinity of its neighbours directly. Whatever interaction between the hæm groups exists must be of a subtle and indirect kind that we cannot yet guess. A few observations of possible significance might be mentioned. Kendrew found that the combination of reduced myoglobin with oxygen involves no structural changes detectable by X-ray analysis[17], both forms being isomorphous with the metmyoglobin normally studied. In the hæmoglobin of horse and of man, on the other hand, the oxygenated and reduced forms are crystallographically different[18–20]. The structure of reduced hæmoglobin is still unknown, but it would not be surprising if loss of oxygen caused the four sub-units to rearrange themselves relative to each other, rather than to change their individual structure to a marked degree.

Riggs has shown that blocking the sulphydryl groups reduces hæm–hæm interaction[21]. As Fig. 8 shows, these groups occupy key positions close to the hæm-linked histidines and to points of contact between two different sub-units. They may well play an important part in the transition between the oxygenated and reduced forms. Incidentally, the cysteine residue should provide a convenient marker for the peptide containing the hæm-linked histidine, and so help to determine the sequence in this important part of the chain.

A full account of this work will be published elsewhere.

We thank the Director and staff of the University of Cambridge Mathematics Laboratory for making their electronic computor *Edsac* II available to us for the many calculations involved in this work. We are

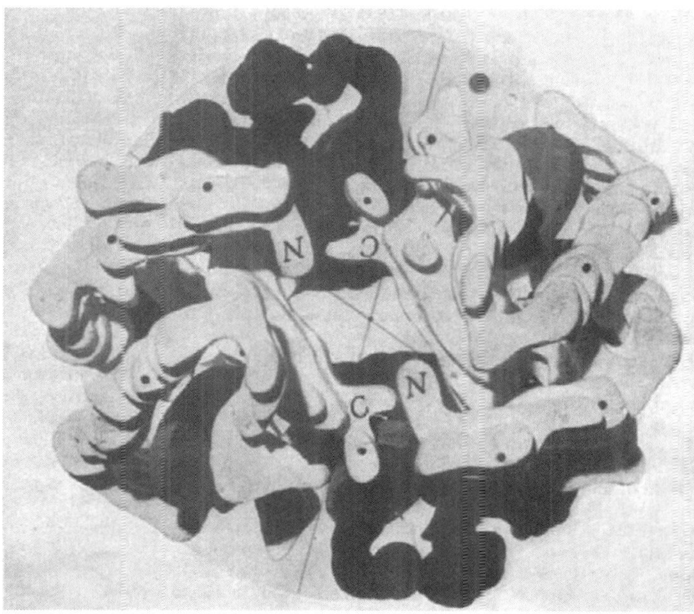

Fig. 10. A view down the *b*-axis. Note the proximity of the *C* and *N* terminal ends which could serve to form links between the two white chains

also deeply grateful to the Rockefeller Foundation for its long-continued financial support and to Sir Lawrence Bragg for his unfailing enthusiasm and encouragement. Finally, we wish to thank Dr. B. R. Baker, of the Stanford Research Institute, for a gift of organic mercurials, Dr. J. Chatt of Imperial Chemical Industries, Ltd., and Mr. A. R. Powell, of Johnson Matthey, Ltd., for gifts of heavy-atom compounds, and Mrs. Margaret Allen, Miss Ann Jury and Miss Brenda Davies for assistance.

[1] Rhinesmith, S. H., Schroeder, W. A., and Pauling, L., *J. Amer. Chem. Soc.*, **79**, 4682 (1957).

[2] Braunitzer, G., *Hoppe-Seyl. Z.*, **312**, 72 (1958).

[3] Wilson, S., and Smith, D. B., *Can. J. Biochem. Physiol.*, **37**, 405 (1959).

[4] Bokhoven, C., Schoone, J. C., and Bijvoet, J. B., *Acta Cryst.*, **4**, 275 (1951).

[5] Green, D. W., Ingram, V. M., and Perutz, M. F., *Proc. Roy. Soc.*, A, **225**, 287 (1954).

[6] Kendrew, J. C., Bodo, G., Dintzis, H. M., Parrish, R. G., Wyckoff, H. W., and Phillips, D. C., *Nature*, **181**, 662 (1958).

[7] Blow, D. M., *Proc. Roy. Soc.*, A, **247**, 302 (1958).

[8] Perutz, M. F., *Proc. Roy. Soc.*, A, **225**, 264 (1954).

[9] Cullis, A. F., Dintzis, H. M., and Perutz, M. F., Conference on Hæmoglobin, National Academy of Sciences, NAS-NRC Publication 557, p. 50 (Washington, 1958).

[10] Rossmann, M. G., *Acta Cryst.* (in the press).

[11] Bodo, G., Dintzis, H. M., Kendrew, J. C., and Wyckoff, H. W., *Proc. Roy. Soc.*, A, **253**, 70 (1959).

[12] Blow, D. M., and Crick, F. H. C., *Acta Cryst.*, **12**, 794 (1959).

[13] Bragg, W. L., and Perutz, M. F., *Proc. Roy. Soc.*, A, **225**, 315 (1954).

[14] Ingram, D. J. E., Gibson, J. F., and Perutz, M. F., *Nature*, **178**, 905 (1956).

[15] Kendrew, J. C., Dickerson, R. E., Strandberg, B., Hart, R. G., Davies, D. R., Phillips, D. C., and Shore, V. C. (see following article).

[16] Scouloudi, H., *Nature*, **183**, 374 (1959).

[17] Kendrew, J. C. (private communication).

[18] Haurowitz, F., *Hoppe-Seyl. Z.*, **254**, 266 (1938).

[19] Jope, H. M., and O'Brien, J. R. P., "Hæmoglobin", 269 (Butterworths, London, 1949).

[20] Perutz, M. F., Trotter, I. F., Howells, E. R., and Green, D. W., *Acta Cryst.*, **8**, 241 (1955).

[21] Riggs, A. F., *J. Gen. Physiol.*, **36**, 1 (1952).

422 NATURE February 13, 1960 VOL. 185

STRUCTURE OF MYOGLOBIN

A THREE-DIMENSIONAL FOURIER SYNTHESIS AT 2 Å. RESOLUTION

By DRS. J. C. KENDREW, R. E. DICKERSON, B. E. STRANDBERG, R. G. HART and D. R. DAVIES*

Medical Research Council Unit for Molecular Biology, Cavendish Laboratory, Cambridge

AND

D. C. PHILLIPS and V. C. SHORE

Davy Faraday Laboratory, The Royal Institution, London

MYOGLOBIN is a conjugated protein consisting of a single polypeptide chain of about 153 amino-acid residues associated with an iron–porphyrin complex, the hæm group; its molecular weight is about 18,000, and the molecule contains some 1,200 atoms (excluding hydrogen). Two years ago a preliminary report of the first stage of an investigation of the three-dimensional structure of sperm-whale myoglobin was published in *Nature*[1] (a detailed account of this work has appeared recently[2]). Several isomorphous crystalline derivatives of myoglobin containing heavy atoms (mercury or gold) at single sites on the molecule were prepared, and by comparing the X-ray diffraction patterns of these crystals with those of the unsubstituted protein, it was possible to deduce the phases of all the reflexions in the X-ray pattern having spacings greater than 6 Å. These phases, together with the observed amplitudes, were used to compute a three-dimensional Fourier synthesis of the electron density in the unit cell (which contains two molecules) at a resolution of 6 Å. In this synthesis the polypeptide chain was visible as a rod of high electron density, folded in a complex pattern (Fig. 5*a*); in addition, the single hæm group with its iron atom, which is much more dense than any other atom in the molecule, could be identified as a disk of high electron density. The shape of the molecule could be determined with some confidence, as could most of the course of the single polypeptide chain, though there were several ambiguities where it turned through a large angle, so that the ends of the chain could not be located with certainty. Thus, the general nature of the tertiary structure of the molecule was revealed, but not the secondary structure of the polypeptide chain, though the results were consistent with a helical configuration.

More recently, Scouloudi[3] used similar methods to obtain a two-dimensional Fourier projection of the unit cell of seal myoglobin, in which the molecular arrangement is entirely different from that in sperm-whale crystals; she was able to show that the myoglobins of these two species have essentially the same tertiary structure in spite of their different amino-acid compositions. Her work, by implication, confirmed the correctness of both analyses, as well as the deductions made about the shape of the molecule. In the accompanying article[4] Perutz *et al.* now describe a three-dimensional analysis of the related protein hæmoglobin, at a slightly greater resolution, and show that each of the four sub-units of which this molecule is composed bears a close structural resemblance to myoglobin. It is apparent, therefore, that sperm-whale myoglobin possesses a structure the significance of which extends beyond a particular species and even beyond a particular protein.

We now present the results of a second stage in the analysis of sperm-whale myoglobin; in this the resolution has been increased to 2 Å., that is to say, not far short of atomic resolution. The resulting Fourier synthesis is very complicated, and a detailed study of it will take many months; in the meantime, our preliminary findings may be of interest.

Methods of X-ray Analysis

In this stage we have simply extended the methods which proved successful in the first stage of the analysis, comparing the diffraction pattern of unsub-

* Visiting scientist from the National Institute of Mental Health, Bethesda, Md.

No. 4711 February 13, 1960 NATURE 423

stituted myoglobin crystals with those of the *p*-chloromercuribenzene sulphonate, mercury diammine and aurichloride derivatives, together with a double derivative containing the first two substituents simultaneously. Whereas myoglobin crystals give 400 reflexions having spacings greater than 6 Å., the number of reflexions with spacings greater than 2 Å. is 9,600, each of which has to be measured not only for the unsubstituted protein but also for each of the derivatives. The very much greater number of data posed many problems, both in recording intensities and in computation, and in this stage we relied much more heavily than before on the use of a high-speed computer; it was fortunate that about the time the work began the *Edsac* Mark I computer used previously was superseded by the very much faster and more powerful Mark II.

The data for each derivative were recorded on twenty-two precession photographs; a separate crystal had to be used for each photograph to keep radiation damage within acceptable limits. The results from the different photographs were scaled together on the computer, the best set of scaling factors being determined by solving an appropriate 22 × 22 matrix[5]. The degree of isomorphism of each derivative was tested, and found adequate, by means of a computer programme which used the $h0l$ reflexions to refine the preliminary values of the heavy-atom parameters, temperature factor, etc., and then compared the values of $\delta F_{obs.}$ and $\delta F_{calc.}$ as a function of sin θ. The co-ordinates of the heavy atoms were further refined using correlation functions[6] computed by means of programmes devised by Dr. M. G. Rossmann, and finally refined again during the process of phase determination itself. The phases were determined by essentially the same method as before, but owing to the very large number of reflexions the determination was carried out on the computer rather than graphically. The 'best' phases and amplitudes[7] were computed and used in the final Fourier synthesis, to which a moderate degree of sharpening was applied. In all, some hundreds of hours of computer time were required, and the Fourier synthesis itself, which was calculated at intervals of about 2/3 Å., took about 12 hr.

The Fourier Synthesis

The electron density distribution was plotted in the form of 96 sections perpendicular to x^* and spaced 2/3 Å. apart, the density in each section being represented by a series of contours. For some purposes this method of representation is unsatisfactory, and we have also constructed models of parts of the structure on a scale of 5 cm. = 1 Å., by erecting vertical steel rods parallel to y in an array corresponding to the grid of points in the xz-plane at which densities were calculated, the value of the electron density at points along the rods being indicated by coloured clips. On this scale the whole molecule is about 6 ft. cube, and about 2,500 rods each 6 ft. high are required (see Fig. 4).

The distance between singly bonded carbon atoms is 1·54 Å.; many of the common types of covalent bond present in protein molecules are shorter than this. We could not expect, therefore, that neighbouring covalently bonded atoms would be resolved in the present Fourier synthesis, and in fact such resolution was not achieved; on the other hand, the degree of resolution actually obtained is very nearly

Fig. 1. (*a*) Cylindrical projection of a helical segment of polypeptide chain, with the α-helix structure superposed: for explanation see text. (*b*) Key to the atomic arrangement in the α-helix. The points marked β and β' are the two alternative projected positions of C_β; β is the position in a right-handed and β' that in a left-handed helix of L-amino-acids

as great as theory would indicate, showing that the errors in the synthesis cannot be very large (see Fig. 2). The distances between neighbouring groups in Van der Waals contact is of the order of 3 Å., however, and as would be anticipated, we find that such groups appear as discrete regions of high density. One factor which complicates the interpretation is that similar groups (for example, peptide bonds) in different parts of the structure have substantially different densities. We attribute these differences to variation in the amplitude of thermal vibration in different parts of the molecule. Thus the terminal amino-end of the chain, which is on the outside of the molecule, has low density and is presumably relatively flexible; on the other hand, structures near the hæm group, which is surrounded by several segments of polypeptide chain running in different directions, possess high density, presumably because this part of the molecule is stabilized by many inter-chain bonds.

Configuration of the Polypeptide Chain

In the 6-Å. Fourier synthesis the polypeptide chain appeared as a solid rod of high density with approximately circular cross-section; segments of the chain were more or less straight, and these were joined by dense regions where the chain turned corners. In the 2-Å. synthesis all the straight segments of chain were found to be hollow cylindrical tubes of high density; projecting at intervals from the cylindrical core are dense regions of various shapes and sizes, which are clearly the amino-acid side-chains. More detailed

424 NATURE February 13, 1960 VOL. 185

(*a*)

(*b*)

Fig. 2. (*a*) Observed electron density distribution in the plane of the hæm group, with the atomic arrangement in the group superposed. One contour interval = 0·5 electron/Å.^{3}. (*b*) Calculated electron density distribution, computed from the known atomic arrangement as described in the text. One contour interval = 0·5 electron/Å.^{3}

examination shows that, in fact, the cylindrical tubes are helices, consisting of a single strand of high density with axial repeat about 5·4 Å. The density in one such helix is shown in Fig. 1*a*. This was obtained by projecting all the density lying within a cylindrical shell of radii 1·6 and 2·3 Å. on to the surface of a cylinder of radius 1·95 Å. and co-axial with the helix; the cylinder was then cut along a line parallel to its axis and unrolled. Superposed on the contours of electron density is the cylindrical projection of an α-helix, with the dimensions given by Pauling and Corey[8]. It will be seen that the observed density follows the configuration of the α-helix with remarkable precision; and it may confidently be asserted that not only this, but also several other lengths of polypeptide chain which have been analysed in the same way, possess the α-helical configuration. It is also found that the side-chains emerge at intervals of 100° around the periphery of the helix, and at axial intervals of 1·5 Å., in conformity with the parameters of the α-helix. Several indirect lines of evidence had led to the conclusion that parts of the polypeptide chain in globular proteins are in the form of α-helixes, and in hæmoglobin Perutz[9] had found traces of an X-ray reflexion characteristic of this configuration; but these results are the first direct proof that α-helices are present and, indeed, enable them to be seen directly for the first time.

The next question is whether the α-helices are left-handed or right-handed; helices of either hand

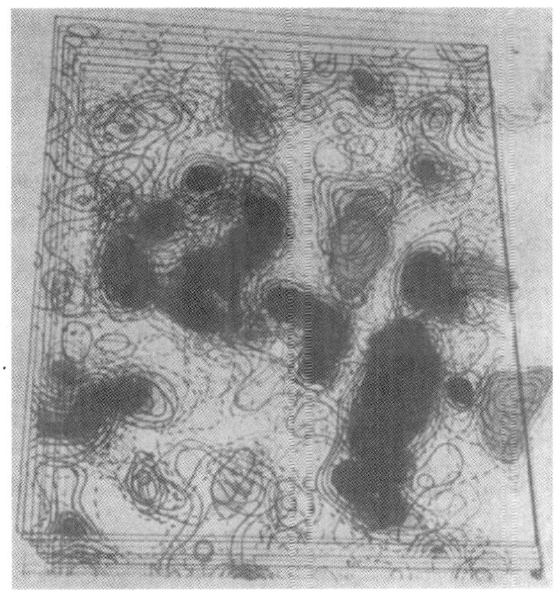

(*a*)

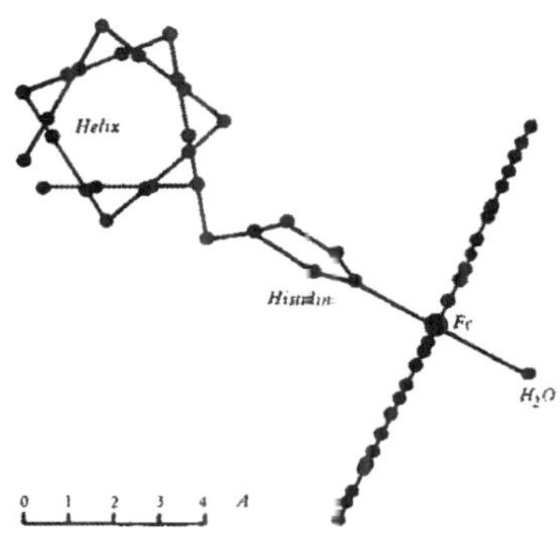

(*b*)

Fig. 3. (*a*) Photograph of a set of sections normal to the plane of the hæm groups showing, from left to right, a helix in cross-section, the histidine residue nearly edge-on, the hæm group edge-on, and a presumed water molecule. (*b*) Sketch showing the atomic arrangement in Fig. 3*a*

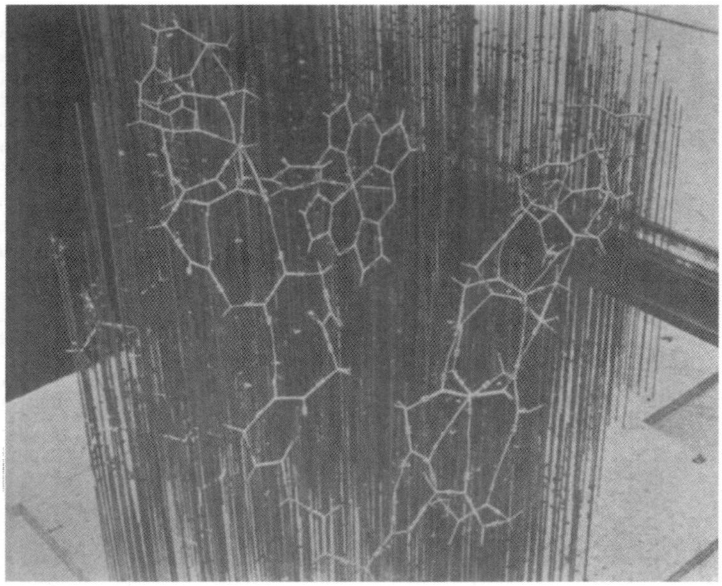

Fig. 4. Photograph of a model of part of the molecule near the hæm group, showing the vertical rods and coloured clips which indicate the electron density at each point of the grid, and atomic models of the hæm group and neighbouring helices. The terminal carboxyl end of the chain is on the extreme left

can be built, and although the right-handed form appears to be marginally the more stable, the difference is not so great as to make it certain that left-handed helices cannot exist. To decide this question, it is necessary to determine the absolute configuration of the Fourier synthesis; this function has no centre or plane of symmetry and might be plotted in two ways, each of which would be the mirror image of the other. The absolute configuration could be settled by making use of the anomalous dispersion of X-rays from the iron atoms or the introduced heavy atoms; we have not in fact made use of this effect, but instead have proceeded in the following way. In a right-handed α-helix composed of L-amino-acids, the first atom of the side-chain (C_β) projects from the main chain in a direction *opposite* to the C—O bonds of the carbonyl groups; in a left-handed helix of L-amino-acids C_β projects in the *same* direction as the carbonyl groups (see Fig. 1*b*). By constructing a second cylindrical projection (not illustrated) of the density near a radius of 3·34 Å., corresponding approximately to the radius at which C_β is found in either case, we found that C_β was systematically on the side of the main chain opposite the oxygen atoms of the carbonyl groups. This shows that the helix must be right-handed, and we were able to plot the Fourier synthesis with the correct absolute configuration by taking account of the known absolute configuration of a L-amino-acid[10]. The molecule was then found to be of the same hand as all the closely similar sub-units in hæmoglobin, the absolute configuration of which was determined by measurements of anomalous dispersion[4]. All the lengths of α-helix in the myoglobin molecule turn out to be right-handed. Finally, it will be clear from Fig. 1 that it is possible to determine by inspection the direction in which the C—O group points, and hence to see which is the terminal carboxyl end of each segment.

The α-helices can also be located by building atomic models into the model of the unit cell made with steel rods; this has been done for all the segments of helix in the molecule, and it is found that the total number of amino-acid residues contained in these segments is 100–110, whereas the number of residues contained in the whole molecule is believed to be 153 (Edmundson, A., unpublished results). Thus, 65–72 per cent of the molecule consists of regular right-handed α-helix, made up of about eight segments, each containing between seven and about twenty residues.

When the chain turns a corner, its regular helical configuration is necessarily disrupted. At the present resolution the precise arrangement of the residues at corners is difficult to determine; we have, nevertheless, built plausible models of several corners, though we do not yet claim to have established their configurations with certainty. Most of the corners take up two or three residues, and in addition there is one region (on the extreme right of Fig. 5*a*), consisting of about 13–18 residues, in which the arrangement is irregular (part of it is helical, but probably not α-helical). Further studies will be devoted to elucidating these problems.

No serious attempt has yet been made to identify the side-chains, and indeed it is doubtful whether this can be done systematically at the present resolution, partly because many of them appear to be subject to considerable thermal vibration, partly because there are often interactions between two side-chains on adjacent turns of the same helix or on adjacent helices, and it is difficult to tell where one ends and the other begins. In special cases, however, identification is easy; thus, in several places two helices approach one another closely, and one or more side-chains must for steric reasons be glycine. On the other hand, some side-chains are so long that they can only be arginine or lysine. It remains to be seen how many positive identifications can be made; but it is clear that if the whole amino-acid sequence of the protein were known, it would be possible at this stage to construct a model of the complete structure with fair precision. Unfortunately, this information is not yet available; but in the meantime it may be possible to correlate and check our findings with the partial results of the sequence determination now being undertaken by Mr. A. Edmundson, of the Rockefeller Institute.

The Hæm Group

The identification of a disk of high electron density in the 6-Å. synthesis as the hæm group is confirmed by the 2-Å. synthesis; but the iron atom is now just resolved from the nitrogens of the porphyrin ring (the iron–nitrogen distance is about 1·9 Å.) and the structure of the group as a whole corresponds closely with theoretical expectation. Fig. 2*a* shows a section through the density distribution, cut in the plane of the hæm group; superposed on the section is a model of the hæm group with the dimensions

426 NATURE February 13, 1960 Vol. 185

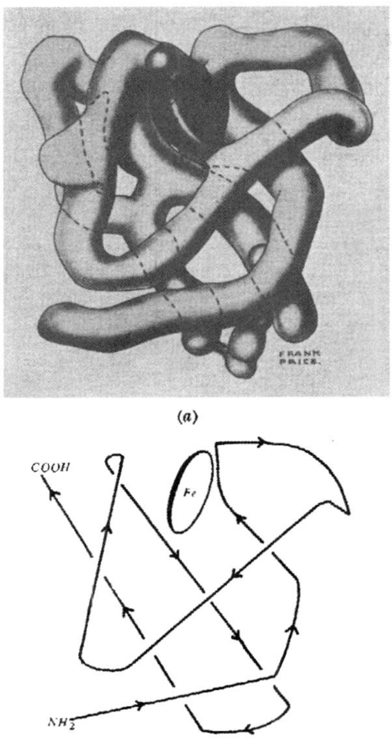

Fig. 5. (*a*) Drawing of the tertiary structure of myoglobin as deduced from the 6-Å. Fourier synthesis. (Note: the plane of the hæm group is indicated incorrectly in this drawing—see text). (*b*) The course of the polypeptide chain, deduced from the 2-Å. Fourier synthesis

found by Crute in his recent structure analysis of nickel etioporphyrin[11]. In the model, only the carbon atoms of the methyl groups and the first carbon atoms of the vinyl and propionic acid groups have been indicated; our analysis has not yet proceeded far enough to identify the latter groups. Fig. 2*b* is a Fourier synthesis made by calculating structure factors for the model structure in a convenient unit cell, applying to these the same temperature factor as that observed in myoglobin, and using them as terms of a Fourier synthesis cut off at 2-Å. spacings. It will be seen that the calculated and observed density distributions agree very closely; furthermore, the observed peak density (8·8 electrons/Å.^{3}) and half-width (1·8 Å.) of the iron atom agree well with the predicted values (8·2 electrons/Å.^{3} peak density, 1·8 Å. half-width). Errors are, of course, inherent in a Fourier synthesis produced from experimental results; the good agreement between the observed and calculated versions of the hæm group, and between the helical polypeptide chain and the α-helix (Fig. 1), indicate that in the present case these errors are fairly small, certainly less than we had anticipated.

In one respect the 6-Å. model now requires correction. Measurements of electron spin resonance[12] had shown that the two hæm groups in the unit cell were tilted out of the *bc* plane through 21° about an axis parallel to *c*, one in one direction and one in the other, but from these measurements one could not decide which direction of tilt was associated with the hæm group of which molecule. In the 6-Å. analysis the wrong choice was made; the new results show that the angle of tilt is indeed very close to 21°, but that it should be in the direction opposite to that shown in earlier models (including Fig. 5*a*).

Chemical studies of hæmoglobin and myoglobin have suggested that the hæm group is attached to the protein by a bond from the iron atom to a nitrogen atom of a histidine residue, and that the sixth co-ordination position of the iron atom is occupied in met-myoglobin (at *p*H's below 7) by a water molecule. Our results are in conformity with these suggestions. Fig. 3*a* is a photograph of a series of sections cut through the hæm group normal to its plane. The hæm group is seen edge-on; to its left is the histidine residue, also nearly edge-on, and to the left of that is the helix to which the histidine is attached, seen in cross-section. To the right of the hæm group is a small isolated peak which we take to be the oxygen atom of the water molecule (see Fig. 3*b*). When this structure is built with atomic models (see Fig. 4), it is found to satisfy all known requirements as to bond-lengths, angles, etc. Thus the iron–oxygen distance is close to 2·1 Å.; that between the iron and one of the histidine nitrogen atoms is 1·9 Å., while the second histidine nitrogen points with its hydrogen directly at a carbonyl group on the helix, presumably forming a hydrogen bond (forked at the carbonyl end); and finally the bonds about the C_α—C_β bond of histidine are almost exactly in the staggered configuration.

General Configuration of the Molecule

In the 2-Å. synthesis the continuity of the polypeptide chain can be followed throughout the molecule; in addition, its direction can be ascertained in each straight segment by examining the direction of the carbonyl groups in the cylindrical projections or by direct model-building. The 'run' of the chain so deduced everywhere agrees with that derived from the 6-Å. synthesis, as well as resolving several ambiguities in that model, and the terminal amino- and carboxyl-ends of the chain can now be identified. The arrangement is shown in Fig. 5 (see also Fig. 5 in the accompanying article by Perutz *et al.*). The overall shape of the molecule is as previously deduced.

Conservative estimates suggest that it may now be possible to locate with some precision at least 43 per cent of all the atoms (excluding hydrogens) in the molecule—namely, those making up the hæm group and the backbone of the regular segments of helix, together with the associated C_β's. We believe that it may be possible, either by straightforward refinement procedures or by heavy-atom methods, to extend the resolution of the synthesis to a point near the limit of the diffraction pattern of sperm-whale myoglobin crystals (about 1·5 Å.), and thus to reveal in any event most of the details of the atomic arrangement of the molecule. Further work is in hand with this end in view.

We wish to thank the Rockefeller Foundation for its generous financial support; the National Science Foundation (U.S.A.) for providing a fellowship for one of us (R. E. D.); to the National Institutes of Health (U.S.A.) for fellowships (R. G. H., R. E. D.); and to the Director and Staff of the Mathematics Laboratory, University of Cambridge, for facilities on the *Edsac* and for their enthusiastic co-operation. We also wish to thank Miss Mary Pinkerton, who was responsible for the organization of the very

No. 4711 February 13, 1960 NATURE 427

extensive programme of data processing, and who helped us in many other ways; a large number of computing assistants, especially Mrs. J. Blows, Mrs. W. Browne, Mrs. V. Cotterill, Mrs. A. Hartley, Miss J. Hughes, Mrs. J. Moss, Mr. P. Mulhall, Mrs. K. Parkes and Miss E. Rowland; Dr. M. G. Rossmann and Dr. E. Alver for making available several of their computer programmes; and Sir Lawrence Bragg for his encouragement at all stages of the work.

[1] Kendrew, J. C., Bodo, G., Dintzis, H. M., Parrish, R. G., Wyckoff, H. W., and Phillips, D. C., *Nature*, **181**, 662 (1958).
[2] Bodo, G., Dintzis, H. M., Kendrew, J. C., and Wyckoff, H. W., *Proc. Roy. Soc.*, A, **253**, 70 (1959).
[3] Scouloudi, H., *Nature*, **183**, 374 (1959).
[4] Perutz, M. F., Rossmann, M. G., Cullis, A. F., Muirhead, H., Will, G., and North, A. C. T. (see preceding article).
[5] Dickerson, R. E., *Acta Cryst.*, **12**, 610 (1959).
[6] Rossmann, M. G., *Acta Cryst.* (in the press).
[7] Blow, D. M., and Crick, F. H. C., *Acta Cryst.*, **12**, 794 (1959).
[8] Pauling, L., and Corey, R. B., *Proc. U.S. Nat. Acad. Sci.*, **37**, 235 (1951).
[9] Perutz, M. F., *Nature*, **167**, 1053 (1951).
[10] Trommel, J., and Bijvoet, J. M., *Acta Cryst.*, **7**, 703 (1954).
[11] Crute, M. B., *Acta Cryst.*, **12**, 24 (1959).
[12] Ingram, D. J. E., and Kendrew, J. C., *Nature*, **178**, 905 (1956).

No. 4443 December 25, 1954 NATURE 1169

X-ray Crystallographic Evidence on the Structure of Vitamin B_{12}

It is not usual in X-ray analysis to present evidence of part only of a crystal structure. We have, however, reached a point in the investigation of the structure of vitamin B_{12} at which we have obtained strong evidence for the existence of a new type of ring system within the molecule, although we cannot yet describe in detail the nature of the groups attached to it.

Vitamin B_{12} has been shown[1] chemically to contain cobalt, a nucleotide-like group (5 : 6-dimethylbenziminazole 1-α-D-ribofuranoside (2′ or 3′)-phosphate), and a CN group, which together account for $C_{15}H_{18}O_7N_3PCo$ out of a total[2] of approximately $C_{61-64}H_{86-92}O_{14}N_{14}PCo$. There is chemical evidence in addition for the existence of groups which give rise to propanolamine on hydrolysis and to various simple acids on oxidation[3], but a major portion of the molecule is at present of largely unknown character. It is for part of this structure that we now have evidence.

We have obtained complete three-dimensional X-ray data for four crystals: air-dried vitamin B_{12} (two independent sets of data obtained at Oxford and at Princeton), wet vitamin B_{12} (the crystal in its mother liquor), vitamin B_{12}-SeCN (having the CN of B_{12} replaced by SeCN), and, very recently, a red degradation product of vitamin B_{12} from which the nucleotide and propanolamine have been removed. In each of these cases, we have found the positions of the heavy atoms in the crystal by direct Patterson methods, and calculated three-dimensional Fourier series using our observed F values and phases based only on the heavy atom positions: cobalt, or cobalt and selenium. The resulting distributions give a crude approximation to the correct electron density series, containing certainly many spurious peaks, but also containing peaks which must correspond to actual atomic positions. We first recognized peaks corresponding to the CN groups in the case of wet and dry vitamin B_{12} and proved our conclusions correct by the substitution of SeCN for CN. We next recognized the general outline of the nucleotide, placed between the cobalt and a second moderately heavy peak—the phosphorus atom—the ribose being linked to the phosphate at the 3-position[4] and one nitrogen atom of the benziminazole nucleus co-ordinated to the cobalt[5]. Between the nucleotide and CN group, and surrounding the cobalt atom, appeared a group of largely planar form, similar in dimensions to a porphyrin nucleus. Attached to this nucleus could be located, though less precisely, various side-chains and possible other ring systems.

1170 **NATURE** December 25, 1954 Vol. 174

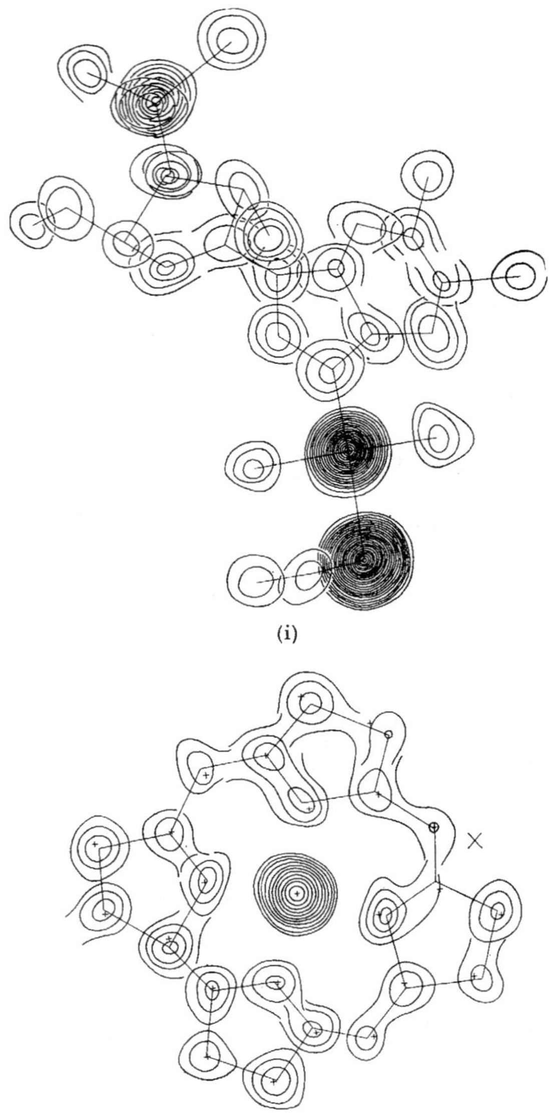

Fig. 1. Positions of certain of the atoms in vitamin B_{12}-SeCN derived from the second calculated three-dimensional electron density distribution

(i) The phosphate, ribose, benziminazole, cobalt and selenocyanide groups viewed along the *b*-axis

(ii) The 'planar' group viewed along the *a*-axis

In this figure, and in Fig. 2, the contours drawn are those showing the electron density in sections in the three-dimensional distribution, passing at or near each individual atomic position

In the next series of calculations on air-dried vitamin B_{12}, made both in Oxford and Princeton, it was assumed that the planar group had in outline the form expected for a normal porphyrin. But more detailed comparison of all three series for vitamin B_{12} later obtained in Oxford suggested that, in fact, no true porphyrin was present. Three of the five-membered rings appeared separated each by a single atom, but the fourth seemed directly linked to the first. At this point, different refinements of the first electron density series were calculated. In those derived at Princeton, for vitamin B_{12} itself (J. W.), atomic positions corresponding to a normal porphyrin were inserted, together with successively increasing numbers of positions postulated for the rest of the molecule. In that obtained at Oxford, based on vitamin B_{12}-SeCN, the modified porphyrin-like structure was used, together with postulated positions for all the atoms in the unit cell. Comparison of results provides strong support for the modified planar group. The region of the directly linked five-membered rings appears most distinctly in the vitamin B_{12}-SeCN series. This is illustrated in Fig. 1, which shows the appearance of the planar group, projected along the *a*-axis (ii) together with that of the benziminazole–ribose–phosphate system viewed in the direction of the *b*-axis (i). The atomic positions in both these systems are individually resolved and of the height expected, except perhaps in the region marked *X*.

The atomic arrangement found in the planar group is so unexpected that we hesitated to accept this until we examined the degradation product isolated by Cannon, Johnson and Todd and described in the preceding communication. In this crystal the nucleotide is absent and is replaced by chlorine. The relative disposition of the planar groups in the crystal is entirely different from that in vitamin B_{12} itself; but, within the molecule, we can still trace the same essential pattern. The planar group here is considerably inclined to all three axes; a linear group, CN, projects on one side, and on the other, a heavy peak, the chlorine atom, completes the octahedron of peaks surrounding the cobalt. Fig. 2 shows the peaks directly traced from the first three-dimensional series, viewed from the *a*-direction.

There are a few points that we should make. While we are reasonably certain that atoms exist bonded essentially as shown in Figs. 1 (ii) and 2, we have no evidence of their chemical nature; it is natural to suppose that the four directly attached to cobalt are nitrogen, and the rest carbon. Secondly, we have evidence that the system as a whole is not completely planar, but our results are not yet precise enough for us to know at which points the rings are reduced. Again, though we have indications of the existence of side-chains attached to the nucleus, their corresponding electron density peaks are usually

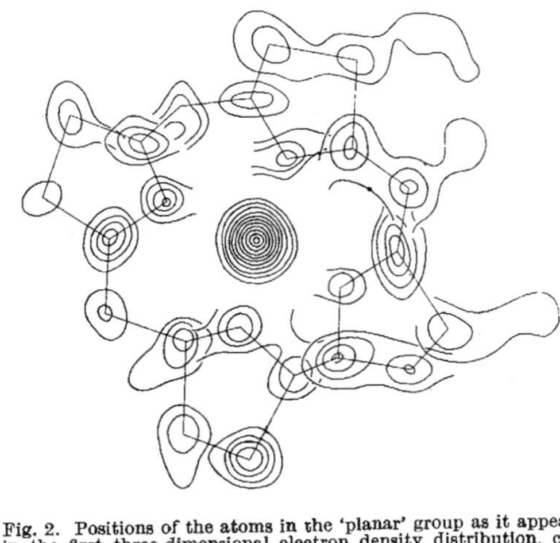

Fig. 2. Positions of the atoms in the 'planar' group as it appears in the first three-dimensional electron density distribution, calculated for the vitamin B_{12} fragment. The structure is viewed along the *a*-axis

No. 4443 December 25, 1954 NATURE 1171

much less clear than those of the nucleus itself and we differ among ourselves in our deduction of their nature. A great deal more work is clearly necessary to discover in detail these remaining features of the vitamin B_{12} structure, but from our present experience we now have little doubt that a complete solution of the crystal structure by X-ray analysis is possible.

We are very grateful for gifts of material, and for invaluable conversations, to Dr. Lester Smith, of Glaxo Laboratories, Ltd., and to Sir Alexander Todd and his colleagues. One of us (J. W.) would like similarly to acknowledge a gift from the Merck Laboratories. We have had considerable assistance in our calculations from Miss Ruth Cox, Mrs. U. Ledwith, Mrs. M. Mackay and Mrs. I. Robertson, using machines supplied to us by the British Tabulating Machines, Ltd. Other computing has been carried out for us by the Scientific Computing Service, the National Physical Laboratory, and by XRAC, through the kindness of Prof. Pepinsky. Our researches in Oxford have been very generously supported by the Nuffield Foundation and by the Rockefeller Foundation. We also gratefully acknowledge the award of a fellowship by the International Federation of University Women (to C. B.) and of a grant from the Department of Scientific and Industrial Research (J. P.).

CLARA BRINK
DOROTHY CROWFOOT HODGKIN
JUNE LINDSEY
JENNY PICKWORTH
JOHN H. ROBERTSON
Laboratory of Chemical Crystallography,
University Museum,
Oxford.

JOHN G. WHITE
Frick Laboratory,
Princeton University,
Princeton, N.J.

Oct. 15.

[1] Rickes, E. L., Brink, N. G., Koniuszy F. R., Wood, T. R., and Folkers, K., *Science*, **107**, 396 (1948); **108**, 134 (1948). Lester Smith, E., *Nature* **161**, 638 (1948); **162**, 144 (1948). Fantes, K. H., Page, J. E., Parker, L. F. J., and Lester Smith, E., *Proc. Roy. Soc.*, B, **136**, 592 (1950).

[2] See review by Todd, A. R., and Johnson, A. W., *Rev. Pure and App. Chem.*, **2**, 23 (1952).

[3] Schmidt, von H., Ebnöther, A., and Karrer, P., *Helv. Chim. Acta*, **36**, 65 (1953).

[4] Kaczka, E. A., and Folkers, K., *J. Amer. Chem. Soc.*, **75**, 6317 (1953).

[5] Beaven, G. H., Holiday, E. R., Johnson, E. A., Ellis, B., and Petrow, V., *J. Pharm. Pharmacol.*, **2**, 944 (1950).

Chemistry 1976

THE BORANES AND THEIR RELATIVES

Nobel Lecture, December 11, 1976
by
WILLIAM N. LIPSCOMB
Harvard University, Cambridge, Massachusetts, USA

This year, 1976, the Nobel Prize in Chemistry has been awarded for research in pure inorganic chemistry, in particular the boranes. May I say that I am most pleased and profoundly grateful. My own orientation to this field has been, as it has in all of my studies, the relationships of the chemical behavior of molecules to their three-dimensional geometrical and electronic structures. The early work on the molecular structures of boranes by X-ray diffraction led to a reasonable basis for a theory of chemical bonding different from that which is typical in carbon chemistry, and yielded an understanding of the pleasing polyhedral-like nature of these compounds. Assimilated by the preparative chemists, the principles helped to establish a large body of a hitherto unknown chemistry, which made a reality of the expectation that boron, next to carbon in the periodic table, should indeed have a complex chemistry.

In these nearly thirty years both the theoretical and experimental methods have been applied by us and others to areas of inorganic, physical, organic and biochemistry. For examples, these areas include low temperature X-ray diffraction techniques, and the theoretical studies of multicentered chemical bonds including both delocalized and localized molecular orbitals. An early example is extended Hückel theory, originally developed for studies of the boranes, and even now one of the most widely applicable approximate methods for theoretical studies of bonding in complex molecules. More soundly based theories are presently in use by my research students for studying how enzymes catalyze reactions, details of which are based on the three-dimensional structures by X-ray diffraction methods. Besides illuminating particular problems, these developments may contribute toward the redefinition of areas of chemistry, and thereby broaden the chemist's view. Our research in the boranes and their related molecular species crosses areas of inorganic, experimental physical, theoretical and organic chemistry, and includes applications in biochemistry. More simply stated, the area is the study of the relationships of molecular structure to function.

BORANES, AND EARLY STRUCTURE STUDIES

By now, large numbers of chemical compounds related to polyborane chemistry exist: boron hydrides, carboranes, metalloboranes, metallocarboranes, mixed compounds with organic moieties, and others. These discoveries of preparative chemists are relatively recent. Long ago, Alfred Stock established borane chemistry. He developed the experimental techniques which were required for the preparation of the volatile and potentially explosive compounds, B_2H_6, B_4H_{10}, B_5H_9, B_5H_{11} and B_6H_{10}, and the relatively stable white crystalline $B_{10}H_{14}$. This work, beautifully summarized in his Baker Lectures,[1] was celebrated in the Third International Meeting on Boron Chemistry this past July in Munich, 100 years after his birth. Sidgwick[2] wrote, "All statements about the hydrides of boron earlier than 1912, when Stock began to work on them, are untrue."

Aside from the classification as either an N+4 or an N+6 series, no simple basis for these unusual formulas was foreseen before their structures were established. A B_6 octahedron (Fig. 1) was known in certain crystalline borides, and the B_{12} icosahedron (Fig. 2) was found in boron carbide, but no one realized that a systematic description of the boron arrangements in these hydrides might be based on fragments of these polyhedra. Most electron diffraction work before 1940 supported more open structures, which Pauling[3] described in terms of resonating one-electron bonds. In 1940-1941 Stitt[4] produced infrared and thermodynamic evidence for the bridge structure of B_2H_6 (Fig. 3). More general realization of the correctness of this structure followed the theoretical studies,[5-8] and especially the infrared work of Price.[9] The three-center bridge BHB bond was clearly formulated by Longuet-Higgins.[8]

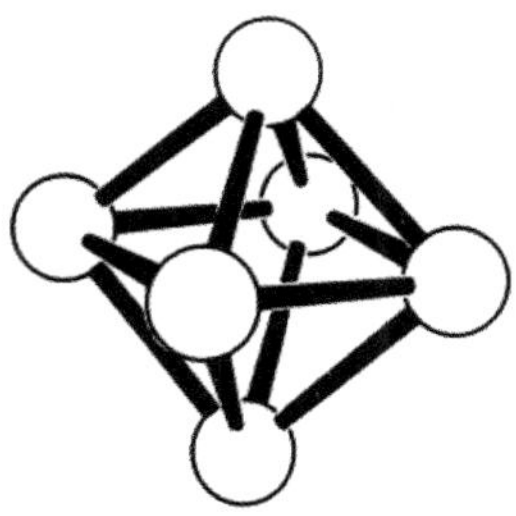

Fig. 1. The B_6 octahedron, which occurs in certain metal borides, in which each boron atom is bonded externally to a boron atom of another octahedron. In $B_6H_6^{-2}$ or $C_2B_4H_6$, a terminal hydrogen atom is bonded externally to each B or C atom.

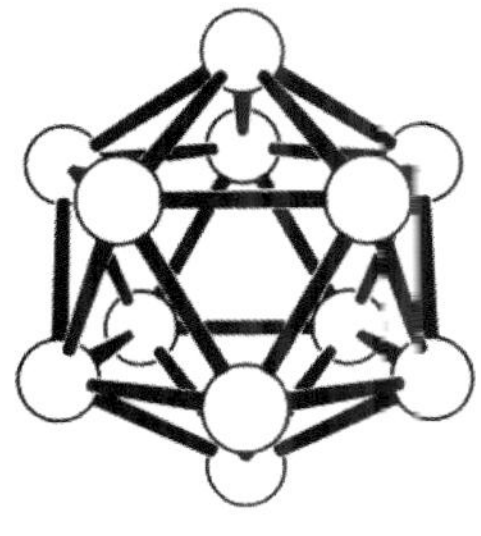

Fig. 2. The B_{12} icosahedron, which occurs in boron carbide $B_{12}C_3$, in elementary boron, and in $B_{12}H_{12}^{-2}$. The three isomers of $C_2B_{10}H_{12}$ also have this icosahedral arrangement in which there is one externally bonded hydrogen on each B or C atom.

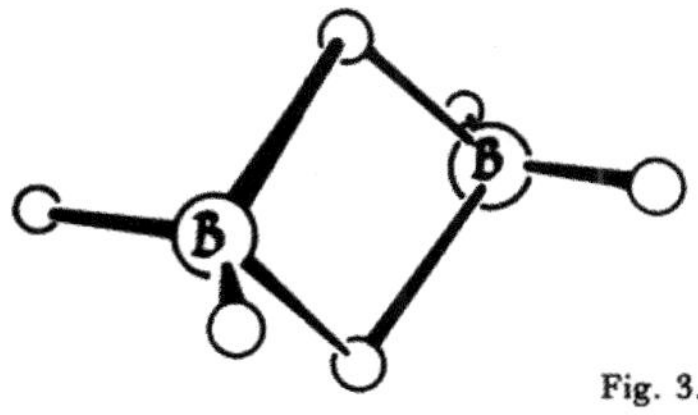

Fig. 3. The geometrical structure of B_2H_6.

The first of the higher hydrides to be structurally characterized was $B_{10}H_{11}$ (Fig. 4). Kasper, Lucht and Harker[10] showed that the boron arrangement was a fragment of a B_{12} icosahedron in which there are four bridge BHB bonds in the open face. Next was B_5H_9,[11, 12] (Fig. 5), a fragment of the B_6, octahedron, and then B_4H_{10}[13,14] (Fig. 6), a B_4 unit from these polyhedra. Both of these structures were established by X-ray diffraction in our laboratory, and by electron diffraction at the California Institute of Technology. Our X-ray diffraction results on B_5H_{11}[15] (Fig. 7) and tetrahedral B_4Cl_4[16] then set the stage for the beginning of the theory of bonding. The B_6H_{10} structure,[17,18] (Fig. 8) completing the compounds found by Stock, was to be one of our later X-ray diffraction studies, which were to include many other boranes and related molecules.

It was actually in 1946 that I decided to enter this area of inorganic chemistry. At that time, no reliable methods were generally available for accumulation of large amounts of X-ray diffraction data at low temperatures. Our methods[19] paralleled those in Fankuchen's laboratory[20] at the then Polytechnic Institute of Brooklyn, but were quite independent. Because of the special difficulties of working with the volatile boranes, I chose among other topics, a series of X-ray diffraction studies of single crystals grown at low temperatures from low melting liquids for which putative residual entropy problems

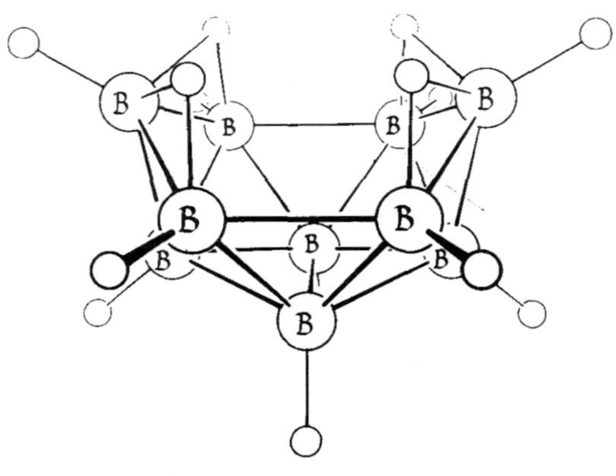

Fig. 4. $B_{10}H_{14}$.

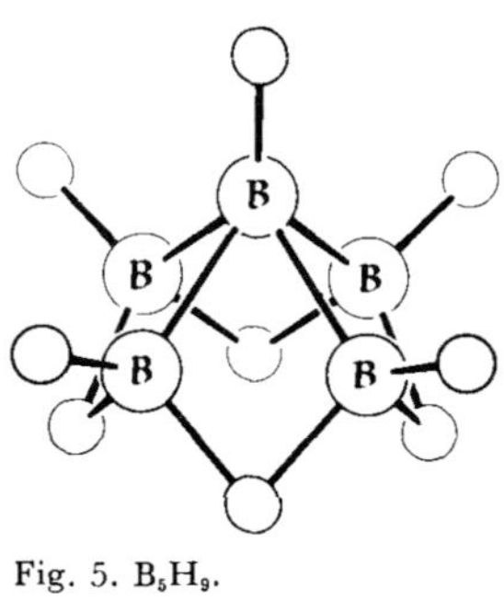

Fig. 5. B_5H_9.

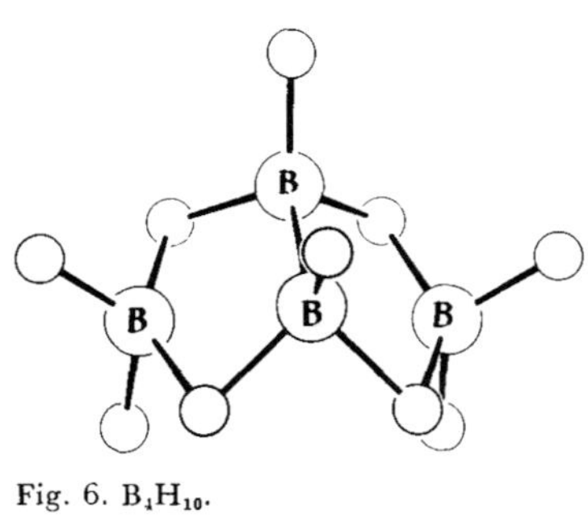

Fig. 6. B_4H_{10}.

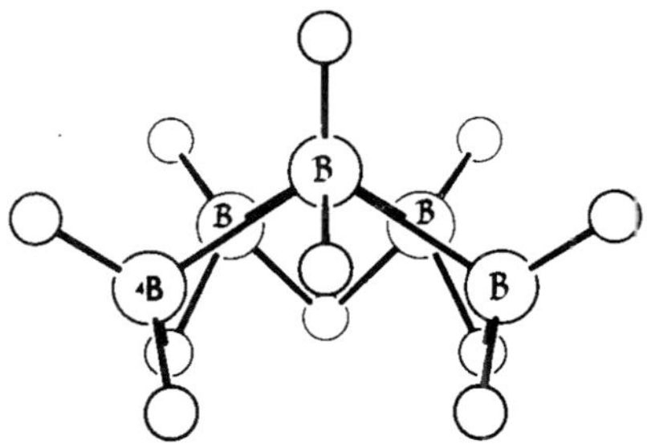

Fig. 7. B_5H_{11}.

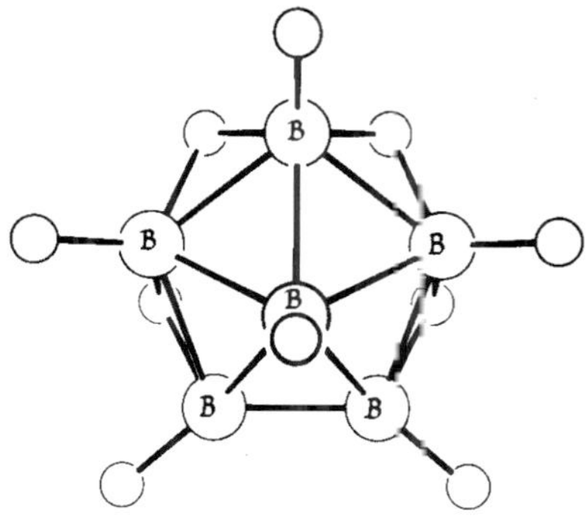

Fig. 8. B_6H_{10}.

existed: N_2O_2, CH_3NH_2, CH_3OH, N_2H_4, $COCl_2$ and H_2O_2. The subsequent low temperature studies of single crystals of these volatile and unstable boranes were not without hazards. Vacuum line techniques were learned as we needed them. Fortunately, no serious injuries were incurred as a result of several explosions resulting from cracks in these vacuum systems. I was relieved, on one occasion, when I had taken Russell Grimes to a hospital in Cambridge after one of these explosions to hear the doctor tell me, "Louis Fieser sends me much more interesting cases than you do." I still have in my office the air gun which I, or my young son, used on a number of occasions to destroy a cracked vacuum system from a safe distance. We also had chemical surprises, for example, when we found a presumed B_8 hydride to be B_9H_{15}.[21] Our only chemical analysis of this compound was the count of the numbers of boron and hydrogen atoms in the electron density map, which was calculated in those days from rough visual estimates of intensities of diffraction maxima on films.

THREE-CENTER BONDS AMONG BORON ATOMS

At the fortunate time in 1953 of W. H. Eberhardt's sabbatical, he, Crawford and I examined[22] the open boron hydrides B_2H_6, B_4H_{10}, B_5H_9, B_5H_{11} and $B_{10}H_{14}$ from the viewpoint of three-center bonds; and we studied B_5H_9, the unknown polyhedral molecule B_4H_4, the then hypothetical ions $B_6H_6^{-2}$ (Fig. 1) and $B_{12}H_{12}^{-2}$ (Fig. 2) from the viewpoint of molecular orbitals. Longuet-Higgins[23] also, independently, formed an early molecular orbital description almost like ours. One of the simple consequences of these studies was that electron deficient molecules, defined as having more valence orbitals than electrons, are not really electron deficient. I mean by this non-sequitur that the three-center two-electron bonds make possible a simple description of these molecules and ions as filled orbital species. Filled molecular orbitals were later extended to closed polyhedral compounds B_nH_n, for all values of n from 5 to 24, all of which have a formal charge of -2, even though hypothetical B_4H_4 is neutral. In fact, all of the experimentally known ions, for $6 \leqslant n \leqslant 12$, do have -2 charge.

in addition to $B_6H_6^{-2}$ (Fig. 1) and $B_{12}H_{12}^{-2}$ (Fig. 2), those polyhedra for $5 \leqslant n \leqslant 10$ are shown in Fig. 9. The isoelectronic series $C_2B_{n-2}H_n$ is known for $5 \leqslant n \leqslant 12$.

Equations for the atom, orbital and electron balance were formulated in our paper of 1954, allowing prediction of many new chemical species. One simple form of these rules is exemplified for a neutral hydride formula B_pH_{p+q} in which each boron atom has at least one terminal hydrogen atom. Define the number of BHB bridges as s, the number of three-center BBB bonds as t, the number of two-center BB bonds as y and the number of extra terminal hydrogens on each BH unit as x. Then

$$s+x = q$$

$$s+t = p$$

$$p = t+y+q/2$$

The first equation is the hydrogen balance. The second comes from the fact that each of the p boron atoms supplies four orbitals and only three electrons, and the extra orbital is utilized in one of the two types of three-center bond. Finally, if each BH unit is recognized as contributing a pair of electrons, these p pairs are used in BBB or BB bonds, or in half of the pairs required for the extra hydrogens, s and x. These rules, and the accompanying valence structures, are especially helpful in describing those polyboron compounds which are open, but they are also useful for closed polyhedral molecules and ions.

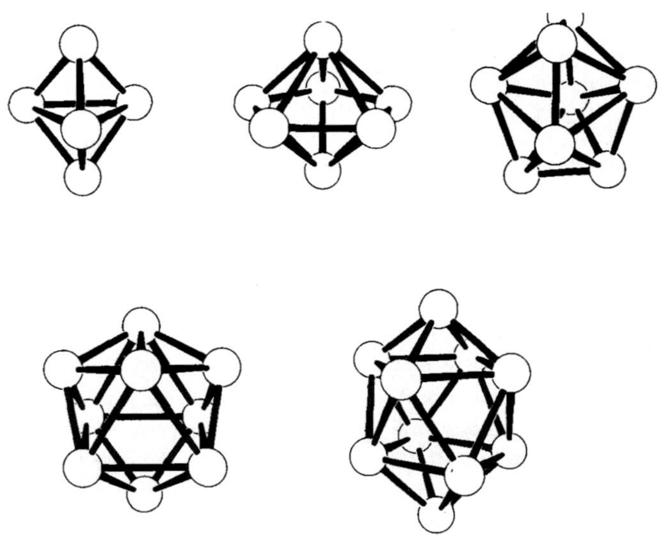

Fig. 9. Structures for the boron and carbon arrangements in some of the polyhedral molecules $C_2B_{n-2}H_n$, which are known for $5 \leqslant n \leqslant 12$. The isoelectronic anions are known for $6 \leqslant n \leqslant 12$.

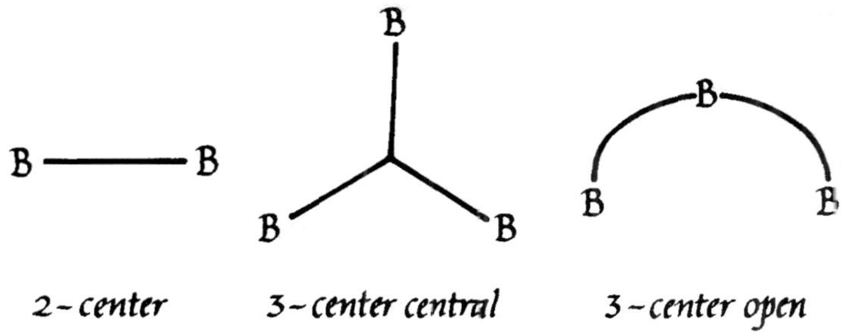

Fig. 10. The three types of bonds within the boron framework of a borane structure, or an equivalent carborane structure, are a BB bond, and the two types of three-center bonds. As shown below, the open three-center bond is known only for BCB bonds, not BBB bonds.

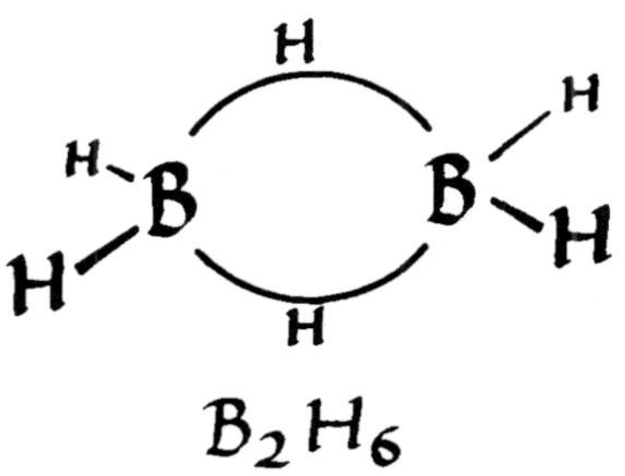

Fig. 11. Bonds in B_2H_6, according to three-center bond theory.

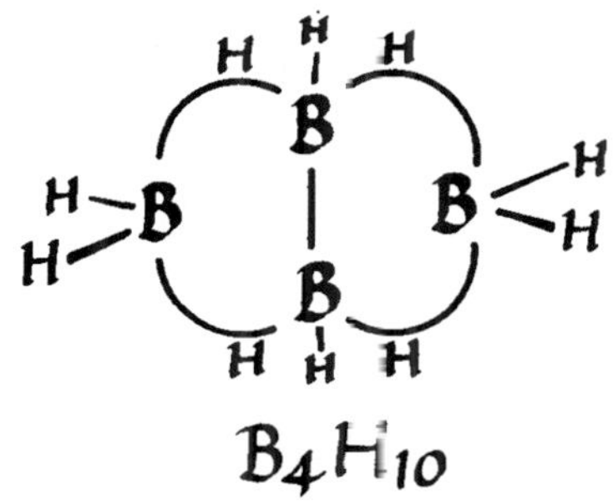

Fig. 12. Bonds in B_4H_{10}.

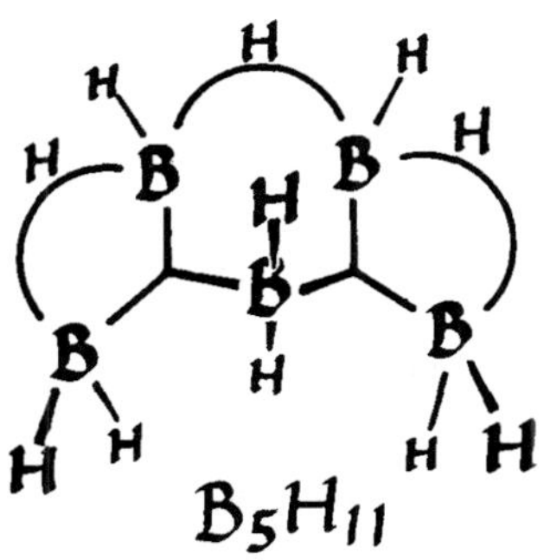

Fig. 13. Bonds in B_5H_{11}.

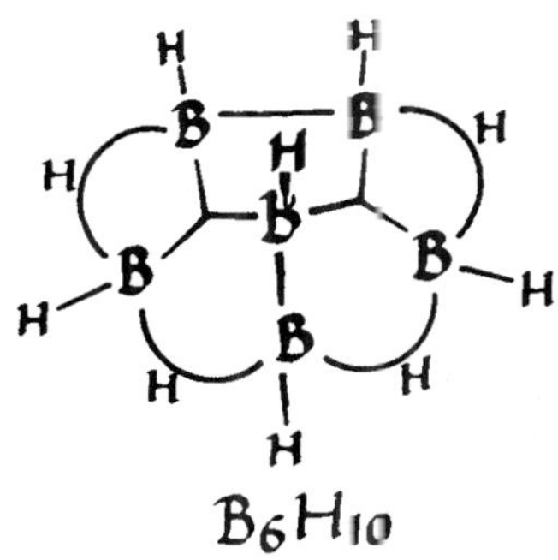

Fig. 14. Bonds in B_6H_{10}.

There are two kinds of three-center bonds among three boron atoms (Fig. 10). The central three-center bond involves positive overlap among hybrid orbitals from each of three boron atoms, while the open three-center bond involves, on the central atom, a π orbital which overlaps in a bonding manner with an orbital from each of the adjacent boron atoms. The less compact B_2H_6, B_4H_{10}, B_5H_{11} and B_6H_{10} structures (Figs. 11-14) are well described with the use of these three-center bonds, omitting open three-center BBB bonds for the moment. However, B_5H_9 (Fig. 15) requires a resonance hybrid of four valence structures, related by the four-fold symmetry; and $B_{10}H_{14}$ (Fig. 16) requires a resonance hybrid of 24 valence structures.

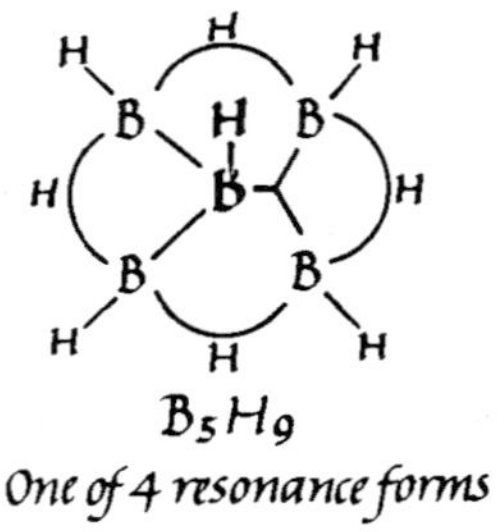

Fig. 15. One of four resonance structures for B_5H_9. The other three are obtained by reorientation of the framework bonds by 90° about the molecular four-fold axis.

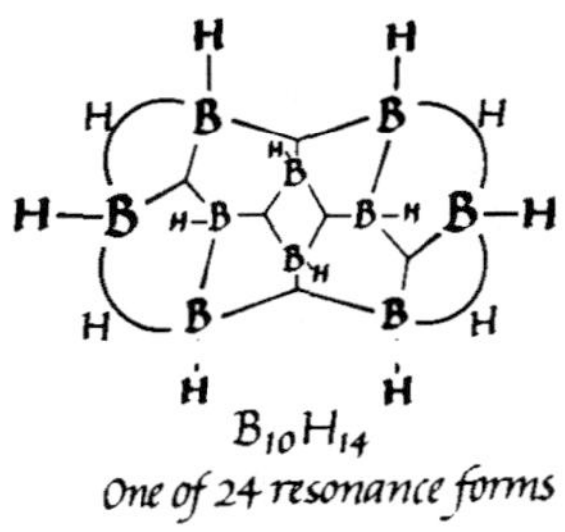

Fig. 16. One of the 24 resonance structures for $B_{10}H_{14}$, omitting open three-center bonds.

Three-center bond theory was further developed in the following 22 years.[24] I realized that the valence rules implied the existence of a large body of boron chemistry, and then ventured predictions, some of which were actually verified experimentally. Dickerson[25] and I formalized my intuitive approach into a theory of connectivity of various bonding patterns within the three-center description. Also, geometrical constraints were introduced in order to avoid overcrowding of hydrogen atoms and to preserve known bonding angles about boron atoms among the boranes.[26] More recently, Epstein[27] and I reformulated the topology using central three-center BBB bonds, to the exclusion of open three-center BBB bonds. The recent work on localized molecular orbitals which has led to the restriction of open three-center bonds, so far, to BCB bonds is exemplified in a later part of this manuscript.

MOLECULAR ORBITAL STUDIES OF BORANES

Molecular orbitals are more appropriate for describing the valence structures of the polyhedral molecules and ions, and the more compact polyhedral fragments. Two simple examples will suffice. In B_5H_9 there are three pairs of electrons in the boron framework, which has four-fold symmetry. The bonding is beautifully described without resonance, by a simple σ and π set of molecular orbitals[22, 24, 28] (Fig. 17). These orbitals are similar to the bonding orbitals, for example, between planar four-fold cyclobutadiene (C_4H_4) and a CH+ placed along the four-fold axis to give the carbonium ion $C_5H_5^+$.

A similar σ, π situation occurs[28] if a BH unit is removed from icosahedral $B_{12}H_{12}^{-2}$ leaving $B_{11}H_{11}^{-2}$ having five atomic orbitals containing four electrons (Fig. 18). This set of five orbitals gives σ, π molecular orbitals which can bond to a similar set of σ, π orbitals from another atom or group of atoms supplying two more electrons. I suggested, conceptually, adding H_3^+ to predict $B_{11}H_{14}^-$, for example. Although recognizing the similarity of this set of orbitals to those in C_5H_5, which was known to form ferrocene $Fe(C_5H_5)_2$, I did not then go quite so far as to suggest bonding of this $B_{11}H_{11}^{-2}$ fragment to a transition metal. Later, Hawthorne did so, using these ideas as a starting point, and thereby created the large family of metalloboranes and metallocarboranes[29] (Fig. 19).

In the early 1960's, when large scale computing facilities became available to us, Roald Hoffmann and Lawrence Lohr independently programmed the

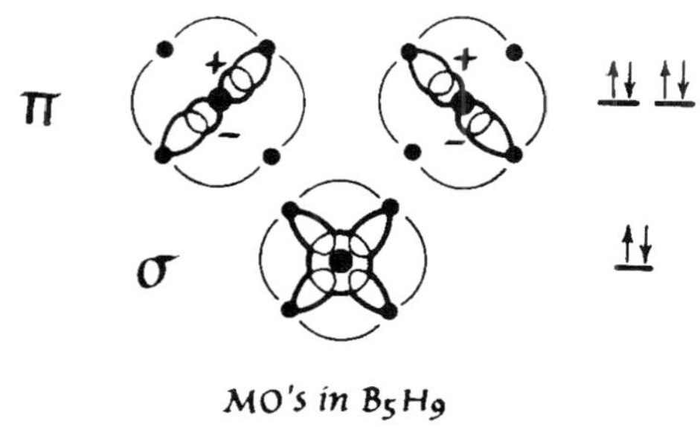

Fig. 17. Symmetry molecular orbitals in the boron framework of B_5H_9. The σ molecular orbital is a five-centered one, and each of the π components is actually an open three-center bond.

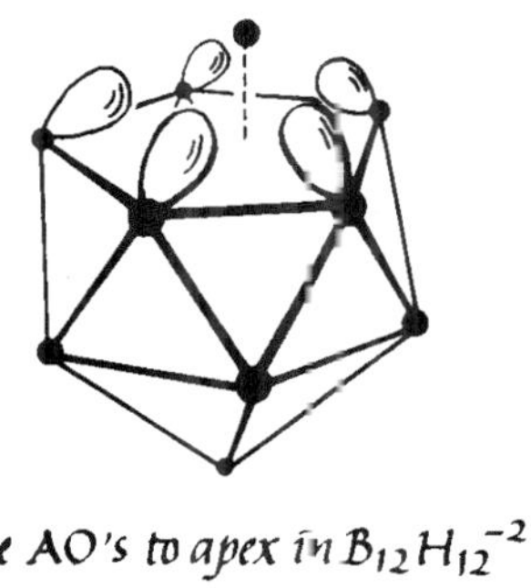

Fig. 18. Five atomic orbitals, containing four electrons remain when a neutral BH unit is removed from the apex of $B_{12}H_{12}^{-2}$, or $C_2B_{10}H_{12}$. These five atomic orbitals form a σ(bonding), π (bonding) and δ (antibonding) set of five molecular orbitals, like those in the π electron system of $C_5H_5^-$.

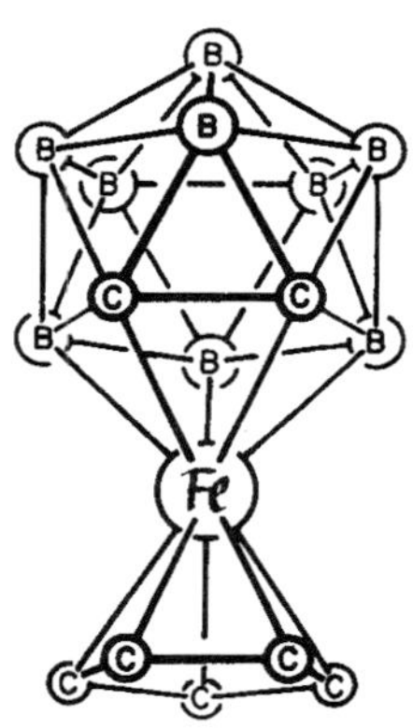

Fig. 19. The structure of $(B_9C_2H_{11})Fe(C_5H_5)$, a metallocarborane derivative.

extended Hückel method in my research group. Its first application was to boron chemistry,[30] where, particularly in the carboranes (compounds of boron, carbon and hydrogen), the charge distributions predicted sites of electrophilic and nucleophilic substitution. One of the rather simple rules which emerged was that nucleophilic attack occurs in a polyhedral carborane at a boron atom closest to carbon, while electrophilic attack was preferred at a boron furthest removed from carbon. Experimental studies of boranes and carboranes in which one or more hydrogens had been replaced by a halogen or by an amine group confirmed these predictions. The extended Hückel method became for a time the most widely used program for the study of molecular theory and reactions in complex organic and inorganic molecules. For example, the Woodward-Hoffmann rules, and related orbital concepts, were tested by their inventors[31] with the use of this theory.

I thought then that progress in structure determination, for new polyborane species and for substituted boranes and carboranes, would be greatly accelerated if the ^{11}B nuclear magnetic resonance spectra, rather than X-ray diffraction, could be used. One approach was empirical,[32] while the other was purely theoretical. This latter approach required the development of a theory for fairly reliable calculation of chemical shifts of ^{11}B from first principles of quantum mechanics.[33] This theory, the coupled Hartree-Fock method for a molecule in a magnetic or electric field, yielded molecular constants good to a few to several percent[34] for many diatomic molecules. A striking result was the prediction of the paramagnetic nature of diatomic BH.[35] However, the application of this method to the complex boranes still lies in the future, even after 13 years of effort. While we understand some of the contributions to chemical shift, diamagnetic and large temperature-independent paramagnetic effects, the use of this method for structure determination of complex polyboranes is still somewhat limited.

These programs yielded accurate self-consistent field molecular orbitals, which were explored in other areas of chemistry, as well as in the boranes and carboranes. One example is the first accurate calculation of the barrier to internal rotation in ethane.[36] At this point, my research students and I set out to explore the gap between extended Hückel theory and self-consistent field theory. We achieved by stages[37,38] a molecular orbital theory[39] still being extended, which was applicable to large polyboranes and other molecules, and which had essentially the accuracy of self-consistent field methods. Molecular orbital methods which do not go beyond symmetry orbitals tend to make each molecule a separate case. Hence we began only a few years ago to explore the connections between these symmetry orbitals and the three-center bonds of the previous section.

LOCALIZED MOLECULAR ORBITALS

Ordinarily, molecular orbitals are classified according to their symmetry types. However, it is possible to make linear combinations of molecular orbitals of different symmetries in such a way that the total electron density of the molecule remains invariant. The most popular methods are those of Edmiston and Ruedenberg[40] who maximize

$$\sum_i \iint \Phi_i(1)\,\Phi_i(1)\,\frac{1}{r_{12}}\,\Phi_i(2)\,\Phi_i(2)\,dV_1 dV_2$$

and of Boys[41] who minimizes

$$\sum \iint \Phi_i(1)\,\Phi_i(1) r^2_{12}\,\Phi_i(2)\,\Phi_i(2)\,dV_1 dV_2$$

These procedures, respectively, maximize the repulsions of those electron pairs within each molecular orbital, and minimize the orbital self-extension of each electron pair. Thus, in slightly different manner, the symmetry orbitals are converted into linear combinations which are a good approximation to the localized electron-pair bond. These objective procedures, without adjustable parameters, have been compared in some detail.[42] They provide support of three-center bond descriptions within a theoretical chemical framework. Also, they test the level at which two-center and three-center bonds require some further delocalization. These studies have also provided preferred descriptions of valence structures, often eliminating or reducing the need for resonance hybrids.

Open three-center and central three-center BBB bonds are almost equivalent descriptions. Including both types there are 111 resonance structures for $B_{10}H_{14}$, of which 24 are based only on central three-center bonds In B_5H_{11} (Fig. 20) two slightly unsymmetrical central three-center BBB bonds are

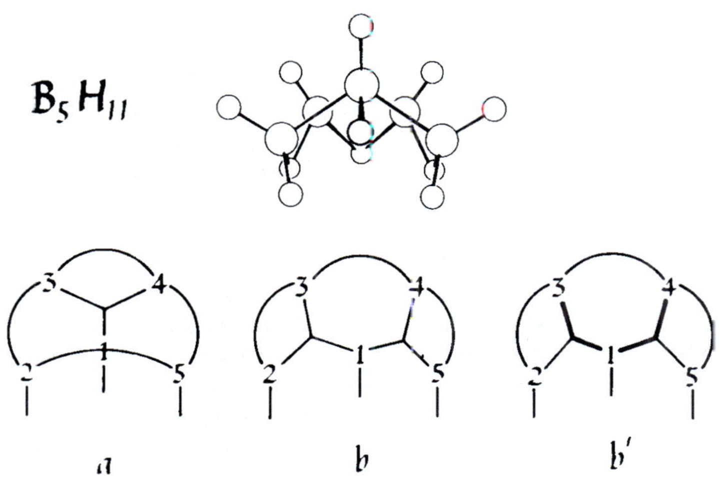

Fig. 20. Three valence structures for B_5H_{11} include (a) the disfavored open three-center bond, (b) the symmetrical central three-center bonds, and (b') the localized slightly unsymmetrical three-center bonds. For clarity, one terminal B-H bond has been omitted at each boron site in (a), (b) and (b').

found[43] by both localization procedures (Fig. 20b'), whereas the open three-center description is not favored. In a series of studies of localized molecular orbitals in all of the boron hydrides, carboranes, and ions of known geometry, we have never found an open three-center BBB bond. However, open three-center BCB bonds do exist. For example, the open three-center BCB bond[44] occurs twice in 1,2-$C_2B_4H_6$ (Fig. 21), when the Edmiston-Ruedenberg procedure is used. A comparison of these results with those of the Boys procedure is given below. In the simplest molecular orbital description of an open three-center bond, the electron pair is distributed as e/2, e, e/2 among these three atoms. It is probable that the extra nuclear charge of carbon stabilizes this distribution, to give open three-center bonds rather than nearly equivalent central three-center bonds, when carbon is the middle atom.

Another new general result is that almost every single bond within a triangulated borane or carborane framework shows some donation to the nearest adjacent atoms (Fig. 22). This tendency for multi-centered bonding usually involves 10 per cent or less of the electron pair. In B_4H_{10}, about 0.2 e is donated[43] from the single BB bond to each of atoms B_1 and B_4, which themselves are relatively electron deficient because of the open three-center bonds of the hydrogen bridges (Fig. 23). This donation then causes these

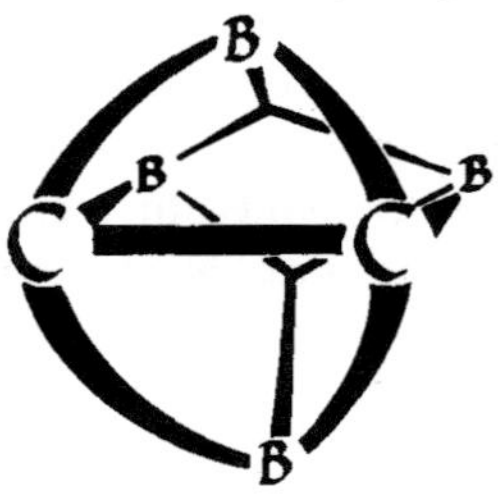

Fig. 21. Localized molecular orbitals are shown in the framework of 1,2-$C_2B_4H_6$. The open three-center bonds go through carbon. An external hydrogen has been omitted at each B or C atom.

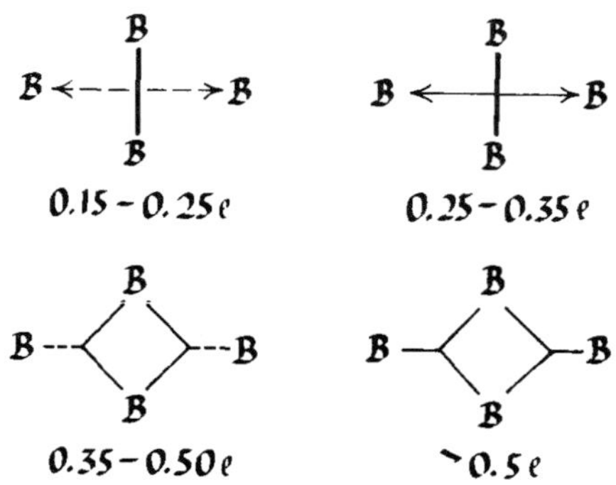

Fig. 22. Notation is shown for amounts of single bond donation to adjacent boron atoms. Unsymmetrical donation may also occur.

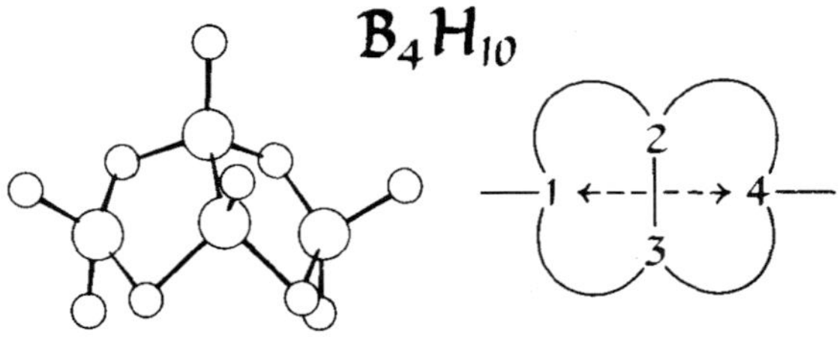

Fig. 23. Single bond donation in B_4H_{10} is about 0.19e from the B_2B_3 bond to each of the outer doubly bridged BH_2 groups of the molecule.

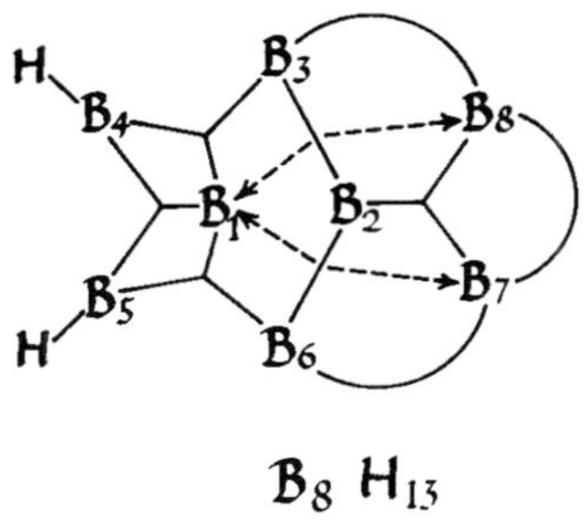

Fig. 24. Single bond donation is shown in $B_8H_{13}^-$.

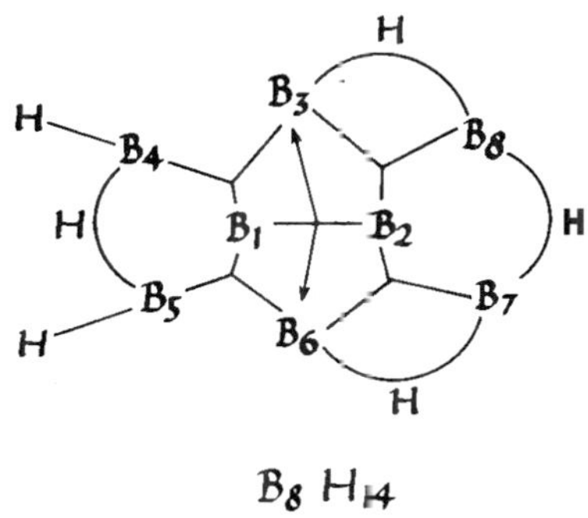

Fig. 25. Donation occurs from the B_1B_2 single bond in B_8H_{14}.

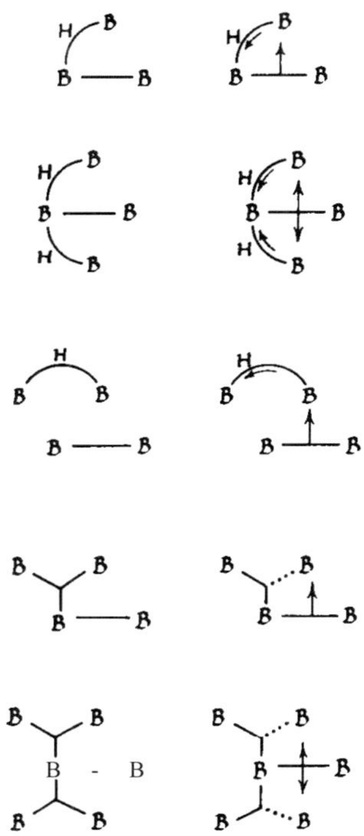

Fig. 26. Some common patterns of single bond donation cause hydrogen atom displacement and electron withdrawal from the atom to which donation occurs.

hydrogen bridges to become unsymmetrical, displaced toward B and B_3, in accord with the results obtained from our X-ray diffraction study. More generally, this donation causes electron withdrawal along other bonds in the molecule, whether or not hydrogen bridges are available for accommodating this electron displacement. Two additional examples are illustrated in $B_8H_{13}^-$ and in the predicted structure of B_8H_{14}, respectively[45] (Fig. 24 and 25). I have abstracted from these examples some typical modes of single bond donation, and accompanying electron withdrawal or hydrogen displacement in Fig. 26. These and similar valence diagrams may be useful in understanding intramolecular distortions and reaction mechanisms.

A new type of bonding, which is conceptually transferable among these molecules, is described as fractional three-center bonds.[46] Its simplest interpretation is the replacement of a pair of resonance structures by a single valence structure (Fig. 27). The fractional use of orbitals, indicated by dotted lines, increases the apparent number of bonds to a given boron above the usual four, in an element in the first row of the periodic table. However, the Pauli exclusion principle is not violated because less than a full atomic orbital is required at a fractional bond. Indeed, the localized molecular orbitals are themselves derived from wavefunctions for which the exclusion principle is rigorously introduced. The use of fractional bonds to reduce the number of resonance hybrids, often to a single preferred structure, is not limited to boranes and their relatives. Our rather extensive studies of bonding among atoms other than boron have indicated that these simplified localized molecular orbitals may be an informative and useful alternative to the 'more conventional valence bond and molecular orbital descriptions in other parts of the periodic table. A very simple example, carboxylate anion,[47] is shown in Fig. 28.

Fractional bonds to atom B_2 in 4,5-$C_2B_4H_8$ are preferred[48] over a resonance hybrid of a single and a central three-center bond, and also are preferred over the open three-center bond (Fig. 29). Also, the bonding of the two carbon atoms in this carborane by both a single bond and a central three-center bond has not been found in boranes, and, in particular, does not occur in B_6H_{10} which

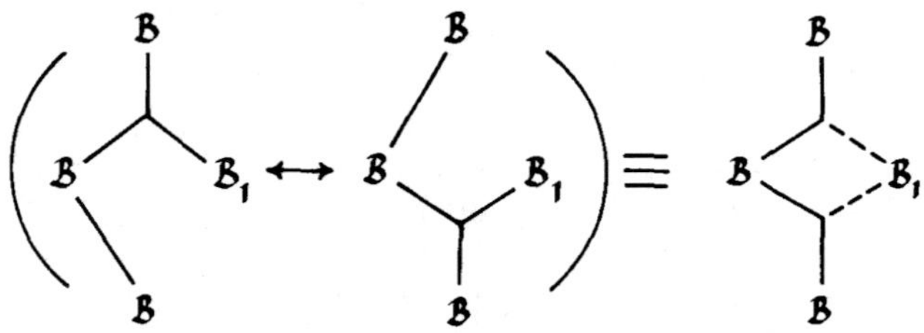

Fig. 27. Equivalence is shown of the two fractional three-center bonds to a resonance hybrid of a single bond and a three-center bond.

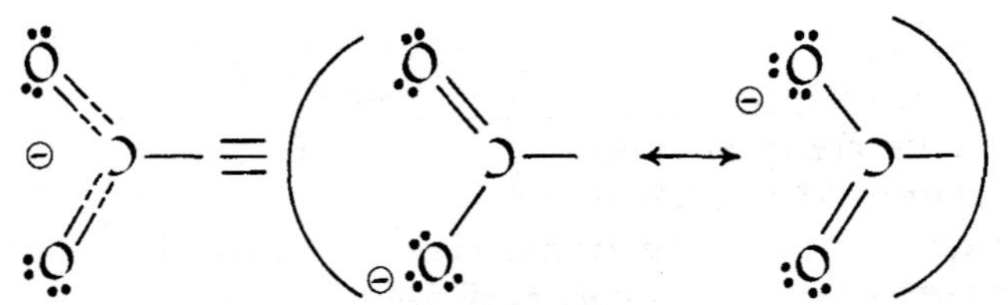

Fig. 28. The four localized molecular orbitals in carboxylate anion have a charge distribution which shows displacement toward oxygen. On the left, thr equivalent resonance hybrid is given.

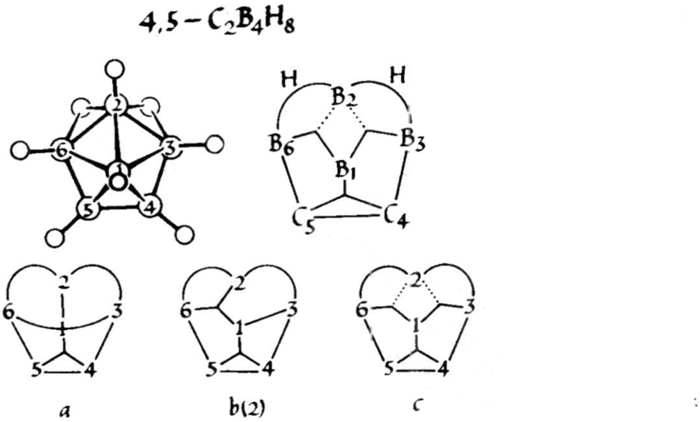

Fig. 29. Localized orbitals in 4,5-$C_2B_4H_8$ show fractional donation to atom B_2 as dotted lines.

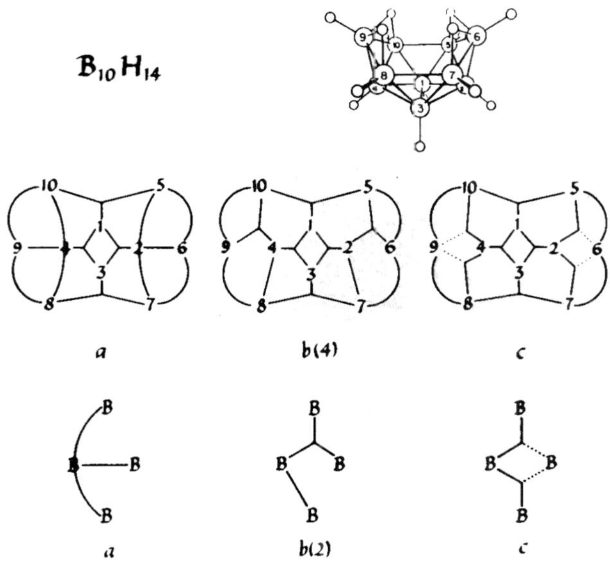

Fig. 30. The pattern of fractional bonds, dotted toward B_6 and B_9, in $B_{10}H_{14}$ can be expected when there is resonance of the type shown in the central diagram, or when one is tempted to draw open three-center **BBB** bonds as shown on the left. The fractional bond description is preferred.

is isoelectronic with 4,5-$C_2B_4H_8$. Fractional bonds give a particularly simple valence description[45] of bonding in $B_{10}H_{14}$ (Fig. 30), where the single valence structure (Fig. 30c) replaces a resonance hybrid of 24 central three-center bond structures. A very similar simplification[45] occurs in $B_{10}H_{14}^{-2}$, where the two pairs offractional bonds are dotted toward atoms $B_2$and B_4, rather than toward $B_6$and B_9.

Similar pairs of fractional bonds are found by Boys' localization procedure in the polyhedral carboranes 1,2-$C_2B_4H_6$(Fig. 31) and 1,7-$C_2B_{10}H_{12}$ (Fig. 32). Perhaps this procedure tends to exaggerate the separation of charge centroids of bonds when they lie at or near a single center. For this reason, we tend to prefer[42] the almost equivalent open three-center bonds, as found by the Edmiston-Ruedenberg procedure in 1,2-$C_2B_4H_6$. When it becomes economically feasible to test this alternative in 1,7-$C_2B_{10}H_{12}$, I would guess that localized open three-center bonds will be found, centered at carbon atoms, in this molecule, like those in 1,2-$C_2B_4H_6$.

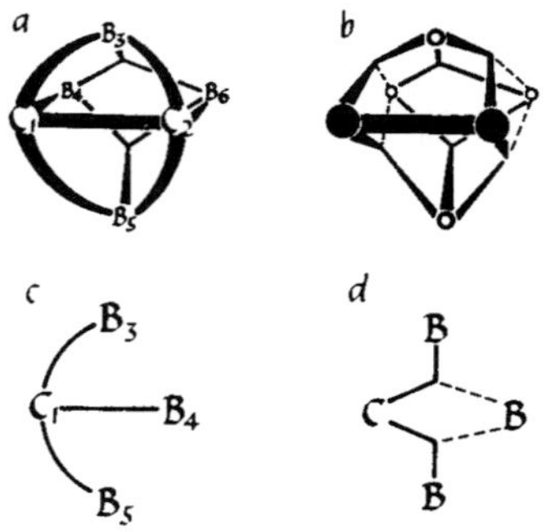

Fig. 31. In 1,2-$C_2B_4H_6$ the Edmiston-Ruedenberg localization yields open three-center bonds at carbon, while the Boys localization gives fractional three-center bonds. The latter procedure gives greater emphasis to separation of orbital centroids when they are on the same atom.

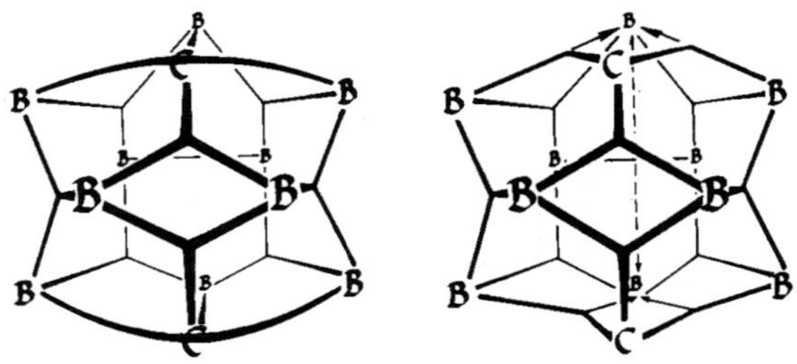

1,7 $C_2B_{10}H_{12}$

Fig. 32. A situation parallel to that of Fig. 31 may occur in 1,7-$C_2B_{10}H_{12}$, where the Boys localization (right side) shows fractional bonds. The open three-center bond description may be found by the Edmiston-Ruedenberg procedure, if and when the calculations can be made.

Fractional three-center bonds are not always unique, particularly in aromatic hydrocarbons, or in those boranes which have a valence pattern similar to that in the aromatics. An example of this orientational ambiguity occurs in the boron framework bonds in B_5H_9(Fig. 33). Actually, the two valence structures represent extremes over a 45° range of orientation angle, about the four-fold axis. There is a continuum of valence structures between these extremes. Moreover, this ambiguity continues throughout the whole 360°, in accord with the four-fold symmetry which the electron density must have in this molecule. All of these valence structures are equally preferred, and each has a total electron density which is consistent with the four-fold molecular symmetry.

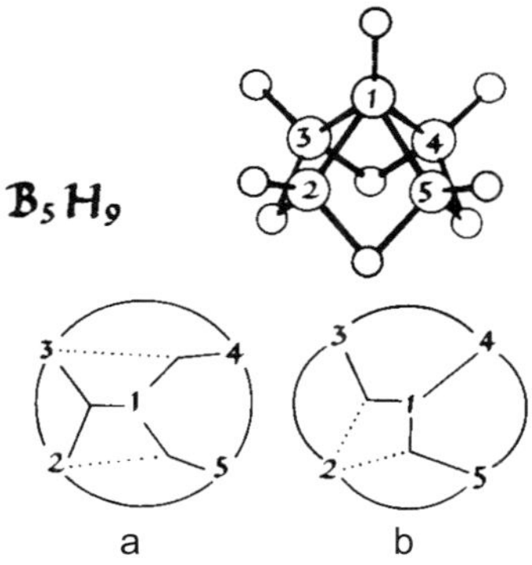

Fig. 33. There is an orientational ambiguity about the four fold axis of B_5H_9 involving a continuum of valence structures between (a) and (b), and extending eventually throughout the complete 360° range.

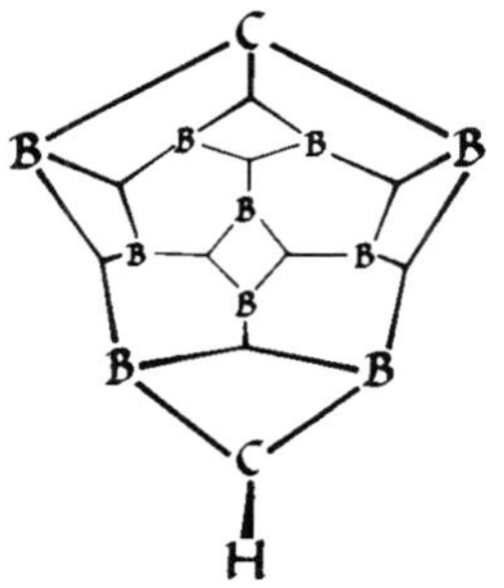

$C_2B_{10}H_{13}^-$

Fig. 34. There is little fractional bonding in the localized orbitals of the nearly polyhedral $C_2B_{10}H_{13}$- ion.

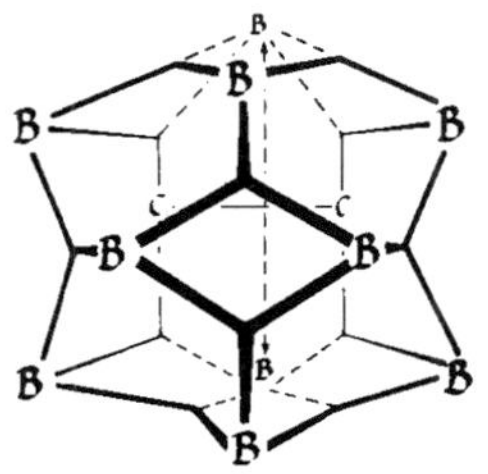

1,2 $C_2B_{10}H_{12}$

Fig. 35. The localized orbitals obtained for 1,2-$C_2B_{10}H_{12}$ by Boys' procedure are the same as those found in 1,7-$C_2B_{10}H_{12}$ in spite of the difference in positions of carbon atoms relative to boron atoms.

In more complex molecules one can find various degrees of simplicity in the bonding patterns of localized molecular orbitals. In $C_2B_{10}H_{13}$-, which has a somewhat open near-icosahedral C_2B_{10} arrangement, the bonding is especially simple,[49] not requiring resonance or appreciable fractional bonding (Fig. 34). In 1,2-$C_2B_{10}H_{12}$, another simple idea occurs: the bonding, as found by Boys' procedure, is to a good approximation just like that in 1,7-C_2B_{10}, H_{12} in spite of the very different positions of the two carbon atoms relative to each other and to their boron neighbors (Fig. 35). Thus there is some tendency for bonding invariance in closely related geometrical structures. In iso-$B_{18}H_{22}$

we find[50] both single bond donation, from B_4B_8 and $B_{14}H_{18}$, and fractional bonds, to B_6 and B_{16} (Fig. 36). Here, the single bond donations are greater toward B_9, which is relatively electron deficient, than toward B_3 and B_{13}. In $B_{20}H_{16}$ there is a remarkable amount of single-bond donation, particularly toward the electron deficient B_9-B_{16} atoms, inclusive (Fig. 37) and the valence structure of each half is clearly like that in $B_{10}H_{14}^{-2}$ (Fig. 38) rather than like that in $B_{10}H_{14}$ (Fig. 30).

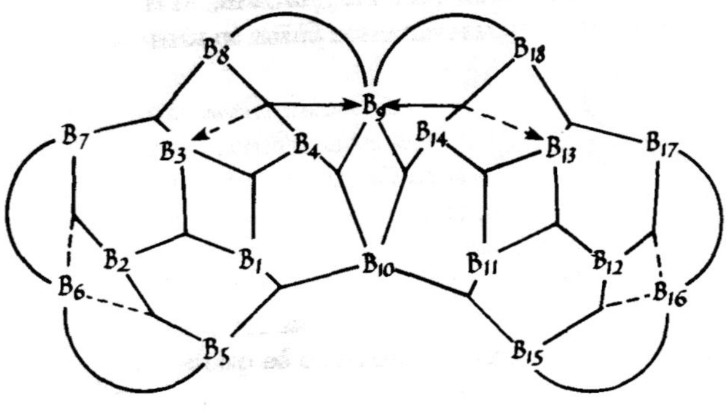

Fig. 36. Single bond donation and fractional bonds occur in iso-$B_{18}H_{22}$, a molecule having a two-fold axis only, and no terminal hydrogen atoms on atoms B_9 and B_{10}.

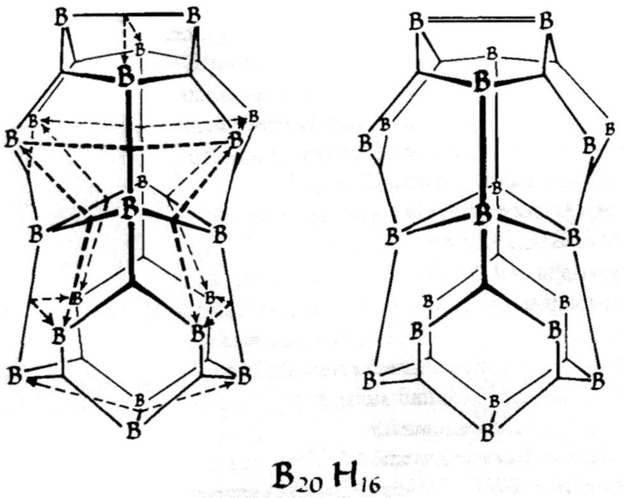

Fig. 37. Donation and non-donation localized orbitals are shown for $B_{20}H_{16}$, in which the four borons nearest the equatorial belt have no terminal hydrogens. Boron atoms B_4B_{16}, inclusive, are particulaly electron deficient, and receive substantial donation from single bonds. Each of these eight boron atoms has only two framework bonds in the figure at the right.

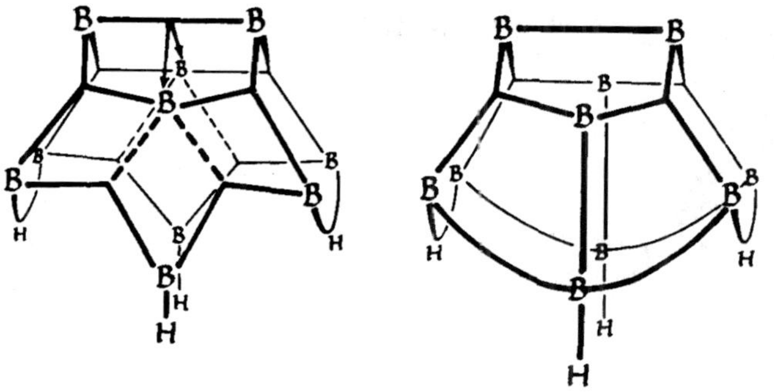

$B_{10}H_{14}^{=}$ (LMO's) $B_{10}H_{14}^{=}$ (open)

Fig. 38. A relationship similar to that in $B_{20}H_{16}$ occurs in $B_{10}H_{14}^{-2}$. The preferred localized orbital structure is shown on the left side, while the open three-center bond structure without donation is an oversimplified valence structure.

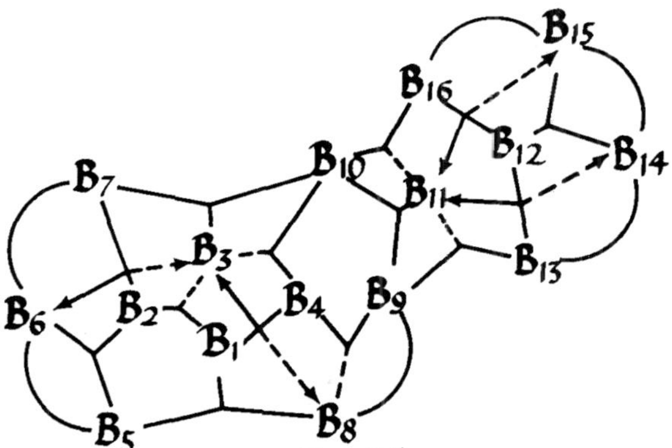

$B_{16}H_{20}$

Fig. 39. A rather complex pattern of localized orbitals is shown for $B_{16}H_{20}$. Bond donations can be withdrawn by the reader in order to discover the preferred valence bond structures of the simpler theory. This single valence structure replaces a resonance hybrid of 216 valence structures based upon central three-center bonds and single bonds. Atoms B_9 and B_{10} do not have terminal hydrogens.

The rather complex localized molecular orbitals in $B_{16}H_{20}$ (Fig. 39) are dominated by the three-center bond approximation; in addition they are modified by single bond donation accompanied by electron withdrawal from atoms to which donation occurs.[50] With recognition that atoms B_9 and B_{10} have no terminal hydrogens attached to them, I offer the reader a challenge to find the close relative valence structures which do not have fractional bonding or single bond donation in this molecule.

Finally, the bonding in reaction intermediates is a new area of study, primarily by purely theoretical methods. In BH_5 (Fig. 40) there is only a very weak donor bond from H_2 to the vacant orbital of BH_3.[51-55] This weakness is probably due to the absence of a pathway for back donation. On the other hand, in the dimerization of two BH_3 molecules to form B_2H_6, donation of electron density from one terminal H to B of the other BH_3 group is balanced by a symmetrically related donation toward the first BH_3 group[56] (Fig. 41). Our other recent progress in theoretical studies of bonding in reaction intermediates, such as B_3H_7 and B_4H_8, shows that the more stable transient species may have a vacant orbital[57] on one boron atom, because of the strain involved when that vacant orbital is filled by converting a terminal BH bond to a

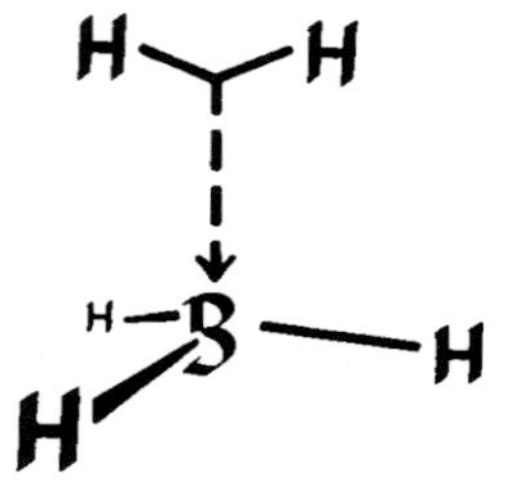

Fig. 40. Fractional donation of a single bond between two hydrogen atoms to the vacant orbital of a slightly pyramidal, nearly planar, BH_3 group. The resulting BH_5 molecule is a very short-lived reaction intermediate.

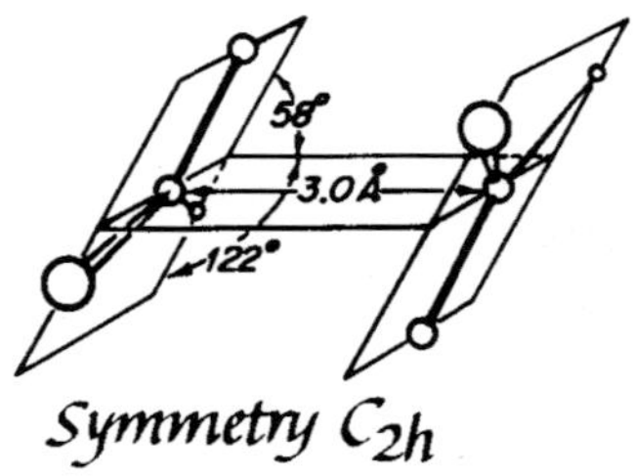

Fig. 41. Approximate geometry of the transition state when two BH_3 molecules form B_2H_6. This geometry is favored over that in an unsymmetrical approach of two BH_3 molecules.

bridge BHB bond. Actually less is known of the detailed reaction mechanisms for reactions of boranes and their related compounds than is known for organic reactions of comparable complexity. Surely, this is a fruitful area for future research.

I shall stop here, omitting descriptions of bonding in large polyhedral borane anions and other related compounds. Also, polyhedral rearrangements, hydrogen atom tautomerism, and particularly the use of bonding theory in bringing some degree of order to chemical transformations of the boranes have been omitted. Attention has thus been concentrated on those aspects of chemical bonding which have been especially illuminated by the molecular and crystal structures that we and others have studied over these many years.

My original intention in the late 1940's was to spend a few years understanding the boranes, and then to discover a systematic valence description of the vast numbers of "electron deficient" intermetallic compounds. I have made little progress toward this latter objective. Instead, the field of boron chemistry has grown enormously, and a systematic understanding of some of its complexities has now begun.

It remains to give credits where they really belong, to my research associates: graduate students, undergraduates, postdoctoral fellows and my other colleagues who have coauthored nearly all of these studies. For the figures of this manuscript, and of the lecture, I thank Jean Evans. I am most grateful to the Office of Naval Research who supported this research during the period from 1948 to 1976, a remarkably long time. I am most aware of the great influence of Linus Pauling on my whole scientific career. Finally, this manuscript is dedicated to the memory of my sister, Helen Porter Lipscomb, composer, teacher and performer.

REFERENCES

1. Stock, A. 1933. *Hydrides of Boron and Silicon.* The George Fisher Baker Non-Resident Lectureship in Chemistry at Cornell University. Cornell University Press, Ithaca. New York.
2. Sidgwick, *N.* V. 1950. *The Chemical Elements and their Compounds.* The Clarendon Press, Oxford, p. 338.
3. Pauling, L. 1940. *The Nature of the Chemical Bond.* The George Fisher Baker Non-Resident Lectureship in Chemistry at Cornell University. Cornell University Press, Ithaca, New York.
4. Stitt, F. 1940. "The Gaseous Heat Capacity and Restricted Internal Rotation of Diborane." J. Chem. Phys. 8, 981-986. Stitt, F. 1941. "Infra-Red and Raman Spectra of Polyatomic Molecules. XV. Diborane." J. Chem. Phys. 9, 780-785.
5. Longuet-Higgins, H. C. and R. P. Bell. 1943. "The Structure of the Boron Hydrides." J. Chem. Soc., 250-255.
6. Pitzer, K. S. 1945. "Electron Deficient Molecules. I. The Principles of Hydroboron Structures." J. Am. Chem. Soc. 67, 1126-1 132.
7. Mulliken, R. S. 1947. "The Structure of Diborane and Related Molecules." Chem. Rev. 41, 207-217.
8. Longuet-Higgins. 1949. "Substances Hydrogenées avec Défaut d'Electrons." J. Chim Phys. 46, 268-275.
9. Price, W. C. 1947. "The Structure of Diborane." J. Chem. Phys. 15, 614.
   Price, W. 1948. "The Absorption Spectrum of Diborane." J. Chem. Phys. 16 894-902.
10. Kasper, J. S., C. M. Lucht and D. Harker. 1950. "The Crystal Structure of Decaborane, $B_{10}H_{11}$." Acta Cryst. 3, 436-455.
11. Dulmage, W. J. and W. N. Lipscomb. 1951. "The Molecular Structure of Pentaborane." J. Am. Chem. Soc. 73, 3539. Dulmage, W. ,J. and W. N. Lipscomb. 1952. "The Crystal and Molecular Structure of Pentaborane." Acta Cryst. 5, 260-264.
12. Hedberg, K., M. E. Jones and V. Schomaker. 1951. "On the Structure of Stable Pentaborane." J. Am. Chem. Soc. 73, 3538-3539.
    Hedberg, K., M. E. Jones and V. Schomaker. 1952. "The Structure of Stable Pentaborane." Proc. Nat. Acad. Sci. 38, G79-685.
13. Nordman, C. E. and W. N. Lipscomb. 1953. "The Molecular Structure of B_4H_{10}." J. Am. Chem. Soc. 75, 4116-4117.
    Nordman, C. E. and W. N. Lipscomb. 1953. "The Crystal and Molecular Structure of Tetraborane." J. Chem. Phys. 21, 1856-1864.
14. Jones, M. E., K. Hedberg and V. Schomaker. 1953. "On the Structure of Tetraborane." J. Am. Chem. Soc. 75, 4116.
15. Lavine, L. and W. N. Lipscomb. 1954. "The Crystal and Molecular Structure of B_5H_{11}" J. Chem. Phys. 22, 614-620.
16. Atoji, M. and W. N. Lipscomb. 1953. "The Molecular Structure of B_4Cl_4." J. Chem. Phys. *21, 172.*

Atoji, M. and W. N. Lipscomb. 1953. "The Crystal and Molecular Structure of B_4Cl_4." Acta Cryst. 6, 547-550.

17. Eriks, K., W. N. Lipscomb and R. Schaeffer. 1954. "The Boron Arrangement in a B_6 Hydride." J. Chem. Phys. 22, 754-755.
18. Hirshfeld, F. L., K. Eriks, R. E. Dickerson, E. L. Lippert, Jr. and W. N. Lipscomb. 1958. "Molecular and Crystal Structure of B_6H_{10}." ,J. Chem. Phys. 28, 56-61.
19. Abrahams, S. C., R. L. Collin, W. N. Lipscomb and T. B. Reed. 1950. "Further Techniques in Single-Crystal X-ray Diffraction Studies at Low Temperatures." Rev. Sci. Instr. 21, 396-397.
20. Kaufman, H. S. and I. Fankuchen. 1949. "A Low Temperature Single Crystal X-ray Diffraction Technique." Rev. Sci. Instr. 20, 733-734.
21. Dickerson, R. E., P. J. Wheatley, P. A. Howell and W. N. Lipscomb. 1957. "Crystal and Molecular Structure of B_9H_{15}." J. Chem. Phys. 27, 200-209.
22. Eberhardt, W. H., B. Crawford, Jr. and W. N. Lipscomb. 1954. "The Valence Structure of the Boron Hydrides." J. Chem. Phys. 22, 989.
23. Longuet-Higgins, H. C. and M. de V. Roberts. 1954. "The Electronic Structure of the Borides MB,." Proc. Roy. Soc. (London) *A224, 336-347.*
Longuet-Higgins, H. C. and M. de V. Roberts. 1955. "The Electronic Structure of an Icosahedron of Boron Atoms." Proc. Roy. Soc. (London) *A230*, 110-l 19.
24. Lipscomb, W. *N., Boron Hydrides.* (W. A. Benjamin, Inc. 1963).
25. Dickerson, R. E. and W. N. Lipscomb. 1957. "Semitopological Approach to Boron-Hydride Structures." J. Chem. Phys. 27, 212-217.
26. Lipscomb, W. N. 1964. "Geometrical Theory of Boron Hydrides." Inorg. Chem. 3, 1683-1685.
27. Epstein, I. R. and W. N. Lipscomb. 1971. "Boron Hydride Valence Structures: A Topological Approach." Inorg. Chem. *10*, 1921-1928.
28. Moore, E. B., Jr., L. L. Lohr, Jr. and W. N. Lipscomb. 1961. "Molecular Orbitals in Some Boron Compounds." J. Chem. Phys. 35, 1329-I 334.
29. Hawthorne, M. F. 1968. "The Chemistry of the Polyhedral Species Derived from Transition Metals and Carboranes." Accounts of Chem. Res. 1, 281-288.
30. Hoffmann, R. and W. N. Lipscomb. 1962. "Boron Hydrides: LCAO-MO and Resonance Studies." J. Chem. Phys. 37, 2872-2883.
Hoffmann, R, and W. N. Lipscomb. 1962. "Theory of Polyhedral Molecules. I. Physical Factorizations of the Secular Equation." J. Chem. Phys. 36, 2179-2189.
Hoffmann, R. and W. N. Lipscomb. 1962. "Theory of Polyhedral Molecules. III. Population Analyses and Reactivities for the Carboranes." J. Chem. Phys. 36, 3489-3493.
31. Woodward, R. B. and R. Hoffmann. 1970. *The Conservation of Orbital Symmetry.* Verlag Chemie GmbH, Academic Press.
32. Eaton, G. R. and W. N. Lipscomb. 1969. *NMR Studies of Boron Hydrides and Related Compounds.* W. A. Benjamin, Inc.
33. Stevens, R. M., R. M. Pitzer and W. N. Lipscomb. 1963. "Perturbed Hartree-Fock Calculations. I. Magnetic Susceptibility and Shielding in the LiH Molecule." J. Chem. Phys. 38, 550-560.
34. Lipscomb, W. N. 1972. "Molecular Properties." MTP International Review of Science, Theoretical Chemistry. Physical Chemistry Series One. Volume 1. Editors: A. D. Buckingham and W. Byers Brown. Butterworths, London, England. pp. 167-196.
35. Hegstrom, R. A. and W. N. Lipscomb. 1968. "Paramagnetism in Closed-Shell Molecules." Rev. Mod. Phys. 40, 354-358.
36. Pitzer, R. M. and W. N. Lipscomb. 1963. "Calculation of the Barrier to Internal Rotation in Ethane." J. Chem. Phys. 39, 1995-2004.
37. Newton, M. D., F. P. Boer and W. N. Lipscomb. 1966. "Molecular Orbital Theory for Large Molecules. Approximation of the SCF LCAO Hamiltonian Matrix." J. Am. Chem. Soc. 88. 2353-2360.

38. Boer, F. P.. M. D. Newton and W. N. Lipscomb. 1966. "Molecular Orbitals for Boron Hydrides Parameterized from SCF Model Calculations." J. Am. Chem. Soc. 88, 2361-2366.

39. Halgren, T. A. and W. N. Lipscomb. 1972. "Approximations to Self-Consistent Field Molecular Wavefunctions." Proc. Nat. Acad. Sci. USA 6.9, 652-656.
Halgren, T. A. and W. N. Lipscomb. 1973. "Self-Consistent Field Wavefunctions for Complex Molecules. The Approximations of Partial Retention of Diatomic Differential Overlap. J. Chem. Phys. 58, 1569-1591.

40. Edmiston, C. and K. Ruedenberg. 1963. "Localized Atomic and Molecular Orbitals." Rev. Mod. Phys. 35, 457-465.

41. Boys, S. F. 1966. *Quantum Theory of Atoms, Molecules and the Solid State.* (P. O Löwdin, Ed.). Academic Press, New York. pp. 253-262.

42. Kleier, D. A., T. A. Halgren, J. H. Hall, Jr. and W. N. Lipscomb. 1974. "Localized Molecular Orbitals for Polyatomic Molecules. I. A Comparison of the Edmiston-Ruedenberg and Boys Localization Methods." J. Chem. Phys. 61, 3905-3919.

43. Switkes, E., W. N. Lipscomb and M. D. Newton. 1970. "Localized Bonds in Self-Consistent-Field Wave Functions for Polyatomic Molecules. II. Boron Hydrides." J. Am. Chem. Soc. 92, 3847-3853.

44. Epstein, I. R., D. S. Marynick and W. N. Lipscomb. 1973. "Localized Orbitals for 1,2- and 1,6-Dicarbahexaborane(6). The Open Three-Center Bond, and Implications for Carborane Topology." J. Am. Chem. Soc. 95, 1760-1766.

45. Hall, J. H., .Jr., D. A. Dixon, D. A. Kleier, T. A. Halgren, L. D. Brown and W. N. Lipscomb. 1975. "Localized Molecular Orbitals for Polyatomic Molecules. II. Structural Relationships and Charge Distributions for Open Boron Hydrides and Ions." J. Am. Chem. Soc. 97. 4202-4212.

46. Marynick, D. S. and W. N. Lipscomb. 1972. "Self-Consistent Field Wave Function and Localized Orbitals for 2,4-Dicarbaheptaborane(7). The Fractional Three-Center Bond." J. Am. Chem. Soc. 94, 8692-8699.

47. Dixon, D. A. and W. N. Lipscomb. 1976. "Electronic Structure and Bonding of the Amino Acids Containing First Row Atoms." J. Biol. Chem. 251, 5992-6000

48. Marynick, D. S. and W. N. Lipscomb. 1972. "A Self-Consistent Field and Localized Orbital Study of 4,5-Dicarbahexaborane(8)." J. Am. Chem. Soc. 94, 8699-8706.

49. Tolpin, E. I. and W. N. Lipscomb. 1973. "Crystal and Molecular Structure of Tetramethylammonium C,C'-Diphenylundecahydrodicarbanido-dodecaborate(1-)."'"OInog. Chem. 12, 2257-2262.

50. Dixon, D. A., D. A. Kleier, T. A. Halgren and W. N. Lipscomb. 1976. "Localized Molecular Orbitals for Polyatomic Molecules. IV. Large Boron Hydrides." J. Am. Chem. Soc. 98, 2086-2096.

51. Kreevoy, M. M. and J. E. C. Hutchins. 1972. " $H_2B\ H_3$ as an Intermediate in Tetrahydridoborate Hydrolysis." .J. Am. Chem. Soc. 94, 6371-6376.

52. Pepperberg, I. M., T. A. Halgren and W. N. Lipscomb. 1976. "A Molecular Orbital Study of the Role of BH_5 in the Hydrolysis of BH_4^-." J. Am. Chem. Soc. 98, 3442--3451.

53. Hoheisel, C. and W. Kutzelnigg. 1975. "Ab Initio Calculation Including Electron Correlation of the Structure and Binding Energy of BH_5 and $B_2H_7^-$." J. Am. Chem. Soc. 97, 6970-6975.

54. Collins, J. B., P. v. R. Schleyer, J. S. Binkley,.J. A. Pople and L. Radom. 1976. J. Am. Chem. Soc., 98, 3436-3441 (1976).

55. Hariharan, P. C., W. A. Latham and J. A. Pople. 1972. "Molecular Orbital Theory of Simple Carbonium Ions." Chem. Phys. Lett. 14, 385-388.

56. Dixon, D. A., I. M. Pepperberg and W. N. Lipscomb. 1974. "Localized Molecular Orbitals and Chemical Reactions. II. A Study of Three-Center Bond Formation in the Borane-Diborane Reaction." J. Am. Chem. Soc. 96, 1325-1333.

57. Dupont, J. A. and R. Schaeffer. 1960. "Interconversion of Boranes. I. A. Kinetic Study of the Conversion of Tetraborane-10 to Pentaborane-11. "J. Inorg. Nucl. Chem. 15, 310-315.

J. Mol. Biol. (1972) **65**, **469–488**

Structure of the Tubular Variants of the Head of Bacteriophage T4 (Polyheads)

I. Arrangement of Subunits in some Classes of Polyheads

D. J. DeRosier† and A. Klug

Medical Research Council Laboratory of Molecular Biology
Hills Road, Cambridge, England

(*Received 1 June 1971, and in revised form 15 November 1971*)

Electron micrographs of tubular variants of the head of bacteriophage T4 arising from a mutation in gene *20* have been analysed by means of optical diffraction and filtering. Four main classes of tube have been observed. The most commonly occurring tube is of width about 800 Å and has a "coarse" structure consisting of a simple near-hexagonal arrangement of capsomeres, which are resolved into six regularly disposed structural units spaced about 33 Å apart. These units can be identified with the major protein component of the head. The edges of the hexamers make a skew angle with respect to the lattice directions. Polyheads of widths greater than 1000 Å which derive from outer layers of originally multi-layered tubes constitute a second class. The basic structure is "coarse-hexagonal" like that of the first class, but the capsomeres are distorted and those in alternate rows have different orientations.

The tubes of the third class have a smooth appearance similar to that of empty capsids. The substructure in this class is closely related to that in the first class and may be derived from it. Subunits are again organised into hexamers but the latter are now spaced 10% further apart and are rotated with respect to the lattice directions so that they fit snugly together. The central region of the hexamer also often appears to contain an additional unit, so that we have a new type of capsomere or morphological unit, the "6+1" unit. The denser packing of hexamers, together with the presence of the central units, leads to the smooth appearance of the tubes.

The tubes of the fourth class have walls of an even more finely grained appearance and are generally filled with some internal substance so that they stain very heavily. Their structure is described in the accompanying paper.

A special instrument is described which allows the rapid survey of electron micrograph plates by optical diffraction. Details are also given of optical filtering procedures.

1. Introduction

(a) *Historical*

The sytems of the conditional mutants of bacteriophage T4 discovered by Epstein & Edgar (Epstein *et al.*, 1963) affords an ideal opportunity for studying morphogenesis in these viruses. Mutations in different genes lead to the formation of incomplete particles or distinct structural components of the particle in limited states of

† Present address: Department of Chemistry, University of Texas, Austin, Texas, U.S.A.

assembly. Through an understanding of the abnormal types of assembly we may learn much about the normal processes of assembly and their regulation. Mutants in one of the head genes (gene *20*) lead to the production of long tubular structures which have a diameter similar to that of normal phage heads and which were originally called polyheads.

The present paper describes the results of an analysis of electron micrographs of these tubular variants by means of optical diffraction and filtering (Klug & Berger, 1964; Klug & DeRosier, 1966). These techniques enable the surface lattices to be determined with precision and enable the arrangement of matter on one side of the tube to be displayed directly, without confusion from the presence of the other side. In the following paper (Yanagida, De Rosier & Klug, 1972) the same techniques are applied to tubular variants arising from a mutation in gene *22*. When these tubes are "aged" a structural transformation takes place *in vitro*, and the rearrangement of matter can be followed by optical filtering of the micrographs.

Kellenberger and his colleagues in Geneva have obtained substantial genetic, immunological and chemical evidence that the polyheads are closely related to the normal head of the phage. The major protein component of the polyhead (Kellenberger, 1968) is probably identical with the product of gene *23* which forms the normal majority head protein. The polyheads have one antigenic determinant in common with the normal head (Laemmli, 1969; see also Laemmli & Kellenberger, 1970).

Since the work described here was completed, it has been shown (Kellenberger & Kellenberger-Van der Kamp, 1970; Laemmli, 1970) that the major protein of the normal head is a cleaved form of the protein 23 found in polyheads. Laemmli (1970) has shown that there is a head precursor particle containing protein 23 but that this and three other protein components of the head are subsequently cleaved during maturation. According to the method of preparation, the polyheads also contain a number of minor protein components, possibly the products of genes *22* and *24*, but when special care was taken to separate polyheads from small quantities of ghosts, only traces of additional components were found.

In a preliminary account (Finch, Klug & Stretton, 1964), there is described the application of the newly developed technique of optical diffraction (Klug & Berger, 1964) to the analysis of the periodicities present in electron micrographs of polyheads. It was shown that the surface lattice of the polyheads was based upon the plane hexagonal lattice (symmetry *p*6) rolled up into a cylindrical surface.

The optical diffraction patterns were interpreted to show that the apparent confusion in the normal electron microscope image arises from the superimposition of the images of the near and far sides of the structure, both of which are shown up by the contrast provided by the staining. The diffraction patterns were analysed to give the distribution of stain in the surface lattice. The procedure is reminiscent of the early days of X-ray crystallography when a model was proposed and checked by comparing its Fourier transform with the observed diffraction pattern. Since that time a further development has taken place in the optical diffraction method of analysis, namely the method of optical filtering (Klug & DeRosier, 1966) which enables the diffracted rays to be sorted into two sets which can be re-imaged to provide a picture of one side of a particle at a time. The distribution of matter in the surface lattice is thus displayed directly. The introduction of this technique essentially parallels the introduction of the Fourier synthesis method in the development of X-ray crystallography.

The first optical transforms were made using a conventional light source, but since then the introduction of a laser as a light source has enabled diffraction patterns to be obtained in fractions of a second rather than the hours necessary formerly. The newer diffraction patterns so obtained have revealed a wealth of spots in the diffraction pattern. The patterns can be indexed but the application of optical filtering to such complicated patterns has had to await the development of an optical filtering system which makes operations on complicated subjects practicable. There has also, of course, been much progress in the physiological, chemical and genetic studies, mainly by the Geneva group, and the more advanced structure investigation described here has been undertaken in parallel with this other work.

(b) *Scope of the present work*

We have used optical diffraction and filtering of electron micrographs to classify polyheads into a number of different types. This classification, of course, is in addition to that which can be made by direct inspection of the more obvious characteristics such as particle width, coarseness of structure, retention of stain within the tubes and so on. The filtering technique has demonstrated a distinct substructure in a class of polyheads which by inspection appear to be so smooth-surfaced as to be entirely lacking in resolved subunits. Because these polyheads are very similar to normal phage heads—on which capsomeres are not visible by negative staining—it seems likely that the two might possess substructures of the same type. The substructure revealed in this "smooth" class of polyheads is closely related to that of the more commonly occurring coarse polyheads in which the hexagonal structure is easily seen.

The aim of the work described in the present and succeeding paper was to investigate the variety of optical diffraction patterns obtained and to interpret them by optical filtering. We are interested in finding out how far the apparent variability within the wide range of tubes called polyheads is a natural one representing different types of tube. The difficulties in making the distinctions arise, of course, through artifacts of the staining, variability of preservation of the specimen on the grid and so on. In this paper we are not concerned with the statistics of the variability within any one class. A study of this kind on the class of coarse, empty polyheads was undertaken by the Geneva workers and the results have now been published (Yanagida, Boy de la Tour, Alff-Steinberger & Kellenberger, 1970).

2. Materials and Methods

(a) *Preparation and electron microscopy*

Polyheads were prepared and the negative staining carried out as described previously (Finch, Klug & Stretton, 1964). Additional experiments were carried out in which the particles were transferred to the grid within minutes of lysis. Likewise various specimens were fixed with formaldehyde or osmium tetroxide.

(b) *Surveying optical diffractometer*

The two operations of optical diffraction and filtering can both be carried out on the same apparatus (e.g. that described below), but it is useful to have a special instrument which allows one to examine or survey many dozens of electron micrographs rapidly. The optical diffractometer shown in Plate XIII has been constructed especially for this purpose. The essential features of the system are an intense light source (a laser) and a folded optical path which permits the instrumental controls to be placed within easy reach of the operator, and a movable carrier stage for the electron micrograph plate. With the laser as a light source, diffraction photographs can be obtained directly from the electron micrograph plates with exposure times of a fraction of a second rather than hours as was the case with a mercury arc source. Since the diffraction pattern is bright enough to be seen, it is possible to inspect visually the optical transforms of a single particle or of a region within a single particle. As a result, the operator can, for example, view the optical transform while continuously varying the region of the image which is diffracting. Such searches were carried out on polyheads in order to find well preserved regions as judged by the diffraction pattern.

The light beam incident on the micrograph is produced by a 3 mW, helium–neon laser (Spectra-Physics model 122) which is fitted with a beam expanding telescope (Spectra-Physics models 131 and 132). A 2-in. extension tube has been fitted between the spatial filter (model 332) and the collimating lens (model 331) so that the micrograph is illuminated with a converging beam rather than a plane parallel beam. The convergence of the beam is adjusted to focus the diffraction pattern onto the photographic film holder. The magnification constant for the optical transform is given by $R = dd^*$, where d is the spacing in the image and d^* is the corresponding spacing in the transform. For this instrument $R = 0{\cdot}6\ (\text{mm})^2$.

The micrograph, positioned just above the collimating lens, is held on a microscope specimen stage (Carl Zeiss stage no. 56) which permits translation and rotation of the micrograph. A simple, four-jaw slit system, placed above the micrograph, allows the operator to mask off unwanted portions of the micrograph. The slit system can be swung out of the optical track when changing micrographs. The mirror, which folds the optical path, is a front-surface, plane-mirror which is optically flat to within $\lambda/8$ (about 3×10^{-6} in). The mirror housing with its tilting controls is mounted on a vertical micrometer in order to facilitate alignment.

The combined recording and viewing system is built around a 35-mm reflex camera (Exakta model no. VX 1000). The viewfinder (Exakta Magnear adapter with top lens magnifier, model no. 308) has been modified to correct for the increase in focal length due to the long wavelength of the red laser light. In addition, a polarizing filter has been mounted in the eyepiece. It allows one to control the intensity of the optical diffraction pattern.

A lens (of focal length 175 mm) mounted in a pivoting assembly has been included in the viewing system so that the micrograph as well as its transform can be inspected. With the lens swung out of the optical path, the operator sees the diffraction pattern, but with the lens in the path, an image of the micrograph is formed instead. The entire recording and viewing system is mounted on an adjustable rack (Exakta "Vielzweck" model no. 155.10).

The alignment of the optical components is not particularly difficult. The mechanical design of the instrument is such that the fine alignment is achieved by adjusting both the mirror and laser system to coincide with the optical axis of the combined viewing and recording system.

The quality of the diffraction patterns is adequate for surveying purposes. Better results could probably be obtained by illuminating the specimen with plane parallel light and using an additional lens just above the slit system to focus the diffraction pattern.

(c) *Filtering optical diffractometer*

The optical apparatus used for filtering is of the type described by Klug & Berger (1964) and Klug & DeRosier (1966) but with extensive modifications which have made filtering of complicated subjects practicable. It would nowadays be described as a coherent optical system for image processing. The introduction of a laser into the system has meant that there is sufficient light intensity in the diffracted rays so that one can align visually a complicated filtering mask. It has also allowed one to use the original electron micrograph rather than an enlarged copy as in our earlier work.

For the purpose of filtering, the diffraction lens is of long focal length (150 cm) and the subject is kept small so that its optical transform is large. This simplifies the construction and alignment of the filtering mask. A typical subject field is of the order of 1 cm across. Because such a small area is used to form the diffraction pattern, the quality of the diffraction lens (focal length 150 cm) is not critical. However, the diffracted rays are widely separated at the diffraction plane so that the lens which recombines them must be of high quality, and we have used a commercial United Kingdom Bausch and Lomb lens 5 cm in diameter with a focal length of 75 cm. The lens was tested for us by Dr E. H. Linfoot, using the Ronchi test, and found to be very satisfactory over an aperture of 4 cm. All lenses used were coated with an anti-reflective layer that greatly reduced artifacts caused by reflection of light from lens surfaces.

The film holder placed at the diffraction plane is also used to hold the filtering mask. The holder is designed to be used with 2·5 in × 3·5 in film sheets. The holder has translational control along 3 orthogonal axes (for focussing and centering) and rotational control (for alignment of the filtering mask). The imaging or reconstruction lens is positioned just after the film holder. This lens holder (similar to that shown on page 15 of Taylor, 1921) has both tilting and translational controls for alignment. Focussing of the filtered image is achieved by translation of the camera which has been mounted on a micrometer for this purpose. The camera consists of a 35-mm camera body with a focal plane shutter. The camera mount has tilting controls so that the film can be set normal to the optic axis.

(d) *Filtering procedures*

We have found that the use of small areas of the micrograph obviates the necessity for immersing the plate or film in cedar wood oil or other liquid having a refractive index equal to that of the gelatine. Both the glass and the average thickness of the emulsion are generally flat over small regions of the photographic plate. We judge this by inspecting the centrosymmetrically related pairs of spots which should have approximately equal intensity in the optical diffraction pattern if the transparency is optically flat.

There is, however, another effect which results when no liquid gate is used. In photographic emulsions there is invariably a relief effect associated with the removal of the unexposed silver halide from the gelatine by the fixing process (Altman, 1966). Thus the transparency represents the image not only by a variation of optical density but also by a variation in the thickness of emulsion. The result is that when illuminated in a coherent optical system the subject will act as a phase object, the degree of retardation being a function of the amount of relief. We have found experimentally that the effect of minimizing such variations in gelatine thickness with oil is to halve the intensity of the diffraction pattern. However, it is worth noting that when oil is not used, one sometimes observes diffracted rays of a higher order than those which are observed with the subject in a liquid gate. This is a result of the non-linear variation of scattering power with the thickness.

In any case these relief effects do not contribute to the filtered image which records only the opacity of the original subject. For this reason we generally do not use a liquid gate in our investigations, and the effect on the final filtered image we believe to be negligible.

For the recording of transforms we find it advantageous to use a film of very high contrast. Ilford Panchroline and Kodak Panmicrofile were used. The high contrast characteristic allows the film to be used as a noise filter. Thus, weak diffraction spots present in longer exposures are completely absent in shorter exposures. This feature gives one a very simple way of picking out the most strongly diffracted beams. Filtered images are generally recorded on Kodak Panmicrofile.

The filter masks were prepared from 12-fold enlargements of the diffraction patterns by means of a punch connected to a 12-fold reducing pantograph. Apertures for diffracted rays common to both sides of the subjects, for example, those in the region of the zero-order diffraction, were covered with a neutral density filter (copper gauze obtained from the Valve Division of EMI, Hayes, Middlesex) transmitting about 50% of the amplitude.

(e) *Variability in the filtered images*

We believe that the variations in the detailed appearance of different parts of a filtered image are genuine because we have used *large* apertures in our filter masks. The linear extensions of the region over which the averaging takes place are inversely proportional to the dimensions of the aperture. We generally used holes large enough to include the whole of a diffraction spot so that the filtered image is not an average over the whole specimen, as would be obtained with very small holes, but a sort of local average of the periodic structure. This displays the variations from one portion of the original image to another. Likewise, we have included a region of fair extent around the zero-order spot, so that the long-range irregularities in the stain and the shape of the structure are included in the final image. Furthermore, when dealing with a new subject, a set of filtered images is taken with slightly different size apertures and with a particular mask moved through slight distances to make sure that the final image is not sensitive to the size and very fine positioning of the apertures. Since the work described here was done we have checked various filterings on polyheads using a computer technique where the positioning and exact size and shape of the "aperture" are completely under control. These computer filterings confirm the results of the optical filtering.

We attribute the variation in the different parts of a filtered image to variations in staining and preservation in different parts of the specimen. Such variation is not unexpected in the case of particles many thousands of ångstroms long. The variations in stain lead to changes in the relative *emphasis* rather than the *position* of structural features, but such differences can of course cause very striking changes in the pattern one observes in the original micrographs. Our whole aim is to find the "best" or most clear, regular examples, but the choice as to which part of the image represents the most faithfully stained region cannot, of course, be made without careful examination of many such images and without some other knowledge of what is expected.

(f) *Conventions in reproductions*

In the final reproductions of the filtered images the negative stain is shown black and the protein white. The net reduction from subject to final filtered image is of the order of a third to a half, so that the reproductions of the filtered images have been enlarged between 20 and 50 times.

The choice of hand of the helical surface lattice, i.e. which of the two filtered images corresponds to the near side of the tube and which to the far side, follows from the results of Yanagida *et al.* (1970) on shadowed polyheads. In negatively stained specimens, one cannot decide which set of diffraction spots is attributable to the upper or lower layer, but since shadowed specimens reveal only the upper layer, the ambiguity can be resolved.

3. Indexing of the Diffraction Patterns

In a preliminary paper Finch, Klug & Stretton (1964) reported the side of the unit cell as being of length 63 Å. It was later reported (Klug, Finch, Leberman & Longley, 1966) that the cell chosen was a pseudo-cell, smaller than the true cell of the surface lattice. This meant that the relation between the morphological units and chemical subunits originally described is in error, although the main conclusion about the hexagonal symmetry of the lattice remains unchanged. The wrong choice of unit cell arose because some relatively weak diffraction spots which corresponded to a larger spacing (about 100 Å) were ignored. When stronger diffraction patterns were obtained it became clear that these spots, though weak, were a constant feature of the patterns and that the true unit cell was correspondingly larger (this possibility was first pointed out to us in personal communications from Drs D. L. D. Caspar and M. F. Moody). However, because the diffraction spots which had been ignored are weak, the distribution of matter in the surface lattice which was deduced by us in 1964 is correct to the first approximation. The way in which the pseudo-cell arises is illustrated in Figure 1. As described by Finch *et al.* (1964), the structure units lie approximately at the vertices of a regular hexagonal honeycomb (Fig. 1(a)). In this type of arrangement there are gaps or apparent holes at both the 6-fold and 3-fold positions of the surface lattice and the two types of holes are very similar. In so far as one cannot resolve the difference between these two types of hole, the lattice can be described by the small pseudo-cell shown at the left of the Figure. In the diffraction pattern shown schematically in

TUBULAR VARIANTS OF THE HEAD OF BACTERIOPHAGE T4 475

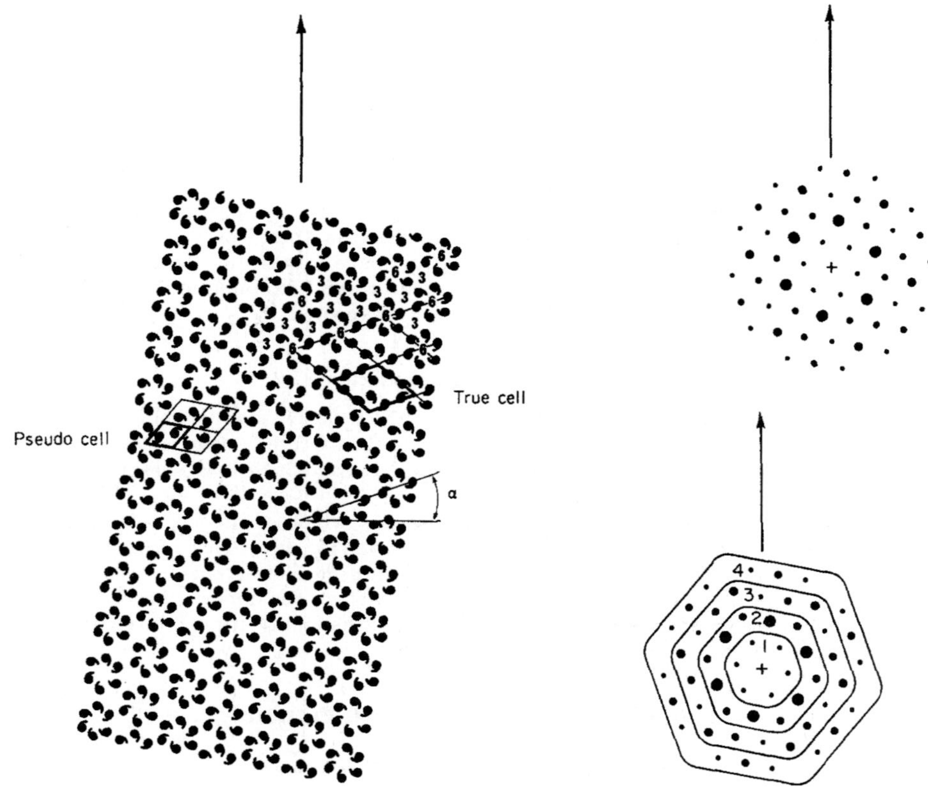

(a) (b)

Fig. 1. A schematic diagram of asymmetric units arranged in a plane *p*6, lattice and the corresponding diffraction pattern. (a) The asymmetric units are represented by commas. Some of the 3-fold and 6-fold rotation axes are marked. The side of the unit cell joins two neighbouring 6-fold positions.

The arrangements of units have been chosen to simulate the appearance of a honeycomb, in which 3-fold and 6-fold axes are no longer distinctly different. If the distinction between the 6-fold and the 3-fold axes is neglected, the structure can be described in terms of a smaller pseudo-cell, the axes of which join a 6-fold position and a 3-fold position. The earliest description of a surface lattice of a polyhead (Finch *et al.*, 1964) was given in terms of this pseudo-cell. When the plane net is rolled into a cylindrical surface, the angular pitch, α, of such a helical surface lattice can be defined by the angle formed by one of the lattice directions with the equatorial plane of the cylinder. Our convention for this pitch angle α is the same as that chosen by Yanagida *et al.* (1970).

(b) The top part of the diagram shows the transform of the arrangement in (a). The innermost set of six diffraction spots is weak because the structure at low resolution approximates to the pseudo-cell shown in (a). For convenience of discussion the diffraction pattern has been divided into four hexagonal bands, referred to in the text as first, second, third and fourth orders. This terminology is useful for classifying the different kinds of polyheads.

Figure 1(b) the pseudo-cell shows itself by a weakness in the innermost hexagonal band of diffraction spots corresponding to a spacing of 100 Å; only the spots lying at a spacing of 60 Å in the second hexagonal band are strong.

A fuller account and justification of the indexing of the various classes of polyhead tube are given in the Appendix.

4. Relation between the Appearance of Different Images and their Diffraction Patterns

The polyheads were classified by their appearance as follows: first the obvious characteristics were noted, such as width of the tube, the presence of some internal substance or the state of flattening on the grid; secondly the detailed appearance was taken into account, namely whether there was an easily visible structure or whether the polyheads seemed to be structureless. The final classification, of course, is made in terms of the diffraction patterns and the distributions of matter revealed by optical filtering.

We have found it useful to use the following terms in discussing the appearance. Coarseness or fineness refers to the scale of detail which is most striking to the eye. A coarse structure is one with prominent detail on the scale of 60 to 100 Å, whereas a fine structure has its most prominent detail in the neighbourhood of 30 to 40 Å. It should, however, be noted that in both cases, interpretable detail extends to spacings of about 25 Å, but the appearance is dominated by different particular spacings in the two cases. Roughness or smoothness refers to the visibility or contrast of the pattern, that is, the strength of the fluctuations above the average. These terms refer, of course, to the distribution of stain, but in so far as the stain distribution smoothly follows the contours of the particle (cf. Klug & Finch, 1968), the same terms may be taken to apply to the distribution of matter as well.

These rather qualitative terms may be formulated precisely in terms of the diffraction pattern, which serves as an objective criterion for distinguishing differences in structure. The fineness of detail apparent in a micrograph is reflected in the relative strengths of different orders in the diffraction pattern. Thus a coarse appearance means that there are strong low orders (first and second hexagonal bands in Fig. 1(b)), whereas a fine appearance means that the stronger spots are to be found in the third and fourth bands. It should be emphasized that the scale of detail in all classes of polyheads is of the same size, which is fixed by the size of the protein subunits. They differ in appearance because the packings of subunits are different, so that different sets of grooves or niches are filled by the stain. Roughness and smoothness refer to the strength of the diffraction spots relative to the strength of the zero-order region in the diffraction pattern. Hence, if the non-zero-order diffraction spots are weak, the polyhead will appear very smooth or structureless, but the method of optical filtering can, of course, be used to enhance these fluctuations relative to the zero order by cutting down the strength of the latter when the diffracted rays are re-imaged to form a filtered image. We would, thus, have four types of appearance; (i) coarse rough, (ii) coarse smooth, (iii) fine rough, and (iv) fine smooth. However, we shall not always use these full terms in a general description. When a particle looks "smooth", the scale of detail cannot be determined without recourse to the diffraction pattern. Similarly,

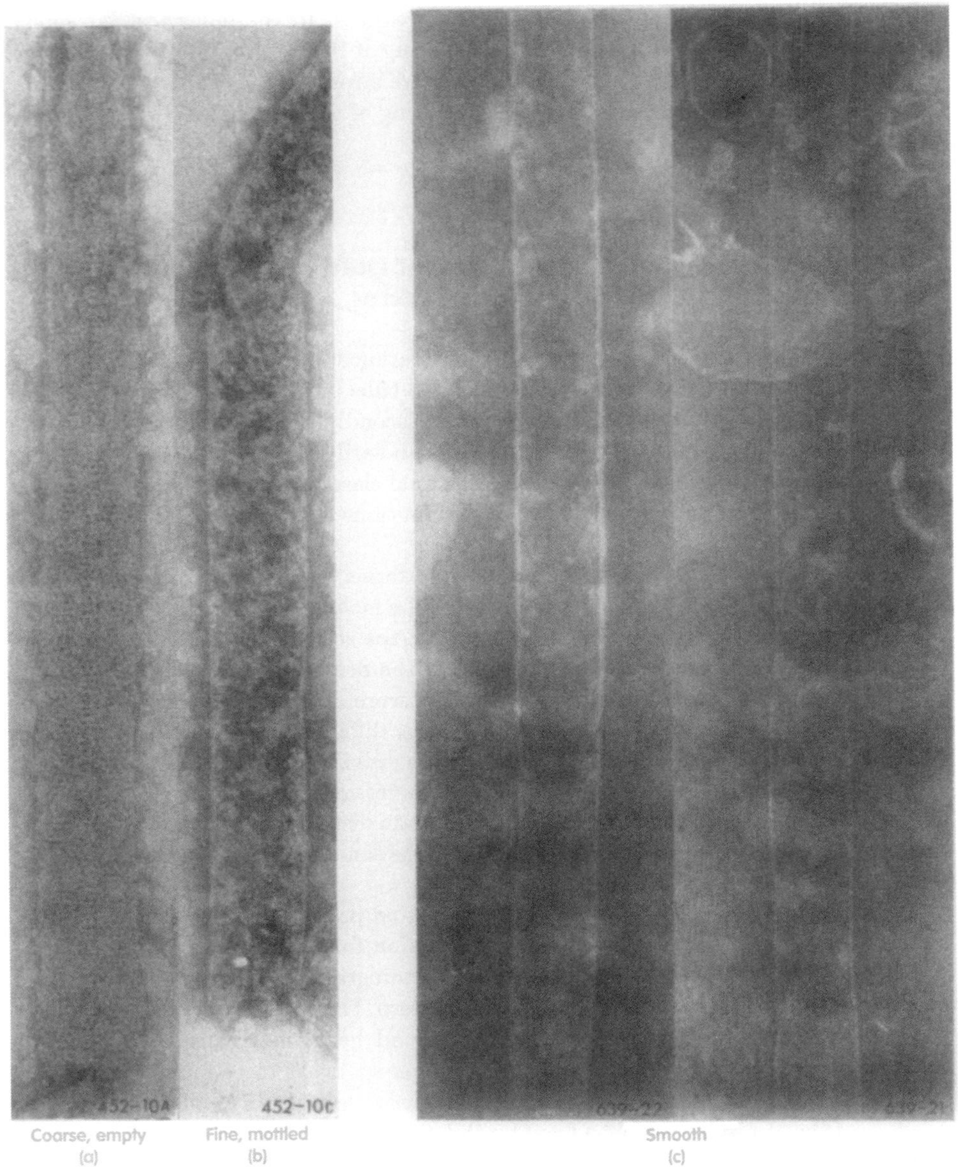

Plate I. Some polyheads classified by appearance. The magnification of the electron micrographs in this and in all subsequent plates is 150,000, unless specifically stated otherwise.

[facing p. 476

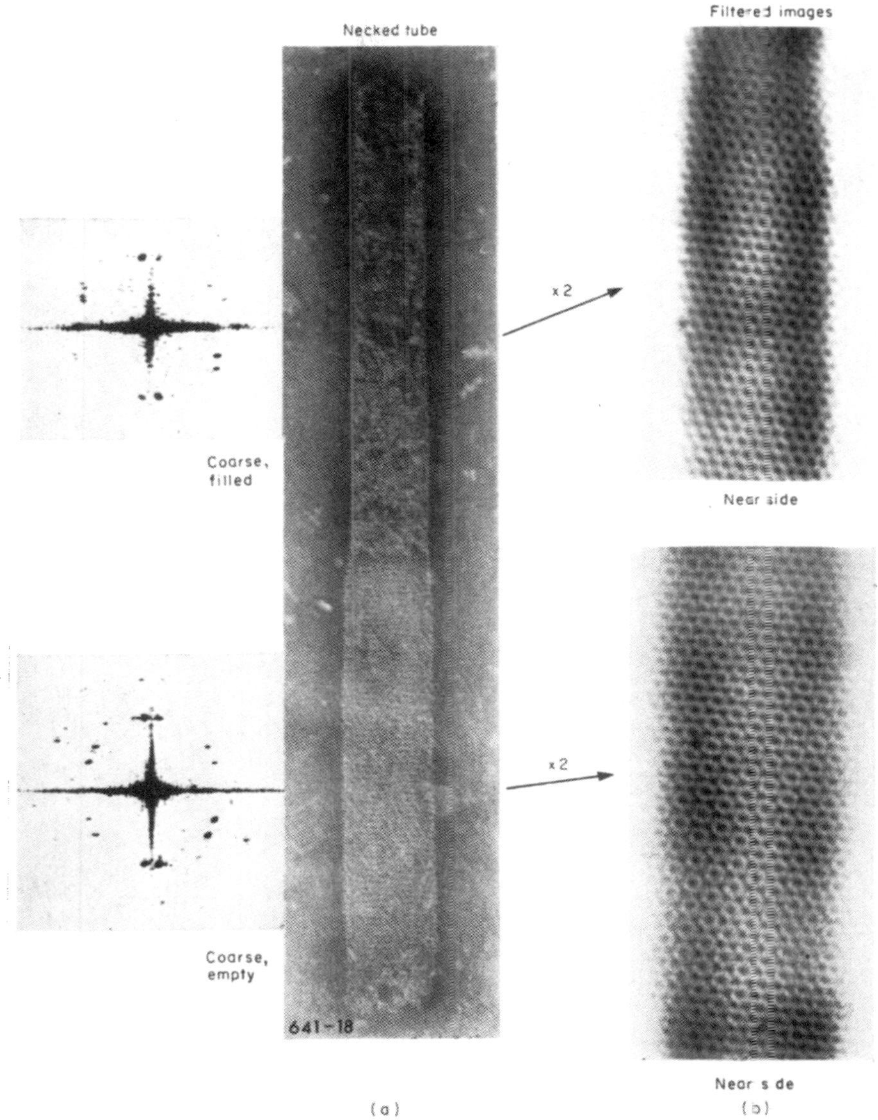

Plate II. (a) A polyhead of the coarse type containing a filled and an empty stretch. The optical transforms of the two stretches are shown alongside. (b) Filtered images of the two stretches corresponding in each case to the near side of the tube. The wall of the filled stretch approximates to a honeycomb structure but near the top distinct capsomeres, circular in outline, can be discerned. In the wall of the flattened, empty stretch hexameric capsomeres are clearly visible.

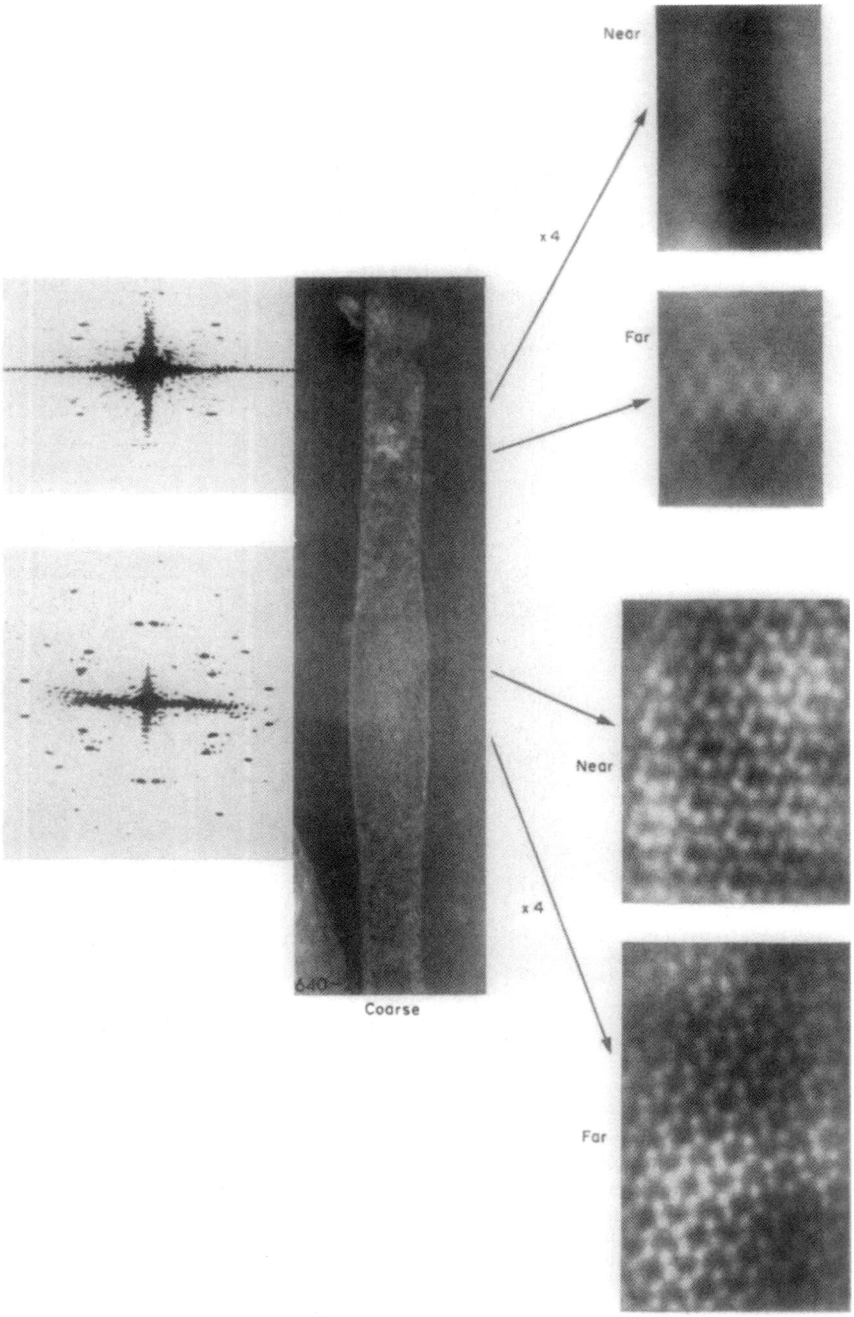

Plate III. A coarse polyhead lying in a sheet of stain spanning a hole in the carbon grid. The particle is much contracted. The middle section of the tube is empty and flattened and the end sections contain some irregularly organized internal substance which preserves the shape. The filtered images of the filled stretch show a honeycomb structure whereas those of the flattened portion show clear hexamers with the individual subunits resolved. There is a difference between the detailed disposition of matter on the near and far sides, presumably due to a difference in environment on drying down.

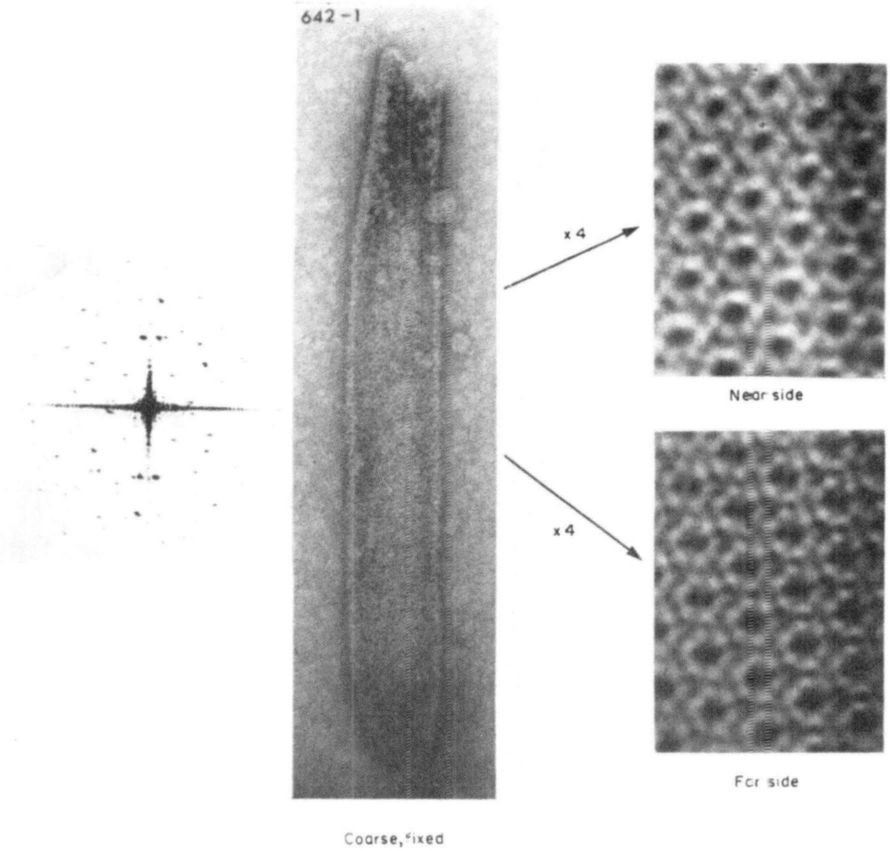

PLATE IV. The effect of fixing the specimen. Despite the appearance of the image, the polyhead belongs to the coarse class, that is, it diffracts most strongly in the 50 Å region (second-order band, Fig. 1 (b)). The fine detail extends to spacings of 25 Å.

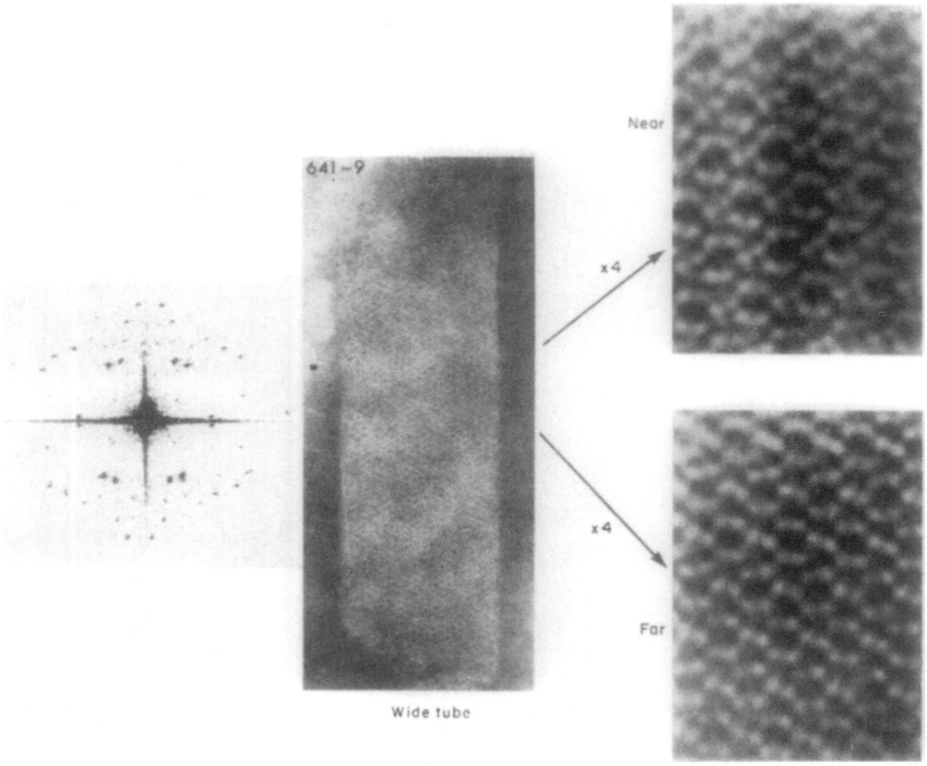

PLATE V. A wide, coarse tube of width about 1000 Å with a pitch angle of 15°. Alternate rows of hexamers (making an angle of 15° with the equatorial line) are distinctly different in shape and orientation.

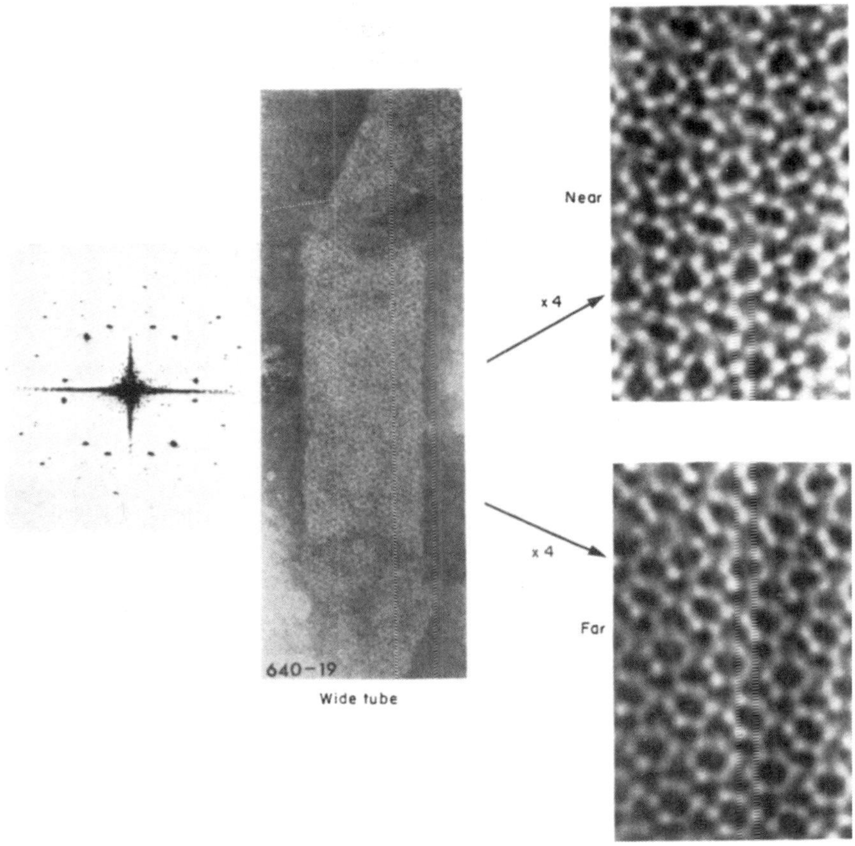

PLATE VI. A very wide tube, diameter 1400 Å, pitch angle 4°. Alternate rows of hexamers are again different as in Plate V, but the differences are not so marked.

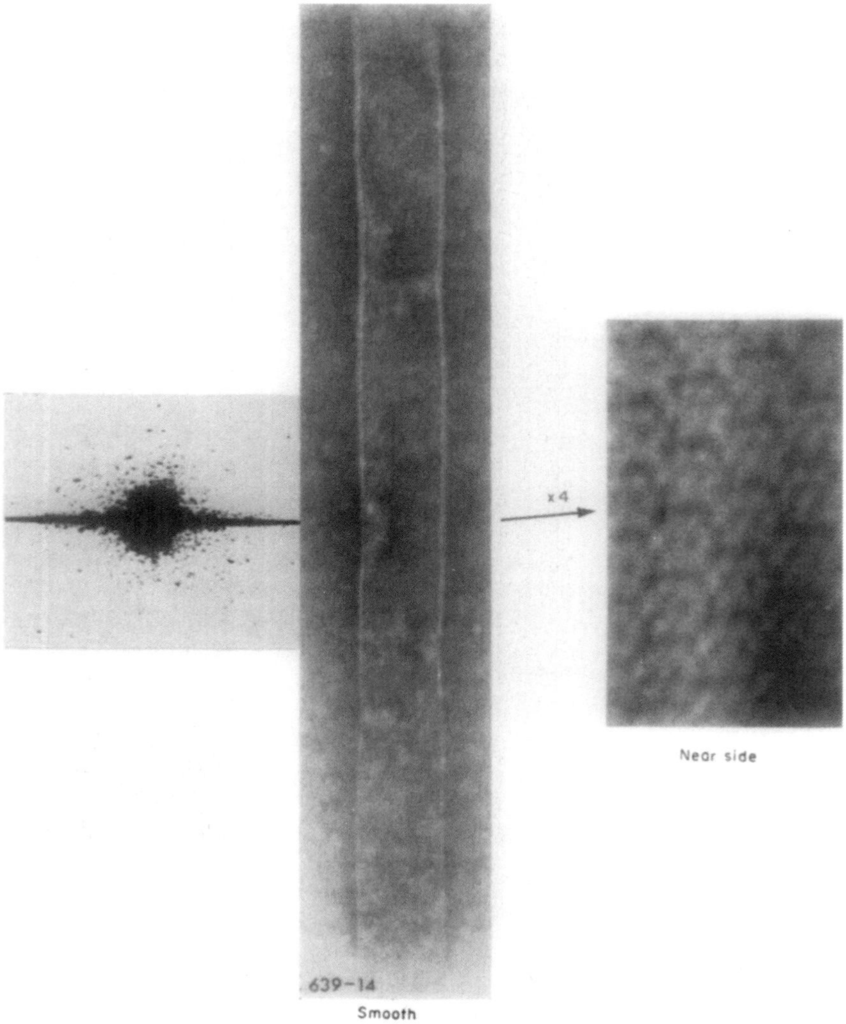

Plate VII. A smooth empty tube together with the diffraction pattern from the middle stretch and the filtered image of the near side. The hexamers are oriented so that their edges are in line, unlike those in Plates III and IV where they are skew to one another. They also possess a central unit so that the morphological units are described as 6 + 1.

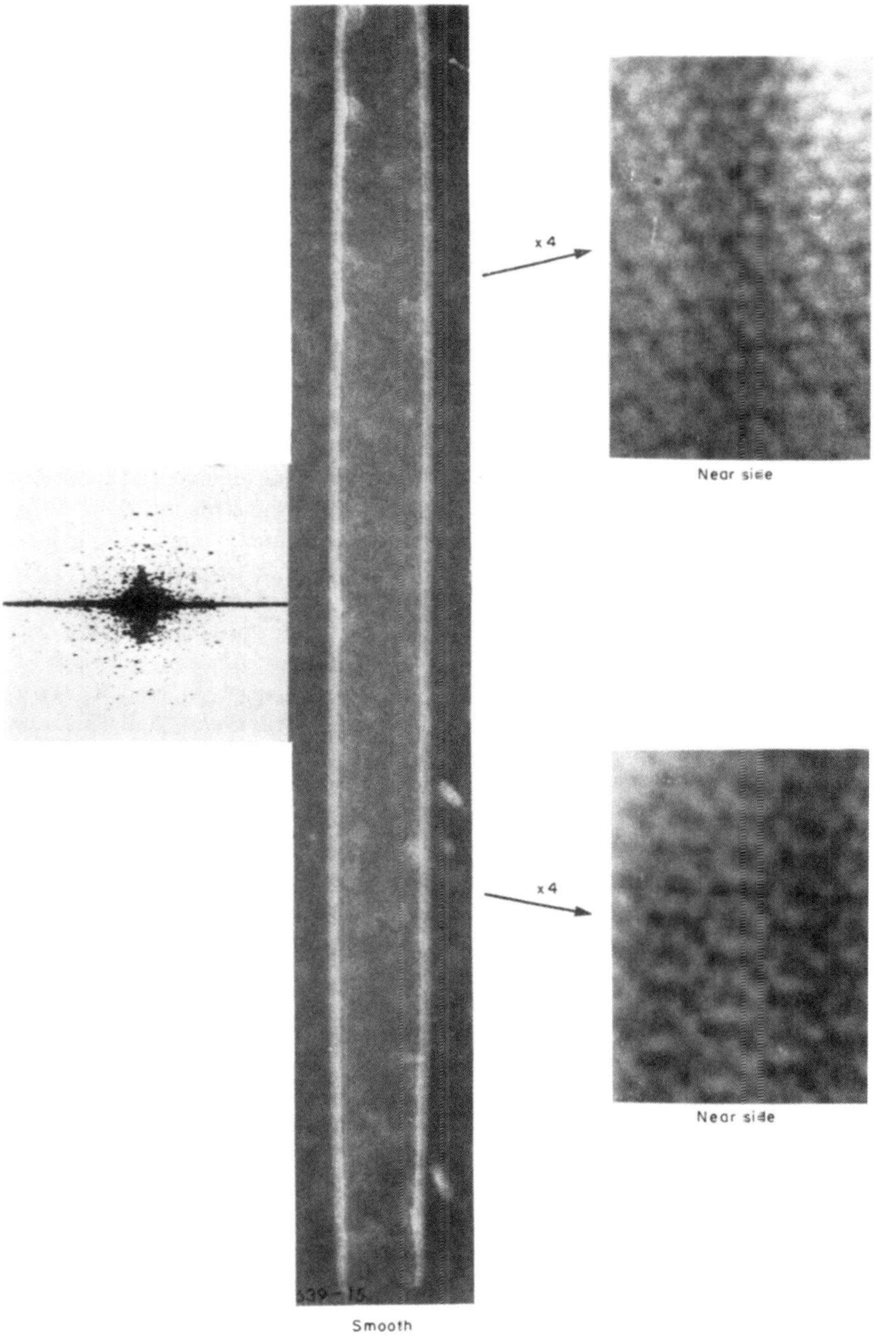

PLATE VIII. Another smooth empty tube with rather thickened walls. Filtered images of two portions are shown. The top portion shows lined-up hexamers with their central regions rather dense, though distinct filling units are not visible. The lower portion shows rather poorly preserved hexamers with their central regions empty.

when we describe a particle as appearing coarse or fine, we obviously mean that we can see the substructure clearly against the uniform background of the average distribution of scattering density and, therefore, it is by implication already rough textured.

5. Results

We have observed four main classes of polyhead in the course of our work and these can in turn be divided into subcategories according to the presence of internal substance or a minor variation in helical parameters.

(a) *Coarse-surface polyheads ("second-order" polyheads)*

The most commonly observed form of polyhead from gene *20* is shown in Plate I(a) and in Plates II, III and IV. This is the class which has been studied and reported in our first communication. The particles are usually empty of internal substance, do not retain any stain within, and lie flattened on the electron microscope grid. The observed width is about 800 Å, corresponding to a true cylindrical diameter of about 500 Å. These give a characteristic diffraction pattern in which the strongest spots lie in the 60 Å region, i.e. in the second band of Figure 1(b). The diffraction pattern is indexed as shown in Plate IX. The extraction by optical filtering of an image of each side of the original two-sided electron microscope image was carried out on the basis of this indexing and the filtered images are shown magnified alongside the original electron micrographs. Distinct morphological units or capsomeres consisting of rings of six structure units (hexamers) at a distance of about 100 Å apart are seen, as expected. The distance between neighbouring subunits in a hexamer is about 35 Å. This distance is only a little less than that between neighbouring subunits of adjacent hexamers, so that, to a crude approximation, the structure looks like a honeycomb. This is particularly evident in one of the filtered images in Plate III. In the best preserved specimens it is clear that the direction of the edge of the hexamer is not parallel to the direction joining the centres of two neighbouring hexamers (i.e. the unit cell edge) so that the arrangement of hexamers is skew. The angle of skewness is about 10°.

Apart from the coarse, empty particles just described (Plates I(a) and IV), a number of particles of similar surface appearance are seen but which appear to contain some matter on the inside and also to be heavily stained in the lumen. The result is that these "coarse, filled" particles have a diameter about 20% less than the coarse, empty particles, and their cylindrical shape seems better preserved. Frequently particles are observed which have adjacent stretches filled and empty. In the example reproduced in Plate II the internal substance is missing from one end and this part of the tube thus appears flattened on the grid. In Plate III an example is shown in which the internal substance is missing from the middle stretch of tube, but both end stretches are filled. The diffraction patterns of the filled stretches are naturally much noisier than those of the empty, because the staining is more irregular, presumably because of the irregular distribution of the internal substance. Nevertheless, it is clear that the diffraction pattern is of the same type as the coarse, empty particle. The corresponding filtered images show either ring-like capsomeres, or, less often, a honeycomb structure in which the individual capsomeres are not resolved. There is, thus, no reason to believe that there is any significant difference in the surface structures of the filled and empty tubes.

It is an open question as to whether this commonly occurring type of coarse polyhead is originally made with a special internal substance or whether such matter as is enclosed within its walls is a random sample of the cytoplasm. Kellenberger and his colleagues (Kellenberger, Eiserling & Boy de la Tour, 1968) have developed a technique of *in situ* lysis which permits the observation of the negatively stained products of single bacterial cells. In some cases, notably in the case of mutants from gene *22*, the polyheads appear to be made with some internal substance, since in fresh lysates they are often observed with rounded and presumably sealed ends and with matter on the inside (Favre, Boy de la Tour, Segre & Kellenberger, 1965). However, Laemmli (1969) has observed products of *in situ* lysis which are very similar in appearance to empty, coarse polyheads of the kind shown in Plates I(a) and IV. Certainly if there is an internal substance it seems to have little influence on the arrangement of subunits once the tube is completed. Measurements made on the necked particles of the type shown in Plate II show that there is a slight shrinkage in the axial direction of about 2 or 3% when the polyhead changes from the filled to the emptied state. This small change could be the result of mechanical deformation or flattening on the grid and is probably insignificant. The large changes in the lateral dimension of the unit cell are more apparent than real. They can be accounted for largely by the effect of flattening a curved surface.

One further point needs to be made about the arrangement of matter in this class of polyhead. It does not possess strict hexagonal rotational symmetry. The lattice spacings in the axial direction tend to be smaller than those in the lateral directions, as though the interactions between adjacent hexamers are different according to the direction of curvature between them. Moreover the hexamers themselves tend to be distorted elliptically, the dimension in the axial direction being some 10% shorter than in the transverse direction. Further information on this point can be found in the accompanying paper.

(b) *Wide polyheads, derived from outer layers*

In addition to the "narrow" polyheads just described, we have observed a number of tubes of similar coarse appearance but with a significantly greater width, 1000 to 1800 Å (Plates V and VI). It is clear by direct inspection of the images that the directions of the lattice vectors are different from those of the narrow tubes since the Moiré patterns are of a different type. The exact lattice directions have been found from the diffraction patterns. These wide polyheads have pitch angles (α in Fig. 1(a)) ranging from 0 to 15°, compared with 20 to 25° for the common polyheads.

These polyheads were at first puzzling objects to us since they were not always observed and it turned out that it was necessary to fix the preparations in order to observe them with regularity. Moreover, we never observed any wide tubes which were filled with the internal substance. It was clear to us that we were dealing with a different population of tubes from the commonly occurring coarse polyheads of width 800 Å. A minor fraction of polyheads with widths around 1000 Å had also been observed by Favre *et al.* (1965), but had not been correlated with structural differences. The final explanation was given by the Geneva workers (Kellenberger *et al.*, 1968; Yanagida *et al.*, 1970) who showed that it was very likely that all the larger empty polyheads observed in bulk lysates were initially outer layers of multi-layered polyheads. When nascent polyheads were observed by the *in situ* lysis technique, about 15% of the polyheads were found to be multi-layered and, moreover, multi-layered polyheads were also observed in thin sections. The process of lysing and agitation of the lysates during the preparation of the grids would lead to a slipping off of the outer concentric tubes. Further evidence on the outer layer polyheads is obtained from the discovery of Laemmli (see Laemmli, Mölbert, Showe & Kellenberger, 1970) that mutants in gene *22* produce more than 60% multi-layered polyheads when observed by the *in situ* lysis technique.

While the optical diffraction patterns of the wide tubes show a variety of pitch angles α, there does seem to be a strong tendency to cluster into two ranges, namely 8 to 15° and 2 to 5°, although intermediates were noticed. This variation in angle seems to be correlated with the width of the tube, the smaller angles being associated with the wider tubes. The frequency of occurrence of different widths and pitch angles in polyheads from genes *20* and *22* has been investigated by Yanagida *et al.* (1970). There appears to be a strong correlation between pitch angle and width, and a tendency to clustering is observed.

Two examples are shown in Plates V and VI. Plate V shows a tube of diameter 1000 Å with a pitch angle of about 8°. Plate VI shows a very wide tube of diameter about 1500 Å with a pitch angle of about 4°. The diffraction patterns are readily indexed (Plate X) following the same scheme as that for the narrow, coarse polyheads (Plate IX), but it is found that there are a few diffraction spots which are not accounted for by the main indexing scheme. These are the spots indicated by filled or empty squares in Plate X. As a check that these additional spots were genuine features of the diffraction pattern of the wide tube, we searched for examples of wide tubes where one side had been torn or broken off and could be observed on its own. The diffraction pattern of the one-sided regions showed the same additional spots (Plate XI). This proves that the additional spots are not artifacts such as could arise from multiple scattering (Finch, Klug & Nermut, 1967) but are produced by the arrangement of matter in the surface lattice. The presence of the additional diffraction spots means that the unit cell is twice as long in the axial direction as it is in the two other, approximately equivalently related, hexagonal directions. In other words, alternate horizontal rows of hexagons in the structure are different. The difference is evident in the filtered images in Plates V and VI. It can be seen that the relative orientations of hexagons in adjacent horizontal rows change by a considerable angle and also that the shapes of the hexamers in alternate rows are different. These differences indicate some kind of strong distortion from the simple hexagonal lattice of the normal, coarse polyhead in which all hexamers are equivalent.

(c) *"Fine mottled" polyheads ("fourth-order" polyheads)*

Most preparations of polyheads contain a small proportion (about 5%) of a distinctive type of long particle which appears to be filled with stain and also contains some rather irregularly arranged internal substance (Plate I(b)). This substance shows itself in the form of irregular patches which seem to be applied to the inner wall of the polyhead. This class of polyhead is quite distinct from the coarse, filled polyheads mentioned above, since the wall of the tube has a rather fine appearance. This is confirmed by optical diffraction which gives patterns having hexagonally arranged spots in the 30 Å region (the fourth-order band in Fig. 1(b)), but none discernible above the noise level in the 50 to 100 Å region. Although it is clear that the surface structure is also based on an hexagonal lattice, the relation of this lattice to the standard lattice could not be obtained since, in the few examples we had of this type of polyhead, no inner spots were observed which could be used to determine the unit cell. To investigate the problem further, it was clearly necessary to obtain better diffraction patterns of this class of polyhead from a larger number of specimens and this called for a new investigation which is described in the accompanying paper. We therefore defer the description of the results to that paper.

This class of fine-mottled, filled polyheads, although present in small amounts, is extraordinarily stable. In a polyhead preparation which was stored at 4°C for a week, the ordinary coarse polyheads disappeared entirely, the only survivors being the smooth, filled polyheads. A possible explanation for this stability is suggested by the results of optical filtering presented in the following paper.

(d) *"Smooth empty" polyheads ("third-order" polyheads)*

In very fresh lysates which were applied within minutes of lysis to the electron microscope grid, we observed a class of tube which resembles in surface appearance the ordinary empty capsids of the bacteriophage (phage ghosts). Examples are shown in Plate I(c). When the specimens were fixed, the appearance changed very little, although there sometimes appeared to be a slight thickening in the wall (Plate VIII). The tubes are so smooth in appearance that substrate and various pieces of cell debris can be seen through them as though they were transparent. We have called this type of tube "smooth, empty". These tubes did not, at first sight, offer much hope as a suitable subject for structural analysis, but after some searching we found stretches which gave diffraction patterns which could eventually be indexed. The task of indexing is made very difficult by the very low signal-to-noise ratio, i.e. the strength of the diffraction maxima lying on a reciprocal lattice compared with the random background in the patterns. By searching for common features in the diffraction patterns of different specimens, we were able to arrive at the consistent and plausible indexing shown in Plate XII. The key to the indexing of these patterns lies in determining whether the spots at the periphery of the patterns are third- or fourth-order spots. A few common inner diffraction spots in different diffraction patterns were used to decide the matter. The indexing arrived at in Plate XII is essentially the same as that found later in a class of variants observed in mutants of gene *22* (accompanying paper, Plate VIII), in which case the signal-to-noise ratio in the patterns is high enough to present no difficulties.

The filtered images of one side of the smooth tubes are shown in Plates VII and VIII. These images are not as clear as those of other classes of polyheads but they leave one in no doubt as to the distribution of matter in the surface lattice of the tubes. The arrangment is again one of hexameric morphological units but there are three differences from the standard lattice of the coarse polyheads. First, the unit cell dimensions are larger by about 10% than in the coarse polyheads and more nearly approximate to hexagonal symmetry—that is the three lattice vectors are nearly equal and the angles are closer to 60°. Second, the edges of adjacent hexagons are parallel so that the hexamers are lined up in a parallel fashion. Third, and perhaps most significantly, the central region of the hexamer does not appear to be penetrated by stain, consistent with there being an additional piece of matter in this position. We call this a 6+1 morphological unit.

In some regions of the filtered images, the boundaries of the morphological units are not visible, and the distinction between the central unit and the six outer units of the large 6+1 unit is lost, so that the appearance is simply one of molecules arranged on a small primitive, pseudo-hexagonal lattice of side 35 Å. This is a very fine distribution of matter and accounts for the weakness of the inner spots in the diffraction patterns.

Since the surface appearance of these tubes is so similar to phage ghosts it is likely that the surface structure of phage ghosts is similar. We have made attempts to obtain optical diffraction patterns of phage ghosts but, although some ghosts did show spots in common with the smooth, empty tubes, their areas are too small to obtain significant patterns.

Whether the apparent central unit in the hexagon represents a new protein subunit or not is a matter for further investigation. The matter in the middle of the hexagon might not represent a distinct molecule at all, but could conceivably arise by a transformation in the structure of the individual subunits of the hexamer, which in turn causes protuberances to point towards the centre of the hexagon. These protuberances could coalesce to give the appearance of a single structure unit. We, therefore, do not call this 6+1 arrangement a heptamer.

6. Discussion

Our analysis of the electron micrographs of polyheads has led to the description of four classes of tube. The most commonly occurring tube is narrow with a width of about 800 Å and the wall structure is coarse or "capsomeric" in type, based on a simple near-hexagonal arrangement of subunits. The tubes appear to be only one layer of subunits thick and this thickness has been estimated to be 94 Å (Yanagida *et al.*, 1970). The filtered images show distinct structural units at a distance of about 33 Å apart arranged in hexamers, or capsomeres. The units can be identified with the molecules of the major protein component of the head, namely P23, the product of gene *23* which has a molecular weight about 50,000. The hexamers are related by approximate 3-fold axes corresponding to the plane group with symmetry *p*6 (Fig. 1). In some specimens the gaps in the 3-fold positions between hexamers are similar in size and shape to the central hole in a hexamer, so that to the first approximation the structure resembles that of a honeycomb. This state of affairs does not, however, seem to be typical of the best preserved specimens, in which the orientation of hexamers in the lattice is skew, so that the distinction between 3-fold and 6-fold axes of symmetry is quite clear.

The hexagonal symmetry of the surface lattice is not exact. The lattice spacing closest to the axial direction is smaller than those in the lateral directions and the hexamers themselves are elliptical rather than circular in shape. If the cylindrical surface is thought of as having been derived from a plane hexagonal net, then deformations are necessary to close the structure. The highest order possible for a rotational symmetry axis perpendicular to the axis of a cylindrical or helical particle is 2-fold (Klug, Crick & Wyckoff, 1958), and this is the true symmetry possessed by the tubes.

The significance of a structure of this type has been pointed out by Caspar & Klug (1962). The plane lattice *p*6 is one which may be folded into a closed surface in such a way that the local bond structure is preserved with a minimum of variation in the local bonding; or in other words, the structure units remain in quasi-equivalent environments. This hexagonal pattern of bonding between structure units can be used to form a spherical shell with icosahedral symmetry, but the same type of bonding can also be incorporated into a surface of lower symmetry, but of similar curvature. This result suggested (Finch *et al.*, 1964) that the polyheads are a tubular variant of an icosahedral type of shell and hence that the T-even bacteriophages themselves may be thought of properly as an elongated version of icosahedral viruses, their lengths being determined by interaction with some other component(s) necessary for normal morphogenesis. Moody (1965) has made careful observations on the head of T-even phage and concluded that its most likely shape is that of an icosahedron elongated along one of its 5-fold axes.

The wide polyheads we have observed, of width 1,000 Å or more, also appear to be one subunit thick, but we know from the work of Kellenberger and his colleagues that they are derived from multi-layer polyheads which consist of layers of protein subunits arranged around the narrow polyhead (what they call a "first-layer polyhead"). These wide tubes from gene *20* show a range of pitch angles α with a tendency to cluster into two ranges, which appear to be correlated with the width of the tube, a smaller angle being associated with a wider tube. The basic structure is hexagonal like that of the narrow, coarse polyhead but there are marked distortions in the packing which leads to alternate rows of the lattice being different. It is conceivable that this distortion arises by a crumpling of the simple hexagonal lattice when an outer layer of the polyhead is formed by growing on an inner layer. The difference would seem to be associated with the pitch angle α of the surface lattice, which is lower for the outer layers than for the innermost layer. This distortion in the hexagonal lattice does not seem to be a necessary feature of the assembly of protein P23 since the outer layers of the multi-layered polyheads derived from mutants in gene *22* have a local structure indistinguishable from that of the narrow tubes or "first layers" (see the accompanying paper).

The statistical analysis of Yanagida *et al*. (1970) shows that the diameters (or radii of curvature) of the first-layer polyheads and the pitch angles (or directions of curvature) are fixed within relatively narrow limits. The higher layers do not have a precisely determined direction of bending but all directions which could give comparable diameter polyheads are not observed with equal frequency. The relatively small variation in the circumferential vector among the first layer polyheads is striking. We are indebted to Dr D. L. D. Caspar for the following remarks. The energetically favoured curvature of the surface lattice may be an intrinsic property of the protein P23 since no mutants have been observed which show a different mode of assembly in isolated polyheads. The only exception is that of the outer-layer polyheads, but these seem to form only when there is a lack of nucleation sites for the assembly of tubes, so that the surplus protein subunits condense on the surface of an already formed tube. The direction of curvature of the higher layers is not related to that of the underlying layer so that the nucleation of outer layers presumably does not involve a special mechanism. In the circumstances it is not surprising that distortions occur when the energetically favoured bending of the surface lattice—a property of the protein—cannot be realised.

The most significant observations that we have made are perhaps those on the "smooth, empty" polyheads which are so similar in appearance to phage ghosts or empty capsids. The optical filtering has revealed a substructure in this class of tubes which is closely related to that of the hexagonal structure of the commonly occurring coarse polyheads and may be derived from it. The subunits are again organized into hexamers but these are now equal in all dimensions, i.e. no longer elliptical, and the hexamers are rotated so that they are all lined up in a parallel fashion and fit snugly against each other. Moreover the central region of the hexamer appears to contain an additional unit, so that we have a new type of capsomers or morphological unit, the 6+1 unit. The fine distribution of matter which results (third-order, Fig. 1(b)) means that the description of the surface in terms of capsomeres which are by definition distinct, may be something of a misnomer.

Only by application of the freeze–etching technique (Bayer & Remsen, 1970) has it been possible to demonstrate capsomeres directly on the surface of a phage head. The capsomeres consist of six units with an arrangement very similar to those revealed in this work and the capsomeres are reported to contain a central depression with a varying degree of filling, which might correspond to the central unit reported here.

Some electron micrographs have been taken of freeze–etched preparations of phage T2 by Professor D. Branton. The pattern observed on polyheads is essentially the same as that described here for coarse polyheads.

The "fine, mottled" polyheads, as will be shown in the accompanying paper may also be described in terms of a morphological unit of the 6+1 type, but in this case additional matter is visible in the 2-fold and 3-fold positions of the surface lattice and this seems to bridge these units. The structure in this case is so fine (fourth-order, Fig. 1(b)) that the description in terms of capsomers becomes physically meaningless, although it may be helpful from a geometrical point of view. The stability of this class of polyhead and also of the third-order class is presumably linked with the presence of the central unit in the hexamer. This structural change may in turn reflect the possibility—and there is some preliminary evidence for this from recent work by Laemmli (personal communication)—that the protein in these classes of tube is already in the cleaved state, the state which has been found by Laemmli (1970) to occur in the mature phage head. It seems likely that the differences in structure between the normal coarse polyheads and the fine and smooth types described here are due to this, or a similar, cleavage having taken place in the tubes. If this is so, the tubular variants may serve as a model system for the maturation of the phage head. Further discussion of this point will be taken up in the following paper.

APPENDIX

Indexing and Filtering of the Diffraction Patterns

(a) *Rationale behind the indexing*

A helical surface lattice such as that of a polyhead can be thought of as arising from the wrapping of a two-dimensional plane lattice on to a cylindrical surface. If the density in a helical surface lattice is projected onto a plane parallel to the axis, the projected patterns of the near and far sides of the surface lattice are mirror images of each other (Fig. A1). In general, the patterns from the two sides will also appear to be rotated with respect to each other, a result of the fact that a helical family which appears to run from the bottom left to the top right on the front side of the tube will appear to run from the bottom right to the top left on the far side. The diffraction pattern of the projection may be thought of as a sum† of the diffraction patterns

† Strictly speaking, the Fourier transform $F(h,k)$ of two juxtaposed layers is not the sum of the transforms of the individual layers, $F_1(h,k)$ and $F_2(h,k)$, but the convolution. There is also an additional Fresnel factor to take account of the fact that the two layers are not simultaneously exactly in focus. The complete expression for successive scattering from two layers a distance z apart is (Cowley & Moodie, 1957):

$$F(h,k) = \sum_{h'} \sum_{k'} F_1(h',k_1') F_2(h-h',k-k') \exp \{-2\pi i \frac{\lambda z}{2} (h'^2 + k'^2)\}. \quad (1)$$

Here λ is the electron wavelength, and (h,k) are co-ordinates in reciprocal space. If the distance z

deriving from the two sides indepedently, since if the diffraction maxima are narrow there will be in general no overlap of the sets of spots deriving from the two sides, i.e. there will be no interference in the sense of physical optics. The separation of the two sets will be the better, the larger the number of unit cells of the lattice wrapped round the circumference. The breadth of a spot is approximately inversely proportional to the width of the whole particle and the distance between reciprocal lattice spots will be inversely proportional to the edge of the unit cell. Thus, the diffraction from a single helical tube will, in the first approximation, be made up of two sets of spots each lying on a reciprocal lattice and related to each other by a mirror line parallel to the axis of the particle (Klug & DeRosier, 1966). This is the basis of the indexing which has been used for the polyhead patterns (Fig. A1).

It follows from this reasoning that, in the diffraction plane of an optical system, the set of rays diffracted by the image of the near side of a helical particle is separated spatially from the set of rays diffracted by the image of the far side of the particle. The two sets can therefore be manipulated separately. This is the basis of the optical filtering method for helical particles proposed by Klug & DeRosier (1966). Once the spots in the diffraction pattern are allocated to one or other side of the particle, separate images of each side can be extracted by filtering from the original two-sided image.

(b) *Polyheads*

The diffraction pattern of a coarse empty polyhead is shown in Plate IX. The strongest maxima in the pattern form a ring of 12 spots in the 60 Å region and, as pointed out in our first paper (Finch *et al.*, 1964), these 12 spots can be separated into two sets of 6 each. If axes are drawn from the origin to the centres of one set of six

is smaller than d^2/λ, where d is the highest spacing in the structure which is diffracting, then the Fresnel term reduces to unity, so that we have a simple convolution. This condition applies here, since $z \sim 100$ Å, and $d \sim 25$ Å. Furthermore, if the diffracted orders are weak relative to the zero order, the convolution reduces to a sum in the first approximation:

$$F(h,k) \simeq F_2(0,0)\,F_1(h,k) + F_1(0,0)\,F_2(h,k). \quad (2)$$

The terms omitted describe the phenomenon of multiple—or in this case double scattering, and when the scattering is strong, so that they are not negligible, extra spots may appear in the diffraction pattern. These can be regarded as arising from the secondary diffraction of a diffracted wave $(h',k')_1$ from the first layer by the set of planes $(h'',k'')_2$ of the second layer. The resulting spot appears at the vector sum of $(h',k')_1$ and $(h'',k'')_2$. When the two layers are rotated relative to one another, these extra spots do not generally fall on the reciprocal lattice points of either layer and can therefore be observed independently of the maxima produced by direct diffraction from each layer. Double scattering of this kind was observed in optical transforms of electron micrographs of negatively stained bacterial cell walls (Finch *et al.*, 1967), but the effect is weak. We have also occasionally observed similar effects in the polyhead transforms, but the secondary spots are again very weak. This justifies our using the approximation of considering the observed electron microscope images as resulting from the simple superposition of two layers. Further experiments relevant to the validity of the approximation are described by Erickson & Klug (1971).

FIG. A1. Indexing of the diffraction pattern of a (flattened) helical structure.

(Top) Schematic diagram of the image of a single layer, the near side of the tube say, together with its diffraction pattern. This would correspond to the diffraction from one of the sides of the tube.

(Centre) Image of the far side and its diffraction pattern.

(Bottom) The image of both layers superimposed and the corresponding diffraction pattern. Whereas the image is confused, and is in fact a Moiré pattern, the diffracted rays from each of the two sides are separated in the diffraction plane. This provides the basis for indexing and filtering a helical particle.

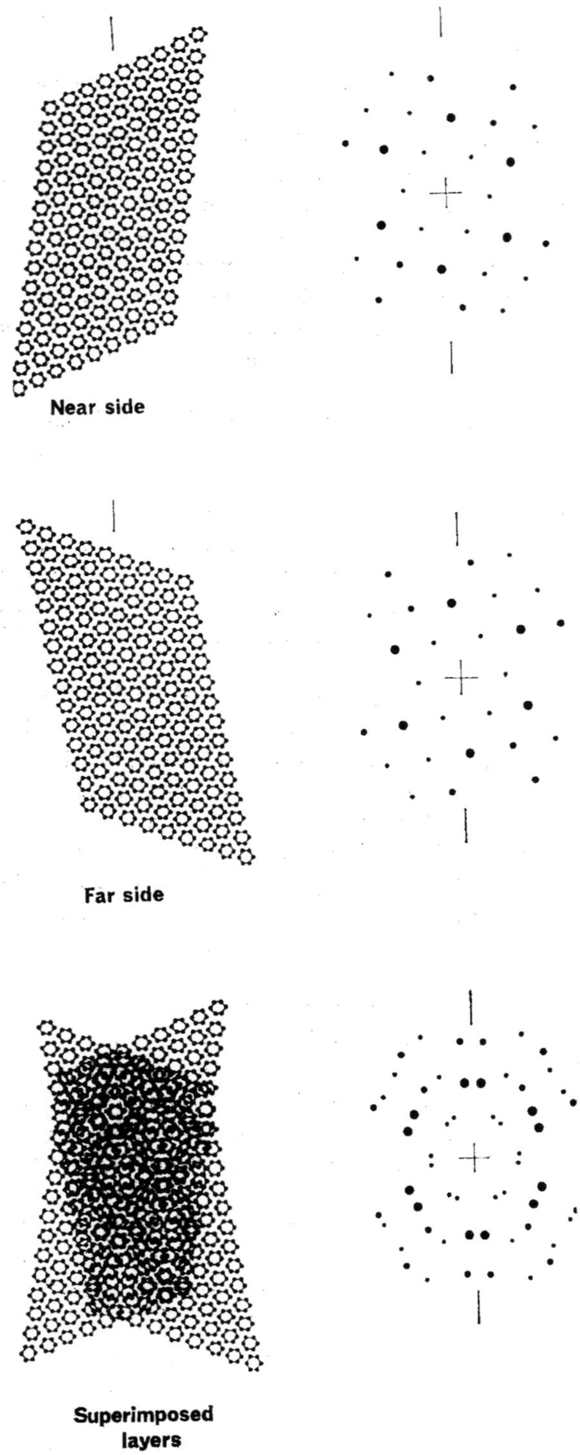

Fig. A1

spots, these can be used as approximate basis vectors for a hexagonal reciprocal lattice, which accounts for many of the other spots observed. However, weaker spots can be seen closer to the zero order which require that the true basic reciprocal lattice vector be drawn at an angle of 30° to the original set and a length of $1/\sqrt{3}$ shorter. It is then found that all the spots observed can be indexed.

In general, it is essential to compare diffraction patterns obtained from many different specimens in order to determine the constant, and therefore presumably genuine, features of the structure and to refer back to the original electron microscope images at each step in the analysis. Similarly a number of filtered images of different particles are examined so that the significance of various features can be judged. Although any single image may show individual distortions and irregularities, it is extremely unlikely that all images will suffer from exactly the same complaint, so that the consistent features can be extracted.

In the case of polyheads, the high symmetry of the reciprocal lattice makes the process of indexing a relatively easy one when the signal-to-noise ratio is high, and once started correctly the process proceeds almost automatically to completion. Details of the indexing may be checked from the information presented in Plate IX. It will be seen that the set of spots allocated to one side form, to a very good approximation, the mirror image of the set allocated to the other side. No other system of indexing we have tried leads to a unique solution which accounts for the great majority of observed spots in all examples and we believe the result to be quite unambiguous.

The diffraction patterns from the wide coarse tubes are basically similar to those from the narrow tubes. An example is shown in Plate X. Here again, there is a set of twelve strong spots at about 60 Å spacing which provides an immediate clue to the indexing and the pattern falls into two hexagonal lattices just as in the first case. There are, however, additional spots which tend to be weaker and which could easily be missed on a first round of indexing. These are the spots indicated by the square boxes and, as discussed in the text, they are genuine (see also Plate XI). The additional spots only appear at high or relatively high orders and are not seen at low angles of diffraction because to a coarse approximation the hexagons in adjacent unit cells are nearly the same. However, the existence of the weak spots shows that alternate rows of hexagons are different from one another (Plates V and VI).

The diffraction patterns of the smooth empty tubes, two examples of which are reproduced in Plate XII, are very difficult to index. The only obvious set of strong, hexagonally arranged spots lies at rather high angles of diffraction. The reasons for the weakness or absence of diffraction maxima corresponding to larger spacings are given in the text but they could only become apparent after the filtering was carried out. However, by comparing the diffraction patterns from different stretches of different specimens of smooth tubes, constant features could be found amongst the noise. Plate XII shows the diffraction patterns from two different particles. It can be seen that the spots chosen as belonging to the reciprocal lattice of particle 639-14 coincide in position with those corresponding to the particle 639-15. In identifying the correct diffraction spots, we have found it useful to obtain diffraction patterns from relatively short stretches of a particle, rather than averaging optically over a long stretch. The latter procedure tends to produce only a blurring of the diffraction spots since slight changes in cell dimensions along the length of a particle on a grid means that the common diffraction spots do not superimpose exactly. In Plate XII we have marked

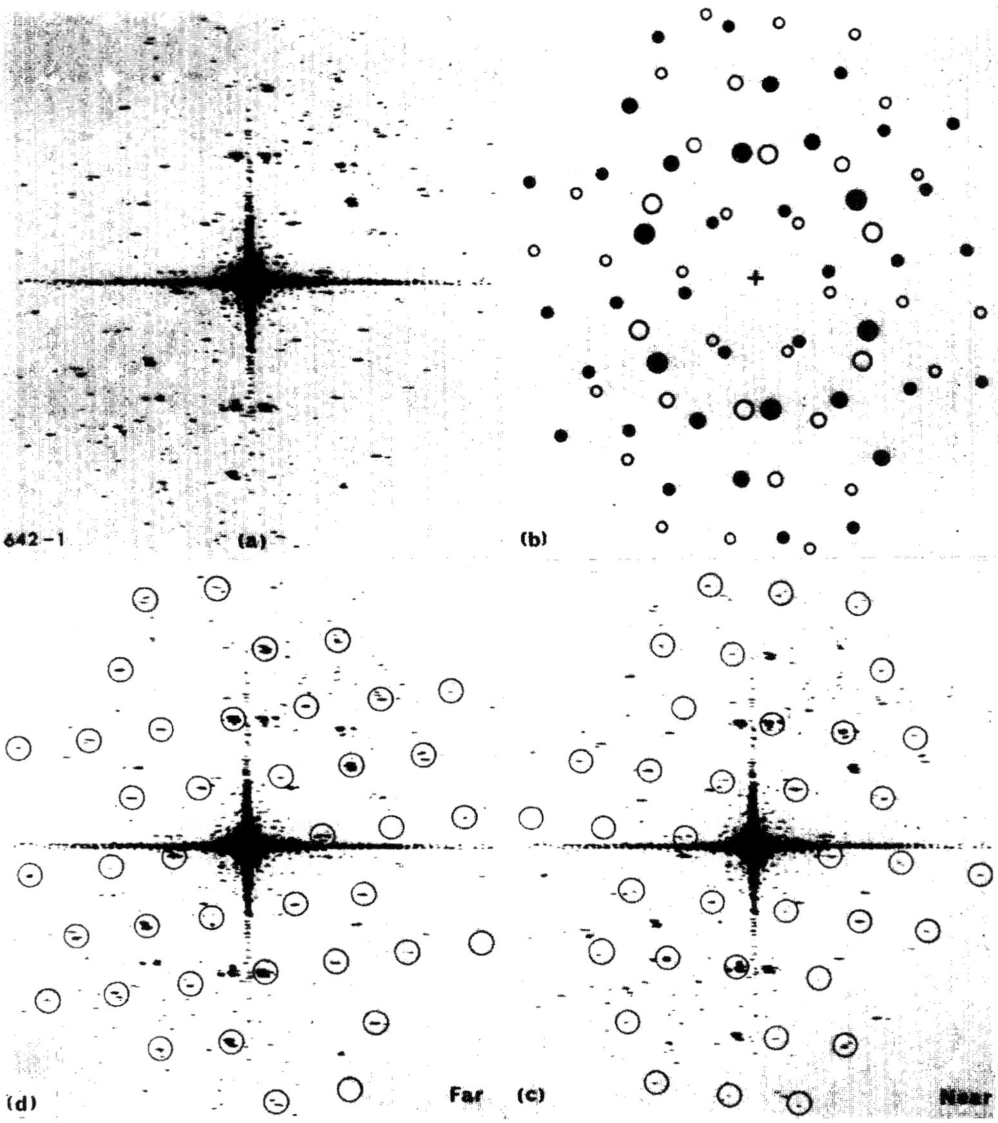

PLATE IX. (a) Optical transform of a coarse narrow tube. (b) The "indexed" diffraction pattern; the majority of the diffracted spots have been sorted into two sets derived respectively from the near and far sides of the tube (open circles and filled circles respectively). (c) and (d) indicate the positions and approximate sizes of the apertures used to produce the filtered images shown in Plate IV. In each case the region around the central beam was covered with a neutral density filter transmitting about 50% of the amplitude.

[facing p. 486

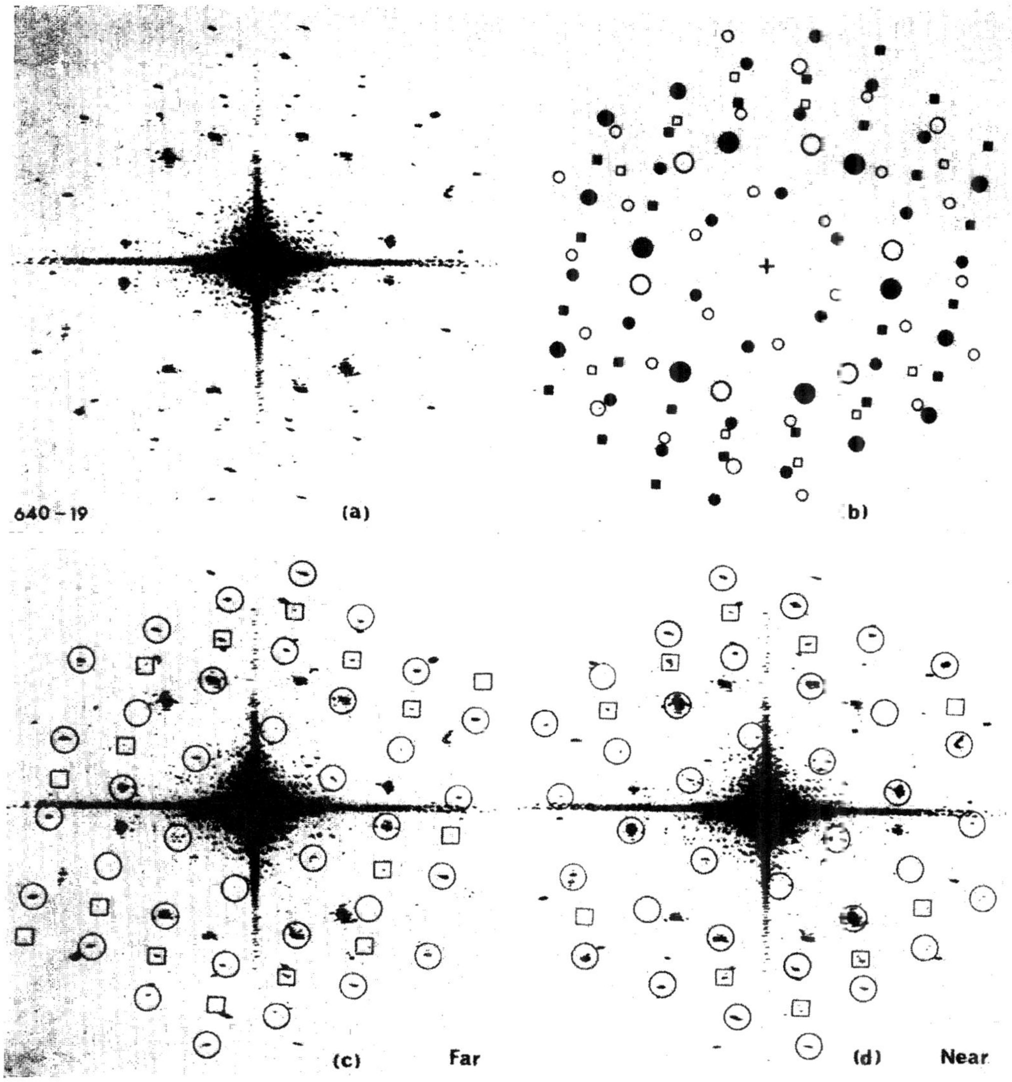

Plate X. (a) Optical transform of a wide tube. (b) The indexed pattern. Contributions from near and far sides are marked by open and closed circles respectively. Additional maxima falling on a reciprocal lattice of half the scale are marked by squares. (c) and (d) Positions and approximate sizes of apertures used to produce the filtered images shown in Plate V.

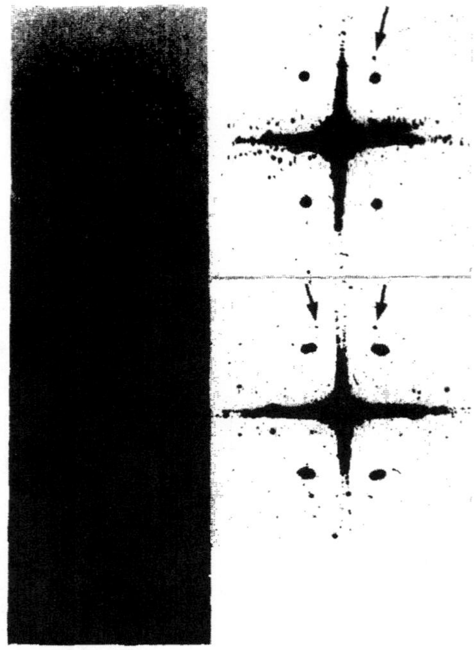

PLATE XI. A wide tube containing a portion from which one of its sides has been torn off. Diffraction patterns from the single layer region and the two layer region are shown alongside. The arrows mark the expected positions for the strongest of the additional diffraction maxima denoted by squares in Plate X. The single layer shows the expected additional diffraction maximum, thus proving that it does not arise as an artifact of the superposition of the two layers.

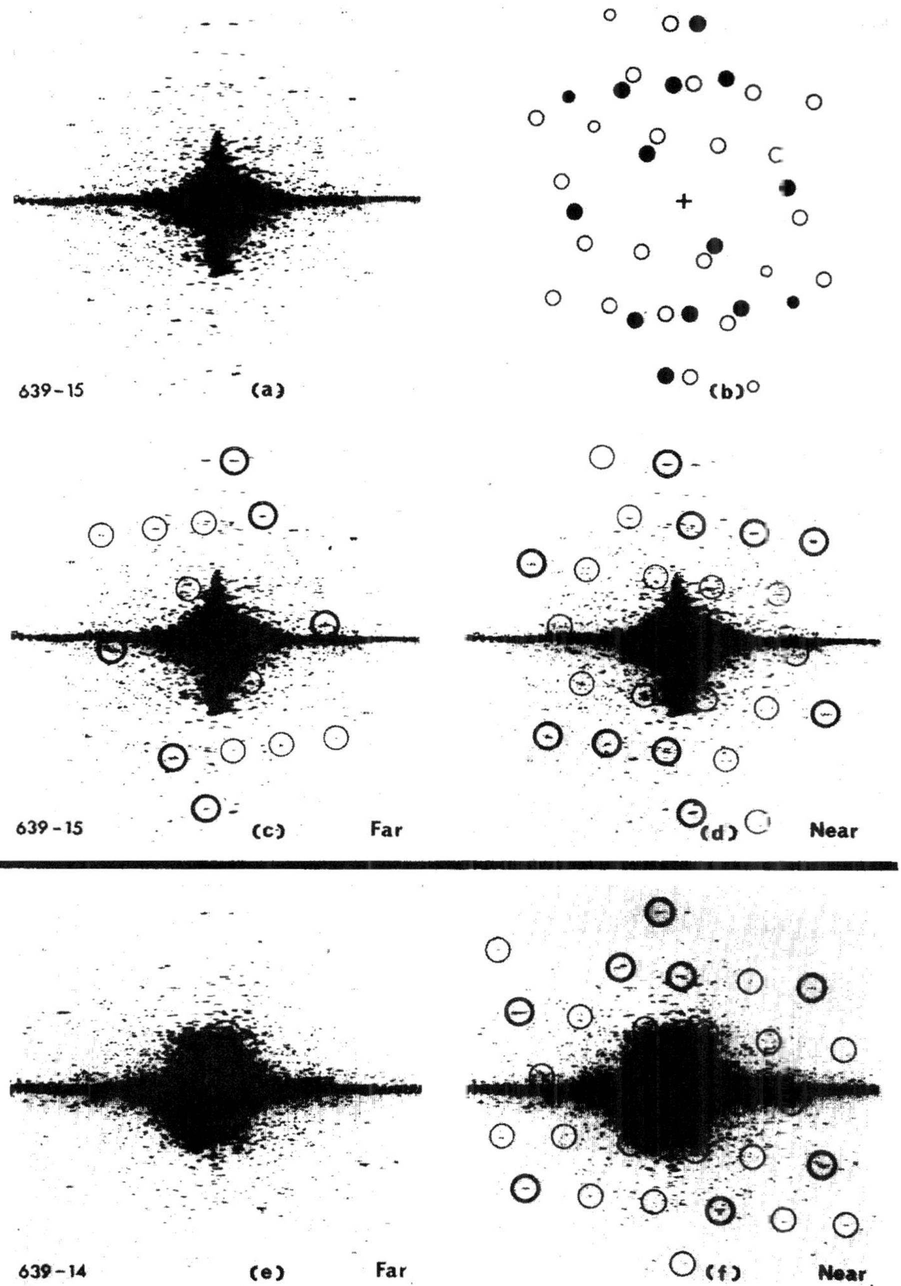

PLATE XII. (a) Optical transform of a smooth empty tube. The signal-to-noise ratio is very low because the signal (the diffraction maxima) is weak relative to the zero order term. (b) The indexed pattern based upon the transforms in (a), (e) and some other examples not shown in this Plate. (c) and (d) Positions of diffraction maxima corresponding to the far and near sides respectively. The circles also mark the apertures used to produce the filtered images in Plate VIII. The double circles mark those spots which can be assigned with reasonable certainty to one or other of the two reciprocal lattices. (e) Another example of a transform of a smooth empty tube. (f) Indexed pattern of the near side and the positions of apertures used to obtain the filtered image in Plate VII. The double circles again mark the strongest and most certain features of the pattern.

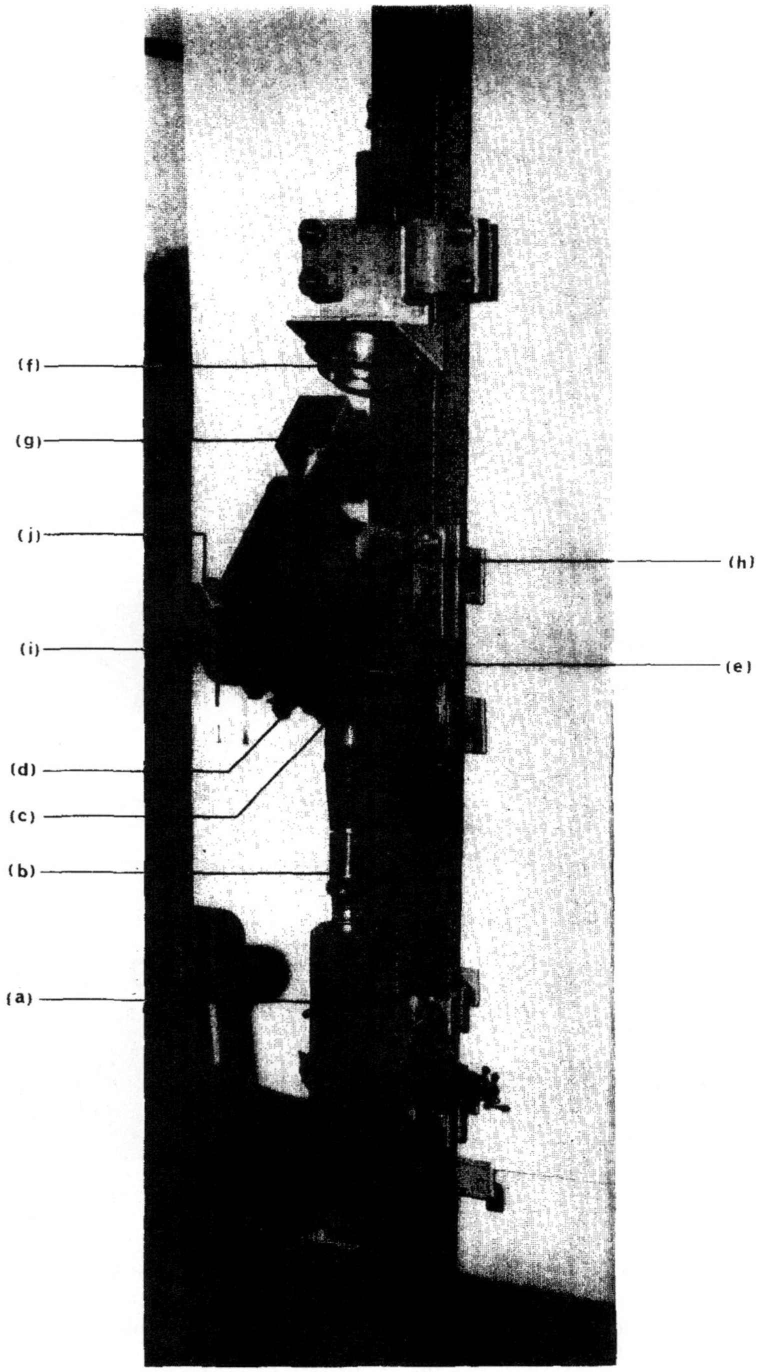

PLATE XIII. The surveying optical diffractometer. (a) Laser; (b) beam expanding telescope; (c) main lens; (d) plate holder; (e) slit system; (f) mirror for folding the optical track; (g) field lens control; (h) adjustable rack; (i) reflex camera for viewing and recording; (j) magnifier and polarizer.

with a double circle those spots the assignment of which we were confident. If the strong spots from different particles are considered together, the problem of identifying the underlying reciprocal lattice can be solved.

(c) *General comments*

The examples given above illustrate the problems and procedures of indexing in a practical case. The majority of the strong diffracted spots have been sorted into two sets derived, respectively, from the near and far sides of the flattened tube. The rationale behind this is that the images of the two sides of the helical structure are related by mirror symmetry when projected on to the plane containing the axis. This is true even when the perfect cylindrical shape has been flattened, providing the flattening has been symmetric. Each such set of spots must lie at the vertices of a lattice. This would be the ordinary reciprocal lattice if the layers were perfectly flat but deviations must be allowed if the layers still retain some of their original cylindrical curvature. These deviations have been calculated and listed by Moody (1967).

The mirror symmetry relating to the two sets of spots is not usually perfect because the staining on the two sides is not always equal—one side of the specimen is attached to the supporting grid, and the other faces the stain only. There are also variations from tube to tube, but the constant features of the diffraction pattern can be found by inspecting a number of different specimens. The highest angle lattice spots we have observed in the polyhead system correspond to spacings of about 25 Å, showing that the long-range order and identity between different unit cells of the structure is preserved to this resolution. All spots which cannot be allocated in this way to the periodic part of the structure are attributed either to the presence of a non-periodic component or to the effects of noise. This noise originates in imperfections and variations in the staining, perturbations of the structure due to imperfect preservation and so on.

It is pertinent here to refer to an example where the indexing (and filtering) proved difficult because of interference between the diffracted rays from the two sides. This is the case of the pentamer tubular variants of papilloma virus (Kiselev & Klug, 1969) where, because the tubes are narrow, the distance between diffraction maxima in the transform is only slightly greater than the width of the maxima. In quantitative terms these tubes contain only three unit cells of the surface lattice around a circumference and hence only one and a half in projection, compared with about 18 and 9, respectively, in the case of the polyheads. The separation of the diffracted rays in the latter case is therefore much easier and is limited only by the chance juxtaposition of two spots deriving from different sides.

(d) *Relation between optical filtering and translational averaging*

The optical filtering method works in Fourier or reciprocal space. To any method in this space there must be a formal counterpart in real or direct space, but the corresponding method in the two spaces may differ considerably in the practicability of implementation. The counterpart of optical filtering is the photographic integration method developed by Markham for enhancing image detail in electron micrographs of objects with a regularly repeating structure (Markham, Hitchborn, Hills & Frey, 1964). In Markham's linear integration technique, an image of one side of the structure is produced by repeatedly translating the electron micrograph in certain directions and making a photographic superposition of the set so obtained. If the translations have been chosen correctly, the structure of the object will be revealed. In the latest form of the method (Hitchborn & Hills, 1968), the directions to be employed are determined by optical diffraction. It can be shown (DeRosier & Klug, manuscript in preparation) that there is an exact relation between averaging by this technique and optical filtering. If translational superposition were carried out over all repeats of *all* the lattice vectors in the subject—something which cannot be done easily in practice—then one would obtain a picture which is equivalent to an optically filtered ideal image, but with a larger noise level. This is because the other side of the structure is not being removed as in filtering, but is being smeared out. The Fourier method would then seem to have the advantage that it requires less work to achieve the elimination of unwanted components in the image.

We thank Dr J. T. Finch for supplying the micrographs on which this work was based and Mrs Marion Iredale for technical assistance. The two optical instruments were constructed in the workshops of the Laboratory of Molecular Biology by Messrs. D. Hart and L. Hayward. One of us (D. J. DeR.) was aided by a grant from the American Cancer Society.

REFERENCES

Altman, J. (1966). *Appl. Opt.* **5**, 1689.
Bayer, M. E. & Remsen, C. C. (1970). *Virology*, **40**, 703.
Caspar, D. L. D. & Klug, A. (1962). *Cold Spr. Harb. Symp. Quant. Biol.* **27**, 1.
Cowley, J. M. & Moodie, A. F. (1957). *Acta Cryst.* **10**, 609.
Epstein, R. H., Bolle, A., Steinberg, C. M., Kellenberger, E., Boy de la Tour, E., Chevalley, R., Edgar, R. S., Sussman, M., Denhardt, G. H. & Lielausis, A. (1963). *Cold Spr. Harb. Symp. Quant. Biol.* **28**, 375.
Erickson, H. P. & Klug, A. (1971). *Phil. Trans. Roy. Soc.* B. **261**, 105
Favre, R., Boy de la Tour, E., Segre, N. & Kellenberger, E. (1965). *J. Ultrastruct. Res.* **13**, 318.
Finch, J. T., Klug, A. & Nermut, M. V. (1967). *J. Cell Sci.* **2**, 587.
Finch, J. T., Klug, A. & Stretton, A. O. W. (1964). *J. Mol. Biol.* **10**, 570.
Hitchborn, J. H. & Hills, G. J. (1968). *Virology*, **35**, 50.
Kellenberger, E. (1968). *Virology*, **34**, 549.
Kellenberger, E., Eiserling, F. A. & Boy de la Tour, E. (1968). *J. Ultrastruct. Res.* **21**, 335.
Kellenberger, E. & Kellenberger-Van der Kamp, C. (1970). *FEBS Letters*, **8**, 140.
Kiselev, N. A. & Klug, A. (1969). *J. Mol. Biol.* **40**, 155.
Klug, A. & Berger, J. E. (1964). *J. Mol. Biol.* **10**, 565.
Klug, A., Crick, F. H. C. & Wyckoff, H. W. (1958). *Acta Cryst.* **11**, 199.
Klug, A. & DeRosier, D. J. (1966). *Nature*, **212**, 29.
Klug, A. & Finch, J. T. (1968). *J. Mol. Biol.* **31**, 1
Klug, A., Finch, J. T., Leberman, R. & Longley, W. (1966). In *Principles of Biomolecular Organisation*, ed. G. E. W. Wolstenholme & M. O. Connor, p. 158. London: Churchill.
Laemmli, U. K. (1969). Thesis, no. 1498, University of Geneva.
Laemmli. U. K. (1970). *Nature*, **227**, 680.
Laemmli, U. K. & Kellenberger, E. (1970). *FEBS Symposium*. **21**, 159.
Laemmli, U. K., Mölbert, E., Showe, M. & Kellenberger, E. (1970). *J. Mol. Biol.* **49**, 99.
Markham, R., Hitchborn, J. H., Hills, G. J. & Frey, S. (1964). *Virology*, **22**, 342.
Moody, M. F. (1965). *Virology*, **26**, 567.
Moody, M. F. (1967). *J. Mol. Biol.* **25**, 167.
Taylor, H. D. (1921). In *The Adjustment and Testing of Telescope Objectives*, p. 15. London: T. Cooke & Sons, Ltd.
Yanagida, M., Boy de la Tour, E., Alff-Steinberger, C. & Kellenberger, E. (1970). *J. Mol. Biol.* **50**, 35.
Yanagida, M., DeRosier, D. J. & Klug, A. (1972). *J. Mol. Biol.* **65**, 489.

Acta Cryst. (1950). **3**, 478

A geometric approach to the crystal structure problem. By H. Hauptman and J. Karle, *U.S. Naval Research Laboratory, Washington, D.C., U.S.A.*

(*Received* 6 *April* 1950 *and in revised form* 17 *April* 1950)

The structure factors for crystals are defined by

$$F_{hkl}=\sum_{j=1}^{N} f_j(\theta_{hkl})\exp\left[-2\pi i(hx_j+ky_j+lz_j)\right], \quad (1)$$

where N is the number of atoms in the unit cell, θ_{hkl} is the angle of scattering for F_{hkl}, $f_j(\theta_{hkl})$ is the scattering factor for the jth atom and x_j, y_j, z_j are its co-ordinates in the unit cell. The vector F_{hkl} may be written

$$F_{hkl}=R_{hkl}\exp\left[i\phi_{hkl}\right] \quad (R_{hkl}\geqslant 0, \ -\pi<\phi_{hkl}\leqslant\pi), \quad (2)$$

where R_{hkl} is the magnitude and ϕ_{hkl} is the phase of F_{hkl}. The crystal structure problem is the problem of solving the system of equations (1) for the atomic co-ordinates x_j, y_j, z_j when, in addition to $f_j(\theta_{hkl})$, only the magnitudes R_{hkl} of the structure factors are known. The problem is completely determinate (assuming the f_j are obtained from tables) if $3(N-1)$ independent magnitudes R_{hkl} are known.

The geometric interpretation of (1) leads to a means for evaluating the atomic co-ordinates.* Equations (1) may be written

$$\sum_{j=1}^{N} f_j\exp\left[-2\pi i(hx_j+ky_j+lz_j)\right]-R_{hkl}\exp\left[i\phi_{hkl}\right]=0. \quad (3)$$

Equation (3) states that a sum of $N+1$ coplanar vectors of known magnitudes is equal to zero, i.e. the vectors form a closed polygon. In this interpretation

$$f_j\exp\left[-2\pi i(hx_j+ky_j+lz_j)\right]$$

is regarded as a vector whose magnitude is f_j and whose angle is $-2\pi(hx_j+ky_j+lz_j)$.

As an illustration of the solution by means of a geometric-mechanical construction we consider the three-atom problem in one dimension. The number of independent structure-factor magnitudes required to determine the one-dimensional problem is $N-1$. In the three-atom problem we use, for example, R_1 and R_2. The defining equations are

$$\left.\begin{aligned} R_1e^{i\phi_1}&=f_1(\theta_1)\,e^{-2\pi ix_1}+f_2(\theta_1)\,e^{-2\pi ix_2}+f_3(\theta_1)\,e^{-2\pi ix_3},\\ R_2e^{i\phi_2}&=f_1(\theta_2)\,e^{-4\pi ix_1}+f_2(\theta_2)\,e^{-4\pi ix_2}+f_3(\theta_2)\,e^{-4\pi ix_3}.\end{aligned}\right\} \quad (4)$$

Given the magnitudes R_1, R_2 and the structure factors f_1, f_2, f_3 at the two different scattering angles, there is a unique construction of the two vector sums with the restriction that the phases of the components in the second sum be twice those of the first sum. This is shown in Fig. 1 which illustrates the vector sums (4). The magnitudes of the vectors are given on the rods which are joined together by free moving links. These permit some degree of relative rotation in the plane. If the criterion is imposed that $\alpha_2=2\alpha_1$, $\beta_2=2\beta_1$, as required by equations (4), there is a unique configuration for the two vector sums. We now may read from this configuration the co-ordinates of the atoms along the line as follows:

$$x_1=0 \text{ (arbitrarily set)}, \quad x_2=-\alpha_1/2\pi, \quad x_3=-(\alpha_1+\beta_1)/2\pi.$$

This represents the solution to the problem of three atoms in one dimension. It should be noted that the phases of R_1 and R_2 may be immediately read if desired from the geometric construction. The point to be emphasized is that the solution is obtained by construction from measured *magnitudes* only and tabulated values of the atomic scattering factors. The coupling restrictions on the angles α and β set by (4) determine the unique configuration. The values of α and β are not *known* previous to the construction.

This geometric procedure generalizes in a straightforward fashion. In order to determine the three-dimensional co-ordinates x_j, y_j, z_j, $3(N-1)$ independent experimental magnitudes, R_{hkl}, are required. There are N vectors corresponding to the N atoms in each of the $3(N-1)$ vector sums whose magnitudes are the atom scattering factors for the particular scattering angle. The coupling restrictions on the angles in the geometrical linkages determine the co-ordinates of the atoms in the unit cell. The angles α, β, ... are, in three dimensions, linear functions of the three atomic co-ordinates.

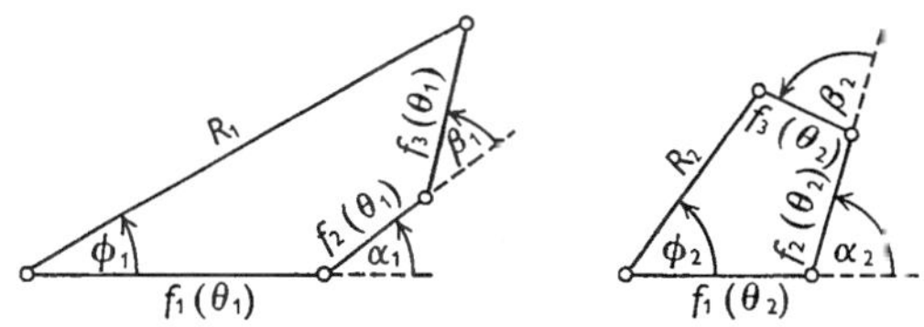

Fig. 1. Illustration of the vector sums of equations (4).

The practical aspects of the geometrical interpretation are apparent. The co-ordinates of the unit cell are obtained immediately by performing a simultaneous set of $3(N-1)$ vector summations subject to the coupling restrictions on the linkage angles (α_1, β_1, ...) given by equation (1). Since alternating currents combine in the same manner as vectors, an electrical model may be found for this construction.

The following pertinent remarks may be made:

(*a*) The geometric construction permits the immediate determination of the co-ordinates of the unit cell from measured magnitudes of the crystal structure factors and tabulated atomic scattering factors.

(*b*) Since more magnitudes are measured than are required for a solution, these afford a basis for adjusting the internal consistency of the data.

(*c*) The phases of the crystal structure factors are obtained immediately from the geometric construction. From this point of view, however, these phases are not required.

(*d*) No symmetry relations are required in these considerations. The formulation is valid for a triclinic crystal.

(*e*) If only relative intensities are measured an additional magnitude is required to determine the solution.

(*f*) The concept of vector summations with coupling restrictions on the angles joining the vectors lends itself to automatic computing techniques for obtaining the atomic co-ordinates.

* A similar interpretation of (1) was presented by Prof. J. M. Robertson at the Computer and Phase Conference, Pennsylvania State College, 7 April 1950.

Acta Cryst. (1950). **3**, 181

The Phases and Magnitudes of the Structure Factors*

By J. Karle and H. Hauptman

U.S. Naval Research Laboratory, Washington, D.C., U.S.A.

(*Received* 19 *May* 1949 *and in revised form* 3 *October* 1949)

Inequalities among the coefficients of a Fourier series representing the electron density in a crystal are derived on the basis that the series represents a positive function. The procedure is formulated for obtaining all inequalities which are based on this characteristic of positiveness, and some of the simpler ones are listed. No symmetry properties are required for deriving the inequalities, but they may be readily introduced into the inequality relationships. It is indicated that application of the linear transformation theory on hermitian forms may prove fruitful in future investigation.

An extensive and fundamental system of inequalities exists among the coefficients of a Fourier series which represents a positive function. The structure factors are the coefficients in the Fourier-series representation of the positive electron density distribution function for crystals. It is the purpose of this paper to derive the fundamental system of inequalities among the structure factors and express them in a useful form.†

By making use of the symmetry characteristics which are found in crystals and the Schwarz inequality, Harker & Kasper (1948) have derived certain useful inequalities among the structure factors. An extension of this work has been made by Gillis (1948), who has applied some additional inequalities of formal mathematical analysis. In both cases it was necessary to resort to symmetry characteristics and certain standard inequalities in analysis. Implicit in their investigations though was the assumption that their distribution function was positive. In recent work on the structure of atoms‡ we have found that the electron distribution about atoms is accurately determined by a limited amount of experimental data since the distribution function is positive. This characteristic of positiveness will be seen to be alone sufficient to yield a system of inequalities which limits the phases and magnitudes of the structure factors for crystals.

Symmetry considerations are not basic to the development of the theory. However, it will be shown how symmetry relations may be introduced into the final results.

Theory

The Fourier coefficient, F_{hkl}, is defined in terms of the electron density distribution function for a crystal, $\rho(x,y,z)$, as follows:

$$F_{hkl}=V\int_0^1\int_0^1\int_0^1 \rho(x,y,z) \times \exp[-2\pi i(hx+ky+lz)]\,dx\,dy\,dz, \quad (1)$$

where V is the volume of the unit cell. We construct from expression (1) useful hermitian forms which will be shown to be non-negative. The forms obtained from (1) are

$$\sum_{hkl}\sum_{h'k'l'}{}_1^m X_{hkl}\overline{X}_{h'k'l'}F_{h-h',\,k-k',\,l-l'}$$
$$=V\int_0^1\int_0^1\int_0^1 \rho(x,y,z)\sum_{hkl}\sum_{h'k'l'}{}_1^m X_{hkl}\overline{X}_{h'k'l'}$$
$$\times\exp\{-2\pi i[(h-h')\,x+(k-k')\,y+(l-l')\,z]\}\,dx\,dy\,dz,$$
$$(m=1,2,\ldots), \quad (2)$$

* Presented at the meeting of the Crystallographic Society of America, Ann Arbor, Michigan, 7 April 1949.

† A system of inequalities for the one-dimensional case has been found by Achyeser & Krein (1934) in their studies of the one-dimensional trigonometric moment problem.

‡ To be published (*Phys. Rev.*, February 1950). Presented at ASXRED Meeting, Columbus, Ohio, December 1948.

where each summation represents the triple sum over each index, X_{hkl} is an independent variable, and $\overline{X}_{hkl}$ is the complex conjugate of X_{hkl}.

Since $\rho(x,y,z)$ is real, $F_{hkl}=\overline{F_{\bar{h}\bar{k}\bar{l}}}$. The sextuple sum on the right side of equation (2) may be expressed as the product of a triple sum and its complex conjugate. The right side may therefore be written

$$V\int_0^1\int_0^1\int_0^1 \rho(x,y,z) \times\left|\sum_{hkl\atop 1}^{m} X_{hkl}\exp[-2\pi i(hx+ky+lz)]\right|^2 dx\,dy\,dz. \quad (3)$$

It is seen that (3) is non-negative since $\rho(x,y,z)$ is a non-negative function. We therefore have from the left side of (2) the non-negative hermitian forms*

$$\sum_{hkl\atop 1}^{m}\sum_{h'k'l'} X_{hkl}\overline{X}_{h'k'l'}F_{h-h',k-k',l-l'}\geqslant 0 \quad (m=1,2,\ldots). \quad (4)$$

Not only are the inequalities (4) a necessary consequence of the positiveness of $\rho(x,y,z)$, but it may also be shown that these inequalities (4) are sufficient to insure that $\rho(x,y,z)$ be non-negative. This is of importance because it implies that any set of inequalities which is derived from $\rho(x,y,z)$ being non-negative is contained in the set which we shall derive from (4). To prove that (4) is sufficient to determine a non-negative Fourier series we assume that (4) is true and shall show that as a consequence $\rho(x,y,z)$ is non-negative.

The proof consists in showing that the function

$$\sum_{hkl\atop -\infty}^{\infty}{}^{\dagger} F_{hkl}X^hY^kZ^l \quad (5)$$

is non-negative in every region $|X|\leqslant 1-\delta$, $|Y|\leqslant 1-\delta$, $|Z|\leqslant 1-\delta$, where $0<\delta<1$. The symbol $\Sigma^\dagger$ means that a negative power of any variable which appears in the sum (5) must be replaced by the conjugate complex of the variable raised to the absolute value of the power, e.g. X^{-h}, $h>0$, is to be replaced by $\overline{X}^h$. Since the density function, $\rho(x,y,z)$, is the limit of (5) as $X\to e^{2\pi ix}$, $Y\to e^{2\pi iy}$ and $Z\to e^{2\pi iz}$, it may be concluded that $\rho(x,y,z)$ is also non-negative.

Since the set of numbers $|F_{hkl}|$ is bounded, the series (5) converges uniformly and absolutely in any region $|X|\leqslant 1-\delta$, $|Y|\leqslant 1-\delta$, $|Z|\leqslant 1-\delta$, where $0<\delta<1$. It may be shown that (5) is equal to

$$\sum_{hkl\atop 0}^{\infty}\sum_{h'k'l'} F_{h-h',k-k',l-l'}X^h\overline{X}^{h'}Y^k\overline{Y}^{k'}Z^l\overline{Z}^{l'} \times(1-X\overline{X})(1-Y\overline{Y})(1-Z\overline{Z}). \quad (6)$$

To demonstrate this we take a typical term of (5), $F_{m,n,-p}X^mY^n\overline{Z}^p$, where $m>0$, $n>0$, $p>0$, and show that it is contained once in (6). The indices of those terms of (6) which contain $F_{m,n,-p}$ as a factor satisfy

$$h-h'=m,\quad k-k'=n,\quad l-l'=-p. \quad (7)$$

* Simple, but relatively weak inequalities on the coefficients may be obtained immediately from (4) by substituting arbitrary complex numbers for the independent variables X_{hkl}.

These relations may be written

$$\begin{aligned} &h=h'+m,\quad k=k'+n,\quad l=l'-p,\\ \text{or}\quad &h'=h-m,\quad k'=k-n,\quad l'=l+p.\end{aligned} \quad (8)$$

In order to obtain all those terms of (6) having $F_{m,n,-p}$ as a factor, it is important to note that only three of the relations (8) may be validly inserted into (6). Otherwise, any of the remaining relations if substituted into (6) would allow certain of the indices to take on negative values since $m,n,p>0$. The relations (8) which are therefore substituted into (6) are

$$h=h'+m,\quad k=k'+n,\quad l'=l+p. \quad (9)$$

We obtain

$$\sum_{h'k'l\atop 0}^{\infty} F_{m,n,-p}X^{h'+m}\overline{X}^{h'}Y^{k'+n}\overline{Y}^{k'}Z^l\overline{Z}^{l+p} \times(1-X\overline{X})(1-Y\overline{Y})(1-Z\overline{Z}), \quad (10)$$

which may be written

$$F_{m,n,-p}X^mY^n\overline{Z}^p(1-X\overline{X})(1-Y\overline{Y})(1-Z\overline{Z}) \times\sum_{h'=0}^{\infty}(X\overline{X})^{h'}\sum_{k'=0}^{\infty}(Y\overline{Y})^{k'}\sum_{l=0}^{\infty}(Z\overline{Z})^l=F_{m,n,-p}X^mY^n\overline{Z}^p, \quad (11)$$

the desired result. If we set

$$\zeta_{hkl}=X^hY^kZ^l,\quad \overline{\zeta}_{h'k'l'}=\overline{X}^{h'}\overline{Y}^{k'}\overline{Z}^{l'}, \quad (12)$$

expression (6) becomes

$$\begin{aligned}&\sum_{hkl\atop 0}^{\infty}\sum_{h'k'l'} F_{h-h',k-k',l-l'}\zeta_{hkl}\overline{\zeta}_{h'k'l'}(1-X\overline{X})(1-Y\overline{Y})(1-Z\overline{Z})\\ &=(1-X\overline{X})(1-Y\overline{Y})(1-Z\overline{Z})\times\lim_{m\to\infty}\sum_{hkl\atop 0}^{m}\sum_{h'k'l'} F_{h-h',k-k',l-l'}\zeta_{hkl}\overline{\zeta}_{h'k'l'}\geqslant 0,\end{aligned} \quad (13)$$

in view of (4) and $X\overline{X}<1$, $Y\overline{Y}<1$, $Z\overline{Z}<1$.

It is assumed that the first partial derivatives of $\rho(x,y,z)$ exist and are continuous at every point. This condition is sufficient to insure that the series

$$\rho(x,y,z)=\sum_{hkl\atop -\infty}^{\infty}F_{hkl}\exp[2\pi i(hx+ky+lz)] \quad (14)$$

converges. In other words the series (5) converges when $X=e^{2\pi ix}$, $Y=e^{2\pi iy}$, $Z=e^{2\pi iz}$, and defines a function of X, Y, Z which is continuous in the closed region $|X|\leqslant 1$, $|Y|\leqslant 1$, $|Z|\leqslant 1$.* It therefore follows from (13) that

$$\rho(x,y,z)=\lim_{\substack{X\to e^{2\pi ix}\\ Y\to e^{2\pi iy}\\ Z\to e^{2\pi iz}}}\sum_{hkl\atop -\infty}^{\infty}{}^{\dagger}F_{hkl}X^hY^kZ^l\geqslant 0. \quad (15)$$

This completes the proof that (4) implies that $\rho(x,y,z)$ is non-negative.

* This assumption, used to complete the proof with mathematical rigor, is no restriction in the physical sense since the infinite sums may be replaced by finite sums each having a sufficient number of terms to insure that the resulting $\rho(x,y,z)$ differs from the true one by an amount far less than that due to the errors arising from experiment. The assumption is then certainly justified.

The necessary and sufficient condition for the hermitian forms (4) to be non-negative is that a system of determinants involving the F_{hkl}'s be non-negative. Before discussing these determinants for the three-dimensional Fourier series there is an advantage in developing in detail the theory of the determinants corresponding to the one-dimensional positive Fourier series. The three-dimensional theory follows readily from the one-dimensional theory, since both are based on the theory of hermitian forms. The major effect of increasing the number of dimensions is to complicate the notation.

The one-dimensional problem

The hermitian forms associated with the one-dimensional problem may be written*

$$\sum_{h}\sum_{h'}{}_{1}^{m} X_h \bar{X}_{h'} F_{h-h'} \geqslant 0 \quad (m=1,2,\ldots), \tag{16}$$

where F_h is a coefficient in a one-dimensional positive Fourier series. The necessary and sufficient condition for (16) to hold is that the following determinants be non-negative (Dickson, 1930, pp. 81–8):†

$$\begin{vmatrix} F_0 & F_{-1} & F_{-2} & \cdots & F_{-n} \\ F_1 & F_0 & F_{-1} & \cdots & F_{-(n-1)} \\ \cdots & \cdots & \cdots & \cdots & \cdots \\ F_n & F_{n-1} & F_{n-2} & \cdots & F_0 \end{vmatrix} \geqslant 0 \quad (n=0,1,2,\ldots). \tag{17}$$

Since any rearrangement may be made on the subscripts of the independent variables of the hermitian forms (16), and since any of these variables may be set equal to zero, we may write the non-negative determinants

$$D=\begin{vmatrix} F_0 & F_{-\epsilon_1} & F_{-\epsilon_2} & \cdots & F_{-\epsilon_n} \\ F_{\epsilon_1} & F_{\epsilon_1-\epsilon_1} & F_{\epsilon_1-\epsilon_2} & \cdots & F_{\epsilon_1-\epsilon_n} \\ \cdots & \cdots & \cdots & \cdots & \cdots \\ F_{\epsilon_n} & F_{\epsilon_n-\epsilon_1} & F_{\epsilon_n-\epsilon_2} & \cdots & F_{\epsilon_n-\epsilon_n} \end{vmatrix} \geqslant 0 \quad (n=0,1,2\ldots), \tag{18}$$

where the subscripts ϵ_i, $i=1,2,\ldots,n$ are different from each other and from zero, but otherwise may take on arbitrary positive or negative values. The distinction between (17) and (18) is that in (17) the subscripts in the first column consist of all the integers from 0 to n arranged in increasing order, whereas in (18), although the subscripts in the first column also start with 0 and are distinct, they are arbitrary otherwise. Evidently (18) includes (17). It is seen that in the determinants (18) the subscripts in the first row are the same as those in the first column, but with opposite signs, and are arranged in the same order. The subscript in the ith row and jth column is obtained by adding the subscript in the ith row and first column to the subscript in the first row and jth column ($\epsilon_{ij}=\epsilon_{i1}+\epsilon_{1j}$).

The determinants (18) may be used to obtain a bound on F_{ϵ_n}. Owing to the general form of (18), this bound may be expressed in terms of some or all of the coefficients whose subscripts precede ϵ_n in numerical order or, if desired, in terms of coefficients having higher subscripts in addition to the previous ones.

In order to obtain the bound on F_{ϵ_n} we apply to (18) the expansion theorem for determinants (Muir & Metzler, 1930, pp. 132–5):

$$\begin{vmatrix} F_{\epsilon_1-\epsilon_1} & F_{\epsilon_1-\epsilon_2} & \cdots & F_{\epsilon_1-\epsilon_{n-1}} \\ F_{\epsilon_2-\epsilon_1} & F_{\epsilon_2-\epsilon_2} & \cdots & F_{\epsilon_2-\epsilon_{n-1}} \\ \cdots & \cdots & \cdots & \cdots \\ F_{\epsilon_{n-1}-\epsilon_1} & F_{\epsilon_{n-1}-\epsilon_2} & \cdots & F_{\epsilon_{n-1}-\epsilon_{n-1}} \end{vmatrix} \times D = \begin{vmatrix} F_0 & F_{-\epsilon_1} & \cdots & F_{-\epsilon_{n-1}} \\ F_{\epsilon_1} & F_{\epsilon_1-\epsilon_1} & \cdots & F_{\epsilon_1-\epsilon_{n-1}} \\ \cdots & \cdots & \cdots & \cdots \\ F_{\epsilon_{n-1}} & F_{\epsilon_{n-1}-\epsilon_1} & \cdots & F_{\epsilon_{n-1}-\epsilon_{n-1}} \end{vmatrix} \times \begin{vmatrix} F_{\epsilon_1-\epsilon_1} & F_{\epsilon_1-\epsilon_2} & \cdots & F_{\epsilon_1-\epsilon_n} \\ F_{\epsilon_2-\epsilon_1} & F_{\epsilon_2-\epsilon_2} & \cdots & F_{\epsilon_2-\epsilon_n} \\ \cdots & \cdots & \cdots & \cdots \\ F_{\epsilon_n-\epsilon_1} & F_{\epsilon_n-\epsilon_2} & \cdots & F_{\epsilon_n-\epsilon_n} \end{vmatrix}$$

$$-\begin{vmatrix} F_{\epsilon_1} & F_{\epsilon_1-\epsilon_1} & \cdots & F_{\epsilon_1-\epsilon_{n-1}} \\ F_{\epsilon_2} & F_{\epsilon_2-\epsilon_1} & \cdots & F_{\epsilon_2-\epsilon_{n-1}} \\ \cdots & \cdots & \cdots & \cdots \\ F_{\epsilon_n} & F_{\epsilon_n-\epsilon_1} & \cdots & F_{\epsilon_n-\epsilon_{n-1}} \end{vmatrix} \times \begin{vmatrix} F_{-\epsilon_1} & F_{\epsilon_1-\epsilon_1} & \cdots & F_{\epsilon_{n-1}-\epsilon_1} \\ F_{-\epsilon_2} & F_{\epsilon_1-\epsilon_2} & \cdots & F_{\epsilon_{n-1}-\epsilon_2} \\ \cdots & \cdots & \cdots & \cdots \\ F_{-\epsilon_n} & F_{\epsilon_1-\epsilon_n} & \cdots & F_{\epsilon_{n-1}-\epsilon_n} \end{vmatrix}. \tag{19}$$

The left side of (19) is the product of D and the determinant formed by omitting the first and last rows and columns of D. The first product on the right side of (19) is formed by multiplying the determinant which arises from omitting the last row and column in D by the determinant which arises from omitting the first row and column in D. The second product on the right side of (19) is formed by multiplying the determinant which arises from omitting the first row and last column in D by the determinant which arises from omitting the first column and last row in D. The rows and columns in the last determinant of (19) have been interchanged without changing its value. The determinants on the left side of (19), as well as the first two on the right side, are non-negative in view of (18). The last two determinants in (19) are complex conjugates since corresponding elements are complex conjugates.

If the determinant on the left side of (19) is called Δ and the determinants on the right side called in order $\Delta_1, \Delta_2, d, \bar{d}$, we get

$$D\Delta=\Delta_1\Delta_2-d\bar{d}. \tag{20}$$

Since D and Δ are non-negative,

$$d\bar{d}\leqslant \Delta_1\Delta_2. \tag{21}$$

By expanding d in terms of the minor of F_{ϵ_n}, we have

$$|d|=|F_{\epsilon_n}\Delta-\Delta'|, \tag{22}$$

* The hermitian forms in (16) may be considered as a subset of the three-dimensional set since the two subscripts k, l may be included in the notation if set equal to zero. It is apparent that there are three such subsets depending upon which two of the subscripts are set equal to zero. There are also three subsets for the two-dimensional problem depending upon which one of the three subscripts is set equal to zero.

† For fixed m, n ranges from zero to $m-1$.

where

$$\Delta' = (-1)^n \begin{vmatrix} F_{\epsilon_1} & F_{\epsilon_1-\epsilon_1} & F_{\epsilon_1-\epsilon_2} & \cdots & F_{\epsilon_1-\epsilon_{n-1}} \\ F_{\epsilon_2} & F_{\epsilon_2-\epsilon_1} & F_{\epsilon_2-\epsilon_2} & \cdots & F_{\epsilon_2-\epsilon_{n-1}} \\ \cdots & \cdots & \cdots & \cdots & \cdots \\ F_{\epsilon_{n-1}} & F_{\epsilon_{n-1}-\epsilon_1} & F_{\epsilon_{n-1}-\epsilon_2} & \cdots & F_{\epsilon_{n-1}-\epsilon_{n-1}} \\ 0 & F_{\epsilon_n-\epsilon_1} & F_{\epsilon_n-\epsilon_2} & \cdots & F_{\epsilon_n-\epsilon_{n-1}} \end{vmatrix}. \tag{23}$$

Since Δ_1 and Δ_2 are non-negative, (21) and (22) imply

$$|F_{\epsilon_n}\Delta - \Delta'| \leqslant \Delta_1^{\frac{1}{2}}\Delta_2^{\frac{1}{2}}. \tag{24}$$

Finally, since Δ is non-negative, this becomes

$$|F_{\epsilon_n} - \delta| \leqslant r, \tag{25}$$

where $\delta = \Delta'/\Delta$ and $r = \Delta_1^{\frac{1}{2}}\Delta_2^{\frac{1}{2}}/\Delta$. *The coefficient F_{ϵ_n} is bounded in terms of others having higher and lower subscripts, in general, within a circle in the complex plane whose center is δ and whose radius is r.*

To complete the proof that (25), $F_0 \geqslant 0$ and $|F_{\epsilon_1}| \leqslant F_0$ include all inequalities based on the positiveness of the distribution function, it must still be shown that, conversely, the system (25), $F_0 \geqslant 0$ and $|F_{\epsilon_1}| \leqslant F_0$ imply the system (18). Assume then that $F_0 \geqslant 0$, $|F_{\epsilon_1}| \leqslant F_0$, and that (25) holds. Our proof is by induction. Since $F_0 > 0$ and $|F_{\epsilon_1}| \leqslant F_0$, (18) is valid when $n=0$ or 1. We make the induction hypothesis that (18) is valid for all values of n less than some positive integer p. In (25) take $n=p$. Then Δ, Δ_1 and Δ_2 are all non-negative and the steps (21)–(25) are all reversible. Therefore (25) implies (21), and (20) becomes

$$D\Delta = \Delta_1\Delta_2 - d\bar{d} \geqslant 0. \tag{26}$$

Hence

$$D \geqslant 0, \tag{27}$$

and (18) is valid for $n=p$. Since p is arbitrary this completes the induction.

The three-dimensional problem

The hermitian forms (4) constitute the basis for formulating the relationships among the Fourier coefficients in the three-dimensional case. In complete analogy with (17), the necessary and sufficient condition for (4) to hold is that an infinite set of determinants be non-negative (Dickson, 1930, pp. 81–8). Their construction will now be described.

For any choice of the value of m in (4) the corresponding determinant is of the order m^3. The indices now consist of number triples, each number being chosen from the set $0, 1, \ldots, m-1$. In the first column, the indices consist of all m^3 triples arranged in 'dictionary' order, i.e. the index α, β, γ precedes α', β', γ' if $\alpha < \alpha'$ or if $\alpha = \alpha'$, $\beta < \beta'$ or if $\alpha = \alpha'$, $\beta = \beta'$, $\gamma < \gamma'$. The indices in the first row are the same and in the same order as those in the first column, but of opposite signs. The index in the ith row and the jth column, as in the one-dimensional case, is obtained by adding the index of the ith row and first column to that of the first row and jth column. As an example we write the determinant when $m=2$:

$$\begin{vmatrix} F_{000} & F_{00\bar{1}} & F_{0\bar{1}0} & F_{0\bar{1}\bar{1}} & F_{\bar{1}00} & F_{\bar{1}0\bar{1}} & F_{\bar{1}\bar{1}0} & F_{\bar{1}\bar{1}\bar{1}} \\ F_{001} & F_{000} & F_{0\bar{1}1} & F_{0\bar{1}0} & \cdots & \cdots & \cdots & \cdots \\ F_{010} & F_{01\bar{1}} & F_{000} & F_{00\bar{1}} & \cdots & \cdots & \cdots & \cdots \\ \cdots & \cdots & \cdots & \cdots & \cdots & \cdots & \cdots & \cdots \\ F_{111} & F_{110} & F_{101} & F_{100} & \cdots & \cdots & \cdots & F_{000} \end{vmatrix} \geqslant 0. \tag{28}$$

This is the three-dimensional analogue of (17) with $n=1$. As before, any rearrangement may be made on the subscripts of the independent variables of the hermitian forms (4), and any of these variables may be set equal to zero, so that we obtain the three-dimensional analogue of (18):

$$\begin{vmatrix} F_{000} & F_{-\epsilon_1,-\eta_1,-\zeta_1} & \cdots & F_{-\epsilon_n,-\eta_n,-\zeta_n} \\ F_{\epsilon_1,\eta_1,\zeta_1} & F_{\epsilon_1-\epsilon_1,\eta_1-\eta_1,\zeta_1-\zeta_1} & \cdots & F_{\epsilon_1-\epsilon_n,\eta_1-\eta_n,\zeta_1-\zeta_n} \\ F_{\epsilon_2,\eta_2,\zeta_2} & F_{\epsilon_2-\epsilon_1,\eta_2-\eta_1,\zeta_2-\zeta_1} & \cdots & F_{\epsilon_2-\epsilon_n,\eta_2-\eta_n,\zeta_2-\zeta_n} \\ \cdots & \cdots & \cdots & \cdots \\ F_{\epsilon_n,\eta_n,\zeta_n} & F_{\epsilon_n-\epsilon_1,\eta_n-\eta_1,\zeta_n-\zeta_1} & \cdots & F_{\epsilon_n-\epsilon_n,\eta_n-\eta_n,\zeta_n-\zeta_n} \end{vmatrix} \geqslant 0. \tag{29}$$

The indices in the first column start with 0, 0, 0 and are distinct, but are arbitrary otherwise. The indices in the first row are the same as those in the first column, but with opposite signs, and are in the same order. The index in the ith row and jth column is determined by

$$\epsilon_{ij} = \epsilon_{i1} + \epsilon_{1j}, \quad \eta_{ij} = \eta_{i1} + \eta_{1j}, \quad \zeta_{ij} = \zeta_{i1} + \zeta_{1j}.$$

By an analysis similar to that given by equations (18)–(25) we obtain a bound for $F_{\epsilon_n,\eta_n,\zeta_n}$,*

$$|F_{\epsilon_n,\eta_n,\zeta_n} - \delta| \leqslant r, \tag{30}$$

where

$$\delta = \frac{\Delta'}{\Delta}, \quad r = \frac{\Delta_1^{\frac{1}{2}}\Delta_2^{\frac{1}{2}}}{\Delta}, \tag{31}$$

and $\Delta_1, \Delta_2, \Delta$ and Δ' are the three-dimensional generalizations of the corresponding one-dimensional determinants in equations (19) and (23). They are formed by adding the additional subscripts η and ζ as is indicated in (29).

Just as in the one-dimensional case, not only are (30), $F_{000} \geqslant 0$, and $|F_{\epsilon_1\eta_1\zeta_1}| \leqslant F_{000}$ necessary consequences of (29), but (30), $F_{000} \geqslant 0$, and $|F_{\epsilon_1\eta_1\zeta_1}| \leqslant F_{000}$ are also sufficient to insure the validity of (29).

It was pointed out in discussion with Dr A. L. Patterson that the inequalities will not in all cases bound the phases of the coefficients, even if the magnitudes of all the structure factors are known. This is easily seen to be true by considering the example in which

$$\sum_{hkl}^{\infty}{}' |F_{hkl}| < 2F_{000}.$$

Here the distribution function is positive regardless of the phases of the F_{hkl}. A situation as unfavourable as

* This general type of inequality applies to all non-negative Fourier series and therefore may find application to syntheses along and projections upon lines and planes.

this is not to be generally expected for electron densities in crystals, because of the high electron densities in the vicinity of the atomic nuclei.

Discussion

The foregoing theory may find application in the direct use of the derived inequalities. In addition, it will be indicated below that the non-negative hermitian forms which represent the Fourier series constitute a working basis from which many new representations of a crystal structure may be developed. The well-known Patterson representation is such an example.

Inequalities of successively increasing complexity may be obtained from (29) or (30). We write down the first four,*†

$$F_{000} \geqslant 0, \qquad (32)$$

$$|F_{hkl}| \leqslant F_{000}, \qquad (33)$$

$$\left| F_{h_1+h_2,\,k_1+k_2,\,l_1+l_2} - \frac{F_{h_1k_1l_1}F_{h_2k_2l_2}}{F_{000}} \right| \leqslant \frac{\begin{vmatrix} F_{000} & F_{\bar{h}_1\bar{k}_1\bar{l}_1} \\ F_{h_1k_1l_1} & F_{000} \end{vmatrix}^{\frac{1}{2}} \begin{vmatrix} F_{000} & F_{\bar{h}_2\bar{k}_2\bar{l}_2} \\ F_{h_2k_2l_2} & F_{000} \end{vmatrix}^{\frac{1}{2}}}{F_{000}}, \qquad (34)$$

$$\left| F_{h_1+h_2+h_3,\,k_1+k_2+k_3,\,l_1+l_2+l_3} + \frac{\begin{vmatrix} F_{h_1k_1l_1} & F_{000} & F_{\bar{h}_2\bar{k}_2\bar{l}_2} \\ F_{h_1+h_2,\,k_1+k_2,\,l_1+l_2} & F_{h_2k_2l_2} & F_{000} \\ 0 & F_{h_2+h_3,\,k_2+k_3,\,l_2+l_3} & F_{h_3k_3l_3} \end{vmatrix}}{\begin{vmatrix} F_{000} & F_{\bar{h}_2\bar{k}_2\bar{l}_2} \\ F_{h_2k_2l_2} & F_{000} \end{vmatrix}} \right|$$

$$\leqslant \frac{\begin{vmatrix} F_{000} & F_{\bar{h}_1\bar{k}_1\bar{l}_1} & F_{\bar{h}_1+\bar{h}_2,\,\bar{k}_1+\bar{k}_2,\,\bar{l}_1+\bar{l}_2} \\ F_{h_1k_1l_1} & F_{000} & F_{\bar{h}_2\bar{k}_2\bar{l}_2} \\ F_{h_1+h_2,\,k_1+k_2,\,l_1+l_2} & F_{h_2k_2l_2} & F_{000} \end{vmatrix}^{\frac{1}{2}} \begin{vmatrix} F_{000} & F_{\bar{h}_2\bar{k}_2\bar{l}_2} & F_{\bar{h}_2+\bar{h}_3,\,\bar{k}_2+\bar{k}_3,\,\bar{l}_2+\bar{l}_3} \\ F_{h_2k_2l_2} & F_{000} & F_{\bar{h}_3\bar{k}_3\bar{l}_3} \\ F_{h_2+h_3,\,k_2+k_3,\,l_2+l_3} & F_{h_3k_3l_3} & F_{000} \end{vmatrix}^{\frac{1}{2}}}{\begin{vmatrix} F_{000} & F_{\bar{h}_2\bar{k}_2\bar{l}_2} \\ F_{h_2k_2l_2} & F_{000} \end{vmatrix}}. \qquad (35)$$

Since the h, k, l's are any integers, positive, negative or zero, it is seen that any of the coefficients may be bounded in terms of sets of other coefficients, and in many different ways. In the general type of inequality of which the last two are examples, the coefficient which is being bounded lies in a circle in the complex plane whose center is given by the second term in the left member and whose radius is the right member.

In addition to limiting the phases of those coefficients whose magnitudes may be obtained from experiment the inequalities also bound the coefficients whose magnitudes are not known. If the inequalities are used to continue the series, the resulting series will be positive only if a new coefficient satisfies a set of inequalities involving all the coefficients whose magnitudes and phases have already been fixed.

The inequalities (4) have been proven to be sufficient to insure that $\rho(x,y,z)$ be non-negative. The set of inequalities (30), (32) and (33) derived from (4), of which (34) and (35) are examples, is necessary and sufficient to insure the validity of (4). Therefore all systems of inequalities based on the non-negative character of $\rho(x,y,z)$ must be included in the set (30), (32) and (33).

It is apparent that symmetry considerations have not been introduced either for the purpose of representing a non-negative Fourier series or for the purpose of deriving the inequalities. Symmetry relationships may be directly introduced into the inequalities. As an example, if the origin is assumed to be a center of symmetry, $F_{hkl} = F_{\bar{h}\bar{k}\bar{l}}$. Since, in general, $\bar{F}_{hkl} = F_{\bar{h}\bar{k}\bar{l}}$, all coefficients in (30) are real. In this case the inequalities (33), (34) and (35) limit the coefficients to intervals on the real axis.* Inequality (34), for example, then becomes

$$\left| F_{h_1+h_2,\,k_1+k_2,\,l_1+l_2} - \frac{F_{h_1k_1l_1}F_{h_2k_2l_2}}{F_{000}} \right| \leqslant \frac{(F^2_{000} - F^2_{h_1k_1l_1})^{\frac{1}{2}}(F^2_{000} - F^2_{h_2k_2l_2})^{\frac{1}{2}}}{F_{000}}. \qquad (36)$$

* The h, k, l are simply related to the ϵ, η, ζ in (30) and may be arbitrary positive or negative integers or zero. For example, in (35)

$$h_i = \epsilon_i - \epsilon_{i-1}, \quad k_i = \eta_i - \eta_{i-1}, \quad l_i = \zeta_i - \zeta_{i-1},$$
$$i = 1, 2, 3, \quad \epsilon_0 = \eta_0 = \zeta_0 = 0.$$

Certain choices of h, k, l should be avoided since they may lead to either (a) a zero denominator, or (b) a trivial identity which would arise if the triple ϵ_i, η_i, ζ_i were equal to the triple ϵ_j, η_j, ζ_j or 0, 0, 0 in (29), making two columns identical in (29), or (c) a bound on F_{hkl} in terms of itself. These choices may be avoided in (35), for example, by means of the following rule: If the numbers α, β, γ are chosen in any way from the sets 0, 1 or 1, 2 except that the combination $\alpha = 2$, $\beta = 1$, $\gamma = 2$ is excluded, then the triples

$$\alpha h_1 + \beta h_2 + \gamma h_3, \quad \alpha k_1 + \beta k_2 + \gamma k_3, \quad \alpha l_1 + \beta l_2 + \gamma l_3$$

are to be different from 0, 0, 0.

† These inequalities may be strengthened by dividing the Fourier coefficients by a function of the atomic scattering factors which effectively concentrates the scattering material about the atomic co-ordinates (see Harker & Kasper, 1948).

* In the case of a center of symmetry, some of the determinants which arise may be factored. See Achyeser & Krein (1935).

For the special case $h_1=h_2=h$, $k_1=k_2=k$, $l_1=l_2=l$, this reduces to

$$\left|F_{2h,2k,2l}-\frac{F^2_{hkl}}{F_{000}}\right|\leqslant\frac{F^2_{000}-F^2_{hkl}}{F_{000}}, \quad (37)$$

which implies

$$\frac{F^2_{hkl}}{F_{000}}-F_{2h,2k,2l}\leqslant F_{000}-\frac{F^2_{hkl}}{F_{000}}, \quad (38)$$

or

$$\left(\frac{F_{hkl}}{F_{000}}\right)^2\leqslant\frac{1}{2}+\frac{1}{2}\frac{F_{2h,2k,2l}}{F_{000}}. \quad (39)$$

This is one of the inequalities derived by Harker & Kasper (1948, p. 72).

Again, if the y axis is a twofold rotation axis then $F_{hkl}=F_{\bar{h}k\bar{l}}$, and (34) may be written

$$\left|F_{h_1\pm h_2,k_1+k_2,l_1\pm l_2}-\frac{F_{h_1k_1l_1}F_{h_2k_2l_2}}{F_{000}}\right|$$
$$\leqslant\frac{(F^2_{000}-|F_{h_1k_1l_1}|^2)^{\frac{1}{2}}(F^2_{000}-|F_{h_2k_2l_2}|^2)^{\frac{1}{2}}}{F_{000}}. \quad (40)$$

In other words, both $F_{h_1+h_2,k_1+k_2,l_1+l_2}$ and $F_{h_1-h_2,k_1+k_2,l_1-l_2}$ lie inside the circle with center at $F_{h_1k_1l_1}F_{h_2k_2l_2}/F_{000}$ and radius $(F^2_{000}-|F_{h_1k_1l_1}|^2)^{\frac{1}{2}}(F^2_{000}-|F_{h_2k_2l_2}|^2)^{\frac{1}{2}}/F_{000}$, so that the distance between them is not greater than the diameter of the circle

$$|F_{h_1+h_2,k_1+k_2,l_1+l_2}-F_{h_1-h_2,k_1+k_2,l_1-l_2}|$$
$$\leqslant\frac{2(F^2_{000}-|F_{h_1k_1l_1}|^2)^{\frac{1}{2}}(F^2_{000}-|F_{h_2k_2l_2}|^2)^{\frac{1}{2}}}{F_{000}}. \quad (41)$$

Taking the special case $h_1=h_2=h$, $k_1=-k_2=k$, $l_1=l_2=l$, and noting that $|F_{hkl}|=|F_{\bar{h}\bar{k}\bar{l}}|$, this reduces to

$$|F_{2h,0,2l}-F_{000}|\leqslant\frac{2(F^2_{000}-|F_{hkl}|^2)}{F_{000}}, \quad (42)$$

whence

$$F_{000}-F_{2h,0,2l}\leqslant\frac{2(F^2_{000}-|F_{hkl}|^2)}{F_{000}}, \quad (43)$$

or

$$\left|\frac{F_{hkl}}{F_{000}}\right|^2\leqslant\frac{1}{2}+\frac{1}{2}\frac{F_{2h,0,2l}}{F_{000}}. \quad (44)$$

This is another inequality derived by Harker & Kasper (1948).

As a final example we take the case in which the y axis is a twofold inversion axis. Then $F_{hkl}=F_{h\bar{k}l}$ and (34) becomes

$$\left|F_{h_1+h_2,k_1\pm k_2,l_1+l_2}-\frac{F_{h_1k_1l_1}F_{h_2k_2l_2}}{F_{000}}\right|$$
$$\leqslant\frac{(F^2_{000}-|F_{h_1k_1l_1}|^2)^{\frac{1}{2}}(F^2_{000}-|F_{h_2k_2l_2}|^2)^{\frac{1}{2}}}{F_{000}}. \quad (45)$$

As before,

$$|F_{h_1+h_2,k_1+k_2,l_1+l_2}-F_{h_1+h_2,k_1-k_2,l_1+l_2}|$$
$$\leqslant\frac{2(F^2_{000}-|F_{h_1k_1l_1}|^2)^{\frac{1}{2}}(F^2_{000}-|F_{h_2k_2l_2}|^2)^{\frac{1}{2}}}{F_{000}}, \quad (46)$$

and the case

$$h_1=-h_2=h,\quad k_1=k_2=k,\quad l_1=-l_2=l,$$

reduces to

$$|F_{0,2k,0}-F_{000}|\leqslant\frac{2(F^2_{000}-|F_{hkl}|^2)}{F_{000}}, \quad (47)$$

and finally to the Harker–Kasper inequality

$$\left|\frac{F_{hkl}}{F_{000}}\right|^2\leqslant\frac{1}{2}+\frac{1}{2}\frac{F_{0,2k,0}}{F_{000}}. \quad (48)$$

Since the Harker–Kasper and Gillis inequalities were derived on the assumption that ρ is non-negative, and since the inequalities (30) include all inequalities which can be obtained on this basis, it was to have been anticipated that the Harker–Kasper and Gillis inequalities would be included in the set (30).

Other types of inequalities using symmetry properties may be obtained by inserting the additional relationships into the determinants (29). For example, if the y axis is a twofold rotation axis, whence

$$F_{hkl}=F_{\bar{h}k\bar{l}},$$

and we choose a fourth-order determinant (29) in which F_{hkl} and $F_{\bar{h}k\bar{l}}$ occur twice along the secondary diagonal, the determinant may be factored to yield*

$$|F_{h_1k_1l_1}\pm F_{h_2k_1l_2}|$$
$$\lesseqgtr(F_{000}\pm F_{h_1+h_2,0,l_1+l_2})^{\frac{1}{2}}(F_{000}\pm F_{h_1-h_2,0,l_1-l_2})^{\frac{1}{2}}, \quad (49)$$

where either $<$ holds for both signs, or $>$ holds for both signs, or $=$ holds for at least one sign. This means that $F_{h_1k_1l_1}$ lies either in the interior of two intersecting circles, or the exterior of both circles or on the boundary of at least one. This type of inequality contains only information which is already included in the set (30). However, it may have more usefulness in application. We have not exploited further the possibility of obtaining new types of inequalities by means of introducing symmetry relations into the determinants (29) in which the coefficient of interest and its complex conjugate occur more than once.

The application of any linear transformations to the variables of the hermitian forms (4) yields other hermitian forms which are evidently non-negative. The determinants associated with these lead to different inequalities which are nevertheless equivalent to the original ones. The inequalities related to these new hermitian forms may be more useful. This may be illustrated by deriving the hermitian forms which represent a Patterson series from those which represent the original series by means of a linear transformation. For example, the hermitian form containing two variables may be transformed to the hermitian form

* As before the h, k, l are simply related to the ϵ, η, ζ in (29) and may be arbitrary positive or negative integers or zero.

corresponding to the Patterson series in accordance with the following scheme:

$$\begin{pmatrix} F_0 & F_{\bar{n}} \\ F_n & F_0 \end{pmatrix} \to \begin{pmatrix} F_0+|F_n| & 0 \\ 0 & F_0-|F_n| \end{pmatrix}$$
$$\to \begin{pmatrix} F_0^2+|F_n|^2 & 0 \\ 0 & F_0^2-|F_n|^2 \end{pmatrix} \to \begin{pmatrix} F_0^2 & |F_n|^2 \\ |F_n|^2 & F_0^2 \end{pmatrix}. \quad (50)$$

The first matrix represents the hermitian form

$$F_0X_1\bar{X}_1+F_nX_1\bar{X}_2+F_{\bar{n}}X_2\bar{X}_1+F_0X_2\bar{X}_2$$

related to the original series, and the last matrix represents the hermitian form corresponding to the Patterson series. The second matrix is the diagonal form of the first and may be derived from it by a linear transformation. The third and fourth matrices are similarly related. Evidently a linear transformation relates the second and third matrices.

The practical significance of this type of transformation is that the inequalities associated with the Patterson series involve only the magnitudes of the Fourier coefficients. These inequalities have the obvious advantage that the quantities contained in them are directly derivable from experiment. Perhaps other intermediate cases occur in which inequalities arise that contain some complex coefficients and the magnitudes of others. Certainly, it is worth while investigating the further implications of linear transformation theory.

References

ACHYESER, N. A. & KREIN, M. (1934). *Commun. Soc. math. Kharkow*, **4**, 9.

ACHYSER, N. A. & KREIN, M. (1935). *Commun. Soc. math. Kharkow*, **11**, 21.

DICKSON, L. E. (1930). *Modern Algebraic Theories*. New York: Sanborn.

GILLIS, J. (1948). *Acta Cryst.* **1**, 76.

HARKER, D. & KASPER, J. S. (1948). *Acta Cryst.* **1**, 70.

MUIR, T. & METZLER, W. H. (1930). *A Treatise on the Theory of Determinants*. Albany: privately published.

Structure of the protein subunits in the photosynthetic reaction centre of *Rhodopseudomonas viridis* at 3Å resolution

J. Deisenhofer, O. Epp, K. Miki, R. Huber & H. Michel*

Abteilungen Strukturforschung II and * Membranbiochemie, Max-Planck-Institut für Biochemie, D-8033 Martinsried, FRG

The molecular structure of the photosynthetic reaction centre from Rhodopseudomonas viridis *has been elucidated using X-ray crystallographic analysis. The central part of the complex consists of two subunits, L and M, each of which forms five membrane-spanning helices. We present the first description of the high-resolution structure of an integral membrane protein.*

The first steps in photosynthesis are mediated by protein-pigment complexes in the photosynthetic membrane. Light energy is absorbed primarily by light-collecting complexes and transferred to reaction centres where it is used with a very high quantum efficiency to transport electrons across the membrane. The reaction centre from the purple bacterium *Rhodopseudomonas viridis* consists of the four protein subunits L, M, H and *c*-type cytochrome[1] with four covalently linked haem groups[2]. Other prosthetic groups associated with the reaction centre are four bacteriochlorophyll *b* (BChl-*b*), two bacteriopheophytin *b* (BPh-*b*), two quinones, one non-haem ferrous iron and carotenoid(s)[1]. Apart from the cytochrome, which has been found as an integral constituent of reaction centre complexes only in some bacteria, it is generally accepted that reaction centres of purple bacteria are organized in a very similar way. This can be concluded from the subunit composition, from homology between the amino-acid sequences of the L and M subunits, from chromophore composition and from similar spectroscopic and electron-transfer properties (for reviews see, for example, refs 3-7).

The reaction centre from *Rps. viridis* was the first to be crystallized[8]; the photochemical activity of the crystalline material has been demonstrated[9]. An X-ray structure analysis of these well-ordered crystals allowed the calculation of an electron-density map at 3 Å resolution; an atomic model of the prosthetic groups was deduced from this map[10]. The arrangement of the pyrrole ring systems of BChl-*b* and BPh-*b* molecules in the central region of the reaction centre (assumed to consist of the subunits L and M) shows an approximate 2-fold symmetry with two closely associated BChl-*bs* near the symmetry axis. These two molecules form the 'special pair' postulated from the results of spectroscopic investigations[11]. The special pair is symmetrically in contact with two BChl-*b* molecules, each of which has a BPh-*b* as a neighbour. These chromophores form two branches that are apparently suitable as electron pathways. The right-hand side branch (see Fig. 4 in ref. 10) ends with a quinone close to the non-haem iron located on the local symmetry axis. The second quinone and the caretenoid are either not present or disordered in the crystal. The four haem of the cytochrome show a roughly linear arrangement with one haem near the special pair[10].

This chromophore model correlates well with the electron-transfer steps determined from spectroscopy[12]. After absorption of light an electron is transferred from the excited primary electron donor (the special pair), possibly via a BChl-*b*[13] to an intermediate acceptor (BPh-*b*), and further to a primary quinone acceptor (Q_A or tightly bound quinone). Q_A reduces a secondary quinone acceptor (Q_B or loosely bound quinone). Q_B, when fully reduced and protonated, is commonly assumed to leave the reaction centre, and to be replaced from a quinone pool in the membrane[14]. The oxidized special pair is reduced by the cytochrome.

Here we report further interpretation of the electron-density map. We describe the folding of the polypeptide chains of the protein subunits in the reaction centre and the spatial relation between protein and prosthetic groups. With this structure we can present the first model of an integral membrane protein determined at nearly atomic resolution.

Electron-density map

The reaction centre crystals belong to the space group $P4_32_12$ with unit cell constants of $a = b = 223.5$ Å, $c = 113.6$ Å, and contain one centre per asymmetrical unit[8,10]. The electron-density map at 3 Å resolution was calculated with phases from multiple isomorphous replacement using five heavy-atom derivatives; this map was improved further by solvent flattening[15]. A detailed report on data collection, map calculation and arrangement of prosthetic groups is described elsewhere[10].

The chromophore model facilitated the identification of the protein subunits of the reaction centre. The haem groups are evidently bound to the cytochrome subunit; BChl-*b*, BPh-*b*, non-haem iron and quinone are known to be associated with the subunits L and M in reaction centres from purple bacteria[4,16,17]. Available amino-acid sequences of 46, 30 and 28 residues[18] at the amino-termini of the subunits L, M and cytochrome, respectively, could be located in the map. The H subunit was identified on the basis of a partial gene sequence corresponding to 68 residues from the amino terminus.

For model building of the protein subunits a minimap (scale 2 mm $Å^{-1}$), and interactive display systems (Vector General 3400 and Evans & Sutherland PS330) driven by the program FRODO[19] were used. Optimum fit of helices and large side chains to the density was achieved with the real-space refinement option in FRODO[20]. Segments with unknown amino-acid sequence were built as polyalanine. Residue numbers were prefixed with the letters L, M, C and H, respectively, for distinction between subunits.

The good quality of the map allowed unambiguous chain tracing even in the absence of amino-acid sequence information. Recently the complete amino-acid sequences of the subunits H[21], L and M (H.M., K. A. Weyer, H. Gruenberg, I. Dunger, D. Oesterhelt and F. Lottspeich, in preparation) became available from the corresponding gene sequences and were incorporated into the model. The amino-acid sequence of the cytochrome, presently being investigated with protein-chemical methods (K. A. Weyer, H.M. and F. Lottspeich, unpublished data), is still incomplete. However, peptide sequences resulting from this work could be located in the electron density and used for model building. Currently, these pieces add up to ~60% of the residues. The absence of a contiguous sequence for the cytochrome leads to an uncertainty in the residue numbering for this subunit caused by possible erroneous omission or insertion of residues during chain tracing.

NATURE VOL. 318 19/26 DECEMBER 1985 ARTICLES 619

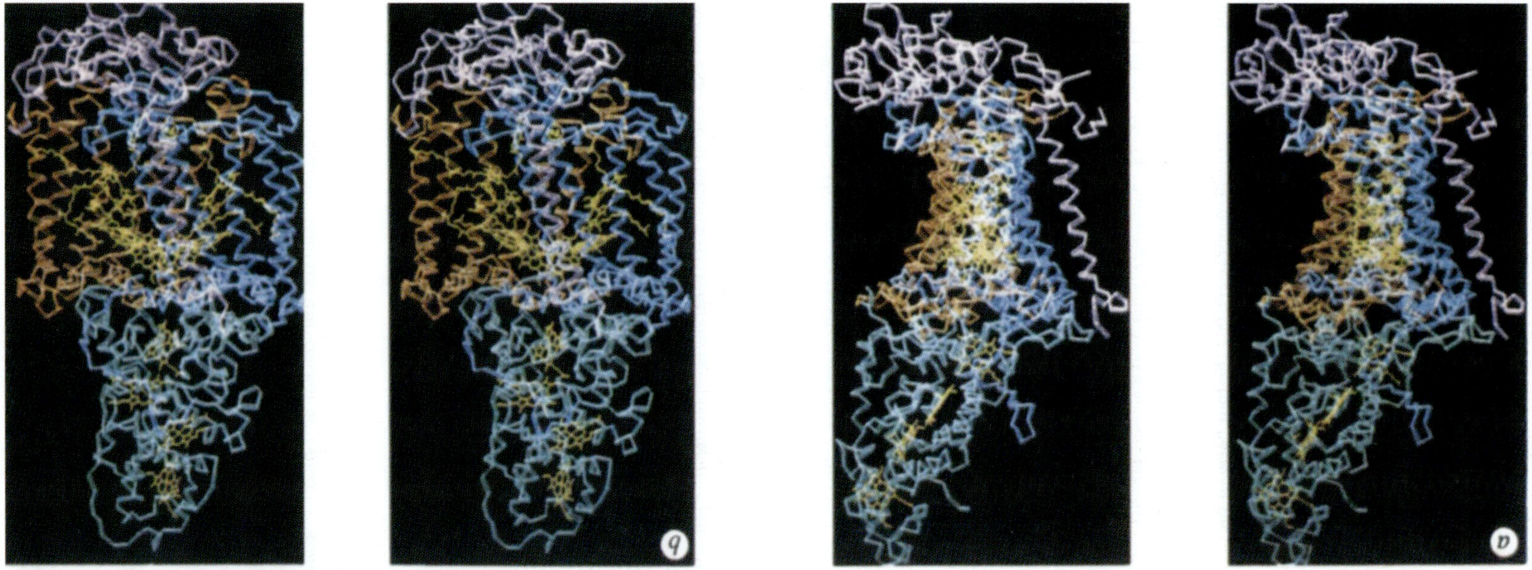

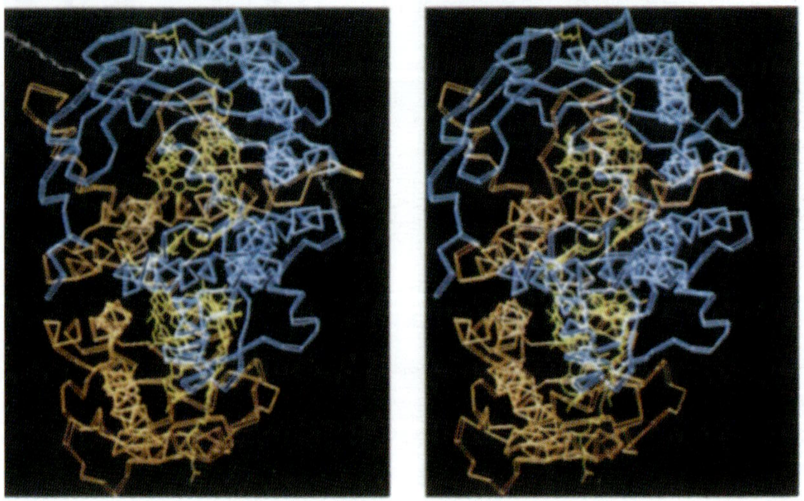

Fig. 1 (Parts *a*, *b* above.) Stereo drawing of the complete reaction centre complex in two orientations. Panel *a* is rotated with respect to *b* by 90° around a vertical axis. Green, cytochrome; orange, L; blue, M; purple, H; yellow, prosthetic groups. Figs 1–5 were produced by a computer program written by A. M. Lesk and K. D. Hardman[50].

Fig. 3 (Left) Stereo drawing of the L–M complex and associated chromophores viewed along the L–M local diad. Protein chains are represented as ribbons. Colours as in Fig. 1.

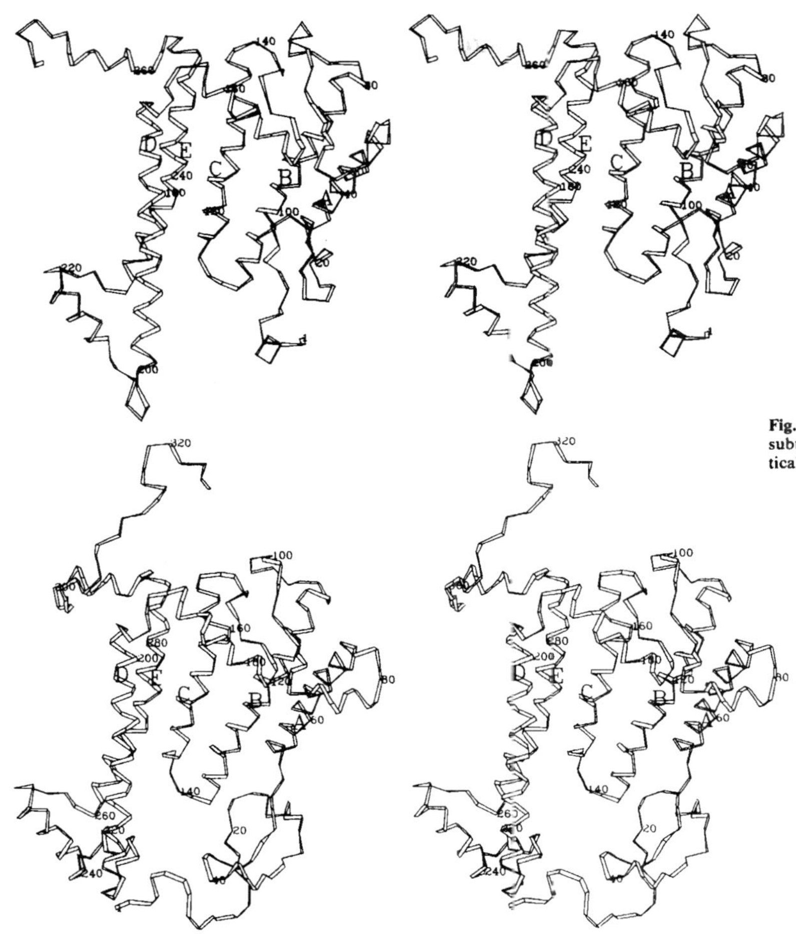

Fig. 2 Stereo drawings of *a*, subunit L; *b*, subunit M represented as ribbons in identical orientations. Transmembrane helices A–E are labelled.

A crystallographic *R*-value ($R = \sum ||F_{obs}| - |F_{calc}|| / \sum |F_{obs}|$) was calculated when the model contained ~70% of the correct sequence. For 46,658 reflections between 7- and 3-Å resolution we obtained $R = 0.359$; this value is exceptionally low for a protein model before crystallographic refinement.

The reaction centre complex

The course of the polypeptide chains and the location of the prosthetic groups of the centre are shown in Fig. 1. The central part consists of the subunits L and M which bind the BChl-*b*, BPh-*b*, the quinone and the non-haem iron. The cytochrome subunit is bound to a flat surface of L–M close to the special pair of BChl-*b*. A large contact area and the carboxy-terminal arm of the M subunit are responsible for the tight binding of the cytochrome to L–M. The major part of the H subunit is in contact with the other flat surface of the L–M complex that is close to the binding sites of the quinone and the non-haem iron. The amino-terminal segment of H forms a helix extending from the cytochrome side along the surface of the M subunit. The longest dimension of the reaction centre complex is ~130 Å between the tips of cytochrome and H subunit.

In our electron-density map there is no clear indication of the presence of ordered lipid or detergent that would provide direct evidence on the position and orientation of the membrane relative to the reaction centre complex. However, the highly hydrophobic surface of the central region of the centre indicates that this part (that is, the subunits L and M and the amino-terminal helix of H) spans the membrane. The same conclusion follows from the fact that with electron transfer from the special pair to the quinone(s) a charge separation across the membrane takes place. Recently, the amino-terminus of the L-subunit was localized at the cytoplasmic side of the photosynthetic membrane from *Rhodospirillum rubrum*[22]. Assuming the same orientation for the reaction centre from *Rps. viridis*, this means that the cytochrome is located at the periplasmic side, and the major part of the H subunit is at the cytoplasmic side of the membrane.

Subunits L and M

Subunit structure. Although L and M differ in length by 50 residues (L, 273 residues; M, 323 residues), their folding is generally very similar. Using the method of Rossmann and Argos[23], 219 C^{α}-positions of M could be rotated onto corresponding positions of L to an r.m.s. distance of 1.25 Å. The rotation axis is defined by the polar angles $\phi = 43°$, $\Psi = 39.4°$, $\kappa = -179.3°$ (where ϕ is the angle between the rotation axis and *y*-axis; ψ is the angle between projection of the rotation axis on *x–z* plane and *x*-axis; and κ is the rotation angle). At the present level of accuracy this L–M local diad is virtually identical to the local diad relating ring atoms of BChl-*b* and BPh-*b* ($\phi = 43°$, $\psi = 37.6°$, $\kappa = -177.1°$)[10]. Figure 2 shows ribbon drawings of subunits L and M in the same orientation. Outstanding structural features of both subunits are five long helical regions (mostly α-helix) numbered A–E (LA–LE in the L subunit.

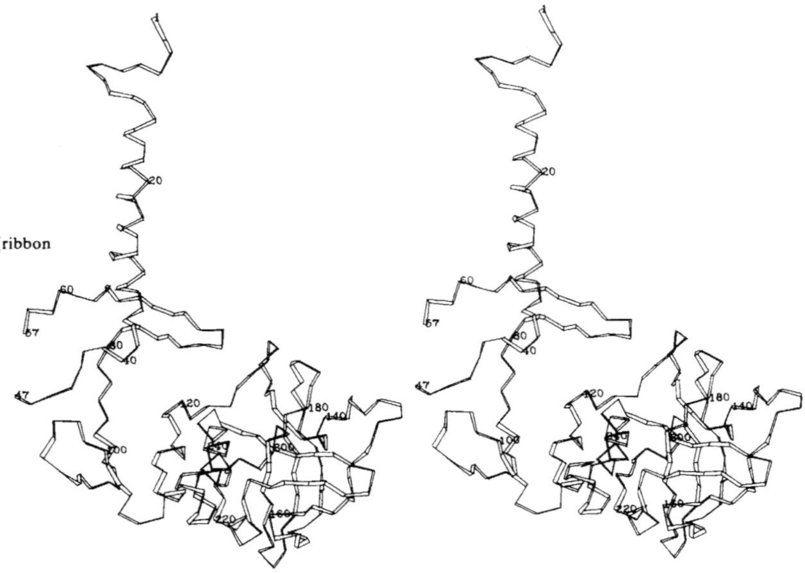

Fig. 4 Stereo drawing of subunit H (ribbon representation).

MA–ME in the M subunit). They consist mainly of hydrophobic residues, many of which are exposed to the environment. Therefore, we assume that these helices span the membrane; their lengths range between 24 and 30 residues (A, L32–L55, M52–M78; B, L84–L112, M110–M139; C, L115–L140, M142–M167; D, L170–L199, M197–M225; E, L225–L251, M259–M285). The helices are arranged side by side in a concave layer in the order A, B, C, E and D, so that helices C and E are neighbours and run approximately parallel, whereas all other neighbouring helix pairs within a subunit run roughly antiparallel. The angles between the helix axes and the L–M local diad are <25° for the helices A, B, C and E and ~38° for helix D.

On both sides of the helical region the remaining segments of L and M form two flat surfaces in intimate contact with the subunits cytochrome and H, respectively. The surface in contact with the H subunit consists of the amino-terminal segments and the helix connections D–E; helix connections B–C are short turns which also come close to this surface. Helix connections A–B and C–D and carboxy-terminal chain segments contribute to the cytochrome contact surface. Apart from as the transmembrane helices, further segments with helical conformation are located in both contact surfaces: L208–L220, M36–M42, M240–M255 at the H contact surface and L149–L164, L259–L269, M80–M88, M178–M192 and M281–M300 at the cytochrome contact surface. Only at the amino termini of L and M two-stranded antiparallel β-sheets are found comprising segments L24–L26, L29–L31 and M27–M35, M44–M51, respectively.

Major structural differences between L and M are observed at or near the contact surfaces. From the amino terminus to the start of transmembrane helix A, M is 20 residues longer than L; these additional residues mainly contribute to the two-stranded antiparallel β-sheet at the H contact surface. The helix connection D–E of M, which is also at the H contact surface, is seven residues longer than in L. At the cytochrome contact surface the helix connection A–B is seven residues longer in M than in L. From the end of helix E to the carboxy terminus M is 16 residues longer than L; the additional residues of M form the arm that folds along the surface of the cytochrome, and appears to contribute to the tight binding of this subunit in the reaction centre complex.

L–M association. The subunits L and M cohere predominantly at or near the H contact surface. Helices D, E and the C–D connection of one subunit intercalate between helices C and E in the other subunit (see Fig. 3). Helices LD, LE, MD and ME come close to the L–M local diad, and are in contact with each other (see Fig. 1). The non-haem iron is bound between these four helices, each of which contributes one ligand to the iron. Further L–M contacts at the H contact surface are contributed by the helix connections D–E and by the amino-terminal segments (Fig. 3).

Near the cytochrome contact surface a gap opens up between helix pairs LD, LE and MD, ME to accommodate the special pair on the L–M local diad. Helices C and E are bent away from the L–M local diad near the cytochrome contact surface, and so increase the space available for the accessory BChl-*b* and the BPh-*b* molecules. Both L and M contribute one half of the cytochrome contact surface with apparently weak lateral contacts between these halves. At the intersection of the L–M local diad and the cytochrome contact surface the side chain of residue Tyr L162 is located between the closest haem group and the special pair; it may be involved in electron transfer.

With its two only partially overlapping layers of transmembrane helices the L–M complex has a remarkably non-globular structure in which the surface area accessible from the environment appears to be significantly larger than in globular water soluble proteins. Its shape suggests that the contact surfaces are parallel and that the L–M local diad is perpendicular to the membrane plane, in agreement with chromophore orientations determined optically in oriented *Rps. viridis* cells[24,25].

Binding of prosthetic groups in L–M. As reported previously[10], protein side chains are bound to the Mg^{2+} ions of BChl-*b*; all four BChl-*b* ligands are histidines as most BChl-*a* ligands found in the BChl-*a* protein from *Prosthecochloris aestuarii*[26,27]. The BChl-*b*s are designated BC_{LP}, BC_{MP}, BC_{LA} and BC_{MA}, where L and M refer to the ligation and P and A indicate 'special pair' and 'accessory', respectively. The Mg^{2+} ligands of BC_{LP} and BC_{MP} are near the start of the transmembrane helices LD (L173), and MD (M200), respectively. The Mg^{2+} ligands for BC_{LA} and BC_{MA} are in the helix connections C–D (L153 and M180, respectively). The BPh-bs are named BP_L and BP_M according to predominant interactions with subunit L or M. Each BPh-*b* binding site is a pocket formed mainly by non-polar residues from both subunits; there are no polar interactions between the protein and the tetrapyrrole nitrogens. BP_L (BP_M) is located near the centres of helices LC, LE (MC, ME) with its ring plane parallel to the axis of helix MD (LD). Accessory BChl-*b* and BPh-*b* are partially accessible from the environment.

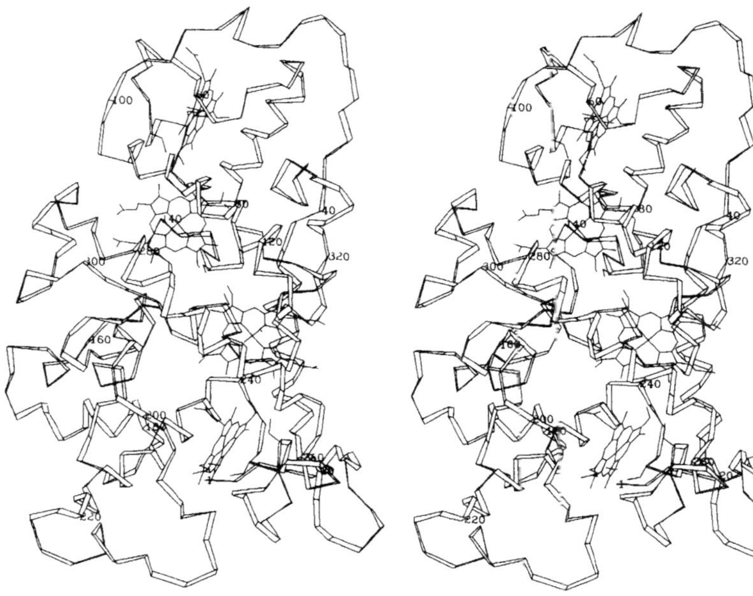

Fig. 5 Stereo drawing of the cytochrome subunit. The protein chain is represented as a ribbon; haem groups and linked cysteine side chains are shown as line drawings.

The non-haem iron is bound by five protein ligands, four histidines and one glutamic acid. The histidines are located in the transmembrane helices LD (L190), LE (L230), MD (M217) and ME (M264), and obey the L–M local 2-fold symmetry. They are arranged approximately in a plane perpendicular to the L–M local diad. The fifth iron ligand, Glu M232, sits on the local L–M diad. The structure near the non-haem iron is in good agreement with conclusions drawn from extended X-ray absorption fine-structure studies[28,29].

As reported previously[10], we believe that the quinone identified in the electron-density map is the tightly bound quinone Q_A (menaquinone-9 in *Rps. viridis*[30]). Q_A is located near BP_L; its head group is bound exclusively to residues of M in the helix connection D–E with close contacts to His M217 (one of the iron ligands), to the peptide nitrogen of M258 and to Trp M250 whose side chain is nearly parallel to the Q_A head group. Its isoprenoid chain is mainly in contact with BP_L and with resdiues of L. It was shown that the loosely bound Q_B (ubiquinone-9 in *Rps. viridis*[30]) can be lost during purification of the reaction centres[14] or during crystallization[30]. To find the Q_B site, the binding of the competitive inhibitors for Q_B *o*-phenanthroline[31], terbutryn (a herbicide)[32] and of ubiquinone-1 in the crystal was studied by soaking these compounds into the crystals (*o*-phenanthroline, 2 mM for 3 days; terbutryn, 1 mM for 3 days, followed by 0.01 mM for 3 days; ubiquinone-1, 1 mM for 3 days), collecting X-ray data and analysing these data using the difference-Fourier technique. These studies showed binding of the three compounds to a pocket formed near the H contact surface by parts of the transmembrane helices LD and LE, and by the LD–LE connecting segment. Additional binding sites were observed for *o*-phenanthroline (two), and ubiquinone-1 (seven). From these results we conclude that the single common binding site of the three compounds is the Q_B binding site. Only this site, which is related to the Q_A site by the L–M local diad, is close to the non-haem iron as required to explain the broadening of electron-spin resonance lines observed for Q_B^- (ref. 33). The Q_B site is in contact to the iron ligand His L190; the iron itself is located midway between Q_A and Q_B. This arrangement strongly suggests that the iron participates in electron tranfer from Q_A to Q_B. However, experiments with iron-depleted reaction centres from *Rps. sphaeroides* showed that electron transfer is only moderately slower in the absence of the iron[34].

The binding sites of Q_A and Q_B are in regions of L and M, where both subunits markedly deviate from the L–M local symmetry. Whereas Q_A is completely shielded from the H subunit by the M chain, the probable Q_B binding site is accessible from H through an opening in the H contact surface. These observations correlate the different binding of Q_A and Q_B to structural differences between L and M; they also suggest that one function of H is related to Q_B binding. A more detailed description of the reaction centre structure near the prosthetic groups will be given elsewhere.

The H subunit

With 258 residues H is the smallest subunit of the reaction centre. The course of its polypeptide chain is shown in Fig. 4. The amino-terminus is near the cytochrome contact surface of L–M; residues H1, H2, H3 and H5 are in contact with residues from the cytochrome. Residues H12 to H37 form a helix which runs along the convex side of the surface formed by the transmembrane helices of M, with contacts to helices ME and MD (see Fig. 1). The amino-acid sequence in this helical region up to residue H32 is also primarily hydrophobic; the carboxy-end of the helix (H33–H37) and further on to H39 consists of charged residues (Arg-Arg-Glu-Asp-Arg-Arg-Glu) which interact partly among themselves, and probably also with the polar head groups of membrane lipids *in vivo*.

Altogether we find 11 transmembrane helices in the reaction centre complex, 6 running from the cytoplasmic side to the periplasmic side and 5 running in the opposite direction. The question of whether the dipole moments of these helices can facilitate electron transfer across the membrane, as proposed recently[35], requires a detailed inspection of the model and will be discussed elsewhere.

The absence of significant electron density for residues H48 to H56 near the end of the transmembrane helix indicates that this segment is disordered in the crystal. The segment H57–H115 lies on the surface of L–M; residues H66–H70 and H74–H78 form a two-stranded antiparallel β-sheet. The remainder of H is folded into a globular domain. Residues H156–H120 form two layers of β-sheet which enclose a hydrophobic core. Both sheets consist of three strands and show a right-handed twist. In sheet 1 the strands (H162–H174, H179–H187 and H191–H197) are antiparallel; in sheet 2 the middle strand (H207–H210) is accompanied by an antiparallel strand (H200–H204) and a parallel strand (H156–H159). Near the carboxy-terminus residues H231–H249 form a helix that runs parallel to the H contact surface of L–M.

Little is known about the function of H in the reaction centre complex. Crosslinking studies[16] indicate that H is close to the light-collecting complexes in the membrane; this observation suggests that H has a role in determining the relative orientation of light-collecting complexes and reaction centre. As mentioned above, H may also play a role in Q_B binding—residue H177 (Glu) and its neighbours have access to the Q_B binding pocket in subunit L. They could be involved in donating protons to the photochemically reduced Q_B. This is in accord with experimental results on L–M complexes of *Rps. sphaeroides*[36] which show abnormal electron transfer from Q_A to Q_B in the absence of H.

The cytochrome

The present model of the cytochrome subunit contains 333 residues. Figure 5 shows the course of the polypeptide chain and the location of the haem groups. Near the amino-terminus, residues C7–C11 and C20–C23 form an antiparallel β-sheet that lies on the L half of the cytochrome contact surface. From there the chain proceeds to the tip of the subunit (around C50); residues C25–C37 form a helix. In the following bulk of the subunit four recurring structural motifs can be observed: a helix runs parallel to the ring plane of each haem group (residues C67–C82. C102–C119, C222–C240 and C262–C286, respectively); the haems are covalently attached near the carboxy termini of these helices where the chains turn back towards the haems. The electron density strongly indicates that these covalent links are thioether bridges, and that the sequences of the haem binding segments are Cys-X-X-Cys-His as in other *c*-type cytochromes; the His residues are the fifth ligands to the haem irons. The haems are numbered according to the order of attachment to the protein; for haem 1 the first covalent link is to residue C87, haem 2, C132, haem 3, C244 and haem 4, C305. All haem irons appear to be six-coordinated; the sixth ligands to the irons of haem 1, 2 and 3 are located in the helices associated with the haems (residues C74, C110 and C233, respectively). The sixth ligand to the haem 4 iron (C124) is in the segment between the helix parallel to haem 2 and the haem 2 attachment site. According to the electron density the sixth ligands to the haem irons could be histidines or methionines.

As reported previously[10], the haem groups are arranged with an approximate 2-fold symmetry: pyrrole ring atoms of haems 3 and 4 can be rotated onto those of haems 1 and 2 with polar angles $\phi = 117.2°$, $\Psi = -70.2°$ and $\kappa = -176.5°$ to an r.m.s. distance of 0.41 Å. Haems 1 and 2 and haems 3 and 4 are related by improper rotations (~105°) with large screw components (~12 Å). Using the same method as for the L–M comparison[23], 66 C^{α} positions from the region C226–C315 could be rotated onto corresponding C^{α} positions from between C67 and C142 to an r.m.s. distance of 1.1 Å. The polar angles of this rotation are $\phi = 116.7°$, $\psi = -68.9°$ and $\kappa = -177.8°$, very close to those for the haem rotation. This internal 2-fold symmetry includes about one third of the cytochrome subunit. More extended structural homologies with 2-fold symmetry in protein molecules have been reported for rhodanese[37] and γ-crystallin[38]. The intramolecular symmetry is broken in the segment connecting the binding sites of haem 2 and 3; this segment contributes to the contact of cytochrome to L–M. The carboxy-terminal segment ends near the helix associated with haem 2. Haem 1 is accessible from the environment at its pyrrole rings I and IV; haem 2 appears to be almost completely buried; and haems 3 and 4 are accessible at their propionic acid groups. Pyrrole rings I and IV of haem 3 are closest to the special pair of BChl-*b* in the L–M complex. Apparently, protein folding in the cytochrome subunit of the reaction centre is different from both cytochromes C2 and cytochromes C3 with known three-dimensional structure (see ref. 39 for review).

Structure of related proteins

Reaction centres from other purple bacteria. Amino-acid sequences are known for the reaction centre subunits L, M and H from *Rps. capsulata*[40], and for L and M from *Rps. sphaeroides*[41,42]. The sequences of L and M from *Rps. virisidis* show between 50 and 60% homology with those of the corresponding subunits from the other two bacteria. On the basis of hydrophobicity five transmembrane helices each were postulated[40–42] in segments of L and M homologous with the transmembrane helix regions in our model. All the ligands to the BChl and to the non-haem iron described above are also conserved, and the polypeptide segments connecting these ligands have the same length in all three organisms. These observations clearly indicate that the structures of L and M from *Rps. capsulata* and *Rps. sphareoides*, and the arrangement of BChls and non-haem iron are very similar to those found for *Rps. viridis*. Significant differences in polypeptide chain length occur exclusively at the carboxy-termini where L from *Rps. viridis* is shorter by 8 residues, and M is longer by 17 residues. These additional residues of M are in contact with the cytochrome in *Rps. viridis*: the reaction centres of *Rps. capsulata* and *Rps. sphaeroides* lack a permanently bound cytochrome subunit.

For the H-subunits from *Rps. capsulata* and *Rps. viridis* the sequence homology is 37% (ref. 21). Common features of these two H-subunits include a hydrophobic region near the amino terminus which corresponds to the transmembrane helix of H in our model. In addition, many of the residues involved in the contact between H and L–M are conserved in both organisms, so that structural similarity can also be expected between the H subunits from *Rps. capsulata* and *Rps. viridis*.

Photosystem II. Recently, amino-acid sequences of polypeptides from photosystem II reaction centres[43–47] have been reported. Sequence homologies were found between L and M of purple bacteria and the D1 polypeptide (also referred to as Q_B protein or herbicide binding polypeptide) from photosystem II of higher plants[40,42] and between D1 and D2 (ref. 45) which is also a constituent of photosystem II. In these plant proteins, histidines which correspond to the ligands of the special pair, and to those of the non-haem iron in our model are conserved. Even a Trp residue corresponding to M250, which forms part of the Q_A binding site in reaction centres of purple bacteria, can be found in the D2 proteins. These observations corroborate our recent proposal[48] that proteins D1 and D2 form the core of the photosystem II reaction centre, similar to the subunits L and M in purple bacteria. Based on functional and sequence homology we believe that D1 and D2 span the membrane five times, contrary to a proposed seven-helix structure[49], where the helix connections A–B and D–E were probably misinterpreted as transmembrane helices.

We thank Professor D. Oesterhelt for support and the Deutsche Forschungsgemeinschaft (SFB 143) for financial help.

Received 1 August; accepted 17 October 1985.

1. Thornber, J. P., Cogdell, R. J., Seftor, R. E. B. & Webster, G. D. *Biochim. biophys. Acta* **593,** 60–75 (1980).
2. Prince, R. C., Leigh, J. S. & Dutton, L. *Biochim. biophys. Acta* **440,** 622–636 (1976).
3. Feher, G. & Okamura, M. Y. in *The Photosynthetic Bacteria* (eds Clayton, R. K. & Sistrom, W. R.) 349–386 (Plenum, New York, 1978).
4. Okamura, M. Y., Feher, G. & Nelson, N. in *Photosynthesis: Energy Conversion by Plants and Bacteria* (ed. Govindjee) 195–272 (Academic, New York, 1982).
5. Parson, W. W. & Ke, B. in *Photosynthesis: Energy Conversion by Plants and Bacteria* (ed. Govindjee) 331–384 (Academic, New York, 1982).
6. Hoff, A. J. in *Light Reaction Path of Photosynthesis* (ed. Fong, F. K.) **80–151,** 322–326 (Springer, Berlin, 1982).
7. Parson, W. W. *A. Rev. Biophys. Bioengng.* **11,** 57–80 (1982).
8. Michel, H. *J. molec. Biol.* **158,** 567–572 (1982).
9. Zinth, W., Kaiser, W. & Michel, H. *Biochim. biophys. Acta* **723,** 128–131 (1983).
10. Deisenhofer, J., Epp, O., Miki, K., Huber, R. & Michel, H. *J. molec. Biol.* **180,** 385–398 (1984).
11. Norris, J. R., Uphaus, R. A., Crespi, H. L. & Katz, J. J. *Proc. natn. Acad. Sci. U.S.A.* **68,** 625–628 (1971).
12. Holten, D., Windsor, M. W., Parson, W. W. & Thornber, J. P. *Biochim. biophys. Acta* **501,** 112–126 (1978).
13. Shuvalov, V. A. & Asadov, A. A. *Biochim. biophys. Acta* **545,** 296–308 (1979).
14. Shopes, R. J. & Wraight, C. A. *Biochim. biophys. Acta* **806,** 348–356 (1985).
15. Wang, B. C. *Lecture Notes for School on Direct Methods and Macromolecular Crystallography*[7] (Medical Foundation of Buffalo, Buffalo, New York, 1983).
16. Peters, J., Welte, W. & Drews, G. *FEBS Lett.* **171,** 267–270 (1984).
17. Wiemken, V. & Bachofen, R. *FEBS Lett.* **166,** 155–159 (1984).

18. Michel, H., Deisenhofer, J., Miki, K., Weyer, K. & Lottspeich, F. in *Structure and Function of Membrane Proteins* (eds Quagliariello, E. & Palmieri, F.) 191–197 (Elsevier, Amsterdam, 1983).
19. Jones, T. A. *J. appl. Crystallogr.* **11**, 268–272 (1978).
20. Jones, T. A. & Liljas, L. *Acta Crystallogr.* **A40**, 50–57 (1984).
21. Michel, H., Weyer, K. A., Gruenberg, H. & Lottspeich, F. *EMBO J.* **4**, 1667–1672 (1985).
22. Brunisholz, R. A., Wiemken, V., Suter, F. & Bachofen, R. *Hoppe-Seyler's Z. Physiol. Chem.* **365**, 689–701 (1984).
23. Rossmann, M. G. & Argos, P. *J. molec. Biol.* **105**, 75–95 (1976).
24. Paillotin, G., Vermeglio, A., Breton, J. *Biochim. biophys. Acta* **545**, 249–264 (1979).
25. Breton, J. in *Proc. Feldafing Workshop 23–25 March* 1985 (ed. Michel-Beyerle, M. E.) (Springer, Berlin, in the press).
26. Matthews, B. W., Fenna, R. E., Bolognesi, M. C. & Schmid, M. F. *J. molec. Biol.* **131**, 259–285 ((1979).
27. Schmid, M. F., Tronrud, D. E. & Matthews, B. W. *Chim. scripta* **21**, 67–69 (1983).
28. Bunker, G., Stern, E. A., Blankenship, R. E. & Parson, W. W. *Biophys. J.* **37**, 539–551 (1982).
29. Eisenberger, P., Okamura, M. Y. & Feher, G. *Biophys. J.* **37**, 523–538 (1982).
30. Gast, P., Michalski, T. J., Hunt, J. E., Norris, J. R. *FEBS Lett.* **179**, 325–328 (1985).
31. Pfister, K., Steinback, K. E., Gardner, G. & Arntzen, C. J. *Proc. natn. Acad. Sci. U.S.A.* **78**, 981–985 (1981).
32. Stein, R. R., Castellvi, A. L., Bogacz, J. P. & Wraight, C. A. *J. cell. Biochem.* **24**, 243–259 (1984).
33. Dismukes, G. C., Frank, H. A., Friesner, R. & Sauer, K. *Biochim. biophys. Acta* **764**, 253–271 (1984).
34. Debus, R. J., Okamura, M. Y. & Feher, G. *Biophys. J.* **47**, 3a (1985).
35. Hol, W. G. J. *Prog. Biophys. molec. Biol.* **45** 149–195 (1985).
36. Debus, R. J., Feher, G. & Okamura, M. Y. *Biochemistry* **24**, 2488–2500 (1985).
37. Ploegman, J. H., Drent, G., Kalk K. H. & Hol, W. G. J. *J. molec. Biol.* **123**, 439–457 (1978).
38. Blundell, T. *et al. Nature* **289**, 771–777 (1981).
39. Mathews, F. S. *Prog. Biophys. molec. Biol.* **45**, 1–56 (1985).
40. Youvan, D. C., Bylina, E. J., Alberti, M., Begusch, H. & Hearst, J. E. *Cell* **37**, 949–957 (1984).
41. Williams, J. C., Steiner, L. A., Ogden, R. C., Simon, M. I. & Feher, G. *Proc. natn. Acad. Sci. U.S.A.* **80**, 6505–6509 (1983).
42. Williams, J. C., Steiner, L. A., Feher, G. & Simon, M. I. *Proc. natn. Acad. Sci. U.S.A.* **81**, 7303–7307 (1984).
43. Zurawski, G., Bohnert, H. J., Whitfield, P. R & Bottomley, W. *Proc. natn. Acad. Sci. U.S.A.* **79**, 7699–7703 (1982).
44. Rasmussen, O. F., Bookjans, G., Stummann, B. M. & Henningsen, K. W. *Pl. Molec. Biol.* **3**, 1984–1984 (1984).
45. Rochaix, J.-D., Dron, M., Rahire, M., Malnoe, P. *Pl. molec. Biol.* **3**, 363–370 (1985).
46. Alt, J., Morris, J., Westhoff, P. & Herrmann R. G. *Curr. Genet.* **8**, 597–606 (1984).
47. Holschuh, K., Bottomley,. W. & Whitfield, P. R. *Nucleic Acids Res.* **12**, 8819–8834 (1984).
48. Michel, H., Deisenhofer, J. in *Encyclopedia of Plant Physiology: Photosynthesis III* Vol. 19 (eds Staehelin, A. C. & Arntzen, C. J.) (Springer, Berlin, in the press).
49. Rao, J. K. M., Hargrave, P. A. & Argos, P. *FEBS Lett.* **156**, 165–169 (1983).
50. Lesk, A. M. & Hardman, K. D. *Science* **216**, 533–540 (1982).

Slow neutron spectroscopy and the grand atlas of the physical world*

Bertram N. Brockhouse
Department of Physics, McMaster University, Hamilton, Ontario, Canada L8S 4M1

On October 12, 1994, telephone communications from Stockholm ensured that I would have the privilege of giving a Nobel lecture, for which I must thank all those involved in arranging for this great and surprising event of my life. But I had to go back some 30 to 45 years, first in memory and then in the entropy of my files and library. The lecture was given on December 8, 1994 (and subsequently on several occasions). This written version covers the same ground, though sometimes in different order, with a little additional material from my preparations, which time prevented including in the spoken lecture.

In August 1950, when I joined the physics division of the Chalk River Nuclear Laboratory, it was 18 years since the identification of the neutron as being emitted in some radioactive processes and 14 years since verification of the supposition that neutrons would exhibit wave-particle duality. A considerable body of theory was available in the open literature—which was still small enough so that one person could have read all that was then available. There had been significant measurements of total cross sections using neutron beams produced by cyclotrons. Self-sustaining reactors employing the neutron-induced fission of uranium had been demonstrated, and full-scale models which emitted potent beams of both fast and slow neutrons had been constructed. Slow neutron beams from reactors were already in use at several laboratories (including Chalk River) for studies of crystals and other forms of matter. The works of E. O. Wollan and C. G. Shull at the Oak Ridge National Laboratory were particularly significant, because they included the first studies of a number of phenomena and because they already presented values of scattering power for neutrons of a substantial number of elements. This history is discussed in the lecture of my senior colleague and co-winner of the 1994 Nobel prize in physics, Clifford Shull.

To 1951, some studies had been made of the elastic scattering of monochromatic slow neutrons by specimens in the form of powdered crystals (the neutron analog of Debye-Scherrer patterns for x rays), which led to improved crystallographic understanding of the substances involved. And there were a few studies of the angular distribution of slow neutrons scattered by certain specimens in the liquid or gaseous form. Most importantly, magnetic scattering of neutrons, by crystalline substances which contained atoms with magnetic moments, had been demonstrated and some simple magnetic structures determined. But the thermal diffuse scattering, expected from theory, had been studied only as the total cross section for incident neutrons of long wavelength (beyond the possibility of Bragg scattering). No spectrometric measurements had been made, though there existed relevant theory, and the phenomenon of neutron slowing down (moderation) gives assurance that energy changes on single scattering must occur.

Figure 1 shows the main floor of the hall of Chalk River with the NRX reactor ("pile"), a very powerful one for the times. A variety of experiments were under way, mostly in nuclear physics, but the double-axis powder diffractometer, constructed by Dr. Donald Hurst and associates, is visible near the centre of the photograph. The large size of the equipment is enjoined by the necessity of shielding against the fast neutrons which accompany the slow neutron beams—one's own beam and one's neighbours'—for low experimental background as well as for biological protection.

It is necessary to remember that many things which a researcher of the present would take for granted were then not available—not available on the market or not available at all. Counting electronics were still under development in-house; recording apparatus was mostly analog and of limited availability as to kind. The first digital computer at Chalk River was delivered in 1954 and occupied a large room with heat-generating electronics (no siliconics!). The single crystals available were mostly of natural origin, though large ingots of NaCl and other alkali halides were produced for use in infrared spectrometry and were available commercially. But opposed to all this sort of thing was the remarkable openness of the possibilities available. If an experiment could be envisaged, it was likely new or at least repeating it would not be merely trivial.

THE NEUTRON AND ITS APPLICATIONS

In 1982 a conference was held at Cambridge in honour of the 50th anniversary of the discovery by James Chadwick of the free neutron. The proceedings [A] deal with the numerous consequences of this event and of their histories. Beyond those military consequences so important in the media and beyond the applications to the public energy supply (nuclear energy), to nuclear medicine, and to nuclear physics and chemistry, there have come into existence a number of new scientific fields.

These fields involve neutrons produced by nuclear reactions (as "fast" neutrons) with energies usually millions of electron volts (MeV) and then allowed to "slow down" by repeated collisions with atomic nuclei of small atomic weight ("moderation") to a near-equilibrium distribution at a temperature a little above that of the moderator (often water at a temperature somewhat above room temperature). A small tube through the reactor shielding admits a beam of the fast, intermediate, an slow neutrons into the experimental hall.

After dealing with the undesired fast neutrons, well-defined beams of the "slow neutrons" are then employed in a variety of ways to study the properties of material speci-

*This lecture was delivered 8 December 1994, on the occasion of the presentation of the 1994 Nobel prize in physics.

FIG. 1. The main floor of the NRX reactor at the Chalk River Laboratory about 1950. The powder diffractometer constructed by Donald Hurst and associates is visible near the centre of the photograph. Most of the other equipment is concerned with nuclear physics or with the physics of the neutron itself. For reasons of space, each apparatus is located at the end of a long tube. (AECL photo.)

mens placed in the neutron beam (or in subsidiary neutron beams derived from the first). The fields of interest here are known generically by the terms (slow) "neutron scattering" or "neutron diffraction." The scattering may be elastic or inelastic; in the latter case, the experiments may involve measurements of the energy changes experienced by the neutrons when scattered by the specimen, in which case the term (slow) "neutron spectrometry" is often used. When the incident neutrons have energies well below what they would have if in equilibrium at room temperature (0.025 eV), they are often termed "cold neutrons."

Slow neutrons have energies from millivolts (MeV) to tenths of an eV and, by the same token, have wavelengths from 10 Å or more to tenths of an angstrom. This fact—that neutrons have, simultaneously, energies of the order of the characteristic quantum energies in condensed matter and wavelengths of the same order as inter-atomic dimensions—is of the first importance to the new subjects. There are several other favourable circumstances: the velocity of the neutrons (hundreds of metres/sec) is suitable for mechanical measurement. For many materials, neutrons penetrate deeply before being captured or scattered; this allows many things to be done which are not possible with other radiations.

The Bibliography gives ample access to the various branches of the subject: neutron optics, neutron diffraction and neutron spectrometry, all possibly involving nuclear/magnetic scatterers. Reference [3], *Fifty Years of Neutron Diffraction,* has many articles from several points of view, by 35 authors (including Dr. Shull and myself), as well as reprints of the 1936 papers of the very first workers in the field. A much larger bibliography, compiled up to 1974 by Larose and Vanderwal [Q] shows the remarkable growth in the number of publications in the subject per year, from the 50 or so in the 1950s to about 800 by 1974 and no doubt many more in recent years.

Thus the fields have grown to the extent that one person, or even a committee, can no longer be fully consonant of what has been done, of what is in the literature. For the larger literature of science in general of physics in general, this is true *a fortiori*; no one—no one at all—can be sure of the nonexistence of some particular results, relevant to his projects. In general this is insurmountable, but physics moves step by step with technology into the future; it is the availability or nonavailability of technologies in the open market which defines the "present" or "state of the art" of research. Neutron spectrometry could not flourish until large neutron sources were for some reason available. The researchers of Dr. Shull and his colleagues, like those of myself and mine, were by-products of programs set in motion for other reasons.

THE GRAND ATLAS

This brings us to the Grand Atlas, viz., the scientistic literature of the state-of-the-art present. There can be histories of the Grand Atlas, but only historical artifacts and theories of history are contained within it. The Grand Atlas includes the "petit atlas"—earthly geography and geology in their most reliable current scientistic guises—and the scientistic, currently accepted "recipes" of physics and chemistry. Thus the Grand Atlas comprises "maps" of the world we live in, metaphorical maps which just might prove to be metaphysical, maps that link percepts (which can be thought of as metaphorical geography) with other percepts, by means of theory (which involves concepts and which can be thought of as metaphorical geology). The percepts are sense data: meter readings, features in a photograph, and the like. The concepts are names involved in the apposite theories. But it does not do to forget that percepts are themselves theory based, that beyond the photograph is the theory of photography—and of the photographer.

The observations (like the theories) are made by reputed scientists; the observers (like the theoreticians) have done their work openly and reported it in ways that permit repetition. The maps of the Grand Atlas can be checked by other scientists; they are scientistic—purportedly backed by men who have a right to an opinion on the matter. Whether the maps are to be called "scientific facts" is to be known only after-the-fact, *a posteriori*, by users of the maps. Similarly, whether a particular map is to be thought of as pertaining to metaphysical reality (or even to actuality now) is a religious and philosophical matter.

For strictly scientific or technological purposes, all this is irrelevant. On a pragmatic view, as on a religious view, theory and concepts are held in faith. On the pragmatic view, the only thing that matters is that the theory is efficacious, that it "works," and that the necessary preliminaries and side issues do not cost too much in time and effort. Beyond that, theory and concepts go to constitute a language in which the scientistic matters at issue can be formulated and discussed.

Of course, we would like much more than that. We would like to know whether it is possible to say of "Alberta 100 million B.C.": "Here be dinosaurs" or, at least, "Here were dinosaurs"; to say of an aluminum crystal: "Here be Al atoms"; to say of the theoretical nuclei of the conceptual Al atoms: "Here be protons and neutrons"; to say of the mathematical Al crystal lattice: "Here be phonons." We would like to be certain of what the Grand Atlas ultimately means. And, of course, once obtained, we might not like that knowledge at all. As it is, each of us takes his own scientismic view of these matters, a more or less conscious aspect of his religious views on "the nature of things."

At a given epoch of the "state-of-the-art," there are applications visible—technological or scientific applications—and also, perhaps, moral implications which go to forbid or enjoin them. And there can be metaphors visible, which modelled upon, may ultimately find places as theory held in faith, in the Grand Atlas. So that metaphors, too, are to be watched for their moral implications; nuclear fission, nuclear fusion are examples. Might it not be better that these notions never have been thought?

Thus concepts which are not immediately testable by experience—which are not themselves scientific—remain always questionable. At any given epoch physics are true by definition; if not true now, the physics in question never were true. The same is true generally of "the scientific." In the world to which the Grand Atlas applies, there is an enduring tension: additional evidence increases the reasonableness of accepting the concepts as actual entities or even as moral, not merely mental, realities—but the burden of proof can always be shifted to the opposite side. Observation or experimentation can suggest an addition or modification to a theory or can support an existing implementation of the theory. Experiment can refute or falsify such an implementation, but a new implementation may be found which agrees sufficiently with experiment to be acceptable. At a deep enough level in the intellectual structure, a theory is very difficult to refute.

SLOW NEUTRON SPECTROMETRY

Theories of the physics of condensed matter involve the most basic aspects of modern physics: the principles of conservation (energy, linear momentum, etc.), the chemical elements in various ionic forms, electrons, neutrons, quantum mechanics. Implementations for a particular substance in a particular setting usually involve drastic approximations if the required quantum statistical calculations are to be possible. Happily, because the nuclear and magnetic interactions between the neutron and atom are (in some sense) weak, the very good "first Born approximation" is applicable, and the neutrons are effectively "decoupled" from the dynamics of the scattering system which can be considered in isolation. The neutron, in being scattered, "causes" transitions between the quantum states of the scattering system, but does not change the states.

In general, the scattering system is modeled, with assistance of experimental information about it. For example, the chemical constitution would be taken as given. An aluminum crystal would be taken to have the face-centered-cubic (fcc) structure indicated from x-ray or neutron diffraction; theoretical calculation would be unlikely to produce this *ab initio*. The dynamics of the so-defined system could be studied *a priori* by simple models, or could be studied using a phenomenological theory with numerous parameters to be fitted. The Born–von Kármán theory [O] of crystal lattice vibrations can host either approach, at least to some degree. From the model, macroscopic properties (optical, thermodynamic, and the like) can be calculated and compared with experiment. Until the advent of neutron spectroscopy, only qualitative comparison was normally possible—the "distance" between the concepts at the atomic level and the percepts at the macroscopic level was simply too great.

Normally neutron energies (with their associated momenta) are inferred from their wavelengths as determined by diffraction from a single crystal through the Bragg law, or from their times of flight over a known distance. In each case the energy is proportional to the inverse square of the measured quantity. Thus, at the theoretical level, spectrometers based on the two methods have the same law of dispersion. (Any measurement of the same spectrum by the two methods is implicitly a test of quantum mechanics.) However, crystal spectrometers have additional dispersion from the Bragg transformation from angle of reflection to wavelength. The two methods are very different in their technical aspects and the associated experimental problems.

The first method considered at Chalk River (in 1951), and (I think) elsewhere, was to employ time of flight to measure both incoming and outgoing neutrons. The "double-chopper" method was to employ electrically phased choppers to select the energy of the incoming neutrons and the time of flight from the second chopper to the detector to infer the energy of the scattered neutrons. (Bernard Jacrot, at Saclay, built and employed for a time [26] a double chopper consisting of two large wheels on a rotating shaft.) It was soon apparent that development of a competent double-chopper instrument would be technologically demanding. So at Chalk River attention shifted to the use of crystal diffraction for both energy measurements—the "triple-axis" neutron crystal spectrometer, which will be discussed here later. The double chopper later received extensive development elsewhere, particularly at the hands of Peter Egelstaff at Harwell. Mixed systems were also developed, of which the "rotating-crystal" spectrometer [43] is probably the best known. This system employs a spinning crystal as initial monochromator and measures the spectrum by time of flight.

There is also an historically important method—the "filter-chopper" or "cold neutron" method [29–32]. The concept originated at Brookhaven, I think. A modified version was developed at Chalk River, largely by my colleague Alec T. Stewart. The method employs a polycrystalline filter of certain materials (notably beryllium metal) with particularly low neutron absorption and "parasitic" scattering, to eliminate all slow neutrons except those of such long wavelength (so slow) that they cannot be Bragg scattered out of the beam by the filter. The scattered neutrons are then analyzed by a chopper-based time-of-flight apparatus. In the Chalk River version, two filters were employed (Be and Pb), in differential fashion.

An inverse of this method, the "filter-detector" or "beryllium detector" method [43], employs crystal diffraction for the monochromator and detects just "cold" neutrons. It has uses in studying those modes of motion in condensed matter of particularly high energy.

Absorption methods

A number of elements exhibit large capture cross sections for slow neutrons. Boron and lithium and others have capture cross sections proportional to the wavelength; cadmium (and some rare earths) have neutron resonances with cross sections steeply varying with energy. When such absorbers intercept the neutrons scattered by a specimen, the measured intensity of the scattered neutrons depends on the neutron spectrum. In fact, the relative transmission of the scattered neutrons, for a sequence of absorbers running the gamut from very thin to very thick, is in essence related to the spectrum by a Laplace transform. The first experiments in neutron spectrometry, done in 1951—52 at A.E.R.E. Harwell by P.A. Egelstaff [21] and R.D. Lowde [23] and at Chalk River by Brockhouse and Hurst [22], employed such methods, but in very different ways. Only the last noted will be discussed here.

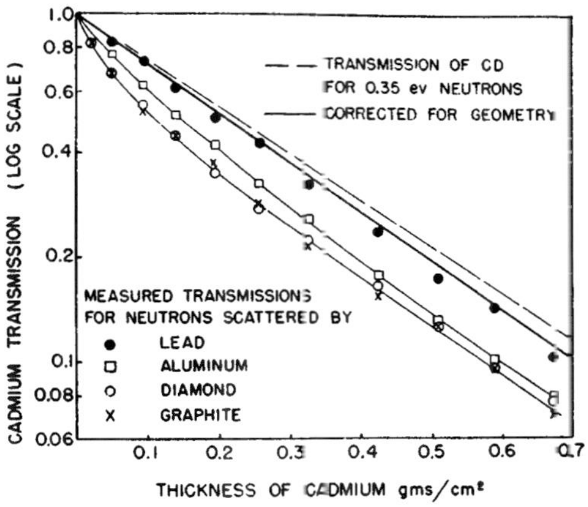

FIG. 2. The relative transmission of neutrons scattered into an annular detector by thin specimens of four materials, for annular Cd absorbers of various thicknesses, as well as the estimated transmission of elastically scattered neutrons. Each point is the result of measurement with a different thickness of absorber [22]. Scattering by the heavy element Pb shows little effect—as expected.

Over the range 0.2 eV to 0.5 eV the total cross section of Cd drops from over 6000 barns to less than 150. For an incident energy of 0.35 eV the transmission of Cd absorbers is quite sensitive to the scattered-neutron spectrum. Figure 2 shows transmission curves of annular Cd absorbers, for neutrons of initial energy 0.35 eV, scattered through a mean angle of 90°, into an annular detector, by satisfactorily "thin" scatterers of lead, aluminum, graphite, and diamond powder. The expected results for elastically scattered neutrons (with corrections for geometric effects) are also shown. (Some small corrections for multiple-event scattering were also made.) As expected, the neutrons scatter from the massive Pb atoms with little change in mean energy. For the lighter atoms Al and C, considerable energy transfer is indicated.

The results were compared with a calculation, by Finkelstein [5], of neutron scattering by a polycrystal, using the Einstein model for the atomic motions. In the Einstein model the atoms are considered to vibrate independently at a determined frequency and thus to be oscillators whose energies are integer multiples of an Einstein quantum energy (E_q), which can be estimated from data on the specific heat. In scattering, the neutron can create or annihilate one or more quanta with consequent change in its energy. For Pb and Al, multiple-quanta transitions predominate; for the two carbon specimens, the scattering is mostly zero quantum (elastic) with about 20% single quantum. The scattered spectra so calculated can be used to generate Cd transmission curves.

For Pb the effects are small; the agreement is satisfactory but not very significant. For aluminum and the two forms of carbon, agreement was obtained with values of E_q which were compatible with specific-heat data. At room temperature, aluminum is in a "classical" regime ($E_q \sim kT$), and

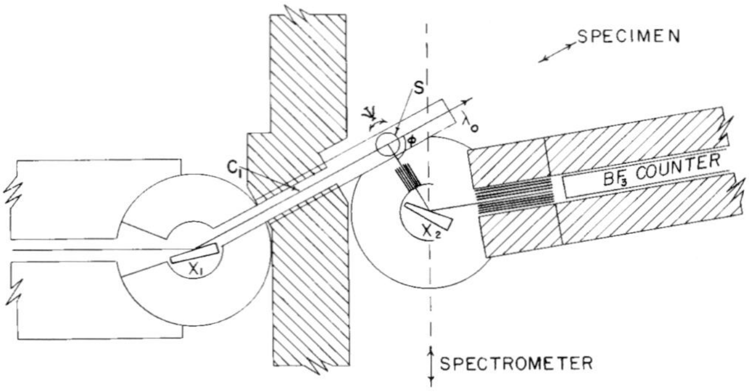

FIG. 3. The first crude version of the triple-axis crystal spectrometer [24–26]. Monoenergetic neutrons are selected by the large single-crystal monochromator ($X1$) and impinge on the specimen (S), which is located on a table whose orientation (ψ) in the horizontal plane can be selected. This table can be moved along the direction of the incident beam as desired. The analyzing spectrometer, which employs crystal $X2$, is a diffractometer (of especially large aperture) which can be translated as a unit; the angle (ϕ) through which the examined neutrons are scattered is determined by triangulation.

graphite and diamond are in a quantum regime ($E_q \gg kT$); so the results here are very satisfactory.

Further calculations were made for the very unrealistic model in which the specimens were considered as ideal gases at room temperature, with atomic masses 12 for C, 27 for Al, and 207 for Pb. The continuous spectra obtained for Pb and Al were very like the histogram spectra for the Einstein case, but the spectrum for the forms of carbon were markedly different. Thus the results suggested that details of the dynamical behaviour were not likely to be visible except where single-quantum scattering dominated.

Had available neutron fluxes been smaller, there might well have developed a small field of absorption neutron spectroscopy, though it is unlikely that I would then have had the opportunity to give this lecture and to write this report. But events soon overtook the development, as will now be discussed.

Triple-axis spectrometer

As recounted earlier, the decision was made in 1951 to develop a neutron spectrometer which employed Bragg reflection by two single crystals as the monochromating units. The beams were to be 2 in. square, larger than with existing instruments. Where distances were short, the angular resolution was to be defined by Soller slits. The concept required low backgrounds and thus heavier shielding than was then employed. Monochromators would clearly have to be more efficient than the existing NaCl ones; so a program for growing large metal crystals was instituted, which focused for the moment on lead and aluminum. Hurst was embarking on a low-temperature program, which looked forward to studies of the rare-gas liquids, especially liquid helium. Chalk River was already well supplied with liquid nitrogen; after some abortive attempts, a "Collins" helium liquefier was eventually installed. In 1952 an experienced low temperature physicist, David G. Henshaw, arrived and took over major responsibilities for the crystal-growing and liquefier programs.

A primitive triple-axis spectrometer was set up at NRX in 1952. The intensities were too low for practical work, though the elastic peak from paraffin wax was seen and probably also that from vanadium. But in the late autumn of 1952 the reactor suffered an accident, and experimental work terminated until the summer of 1954. Meanwhile Henshaw and Jack Freeborn had been successful in growing large single-crystal ingots of aluminum and lead. So that when work resumed at NRX, both the powder diffractometer and the triple-axis instruments were empowered by much superior monochromators. Hurst and Henshaw embarked on studies of the rare-gas liquids, instituting an ongoing program which, in successive forms and hands, continued for over 40 years.

The initial form taken by the spectrometer is shown in Fig. 3. The specimen is mounted on an indexed table which translates along the direction of the "monochromatic" neutron beam from the initial monochromator. Another single-axis spectrometer [7], which views the specimen through Soller slits, translates along the rails parallel to the reactor face. The angle of scattering is calculated by triangulation and hence is changed only on occasion.

The triple-axis program met with almost immediate success. Some *ad hoc* additions to shielding were added, but otherwise the apparatus performed as expected (see Fig. 3). A variety of experiments were performed, some guided by existing theory and some not. A contributed paper, which turned into an invited paper [24], was given at the annual meeting of the American Physical Society in New York City in January 1955. Preliminary results were presented for scattering from liquid and solid Pb, water, heavy water, vanadium, and from certain paramagnetic materials with differing characteristics, including $MnSO_4$, Mn_2O_3, and MnO.

Several of these projects were continued and eventually reached publication, the first being the results for the two paramagnetics $MnSO_4$ and Mn_2O_3 [25]. $MnSO_4$ has a susceptibility fairly obedient to the Curie law and hence it is expected that the interactions between the Mn^{++} ions are small; for the same reasons, it was expected that the diffuse magnetic scattering should be almost elastic. In fact, the scattering at angles smaller than the Debye-Scherrer lines had virtually the same energy width as the incoherent elastic scattering from vanadium. For Mn_2O_3, on the other hand, there was substantial inelastic scattering with energy width consistent with a calculation of Van Vleck [15]. Similar results were obtained for MnO, in which the energy widths correlated with the short-range-order indications in the diffraction pattern. These latter results, continued successively with P.K. Iyengar and R.S. Weaver, did not get properly published, but were reported in review papers.

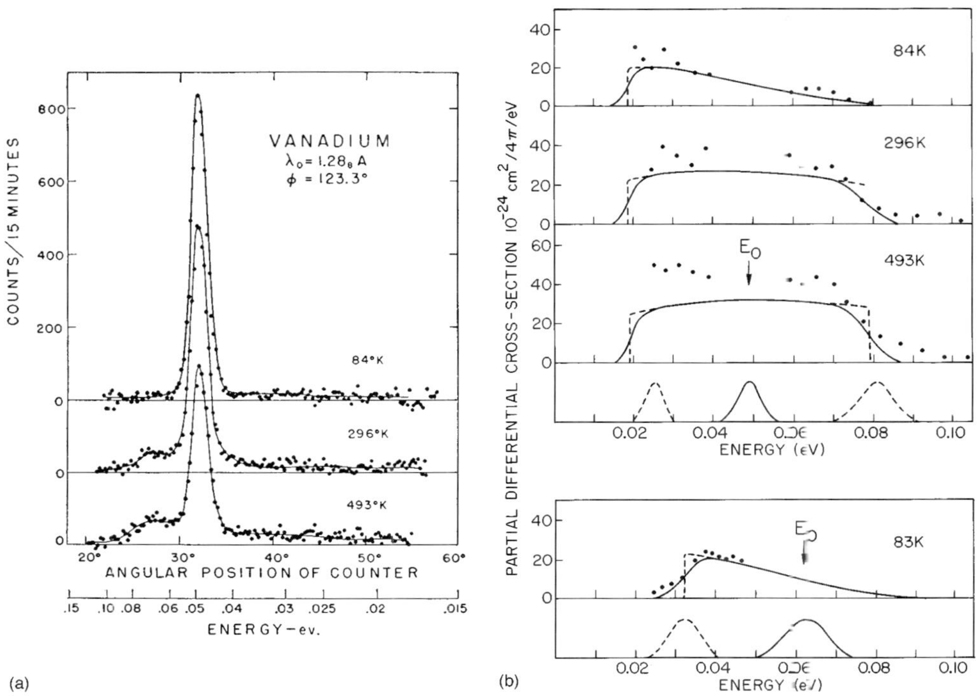

FIG. 4. (a) Measured energy distributions of neutrons scattered by vanadium as functions of the angular settings of the analyzing spectrometer, for three temperatures, with backgrounds subtracted and intensities normalized [27]. The intensity of the elastic peak decreases with temperature in about the manner expected from Debye-Waller theory. The inelastic component is affected by the extreme dispersion of neutron spectrometers and requires transformation to a true energy distribution. (b) The transformed energy distributions (points) of the inelastic component in vanadium [Fig. 4(a)]. The curves are calculated (one-phonon) cross sections for a Debye frequency distribution for the lattice vibrations. The elastic component provides a measure of the resolution function at the incident energy.

From the beginning our program had been directed towards study of the frequency/wave-number (energy/momentum) dispersion relation of the normal modes (lattice vibrations) in crystals, by means of study of the one-phonon scattering. (By analogy, it was assumed that spin waves in magnetic crystals could also be studied, as turned out to be the case.) In 1954 a theoretical paper by Placzek and Van Hove [9] discussed the phonon experiment in detail. More importantly, for us, they proposed a different kind of experiment—to determine the frequency distribution of the normal modes from the incoherent one-phonon scattering by a cubic crystal with one atom per unit cell. From a practical point of view, vanadium is the only natural element which is a candidate for this experiment, which was added immediately to our program.

Such an experiment was carried out [27] on the primitive triple-axis installation of Fig. 3, with the results shown in Fig. 4(a). A strong elastic peak, whose intensity decreases with increasing temperature, is visible—as expected from a Debye-Waller thermal cloud of increasing size. A weak inelastic component, whose intensity increases with increasing temperature, is also visible; it is this in which we are mainly interested. When this component is transformed from the distribution in (analyzer) angle to a distribution in energy, the curves of Fig. 4(b) are obtained. An Einstein model can be ruled out by inspection. One-phonon distributions calculated on a Debye model, with Debye temperature consistent with the Debye-Waller behaviour of the elastic peak, are also shown. (The additional scattering can reasonably be attributed to multiphonon processes.) Thus the experiments supported the structure of the theory, but the resolution did not permit improving upon the Debye model for the frequency distribution of the normal modes. Later experiments, using the filter-chopper method, by Stewart *et al.* [32] at Chalk River and by Eisenhauer *et al.* [31] at Brookhaven, showed plainly structure in the frequency distribution, indicating [see M–O] realistic structure for the lattice vibrations beyond the capabilities of a Debye model.

Least guided by theory and arguably the most important, were the experiments on liquids, particularly those on light and heavy water [33]. Here quasielastic scattering was found, whose energy widths could be qualitatively related to the manner in which diffusion went on in the liquid. Analogs to the Debye-Waller factor for the quasielastic component and the energy widths of the inelastic components in the scattering could be related to more general aspects of the thermal motions of the molecules. In the later work, by the author and by Sakamoto *et al.* [33], a high-resolution rotating-

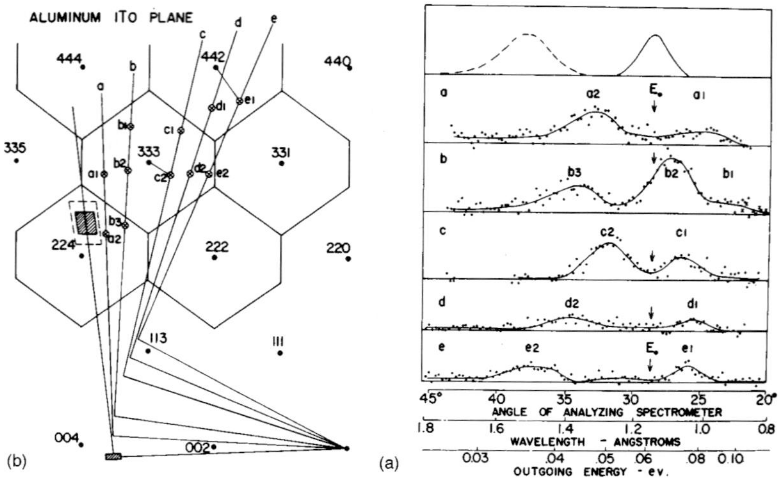

FIG. 5. Phonons in aluminum at room temperature [26]. (a) Some measured distributions in the (1,1,0) plane of a single crystal at a fixed angle of scattering of 95.1°. (b) The reciprocal lattice for the (110) plane with the crystal momenta plotted for the experiments in (a). One typical case is shown in detail as well as a crude depiction of the resolution in momentum.

crystal spectrometer was employed, as well as the triple-axis instrument. The developments by Van Hove [13], of generalized (space-time vs momentum-energy) correlation functions, contributed to thinking about this notoriously difficult area, the physics of liquids.

The phonon experiment envisaged from the beginning—almost the *raison d'etre* of the program—was carried out on a large single crystal of aluminum, by Brockhouse and Stewart [28]. This material was selected for technical reasons: Al has little neutron capture and incoherent scattering, and the coherent scattering cross section is small; so the physical size can be large without leading to large multiple-scattering effects. And we had the crystal! The first results were obtained on the early crystal spectrometer with its fixed angle of scattering. The crystal was oriented with its (110) plane in the plane of the spectrometer and was turned a few degrees about an axis normal to that plane, to carry out successive experiments in successive orientations, as shown in Fig. 5. Neutron groups, with the expected properties of one-phonon groups, are observed in the different experiments, at different settings of the analyzing spectrometer and with different intensities. (Multiphonon scattering would show a continuum.)

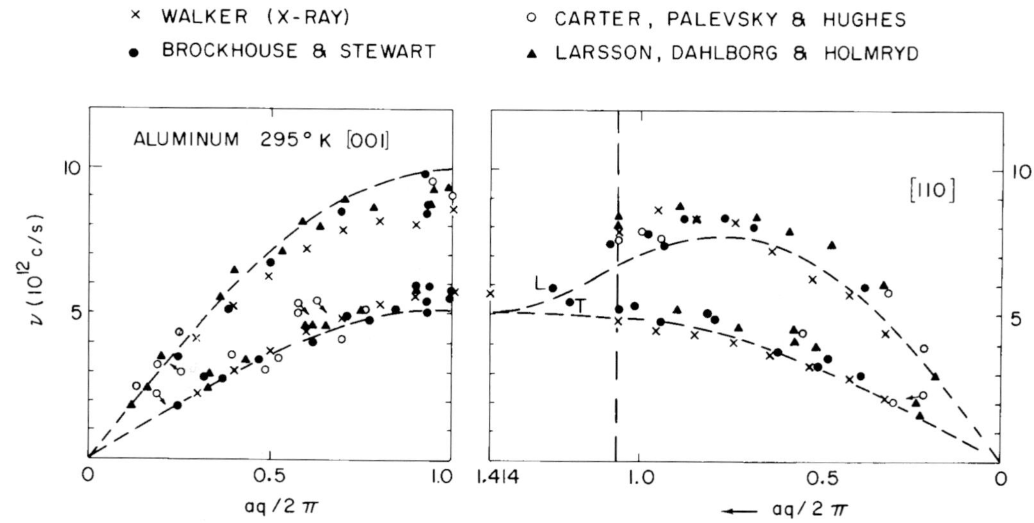

FIG. 6. Dispersion curves for aluminum at room temperature, for two symmetry directions in the crystal (collected results [28–30]). The neutron measurements from three laboratories are consistent with each other within the errors and are almost so with those inferred [18] from intensities of scattered x rays. Nevertheless, the discrepancies would be very important in attempts to extract interatomic forces from the data.

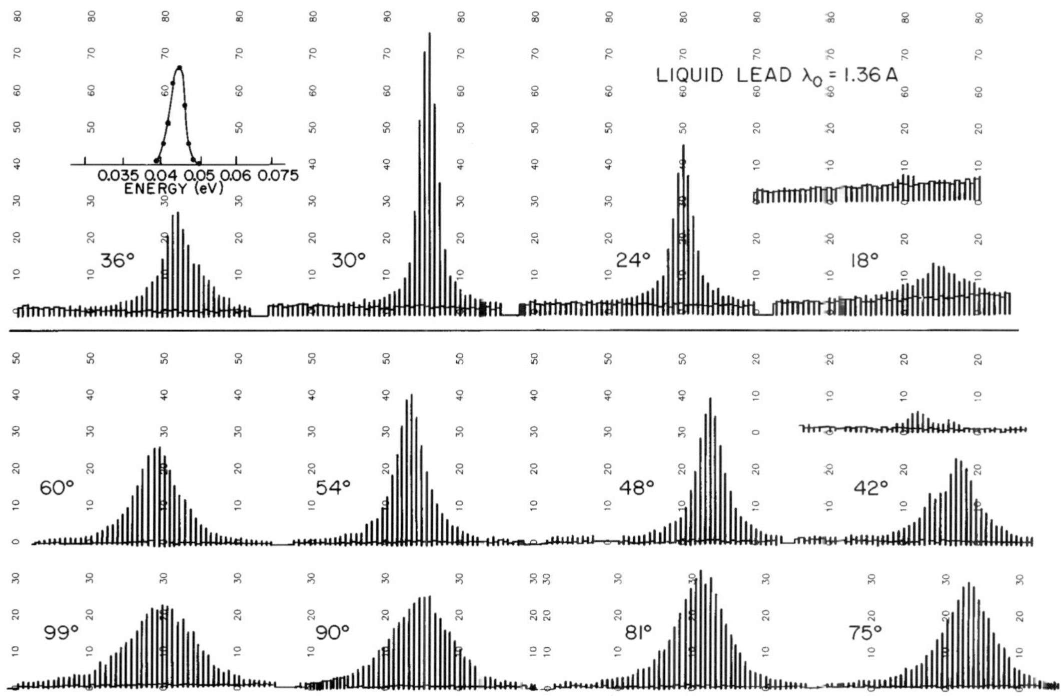

FIG. 7. Histograms [43] of energy-analyzed distributions of neutrons scattered by liquid lead at a series of scattering angles [34]. The scattering angles are roughly related to the momentum transfer Q; the abscissa angles, to the energy transfer. Appropriate measured backgrounds are also given. The resolution function measured with a vanadium scatterer is shown at the upper left.

From the momentum diagram, the wave vectors (**q**) of a phonon can be deduced; and from energy conservation, its frequency (ν). From a number of phonons with **q** in the same direction, it can be made plausible (not shown) that the phonons obey a frequency/wave-vector dispersion relation.

The experiments were continued on the differential filter-chopper as well as on the triple-axis apparatus, some 210 phonons distributed over two symmetry planes of the crystal's momentum space (reciprocal lattice). The results permitted refutation of the simple models for the interatomic forces then extant. The Born–von Kármán theory itself can be taken to be phenomenological with innumerable possibilities for parameters, and thus hardly to be capable of refutation. If taken literally, as involving forces between ions in the crystal, then already these early results show surprisingly long-range behaviour for the interatomic force system. In the work, intensities were used to assign polarization to phonons in symmetry directions. Henceforth measurements were made in symmetry directions as far as possible, in order to reduce the difficulties of analysis.

Dispersion curves for the phonons in two symmetry directions of single crystalline aluminum are shown in Fig. 6, as determined in the above experiments and by Carter, Palevsky, and Hughes at Brookhaven [29] and by Larsson, Dahlborg, and Holmryd at Stockholm [30], both groups using the beryllium filter-chopper method. The various experiments agree within their still substantial uncertainties. Dispersion curves deduced by Walker [18] from his x-ray intensity measurements are also in substantial, though not complete, agreement. Because the frequencies from neutron experiments are deduced directly from the conservation rules, while those from x rays require difficult calibrations and corrections, the neutron frequencies are believed to be the more reliable in principle.

Experiments were carried out to study spin waves ("magnons") in single-crystal magnetite (Fe_3O_4). This cubic ferrite ("lodestone") was selected because large single crystals of natural origin were available. It was expected that a more or less isotropic, quadratic dispersion curve would be found at small wave vectors, close to positions in reciprocal space where strong Bragg scattering occurred. This was indeed found [39] to be the case. In a later experiment [39], it was verified that the neutron groups concerned showed the change of intensity expected when a suitable magnetic field was applied, and thus that the quanta concerned were indeed magnons and not phonons.

Early results for liquid lead were not easily analyzable. Because of the large mass of the Pb atom, the energy transfers are small. The narrow inelastic component would not be separable from a quasielastic component, if the latter existed, because of the instrumental energy resolution. So the early results were never published, though they were presented at

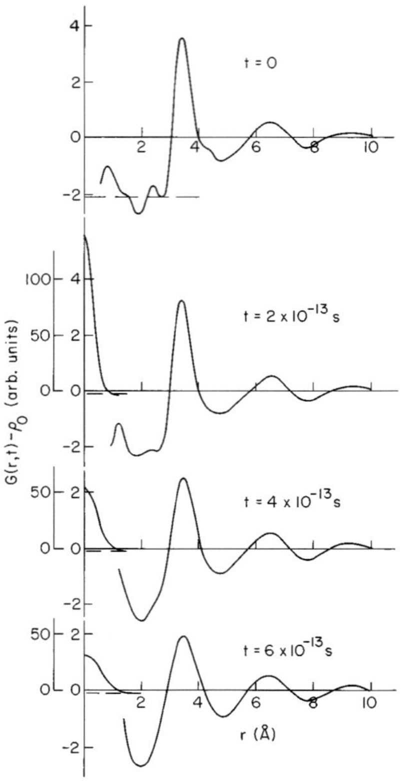

FIG. 8. Van Hove transformations of the data [34]. The distribution for time $t=0$ is essentially the familiar pair distribution of x-ray scattering experiments. The peak (δ function) at the origin gradually diffuses outwards with elapsed time t, as do the first and subsequent neighbor peaks, in a sort of moving picture of what goes on in the liquid.

conferences. But with much improved facilities available, new measurements (Fig. 7) with Noel K. Pope [34] were analyzed using the Van Hove [13] transformations to give directly the self- and pair-correlation functions of the atoms in liquid lead (Fig. 8). The temporal development of these functions, from the defined origin at $t=0$ to the vanishing of correlation some 10 picoseconds later, provides a sort of statistical "movie" of the atomic motions in the liquid.

In a wide-ranging memoir such as this, it would not be right to omit mention of the work [35–38] on liquid helium, even though I had little part in it. From early on, Henshaw and Hurst had looked forward to including study of the inelastic scattering of neutrons by liquid He as part of their program on the rare-gas liquids. But the experiment looked formidable, as to both execution and analysis of results. This appraisal changed abruptly with the publication of a theoretical paper by Cohen and Feynman [35], who predicted a "line" spectrum instead of the continuous spectrum of unknown behavior which had been assumed. Preparatory work was immediately started. Palevsky, Otnes, and Larsson [36], at Stockholm, found the predicted line ("roton") spectrum, using the beryllium-chopper method. This was followed closely by work by Yarnell *et al.* [37] and by Henshaw [38], improving and extending the Stockholm results. At Chalk River, work has continued on various aspects of liquid-helium physics almost to this day.

Developments in apparatus and methods

In 1956 the instrument had been completed by construction (by the Engineering Design Branch at Chalk River) of a proper analyzing spectrometer which was set upon the arm of the original [7] single-axis instrument, whose "half-angling" table now carried the specimen. With this new triple-axis instrument, all variables were motor-driven except the initial monochromator on the first (fixed) axis. Thus the initial energy was changeable only by changing the monochromator. When the spectrometer was moved to the NRU reactor in 1959, the first axis was also engineered (by W. McAlpin) so that the initial energy (E_0) was also controllable within an experiment. Figure 9 shows the spectrometer in place at hole C5 of the NRU reactor. Figure 10 shows a schematic diagram of the arrangement. It might be noted that the relative arrangement of the incident and outgoing beams has importance because of correlations which affect the resolution in momentum space.

In all the work the intensity of the incoming neutron beam is monitored by means of a "thin" neutron detector of known efficiency. In the experiments so far discussed, it was the analyzing spectrometer which was changed and, consequently, the outgoing energy (E') which was varied. Corrections to the intensities for the efficiency of the analyzer had to be made. For a crystal spectrometer, these are not readily calculable; so suitable auxiliary measurements were often required. If the analyzer is kept fixed, its efficiency does not enter the problem unless absolute partial differential cross sections are desired. So from 1959 on, distributions were often measured at constant outgoing energy.

It had been recognized in crystallography that wave-vector (momentum) transfers, not the angles of scattering and crystal orientation, were the "natural variables." This had shown up also in our original consideration, in 1950–51, of the theoretical analysis by Weinstock [19] and others, of the scattering by phonons in crystals. Over the years it came to seem obvious that the neutron, in scattering, conserved energy and momentum with the specimen scatterer—the transferred momentum ($\mathbf{Q}$) might be carried by the specimen as a whole or by entities within it or by a combination of the two. The specimen being so large, the energy transferred ($\omega=E_0-E'$) is inevitably carried by the quasiparticle entities within it. When no such entities are involved, the scattering is "elastic." So, in thinking about an experiment, one can start with energy-momentum ($\omega,\mathbf{Q}$) conservation, not have to reach it at the end of a calculation. This point of view was reenforced by the evocative Van Hove transformations. The intensities are then governed by a generalized structure factor analogous to the structure factors of neutron and x-ray crystallography. Of course, as for quasielastic scattering, some care must be used, in such a semiclassical approach, that the necessary internal entities are considered.

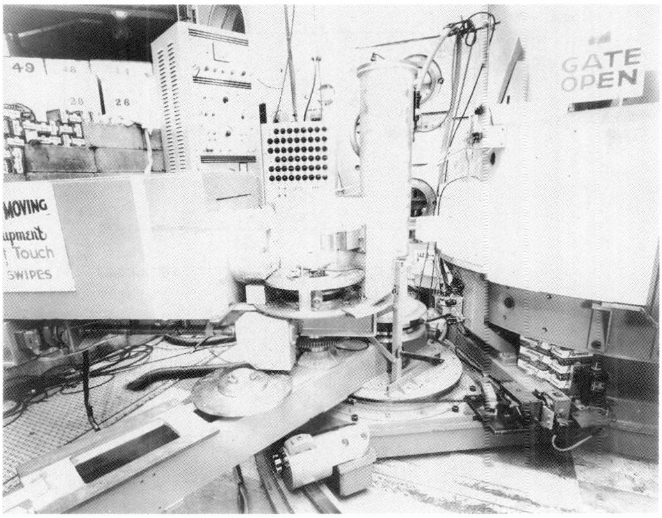

FIG. 9. Photograph (1959) of the original triple-axis spectrometer at the NRU reactor at Chalk River. The bank of 52 rotary switches, preset to go through an energy scan of up to 26 points, can be seen in the upper centre. One of the three variables involved would traverse linearly through the domain of settings desired, while the other two advance nonlinearly according to the settings of the appropriate switches. (AECL photo.)

This was the era of the early development of digital computers. Chalk River obtained such a facility about 1954; programs written in a predecessor of FORTRAN could be run on the Datatron to print out "hard copy" results of a sequence of calculations. The calculations could provide a sequence of angular increments of the variables of a triple-axis spectrometer: the angles of the analyzer and the initial monochromator, the angle of scattering, and the angular setting of the specimen in the plane of the spectrometer. Figure 10 shows the schematic diagram.

In operation, one variable would be held constant; one variable moved linearly in constant increments. The other two variables would move nonlinearly in the increments provided by the computer. Initially these were set by means of

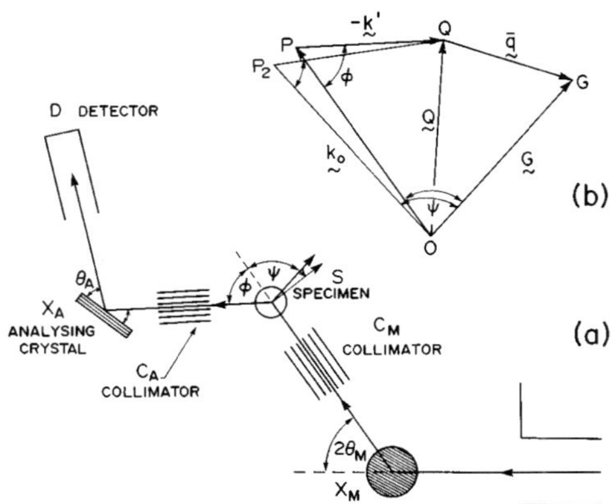

FIG. 10. Schematic drawing of a triple-axis spectrometer together with the momentum space diagram OPQ in the same coordinate system. The second diagram OP_2Q indicates a second point on a "constant-**Q**" curve at fixed incoming energy [43]. The massive shielding required, particularly around the monochromator, is not shown.

52 rotary switches, which can be seen in the upper centre of Fig. 9. Thus an experiment ("scan") could involve up to 26 points. A stepping switch would accept the two instructions in sequence and match them with the number of microswitch closings by the angular variables concerned. The overall angular traverse was set by the number of increments traversed by the angular variable using fixed increments. This system, largely designed by E. A. Glaser, was brought into operation in 1959. A bit later a system designed by W. D. Howell, which employed punch tape put out by the computer to transport the data to the spectrometer, was in turn installed. Figure 10 shows the initial arrangement for a scan in which the analyzer setting would be changed linearly and the angular variables set to give the desired momentum transfer. A second setting is indicated for a second point on the scan. And so on.

The method can be thought of in optical analog. There is an incoming beam of "green" neutrons and the analyzer goes through a "rainbow" sequence, the angular variables meanwhile being changed sequentially to keep the momentum transfer **Q** constant. Alternatively, the analyzer is set to detect "green" neutrons, and the incident neutrons go through the "rainbow" sequence. In the simple theory the results are equivalent, but technical matters may favor one over the other.

The variables, momentum transfer **Q** and energy transfer ω, are often given in units of Planck's constant $\hbar$ and are then referred to as, respectively, wave vector and (angular) frequency. The (cyclic) frequency ν is then equal to $\omega/2\pi$.

Subject to some geometrical limitations, an experiment could in principle be carried out over any track in **Q**-ω space. But the most useful has been the "constant-**Q**" or constant momentum transfer method. (The method of constant energy transfer has also been used, especially for steep dispersion curves.) The major advantage of these methods [43] is, of course, that the experimenter can get the data wanted for the analysis proposed, rather than having the experiment turn up what might be a great deal of unwanted

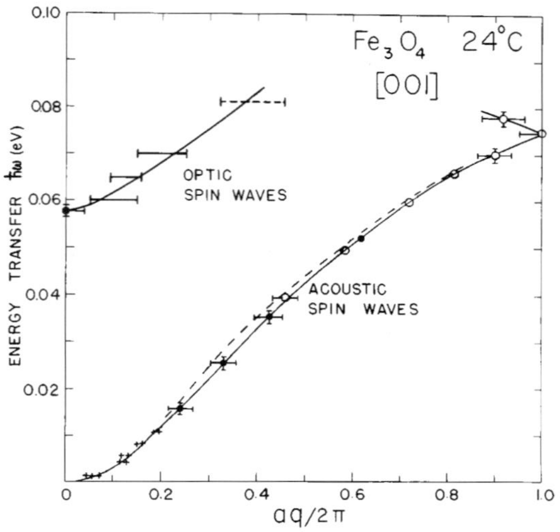

FIG. 11. Dispersion curves for spin waves in a magnetite (Fe_3O_4) crystal [40], showing the expected initial quadratic behaviour and the initial part of an optical mode.

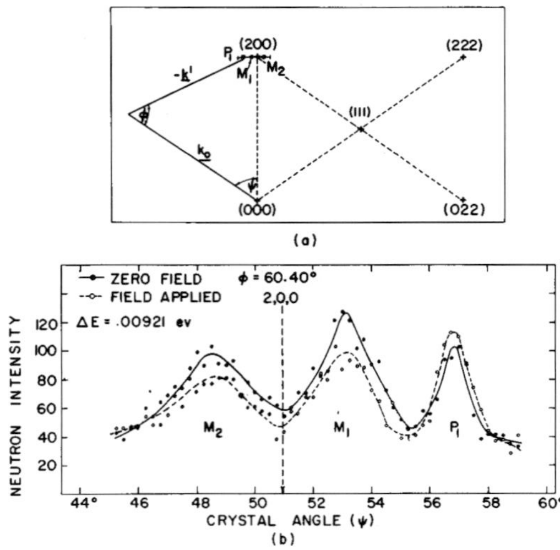

FIG. 12. Momentum diagram (a) and neutron groups (b) for a crystal of (fcc) Co (8% Fe). The magnetic signatures on the intensities indicate that M_1 and M_2 are groups for spin wave quanta (magnons) and that P_1 is a phonon group [41].

data. Of course, also, there is a downside to this—something unexpected might be missed. Good practice thus involves giving in any report the value of the fixed variable, usually the incident or outgoing energy, even though it does not appear explicitly in the analysis. (It may affect contaminant scattering.) There is another important virtue of the constant-**Q** method: for important technical reasons, the accuracy is greater and the intensities are more easily interpretable than with ordinary neutron spectrometry. (These facts are connected with the occurrence of the Waller-Froman Jacobian [17] in the theory.)

With the advent of the constant-**Q** ability and the considerable improvements brought about by the move to the new NRU reactor, experiments with much improved resolution and precision were possible. Technical advances were now concerned with making more precise certain "rules of thumb" regarding selection of points in momentum space for optimum resolution, discrimination between different symmetry types of neutron groups, and so on. Of course, as new components became available, such as ^{3}He proportional counters and better monochromators, these were eagerly adopted.

Using the new methods and the improved facilities at NRU, Watanabe and Brockhouse [40] were able to extend considerably the earlier work [39] on the spin waves in magnetite (see Fig. 11). The complete "acoustic" dispersion curve was obtained as well as part of an "optical" branch near the zone centre, both in qualitative agreement with theoretical expectations [4,40].

At the time it was still an open question whether spin waves existed in metals. Sinclair and Brockhouse [41] studied the neutron scattering in a fcc cobalt-iron alloy (8% Fe). (This crystal was available for use as a neutron polarizer. The magnetic and nuclear scattering are comparable, and this fact overrode the considerable objection that Co has a large capture cross section for neutrons.) The rather steep spin-wave dispersion curve could be measured only at small wave vectors (to about 20% of the zone centre) (see Fig. 12). It was established from the "signature" under an applied magnetic field that the neutron groups indeed represented spin-wave quanta (magnons).

Experiments of other types were carried out. The crystal-field spectra of a number of rare-earth oxides were studied [42]; but analysis proved intractable, and these will not be discussed here. Watanabe and Brockhouse [40] studied the exchange field splitting in ytterbium iron garnet. And a visiting colleague, L. N. Becka, studied rotational modes in a

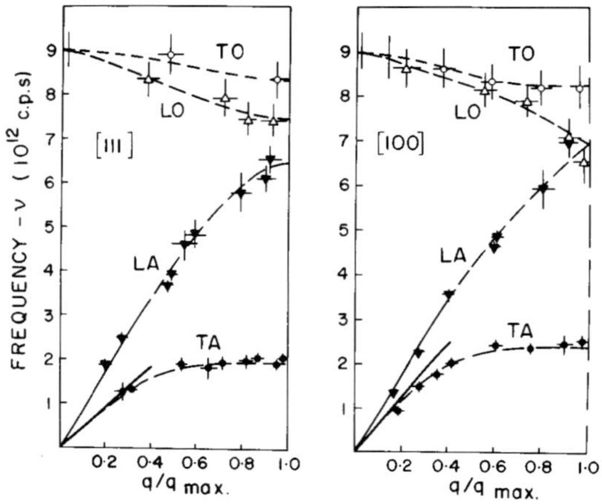

FIG. 13. Dispersion curves for germanium at room temperature, showing optical and acoustic branches in two symmetry directions [45].

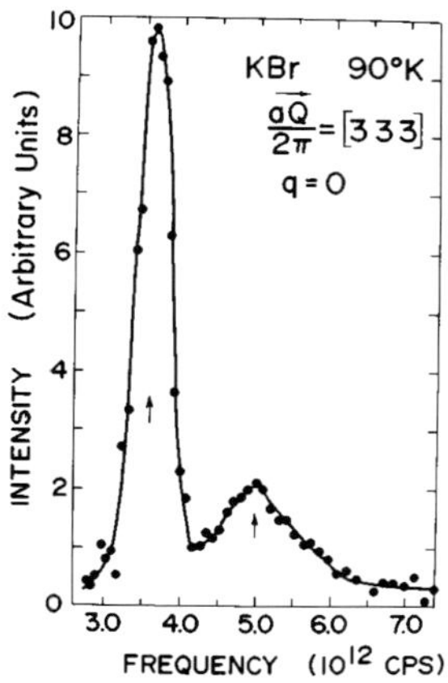

FIG. 14. The zone-centre phonons in a crystal of KBr [50]. The transverse optical mode is active in infrared spectroscopy, but the longitudinal optical mode is not. Both are visible in this neutron pattern, and the ratio of their two frequencies is about as predicted from theory [O].

number of organic compounds. But the major target of my program from here on was phonon physics.

Phonons, dispersion curves, and interatomic forces

Already, before the major acquisitions of 1959, there had been phonon studies which, though desirous of improvement in resolution and precision, were fully competent to provide important instruction on the physics of the specimens concerned. The first of these was a study of the lattice vibrations in two symmetry directions of germanium, by Brockhouse and Iyengar [45] (see Fig. 13). The dispersion curves for the acoustic and optical branches were well delineated and were analyzed in terms of simple interatomic force-constant models from the literature, but without complete success, indicating some missing factors in the models. Some elegant connections with far-infrared spectra were brought out, and the accepted placements of the maximum of the valence band and the minimum in the conduction band were supported. Further experiments were concerned with the phonons in silicon [46], and the homology found to exist with those in germanium, and with the temperature dependence in the phonons in germanium [47]. It should be noted that some results from Brookhaven (using the filter-chopper method) are noted in [45].

An extensive program was undertaken on alkali halide crystals. In selecting which of these to study, attention was paid to the neutron properties and to the mass ratio between the two component atoms. Two crystals were studied extensively, sodium iodide (NaI), initially [49], and then potassium bromide (KBr) [50]. In the course of this work our visiting colleague, William Cochran, invented his famous and useful "shell model" for polarizable ions in crystals. One of the results of these studies was the verification of the well-known Lyddane-Sachs-Teller formula [O] for the zone-centre optical phonons in terms of the dielectric constants (see Fig. 14).

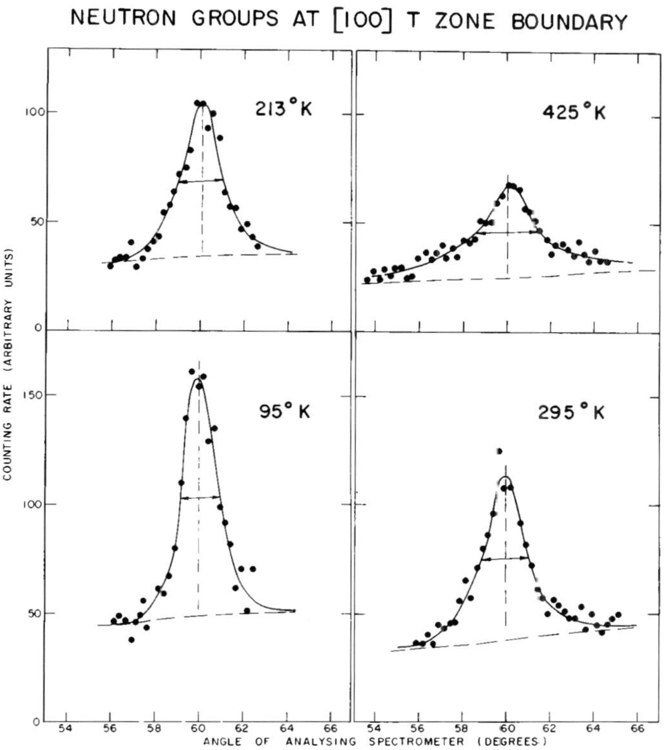

FIG. 15. Neutron groups for a symmetry phonon in a crystal of Pb at four temperatures showing linewidths increasing with temperature [60]. This received a natural interpretation as a reduction with temperature of the lifetime and coherence length of the phonon involved, because of interaction with other phonons.

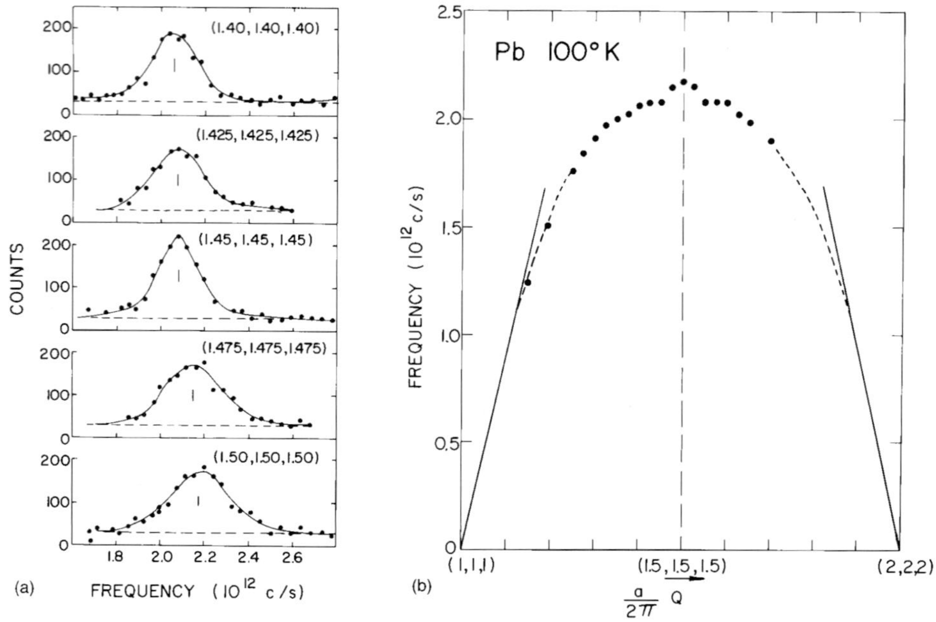

FIG. 16. Kohn effect in dispersion curves of Pb [53]. (a) A series of closely spaced phonons show anomalous variation. (b) In the corresponding dispersion relation $v(q)$, in this case at a caliper dimension of the Fermi surface in the $[\zeta, \zeta, \zeta]$ direction.

An extensive program on metals was carried out, with special emphasis on the metal (sodium) then considered to be the "simplest" and on a metal (lead) in which the phonon-electron interaction was known to be appreciable (from the fact that it is a superconductor with high transition temperature). And, as always, an important consideration was the availability of suitable single-crystal specimens. We had lead crystals from Henshaw's work on monochromators, and Raymond Bowers of Cornell University was able to grow the large crystals of sodium required.

Measurements were made on Pb [51] and Na [54] at a low (liquid-nitrogen-cooled) temperature for all the symmetry directions available, including some on the surface of the Brillouin zone. In both cases measurements were made also at higher temperatures, to study thermal broadening and shifts in the phonons (see Fig. 15). Substantial effects were observed, which proved difficult to interpret quantitatively in terms of atomic forces. For both metals at high temperatures, the phonon lifetimes and mean free paths became remarkably short at wave vectors far from the zone centre. Similar results for Al were obtained independently by Larsson *et al.* [30], using the filter-chopper method. Our high-temperature results received only incomplete or late publication.

Analysis of the results for Pb [51] displayed an interatomic force system of long range and great complexity. Calculations for alkali metals by Walter Kohn [52] suggested that anomalies in the phonon-dispersion curves, arising from electron-phonon interaction, might be found at wave vectors related to the Fermi surface of the metal. A series of measurements were made to search for such anomalies in the dispersion curves of Pb. The Kohn anomalies were indeed found [53] in repeated experiments at more or less the expected phonon wave vectors (see Fig. 16). A search was made [54] also in Na (for which the electron phonon is believed small) without positive result, supporting to some degree the assignment for Pb.

The results of Woods *et al.* [54] for sodium could be analyzed in terms of a Born–von Kármán model of much shorter range than for lead. Dixon *et al.* [55] used the model

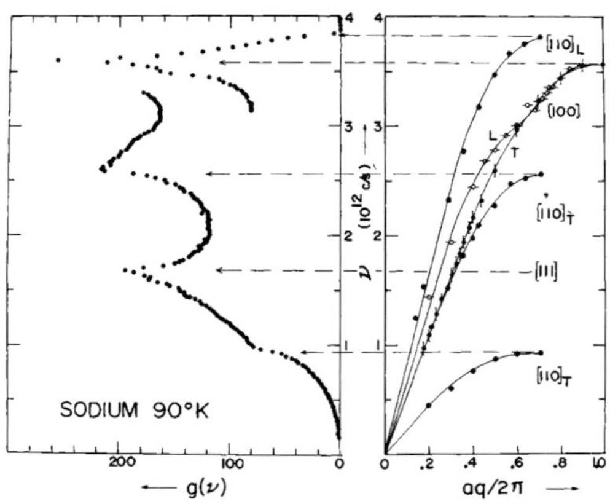

FIG. 17. The frequency distribution $g(v)$ for Na at 90 K [55], calculated from a force-constant model fitting the dispersion curves $v(q)$ also shown [54]. The relation between the frequencies at symmetry points and the Van Hove [14] singularities in $g(v)$ are exhibited.

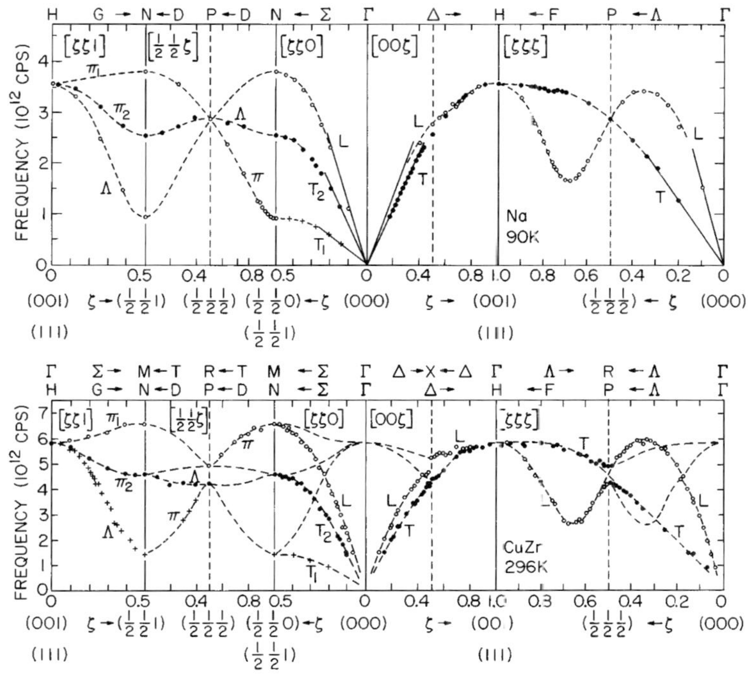

FIG. 18. Complete dispersion curves for the symmetry directions in Na [54] and in ordered CuZn [56]. The group-theoretical notation for the various branches are given as letters. Dashed lines without points represent measurements in symmetry-related directions of the zone.

to calculate the frequency distribution of the normal modes, which gave excellent agreement with specific-heat results from the literature (see Fig. 17). The computer calculation had a resolution capable of showing the topologically significant Van Hove critical points [14] with precision.

Finally, Fig. 18 shows a comparison of the complete results for (body-centered-cubic) Na and those obtained, on the same apparatus by Gilat and Dolling [56], for β-brass (CnZn) ordered with copper mainly on the cube corners and zinc on the body centres. It will be seen how much of the general appearance comes about from symmetry; the effects of what must be considerably different interatomic force systems are for the most part hidden in the fine details. [This is not always the case—Na and CuZn are both (probably) nearly-free-electron metals. Drastically different binding schemes can produce much greater effects.] The curves of Fig. 18 point up the effects of the symmetry breaking by the difference introduced by the ordering of the Cu and Zn atoms on the crystographically equivalent cube corners and body centres.

The results of Figs. 13–18 illustrate a basic fact about neutron spectroscopy of crystals. Despite the considerable amount of data shown, it should be realized that vastly more is potentially available in the off-symmetry frequencies and in the intensities of the neutron groups. But presentation and analysis are alike difficult: Presentation, because our intuition and visualization processes are limited to essentially three spatial dimensions and counted time; the data for crystals with one atom per cell require five-dimensional presentation and, for more complicated crystals, still more. Analysis, because it is not possible generally to assign a particular frequency or intensity to a particular formula—there is sorting out to be done first. Several of the works cited go into this in a small way, via calculations and experimental determinations of dynamic structure factors.

These works and others done over the next few years were extensively reported in review articles [60–63]. In these years, also, comparable results began to appear [48,57] from other laboratories. The field was now a field in being and not merely in development. My own program, now centered at McMaster University, continued to focus on phonon physics with emphasis on metals and alloys. Reference is made to work [44] by two of my first students there, Eric Svensson and Michael Rowe, on phonons in crystals with point mass defects. The initial experiments were on single crystals of an alloy of copper with 9.3 at. % of gold. The fraction of defects was larger than might be wished, but interesting results were obtained, which Svensson and Rowe subsequently improved upon.

Nature, neutron spectroscopy, and the grand atlas

The result of human action is the reward for that action. In physics experimentation we are guided in our choice of action by the existing theory and by calculations based upon that theory. If things turn out as expected, the reward is additional confidence in that theory. If the result is not as expected, then new possibilities for experimentation and theory construction are opened up. The early work on neutron physics resulted in little to challenge the basic theories of physics (quantum mechanics, in particular), but did, I think, add somewhat to the confidence with which that theory was held. With the models to be found in the literature of the time, the situation was different. I cannot recall a single instance in which experimental results were in quantitative agreement with a preexisting model calculation. Indeed, the results were usually qualitatively different, but in a manner in which the differences could indicate the ways in which the model needed modification if it were to describe the experimental results. The existing theory normally functioned well as a language in which to discuss the experiments, but had to be extended if it were to describe the results satisfactorily. And only for the phenomenological theory, with the possibility of numerous fittable parameters (as the Born–von Kármán theory), would the extension be "trivial."

The fact that theory functioned satisfactorily as a language meant that sometimes there was a possibility to compare directly neutron results with those from other types of experiment (such as infrared absorption) and thus, in a sense, to give an interpretation of the latter. But the major importance of the neutron experiments derives from their influence on the introduction of new microscopic models which can be applied to calculations of other quantities as well. With the vast numbers of "universes of discourse" (physical systems at a particular temperature, under a particular applied pressure, and so on), the question arises: how to decide what universes of discourse should be included in the Grand Atlas and to what extent. Chemical and crystallographic structure seem already to have been admitted for any and all such "universes." Are phonons and their dispersion curves and decay descriptions (their "laws"), magnons and their "laws," the numerous other quantum entities and their "laws" to be admitted? Are time-dependent distribution functions also to be admitted? Or are we to act as if the basic theories fully contain all these and that only those quantities are to be considered for inclusion which go to characterize the system? I have had opinions on these matters, but that is another story.

ACKNOWLEDGMENTS

In addition to the numerous colleagues who contributed long ago to the work discussed, many of whom are acknowledged in the text and references, I should thank also Atomic Energy of Canada Ltd. and McMaster University and several of their members, for help in meeting the many demands arising from this remarkable event. In particular, Mrs. Pat Carter gave assistance with this manuscript and other secretarial necessities. Dr. William Buyers and Dr. Gerald Dolling were helpful in many ways, including provision of quality copies of several figures, especially Figs. 1 and 9, the two photographs. Except for Fig. 10, the illustrations were taken from slides of the era (mostly from the papers cited).

BIBLIOGRAPHY

Sources

[A] *The Neutron and its Applications, 1982*, edited by P. Scholfield [The Institute of Physics, Bristol and London (Conference Series Number 64), 1983]. Also: *Pile Neutron Research in Physics* (International Atomic Energy Agency, Vienna, 1962).

[B] *Fifty Years of Neutron Diffraction*, edited by G. E. Bacon (Adam Hilger, Bristol, 1986).

[C] *Inelastic Scattering of Neutrons in Solids and Liquids* (International Atomic Energy Agency, Vienna, 1961).

[D] *Inelastic Scattering of Neutrons in Solids and Liquids* [International Atomic Energy Agency, Vienna; Vols. I, II (1963); I, II (1965); I, II (1968, 1972)].

[E] *Proceedings of the International Conference on Magnetism and Crystallography,* Vol. II, III, 1961 (J. Phys. Soc. Jpn. **17**: Suppl. B-II, B-III, 1962).

[F] *The Interaction of Radiation with Solids*, edited by R. Strumane, J. Nihoul, R. Givers, and S. Amelinckx (North-Holland, Amsterdam, 1964).

[G] *Neutron Physics*, edited by M. L. Yeater (Academic, New York, 1962).

[H] *Solid State Physics*, edited by F. Seitz and D. Turnbull (Academic, London and New York, 1955–1994).

[I] *Phonons and Phonon Interactions*, edited by Thor A. Bak (Benjamin, New York and Amsterdam, 1964).

[J] *Phonons in Perfect Lattices and in Lattices with Point Imperfections,* edited by R. W. H. Stevenson (Oliver and Boyd, Edinburgh and London, 1966).

[K] *Thermal Neutron Scattering*, edited by P. A. Egelstaff (Academic, London and New York, 1965).

[L] *Magnetic and Inelastic Scattering of Neutrons by Metals*, Metallurgical Society Conferences, Vol. 43, edited by T. J. Rowland and P. A. Beck (Gordon and Breach, New York, 1968).

Monographs and texts

[M] *Properties of Metals and Alloys,* by N. F. Mott and H. Jones (Oxford University Press, 1936).

[N] *The Modern Theory of Solids*, by F. Seitz (McGraw-Hill, New York and London, 1940).

[O] *Dynamical Theory of Crystal Lattices*, by Max Born and Kun Huang (Oxford University Press, 1954).

[P] *Neutron Diffraction*, by G. E. Bacon (Oxford University Press, Editions 1955, 1962, 1975).

[Q] *Scattering of Thermal Neutrons* (A Bibliography 1932–1974), Solid State Literature Guides Vol. 7, edited by A. Larose and J. Vanderwal (IFI/PLENUM, New York and London, 1974).

REFERENCES

[1] Brockhouse, B. N., "Slow neutron spectroscopy: an historical account over the years 1950–1977." Source [A]: 193–198 (1983).

[2] Cassels, J. M., "The Scattering of Neutrons by Crystals," edited by O. R. Frisch, Prog. Nucl. Phys. **1**, 185–215 (1950).

[3] DeGennes, P. G., "Liquid dynamics and inelastic scattering of neutrons," Physica **25**, 825–839 (1959).

[4] Elliott R. J. and R. D. Lowde, "The inelastic scattering of neutrons by magnetic spin waves," Proc. R. Soc. London Ser. A **230**, 46–73 (1955).

[5] Finkelstein, R. J., "Scattering of neutrons in polycrystals," Phys. Rev. **72**, 907–913 (1947).

[6] Halpern, O., and M. H. Johnson, "On the magnetic scattering of neutrons," Phys. Rev. **55**, 898–923 (1938).

[7] Hurst, D. G., A. J. Pressesky, and P. R. Tunnicliffe, "The Chalk River single-crystal neutron spectrometer," Rev. Sci. Instrum. **21**, 705–712 (1950).

[8] Moorhouse, R. G., "Slow neutron scattering by ferromagnetic crystals," Proc. Phys. Soc. London Sect. A **64**, 1097–1107 (1951).

[9] Placzek, G., and L. Van Hove, "Crystal dynamics and inelastic scattering of neutrons," Phys. Rev. **93**, 1207–1214 (1954).

[10] Seeger, R. J., and E. Teller, "On the inelastic scattering of neutrons by crystal lattices," Phys. Rev. **62**, 37–40 (1942).

[11] Shull, C. G., "Wave properties of the neutron." Source [A]: 157–168 (1983); "Early neutron diffraction technology." Source [B]: 19–25 (1986).

[12] Squires, G. L., "Multi-oscillator processes in the scattering of neutrons by crystals," Proc. R. Soc. London Ser. A **212**, 192–206 (1952).

[13] Van Hove, L., "Correlations in space and time and Born approximation scattering in systems of interacting particles," Phys. Rev. **95**, 249–262 (1954); "Time-dependent correlations between spins and neutrons scattering in ferromagnetic crystals," Phys. Rev. **95**, 1374–1384 (1954).

[14] Van Hove, L., "The occurrence of singularities in the elastic frequency distribution of a crystal," Phys. Rev. **89**, 1189–1193 (1953).

[15] Van Vleck, J. H., "On the theory of the forward scattering of neutrons by paramagnetic media," Phys. Rev. **55**, 924–930 (1939).

[16] Vineyard, G. H., "Scattering of slow neutrons by a liquid," Phys. Rev. **110**, 999—1010 (1958).

[17] Waller, I., and P. O. Fröman, "On neutron diffraction phenomena according to the kinematic theory. I," Ark. Fys. **4**, 183–189 (1952); P. O. Froman, *ibid.* II, **4**, 191–202 (1952).

[18] Walker, C. B., "X-ray study of lattice vibrations in aluminum," Phys. Rev. **103**, 547–557 (1956).

[19] Weinstock, R., "Inelastic scattering of slow neutrons," Phys. Rev. **65**, 1–20 (1944).

[20] Brockhouse, B. N., "A childhood of slow neutron spectroscopy." Source [B]: 35–46 (1986).

[21] Egelstaff, P. A., "Inelastic scattering of cold neutrons," Nature **168**, 290 (1951).

[22] Brockhouse, B. N., and D. G. Hurst, "Energy distribution of slow neutrons scattered from solids," Phys. Rev. **88**, 542–547 (1952).

[23] Lowde, R. D., "Diffuse reflection of neutrons from a single crystal," Proc. Phys. Soc. London Sect. A **65**, 857–858 (1952). "On the diffuse reflexion of neutrons by a single crystal," Proc. R. Soc. London Sect. A **221**, 206–223 (1954).

[24] Brockhouse, B. N., "A new tool for the study of energy levels in condensed systems" (A) and "Study of energy levels in solids, liquids, and magnetic materials by scattered slow neutrons" (T), Phys. Rev. **98**, 1171 (1955).

[25] Brockhouse, B. N., "Energy distribution of neutrons scattered by paramagnetic substances," Phys. Rev. **99**, 601–603(L) (1955).

[26] Jacrot, B., "Mesure de l'energie de neutrons tres lents apres une diffusion inelastique par des polycrystaux et des monocrystaux," Compt. Rend. **240**, 745–747 (1955).

[27] Brockhouse, B. N., "Neutron scattering and the frequency distribution of the normal modes of vanadium metal," Can. J. Phys. **33**, 889–891 (1955).

[28] Brockhouse, B. N., and A. T. Stewart, "Scattering of neutrons by phonons in an aluminum single crystal," Phys. Rev. **100**, 756–757 (1955); "Normal modes of aluminum by neutron spectrometry," Rev. Mod. Phys. **30**, 236–249 (1958).

[29] Carter, R. S., H. Palevsky, and D. J. Hughes, "Inelastic scattering of slow neutrons by lattice vibrations in aluminum," Phys. Rev. **106**, 1168–1174 (1957).

[30] Larsson, K.-E., U. Dahlborg, and S. Holmryd, "A study of some temperature effects on the phonons in aluminum by use of cold neutrons," Ark. Fys. **17**, 369–392 (1960).

[31] Eisenhauer, C. M., I. Pelah, D. J. Hughes, and H. Palevsky, "Measurement of lattice vibrations in vanadium by neutron scattering," Phys. Rev. **109**, 1045–1051 (1958).

[32] Stewart, A. T., and B. N. Brockhouse, "Vibration spectra of vanadium and a Mn-Co alloy by neutron spectrometry," Rev. Mod. Phys. **30**, 250–255 (1958)

[33] Brockhouse, B. N., "Structural dynamics of water by neutron spectrometry," Acta Crystallogr. **10**, 827-828 (1957), Nuovo Cimento Suppl. **9**, 45–71 (1958), "Diffusive motions in liquids and neutron scattering," Phys. Rev. Lett. **2**, 287–289 (1959), Sakamoto, M., B. N. Brockhouse, R. G. Johnson, and N. K. Pope, "Neutron inelastic scattering study of water," Source [E], Suppl. B-II:370–372.

[34] Brockhouse, B. N., and N. K. Pope, "Time-dependent pair correlations in liquid lead," Phys. Rev. Lett. **3**, 259–262 (1959).

[35] Cohen, M., and R. P. Feynman "Theory of inelastic scattering of cold neutrons from liquid helium," Phys. Rev. **107**, 13–24 (1957).

[36] Palevsky, H., K. Otnes, K. E. Larsson, R. Pardi, and R. Stedman, "Excitation of rotons in Helium II by cold neutrons," Phys. Rev. **108**, 1346–1347 (1957); H. Palevsky, K. Otnes, and K. E. Larsson, Phys. Rev. **112**, 11–18 (1958).

[37] Yarnell, J. L., G. P. Arnold, P. J. Bendt, and E. C. Kerr, "Energy vs momentum relations for excitations in liquid helium," Phys. Rev. Lett. **1** 9–11 (1958) "Excitations in liquid helium: neutron scattering measurements," Phys. Rev. **113**, 1379–1386 (1959).

[38] Henshaw, D. G., "Energy-momentum relation in liquid helium by inelastic scattering of slow neutrons," Phys. Rev. Lett. **1**, 127–129 (1958); Henshaw, D. G., and A. D. B. Woods, "Modes of atomic motions in liquid helium by inelastic scattering of neutrons," Phys. Rev. **121**, 1266–1274 (1961).

[39] Brockhouse, B. N., "Scattering of neutrons by spin waves in magnetic," Phys. Rev. **106**, 859–864 (1957); "Field dependence of neutron scattering by spin waves," Phys. Rev. **111**, 1273–1274 (1958).

[40] Watanabe, H., and B. N. Brockhouse, "Observation of optical and acoustical magnons in magnetite," Phys. Lett. **1**, 189–190 (1962), "Exchange field splitting in ytterbium iron garnet," Phys. Rev. **128**, 67 (1962).

[41] Sinclair, R. N., and B. N. Brockhouse, "Dispersion relation for spin waves in a fcc cobalt alloy," Phys. Rev. **120**, 1638–1640 (1960).

[42] Brockhouse, B. N., L. N. Becka, K. R. Rao, R. N. Sinclair, and A. D. B. Woods, "Crystal field spectra in rare earth oxides" Source [E]: Suppl. B-III, 63–66 (1962).

[43] Brockhouse, B. N., "Methods for neutron spectrometry" Source [C]: 113–150 (1961).

[44] Svensson, E. C., B. N. Brockhouse, and J. M. Rowe, " 'In-Band' modes of vibration of a dilute disordered alloy-Cu(Au)," Solid State Commun. **3**, 245–249 (1965).

[45] Brockhouse, B. N., and P. K. Iyengar, "Normal vibrations of germanium by neutron spectrometry," Phys. Rev. **108**, 894–895 (1957), "Normal modes of germanium by neutron spec-

trometry," Phys. Rev. **111**, 747–754 (1958).

[46] Brockhouse, B. N., "Lattice vibrations in semiconductors by neutron spectrometry," J. Phys. Chem. Solids **8**, 400–405 (1959); "Lattice vibrations in silicon and germanium," Phys. Rev. Lett. **2**, 256–258 (1959).

[47] Brockhouse B. N., and B. A. Dasannacharya, "Temperature effects on lattice vibrations in germanium," Solid State Commun. **1**, 205–209 (1963).

[48] Yarnell, J. L., J. L. Warren, and R. G. Wenzel, "Lattice vibrations in diamond," Phys. Rev. Lett. **13**, 13–15 (1964).

[49] Woods, A. D. B., W. Cochran, and B. N. Brockhouse, "Lattice dynamics of alkali halide crystals," Phys. Rev. **119**, 980–999 (1957).

[50] Woods, A. D. B., B. N. Brockhouse, R. A. Cowley, and W. Cochran, "Lattice dynamics of alkali halide crystals. II. Experimental studies of KBr and NaI," Phys. Rev. **131**, 1025–1029 (1963); Cowley, R. A., W. Cochran, B. N. Brockhouse, and A. D. B. Woods, "III. Theoretical," Phys. Rev. **131**, 1030–1039 (1963).

[51] Brockhouse, B. N., T. Arase, G. Caglioti, K. R. Rao, and A. D. B. Woods, "Crystal dynamics of lead. I. Dispersion curves at 100 K," Phys. Rev. **128**, 1099–1111 (1962).

[52] Kohn, W., "Image of the Fermi surface in the vibration spectrum of a metal," Phys. Rev. Lett. **2**, 393–394 (1959).

[53] Brockhouse, B. N., K. R. Rao, and A. D. B. Woods, "Image of the Fermi surface in the lattice vibrations of lead," Phys. Rev. Lett. **7**, 93–95 (1961).

[54] Woods, A. D. B., B. N. Brockhouse, R. H. March, A. T. Steward, and R. Bowers, "Crystal dynamics of sodium at 90° K," Phys. Rev. **128**, 1112–1120 (1962).

[55] Dixon, A. E., A. D. B. Woods, and B. N. Brockhouse, "Frequency distribution of the lattice vibrations in sodium," Proc. Phys. Soc. London **81**, 973–974 (1963).

[56] Gilat, G., and G. Dolling, "Normal vibrations of β brass," Phys. Rev. **138**, A1053–A1065 (1963).

[57] Stedman, R., and S. Nilsson, "Phonons in aluminum at 80 K." Source [D]: Vol. I, 211–223 (1965).

Reviews

[58] Shull, C. G., and E. O. Wollan, "Applications of neutron diffraction to solid state problems." Source [H]: Vol. 2, 138–217 (1956).

[59] Kothari, L. S., and K. S. Singwi, "Interaction of thermal neutrons with solids." Source [H]: Vol. 8, 110–190 (1959).

[60] Brockhouse, B. N., "Crystal and liquid dynamics from neutron energy distributions," Source [G]: 129–144 (1962); "Phonons and neutron scattering," Source [I]: 221–275 (1964); "Neutron scattering by phonons," Source [J]: 110–152 (1966).

[61] Brockhouse, B. N., S. Hautecler, and H. Stiller, "Inelastic scattering of slow neutrons." Source [F]: 580–642 (1964).

[62] Dolling, G., and A. D. B. Woods, "Thermal vibrations of crystal lattices." Source [K]: 193–249 (1965).

[63] Lomer, W. M., and G. G. Low, "Introductory Theory." Source [K]: 1–52 (1965). Other articles by R. M. Brugger, P. K. Iyengar, S. J. Cocking, F. J. Webb, G. Dolling, A. D. B. Woods, B. Jacrot, T. Riste, A. Sjolander, K. E. Larsson, J. A. Janik, and A. Kowalska provide rather complete reports on the state of the field by 1965.

Early development of neutron scattering*

Clifford G. Shull

Department of Physics, Massachusetts Institute of Technology, Cambridge, Massachusetts 02139

Neutrons were discovered by Chadwick in 1932 when he observed a penetrating form of radiation emanating from beryllium metal when activated by alpha particles from a radium source. Further study showed this to be neutral particle radiation which could be degraded in kinetic energy to thermal energy upon successive inelastic scattering by light atoms in a medium. With this Fermi thermalization, it was realized that thermal neutrons, because of the wave-particle duality principle, should exhibit a wave character with de Broglie wavelength comparable to the atom spacing in solids. Thus diffraction effects should be expected in the scattering of neutrons by crystals, just as with x radiation, and early experiments in 1936 showed in a crude way that this was true.

At this time, x-ray diffraction had been developed to become an important tool in establishing the structure and atomic interactions in materials, and it seemed unlikely that neutron diffraction would develop as a useful tool because of very low source intensity. This limitation changed dramatically in the early war years of 1939–1943 with the discovery of nuclear fission by Hahn and Meitner and the subsequent demonstration by Fermi of a self-sustaining, neutron chain-reacting assembly. Following this Fermi demonstration, immediate design and construction of a full-scale nuclear reactor, or "pile" as it was called then, was effected at Oak Ridge, Tennessee.

This reactor, then called the Clinton Pile, was meant to produce the first measurable quantities of plutonium and to serve as a pilot plant for the much larger production reactors which were being designed for construction in the state of Washington. The Clinton Pile was a graphite-moderated, air-cooled assembly which operated at a power level of about 3.5 MW, thereby producing a slow neutron flux density of about 10^{12} neutrons/cm^2 sec. During this same period, another assembly was designed and constructed at Chicago with the feature of heavy water moderation, the CP-3 assembly. I show in Fig. 1 the start-up dates of these early and postwar reactors. Both the Clinton and the CP-3 pile assemblies operated through the war years producing man-made elements and isotopes, and, to a limited extent, their neutron radiation was used to obtain some critically needed cross section data.

Figure 2 shows an early photograph of the loading face of the Clinton Pile with operators performing some function on a movable platform in front of the assembly. The pile is cubic in form with edge length about 7 m and consists of a nestled collection of graphite "stringers," rectangular blocks of square cross sections about 10 cm on a side and length 150 cm. Pure uranium metal serves as the fissionable fuel for the assembly and is positioned in a lattice-array of channels in the graphite extending from the loading face to the back, discharge face. The uranium metal is in the form of cylindrical rods, about 2 cm in diameter and 10 cm in length, and each is carefully sealed in an aluminum can. Cooling air is drawn through the channels to the back face by large suction fans, and control of the nuclear reaction level is effected by the positioning of various neutron-absorbing control rods within the assembly. Surrounding the cubic graphite pile on all sides is a radiation protection shield of concrete of thickness about 1.5 m. With foresight, the designers had arranged some channels through the concrete shield for access to the active volume of the assembly.

During the war period, a single-axis spectrometer had been used at one of these channels to supply monoenergetic neutrons for transmission cross section measurements. At the end of the war, Ernest Wollan decided to set up a two-axis spectrometer for full diffraction pattern study with the first axis providing a monochromatic beam, hopefully of enough intensity to permit second-axis diffraction patterns to be studied. Wollan had arranged for delivery to Oak Ridge of a base spectrometer that he had used in his earlier thesis work on x-ray gas scattering under Arthur Compton at Chicago. This spectrometer, with coaxial control of specimen and detector positions, had to be modified to support the larger loading of a neutron detector with its necessary shielding, and this was done with support cables running to a swivel-bearing in the ceiling directly above the spectrometer axis. Figure 3 is a photograph of this assembly taken during the first year of use, after some improvements in the shielding around the detector and the monochromating crystal had been made.

Operation of the spectrometer was entirely by hand, and it was a time-consuming chore for Wollan and his early colleague, R. B. Sawyer, to collect what were the first neutron diffraction patterns of polycrystalline NaCl and light and heavy water in the early months of 1946. I was shown these patterns on a visit that I made to Oak Ridge in the spring of

Start-up Dates of Early "Piles"

1942 (Dec.)	Fermi Demonstration (Chicago)
1943 (Nov.)	Clinton Pile, graphite, (3 MW)
1944	Argonne CP-3, heavy water
1947	NRX - Chalk River
1948	BEPO - Harwell
1950	LITR - Oak Ridge (enriched U) BGRR - Brookhaven

FIG. 1. Dates of operation of early nuclear reactors.

*This lecture was delivered 10 December 1994, on the occasion of the presentation of the 1994 Nobel Prize in Physics.

FIG. 2. Photograph of the loading face of the Clinton Pile.

that year prior to my relocating there and joining Wollan in further work in June 1946. I show in Fig. 4 a portion of the NaCl pattern and, although the background level is high, it is seen that the diffraction peaks can be measured and interpreted in a quantitative way. This was important because Wollan had hoped, by using powder-diffraction patterns, to avoid the uncertainties in single-crystal intensities caused by extinction effects. The light and heavy water patterns were taken to see if there was any suggestion of hydrogen (deuterium) atom contribution to the diffuse, liquid patterns as taken with neutrons. Some differences in the coherent features of the two patterns were to be seen, but no attempt to analyze this was made.

During the early period of study with this spectrometer, the emphasis was centered on understanding the intensity that was being measured in the diffraction pattern, both within the coherent diffraction peaks and in the diffuse scattering. At that time, the only pertinent, quantitative information that was available were values of the total scattering and absorption cross sections as measured with thermal neutrons for many elements. Although these quantities might be expected to establish the total intensity in all of the diffraction pattern, the existence of isotopes for some elements, the possible presence of different spin state scattering amplitudes for nuclei with spin, and the presumed thermal disorder or excitation scattering were all factors that had to be considered case by case. Progress in isolating and defining these various factors came very slowly because of the choresome mode of data collection on the instrument, which was entirely by hand operation.

Towards the end of the first year, we were joined by a young technician, Milton Marney, who assisted in the tedious data collection, and by two volunteer physicist-scientists from the Oak Ridge Training School, William Davidson and

FIG. 3. Photograph of the first double-axis neutron spectrometer used to take powder-diffraction patterns at the Clinton Pile.

George Morton. The Training School was organized in late 1946 and was meant to bring a blue-ribbon group of established scientists from academia, industry, and government to Oak Ridge for an extended period so as to attend lectures, collaborate with research groups and generally to learn the new technology and science that had been developed during the war period. Davidson and Morton took interest in our neutron-scattering work and helped with the data collection; however, Morton, who was skilled in instrumentation, soon tired of this and decided to design an automatic control system that would permit step-scanning with intensity measurement and its recording. This took the form of motor drives, cam wheels and microswitches, and, most importantly, a printout recorder of the neutron counter measurement. These changes took several months to effect and, once operational, it permitted continuous, uninterrupted data collection; and this represented a very important step in the progress of the program.

Along with these improvements in the instrumentation, regular changes were being made in the shielding around the monochromating crystal and the movable detector on the outer spectrometer. Initially, the background intensity in the diffraction patterns was comparable to the peak intensity, as seen in Fig. 4, and about as much time was spent in measuring background as in studying the pattern itself. Regular background intensity measurements through insertion of a solenoid-activated cadmium filter into the incident beam were a part of the initial automatic program system.

With this new system of data accumulation, we returned to our initial program of classifying various elements according to their coherent scattering amplitudes and of exploring various novel cases where neutron diffraction might supply useful information beyond that obtained with x-ray scattering. In

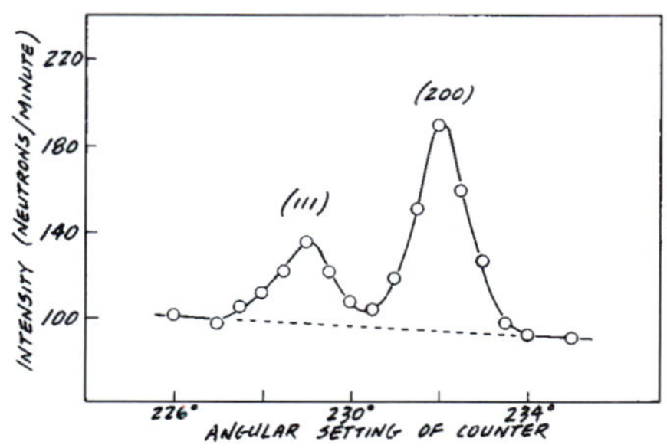

FIG. 4. Portion of the first powder-diffraction pattern of NaCl taken at the Clinton Pile.

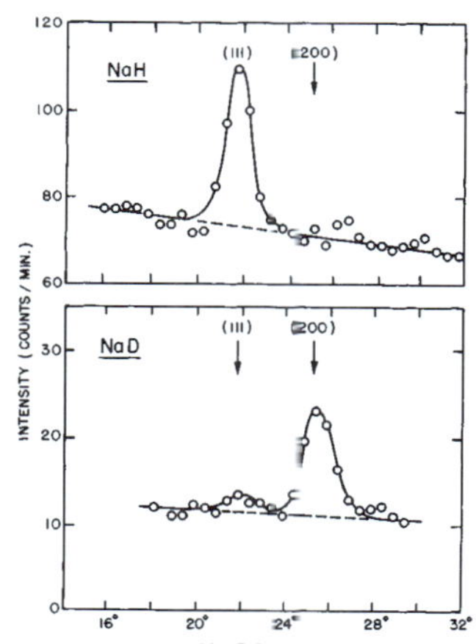

FIG. 5. Patterns for NaH and and showing that hydrogen contributes normally to the pattern.

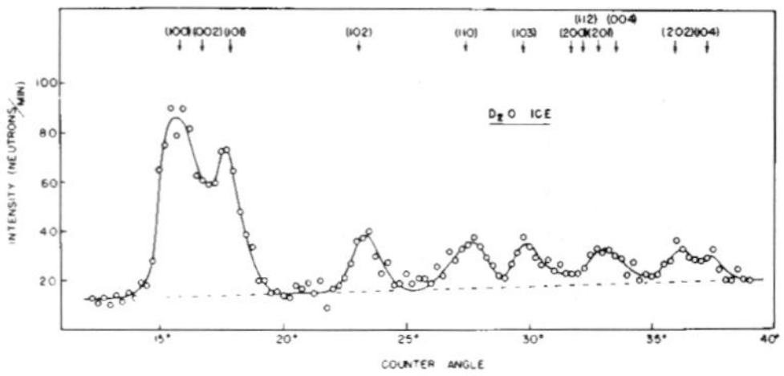

FIG. 6. Pattern taken for a sample of powdered D_2O ice.

the special case of a monoisotopic element with zero nuclear spin (nearly satisfied in oxygen and carbon), the coherent scattering amplitude, which defines the coherent peak intensity in the diffraction pattern, can be obtained from the total scattering cross section. This can then be used as an internal standard in quantifying the diffractive properties of other elements when combined with the standard element. This program of classification went on for several years, continuing into the period when separated isotopes became available.

Rather early in the program, a puzzling feature of the intensity distribution was noted, namely, that there was larger than expected diffuse scattering in the pattern. Careful intensity measurements on an absolute scale showed the diffuse intensity to be larger than that caused by thermal diffuse scattering or isotopic and nuclear-spin incoherence. This was eventually identified as resulting from multiple scattering within the scattering samples, a novel effect characteristic of neutron scattering and never encountered in equivalent x-ray scattering technology.

With these improvements in instrumentation and a better understanding of the components that were being seen in the patterns, attention was then turned to the important crystal-

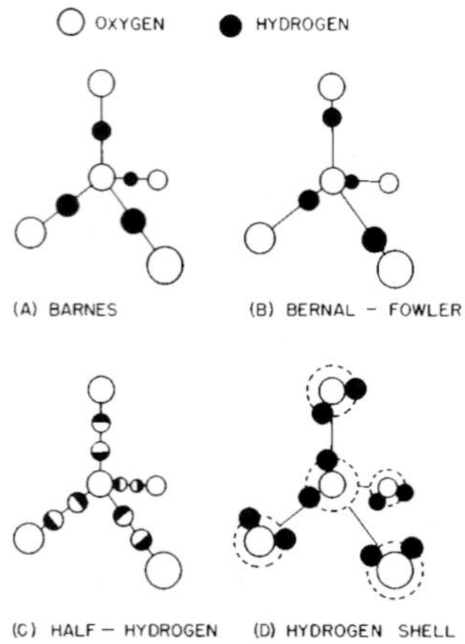

FIG. 7. Suggested models for the hydrogen atom distribution in ice.

lographic case of hydrogen-containing crystals. It was known at the time that the coherent scattering amplitude of hydrogen for neutrons must be small, since that quantity could be derived from measurements of total scattering of neutrons by ortho- and parahydrogen. However, the precision of those measurements did not specify either its finiteness or its algebraic sign. There was also the additional suspicion that hydrogen atoms might display anomalously large thermal motion in crystals and that this would wipe out coherent scattering. This all became settled when powder patterns of NaH and NaD were obtained, with parts being shown in Fig. 5. Analysis of the intensities in these full patterns yielded the coherent scattering amplitudes for hydrogen and deuterium (they are of opposite sign) and showed the hydrogen centers

FIG. 8. Later neutron diffraction spectrometer with Ernest Wollan (kneeling) and the author. This has many of the features of present-day spectrometers.

to possess rather normal thermal oscillation. The large diffuse scattering seen for NaH arises from anomalously large nuclear-spin incoherent scattering from hydrogen centers.

Another interesting hydrogen structure that attracted our attention was that of common ice. It was known from x-ray diffraction studies that the basic structure was formed by interlocking oxygen atom tetrahedra, but the bonding location of the hydrogen atoms was unknown, and several models for this had been proposed. Figure 6 shows a powder pattern that was taken with a deuterated sample of ice, and the analysis of it shows clearly the validity of the "half-hydrogen" or "Pauling" model of the structure as illustrated in Fig. 7. Preparing a powder sample of D_2O ice presented new problems, and a new specimen cryostat (with dry-ice–acetone coolant) was built, which some people called an "ice-ball machine." D_2O vapor was frozen on the inside surface of a cylindrical specimen tube, followed by reaming and light packing of successive layers without contamination from normal atmospheric moisture. Other hydrogen structures were studied in the next two years, mostly transition metal and heavy atom hydrides including the interesting extreme case of uranium hydride.

In mid-1949 a second spectrometer was installed at a beam port adjoining the early one. This spectrometer, again a two-axis unit, was really the first prototype of present-day neutron spectrometers, having components that were built for specific neutron use rather than being improvised from x-ray scattering units. An elegant cylindrical shield (A) surrounded the monochromating crystal, and this was supported on a large bearing which permitted easy change of beam exiting angle or beam wavelength. The specimen spectrometer (B) was supported from the monochromator shield and was now sturdy enough to support the counter arm without the use of support cables that characterized the earlier unit. Figure 8 shows this spectrometer, with Ernest Wollan (kneeling) and the author, in an early photograph.

Along with the use of these spectrometers in which neutron-sensitive counting tubes were used to detect diffraction intensity effects, we were interested in exploring photographic detection as used effectively in x-ray technology. It was realized that photographic film would be very insensitive to neutron radiation and that a converter screen placed next to the emulsion would be required. After some tests, it was found that an indium sheet would serve this purpose: indium has a sizable absorption cross section for thermal neutrons and emits beta particles in its decay which will produce photographic action. This screen-film detector was used to record the first neutron Laue pattern, taken for a NaCl crystal with an exposure time of about 10 h as shown in Fig. 9. A direct, full spectrum beam was taken from the reactor through a long shielding collimator and this passed on to the crystal with the film positioned about 10 cm behind

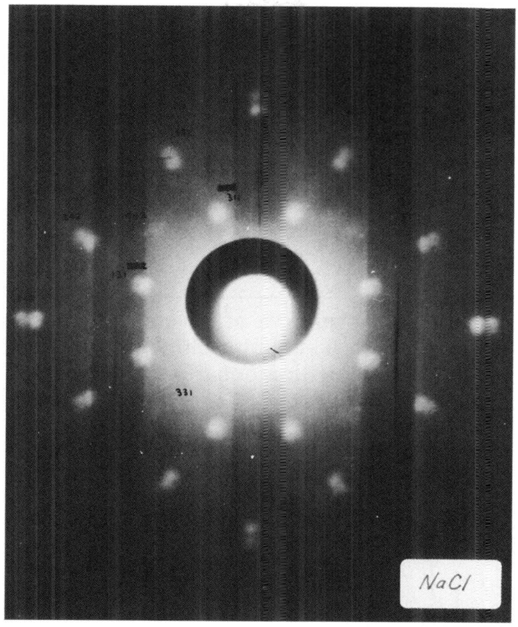

FIG. 9. Neutron Laue pattern taken with NaCl crystal.

the crystal. Among the features to be seen in this first pattern is the apparent double structure of the Laue spots. This puzzled us at first, until later investigation showed it to be an artifact caused by twinning in our first selected crystal. One can notice also in the background a series of narrow band images running from top to bottom. This results from the presence of Scotch Tape ribbon holding adjacent strips of indium sheet together. So not only is this the first neutron Laue pattern, but, as well, it is the first neutron radiograph!

My abbreviated description of the developments that occurred during the first few years of neutron-scattering investigation at Oak Ridge has failed to mention that many other colleagues and collaborators beyond those specifically mentioned were involved in the effort. My heartfelt appreciation is due them as well as to other members of the Physics Division and the Oak Ridge laboratory administration for unfailing support. But mostly I appreciate the association, collaboration and close friendship, extending for nearly a decade, that I enjoyed with Ernest Wollan, who first guided me to the wonders of neutron-scattering research. I regret that he did not live long enough to share in the honors that have come to me.

ARTICLES

Structure at 2.8 Å resolution of F_1-ATPase from bovine heart mitochondria

Jan Pieter Abrahams, Andrew G. W. Leslie, René Lutter* & John E. Walker†

Medical Research Council Laboratory of Molecular Biology, Hills Road, Cambridge CB2 2QH, UK
* Present address: Academic Medical Centre, Department of Pulmonology, University of Amsterdam, Meibergdreef 9, 1105 AZ Amsterdam, The Netherlands
† To whom correspondence should be addressed

In the crystal structure of bovine mitochondrial F_1-ATPase determined at 2.8 Å resolution, the three catalytic β-subunits differ in conformation and in the bound nucleotide. The structure supports a catalytic mechanism in intact ATP synthase in which the three catalytic subunits are in different states of the catalytic cycle at any instant. Interconversion of the states may be achieved by rotation of the $\alpha_3\beta_3$ subassembly relative to an α-helical domain of the γ-subunit.

ADENOSINE triphosphate synthase (ATP synthase, F_1F_o-ATPase) is the central enzyme in energy conversion in mitochondria, chloroplasts and bacteria (reviewed in refs 1–6). It uses a protonmotive force, generated across the membrane by electron flow, to drive the synthesis of ATP from ADP and inorganic phosphate[7]. The multisubunit assembly consists of a globular domain, F_1, and an intrinsic membrane domain, F_o, linked by a slender stalk about 45 Å long. The F_1 domain is an approximate sphere 90–100 Å in diameter, and contains the catalytic binding sites for the substrates ADP and inorganic phosphate. Energy released by proton flux through F_o is relayed to the catalytic sites in the F_1 domain, probably by conformational changes through the stalk. About three protons flow through the membrane per ATP synthesized.

Disruption of the stalk releases the water soluble enzyme, F_1-ATPase, a complex of five different proteins with the stoichiometry $3\alpha:3\beta:1\gamma:1\delta:1\varepsilon$. In bovine mitochondria they contain 510, 482, 272, 146 and 50 amino acids, respectively[8], giving a relative molecular mass of 371,000. The sequences of the α- and β-subunits are homologous (20% identical), including the P-loop nucleotide-binding motif[9]. The catalytic sites are in the β-subunits, whereas the function of nucleotides in the α-subunits, which do not exchange during catalysis, is obscure. According to a binding change mechanism of ATP synthesis[1,10], the structures of the three catalytic sites are always different, but each passes through a cycle of 'open', 'loose' and 'tight' states. The mechanism suggests that F_1-ATPase is an inherently asymmetric assembly, as clearly indicated by subunit stoichiometry, electron microscopy[11,12] and low-resolution X-ray crystallography[13]. However, in one model proposed for rat F_1-ATPase based on analysis of a crystal form that implies three-fold symmetry, the three different α- and β-subunits are indistinguishable[14].

To obtain a fuller understanding of the mechanism of F_1 and of the non-equivalence of the catalytic sites, we have determined the structure of F_1-ATPase from bovine heart mitochondria by X-ray diffraction at 2.8 Å resolution. Bovine F_1-ATPase is the largest asymmetric structure solved at atomic resolution so far. The α- and β-subunits, which have a similar fold, are arranged alternately like the segments of an orange around a central α-helical domain containing both the N- and C-terminal regions of the γ-subunit. As the three β-subunits contain different nucleotides and have different conformations, the structure as found in the crystal is compatible with one of the states to be expected in the cyclical binding change mechanism. It is likely that the central helical domain is part of a mechanism for transferring energy from the F_o domain into the catalytic sites, and that in the process the 3α–3β assembly (referred to below as $\alpha_3\beta_3$) may rotate about the central α-helical domain of the γ-subunit.

Structure determination

Crystallization of bovine F_1 and data collection procedures have been described[13,15]. The quality of data and initial phases determined by single isomorphous replacement and anomalous scattering are summarized in Table 1, and a representative sample of the electron density map is shown in Fig. 1. The model (Fig.

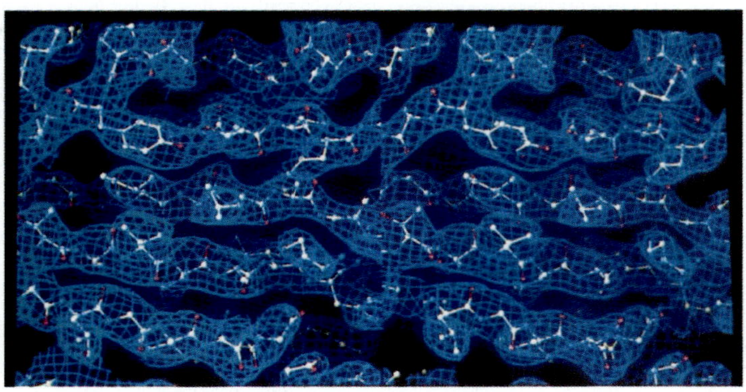

FIG. 1 Stereo illustration of part of the 3.2 Å-resolution electron density map used to build the initial model of bovine F_1-ATPase, with the refined model coordinates superimposed. The region shown corresponds to part of the β-sheet in the nucleotide-binding domain of subunit β_{DP}. The contour level is 1.2 times the r.m.s. density of the map.

TABLE 1 Crystallographic data

	Data set 1		Data set 2		Data set 3	
	Native	0.2 mM methylmercury	Native	1 mM methylmercury	Native	0.2 mM methylmercury
Resolution	2.85 Å	3.2 Å	3.5 Å	3.5 Å	3.5 Å	3.5 Å
R_{merge}*	0.098 (0.19)	0.077 (0.145)	0.059 (0.106)	0.067 (0.111)	0.114 (0.255)	0.086 (0.210)
Completeness	97.5% (92%)	96.3% (95.6%)	97.7% (87.7%)	97.4% (90.4%)	97.5% (99.8%)	93.5% (91.5%)
Multiplicity	5.4 (2.9)	3.7 (3.5)	3.2 (2.8)	7.5 (6.4)	8.9 (7.8)	2.4 (2.0)
No. of crystals	17	6	1	6	12	3
Rejected outliers†	0.37%	0.02%	0.56%	0.4%	0.3%	0.4%
Mean $\langle F/\sigma(F)\rangle$	30.9 (7.8)	28.8 (11.8)	46.9 (24.8)	53.8 (30.0)	32.6 (15.1)	19.2 (8.9)
R_{deriv}‡		0.114 (0.164)		0.120 (0.132)		0.170 (0.215)
No. of mercury sites		4		9		4
Total occupancy§		2.4		3.0		2.4
R_{cullis}‖		0.60 (0.80)		0.56 (0.71)		0.72 (0.87)
Phasing power¶		1.37 (0.95)		2.0 (1.6)		1.08 (0.78)

Diffraction data. As native crystals were sometimes non-isomorphous, every time a derivative data set was collected, a corresponding native data set was collected too. Data sets 1 and 2 were collected on stations 9.6 and 9.5 respectively at the Synchrotron Radiation Source, Daresbury, UK. Data set 3 was collected using an Elliot GX11 generator with a 100 μ focal spot, collimated by a Supper double-mirror system. Data sets were recorded on MAR Research image-plate detectors, with a diameter of 300 mm for data sets 1 and 3, or 180 mm for data set 2. Data were processed with MOSFLM[43] and programs from the CCP4 suite[44]. Values given in parenthesis are for the highest resolution bin. The R_{deriv} between native data sets 1 and 3 was 0.074, that between data sets 1 and 2 was 0.157, and that between data sets 2 and 3 was 0.162, indicating that data set 2 was not isomorphous with data sets 1 and 3, and that data sets 1 and 3 were not isomorphous to a lesser degree. Data from derivatized crystals were merged only if refinement of the heavy-atom parameters of the individual crystals indicated that this was allowed; otherwise they were treated separately. The phase-probability distributions of different data sets were combined, but only the structure-factor amplitudes of the high-resolution data set were used in the structure determination. **Solvent density modification.** Phases to 3.2 Å were improved by a new type of solvent density modification, to be described elsewhere. The solvent region (50% of the crystal volume) was determined from the local variation in electron density within a sphere of radius 3.7 Å. Features in the solvent region were inverted with respect to the mean solvent density. The phases of the resulting map were combined with the original phase probability distribution using σ_A weighting[45] and a new map was calculated. Iteration of this process decreased the mean phase error from 70 to 40°. Phase improvement was monitored by measuring the reduction in variation of the electron density within an unmodified spherical patch of solvent with a radius of 8 Å. **Model building and refinement.** The atomic model was built using program O (ref. 46). The few ambiguities in following the path of the polypeptide backbone of the α- and β-subunits were resolved by comparing the electron density of different copies of these subunits. Conventional refinement using XPLOR-3.1 (ref. 47) resulted in a model with an R-factor of 19.4% for 95% of the data between 6.0 and 2.86 Å. The free R-factor[48] for the remaining 5% of the data within this resolution range was 28.5%. The r.m.s. deviations from standard values of bond lengths and angles are 0.016 Å and 2.1°.

* $R_{merge} = \sum\sum_i |I(h) - I(h)_i| / \sum\sum_i I(h)_i$, where $I(h)$ is the mean intensity after rejections. The contributing reflections were weighted by their standard deviations, which were determined by adjusting the measured standard deviation to reflect the observed differences between symmetry related reflections.

† Reflections with intensities differing more than 3.5 $\sigma(I)$ from the weighted mean were rejected.

‡ $R_{deriv} = \sum ||F_{PH}| - |F_P|| / \sum |F_P|$, where F_{PH} is the structure factor of the derivative and F_P is that of the native data.

§ The total refined occupancy of the sites, normalized to the site with the highest occupancy after setting this to 1.

‖ $R_{cullis} = \sum ||F_{PH} - F_P| - |F_{H(calc)}|| / \sum |F_{PH} - F_P|$; F_P and F_{PH} are defined above, $F_{H(calc)}$ is the calculated heavy-atom structure factor. The summation is over centric reflections only.

¶ R.m.s. isomorphous difference divided by the r.m.s. residual lack of closure.

2) contains 2,983 amino acids in the following sequences: α_{DP} 19–510, α_{TP} 25–510, α_E 24–510, 9–474 in all three β-subunits, γ1–45, γ73–90 and γ209–272 (see Fig. 2 legend for subunit nomenclature). Four AMP-PNP molecules and one ADP are also present (where AMP-PNP is 5′-adenylyl-imidodiphosphate). No water molecules were modelled into the structure. The δ- and ε-subunits and 145 residues of the γ-subunit were not located in the electron density map.

It was possible to solve the structure with data from a single heavy-atom derivative as a result of recent advances in detector technology. Apart from the new solvent density modification method (Table 1), standard techniques were used.

Architecture of F_1-ATPase

F_1 is a flattened sphere 80 Å high and 100 Å across. The three α- and three β-subunits are arranged alternately like the segments of an orange around a central α-helix 90 Å long formed by the C-terminal amino acids γ209–272 (Fig. 2). The C-terminal residue of this helix emerges into a dimple 15 Å deep in the surface at the top of the assembly. The lower half of the helix forms a left-handed antiparallel coiled coil with a second α-helix composed of amino acids γ1–45. This helical structure protrudes about 30 Å from the bottom of the main body in a stem, almost certainly a vestige of the 45 Å stalk in ATP synthase. An increased local variation of the electron density suggests that the δ- and ε-subunits are probably in the stem region, but are not well ordered. However, it cannot be excluded that the δ- and ε-subunits are sub-stoichiometric in the crystals. The γ-subunit passes through a large internal cavity connected to the outside by crevices between neighbouring α- and β-subunits. In the low-resolution structure, this cavity was interpreted to be a pit[13]. The C-terminal helix of the γ-subunit is twisted (Fig. 2). At the top

FIG. 2 The three-dimensional structure of bovine F_1-ATPase. In this figure and Fig. 3, parts *a–c* were prepared with MOLSCRIPT[49]. The α-, β- and γ-subunits are red, yellow and blue, respectively, and nucleotides are black, in a 'ball-and-stick' representation. The axis of pseudo-symmetry is vertical. AMP-PNP is bound to the three α-subunits, and to the β-subunit defined as β_{TP}. Subunit β_{DP} has bound ADP and subunit β_E has no associated nucleotide. The relationships of the various α- and β-subunits to each other is summarized in the top left or right corner (*a–d*). In *b–d*, the shaded part of the diagram indicates which subunits are depicted. Subunit α_{TP} contributes to the nucleotide-binding site of β_{TP}, and similarly for α_{DP} and α_E. Subunits α and γ are numbered from 1–510 and 1–272, respectively. By convention, the fifth amino acid (serine) in subunit β is residue 1 and the first four amino acids (AlaAlaGlnAla) are referred to as residues −1 to −4. The C-terminal amino acid is residue 478. *a*, A view of the entire F_1 particle in which subunits α_E and β_E point towards the viewer, revealing the antiparallel coiled coil of the N- and C-terminal helices of the γ-subunit through the open interface between them. The bar is 20 Å long. *b*, Subunits α_{TP}, γ and β_{DP}. The N and C termini of the β- and γ-subunits are shown. The

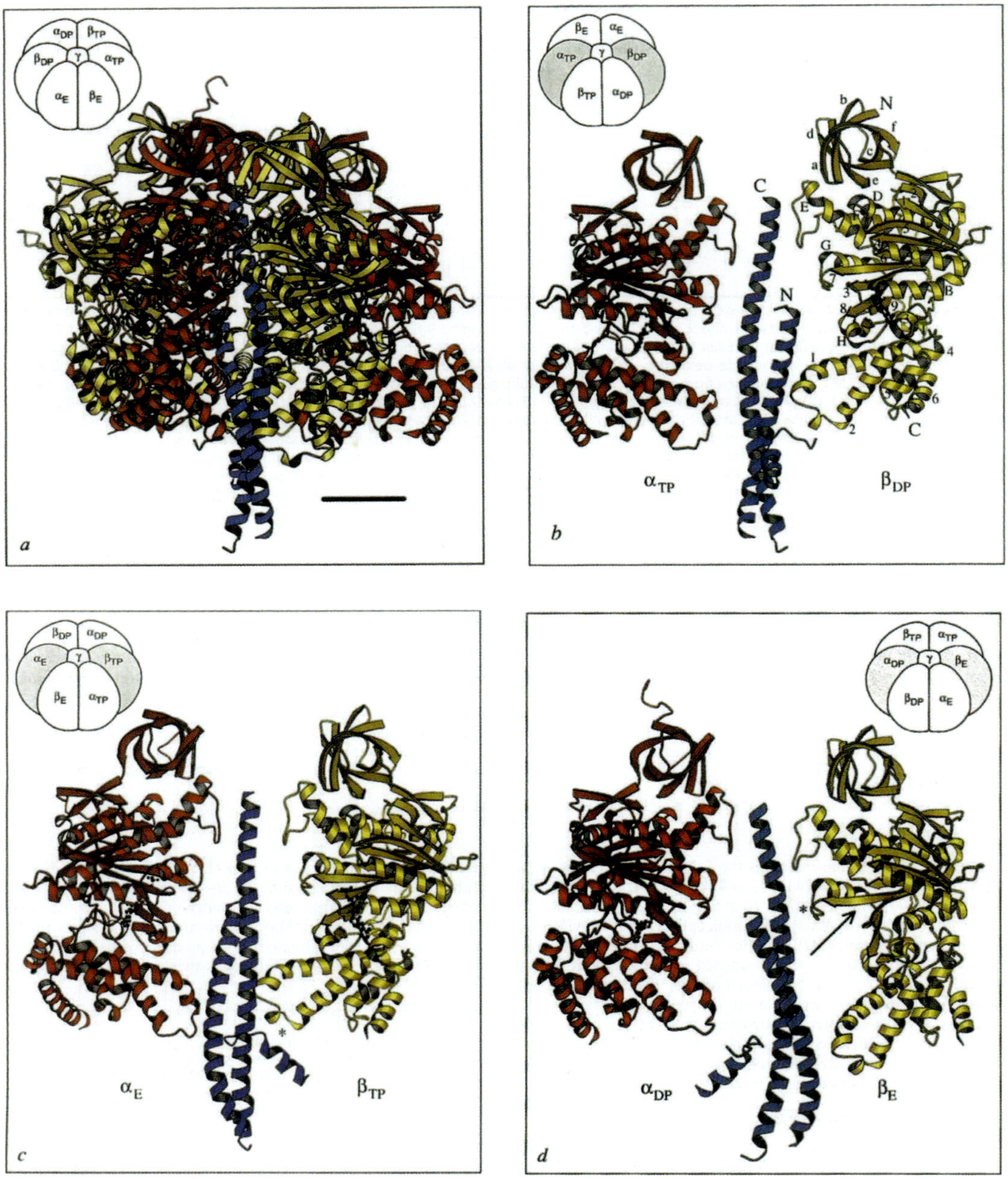

β-sheets in the N-terminal β-barrel are labelled a–f, consecutively. The order of helices (A–I) and the β-sheets (1–9) in the nucleotide-binding domain is as follows: 1, 2, A (not visible, hidden by helix 3 from the C-terminal domain), 3, B, 4, C, 5, D, 6, E, F, 7, G, 8, H, 9, I. The helices in the C-terminal domain are labelled 1–6, consecutively. The α-subunits have the same fold, with a minor deviation in the C-terminal domain, where the equivalent of helix 3 in the β-subunit is missing, and where two more helices are present after the equivalent of helix 6. *c*, Subunits α_E, γ and β_{TP}. The asterisk indicates a 'catch' interaction of the loop containing the DELSEED sequence and the γ-subunit. *d*, Subunits α_{DP}, γ, and β_E. The arrow indicates the disruption of the β-sheet in the nucleotide-binding domain. β_E-Asp 316, β_E-Thr 318 and β_E-Asp 319 in a loop of the nucleotide-binding domain hydrogen-bonds with residues γ-Arg 254 and γ-Gln 255 from the C-terminal helix, 6 Å below the hydrophobic sleeve. The asterisk indicates the loop that makes a 'catch' with the C-terminal part of the γ-subunit. *e, f*, Stereo α-carbon backbone drawings of subunits α_{TP} and β_{DP}, respectively, in the same orientations as in the β-subunit in *b*.

of the assembly, its axis is coincident with the pseudo-threefold axis relating the α- and β-subunits, but at the lower end of the stem it is displaced from the axis by 7 Å. A third segment of the γ-subunit (γ73–90) forms a small α-helix and loop in the region of the stem (Fig. 2).

The folds of the α- and β-subunits are almost identical, so that elements of three- and six-fold symmetry are present in the structure. Each subunit consists of an N-terminal six-stranded β-barrel (α19–95, β9–82), a central α–β domain containing the nucleotide-binding site (α96–379, β83–363), and a C-terminal bundle (α380–510, β364–474) of 7 and 6 helices, respectively, in the α- and β-subunits (Fig. 2a). In each subunit, the β-barrel and nucleotide-binding domains are separated, whereas the nucleotide-binding domain and the helical bundle interdigitate to a minor extent. The extent of contact between neighbouring α- and β-subunits varies (Fig. 3). The β-barrel domains crown the particle with a continuous 24-stranded β-sheet (Fig. 3a), whereas the shortest distance between side chains of some of the helical bundles (Fig. 3c) is 10 Å. Between 18 and 24 amino acids at the N termini of α-subunits and amino acids β(−4 to 8) are disordered. In F_1-ATPase and in ATP synthase, these regions are protease-sensitive[8,16].

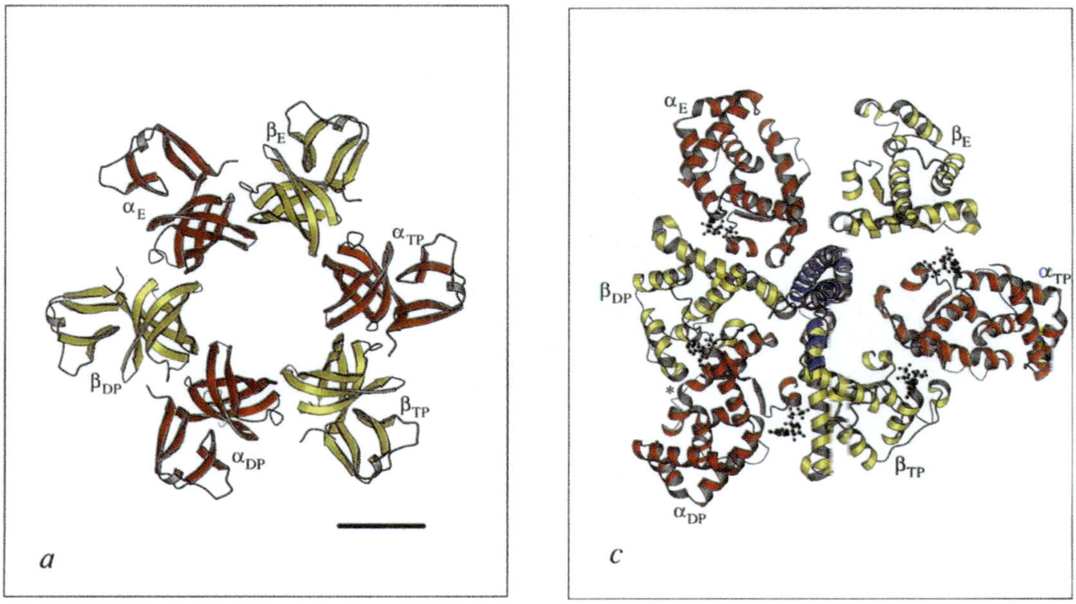

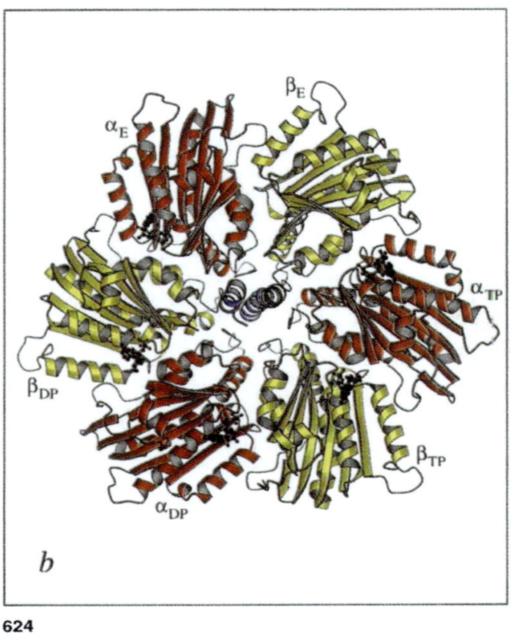

FIG. 3 Asymmetrical features in bovine F_1-ATPase and the binding change mechanism; *a–c* are viewed from the membrane side, the three views being drawn to the same scale and in the same orientation (rotated by 90° relative to Fig. 2*a*). *a*, The N-terminal domains and part of the nucleotide-binding domains of α- and β-subunits. The N-terminal domains make a 24-stranded β-barrel with the six a-strands hydrogen-bonded to d-strands of the neighbouring subunit (Fig. 2*b*). Scale bar, 20 Å. *b*, The nucleotide-binding domains. The β-sheet of subunit β_E is disrupted close to the γ-subunit. *c*, The C-terminal domains. The deviations from perfect symmetry (most apparent in this view) result in the characteristic features of the individual nucleotide-binding sites. *d*, An energy-dependent binding change mechanism of ATP synthesis. The catalytic sites in β-subunits interact and interconvert between three forms: O, open site with very low affinity for ligands and catalytically inactive; L, loose site, loosely binding ligands and catalytically inactive; T, tight site, tightly binding ligands and catalytically active. The proton-induced conformational changes convert a T-site with bound ATP to an O-site, releasing the bound nucleotide. Concomitantly, an L-site with loosely bound ADP and phosphate is converted to a T-site, where the substrates are tightly bound and ATP forms. Fresh substrates bind to an O-site, converting it to an L-site, and so on. The energy-requiring steps are substrate binding and product release. One third of the catalytic cycle is shown. (Reproduced with permission from ref. 1).

The nucleotide-binding domain contains a nine-stranded β-sheet with nine associated α-helices. In both α- and β-subunits, β-strands 3–9 and helices A–G have the same topology as β strands 1–7 and associated helices in RecA[17]. Within this region, between 113 and 145 α-carbons superimpose on RecA with an rms separation of 1.9–2.0 Å (depending on the subunit). The nucleotide pyrophosphate group binds in the same way in both structures, but the orientation of the adenosine moiety is very different. In adenylate kinase[18], often used as a model for F_1-ATPase, the corresponding β-sheet has five strands with different connectivity.

d

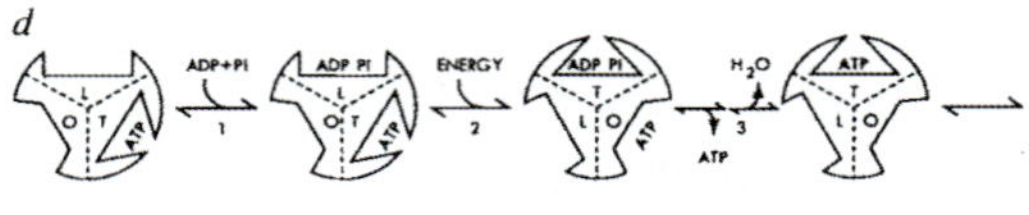

e

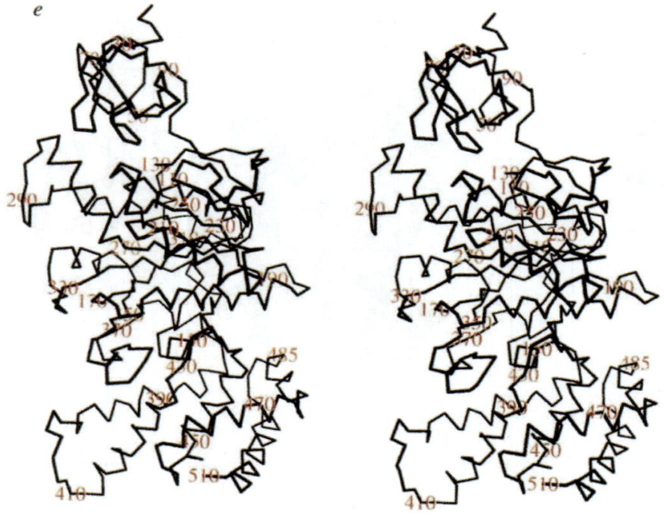

f

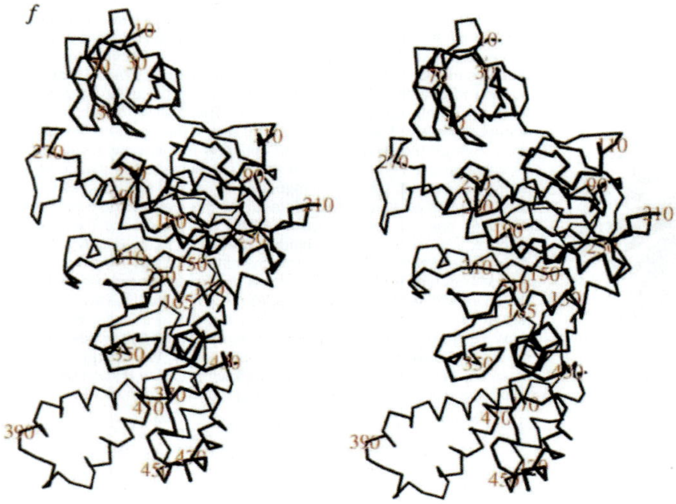

Nucleotide-binding sites

The nucleotide-binding sites are at the interfaces between α- and β-subunits. The catalytic sites are predominantly in β-subunits, with some contributions from side chains in the α-subunits, and conversely with the non-catalytic sites. Most notably, β-Tyr 368, a site of reaction of both photoactivated 2-azido-ADP[19] and of the nucleotide analogue 5′-p-fluorosulphonylbenzoyladenosine[20], protrudes into the α-subunit nucleotide site (Fig. 4). The average distance between adjacent catalytic and non-catalytic sites is 27 Å (between β-phosphates). In both α- and β-subunits, one face of the ribose and phosphate moieties is accessible to external solvent via a conical tunnel. In the crystals, the subunit denoted β_{TP} and each of the three α-subunits bind one molecule of the nucleotide triphosphate analogue AMP-PNP, whereas the diphosphate ADP is bound to β_{DP}. In subunit β_E, the nucleotide-binding site is empty and distorted, despite the presence of 250 μM AMP-PNP and 5 μM ADP in the solvent[15]. Inorganic phosphate is not present in the ADP site, and so the structure is probably that of the ADP-inhibited state[10]. The three α-subunits are defined in relation to β_{TP}, β_{DP} and β_E. Subunit α_{TP} forms the interface with β_{TP} that contains the β_{TP} nucleotide site, and similarly for α_{DP} and α_E (Fig. 2).

The amino acids involved in nucleotide binding are summarized in Fig. 4. Those that define the adenine- and ribose-binding pocket in the β-subunits include β-Tyr 345, which forms a crosslink with photoactivated 2-azido-ATP[21]. In contrast, β-Tyr 311, which reacts with photoactivated 8-azido-ATP[22,23], is 5.5 Å from the terminal phosphate at the opposite end of the binding site to the adenine pocket. The reaction appears to require non-productive binding of the nucleotide. The most important phosphate-binding amino acids are in the P-loop, which can be seen between strand 3 and helix B in subunit β_{DP} in Fig. 2*b*. This loop contains the conserved nucleotide-binding motif, GXXXXGKT/S, first recognized in the sequences of the α- and β-subunits of F_1-ATPase[9]. The motif has been used to predict the nucleotide-binding sites of many other proteins. The crystal structures of p21^{ras} (ref. 24), adenylate kinase[25], RecA[26], elongation factor Tu[27,28] and transducin-α[29], for example, show that the motif binds the di- or triphosphate moiety, and this is also true in the α- and β-subunits of F_1-ATPase (Fig. 4).

The crystal structure provides new details of the chemical mechanism of ATP synthesis. In β_{TP}, at a distance of 4.4 Å from the γ-phosphate of AMP-PNP, there is clear density for a bound water molecule, hydrogen-bonded to the carboxylate of β-Glu 188. This residue is appropriately positioned to activate the water molecule, promoting an in-line nucleophilic attack on the terminal phosphate. The guanidinium of α-Arg 373 could help to stabilize the negative charge that develops on the terminal phosphate in a pentacoordinate transition state. A similar arrangement of a glutamic acid and an arginine side chain is found in the active site of transducin-α complexed with GTP-γS[29]. The structure also indicates why the α-subunits

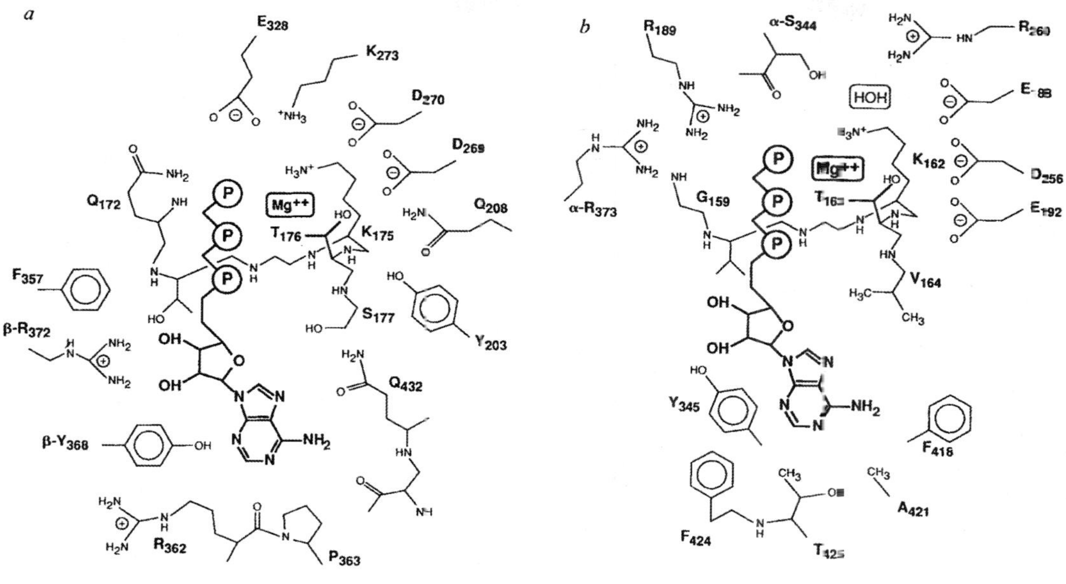

FIG. 4 The nucleotide-binding sites in the α- and β-subunits of bovine F_1-ATPase. *a*, Subunit α_E. All of the amino acids forming the pocket are in the α-subunit, except for the neighbouring β_{DP}-Arg 372 and β_{DP}-Tyr 368. Residues α172–177 are part of the P-loop. The nucleotide-binding sites of α_{DP} and α_{TP} are very similar, except that β-Arg 372 does not contribute to either binding site, and β-Tyr 368 only contributes to nucleotide binding in α_{DP}. *b*, Subunit β_{TP}. Except for α_{TP}-Arg 373 and the main-chain carbonyl and the side chain of α_{TP}-Ser 344, all of the amino acids are in β_{TP}. Residues 159–164 are part of the P-loop. In β_{DP}, there is density for an ordered water molecule replacing the terminal phosphate, and Arg 373 moves closer to the diphosphate moiety. Except for β-Arg 372 in the α-subunit nucleotide-binding site, all residues forming the pockets are widely conserved. The ligands to magnesium identified at this stage are as follows: in the α-subunits, O_γ-Thr 176, two oxygens from the terminal phosphate and one from the β-phosphate; in β_{TP}, O_γ-Thr 163 and one oxygen atom from each of the β- and terminal phosphates; in β_{DP}, O_γ-Thr 163 and one oxygen from the β-phosphate. The remaining ligands are presumably water molecules.

are non-catalytic. In the α-subunits, there is no spatial equivalent of the carboxylate of β-Glu 188, which is replaced in α-subunits by Gln 208, with the side chain pointing away from the terminal phosphate.

The binding of adenine by the α-subunit involves several hydrogen bonds, whereas in the β-subunit the adenine is in contact with a hydrophobic interface. Together with the presence of β-Tyr 368 close to the 2-position of the adenine ring in the α-subunit, this may explain the specificity of the α-subunit nucleotide sites for adenine nucleotides, and the ability of the β-subunit nucleotide sites to accommodate GTP and ITP as well as ATP[6,30]. The roles of the non-exchangeable nucleotides in the α-subunits remain obscure.

The site of reaction of the inhibitor dicyclohexylcarbodiimide (β-Glu 199; ref. 31) points into the conical tunnel leading to the nucleotide sites. The presence of the two bulky cyclohexyl moieties in the conical tunnel could impede nucleotide binding and the conversion of the catalytic sites between conformational states.

Asymmetrical features

With the exception of the nucleotide-binding domain of β_E, the domains of the three α- and three β-subunits superimpose with a root-mean-squared separation of less than 1 Å between Cα atoms, but because of differences in the relative orientations of the domains and the interactions with the unique γ-subunit, the overall structure of F_1 is strikingly asymmetric (Fig. 3).

In the β_E nucleotide-binding domain, the interaction between strands 3 and 7 is disrupted (above and below the arrow in Fig. 2*d*). The lower part of this domain and the α-helical domain are rotated by 30° with respect to their positions in β_{TP} and β_{DP}, resulting in movements by up to 20 Å. At the C terminus of helix B following the P-loop (below the arrowhead in Fig. 2*d*), the disruption of the β-sheet is accompanied by relaxation of a few residues into an α-helical conformation. A number of other hydrogen-bonding rearrangements occur throughout the rest of β_E. Also, β_E-Asp 316, β_E-Thr 318, and β_E-Asp 319 in the loop following strand 7 (indicated by the asterisk in Fig. 2*d*) hydrogen-bond with γ-Arg 254 and γ-Gln 255 in the C-terminal helix. Close by, α_E-Asp 333 makes a salt bridge with γ-Arg 252. These H-bonds form a 'catch', and are the major specific interactions between the two long helices of the γ-subunit and the subassembly $\alpha_3\beta_3$.

A second 'catch' is found in the C-terminal domain of subunit β_{TP}, but not in β_{DP} or β_E. Additional hydrogen bonds link γ-Lys 87, γ-Lys 90 and the backbone amide of γ-Ala 80 (all in the short helix, which is inclined at about 45° to the axis of pseudosymmetry) with β_{TP}-Asp 394 and β_{TP}-Glu 398 in a loop between helices 1 and 2. This sequence (DELSEED, β 394–400) is part of the binding site of amphipathic cationic inhibitors[32,33] and possibly of the ATPase inhibitor protein[34] in one or other of the β-subunits.

Asymmetry is also evident in α-subunits. The non-catalytic nucleotide-binding sites in α_{DP} and α_E are very similar, but in α_{TP} the site is more open because the lower half of β_E (which contributes to this binding site) is hinged out. Therefore, β_E-Tyr 368 no longer interacts with the N3 atom of the adenosine. It is likely that α_{TP} binds the exchangeable non-catalytic nucleotide observed in nucleotide-binding studies[10].

Asymmetry is also manifest in the coiled coil of the γ-subunit. Its C terminus (γ265–272) passes through a hydrophobic sleeve

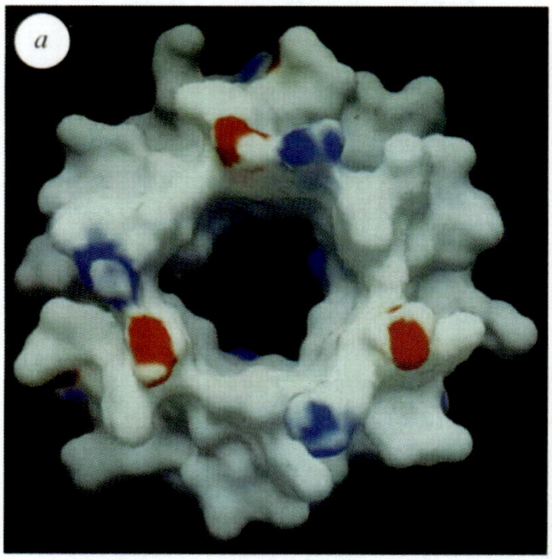

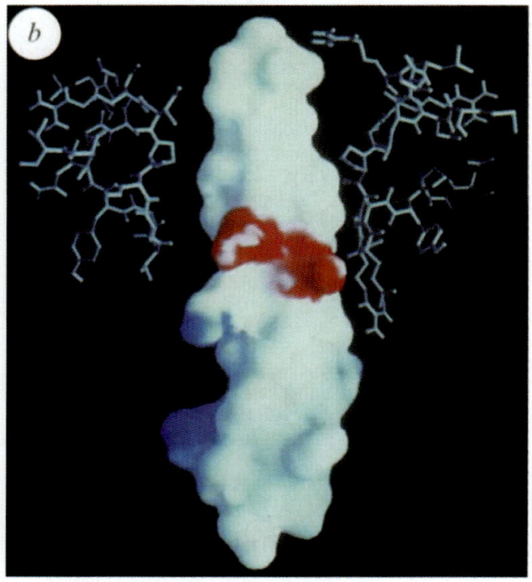

FIG. 5 Features of the structure of F_1-ATPase that are compatible with rotational catalysis. *a*, The electric surface potential of the hydrophobic sleeve surrounding the C-terminal tip of the γ-subunit, viewed from inside the assembly. The sleeve is made from the regions of the 3α and 3β subunits immediately below the crown of β-barrels. The surface was calculated with GRASP[50] from the coordinates of α287–294 and of β274–281. Negative and positive electric potentials of the side chains are red and blue, respectively; white is neutral. *b*, The electric surface potential of the C-terminal helical domain of the γ-subunit (residues γ253–272) shown with a 'ball-and-stick' representation of the amino acids from two diametrically opposed loops making the sleeve in *a*. Because no charged residues are present in the sleeve, the assembly could act as a molecular bearing, lubricated through a hydrophobic interface. *c*, Cross-section of F_1-ATPase showing the central cavity (black) traversed by the γ-subunit. The cross-section through an electron density map was generated from the atomic model and contoured at a level corresponding to the van der Waal's surface. The density for the α- and β-subunits is orange, and that for the γ-subunit is blue. The α-carbon backbones of the α-, β- and γ-subunits (red, yellow and yellow, respectively), and a CPK model of AMP-PNP are presented.

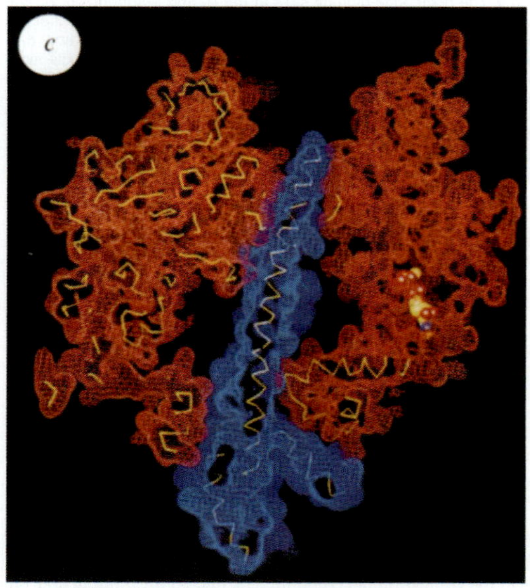

formed by six proline-rich loops (α287–294, β274–281; following helix E in β_{DP} in Fig. 2*b*, and the related loop in α_{TP}). The remaining interactions with the γ-subunit take place through hydrophobic loops, some of them poorly ordered, in the C-terminal domains of the α- and β-subunits, and helix 1 in β_{DP}. In the crystal structure, the γ-subunit seems to prevent β_E from adopting a nucleotide-binding conformation by obstructing the rotation of its lower half towards the central axis. In consequence, the lower halves of β_E and α_{TP} do not interact, and the middle and C-terminal domains of α_{TP} have moved away from the central axis (Fig. 3*c*). Therefore, each β-subunit adopts a different orientation towards the α-subunit that contributes to its active site, and the active site interfaces of β_{DP}, β_{TP} and β_E appear from a structural point of view to be tightly closed, more open, and fully open, respectively. This suggests that β_{DP} contains the non-exchangeable catalytic site[10]. The closure of the active site interface is accompanied by a distortion of helix 1 in the C-terminal domain of the α-subunits at Gly 388 (Fig. 3*c*). The preference of β_{TP} and β_{DP} for AMP-PNP and ADP, respectively, is probably caused by small conformational differences near to their P-loops, the most significant being the shift of α_{DP}-Arg 373 by 1.5 Å towards the β-phosphate moiety in β_{DP}.

Rotational catalysis

The asymmetry required by the binding change mechanism of ATP synthesis (Fig. 3*d*), particularly in the region of the active

ARTICLES

sites, is evident in the crystal structure. The three different conformations of the subunits β_E, β_{TP} and β_{DP} could be identified on the structural grounds described above as representing the open, loose and tight states, respectively. This asymmetry is associated with the asymmetric positioning of the γ-subunit relative to the $\alpha_3\beta_3$ subassembly.

The nucleotides that are found in the catalytic sites in the crystals appear at first sight to be inconsistent with this proposal, as the tight site (β_{DP}) should be occupied by ATP, and the loose site (β_{TP}) by ADP or ADP plus phosphate. However, if, as we propose, the structure in the crystal is that of the ADP-inhibited form of the enzyme, ADP (but not P_i) would be bound at the tight catalytic site. The presence of AMP-PNP in the loose site may be due to its 50-fold higher concentration over ADP, and to the absence of P_i from the crystallization medium. Alternatively, it is possible that the sequence of events during catalysis differs from that shown in Fig. 1, and that the 'tight' site of newly synthesized ATP first becomes 'loose' and then 'open'. Crystals of F_1 grown in the presence of P_i and alternative nucleotides (ATP, ATP-γS) could help to clarify this ambiguity.

The transition between the different catalytic states implied by the cycle of binding changes could be achieved by rotation of the catalytic subunits relative to the rest of the ATP synthase. This hypothesis cannot be tested by a single 'snapshot' of F_1-ATPase provided by the crystal structure, but several features of the structure are compatible with rotation of the $\alpha_3\beta_3$ subassembly relative to the γ-subunit (and presumably the remainder of the intact synthase). First, the sleeve surrounding the C terminus of the γ-subunit looks like a molecular bearing, lubricated by a hydrophobic interface (Fig. 5*a* and *b*). The crown of β-barrels above it, which form a continuous β-sheet linking all six subunits, holds the structure together, and the top of the nucleotide domain also contributes to its stability. Second, the central cavity would permit passage of the long N-terminal helix of the γ-subunit within the core of F_1 during rotation, with minimal contact between the γ-subunit and the rest of the interior of the particle (Fig. 5*c*). Third, the differences between the three catalytic sites are correlated with the asymmetric position of the γ-subunit. If there is to be an interconversion of the three sites, as the binding-change mechanism suggests, the most obvious way to achieve this is by rotation of $\alpha_3\beta_3$ relative to the γ-subunit. The nature of the interactions between the γ-subunit and $\alpha_3\beta_3$ suggest that the energy barrier opposing rotation is quite small. It could be overcome with energy from conformational changes induced by proton transport and relayed to the active sites via subunit γ and other stalk components. The V_{max} of the bovine ATP synthase is about 400 per second, and twice that value in the chloroplast enzyme[10]. Therefore, the rate of rotation in ATP synthases would be about 130–270 Hz.

Rotation of F_1 in ATP synthase and of the hydrophobic subunits in F_o have been considered in the past[10,35–37]. Attempts to demonstrate rotation have given negative results[10], but the possibility of rotation should be re-examined in the light of the structure. In the F_1-ATPase from *Escherichia coli*, nucleotide-dependent movements of the minor subunits (γ, δ and ε) in the central space between the hexameric arrangement of α- and β-subunits have been observed by cryoelectron microscopy[12] and by changes in the properties of fluorescent labels[38]. In the chloroplast enzyme, there is also evidence of movement of a central mass[39]. The proposed rotation within ATP synthase may be analogous to the rotation of the bacterial flagellum, which is also driven by the transmembrane protonmotive force[40]. A homologue of the β-subunit of ATPase is associated with the flagella complex, and may be involved in its assembly[41]. Assessment of the validity of the analogy between ATPase and flagellae requires appropriate biophysical examination of the ATP synthase based on the structure of F_1 and further detailed structural investigation of both ATP synthase and the flagellar motor. In the ATP synthase, it will be necessary to know more about the coupling mechanism (of which the γ-subunit and probably δ and ε are part) linking the catalytic sites with proton transport in the F_0 domain. The characterization of the proteins in the bovine stalk[16,42] and the *in vitro* assembly of an F_1/stalk complex[42] are necessary steps towards that goal. Little is known about the structure of the intrinsic membrane domain, F_o, and determining its high-resolution structure and understanding the structural basis of the coupling of proton transport to ATP synthesis are still challenges for the future. □

Received 21 June; accepted 19 July 1994.

1. Cross, R. L. *A. Rev. Biochem.* **50,** 681–714 (1981).
2. Senior, A. E. *Physiol. Rev.* **68,** 177–231 (1988).
3. Fillingame, R. H. *The Bacteria* **12,** 345–391 (1990).
4. Walker, J. E., Fearnley, I. M., Lutter, R., Todd, R. J. & Runswick, M. J. *Phil. Trans. R. Soc.* **326,** 367–378 (1990).
5. Penefsky, H. S. & Cross, R. L. *Adv. Enzym.* **64,** 173–214 (1991).
6. Issartel, J. P., Dupuis, A., Garin, J., Lunardi, J., Michel, L. & Vignais, P. V. *Experientia* **48,** 351–362 (1992).
7. Mitchell, P. *Biol. Rev.* **41,** 455–502 (1966).
8. Walker, J. E. *et al. J. molec. Biol.* **184,** 677–701 (1985).
9. Walker, J. E., Saraste, M., Runswick, M. J. & Gay, N. J. *EMBO J.* **1,** 945–951 (1982).
10. Boyer, P. D. *Biochim. biophys. Acta* **1140,** 215–250 (1993).
11. Boekema, E. J., Berden, J. A. & van Heel, M. G. *Biochim. biophys. Acta* **851,** 353–360 (1986).
12. Wilkens, S. & Capaldi, R. A. *Biol. Chem. Hoppe-Seyler* **375,** 43–51 (1994).
13. Abrahams, J. P. *et al. EMBO J.* **12,** 1775–1780 (1993).
14. Bianchet, M., Ysern, X., Hullihen, J., Pedersen, P. L. & Amzel, M. *J. biol. Chem.* **266,** 21197–21201 (1991).
15. Lutter, R. *et al. J. molec. Biol.* **229,** 787–790 (1993).
16. Collinson, I. R., Fearnley, I. M., Runswick, M. J., Skehel, J. M. & Walker, J. E. *Biochem. J.* (in the press).
17. Story, R. M., Weber, I. T. & Steitz, T. A. *Nature* **355,** 318–325 (1992).
18. Pai, E. F., Sachsenheimer, W., Schirmer, R. H. & Schulz, G. E. *J. molec. Biol.* **114,** 37–45 (1977).
19. Cross, R. L. *et al. Proc. natn. Acad. Sci. U.S.A.* **84,** 5715–5719 (1987).
20. Esch, F. S. & Allison, W. S. *J. biol. Chem.* **253,** 6100–6106 (1978).
21. Garin, J., Boulay, F., Issartel, J. P., Lunardi, J. & Vignais, P. V. *Biochemistry* **25,** 4431–4437 (1986).
22. Hollemans, M., Runswick, M. J., Fearnley, I. M. & Walker, J. E. *J. biol. Chem.* **258,** 9307–9313 (1983).
23. Garin, J., Vinçon, M., Gagnon, J. & Vignais, P. *Biochemistry* **33,** 3772–3777 (1994).
24. Milburn, M. V. *et al. Science* **247,** 939–945 (1990).
25. Mueller, C. W. & Schultz, G. E. *J. molec. Biol.* **224,** 159–177 (1992).
26. Story, R. M. & Steitz, T. A. *Nature* **355,** 374–376 (1992).
27. Berchtold, H. *et al. Nature* **365,** 126–132 (1993).
28. Kjeldgaard, M., Nissen, P. O., Thirup, S. & Nyborg, J. *Structure* **1,** 35–50 (1993).
29. Noel, J. P., Hamm, H. E. & Sigler, P. B. *Nature* **366,** 654–663 (1993).
30. Perlin, D. S., Latchney, L. R., Wise, J. G. & Senior, A. E. *Biochemistry* **23,** 4998–5003 (1984).
31. Esch, F. S., Böhlen, P., Otsuka, A. S., Yoshida, M. & Allison, W. S. *J. biol. Chem.* **256,** 9084–9089 (1981).
32. Bullough, D. A., Ceccarelli, E. A., Roise, D. & Allison, W. S. *Biochim. biophys. Acta* **975,** 377–383 (1989).
33. Zhuo, S., Paik, S. R., Register, J. A. & Allison, W. S. *Biochemistry* **32,** 2219–2227 (1993).
34. Jackson, P. J. & Harris, D. A. *Biochem. J.* **235,** 577–583 (1986).
35. Cox, G. B., Jans, D. A., Fimmel, A. L., Gibson, F. & Hatch, L. *Biochim. biophys. Acta* **768,** 201–208 (1984).
36. Cox, G. B., Fimmel, A. L., Gibson, F. & Hatch, L. *Biochim. Biophys Acta* **849,** 62–69 (1986).
37. Mitchell, P. *FEBS Lett.* **182,** 1–7 (1985).
38. Capaldi, R., Aggeler, R., Turina, P. & Wilkens, S. *Trends biochem. Sci.* **19,** 284–289 (1994).
39. Boekema, E. J. & Böttcher, B. *Biochim. biophys. Acta* **1098,** 131–143 (1992).
40. Macnab, R. M. *A. Rev. Genet.* **26,** 131–158 (1992).
41. Dreyfus, G., Williams, A. W., Kawagishi, I. & Macnab, R. M. *J. Bact.* **175,** 3131–3138 (1993).
42. Collinson, I. R. *et al. J. molec. Biol.* (in the press).
43. Leslie, A. G. W. in *Joint CCP4 and ESF-EACMB Newsletter on Protein Crystallography* No. 26 (Daresbury Laboratory, Warrington, UK, 1992).
44. *CCP4: A Suite of Programs for Protein Crystallography* (SERC Collaborative Computing Project No. 4, Daresbury Laboratory, Warrington, UK, 1979).
45. Read, R. J. *Acta crystallogr.* A **42,** 140–149 (1986).
46. Jones, T. A., Zou, J. Y., Cowan, S. W. & Kjeldgaard, M. *Acta crystallogr.* A **47,** 110–119 (1991).
47. Brünger, A. T. *XPLOR Version 3.1 Manual* (Yale Univ., USA, 1993).
48. Brünger, A. T. *Nature* **335,** 472–475 (1992).
49. Kraulis, P. J. *J. appl. Crystallogr.* **24,** 946–950 (1991).
50. Nicolls, A., Sharp, K. A. & Honig, B. *Proteins* **11,** 281–296 (1991).

ACKNOWLEDGEMENTS. We thank R. Henderson, A. Klug and M. F. Perutz for discussions, the staff of the Synchrotron Radiation facility at Daresbury for their help, and M. J. van Raaij and R. Noel for assistance. J.P.A. was supported during part of this work by an EMBO Fellowship. After completion of the current refinement, the coordinates will be deposited in the Brookhaven Protein Databank where they will be held for one year before release. We dedicate this paper to M. F. Perutz on the occasion of his 80th birthday

RESEARCH ARTICLES

The Structure of the Potassium Channel: Molecular Basis of K^+ Conduction and Selectivity

Declan A. Doyle, João Morais Cabral, Richard A. Pfuetzner, Anling Kuo, Jacqueline M. Gulbis, Steven L. Cohen, Brian T. Chait, Roderick MacKinnon*

The potassium channel from *Streptomyces lividans* is an integral membrane protein with sequence similarity to all known K^+ channels, particularly in the pore region. X-ray analysis with data to 3.2 angstroms reveals that four identical subunits create an inverted teepee, or cone, cradling the selectivity filter of the pore in its outer end. The narrow selectivity filter is only 12 angstroms long, whereas the remainder of the pore is wider and lined with hydrophobic amino acids. A large water-filled cavity and helix dipoles are positioned so as to overcome electrostatic destabilization of an ion in the pore at the center of the bilayer. Main chain carbonyl oxygen atoms from the K^+ channel signature sequence line the selectivity filter, which is held open by structural constraints to coordinate K^+ ions but not smaller Na^+ ions. The selectivity filter contains two K^+ ions about 7.5 angstroms apart. This configuration promotes ion conduction by exploiting electrostatic repulsive forces to overcome attractive forces between K^+ ions and the selectivity filter. The architecture of the pore establishes the physical principles underlying selective K^+ conduction.

Potassium ions diffuse rapidly across cell membranes through proteins called K^+ channels. This movement underlies many fundamental biological processes, including electrical signaling in the nervous system. Potassium channels use diverse mechanisms of gating (the processes by which the pore opens and closes), but they all exhibit very similar ion permeability characteristics (*1*). All K^+ channels show a selectivity sequence of $K^+ \approx Rb^+ > Cs^+$, whereas permeability for the smallest alkali metal ions Na^+ and Li^+ is immeasurably low. Potassium is at least 10,000 times more permeant than Na^+, a feature that is essential to the function of K^+ channels. Potassium channels also share a constellation of permeability characteristics that is indicative of a multi-ion conduction mechanism: The flux of ions in one direction shows high-order coupling to flux in the opposite direction, and ionic mixtures result in anomalous conduction behavior (*2*). Because of these properties, K^+ channels are classified as "long pore channels," invoking the notion that multiple ions queue inside a long, narrow pore in single file. In addition, the pores of all K^+ channels can be blocked by tetraethylammonium (TEA) ions (*3*).

D. A. Doyle, R. A. Pfuetzner, A. Kuo, and R. MacKinnon are in the Laboratory of Molecular Neurobiology and Biophysics and the Howard Hughes Medical Institute, Rockefeller University, 1230 York Avenue, New York, NY 10021, USA. J. M. Cabral and J. M. Gulbis are in the Laboratory of Molecular Neurobiology and Biophysics, Rockefeller University, 1230 York Avenue, New York, NY 10021, USA. S. L. Cohen and B. T. Chait are in the Laboratory of Mass Spectrometry and Gaseous Ion Chemistry, Rockefeller University, 1230 York Avenue, New York, NY 10021, USA.

*To whom correspondence should be addressed. E-mail: mackinn@rockvax.rockefeller.edu

Molecular cloning and mutagenesis experiments have reinforced the conclusion that all K^+ channels have essentially the same pore constitution. Without exception, all contain a critical amino acid sequence, the K^+ channel signature sequence. Mutation of these amino acids disrupts the channel's ability to discriminate between K^+ and Na^+ ions (*4*).

Two aspects of ion conduction by K^+ channels have tantalized biophysicists for the past quarter century. First, what is the chemical basis of the impressive fidelity with which the channel distinguishes between K^+ and Na^+ ions, which are featureless spheres of Pauling radius 1.33 Å and 0.95 Å, respectively? Second, how can K^+ channels be so highly selective and at the same time, apparently paradoxically, exhibit a throughput rate approaching the diffusion limit? The 10^4 margin by which K^+ is selected over Na^+ implies strong energetic interactions between K^+ ions and the pore. And yet strong energetic interactions seem incongruent with throughput rates up to 10^8 ions per second. How can these two essential features of the K^+ channel pore be reconciled?

Potassium Channel Architecture

The amino acid sequence of the K^+ channel from *Streptomyces lividans* (KcsA K^+ channel) (*5*) is similar to that of other K^+ channels, including vertebrate and invertebrate voltage-dependent K^+ channels, vertebrate inward rectifier and Ca^{2+}-activated K^+ channels, K^+ channels from plants and bacteria, and cyclic nucleotide-gated cation channels (Fig. 1). On the basis of hydrophobicity analysis, there are two closely related varieties of K^+ channels, those containing two membrane-spanning segments per subunit and those containing six. In all cases, the functional K^+ channel protein is a tetramer (*6*), typically of four identical subunits (*7*). Subunits of the two membrane-spanning variety appear to be shortened versions of their larger counterparts, as if they simply lack the first four membrane-spanning segments. Although the KcsA K^+ channel is a two membrane-spanning K^+ channel, its amino acid sequence is actually closer to those of eukaryotic six membrane-spanning K^+ channels. In particular, its sequence in the pore region, located between the membrane-spanning stretches and containing the K^+ channel signature sequence, is nearly identical to that found in the *Drosophila* (*Shaker*) and vertebrate voltage-gated K^+ channels (Fig. 1). In an accompanying paper, through a study of the KcsA K^+ channel interaction with eukaryotic K^+ channel toxins, we confirm that the KcsA pore structure is indeed very similar to that of eukaryotic K^+ channels and that its structure is maintained when it is removed from the membrane with detergent (*8*).

We have determined the KcsA K^+ channel structure from residue position 23 to 119 by x-ray crystallography (Table 1). The cytoplasmic carboxyl terminus (residues 126 to 158) was removed in the preparation and the remaining residues were disordered. The KcsA K^+ channel crystals are radiation-sensitive and the diffraction pattern is anisotropic, with reflections observed along the best and worst directions at 2.5 Å and 3.5 Å Bragg spacings, respectively. By data selection, anisotropy correction, introduction of heavy atom sites by site-directed mutagenesis, averaging, and solvent flattening, an interpretable electron density map was calculated (Fig. 2, A through C). This map was without main chain breaks and showed strong side chain density (Fig. 2C). The model was refined with data to 3.2 Å (the data set was 93 % complete to 3.2 Å with 67% completeness between 3.3 Å and 3.2 Å), maintaining highly restrained stereochemistry and keeping tight noncrystallographic symmetry restraints. The refinement procedure was monitored by minimizing the value *R*-free (29.0%) and its separation from *R*-crystallographic (28.0%). The presence of four molecules (subunits) in the asymmetric unit of the crystal provides a very significant enhancement of the accuracy of the crystallographic analysis; first, by enabling averaging of the electron density over four crystallographically independent regions of the multiple isomorphous replacement (MIR) map, and second, by providing a powerful set of constraints on the atomic model during refinement (*9*).

The K^+ channel is a tetramer with fourfold symmetry about a central pore (Fig. 3, A and B). Like several other membrane proteins, it has two layers of aromatic amino acids positioned to extend into the lipid bilayer, presumably near the membrane-water interfaces (Fig. 3C) (*10*). Each subunit has two transmembrane α-helices connected by the roughly 30 amino acid pore region, which consists of the turret, pore helix, and selectivity filter (Fig. 3, A and B). A subunit is inserted into the tetramer such that one transmembrane helix (inner helix) faces the central pore while the other (outer helix) faces the lipid membrane. The inner helices are tilted with respect to the membrane normal by about 25° and are slightly kinked, so that the subunits open like the petals of a flower facing the outside of the cell. The open petals house the structure formed by the pore region near the extracellular surface of the membrane. This region contains the K^+ channel signature sequence, which forms the selectivity filter (*4*). The essential features of subunit packing can be appreciated by viewing the relation between the four inner helices and the four pore helices (Fig. 3D). The four inner helices pack against each other as a bundle near the intracellular aspect of the membrane, giving the appearance of an inverted teepee. The pore helices are slotted in between the poles of the teepee and are directed, with an amino-to-carboxyl sense, toward a point near the center of the channel (Fig. 3, A, B, and D). This pore helix arrangement provides many of the intersubunit contacts that hold the tetramer together and, as discussed below, is also critical in the operation of the ion conduction pore.

Sequence conservation among K^+ channels (including ones with two and six membrane-spanning segments), as well as cyclic nucleotide-gated cation channels, is strongest for the amino acids corresponding to the pore region and the inner helix. Even Na^+ and Ca^{2+} channels show distant relatedness over these segments. The teepee architecture of the K^+ channel pore likely will be a general feature of all of these cation channels, with four inner helices arranged like the poles of a teepee, four pore helices, and a selectivity filter—tuned to select the appropriate cation—located close to the extracellular surface.

This structure of the KcsA K^+ channel is in excellent agreement with results from functional and mutagenesis studies on *Shaker* and other eukaryotic K^+ channels (Fig. 4). The pore-region of K^+ channels was first defined with pore-blocking scorpion toxins (*11*). These inhibitors interact with amino acids (Fig. 4, white) comprising the broad extracellular-facing entryway to the pore (*12*). The impermeant organic cation TEA blocks K^+ channels from both sides of the membrane at distinct sites (*13*). Amino acids interacting with externally and internally applied TEA are located just external to (Fig. 4, yellow) and internal to (Fig. 4, mustard) the structure formed by the signature sequence amino acids (*14, 15*). Alteration of the signature sequence amino acids (Fig. 4, red main chain atoms) disrupts K^+ selectivity (*4*). Amino acids close to the intracellular opening on the *Shaker* K^+ channel map to the inner helix on the KcsA K^+ channel (*16*). Interestingly, exposure to the cytoplasm of the region above the inner helix bundle (Fig. 4, pink side chains) requires an open voltage-dependent gate, whereas the region at or below the bundle (Fig. 4, green side chains) is exposed whether or not the gate was open. The correlation between the transition zone for gate-dependent exposure to the cytoplasm in the *Shaker* K^+ channel and the inner helix bundle in this structure has implications for mechanisms of gating in K^+ channels.

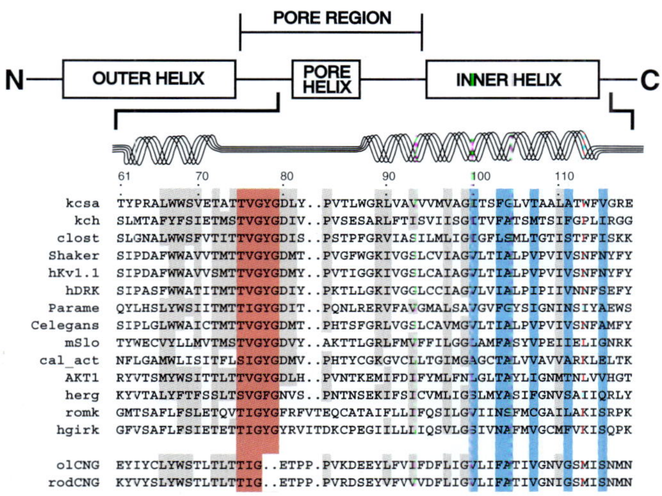

Fig. 1. Sequence alignment of selected K^+ channels and cyclic nucleotide-gated channels. The numbering and secondary structural elements for the *Streptomyces lividans* K^+ channel (KcsA) is given above the sequences. Selectivity filter, red; lining of the cavity and inner pore, blue; residues in which the nature of the side chain is preserved (>50% similarity), grey. The sequences are: KcsA, *Streptomyces lividans*, accession number (acc) PIR S60172; kch, *Escherichia coli*, acc GenBank U24203; clost, *Clostridium acetobutylicum* (Genome Therapeutics Corp.); Shaker, *Drosophila melanogaster*, acc PIR S00479; hKv1.1, *Homo sapiens*, acc Swissprot Q09470; hDRK, *H. sapiens*, acc PIR S31761; Parame, *Paramecium tetraaurelia*, acc GenBank U19908; Celegans, *Caenorhabditis elegans*, acc GenBank AF005246; mSlo, *Mus musculus*, acc PIR A48206; cal_act, *H. sapiens*, acc GenBank AF031815; AKT1, *Arabidopsis thaliana*, acc PIR S62694; herg, *H. sapiens*, acc PIR I38465; romk, *Rattus norvegicus*, acc Swissprot P35560; hgirk, *H. sapiens*, acc GenBank S78684; olCNG, *H. sapiens*, acc Swissprot Q16280; rodCNG, *H. sapiens*, acc PIR A42161. The last two sequences, separate from the rest, are from cyclic nucleotide–gated channels, which are not K^+ selective.

General Properties of the Ion Conduction Pore

As might have been anticipated for a cation channel, both the intracellular and extracellular entryways are negatively charged by acidic amino acids (Fig. 5A, red), an effect that would raise the local concentration of cations while lowering the concentration of anions. The overall length of the pore is 45 Å, and its diameter varies along its distance (Fig. 5B). From inside the cell (bottom) the pore begins as a tunnel 18 Å in length (the internal pore) and then opens into a wide cavity (~10 Å across) near the middle of the membrane. A K^+ ion could move throughout the internal pore and cavity and still remain mostly hydrated. In contrast, the selectivity filter separating the cavity from the extracellular solution is so narrow that a K^+ ion would have to shed its hydrating waters to enter. The chemical composition of the wall lining the internal pore and cavity is predominantly hydrophobic (Fig. 5A, yellow). The selectivity filter, on the other hand, is lined exclusively by polar main chain atoms belonging to the signature sequence amino acids. The distinct mechanisms operating in the cavity and internal pore versus the selectivity filter will be discussed below, but first we introduce the determination of K^+ ion positions in the pore.

Potassium channels exclude the smaller alkali metal cations Li^+ (radius 0.60 Å) and Na^+ (0.95 Å) but allow permeation of the larger members of the series Rb^+ (1.48 Å) and Cs^+ (1.69 Å). In fact, Rb^+ is nearly a perfect K^+ (1.33 Å) analog because its size and permeability characteristics are very similar to those of K^+. Because they are more electron dense than K^+, Rb^+ and Cs^+ allow visualization of the locations of permeant ions in the pore. By difference electron density maps calculated with data from crystals transferred into Rb^+-containing (Fig. 6a) or Cs^+-containing (Fig. 6b) solutions, multiple ions are well defined in the pore. The selectivity filter contains two ions (inner and outer ions) located at opposite ends, about 7.5 Å apart (center to center). In the Rb^+ difference map, there actually are two partially separated peaks at the inner aspect of the selectivity filter. These peaks are too close to each other (2.6 Å) to represent two simultaneously occupied ion binding sites. We suspect that they represent a single ion (on average) in rapid equilibrium between adjacent sites. The single inner ion peak in the Cs^+ difference map undoubtedly reflects the lower resolution at which the map was calculated (to 5 Å for Cs^+ versus 4.0 Å for Rb^+), because the Rb^+ difference map, when calculated at the same lower resolution, also shows only a single peak at the Cs^+ position. The Rb^+ positions correspond to strong peaks (pre-

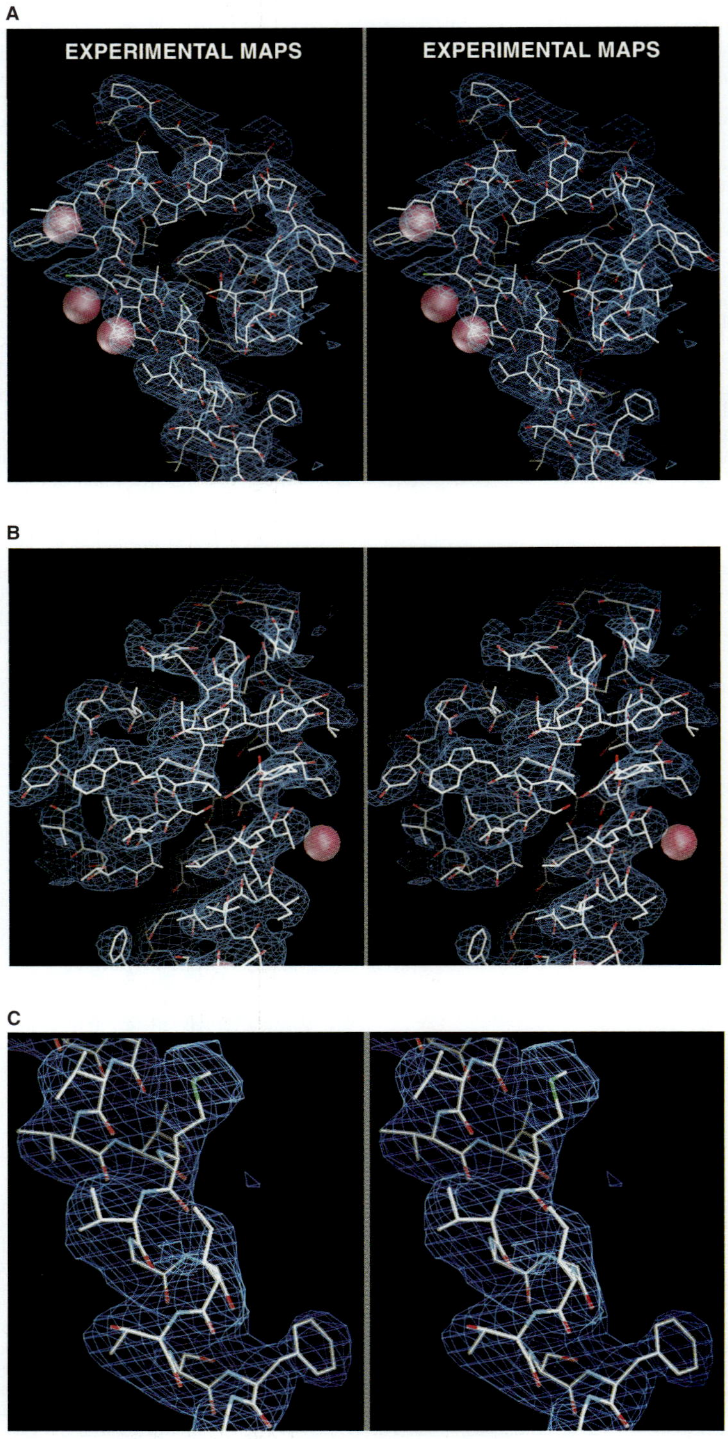

Fig. 2. Experimental electron density map. Stereoviews of the experimental electron-density map contoured at 1 σ covering nearly an entire subunit (removed from the tetramer) of the final model. The map was calculated at 3.2 Å resolution with the following Fourier coefficients: native-sharpened amplitudes and MIR solvent flattened averaged phases. (**A**) Foreground: map showing inner helix, loop structures and selectivity filter; background: the pore helix and outer helix. CPK spheres show positions of mercury atoms used as residue markers (from the top, marked residues are Leu[86], Leu[90] and Val[93]). (**B**) Alternative view. Foreground: pore helix and part of outer helix; background: selectivity filter and turret. CPK sphere marks position of Ala[42]. (**C**) Close up view of electron density.

sumably, K^+ ions) in a high contour native electron density map (not shown). Thus, the selectivity filter contains two K^+ ions. A third weaker peak is located below the selectivity filter at the center of the large cavity in the Rb^+ difference map (Fig. 6a, cavity ion) and in the Cs^+ difference map at a lower contour (not shown). Electron density at the cavity center is prominent in MIR maps, even prior to averaging (Fig. 6c, lower diffuse peak). The difference electron density maps show this to be related to the presence of one or more poorly localized cations situated at least 4 Å away from the closest protein groups.

The Cavity and Internal Pore

Why is there a 10 Å diameter cavity in the center of the channel with an ion in it (Fig. 5B and Fig. 6)? Electrostatic calculations show that when an ion is moved along a narrow pore through a membrane, it must cross an energy barrier that is maximum at the membrane center (*17*). The electrostatic field emanating from a cation polarizes its environment, bringing the negative ends of dipoles closer to it and thereby stabilizing it. At the bilayer center, the polarizability of the surround-

Table 1. Summary of data collection and refinement statistics. Crystals (space group C2: a = 128.8 Å, b = 68.9 Å, c = 112.0 Å, β = 124.6°) were flash-frozen by being transferred directly from the crystal mother liquor to a stream of boiled-off nitrogen (*24*). Because crystals of the mutant L90C diffracted significantly better than wild-type protein crystals, the former were used for native data collection. Data were collected from multiple crystals, and six sets were selected and merged to form the native data set used for structure determination. Mercury derivatives were obtained by direct addition of methyl mercury to the crystallization solution of cysteine mutant crystals. MALDI-TOF mass spectrometry confirmed 60 to 90% derivatization of crystals prior to data collection. All data were collected at Cornell High Energy Synchrotron Source (CHESS), station A1, with the Princeton 2K CCD (*25*). Data were processed with DENZO and SCALEPACK (*26*) and the CCP4 package (*27*). Heavy atom positions were determined with SHELX-97 (*28*) and cross-difference Fourier analysis. These positions confirmed the fourfold noncrystallographic symmetry observed in the self-rotation Patterson function and allowed the determination of initial orientation matrices. An initial model (90% complete) was built into a solvent flattened (64% solvent content), fourfold averaged electron density map with the program O (*29*). The tracing of the model was facilitated by the use of the mercury atom positions as residue markers. L86C was used solely for this purpose. After torsional refinement (with strict fourfold noncrystallographic symmetry constraints) with X-PLOR 3.851 (*30*), this model was used in the anisotropic scaling [sharpening (*31*)] of the native data with X-PLOR. The structure factor sigma values were also rescaled appropriately, and the corrected data were used for all subsequent procedures. Fourfold averaging, solvent flattening, and phase extension were applied in DM (*32*), resulting in a marked improvement of the electron density that allowed correction of the model and the building of additional residues. Refinement consisted of rounds of positional (in the initial stages phase information was also included as a restraint) and grouped B-factor refinement in X-PLOR. Fourfold noncrystallographic symmetry was highly restrained with the force constant for positional restraints set as 1000 kcal/mol/Å². The diffuse ion cloud described in the text was initially modeled as one or more K^+ ions and several water molecules; however, the results were unsatisfactory. Therefore, this and other strong unmodelled density present in solvent-flattened maps (no averaging included) was Fourier back-transformed, scaled, and included in the refinement procedure as partial structure factors. The final model includes amino acids 23 to 119 of each chain. The following residues were truncated: Arg[27] to Cβ, Ile[60] to Cγ, Arg[64] to Cβ, Glu[71] to Cβ, and Arg[117] to Nε. The stereochemistry is strongly restrained, with no outliers on the Ramachandran plot. The high B-factor values reflect the intensity decay of the data beyond 4 Å.

			Data collection and phasing			
Data set	Resolution (Å)	Redundancy	Completeness overall/outer (%)	R_{merge}*	Phasing power†	R-Cullis‡
L90C-a	15.0–3.7	3.5	91.3/93.3	0.071	1.61	0.70
L90C-b	15.0–3.7	7.0	91.5/94.1	0.083	1.87	0.50
V93C	15.0–3.7	4.1	98.3/99.1	0.075	1.35	0.63
A32C	15.0–4.0	2.3	84.1/83.8	0.076	1.45	0.66
A29C	15.0–5.0	2.7	73.9/74.0	0.063	1.03	0.85
A42C	15.0–6.5	2.0	90.7/90.3	0.057	0.97	0.81
L86C	30.0–6.0	2.3	58.7/58.9	0.057	—	—
					$I/\sigma I$	% of measured data with $I/\sigma I$ > 2
Native	30.0–3.2	6.1	93.3%	0.086	15.8	75
Outer Shell	3.3–3.2	2.3	66.6%	0.286	3.9	50

	Anisotropic correction	
	Average F.O.M‖ (30.0–3.2 Å)	Average F.O.M‖ (3.4–3.2 Å)
Before sharpening¶	0.76	0.55
After sharpening¶	0.83	0.64

Refinement		*Root-mean-square deviation of*	
Resolution:	10.0–3.2 Å	Bond angles:	1.096°
R-cryst.§:	28.0%	Bond lengths:	0.005 Å
R-free§:	29.0%	Ncs related atoms:	0.006 Å
No. of reflections with $\|F\|/\sigma\|F\|$ > 2:	12054	B-factor for ncs related atoms:	10 Å²
No. of protein atoms:	710 per subunit	B-factor for non-bonded atoms:	36 Å²
No. of ligand atoms:	1 water, 3 K^+ ions		
Mean B-factor for main-chain atoms:	90 Å²		
Mean B-factor for side-chain atoms:	110 Å²		

*$R_{merge} = \Sigma\Sigma|I-I_j|/\Sigma I$. †Phasing power = $\langle|F_h|\rangle/\langle E\rangle$. ‡$R$-Cullis = $\Sigma||F_{ph} \pm F_p| - |F_{hc}||/\Sigma|F_{ph} \pm F_p|$, only for centric data. §R-cryst. = $\Sigma|F_p - F_{p(calc)}|/\Sigma|F_p|$, R-free the same as R-cryst, but calculated on 10% of data selected in thin resolution shells and excluded from refinement. ‖Figure of merit. ¶In both cases, fourfold averaging and solvent flattening were applied; I_j is the observed intensity, I is the average intensity, F_h is the root-mean-square heavy-atom structure factor, E is the lack of closure error, F_{ph} is the structure factor for the derivative, F_p is the structure factor for the native, F_{hc} is the calculated structure factor for the heavy atom, and $F_{p(calc)}$ is the calculated native structure factor

ing medium is minimal and therefore the energy of the cation is highest. Thus, simple electrostatic considerations allow us to understand the functional significance of the cavity and its strategic location. The cavity overcomes the electrostatic destabilization resulting from the low dielectric bilayer by simply surrounding an ion with polarizable water. A second feature of the K^+ channel structure also stabilizes a cation at the bilayer center. The four pore helices point directly at the center of the cavity (Fig. 3, A, B, and D). The amino to carboxyl orientation of these helices will impose a negative electrostatic (cation attractive) potential via the helix dipole effect (*18*). The ends of the helices are rather far (~8 Å) from the cavity center, but all four contribute to the effect. Therefore, two properties of the structure, the aqueous cavity and the oriented helices, help to solve a fundamental physical problem in biology—how to lower the electrostatic barrier facing a cation crossing a lipid bilayer. Thus, the diffuse electron density in the cavity center (Fig. 6C) likely reflects a hydrated cation cloud rather than an ion binding site (Fig. 7).

Alternatively, the channel could have overcome the destabilizing electrostatic effects of the bilayer center by lining the entire pore with a polarizable surface, putting ion binding sites along its entire length. But the structure shows that, with the exception of the selectivity filter, the pore lining is mainly hydrophobic, a general property of K^+ channels (Fig. 1). This conclusion was anticipated by the landmark experiments of Armstrong, which showed that hydrophobic cations bind in

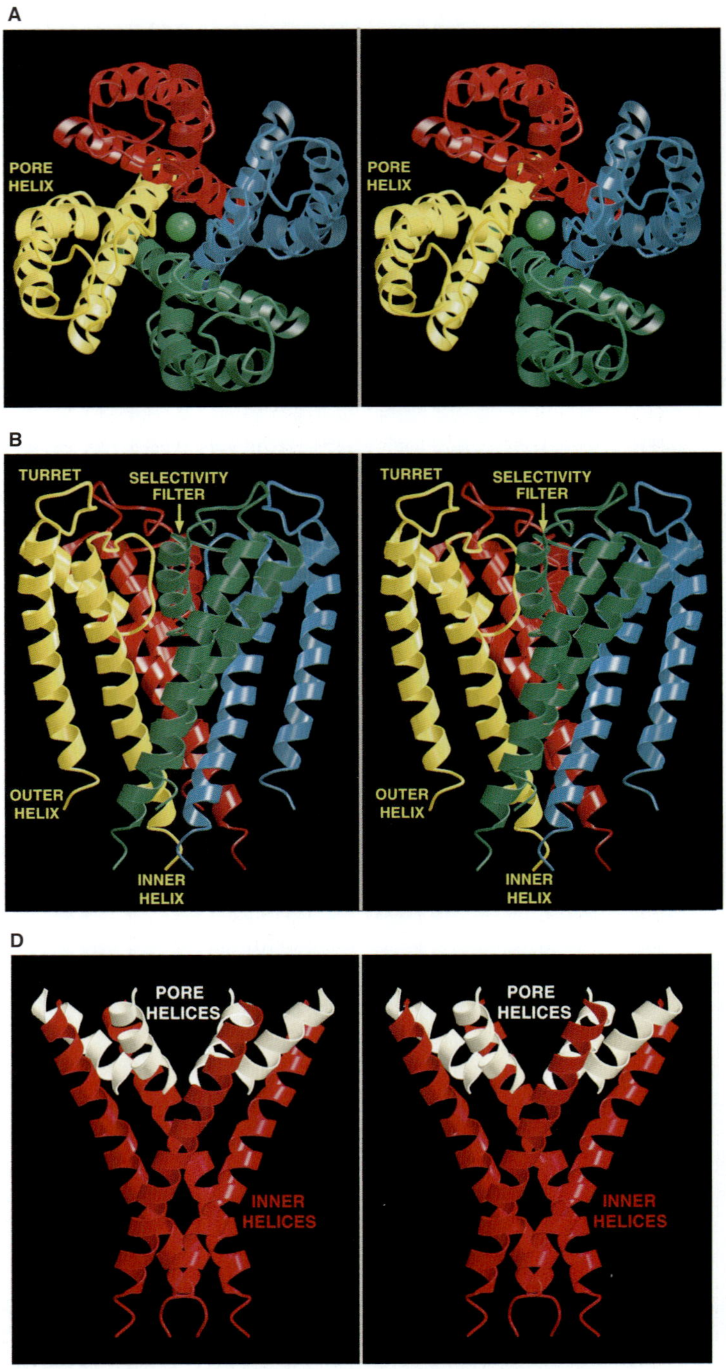

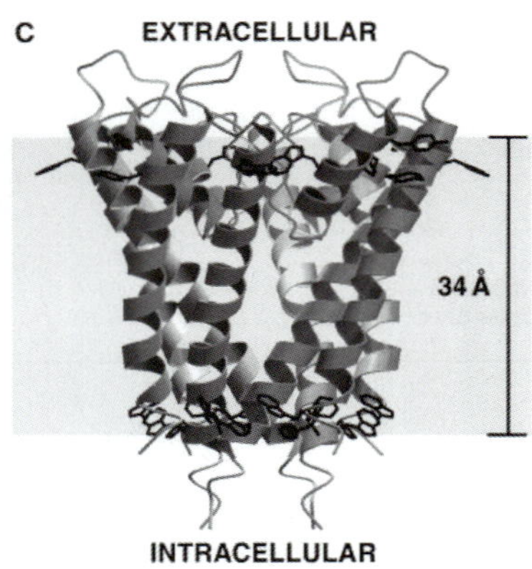

Fig. 3. Views of the tetramer. (**A**) Stereoview of a ribbon representation illustrating the three-dimensional fold of the KcsA tetramer viewed from the extracellular side. The four subunits are distinguished by color. (**B**) Stereoview from another perspective, perpendicular to that in (A). (**C**) Ribbon representation of the tetramer as an integral-membrane protein. Aromatic amino acids on the membrane-facing surface are displayed in black. (**D**) Inverted teepee architecture of the tetramer. These diagrams were prepared with MOLSCRIPT and RASTER-3D (*33*).

the pore of K^+ channels (*3*). What is the significance of the hydrophobic lining? We suggest that it would be counterproductive to achieving a high throughput of K^+ ions were the lining of the channel to interact strongly with ions outside of the selectivity filter. The hydrophobic lining presents a relatively inert surface to a diffusing ion over most of the length of the pore.

In summary, the inner pore and cavity lower electrostatic barriers without creating deep energy wells. The structural and chemical design of this part of the pore ensure a low resistance pathway from the cytoplasm to the selectivity filter, facilitating a high throughput. Functional experiments on K^+ channels support this conclusion. When TEA from the cytoplasm migrates to its binding site at the top of the cavity [>50% of the physical distance across the membrane (Figs. 4 and 5)], it traverses only about 20% of the transmembrane voltage difference (*15*). Thus, 80% of the transmembrane voltage is imposed across the relatively short selectivity filter. The rate limiting steps for a K^+ ion traversing the channel are thereby limited to this short distance. In effect, the K^+ channel has thinned the relevant transmembrane diffusion distance to a mere 12 Å.

The Selectivity Filter

The atomic model for the K^+ channel selectivity filter was based on the experimental electron density map, which showed a continuous ridge of electron density attributable to the main chain, as well as strong Val and Tyr side chain density directed away from the pore (Fig. 8A). We also used K^+ ion positions defined by difference Fourier analysis (Figs. 6 and 8A, yellow density) and our knowledge of alkali metal cation coordination in small molecules. The side chain locations preclude their direct participation in ion coordination, leaving this function to the

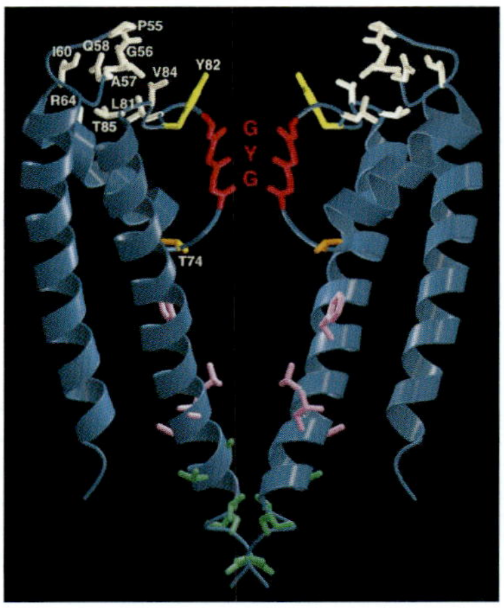

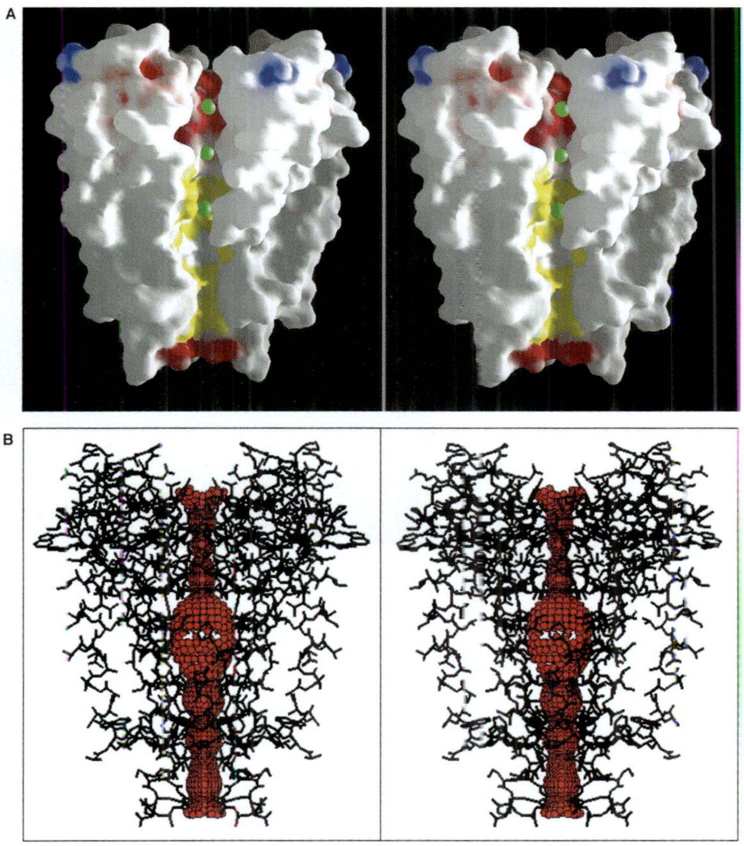

Fig. 4 (**above**). Mutagenesis studies on *Shaker*: Mapping onto the KcsA structure. Mutations in the voltage-gated *Shaker* K^+ channel that affect function are mapped to the equivalent positions in KcsA based on the sequence alignment. Two subunits of KcsA are shown. Mutation of any of the white side chains significantly alters the affinity of agitoxin2 or charybdotoxin for the *Shaker* K^+ channel (*12*). Changing the yellow side chain affects both agitoxin2 and TEA binding from the extracellular solution (*14*). This residue is the external TEA site. The mustard-colored side chain at the base of the selectivity filter affects TEA binding from the intracellular solution [the internal TEA site (*15*)]. The side chains colored green, when mutated to cysteine, are modified by cysteine-reactive agents whether or not the channel gate is open, whereas those colored pink react only when the channel is open (*16*). Finally, the residues colored red (GYG, main chain only) are absolutely required for K^+ selectivity (*4*). This figure was prepared with MOLSCRIPT and RASTER-3D. **Fig. 5** (**right**). Molecular surface of KcsA and contour of the pore. (**A**) A cutaway stereoview displaying the solvent-accessible surface of the K^+ channel colored according to physical properties. Electrostatic potential was calculated with the program GRASP, assuming an ionic strength equivalent to 150 mM KCl and dielectric constants of 2 and 80 for protein and solvent, respectively. Side chains of Lys, Arg, Glu, and Asp residues were assigned single positive or negative charges as appropriate, and the surface coloration varies smoothly from blue in areas of high positive charge through white to red in negatively charged regions. The yellow areas of the surface are colored according to carbon atoms of the hydrophobic (or partly so) side chains of several semi-conserved residues in the inner vestibule (Thr^{75}, Ile^{100}, Phe^{103}, Thr^{107}, Ala^{108}, Ala^{111}, Val^{115}). The green CPK spheres represent K^+ ion positions in the conduction pathway. (**B**) Stereoview of the entire internal pore. Within a stick model of the channel structure is a three-dimensional representation of the minimum radial distance from the center of the channel pore to the nearest van der Waals protein contact. The display was created with the program HOLE (*34*).

main chain atoms. The precise orientation of individual carbonyl oxygens cannot be discerned at the resolution of this x-ray analysis, but we propose that they are directed inward to account for K^+ ion coordination (Fig. 8B). A single water molecule (the only one modeled in the structure) located between the two K^+ ions in the selectivity filter was justified by the presence of a strong electron density peak in the experimental map, which was never associated with an ion peak in the difference Fourier maps (*19*).

The structure of the selectivity filter exhibits two essential features. First, the main chain atoms create a stack of sequential oxygen rings and thus afford numerous closely spaced sites of suitable dimensions for coordinating a dehydrated K^+ ion. The K^+ ion thus has only a very small distance to diffuse from one site to the next within the selectivity filter. The second important structural feature of the selectivity filter is the protein packing around it. The Val and Tyr side chains from the V-G-Y-G sequence point away from the pore and make specific interactions with amino acids from the tilted pore helix. Together with the pore helix Trp residues, the four Tyr side chains form a massive sheet of aromatic amino acids, twelve in total, that is positioned like a cuff around the selectivity filter (Fig. 8C). The hydrogen bonding, for example between the Tyr hydroxyls and Trp nitrogens, and the extensive van der Waals contacts within the sheet, offer the immediate impression that this structure behaves like a layer of springs stretched radially outward to hold the pore open at its proper diameter.

How does the K^+ channel structure account for its prodigious ion selective properties? When an ion enters the selectivity filter, it evidently dehydrates (nearly completely). To compensate for the energetic cost of dehydration, the carbonyl oxygen atoms must take the place of the water oxygen atoms, come in very close contact with the ion, and act like surrogate water (*20, 21*). The structure reveals that the selectivity filter is held open as if to prevent it from accommodating a Na^+ ion with its smaller radius. We propose that a K^+ ion fits in the filter precisely so that the energetic costs and gains are well balanced. The structure of the selectivity filter with its molecular springs holding it open prevents the carbonyl oxygen atoms from approaching close enough to compensate for the cost of dehydration of a Na^+ ion.

In about 150 mM K^+, the selectivity filter contains two K^+ ions (Figs. 6 and 8). The ions are located at opposite ends of the selectivity filter, separated by about 7.5 Å, roughly the average distance between K^+ ions in a 4 M KCl solution, and in the selectivity filter there are no intervening Cl^- anions to balance the charge. We therefore conclude that the selectivity filter attracts and concentrates K^+ ions. But how does such a selectivity filter ever conduct ions? The structure implies that a single K^+ ion would be held very tightly, but that the presence of two K^+ ions results in mutual repulsion, hence their locations near opposite ends of the selectivity filter. Thus, when a second ion enters, the attractive force between a K^+ ion and the selectivity filter becomes perfectly balanced by the repulsive force between ions, and this is what allows conduction to occur. This picture accounts for both a strong interaction between K^+ ions and the selectivity filter and a high throughput mediated by electrostatic repulsion. On the basis of functional measurements, the same concept of destabilization by multiple ion occupancy has been proposed for Ca^{2+} channels (*22*) and for K^+ channels (*23*) and may be a general property of all selective ion channels.

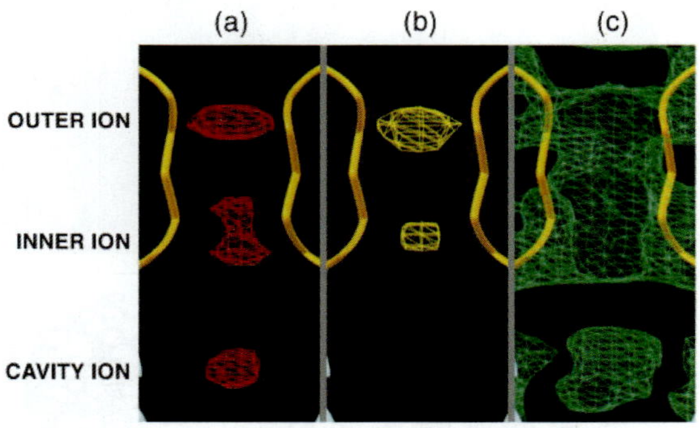

Fig. 6. Identification of permeant ion positions in the pore. (**a**) A Rb^+ difference Fourier map calculated to 4.0 Å and contoured at 6 σ identifies two strong peaks corresponding to ions in the selectivity filter (inner and outer ions) and a weaker peak corresponding to ions in the cavity (cavity ion). The inner ion density has two closely spaced peaks. (**b**) A Cs^+ difference Fourier map calculated to 5.0 Å and contoured at 6 σ shows the inner and outer ion peaks in the selectivity filter. Both difference Fourier maps were calculated with Fourier coefficients: F(soak) − F(native-unsharpened) and MIR phases. (**c**) Electron density map contoured at 1 σ showing diffuse density at the cavity ion position. This map was calculated with the following Fourier coefficients: unsharpened native amplitudes and MIR solvent-flattened phases (no averaging information was included).

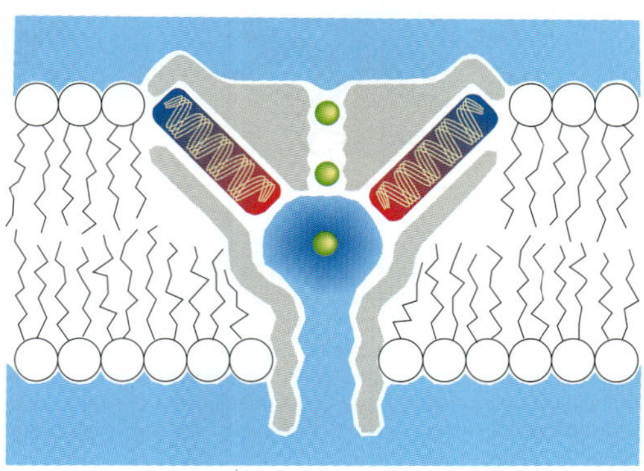

Fig. 7. Two mechanisms by which the K^+ channel stabilizes a cation in the middle of the membrane. First, a large aqueous cavity stabilizes an ion (green) in the otherwise hydrophobic membrane interior. Second, oriented helices point their partial negative charge (carboxyl end, red) towards the cavity where a cation is located.

Summary

We propose that the following principles underlie the structure and operation of K^+ channels. (i) The pore is constructed of an inverted teepee, with the selectivity filter held at its wide end. This architecture also describes the pore of cyclic nucleotide–gated channels and probably Na^+ and Ca^{2+} channels as well. (ii) The narrow selectivity filter is only 12 Å long, whereas the remainder of the pore is wider and has a relatively inert hydrophobic lining. These structural and chemical properties favor a high K^+ throughput by minimizing the distance over which K^+ interacts strongly with the channel. (iii) A large water-filled cavity and helix dipoles help to overcome the high electrostatic energy barrier facing a cation in the low dielectric membrane center. (iv) The K^+ selectivity filter is lined by carbonyl oxygen atoms, which provide multiple closely spaced sites. The filter is constrained in an optimal geometry so that a dehydrated K^+ ion fits with proper coordination but the Na^+ ion is too small. (v) Two K^+ ions at close proximity in the selectivity filter repel each other. The repulsion overcomes the otherwise strong interaction between ion and protein and allows rapid conduction in the setting of high selectivity.

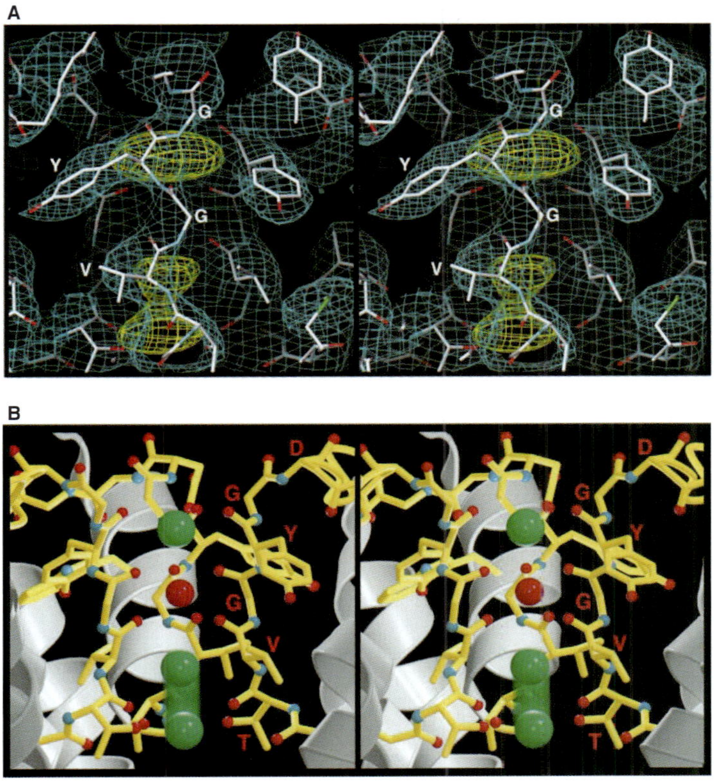

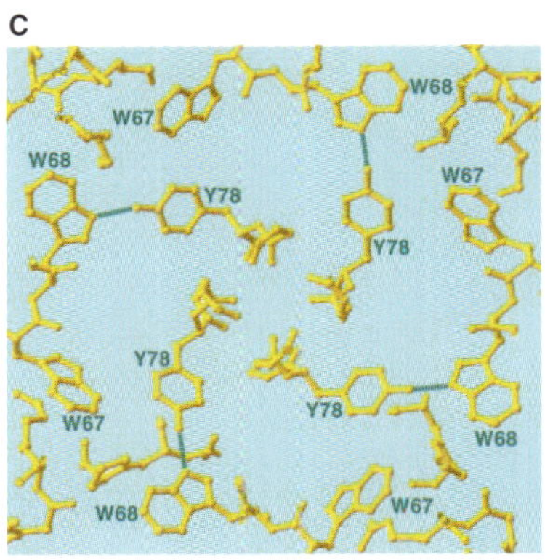

Fig. 8. Detailed views of the K^+ channel selectivity filter. (**A**) Stereoview of the experimental electron density (green) in the selectivity filter. The map was calculated with native-sharpened amplitudes and MIR-solvent-flattened-averaged phases. The selectivity filter of three subunits is shown as a stick representation with several signature sequence residues labeled. The Rb^+ difference map (yellow) is also shown. (**B**) Stereoview of the selectivity filter in a similar orientation to (A) with the chain closest to the viewer removed. The three chains represented are comprised of the signature sequence amino acids Thr, Val, Gly, Tyr, Gly running from bottom to top, as labeled in single-letter code. The Val and Tyr side chains are directed away from the ion conduction pathway, which is lined by the main chain carbonyl oxygen atoms. Two K^+ ions (green) are located at opposite ends of the selectivity filter, roughly 7.5 Å apart, with a single water molecule (red) in between. The inner ion is depicted as in rapid equilibrium between adjacent coordination sites. The filter is surrounded by inner and pore helices (white). Although not shown, the model accounts for hydrogen bonding of all amide nitrogen atoms in the selectivity filter except for that of Gly^{77}. (**C**) A section of the model perpendicular to the pore at the level of the selectivity filter and viewed from the cytoplasm. The view highlights the network of aromatic amino acids surrounding the selectivity filter. Tyrosine-78 from the selectivity filter (Y78) interacts through hydrogen bonding and van der Waals contacts with two Trp residues (W67, W68) from the pore helix.

REFERENCES AND NOTES

1. B. Hille *Ionic Channels of Excitable Membranes* (Sinauer, Sunderland, MA, ed. 2, 1992).
2. A. L. Hodgkin and R. D. Keynes, *J. Physiol. (London)* **128**, 61 (1955); S. Hagiwara, S. Miyazaki, S. Krasne, S. Ciani, *J. Gen. Physiol.* **70**, 269 (1977); B. Hille and W. Schwartz, *ibid.* **72**, 409 (1978); J. Neyton and C. Miller, *ibid.* **92**, 549 (1988).
3. C. M. Armstrong and L. Binstock, *J. Gen. Physiol.* **48**, 859 (1965); C. M. Armstrong, *ibid.* **50**, 491 (1966); *ibid.* **54**, 553 (1969); *ibid.* **58**, 413 (1971).
4. L. Heginbotham, T. Abramson, R. MacKinnon, *Science* **258**, 1152 (1992); L. Heginbotham, Z. Lu, T. Abramson, R. MacKinnon, *J. Biophys.* **66**, 1061 (1994).
5. H. Schrempf *et al.*, *EMBO J.* **14**, 5170 (1995); L. Heginbotham, E. Odessey, C. Miller, *Biochemistry* **36**, 10335 (1997); D. Marien Cortes and E. Perozo, *ibid.*, p. 10343.
6. R. MacKinnon, *Nature* **350**, 232 (1991).
7. Certain K^+ channels contain the equivalent of two subunits in a single open reading frame. These are thought to form the tetramer through the assembly of two dimer subunits [K. A. Ketchum *et al.*, *Nature* **376**, 690 (1995)].
8. R. MacKinnon *et al.*, *Science* **280**, 106 (1998).
9. G. J. Kleywegt and R. J. Read, *Structure* **5**, 1557 (1998).
10. J. Deisenhofer *et al.*, *Nature* **318**, 618 (1985); S. W. Cowan *et al.*, *ibid.* **358**, 727 (1992); A. Kreusch and G. E. Schulz, *J. Mol. Biol.* **243**, 891 (1994).
11. R. MacKinnon and C. Miller, *Science* **245**, 1382 (1989).
12. R. MacKinnon, L. Heginbotham, T. Abramson, *Neuron* **5**, 767 (1990); M. Stocker and C. Miller, *Proc. Natl. Acad. Sci. U.S.A.* **91**, 9509 (1994); S. A. N. Goldstein, D. J. Pheasant, C. Miller, *Neuron* **12**, 1377 (1994); P. Hidalgo and R. MacKinnon, *Science* **268**, 307 (1995); J. Aiyar *et al.*, *Neuron* **15**, 1169 (1995); D. Naranjo and C. Miller, *ibid.* **16**, 123 (1996); R. Ranganathan, J. H. Lewis, R. MacKinnon, *ibid.*, p. 131; A. Gross and R. MacKinnon, *ibid.*, p. 399.
13. C. M. Armstrong and B. Hille, *J. Gen. Physiol.* **59**, 388 (1972).
14. R. MacKinnon and G. Yellen, *Science* **250**, 276 (1990).
15. G. Yellen, M. E. Jurman, T. Abramson, R. MacKinnon, *ibid.* **251**, 939 (1991).
16. Y. Liu, M. Holmgren, M. E. Jurman, G. Yellen, *Neuron* **19**, 175 (1997).
17. V. A. Parsegian, *Ann. N.Y. Acad. Sci.* **264**, 161 (1975).
18. D. Sali, M. Bycroft, A. R. Fersht, *Nature* **335**, 740 (1988); J. Aqvist, H. Luecke, F. A. Quiocho, A. Warshel, *Proc. Natl. Acad. Sci. U.S.A.* **88**, 2026 (1991); D. J. Lockhart and P. S. Kim, *Science* **257**, 947 (1992); *ibid.* **260**, 198 (1993).
19. The temperature factors for Val^{76} and Gly^{77} main chain atoms (but not side chain atoms) refined to higher values than for neighboring atoms. This result can be explained by the difference Fourier analysis which shows alternative positions of the inner K^+ ion in the selectivity filter and therefore, by inference, alternative conformations of the coordinating main chain atoms, depending on the location of the K^+ ion.
20. F. Bezanilla and C. M. Armstrong, *J. Gen. Physiol.* **60**, 588 (1972).
21. B. Hille, *ibid.* **61**, 669 (1973).
22. W. Almers and E. W. McCleskey, *J. Physiol. (London)* **353**, 585 (1984); P. Hess and R. W. Tsien, *Nature* **309**, 453 (1984).
23. J. Neyton and C. Miller, *J. Gen. Physiol.* **92**, 569 (1988).
24. The KcsA gene was subcloned into pQE60 (Qiagen) vector and expressed in *E. coli* XL-1 Blue cells upon induction with 1-β-D-thiogalactopyranoside. The carboxy-terminal histidine tagged protein was extracted by homogenization and solubilization in 40 mM decylmaltoside (Anatrace). The KcsA channel

was purified on a cobalt affinity column. Thirty-five carboxyl terminal amino acids were cleaved by chymotrypsin proteolysis. The truncated channel was purified to homogeneity by gel filtration and the detergent exchanged in a final dialysis step against 5 mM N,N-dimethyldodecylamine-N-oxide (LDAO). Crystals were grown at 20°C with the sitting drop method by mixing equal volumes of protein solution [5 to 10 mg/ml, 150 mM KCl, 50 mM tris (pH 7.5), and 2 mM dithiothreitol] with reservoir mixture [200 mM $CaCl_2$, 100 mM Hepes (pH 7.5), and 48% PEG 400]. Through the entire preparation, the channel protein was maintained in solutions containing 150 mM KCl. For definition of K^+ sites, crystals were transferred into solutions where 150 mM KCl was replaced by 150 mM RbCl or 150 mM CsCl.
25. M. W. Tate *et al.*, *J. Appl. Crystallogr.* **28**, 196 (1995); D.J. Thiel *et al.*, *Rev. Sci. Instrum.* **67**, 1 (1996).
26. Z. Otwinowski, in *Data Collection and Processing*, L. Sawyer and S. Bailey, Eds. (Science and Engineering Research Council Daresbury Laboratory, Daresbury, UK, 1993), pp. 56–62.
27. Collaborative Computational Project 4 (CCP4), *Acta Crystallogr.* **D50**, 760 (1994).
28. G. M. Sheldrick, *ibid.* **46**, 467 (1990).
29. T. A. Jones, J. Y. Zou, J. Y. Cowan, M. Kjeldgaard, *ibid.* **A47**, 110 (1991).
30. A. T. Brunger, X-PLOR (Version 3.851) Manual (The Howard Hughes Medical Institute and Department of Molecular Biophysics and Biochemistry, Yale University, New Haven, CT).
31. S. J. Gamblin, D. W. Rodgers, T. Stehle, Proceedings of the CCP4 study weekend, Daresbury Laboratory (1996), pp. 163–169.
32. K. Y. J. Zhang and P. Main, *Acta Crystallogr.* **A46**, 377.
33. P. J. Kraulis, *J. Appl. Crystallogr.* **24**, 946 (1991).
34. O. S. Smart, J. G. Neduvelil, X. Wang, B. A. Wallace, M. P. Sansom, *J. Mol. Graphics* **14**, 354 (1996).
35. We thank D. Thiel, S. Gruner, and members of the MacCHESS staff for support and assistance in data collection at A1; J. Kuriyan, S. K. Burley, S. Harrison, P. Kim, E. Gouaux, and D. Wang for helpful discussions; Y. Jiang for help in data collection; D. Gadsby and J. Kuriyan for comments on the manuscript; and T. Rahman for patience and support. R.M. is forever grateful to T. Wiesel and A. L. MacKinnon for making this project possible. R.M. is an investigator of the Howard Hughes Medical Institute.

23 February 1998; accepted 13 March 1998

RESEARCH ARTICLES

Structural Basis of Transcription: RNA Polymerase II at 2.8 Ångstrom Resolution

Patrick Cramer,* David A. Bushnell, Roger D. Kornberg†

Structures of a 10-subunit yeast RNA polymerase II have been derived from two crystal forms at 2.8 and 3.1 angstrom resolution. Comparison of the structures reveals a division of the polymerase into four mobile modules, including a clamp, shown previously to swing over the active center. In the 2.8 angstrom structure, the clamp is in an open state, allowing entry of straight promoter DNA for the initiation of transcription. Three loops extending from the clamp may play roles in RNA unwinding and DNA rewinding during transcription. A 2.8 angstrom difference Fourier map reveals two metal ions at the active site, one persistently bound and the other possibly exchangeable during RNA synthesis. The results also provide evidence for RNA exit in the vicinity of the carboxyl-terminal repeat domain, coupling synthesis to RNA processing by enzymes bound to this domain.

RNA polymerase II (Pol II) is responsible for all mRNA synthesis in eukaryotes. Pol II transcription is the first step in gene expression and a focal point of cell regulation. It is a target of many signal transduction pathways, and a molecular switch for cell differentiation in development.

Pol II stands at the center of a complex machinery, whose composition changes in the course of gene transcription. As many as six general transcription factors assemble with Pol II for promoter recognition and melting (*1–3*). A multiprotein Mediator transduces regulatory information from activators and repressors (*4*). Additional regulatory proteins interact with Pol II during RNA chain elongation (*5, 6*), as do enzymes for RNA capping, splicing, and cleavage/polyadenylation (*7, 8*).

Pol II comprises 12 subunits, with a total mass of >0.5 MD. A backbone model of a 10-subunit yeast Pol II (lacking two small subunits dispensable for transcription) was previously obtained by x-ray diffraction and phase determination to ~3.5 Å resolution (*9*). The model revealed the general architecture of the enzyme and led to proposals for interactions with DNA and RNA in a transcribing complex. DNA was suggested to enter a cleft down the middle of the enzyme, passing between a pair of mobile elements termed "jaws." Beyond the active site, marked by a Mg^{2+} ion in the floor of the cleft, the DNA path was blocked by a protein "wall." DNA-RNA hybrid emanating from the active site would have to pass up the wall, at nearly right angles to the incoming DNA in the cleft. Both DNA and RNA would be held in place by a massive "clamp" swinging over the cleft and active-center region. A hole in the floor of the cleft beneath the active site ("pore 1") would allow entry of substrate nucleoside triphosphates and would also allow exit of RNA during retrograde movement of the polymerase on the DNA.

The Pol II backbone model thus revealed the gross structural elements of the enzyme and their likely roles in transcription. To understand how the elements perform their roles, an atomic structure is required. Here we present atomic structures determined from the previous crystal form at 3.1 Å resolution and from a new crystal form, containing the enzyme in a different conformation, at 2.8 Å resolution. The structures illuminate the transcription mechanism. They provide a basis for understanding both transcription initiation and RNA chain elongation. They permit the identification of protein features and amino acid residues crucial in the structure of an actively transcribing complex (*10*).

Atomic structures of Pol II. The Pol II crystals from which the previous backbone model was derived were grown and then shrunk by transfer to a solution of different composition (*9*). Shrinkage reduced the **a** axis of the unit cell by 11 Å and improved the diffraction from about 6.0 to 3.0 Å resolution (crystal form 1). We subsequently found that addition of Mn^{2+}, Pb^{2+}, or other metal ions induced a further shrinkage by 8 Å along the same unit cell direction and improved diffraction to 2.6 Å resolution in favorable cases (crystal form 2, Table 1) (*11*). Differences in Pol II conformation between form 1 and form 2, as well as atomic details most visible in form 2, led to the conclusions reported here.

An atomic model was initially built in electron density maps from crystal form 1, for which phase information from multiple isomorphous heavy-atom derivatives was available (*9*). Model building was facilitated by the use of sequence markers (*12*), especially 94 selenomethionine residues, and maps were gradually improved by phase combination (*13*). The model was refined at 3.1 Å resolution by classical positional and B-factor minimization, alternating with manual rebuilding (*14*). The resulting structure was placed in crystal form 2 and further refined at 2.8 Å resolution to a free *R* factor of 28.2% (Table 1) (*15*). Electron density maps at that resolution revealed side-chain conformations and the orientations of backbone carbonyl groups (Fig. 1A).

Both form 1 and form 2 structures contain over 3500 amino acid residues, with more than 28,000 nonhydrogen atoms and 8 Zn^{2+} ions (Table 1). The Mg^{2+} ion in form 1 is replaced by a Mn^{2+} ion in form 2, and several additional loops, as well as 78 structural water molecules, are also seen in form 2. The stereochemical quality of the structures is high (Table 1), with 98.0% of the residues in form 2 in allowed regions of the Ramachandran plot (*16*), and all residues in disallowed regions located in mobile loops for which only main-chain density was observed (*17*). Disordered regions in the structures are limited to the COOH-terminal repeat domain (CTD) of the largest subunit, Rpb1, to the nonconserved NH_2-terminal tails of Rpb6 and Rpb12, and to several short exposed loops in Rpb1, Rpb2, and Rpb8 (*18*).

Over 53,000 $Å^2$ of surface area is buried in subunit interfaces (Fig. 1B and Table 2), about a third of it between Rpb1 and Rpb2, accounting for the high stability of Pol II (*19*). Many salt bridges and hydrogen bonds, and some structural water molecules, five at 2.8 Å resolution, are observed in the interfaces. There are seven instances of a "β-addition motif," in which a strand from one subunit is added to a β sheet of another. The COOH-terminal region of Rpb12, which bridges between Rpb2 and Rpb3, participates in two such β-addition motifs (Table 2). The importance of one of these motifs is shown by deletion of two residues from the COOH-terminus of Rpb12, which confers a lethal phenotype (*20*). Termini of Rpb10 and Rpb11 also play structural roles (*21*), whereas the remaining 17 subunit termini extend outwards into solvent.

For ease of display and discussion, we represent all Pol II subunits as arrays of domains or domainlike regions, named according to their locations or presumed functional roles (Figs. 2 to 5). In many cases, however, these domains and regions do not appear to be independently folded. For exam-

Department of Structural Biology, Stanford University School of Medicine, Stanford, CA 94305–5126, USA.

*Present address: Institute of Biochemistry, Gene Center, University of Munich, 81377 Munich, Germany.

†To whom correspondence should be addressed. E-mail: kornberg@stanford.edu

ple, the "active site" region of Rpb1 and the "hybrid-binding" region of Rpb2 combine in a single fold that forms the active center of the enzyme (Figs. 1B, 2, and 3). None of the folds in Rpb1 and Rpb2 could be found in the protein structure database and so all are evidently unique (*22*). Rpb3, Rpb5, and Rpb9 each consist of two independent domains, whereas the remaining small subunits form single domains (Figs. 4 and 5).

The surface charge of Pol II is almost entirely negative, except for a uniformly positively charged lining of the cleft, the active center, the wall, and a "saddle" between the clamp and the wall (Fig. 6). This strongly asymmetric charge distribution accords with previous proposals for the paths of DNA and RNA in a transcribing complex. It is also consistent with previous evidence for an electrostatic component of the polymerase-DNA interaction (*23*). The positively charged environment of the cleft may help to localize DNA without restraining movement toward the active site for transcription. The positive charge on the saddle supports the proposal that it serves as an exit path for RNA (*9*). Homology modeling of human Pol II reveals that the overall surface charge distribution is well conserved (*24*).

Four mobile modules. Comparison of the form 1 and form 2 structures reveals a division of the polymerase into four mobile modules (Fig. 7 and Table 3). Half the mass of the enzyme lies in a "core" module, containing the regions of Rpb1 and Rpb2 that form the active center and subunits Rpb3, Rpb10, Rpb11, and Rpb12, which have been implicated in Pol II assembly. Three additional modules, whose positions relative to the core module change between form 1 and form 2, lie along the sides of the DNA-binding cleft, before the active center. The "jaw-lobe" module contains the "upper jaw" (*9*), made up of regions of Rpb1 and Rpb9, and the "lobe" of Rpb2 (Figs. 3 and 4). The "shelf" module contains the "lower jaw" (*9*) (a domain of Rpb5), the "assembly" domain of Rpb5, Rpb6, and the "foot" and "cleft" regions of Rpb1 (Figs. 3 and 4). The remaining module, the "clamp," was originally identified as a mobile element in a Pol II map at 6 Å resolution (*25*).

The changes observed between form 1 and form 2 structures are small rotations of the jaw-lobe and shelf modules about axes roughly parallel to the cleft (perpendicular to the plane of the page in Fig. 7B), producing movements of individual amino acid residues of up to 4 Å, and a larger swinging motion of the clamp, resulting in movements of as much as 14 Å (Table 3). The mobility of the clamp is also evidenced by its high overall temperature factor (Table 4). Rotations of the jaw-lobe and shelf modules might contribute to a helical screw rotation of the DNA as it advances toward the active center.

The clamp and transcription bubble interaction. The swinging motion of the clamp produces a greater opening of the cleft in form 2 than form 1, which may permit the entry of promoter DNA for the initiation of transcription (see below). Features seen in the form 2 structure suggest that, upon closure in a transcribing complex, the clamp serves as a multifunctional element, sensing the DNA-RNA hybrid conformation and separating DNA and RNA strands at the upstream end of the transcription bubble. The unique clamp fold is formed by NH_2- and COOH-terminal regions of Rpb1 and the COOH-terminal region of Rpb2 (*26*). It is stabilized by three Zn^{2+} ions, two within the "clamp core" and one underlying a distinct region at the upper end, termed the "clamp head" (*27*). Mutations of the Zn^{2+}-coordinating cysteine residues in the clamp confer a lethal phenotype (*28*, *29*). At its base, the clamp is connected to the "cleft" region of Rpb1, to the "anchor" region of Rpb2, and to Rpb6 through a set of "switch" regions that are flexible and enable clamp movement (Figs. 2 and 3). Whereas the shorter switches (4 and 5) are well ordered, the longer switches are poorly ordered (switches 1 and 2) or disordered (switch 3). All five switches undergo conformational changes in the transition to a transcribing complex, and switches 1, 2, and 3 contact the DNA-RNA hybrid in the active center (*10*). The switches therefore couple closure of the clamp to the presence of the DNA-RNA hybrid, which is key to the processivity of transcription. Interaction with the DNA-RNA hybrid may also be instrumental in the readout of the template DNA sequence in the active center (*10*).

Weak electron density is seen for three loops extending from the clamp that may interact with DNA and RNA upstream of the

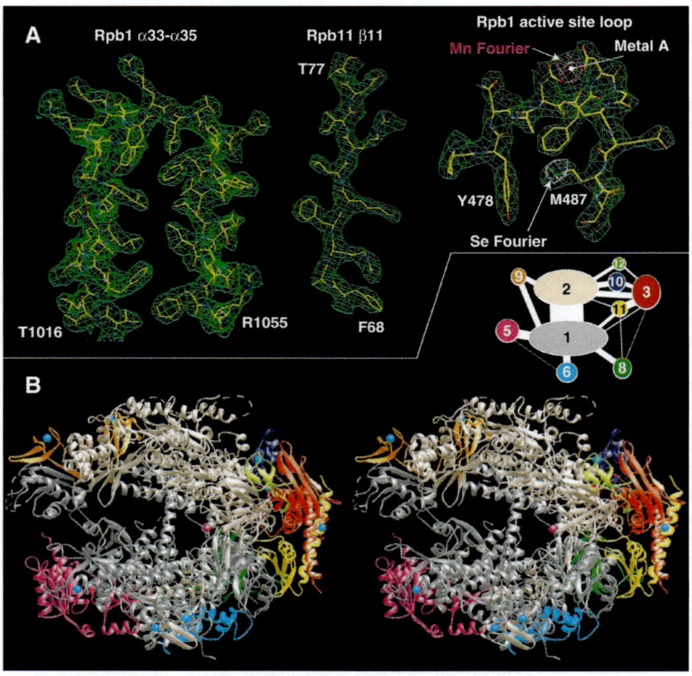

Fig. 1. Refined Pol II structure. (**A**) σ_A-weighted $2mF_{obs} - DF_{calc}$ electron density at 2.8 Å resolution (green) superimposed on the final structure in crystal form 2. Three areas of the structure are shown: the packing of α helices in the foot region of Rpb1, a β strand in Rpb11, and the active-site loop in Rpb1. Backbone carbonyl oxygens are revealed in the map. An anomalous difference Fourier of the Mn^{2+}-soaked crystal reveals the location of the active-site metal A (magenta, contoured at 10σ). An anomalous difference Fourier of a crystal of partially selenomethionine-substituted polymerase reveals the location of the S atom in residue M487 (white, contoured at 2.5σ). This figure was prepared with O (*75*). (**B**) Stereoview of a ribbon representation of the Pol II structure in form 2. Secondary structure was assigned by inspection. The diagram in the upper right corner is a key to the color code and an interaction diagram for the 10 subunits. The thickness of the connecting lines corresponds to the surface area buried in the corresponding subunit interface. This figure and others were prepared with RIBBONS (*81*).

active-center region. The loop nearest the active center corresponds to a "rudder" previously noted in the structure of bacterial RNA polymerase and suggested to participate in the separation of RNA from DNA and maintenance of the upstream end of the RNA-DNA hybrid (*30–32*). The second and third loops, here termed "lid" and "zipper" (Fig. 2D, "Clamp core, Linker," viewed in stereo), may be involved in these processes as well. Although disordered in the bacterial polymerase structure, both lid and zipper are apparently conserved (*33*). They lie 10 to 20 Å, corresponding to roughly three to six nucleotides, beyond the rudder. We speculate that the rudder and lid are involved in the separation of RNA from DNA, whereas the lid and zipper maintain the upstream end of the transcription bubble (*10*). In keeping with this idea, a region in the largest subunit of the *Escherichia coli* enzyme containing residues corresponding to the zipper has been cross-linked to the upstream end of the bubble (*32*). A disordered loop on top of the wall, termed the "flap loop" (Fig. 3) (*34*), may cooperate with the lid and zipper in the maintenance of the bubble.

Two metal ions at the active site. A Mg^{2+} ion, bound by the invariant aspartates D481, D483, and D485 of Rpb1, identifies the active site of Pol II (*9*) and is here referred to as metal A. At the corresponding position in the structure of a bacterial RNA polymerase, a metal ion was previously detected as well (*30*). The presence of only a single metal ion was unexpected, because a two-metal-ion mechanism had been proposed for all nucleic acid polymerases on the basis of x-ray studies of single-subunit enzymes (*35–40*). We now present evidence at the higher resolution of the form 2 data for a second metal ion in the Pol II active site. A difference Fourier map computed with only the protein structure and no metals contained two peaks, one at 21.0σ, owing to metal A, and a second at 4.6σ, designated metal B (Fig. 8). Peaks with comparable relative intensities were observed at the same locations in anomalous difference Fourier maps computed for the Mn^{2+}-soaked crystal. Metal B was not included in the structure because of its low occupancy

Three observations suggest that metal B is part of the active site and that it corresponds to the second metal ion of single-subunit polymerases. (i) Metal B is in the vicinity of metal A, at a distance of 5.8 Å, compared with about 4 Å in the single-subunit polymerases (*36*, *41*). (ii) Metal B is located near three invariant acidic residues—D481 in Rpb1, and E836 and

Table 1. Crystallographic data and structure statistics.

Crystal form	1	2
	*Data collection**	
Space group	I222	I222
Unit cell dimensions (Å)	130.7 by 224.8 by 369.4	122.7 by 223.0 by 376.1
Wavelength (Å)	1.283†	1.291†
Resolution (Å)	40–3.1 (3.2–3.1)‡	40–2.8 (2.9–2.8)‡
Unique reflections	98,315 (9,073)‡	125,251 (12,023)‡
Completeness (%)	99.2 (92.7)‡	99.0 (96.2)‡
Redundancy	4.7	3.6
Mosaicity (°)	0.44	0.36
R_{sym} (%)§	8.4 (29.8)‡	5.8 (34.4)‡
	Refinement	
Nonhydrogen atoms	28,173	28,379
Protein residues	3543	3559
Water molecules	0	78
Metal ions	8 Zn^{2+}, 1 Mg^{2+}	8 Zn^{2+}, 1 Mn^{2+}
Anisotropic scaling (B_{11}, B_{22}, B_{33})	−7.9, 11.3, 6.7	−14.2, 4.3, 9.9
rmsd bonds (Å)	0.008	0.007
rmsd angles (°)	1.50	1.43
Reflections in test set (%)	4,778 (4.8)	3,800 (3.0)
R_{cryst}/R_{free}‖	22.9/28.3	22.9/28.2

*Data for form 1 are from (*9*). Data collection for form 2 was carried out at 100 K as described (*9*) with an ADSC Quantum 4 charge-coupled device detector at beamline 9-2 of SSRL. Diffraction data were processed with DENZO and SCALEPACK (*79*). †Data for form 1 were collected at the Zn^{2+} anomalous peak to reveal native Zn^{2+} sites. Data for form 2 were collected below the Zn^{2+} anomalous peak energy to localize the Mn^{2+} ion at the active center. ‡Values in parentheses correspond to the highest resolution shells. §$R_{sym} = \Sigma_{i,h}|I(i,h) - \langle I(h)\rangle| / \Sigma_{i,h}|I(i,h)|$, where $\langle I(h)\rangle$ is the mean of the I observations of reflection h. R_{sym} was calculated with anomalous pairs merged; no σ cut-off was applied. ‖$R_{cryst/free} = \Sigma_h||F_{obs}(h)| - |F_{calc}(h)||/\Sigma_h|F_{obs}(h)|$. R_{cryst} and R_{free} were calculated from the working and test reflection set, respectively.

Table 2. Subunit interactions.

Subunit interface	Buried surface area (Å²)*	Salt bridges†	Hydrogen bonds‡	β-addition motifs§
Rpb1-Rpb2	17,178	6	58	Rpb2-β41-Rpb1-β7; Rpb2-β45-Rpb1-β1
Rpb1-Rpb3	608	1	3	-
Rpb1-Rpb5	4,768	5	19	-
Rpb1-Rpb6	3,797	3	12	Rpb1-β35-Rpb6-β3
Rpb1-Rpb8	3,056	3	6	Rpb8-β6-Rpb1-β18
Rpb1-Rpb9	3,011	2	21	Rpb9-β4-Rpb1-β28
Rpb1-Rpb11	1,913	-	8	-
Rpb2-Rpb3	3,070	5	26	-
Rpb2-Rpb9	2,705	1	5	-
Rpb2-Rpb10	2,941	1	11	-
Rpb2-Rpb11	608	1	2	-
Rpb2-Rpb12	1,923	4	14	Rpb12-β3-Rpb2-β32
Rpb3-Rpb8	333	1	1	-
Rpb3-Rpb10	2,175	4	15	-
Rpb3-Rpb11	3,899	4	6	-
Rpb3-Rpb12	993	3	7	Rpb12-β4-Rpb3-β3
Rpb5-Rpb6	204	1	3	-
Rpb8-Rpb11	396	-	-	-
Total	53,578	45	217	7 instances

*Calculated with programs AREAIMOL and RESAREA (*73*) with a standard probe radius of 1.4 Å. †A conservative distance cut-off of 3.6 Å was used [program CONTACT (*73*)]. ‡Potential hydrogen bonds with a donor-acceptor distance below 3.3 Å were included. §The order of strands in a β-addition motif is added β strand–accepting strand of a β sheet. Biochemical mapping (*80*) suggests that the β-addition motif formed by Rpb1 and Rpb9 may be largely responsible for the interaction of these subunits. The β-addition motif formed between Rpb1 and Rpb6 restrains clamp mobility.

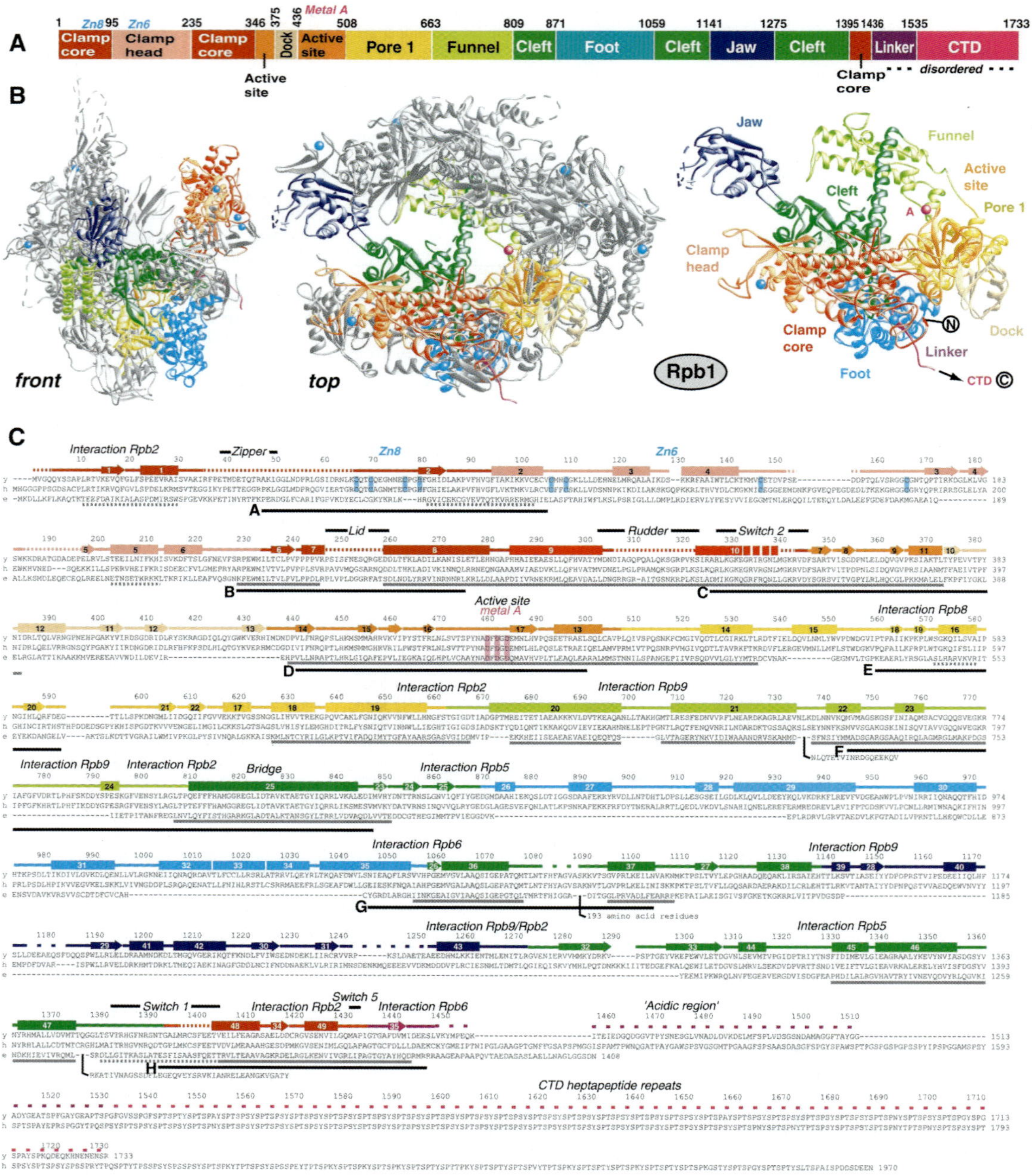
A
B
C
Clamp core
Clamp head
Dock
Active site
Pore 1
Funnel
Cleft
Foot
Jaw
Linker
CTD
Zn8
Zn6
Metal A
disordered
front
top
Rpb1
Interaction Rpb2
Zipper
Lid
Rudder
Switch 2
Active site
Interaction Rpb8
Interaction Rpb9
Bridge
Interaction Rpb5
Interaction Rpb6
Interaction Rpb9/Rpb2
Switch 1
Switch 5
'Acidic region'
CTD heptapeptide repeats
193 amino acid residues

Fig. 2.

RESEARCH ARTICLES

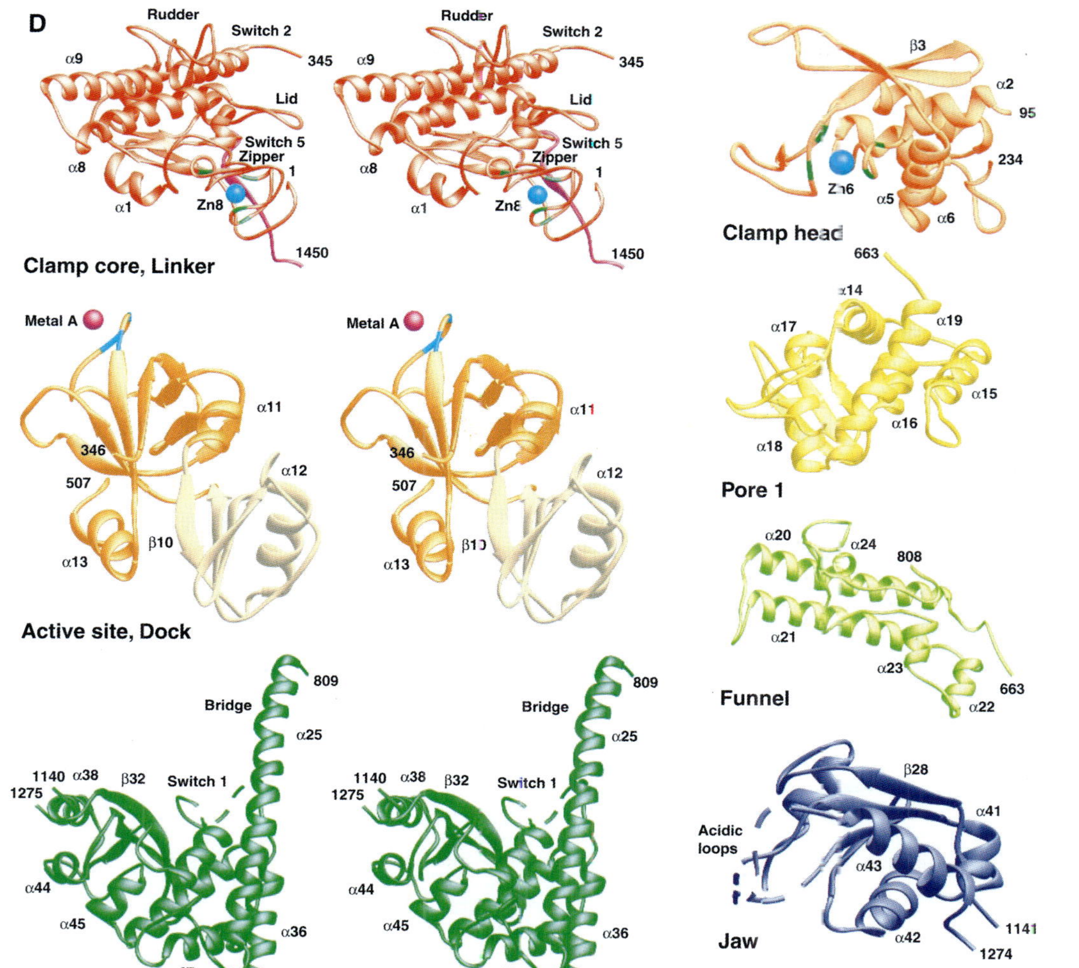

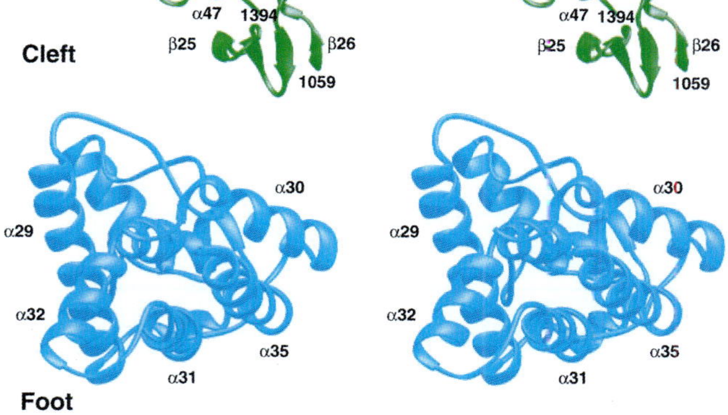

Fig. 2. Structure of Rpb1. (**A**) Domains and domainlike regions of Rpb1. The amino acid residue numbers at the domain boundaries are indicated. (**B**) Ribbon diagrams, showing the location of Rpb1 within Pol II ["front" and "top" views of the enzyme, as previously designated (*9*)], and Rpb1 alone. Locations of NH_2- and COOH-termini are indicated. Color-coding as in (A). (**C**) Secondary structure and amino acid sequence alignment. Yeast amino acid residue numbers are indicated above the sequence. Secondary structure elements were identified by inspection and are indicated and numbered above the sequence (boxes for α helices, arrows for β strands). Solid, dotted, and dashed lines above the sequences indicate ordered, partially ordered, and disordered loops, respectively. Alignment of Rpb1 from yeast (y) with human Rpb1 (h) and *E. coli* subunit β′ (e) was initially carried out with CLUSTALW (*82*) and then edited by hand. Alignment of the *E. coli* sequence is based on the structure of the bacterial enzyme (*30*). Regions for which the polypeptide backbones follow the same course are indicated by gray bars below the sequences (dotted when uncertain). The remaining regions could not be aligned because of disorder or because they differ in structure so that alignment is meaningless. Sequence homology blocks A to H (*70*) are indicated below the sequences by black bars. Important structural elements and prominent regions involved in subunit interactions are also noted. Residues involved in Zn^{2+} and Mg^{2+} coordination are highlighted in blue and pink, respectively. (**D**) Views of the domains and domainlike regions of Rpb1 (stereo on the left, mono on the right). These views reveal the entire course of the polypeptide chain from NH_2- to COOH-terminus and the locations of all secondary structure elements.

D837 in Rpb2 (Fig. 8), with aspartate D481 located between the two metals—resembling the situation in several single-subunit polymerases. The distance from metal B to the acidic residues, 3 to 4 Å, is too great for coordination, but may change during transcription (see below). (iii) The general organization of the active center resembles that of T7 RNA polymerase (*42–44*) and DNA polymerases of various families (*36, 37, 45, 46*). The two metal ions in Pol II are accessible to substrates from one side, and the Rpb1 helix bridging the cleft to Rpb2 is in about the same location relative to the metal ions as a helix in several single-subunit polymerases, generally referred to as the "O-helix."

The location of the two metals is consistent with the geometry of substrate binding inferred from structures of a Pol II transcription elongation complex (*10*) and of some single-subunit polymerases (*35–40*). In the single-subunit structures, metal A coordinates the 3′-OH group at the growing end of the RNA and the α-phosphate of the substrate nucleoside triphosphate, whereas metal B coordinates all three phosphate groups of the triphosphate. Both metals stabilize the transition state during phosphodiester bond formation. In Pol II, only metal A is persistently bound, at the upper edge of pore 1, whereas metal B, located further down in the pore, may enter with the substrate nucleotide. Orientation of the nucleotide by base pairing with the template may enable complete coordination of metal B, leading to phosphodiester bond formation.

Possible structural changes during translocation. A central mystery of all processive enzyme-polymer interactions is how the enzyme translocates along the polymer between catalytic steps without dissociation. Comparison of the Pol II structure with that of bacterial RNA polymerase has given unexpected insight into this aspect of the transcription mechanism. The bridge helix, highly conserved in sequence, is straight in Pol II but bent and partially unfolded in the bacterial polymerase structure (*30*). The bridge

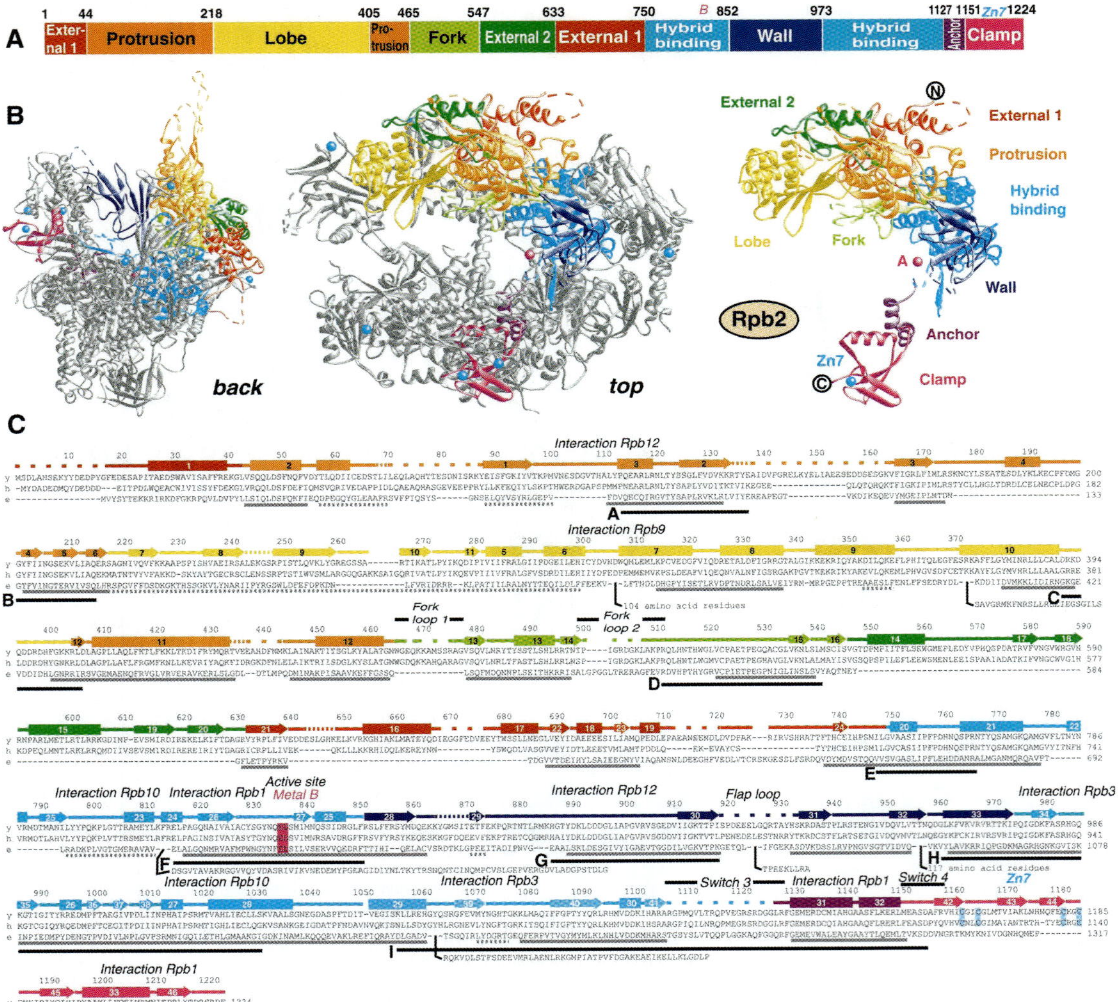

Fig. 3.

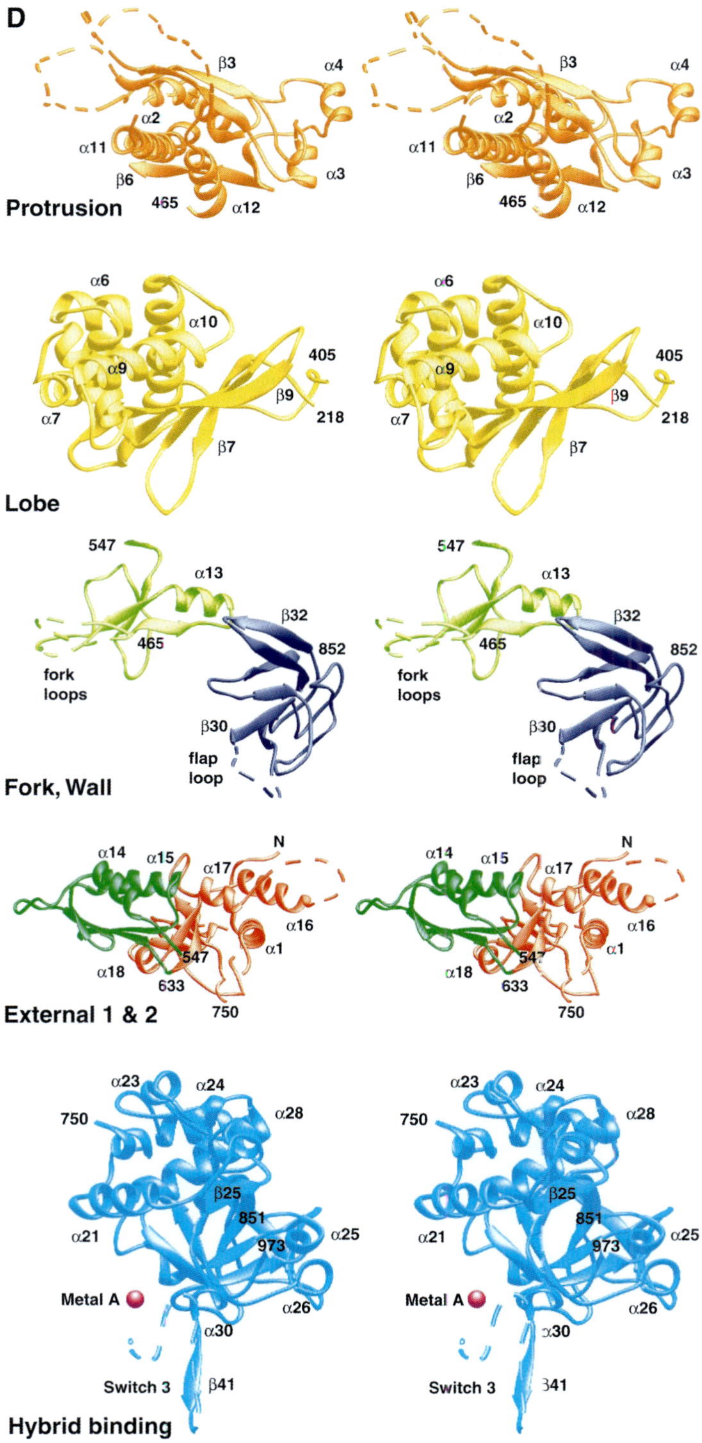

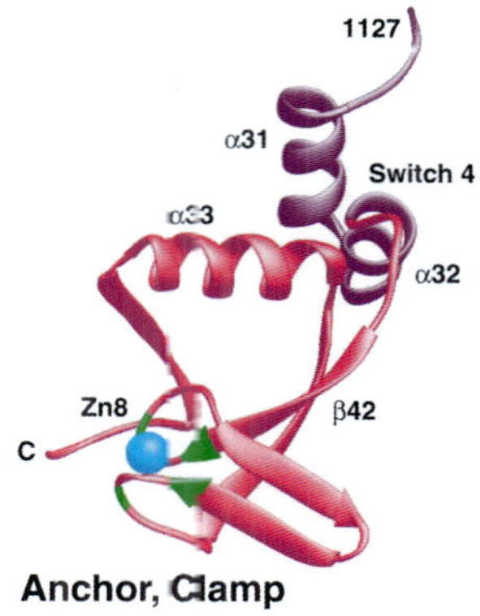

Fig. 3. (**A** to **D**) Structure of Rpb2. Organization and notation as in Fig. 2, except that the sequence alignment in (C) is with *E. coli* subunit β and its homology blocks A to I (*71*).

helix contacts the end of the DNA-RNA hybrid in a Pol II transcription elongation complex (*10*), and we speculate that bending of the helix is important for maintaining nucleic acid–protein interaction during translocation.

RNA exit, the CTD, and coupling of transcription to RNA processing. Two grooves in the Pol II surface were previously noted as possible paths for RNA exiting from the active-center region: "groove 1," at the base of the clamp, and "groove 2," passing alongside the wall (Fig. 9A) (*9*). The atomic structure, together with a result from RNA-protein cross-linking, argue in favor of groove 1. A cross-link is formed to the NH_2-terminal region of β′, the homolog of Rpb1, in an *E. coli* transcription elongation complex (*32*, *47*). The corresponding residues in Rpb1 are located on the side of the clamp core above the beginning of groove 1 (Fig. 9A). The length of RNA in groove 1 may be short, because it enters at about residue 12 and becomes accessible to nuclease digestion at about residue 18 in Pol II (*48*) and at about residue 15 in the bacterial enzyme (*49*). RNA in this part of groove 1 would lie on the saddle, beneath the Rpb1 lid and Rpb2 "flap loop." As noted above, the surface of the saddle is positively charged, appropriate for nucleic acid interaction.

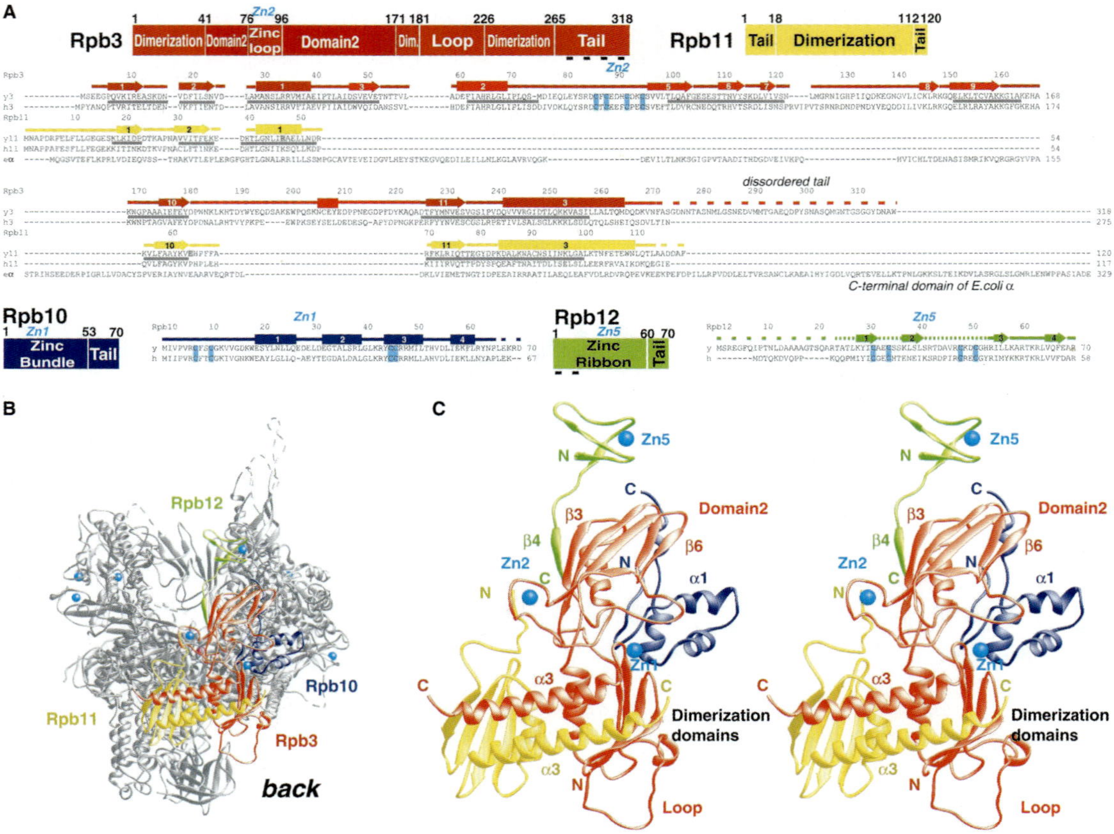

Fig. 4. Structure and location of the Rpb3/10/11/12 subassembly. (**A**) Domain structure and sequence alignments. Rpb3 and Rpb11 from yeast (y3, y11) and human (h3, h11) were aligned with *E. coli* subunit α (eα) on the basis of comparison with the bacterial structure (*30*). Regions for which the polypeptide backbones follow the same course are indicated by gray bars. Rpb10 and Rpb12 from yeast (y) were aligned with the human subunits (h). See Fig. 2 for details. (**B**) Location of the Rpb3/10/11/12 subassembly in Pol II ["back" view, of the enzyme, as previously designated (*9*)]. (**C**) Stereoview of the subassembly from the same direction as in (B).

Table 3. Mobile modules.

Module	Subunits and regions	Percentage of total mass	Maximum Cα atom displacement (Å) (residue number)
Core	All except other three modules	57	-
Shelf	Rpb1 cleft, Rpb1 foot, Rpb5, Rpb6	21	3.3 (N903 of Rpb1)
Clamp	Rpb1 clamp core and clamp head, Rpb2 clamp	12	14.2 (D193 of Rpb1); 14.4 (G283 of Rpb1)
Jaw-lobe	Rpb1 jaw, Rpb9 jaw, Rpb2 lobe	10	4.3 (K347 of Rpb2)

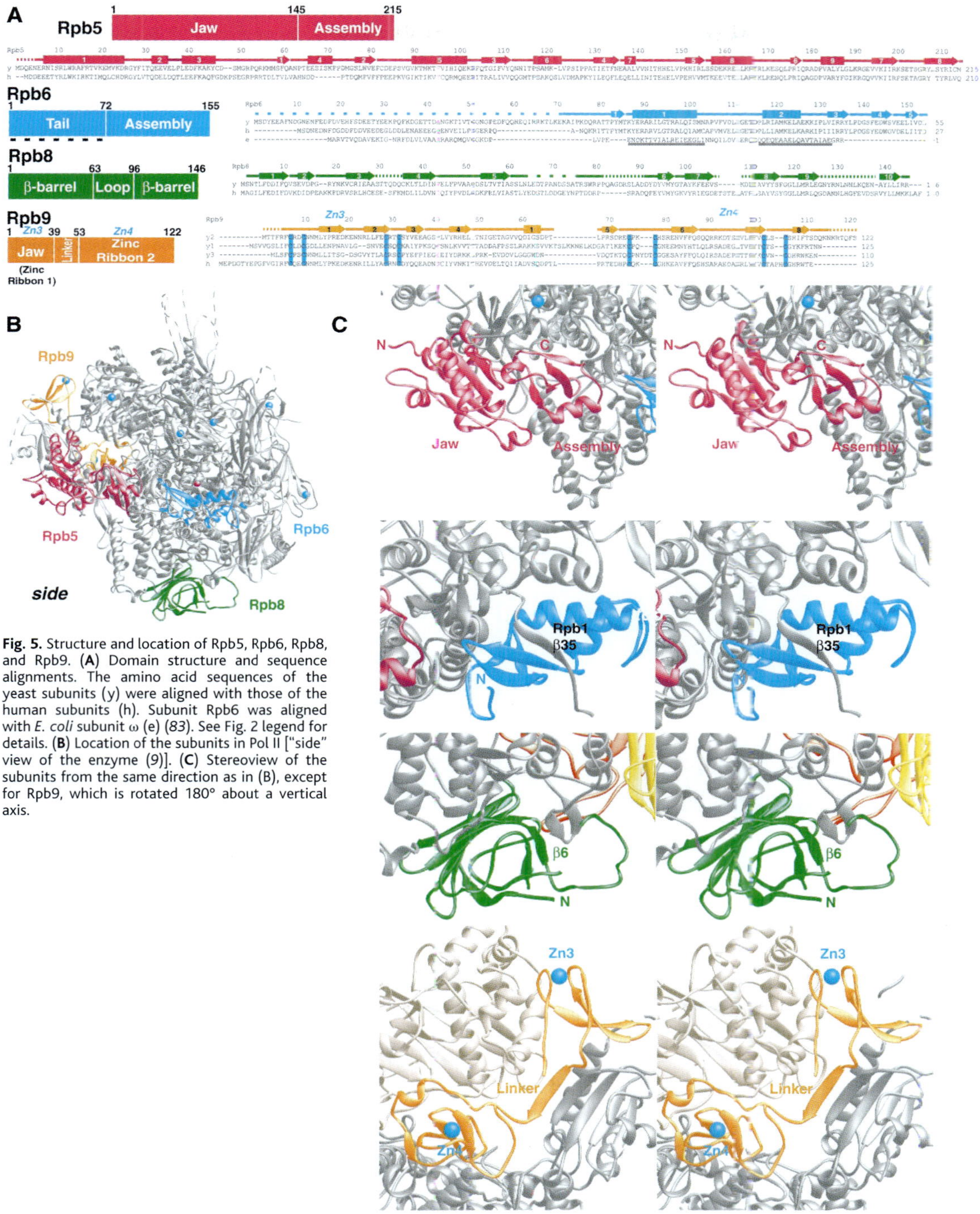

Fig. 5. Structure and location of Rpb5, Rpb6, Rpb8, and Rpb9. (**A**) Domain structure and sequence alignments. The amino acid sequences of the yeast subunits (y) were aligned with those of the human subunits (h). Subunit Rpb6 was aligned with *E. coli* subunit ω (e) (*83*). See Fig. 2 legend for details. (**B**) Location of the subunits in Pol II ["side" view of the enzyme (*9*)]. (**C**) Stereoview of the subunits from the same direction as in (B), except for Rpb9, which is rotated 180° about a vertical axis.

RESEARCH ARTICLES

Soon after exiting from the polymerase, RNA must be available for processing, because capping occurs upon reaching a length of about 25 residues (*50*, *51*). Consistent with this requirement, the exit from groove 1 is located near the last ordered residue of Rpb1, L1450, at the beginning of the linker to the CTD (Fig. 9B), and capping and other RNA processing enzymes interact with the phosphorylated form of the CTD (*7*, *8*, *52*). It may be argued that the length of the linker would allow the CTD to reach any point on the Pol II surface (Fig. 9B), and nuclear magnetic resonance (NMR) and circular dichroism studies have demonstrated a disordered state of a free, unphosphorylated CTD-derived peptide (*53*). The absence of electron density in Pol II maps owing to the linker and CTD provides evidence of motion or disorder, but even if disordered, the linker and CTD are unlikely to be in an extended conformation. The linker and CTD regions of four neighboring Pol II molecules share a space in the crystal sufficient to accommodate them only in a compact conformation (Fig. 9B).

Whereas the 5′ end of the RNA exits through groove 1 during RNA synthesis and forward movement of Pol II, the 3′ end of the RNA is extruded during retrograde movement of the enzyme. The previous backbone model suggested extrusion through pore 1 into a "funnel" on the back side of the enzyme (*9*). Transcription factor TFIIS, which provokes cleavage of extruded RNA, was thought to bind in the funnel as well (*9*). The atomic structure of Pol II lends support to these previous suggestions. A fragment of the largest bacterial polymerase subunit that can be cross-linked to the end of extruded RNA (*54*) is located in the funnel (Fig. 6). Further, Rpb1 residues that interact either physically or genetically with TFIIS (*55*, *56*) cluster on the outer rim of the funnel (Fig. 6). The Gre proteins, bacterial counterparts of TFIIS, also bind to the rim of the funnel (*32*). A cluster of mutations that cause resistance to the mushroom toxin α-amanitin is located in the funnel as well (Fig. 6) (*57*, *58*).

Implications for the initiation of transcription. The previous Pol II backbone model posed a problem for initiation because DNA entering the cleft and passing through the model would have to bend at the wall, whereas promoter DNA around the start site of transcription must be essentially straight (before binding to the enzyme and melting to form a transcription bubble). The only apparent solution to the problem, passage of promoter DNA over the wall, was unappealing because the DNA would be suspended over the cleft, far above the active center. A large movement of the DNA would be required for the initiation of transcription.

The form 2 structure suggests a new and more plausible solution of the initiation problem. In form 2, the clamp has swung further away from the active-center region, opening a wider gap than in form 1. A path is created for straight duplex DNA through the cleft from one side of the enzyme to the other (Fig. 10) (*59*). Following this path, the DNA contacts the jaw domain of Rpb9, fits into a concave surface of the Rpb2 lobe, and passes over the saddle, where it is surrounded by switch 2, switch 3, the rudder, and the flap loop. These surrounding elements probably do not impede entry of DNA, because they are all poorly ordered or disordered.

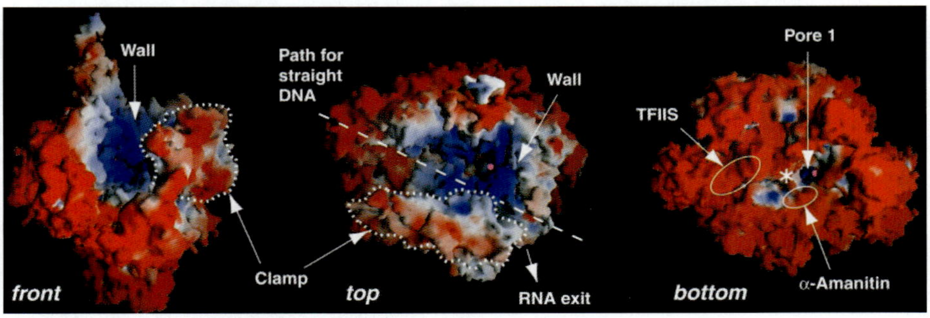

Fig. 6. Surface charge distribution and factor binding sites. The surface of Pol II is colored according to the electrostatic surface potential (*84*), with negative, neutral, and positive charges shown in red, white, and blue, respectively. The active site is marked by a pink sphere. The asterisk indicates the location of the conserved start of a fragment of *E. coli* RNA polymerase subunit β′ that has been cross-linked to an extruded RNA 3′ end (*54*).

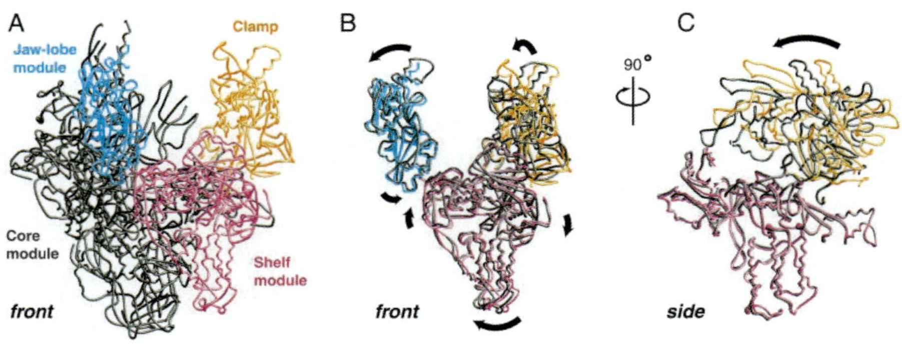

Fig. 7. Four mobile modules of the Pol II structure. (**A**) Backbone traces of the core, jaw-lobe, clamp, and shelf modules of the form 1 structure, shown in gray, blue, yellow, and pink, respectively. (**B**) Changes in the position of the jaw-lobe, clamp, and shelf modules between form 1 (colored) and form 2 structures (gray). The arrows indicate the direction of charges from form 1 to form 2. The core modules in the two crystal forms were superimposed and then omitted for clarity. (**C**) The view in (B) rotated 90° about a vertical axis. The core and jaw-lobe modules are omitted for clarity. In form 2, the clamp has swung to the left, opening a wider gap between its edge and the wall located further to the right (not shown).

Genetic evidence supports the proposed path for straight DNA during the initiation of transcription. A Pol II mutant lacking Rpb9 is defective in transcription start site selection (*60*), and complementation of the mutant with the Rpb9 jaw domain relieves the defect (*61*). Mutations in Rpb1 and Rpb2 affecting start site selection or otherwise altering initiation (*62*, *63*) lie along the proposed path as well (Fig. 10). Some of these mutations are in residues that could contact the DNA, whereas others are in residues that may interact with general transcription factors.

Previous biochemical studies have suggested that the general transcription factor TFIIB bridges between the TATA box of the promoter and Pol II during initiation (*64*). Structural studies led to the suggestion that TFIIB brings a TFIID-TATA box complex to a point on the Pol II surface from which the DNA can run straight to the active center (*65*). A conserved

Table 4. Crystallographic temperature factors.

Selection of model atoms	Average atomic *B* factor (Å^2)	
	Crystal form 1	Crystal form 2
Rpb1	71.8	64.0
Rpb2	70.4	61.5
Rpb3	59.1	59.5
Rpb5	78.6	69.1
Rpb6	59.5	51.8
Rpb8	101.7	100.0
Rpb9	75.1	67.6
Rpb10	57.6	51.2
Rpb11	56.2	62.0
Rpb12	108.0	97.7
Clamp	113.3	81.6
Water molecules	-	39.4
Active-site metal A	58.4 (Mg^{2+})	40.7 (Mn^{2+})
Zn^{2+} ions	119.1	84.9
Overall	71.5	64.5

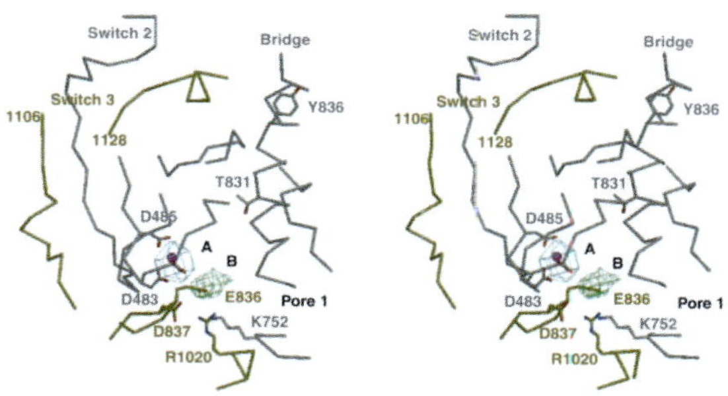

Fig. 8. Active center. Stereoview from the Rpb2 side toward the clamp. Two metal ions are revealed in a σ_A-weighted $mF_{obs} - DF_{calc}$ difference Fourier map (shown for metal B in green, contoured at 3.0σ) and in a Mn^{2+} anomalous difference Fourier map (shown for metal A in blue, contoured at 4.0σ). This figure was prepared with BOBSCRIPT (*85*) and MOLSCRIPT (*86*).

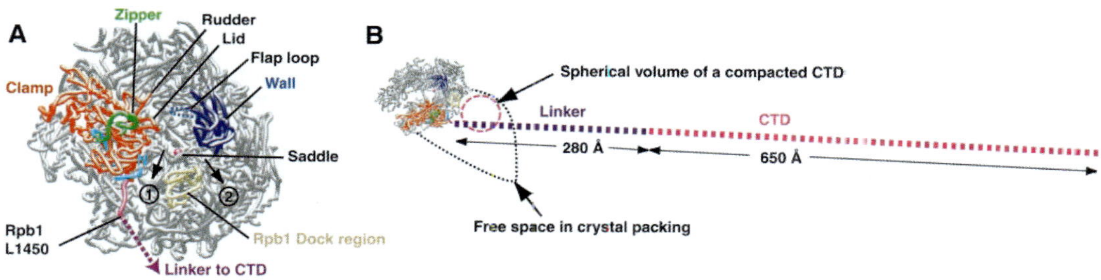

Fig. 9. RNA exit and Rpb1 COOH-terminal repeat domain (CTD). (**A**) Previously proposed RNA exit grooves 1 and 2 (*9*). The two grooves begin at the saddle between the clamp and wall and continue on either side of the Rpb1 dock region. The last ordered residue in Rpb1 (L1450) is indicated. The NH_2-terminal 25 residues of Rpb1 are highlighted in blue and correspond to an *E. coli* RNA polymerase fragment that was cross-linked to exiting RNA (*32*, *47*). The next 30 residues of Rpb1, which form the zipper, are highlighted in green and likely mark the location of *E. coli* residues that have been cross-linked to exiting RNA (*47*) and to the upstream end of the transcription bubble (*32*). (**B**) Size and location of the CTD. The space available in the crystal lattice for the CTDs from four neighboring polymerases is indicated. The dashed line represents the length of a fully extended linker and CTD. The pink dashed circle indicates the size of a compacted random coil with the mass of the CTD.

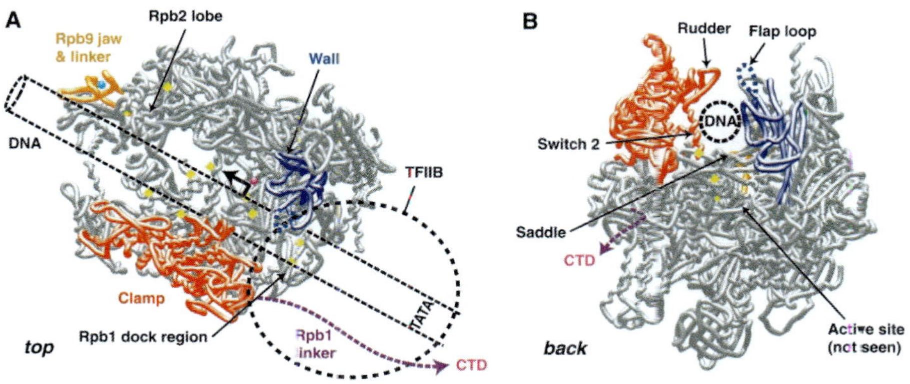

Fig. 10. Proposed path for straight DNA in an initiation complex. (**A**) Top view. A B-DNA duplex was placed as indicated by the dashed cylinder. Rpb9 regions involved in start site selection are shownin orange. The location of mutations that affect initiation or start site selection are marked in yellow. The presumed location of general transcription factor TFIIB in a preinitiation complex is indicated by a dashed circle. (**B**) Back view. DNA may pass through the enzyme over the saddle between the wide open clamp (red) and the wall (blue). The circle corresponds in size to a B-DNA duplex viewed end-on.

spacing of about 25 base pairs between the TATA box and transcription start site in Pol II promoters would correspond to the straight distance to the active center. This hypothesis for transcription start site determination is consistent with the path for straight DNA proposed here. There is space appropriate for a protein the size of TFIIB between a TATA box some 25 base pairs (85 Å) from the active center and the Pol II surface (Fig. 10). TFIIB in this location would contact a region of Pol II around the Rpb1 "dock" domain that is not conserved in the bacterial polymerase sequence or structure (*66*). Binding of TFIIB in this area would also explain its interaction with an acidic region of Rpb1 that includes the adjacent "linker" (*67*).

Once bound to Pol II, promoter DNA must be melted for the initiation of transcription by the adenosine 5′-triphosphate–dependent helicase activity of general transcription factor TFIIH. The region to be melted, extending from the transcription start site about half way to the TATA box, passes close to the active center and across the saddle. As the template single strand emerges, it can bind to nearby sites in the active center, on the floor of the cleft and along the wall, where it is localized in a transcribing complex (*10*). The transition from duplex to melted promoter would thus be effected with minimal movement of protein and DNA. The transition would also remove duplex DNA from the saddle, clearing the way for RNA, whose exit path crosses the saddle.

Conservation of RNA polymerase structure. All 10 subunits in the Pol II structure are identical or closely homologous to subunits of RNA polymerases I and III (*68*). Pol II is also highly conserved across species. Yeast and human Pol II sequences exhibit 53% overall identity, and the conserved residues are distributed over the entire structure (Fig. 11A). The yeast Pol II structure is therefore applicable to all eukaryotic RNA polymerases.

Some of the amino acid differences between Pol I, Pol II, and Pol III may relate to the specificity of assembly. A complex of Rpb3, Rpb10, Rpb11, and Rpb12 anchors Rpb1 and Rpb2 in Pol II and appears to direct their assembly (*19*, *20*, *69*). Rpb10 and Rpb12 are also present in Pol I and Pol III, together with homologs of Rpb3 and Rpb11, designated AC40 and AC19. Residues that interact with the common subunits Rpb10 and Rpb12 are conserved between the three polymerases. Most residues in the interface between Rpb3 and Rpb11 differ in the homologs, accounting for the specificity of heterodimer formation (*24*). Moreover, an important part of the Rpb2-Rpb3 interface (strand β10 of Rpb2 and "loop" region of Rpb3) is not conserved, which may account for the specificity of AC40 (Rpb3 homolog) interaction with the second largest subunits of Pol I and Pol III.

Sequence conservation between yeast and bacterial RNA polymerases is far less than for yeast and human enzymes. Identical residues are scattered throughout the structure (Fig. 11B). Regions of sequence homology between eukaryotic and bacterial RNA polymerases (*70*, *71*), however, cluster around the active center (Fig. 12A). Structural homology, determined by comparison of the Pol II protein folds with the bacterial RNA polymerase structure (*30*), is even more extensive (Fig. 12B). Yeast Pol II evidently shares a core structure, and thus a conserved catalytic mechanism, with the bacterial enzyme, but differs entirely in peripheral and surface structure, where interactions with other proteins, such as general transcription factors and regulatory factors, take place.

Conclusions and prospects. The immediate implications of the atomic Pol II structure are for understanding the transcription mechanism. The structure has given insight into the formation of an initiation complex, the transition to a transcribing complex, the mechanism of the catalytic step in transcription, a possible structural change accompanying the translocation step, the unwinding of RNA and rewinding of DNA, and the coupling of transcription to RNA processing. No less important are the implications for future genetic and biochemical studies of all RNA polymerases. The atomic structure provides a basis for interpretation of available data and the design of experiments to test hypotheses, such as those advanced here, for the transcription mechanism. Amino acid residues of structural elements such as the bridge helix, rudder, lid, zipper, and so forth may be altered by site-directed mutagenesis to assess their roles. Homology modeling of human RNA polymerase II may enable structure-based drug design.

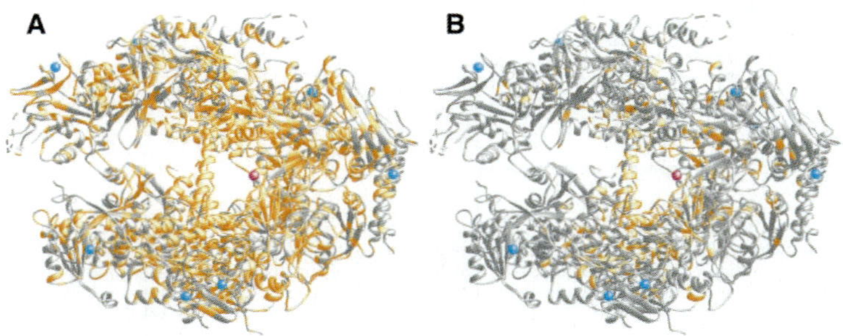

Fig. 11. Sequence identity between RNA polymerases. (**A**) Residues identical in yeast and human Pol II sequences are highlighted in orange. (**B**) Residues identical in the corresponding yeast and *E. coli* sequences are highlighted in orange.

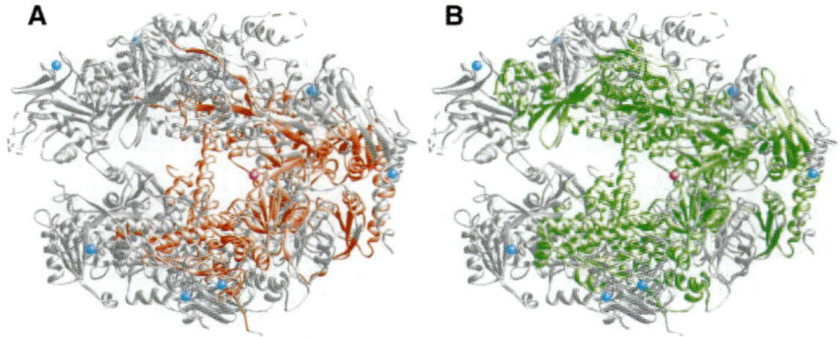

Fig. 12. A conserved RNA polymerase core structure. (**A**) Blocks of sequence homology between the two largest subunits of bacterial and eukaryotic RNA polymerases (*70*, *71*) are in red. (**B**) Regions of structural homology between Pol II and bacterial RNA polymerase (*30*), as judged from a corresponding course of the polypeptide backbone, are in green.

References and Notes

1. R. G. Roeder, *Trends Biochem. Sci.* **21**, 327 (1996); R. Conaway, J. Conaway, *Prog. Nucleic Acid Res. Mol. Biol.* **56**, 327 (1997).
2. R. D. Kornberg, *Trends Cell Biol.* **9**, M46 (1999).
3. T. I. Lee, R. A. Young, *Annu. Rev. Genet.* **34**, 77 (2000).
4. L. C. Myers, R. D. Kornberg, *Annu. Rev. Biochem.* **69**, 729 (2000).
5. J. W. Conaway, R. C. Conaway, *Annu. Rev. Biochem.* **68**, 301 (1999).
6. A. Shilatifard, *FASEB J.* **12**, 1437 (1998).
7. Y. Hirose, J. L. Manley, *Genes Dev.* **14**, 1415 (2000).
8. N. Proudfoot, *Trends Biochem. Sci.* **25**, 290 (2000).
9. P. Cramer *et al.*, *Science* **288**, 640 (2000).
10. A. L. Gnatt, P. Cramer, J. Fu, D. A. Bushnell, R. D. Kornberg, *Science* **292**, 1876 (2001); published online 19 April 2001 (10.1126/science.1059495).

11. Addition of 1 to 10 mM Mg^{2+}, Mn^{2+}, Pb^{2+}, or lanthanide ions led to further shrinkage. The resulting form 2 crystals had a slightly lower solvent content

and lower mosaicity. Shrinkage of form 1 to form 2 results in additional crystal contacts of the mobile clamp and jaw-lobe module (see below), which may account for the improvement in diffraction.

12. A total of 141 amino acid residues were located by sequence markers. Out of 103 methionine residues in the final structure, 94 were revealed as peaks of greater than 3.3σ in a 4 Å anomalous difference Fourier map calculated with data from partially selenomethionine-substituted Pol II and with experimental multiple isomorphous replacement with anomalous scattering (MIRAS) phases (*72*). The few remaining methionines are located in poorly ordered regions. In the selenomethionine-substituted Pol II map, three cysteine residues, C520 and C1400 in Rpb1 and C207 in Rpb3, also showed peaks. Eight Zn^{2+} ions (*9*) confirmed the location of 31 cysteine residues and one histidine residue (Figs. 2 to 5). The active-site metal A is coordinated by three invariant aspartate residues in Rpb1 (Fig. 2). Two different Hg derivatives (*9*) revealed the location of 10 surface cysteine residues (Rpb1, C1400, C1421; Rpb2, C64, C302, C388, C533; Rpb3, C207; Rpb5, C83; Rpb8, C24, C36).
13. MIRAS phases were combined by the program SIGMAA (*73*) with phases from the initial polyalanine model. Phase combination was followed by solvent flattening with DM (*74*). This led to an electron density map at 3.1 Å resolution in which many side chains were visible. Improved maps were obtained by combination of the MIRAS phases with improved phases from combined polyalanine and atomic models in an iterative process.
14. Model building was carried out with the program O (*75*), and refinement, with the program CNS (*76*). After bulk solvent correction and anisotropic scaling, the model was subjected to positional minimization in CNS with experimental phase restraints (MLHL target). After several rounds of model building into the resulting σ_A-weighted electron density maps and subsequent further refinement, the maximum likelihood target function (MLF) was used and restrained atomic B-factor refinement was carried out. With the resulting phase-combined maps, poorly ordered regions such as parts of the clamp and the Rpb2 lobe region could be built. Extensive rebuilding and refinement of atomic positions and *B* factors lowered the free *R* factor to 29.8%. Inclusion in the form 1 structure of fine stereochemical adjustments that were achieved in refinement of the form 2 structure (*15*) lowered the free *R* factor to 28.3%.
15. The form 1 structure was manually placed with experimental Zn^{2+}-ion positions and the positon of the active-site metal in form 2. The clamp was adjusted to its new position relative to the rest of Pol II. After initial rigid body refinement of the entire polymerase in CNS, σ_A-weighted difference electron density maps revealed regions that had moved. Manual adjustment of these regions was followed by rigid body refinement in groups and positional and atomic *B*-factor refinement. The structure in form 2 was further confirmed with the use of sequence markers, including selenomethionine (*24*). After several rounds of fine adjustment of the model stereochemistry and further refinement, 78 water molecules could be included.

16. R. A. Laskowski, M. W. MacArthur, D. S. Moss, J. M. Thornton, *J. Appl. Crystallogr.* **26**, 283 (1993).
17. Regions showing only main-chain electron density: Rpb1, amino acids 1 to 4, 36 to 66, 154 to 157, 186 to 197, 248 to 266, 307 to 323, 330 to 338, 1388 to 1403; Rpb2, 69 to 70, 133 to 138, 241 to 251, 434 to 437, 643 to 649, 864 to 872, 915 to 919, 933 to 935, 1104 to 1110; Rpb5, 1 to 5; Rpb8, 29 to 35, 82 to 91, 107 to 113, 127 to 139; Rpb9, 1 to 4, 116 to 122; Rpb12, 24 to 53.
18. Disordered regions: Rpb1, amino acids 1082 to 1091, 1177 to 1186, 1244 to 1253, 1451 to 1733; Rpb2, 1 to 17, 71 to 88, 139 to 163, 438 to 445, 468 to 476, 503 to 508, 669 to 677, 713 to 721, 920 to 932, 1111 to 1126; Rpb3, 1 to 2, 269 to 318; Rpb6, 1 to 71; Rpb8, 1, 64 to 75; Rpb10, 66 to 70, Rpb11, 115 to 120; Rpb12, 1 to 23.
19. M. Kimura, A. Ishiguro, A. Ishihama, *J. Biol. Chem.* **272**, 25851 (1997).
20. L. Rubbi, S. Labarre-Mariotte, S. Chedin, P. Thuriaux, *J. Biol. Chem.* **274**, 31485 (1999).
21. The NH_2-terminal methionine of Rpb10 is inserted in a hydrophobic pocket lined by Rpb2, Rpb3, and Rpb11. The NH_2-terminus of Rpb11 binds in the previously proposed RNA exit groove 2 (*9*). The charge of its terminal amino group is neutralized by the conserved residue D1100 of Rpb2. The COOH-terminal residue R70 of Rpb12 is linked by a salt-bridge to the conserved residue E166 of Rpb3, whereas the charge of its carboxylate is neutralized by the conserved residue R852 of Rpb2.
22. Domains and domainlike regions of Rpb1 and Rpb2 did not produce any significant matches when submitted to the DALI server (*77*). The unique folds of the large subunits appear to depend on extensive contacts with small subunits on the periphery (Table 2).
23. P. H. von Hippel, D. G. Bear, W. D. Morgan, J. A. McSwiggen, *Annu. Rev. Biochem.* **53**, 389 (1984).
24. P. Cramer, D. A. Bushnell, R. D. Kornberg, data not shown.
25. J. Fu *et al.*, *Cell* **98**, 799 (1999).
26. At the base of the clamp, these regions are held together in a β sheet made up of one strand from each region (Rpb1 β1, Rpb1 β34, and Rpb2 β46). Not included at the base of the clamp is the NH_2-terminal tail of Rpb6, the only change in subunit assignment of a density feature between the atomic structures and the previous backbone model. Incorporation of the Rpb6 tail in the backbone model was based on early electron density maps and the NMR structure of free Rpb6 (*78*). Several residues in the NH_2-terminal tail form an outer strand of a β sheet in the NMR structure. In the course of building the previous Pol II backbone model, the NMR structure was placed in the available electron density and the outer strand of the Rpb6 β sheet was extended toward the NH_2-terminus, following continuous density into the base of the clamp. The current, improved maps and sequence markers show that the continuous density near the base of the clamp instead corresponds to part of conserved region H of Rpb1, and that the NH_2-terminal tail of Rpb6 is disordered.
27. Zinc ions Zn7 and Zn8 in the clamp core are bound by residues in the common motif $CX_2CX_nCX_2C/H$ (where X is any amino acid). Zinc ion Zn6 shows an unusual coordination that underlies the clamp head fold (Fig. 2).
28. I. M. Donaldson, J. D. Friesen, *J. Biol. Chem.* **275**, 13780 (2000).
29. M. Werner, S. Hermann-Le Denmat, I. Treich, A. Sentenac, P. Thuriaux, *Mol. Cell. Biol.* **12**, 1087 (1992).
30. G. Zhang *et al.*, *Cell* **98**, 811 (1999).
31. The rudder, corresponding to Rpb1 residues 304 to 324, was not detected in early electron density maps of Pol II and so is absent from the previous backbone model of Pol II (*9*). Main-chain density for the rudder is clearly revealed in the improved, phase-combined electron density maps reported here.
32. N. Korzheva *et al.*, *Science* **289**, 619 (2000).
33. The lid and zipper are located in sequence homology blocks B and A, respectively. The lid is also flanked by regions of conserved structure.
34. The region termed the "wall" in Pol II (*9*) corresponds to a feature referred to as the "flap" in the bacterial RNA polymerase structure (*30*). The "flap loop" extending from the top of the wall, disordered in Pol II, corresponds to a loop six residues longer in *E. coli* that is ordered in the bacterial polymerase structure.
35. L. S. Beese, T. A. Steitz, *EMBO J.* **10**, 25 (1991).
36. S. Doublie, S. Tabor, A. M. Long, C. C. Richardson, T. Ellenberger, *Nature* **391**, 251 (1998).
37. J. R. Kiefer, C. Mao, J. C. Braman, L. S. Beese, *Nature* **391**, 304 (1998).
38. M. R. Sawaya, R. Prasad, S. H. Wilson, J. Kraut, H. Pelletier, *Biochemistry* **36**, 11205 (1997).
39. T. A. Steitz, *Curr. Opin. Struct. Biol.* **3**, 31 (1993).
40. ———, *Nature* **391**, 231 (1998).
41. J. F. Davies, R. J. Almassy, Z. Hostomska, R. A. Ferre, Z. Hostomsky, *Cell* **76**, 1123 (1994).
42. G. M. Cheetham, T. A. Steitz, *Science* **286**, 2305 (1999).
43. G. M. Cheetham, D. Jeruzalmi, T. A. Steitz, *Nature* **399**, 80 (1999).
44. D. Jeruzalmi, T. A. Steitz, *EMBO J.* **17**, 4101 (1998).
45. K. P. Hopfner *et al.*, *Proc. Natl. Acad. Sci. U.S.A.* **96**, 3600 (1999).
46. A. C. Rodriguez, H. W. Park, C. Mao, L. S. Beese, *J. Mol. Biol.* **299**, 447 (2000).
47. E. Nudler, I. Gusarov, E. Avetissova, M. Kozlov, A. Goldfarb, *Science* **281**, 424 (1998).
48. W. Gu, M. Wind, D. Reines, *Proc. Natl. Acad. Sci. U.S.A.* **93**, 6935 (1996).
49. N. Komissarova, M. Kashlev, *Proc. Natl. Acad. Sci. U.S.A.* **95**, 14699 (1998).
50. Y. Furuichi, A. J. Shatkin, *Adv. Virus Res.* **55**, 135 (2000).
51. S. Shuman, *Prog. Nucleic Acid Res. Mol. Biol.* **66**, 1 (2000).
52. E. J. Steinmetz, *Cell* **89**, 491 (1997).
53. P. M. Cagas, J. L. Corden, *Proteins* **21**, 149 (1995).
54. V. Markovtsov, A. Mustaev, A. Goldfarb, *Proc. Natl. Acad. Sci. U.S.A.* **93**, 3221 (1996).
55. J. Archambault, F. Lacroute, A. Ruet, J. D. Friesen, *Mol. Cell. Biol.* **12**, 4142 (1992).
56. J. Wu, D. E. Awrey, A. M. Edwards, J. Archambault, J. D. Friesen, *Proc. Natl. Acad. Sci. U.S.A.* **93**, 11552 (1996).
57. M. S. Bartolomei, J. L. Corden, *Mol. Gen. Genet.* **246**, 778 (1995).
58. ———, *Mol. Cell. Biol.* **7**, 586 (1987).
59. The path for straight DNA is offset by 20° to 30° from the path of DNA entering a transcribing complex (*10*). Movement of DNA to this extent in the transition from an initiating to a transcribing complex seems plausible, because the DNA in this region is loosely held in the transcribing complex; the jaws, lobe, and clamp surrounding it are mobile; and a far larger movement of upstream DNA occurs upon promoter melting.
60. W. W. Hull, K. McKune, N. A. Woychik, *Genes Dev.* **9**, 481 (1995).
61. S. A. Hemming *et al.*, *J. Biol. Chem.* **275**, 35506 (2000).
62. R. W. Berroteran, D. E. Ware, M. Hampsey, *Mol. Cell. Biol.* **14**, 226 (1994).
63. D. S. Hekmatpanah, R. A. Young, *Mol. Cell. Biol.* **11**, 5781 (1991).
64. S. Buratowski, H. Zhou, *Proc. Natl. Acad. Sci. U.S.A.* **90**, 5633 (1993).
65. K. K. Leuther, D. A. Bushnell, R. D. Kornberg, *Cell* **85**, 773 (1996).
66. The proposed site of interaction with TFIIB in the vicinity of the "dock" domain, is unrelated to a site seen previously in a difference Fourier map of a two-dimensional TFIIB-Pol II cocrystal (*65*). The difference peak attributed to TFIIB was small and may have been misleading.
67. H. Xiao, J. D. Friesen, J. T. Lis, *Mol. Cell. Biol.* **14**, 7507 (1994).
68. A. Sentenac *et al.*, *Yeast RNA Polymerase Subunits and Genes. Transcriptional Regulation* (Cold Spring Harbor Laboratory Press, Cold Spring Harbor, NY, 1992); E. P. Geiduschek, M. S. Bartlett, *Nature Struct. Biol.* **7**, 437 (2000).
69. D. Lalo, C. Carles, A. Sentenac, P. Thuriaux, *Proc. Natl. Acad. Sci. U.S.A.* **90**, 5524 (1993).
70. R. S. Jokerst, J. R. Weeks, W. A. Zehring, A. L. Greenleaf, *Mol. Gen. Genet.* **215**, 266 (1989).
71. D. Sweetser, M. Nonet, R. A. Young, *Proc. Natl. Acad. Sci. U.S.A.* **84**, 1192 (1987).
72. D. A. Bushnell, P. Cramer, R. D. Kornberg, *Structure* **9**, R11 (2001).
73. CCP4, *Acta Crystallogr.* **D50**, 760 (1994).
74. K. Cowtan, *Joint CCP4 ESF-EACBM Newslett. Protein Crystallogr.* **31**, 34 (1994).
75. T. A. Jones, J. Y. Zou, S. W. Cowan, M. Kjeldgaard, *Acta Crystallogr.* **A47**, 110 (1991).
76. A. T. Brünger, *et al.*, *Acta Crystallogr.* **D54**, 905 (1998).
77. L. Holm, C. Sander, *Trends Biochem. Sci.* **20**, 478 (1995).
78. F. del Rio-Portilla, A. Gaskell, D. Gilbert, J. A. Ladias, G. Wagner, *Nature Struct. Biol.* **6**, 1039 (1999).
79. Z. Otwinowski, W. Minor, *Methods Enzymol.* **276**, 307 (1996).
80. S. A. Hemming, A. M. Edwards, *J. Biol. Chem.* **275**, 2288 (2000).
81. M. Carson, *Methods Enzymol.* **277**, 493 (1997).

82. J. D. Thompson, D. G. Higgins, T. J. Gibson, *Nucleic Acids Res.* **22**, 4673 (1994).
83. L. Minakhin *et al.*, *Proc. Natl. Acad. Sci. U.S.A.* **98**, 892 (2001).
84. A. Nicholls, K. A. Sharp, B. Honig, *Proteins* **11**, 281 (1991).
85. R. M. Esnouf, *J. Mol. Graph.* **15**, 132 (1997).
86. P. J. Kraulis, *J. Appl. Crystallogr.* **24**, 946 (1991).
87. We thank P. Ellis, P. Kuhn, T. McPhillips, and M. Soltis for assistance at beamline 9-2 of the Stanford Synchrotron Radiation Laboratory (SSRL). This research is based in part on work done at SSRL, which is funded by the U.S. Department of Energy, Office of Basic Energy Sciences. The structural biology program is supported by the NIH National Center for Research Resources Biomedical Technology Program and the U.S. Department of Energy, Office of Biological and Environmental Research. We thank COMPAQ for providing a Unix workstation, N. Thompson and R. Burgess for providing antibody for protein purification, J. Puglisi and members of the Kornberg laboratory for comments on the manuscript, and K. Westover and M. Levitt for homology modeling of human Pol II. Supported by a postdoctoral fellowship from the Deutsche Forschungsgemeinschaft (P.C.), postdoctoral fellowship PF-00-014-01-GMC from the American Cancer Society (D.A.B.), and NIH grant GM49985 (R.D.K.). Coordinates have been deposited at the Protein Data Bank (accession codes 1I3Q and 1I50 for the form 1 and form 2 structures, respectively).

1 February 2001; accepted 28 March 2001
Published online 19 April 2001;
10.1126/science.1059493
Include this information when citing this paper.

Structural Basis of Transcription: An RNA Polymerase II Elongation Complex at 3.3 Å Resolution

Averell L. Gnatt,* Patrick Cramer,† Jianhua Fu,‡ David A. Bushnell, Roger D. Kornberg§

The crystal structure of RNA polymerase II in the act of transcription was determined at 3.3 Å resolution. Duplex DNA is seen entering the main cleft of the enzyme and unwinding before the active site. Nine base pairs of DNA-RNA hybrid extend from the active center at nearly right angles to the entering DNA, with the 3′ end of the RNA in the nucleotide addition site. The 3′ end is positioned above a pore, through which nucleotides may enter and through which RNA may be extruded during back-tracking. The 5′-most residue of the RNA is close to the point of entry to an exit groove. Changes in protein structure between the transcribing complex and free enzyme include closure of a clamp over the DNA and RNA and ordering of a series of "switches" at the base of the clamp to create a binding site complementary to the DNA-RNA hybrid. Protein–nucleic acid contacts help explain DNA and RNA strand separation, the specificity of RNA synthesis, "abortive cycling" during transcription initiation, and RNA and DNA translocation during transcription elongation.

The recent structure determination of yeast RNA polymerase II at 2.8 Å resolution and of bacterial RNA polymerase at 3.3 Å resolution has led to proposals for polymerase-DNA and -RNA interactions (*1–3*). A DNA duplex was suggested to enter a positively charged cleft between the two largest subunits and to make a right angle bend at the active center, where the DNA strands are separated and from which a DNA-RNA hybrid emerges. Avenues for entry of substrate nucleoside triphosphates and for exit of RNA could also be surmised. Although consistent with results of cross-linking experiments (*4–8*), these general proposals for polymerase–nucleic acid interaction have not been proven, and they do not address key questions about the transcription process: How is an unwound "bubble" of DNA established and maintained in the active center? Why does the enzyme initiate repeatedly, generating many short transcripts, before a transition is made to a stable elongating complex? What is the nature of the presumptive DNA-RNA hybrid duplex? How are DNA and RNA translocated across the surface of the enzyme, forward and backward, during RNA synthesis and back-tracking? Here we report the crystal structure determination of yeast RNA polymerase II in the form of an actively transcribing complex, from which answers to these questions and additional insights into the transcription mechanism are derived.

The main technical challenge of this work was the isolation and crystallization of a transcribing complex. Initiation at an RNA polymerase II promoter requires a complex set of general transcription factors and is poorly efficient in reconstituted systems (*9, 10*). Moreover, most preparations contain many inactive polymerases, and the transcribing complexes obtained would have to be purified by mild methods to preserve their integrity (*11*). The initiation problem was overcome with the use of a DNA duplex bearing a single-stranded "tail" at one 3′-end (Fig. 1A) (*12, 13*). Pol II starts transcription in the tail, two to three nucleotides from the junction with duplex DNA, with no requirement for general transcription factors. All active polymerase molecules are converted to transcribing complexes, which pause at a specific site when one of the four nucleoside triphosphates is withheld. The problem of contamination by inactive polymerases was solved by passage through a heparin column (*13*); inactive molecules were adsorbed, whereas transcribing complexes flowed through, presumably because heparin binds in the positively charged cleft of the enzyme, which is occupied by DNA and RNA in transcribing complexes. The purified complexes formed crystals diffracting anisotropically to 3.1 Å resolution (*14*).

Structure of a pol II transcribing complex. Diffraction data complete to 3.3 Å resolution were used for structure determination by molecular replacement with the 2.8 Å pol II structure (*15*). A native zinc anomalous difference Fourier map showed peaks coinciding with five of the eight zinc ions of the pol II structure, confirming the molecular replacement solution (*16*). The remaining three zinc ions were located in the clamp, a region shown previously to undergo a large conformational change between different pol II crystal forms (*3*). The locations of the three zinc ions served as a guide for manual repositioning of the clamp in the transcribing complex structure. An initial electron density map revealed nucleic acids in the vicinity of the active center. After adjustment of the protein model, the nucleic acid density improved and nine base pairs of DNA-RNA hybrid could be built (*17*). Additional density along the DNA template strand allowed another three nucleotides downstream and one nucleotide upstream to be built. Modeling of the nucleic acids assumed the 3′-end of the RNA at the biochemically defined pause site (Fig. 1A), because the nucleic acid se-

Department of Structural Biology, Stanford University School of Medicine, Stanford, CA 94305–5126, USA.

*Present address: Department of Pharmacology and Experimental Therapy, University of Maryland, 655 West Baltimore Street, HH403, Baltimore, MD 21201, USA.

†Present address: Institute of Biochemistry, Gene Center, University of Munich, 81377 Munich, Germany.

‡Present address: Department of Molecular Biology and Genetics, Cornell University, 223 Biotechnology Building, Ithaca, NY 14853, USA.

§To whom correspondence should be addressed. E-mail: kornberg@stanford.edu

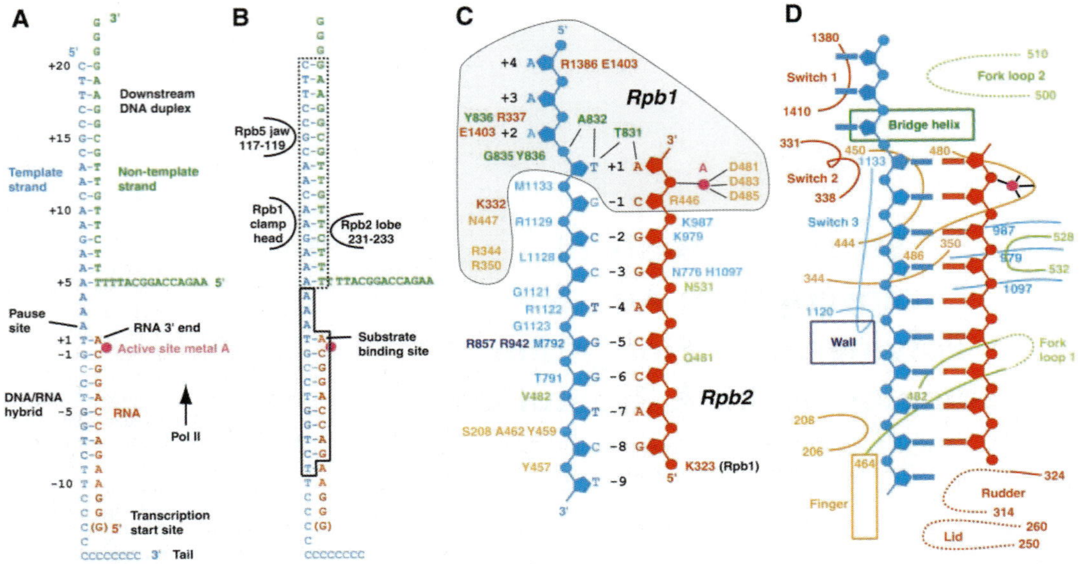

Fig. 1. Nucleic acids in the transcribing complex and their interactions with pol II. (**A**) DNA ("tailed template") and RNA sequences. DNA template and nontemplate strands are in blue and green, respectively, and RNA is in red. This color scheme is used throughout. (**B**) Ordering of nucleic acids in the transcribing complex structure. Nucleotides in the solid box are well ordered. Nucleotides in the dashed box are partially ordered, whereas those outside the boxes are disordered. Three protein regions that abut the downstream DNA are indicated. (**C**) Protein contacts to the ordered nucleotides boxed in (B). Amino acid residues within 4 Å of the DNA are indicated, colored according to the scheme for domain or domainlike regions of Rpb1 or Rpb2 (*3*). Ribose sugars are shown as pentagons, phosphates as dots, and bases as single letters. Amino acid residues listed beside phosphates contact only this nucleotide. Amino acid residues listed beside riboses contact this nucleotide and its 3′-neighbor. Single-letter abbreviations for the amino acid residues are as follows: A, Ala; D, Asp; E, Glu; G, Gly; H, His; K, Lys; L, Leu; M, Met; N, Asn; Q, Gln; R, Arg; S, Ser; T, Thr; V, Val; and Y, Tyr. (**D**) Schematic representation of protein features participating in the detailed interactions shown in (C). Same notation as in (C), except that bases are shown as thick bars.

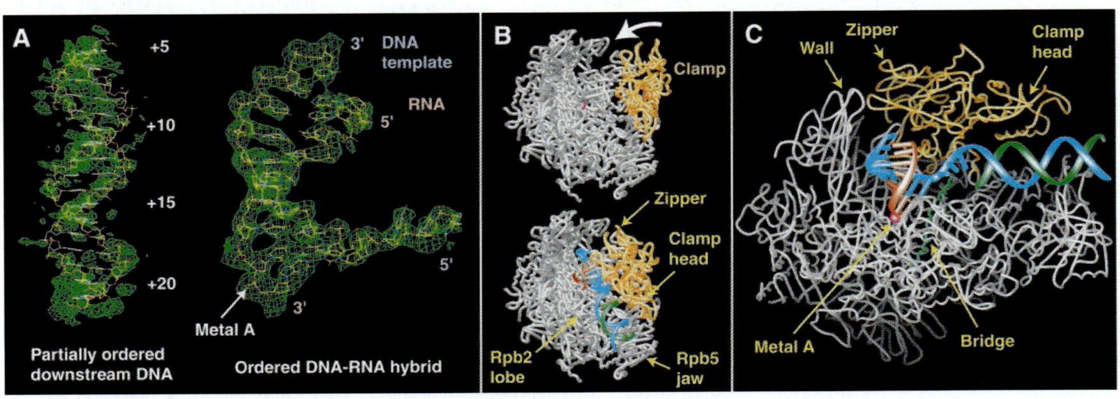

Fig. 2. Crystal structure of the pol II transcribing complex. (**A**) Electron density for the nucleic acids. On the left, the final sigma-weighted $2mF_{obs} - DF_{calc}$ electron density for the downstream DNA duplex (dashed box in Fig. 1B) is contoured at 0.8σ (green). At this contour level, the surrounding solvent region shows only scattered noise peaks. A canonical 16–base pair B-DNA duplex was placed into the density. On the right, the final model of the DNA-RNA hybrid and flanking nucleotides (boxed in Fig. 1B) is superimposed on a simulated-annealing $F_{obs} - F_{calc}$ omit map, calculated from the protein model alone with CNS (*45*) (green, contoured at 2.6σ). The location of the active site metal A is indicated. (**B**) Comparison of structures of free pol II (top) and the pol II transcribing complex (bottom). The clamp (yellow) closes on DNA and RNA, which are bound in the cleft above the active center. The remainder of the protein is in gray. (**C**) Structure of the pol II transcribing complex. Portions of Rpb2 that form one side of the cleft are omitted to reveal the nucleic acids. Bases of ordered nucleotides (boxed in Fig. 1B) are depicted as cylinders protruding from the backbone ribbons. The Rpb1 bridge helix traversing the cleft is highlighted in green. The active site metal A is shown as a pink sphere.

quences could not be inferred from the crystallographic data (*18*).

The final model contains 3521 amino acid residues, 22 nucleotides, eight Zn^{2+} ions, and one Mg^{2+} ion and has a free *R* factor of 29.8% (*R* factor 25.0%, 40 to 3.3 Å) (Fig. 2). A simulated-annealing omit map computed from a model of the protein alone revealed the phosphate groups and most bases in the DNA-RNA hybrid region, confirming the modeling of the nucleic acids (Fig. 2A). Density for DNA in the downstream region was very weak and discontinuous but revealed the major groove, allowing a canonical B-DNA duplex to be approximately placed [not included in the model (*19*)]. Numbering of nucleotides in the DNA begins with +1 immediately downstream and −1 upstream of the Mg^{2+} ion (Fig. 1A).

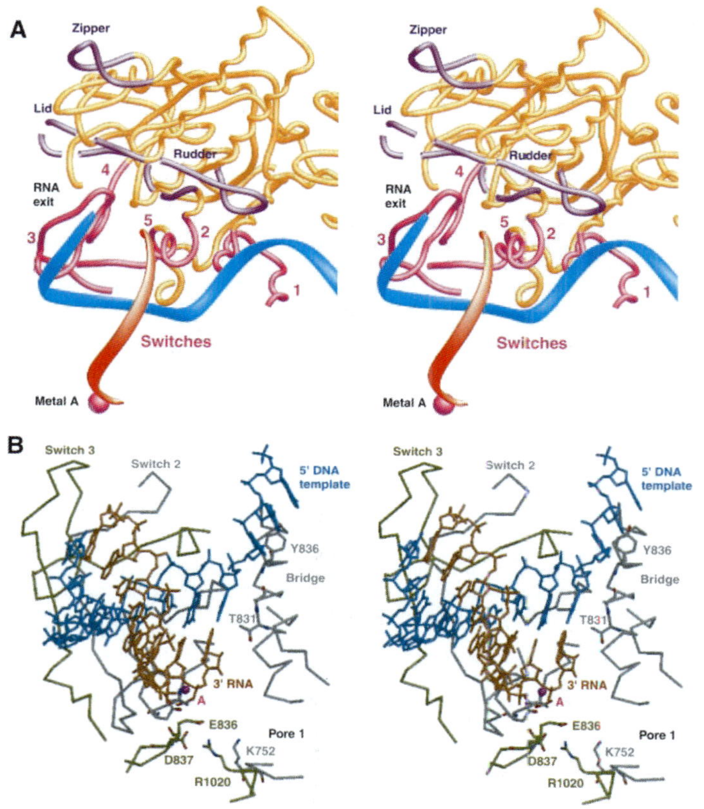

Fig. 3. Switches, clamp loops, and the hybrid-binding site. (**A**) Stereoview of the clamp core (1, yellow) and the DNA and RNA backbones. The view is as in Fig. 2C. The five switches are shown in pink and are numbered. Three loops, which extend from the clamp and may be involved in transactions at the upstream end of the transcription bubble, are in violet. Major portions of the protein are omitted for clarity. (**B**) Stereoview of nucleic acids bound in the active center.

Table 1. Switch regions.

Switch	Subunit	Domain	Residues	DNA contact	Structural changes upon clamp closure
1	Rpb1	Cleft-clamp core	1384–1406	+1 to +4	Two short helices formed (α47a, α47b)
2	Rpb1	Clamp core	328–346	−2, −1, +2	Helical turn flipped out
3	Rpb2	Hybrid-binding anchor	1107–1129	−5 to −1	Loop becomes ordered
4	Rpb2	Clamp	1152–1159	–	One turn added to helix α32 in the anchor region
5	Rpb1	Clamp core	1431–1433	–	Hinge-like bending

Closure of the clamp. The structures of free and transcribing pol II differ mainly in the position of the clamp (Fig. 2B). As previously suggested (*1*), and now demonstrated, the clamp swings over the cleft during formation of the transcribing complex, trapping the template and transcript. The clamp rotates by about 30° with a maximum displacement of over 30 Å at external sites (at the Rpb1 "zipper"). Although most of the clamp moves as a rigid body, five "switch" regions undergo conformational changes and folding transitions (Table 1). Switches 1, 2, 4, and 5 form the base of the clamp (Fig. 3). Switches 1 and 2 are poorly ordered and switch 3 is disordered in free pol II; all three switches become well ordered in the transcribing complex. Ordering is likely induced by binding of the switches to DNA downstream and within the DNA-RNA hybrid (see below). Binding to the hybrid may help couple clamp closure to the presence of RNA. The conformational changes of the switch regions may be concerted, because the switches interact with one another. The conformational changes are accompanied by changes in a network of salt linkages to the "bridge" helix across the cleft (Rpb1 residues Arg^{839}, Arg^{840}, and Lys^{843}).

Downstream DNA mobility. Downstream DNA lies in the cleft between the clamp and Rpb2 (Figs. 1B and 2, B and C), consistent with results from electron crystallography of the transcribing complex (*20*) and results of DNA-protein cross linking (*4–8*). The DNA contacts the Rpb5 "jaw" domain at a loop containing proline residue Pro^{118}, as previously suggested (*1*), and then passes between the Rpb2 "lobe" region and the Rpb1 "clamp head." The sequence of the Rpb2 lobe is divergent between yeast and bacteria, but the fold is conserved, whereas the clamp head is not conserved.

Details of downstream DNA–pol II interaction are lacking because the electron density is weak, indicative of mobility of the DNA. Furthermore, downstream DNAs from neighboring transcribing complexes in the crystal interact end to end, stacking on one another, so the precise location of the DNA may be determined by crystal packing forces. This could be the reason why there is no apparent contact between downstream DNA and the upper jaw. In addition, the length of DNA used here is possibly too short for passage all the way through the jaws.

Transcription bubble. The downstream

edge of the transcription bubble lies between the poorly ordered downstream duplex DNA and the first ordered nucleotide of the template strand at position +4, three nucleotides before the beginning of the RNA-DNA hybrid (Fig. 3B). The nucleotide at position +4 in the nontemplate strand and the remainder of this strand are disordered. The template strand follows a path along the bottom of the clamp and over the "bridge" helix. Template nucleotides +4, +3, and +2 are stacked in the manner of right-handed B-DNA. The base of nucleotide +1 is flipped with respect to that of nucleotide +2 by a left-handed twist of 90°. The base at +1 therefore points downward into the floor of the cleft for readout at the active site, whereas the base at +2 is directed upward into the opening of the cleft. This unusual conformation of the DNA results from binding to switches 1 and 2, as well as to the bridge helix (Figs. 1, C and D, and 3). Invariant bridge helix residues Ala^{832} and Thr^{831} position the coding nucleotide through van der Waals interactions, whereas Tyr^{836} binds nucleotide +2 and may correspond to a tyrosine in the "O-helix" of some single subunit DNA polymerases (*21*, *22*).

Maintenance of the downstream edge of the transcription bubble may be attributed not only to the binding of nucleotides +2, +3, and +4 but also to Rpb2 "fork loop" 2 (Figs. 1D and 4). Although this loop includes several disordered residues (*23*), it would likely clash with the nontemplate strand at position +3 if the nontemplate strand was still base paired with the template strand. A corresponding loop in the bacterial enzyme ("βD loop I"), four residues longer than that in yeast, was previously suggested to play such a role (*5*). Rpb2 fork loop 1 may help maintain the transcription bubble further upstream (Figs. 1D and 4). This loop is absent from the bacterial enzyme, perhaps reflecting a difference in promoter melting between eukaryotes, which require general transcription factors for the process, and bacteria, which do not. Both fork loops, although exposed, are highly conserved between yeast and human polymerases.

DNA-RNA hybrid. The base in the template strand at position +1 forms the first of nine base pairs of DNA-RNA hybrid, located between the bridge helix and Rpb2 "wall" (Figs. 1D and 4). The length of the hybrid corroborates the value of eight to nine base pairs determined biochemically (*24*, *25*). The hybrid heteroduplex adopts a nonstandard conformation, intermediate between those of standard A- and B-DNA (Fig. 5), and is underwound (*26*), in comparison with the crystal structure of a free DNA-RNA hybrid, which is closely related to the A-form (*27*).

The electron density for the hybrid is strongest in the downstream region around the active center, indicative of a high degree of order, important for the high fidelity of transcription. The electron density remains strong for the DNA template strand further upstream, but the density for the RNA strand becomes weaker (Fig. 2A). This gradual loss of density reflects a diminution in the number of RNA-protein contacts. The template DNA strand is bound by protein over the entire length of the hybrid, whereas RNA contacts are limited to the downstream region (Fig. 1C). The five upstream ribonucleotides are held mainly through base pairing with the template DNA.

Contacts to the downstream and upstream parts of the hybrid are made by Rpb1 and Rpb2, respectively (Fig. 1C). Fifteen protein regions are involved, with a substantial portion of the contacts arising from the ordering of Rpb1 switches 1, 2, and 3 upon nucleic acid binding. The entire set of protein contacts forms an extended, highly complementary binding surface. A surface area of 3400 $Å^2$ is buried in the protein–nucleic acid interface, comparable to values for transcription factors bound specifically to DNA sites of similar size. Biochemical studies have shown the binding interaction contributes substantially to the stability of a transcribing complex and thus to the high processivity of transcription (*25*, *28*).

Although a strong pol II–nucleic acid interaction is important for the ordering of nucleic acids in the active center region and for the stability of a transcribing complex, the interaction must not interfere with the translocation of nucleic acids during transcription. Indeed, the nucleic acids in the transcribing complex are mobile, as shown by the partial order of the downstream DNA (see above) and by a high overall crystallographic temperature factor of the hybrid, which appears to reflect mobility rather than static disorder (*29*). The conflicting requirements of tight binding and mobility may be reconciled in at least three ways. First, almost all protein contacts are to the sugar-phosphate backbones of the DNA and RNA. There are no contacts with the edges of the bases, so there is no base specificity. A large open space between pol II and the major groove of the hybrid is a prominent feature of the structure. Second, several side chains interact with two phosphate groups along the backbone simultaneously (Fig. 1C), which may reduce the activation barrier for translocation. Finally, about 20 positively charged side chains form a "second shell" around the hybrid at a distance of 4 to 8 Å, which may attract the hybrid without restraining its movement across the enzyme surface (*30*).

RNA synthesis. The active site metal ion in the transcribing complex structure corresponds to one of two metal ions in the 2.8 Å pol II structure, referred to as metal A (*3*). The location of this metal in the transcribing complex is appropriate for binding the phosphate group between the nucleotide at the 3′-end of the RNA and the adjacent nucleotide, designated +1 and –1, respectively (Fig.

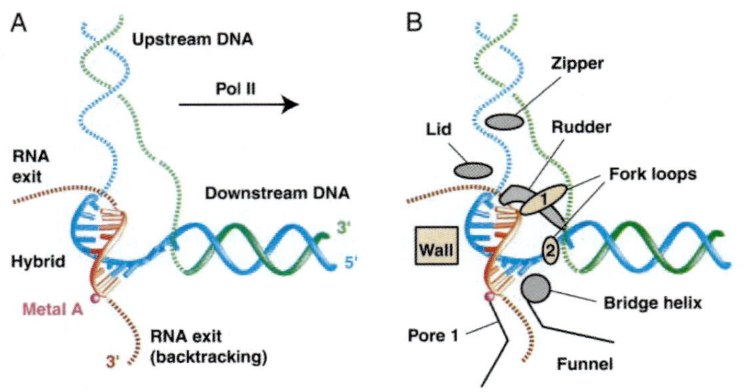

Fig. 4. Maintenance of the transcription bubble. (**A**) Schematic representation of nucleic acids in the transcribing complex. Solid ribbons represent nucleic acid backbones from the crystal structure. Dashed lines indicate possible paths of nucleic acids not present in the structure. (**B**) Protein elements proposed to be involved in maintaining the transcription bubble. Protein elements from Rpb1 and Rpb2 are shown in silver and gold, respectively.

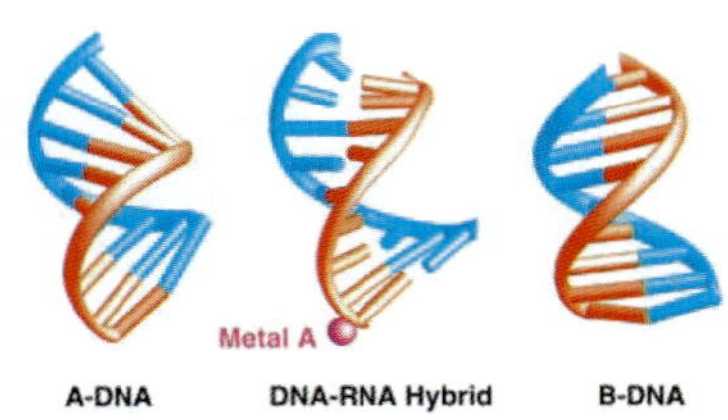

Fig. 5. DNA-RNA hybrid conformation. The view is similar to that in Fig. 2C. The conformation of the DNA-RNA hybrid is intermediary between canonical A- and B-DNA. DNA, blue; RNA, red.

1C). In the two-metal–ion mechanism proposed for single subunit polymerases, metal A contacts the α-phosphate of the incoming nucleoside triphosphate and metal B binds all three phosphates (*21*, *31–35*). Metal B may be absent from the transcribing complex structure because it has left with the pyrophosphate after nucleotide addition. On this basis, position +1 in the transcribing complex would be that of a nucleotide just added to the growing RNA, before translocation to bring the next template base into position opposite an empty nucleotide-binding site at the end of the RNA (*36*) (Fig. 6).

The ribonucleotide in position +1 lies in the entrance to the previously noted "pore 1," which extends from the floor of the cleft through to the backside of the enzyme. This location and orientation of the 3′-end of the RNA lend strong support to the previous proposal that nucleoside triphosphates enter through the pore during RNA synthesis and that RNA is extruded through the pore during back-tracking (*1*). The close fit of the DNA-RNA hybrid to the surrounding protein leaves no alternative to the pore for access of nucleotides to the active site. (Major conformational changes creating access are unlikely, because they would disrupt protein–nucleic acid contacts important for the fidelity and processivity of transcription.)

Specificity for ribo- rather than deoxyribonucleotides may be attributed to recognition of both the ribose sugar and the DNA-RNA hybrid helix. The 2′-hydroxyl group of a ribonucleotide in the substrate binding site (position +1) is 5 Å from the side chain of the highly conserved Rpb1 residue Asn[479]. Although this distance is too great for specific interaction, a slightly different positioning of an incoming nucleoside triphosphate might permit hydrogen bonding and discrimination of the ribose sugar. Different positioning of the nucleoside triphosphate could result from chelation by metal B, bound at a site in the structure of free pol II (*3*). RNA 2′-hydroxyl groups at positions −1, −3, and −5 are at hydrogen bonding distance from the side chains of Rpb1 residue Arg[446] and Rpb2 residues His[1097] and Gln[481]. The nucleic acid binding site is, furthermore, highly complementary to the nonstandard conformation of the hybrid helix and not to the standard conformation of a DNA double helix. Such indirect discrimination was previously suggested to contribute to the specificity of T7 RNA polymerase transcription (*37*).

Recognition of RNA in the transcribing complex from positions –1 to –5, by both hydrogen bonding and indirect discrimination, can contribute to the specificity of RNA synthesis through proofreading. The presence of a deoxyribonucleotide or of an incorrect base anywhere in this region of the RNA will be destabilizing. A back-tracked complex, with previously correctly synthesized RNA in the hybrid region and with the RNA containing the misincorporated nucleotide extruded at the 3′-end, will be favored. The extruded RNA can be removed by cleavage at the active site, through the action of transcription factor TFIIS.

Key nonspecific (van der Waals) contacts to the nucleotide base at the end of the hybrid region, in position +1, are made by residues Thr[831] and Ala[832] from the Rpb1 bridge helix, as mentioned above. Although highly conserved, the bridge helix is essentially straight in the pol II structures so far determined but bent in the bacterial enzyme structure in the vicinity of the residues corresponding to Thr[831] and Ala[832] (*1*, *2*). The bend would produce a movement of this region of the bridge helix by 3 to 4 Å, resulting in a clash with the nucleotide at position +1 (Fig. 6). Modeling of a bacterial transcribing complex resulted in such a clash (*5*). We speculate that the bridge helix oscillates between straight and bent states and that this movement accompanies the translocation of nucleic acids during transcription: Addition of a nucleotide at position +1 would occur in the straight state; translocation to position –1 and movement of nucleic acids through the distance between base pairs, about 3.2 Å, would be accompanied by a conformational change to the bent state; and reversion to the straight state without movement of nucleic acids would create an empty site at position +1 for entry of the next nucleotide, completing a cycle of nucleotide addition during RNA synthesis (Fig. 6).

Protein-RNA contacts are of special importance at the very beginning of transcription. Nucleoside triphosphates must be held in positions +1 and –1 for the synthesis of the first phosphodiester bond. After translocation to positions –1 and –2, the dinucleotide product must still be held by protein-RNA contacts, as the energy of base-pairing alone is insufficient for retention in the complex. Indeed, RNA is deeply buried in the transcribing complex as far as position –3 (Fig. 1C). Di- and trinucleotides are nevertheless occasionally released, and transcription must restart, resulting in "abortive cycling" (*38*). RNA is exposed at position –4 and beyond, with no direct protein contacts except for the hydrogen bond at position –5 mentioned above. Coincident with exposure of the RNA, biochemical studies reveal a transition in stability at a transcript length of four residues, beyond which the RNA is generally retained (*39*). Although the direct protein-RNA contacts observed up to this point may be largely responsible for retention, long-range interactions also play a role. For example, a highly conserved arginine makes long-range electrostatic interactions with the RNA around position –4 (Arg[497] in Rpb2, Arg[529] in *Escherichia coli* β), and mutation of this residue results in the overproduction of abortive transcripts (*40*).

RNA exit. Abortive cycling yields an abundance of two- to three-residue transcripts, as well as transcripts of up to 10 residues (*41*). An initiating complex evidently undergoes a second transition when the transcript reaches 10 residues in length. At this point, the newly synthesized RNA must separate from the DNA-RNA hybrid and enter an exit channel on the surface of the enzyme, where it remains protected from nuclease attack for about six more residues (*42*). Three loops extending from the clamp, termed "rudder," "lid," and "zipper," have been suggested to play roles in hybrid disso-

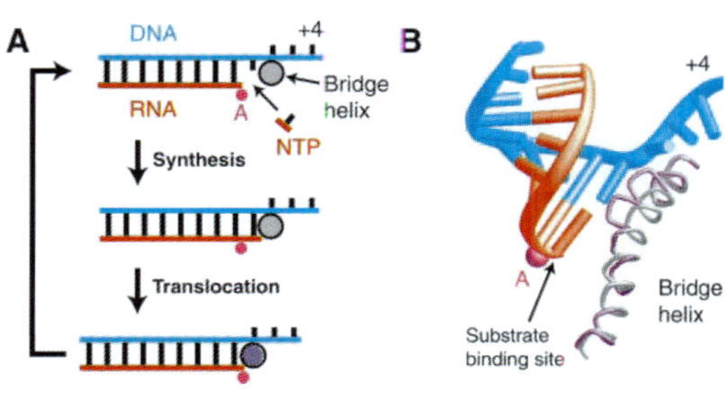

Fig. 6. Proposed transcription cycle and translocation mechanism. (**A**) Schematic representation of the nucleotide addition cycle. The nucleotide triphosphate (NTP) fills the open substrate site (top) and forms a phosphodiester bond at the active site ("Synthesis"). This results in the state of the transcribing complex seen in the crystal structure (middle). We speculate that "Translocation" of the nucleic acids with respect to the active site (marked by a pink dot for metal A) involves a change of the bridge helix from a straight (silver circle) to a bent conformation (violet circle, bottom). Relaxation of the bridge helix back to a straight conformation without movement of the nucleic acids would result in an open substrate site one nucleotide downstream and would complete the cycle. (**B**) Different conformations of the bridge helix in pol II and bacterial RNA polymerase structures. The view is the same as in Fig. 2C. The bacterial RNA polymerase structure (*2*) was superimposed on the pol II transcribing complex by fitting residues around the active site. The resulting fit of the bridge helices of pol II (silver) and the bacterial polymerase (violet) is shown. The bend in the bridge helix in the bacterial polymerase structure causes a clash of amino acid side chains (extending from the backbone shown here) with the hybrid base pair at position +1.

ciation, RNA exit, and maintenance of the upstream end of the transcription bubble (*2, 3*) (Fig. 4). Modeling of the DNA-RNA hybrid beyond the nine base pairs seen in the transcribing complex structure would produce a clash with the rudder. Extension of the RNA from the last hybrid base pair leads beneath the rudder to the previously proposed "exit groove 1." Continuation of this RNA path also leads beneath the lid, whose role may be to maintain the separation of RNA and template DNA strands. The zipper may play a similar role in separating template and nontemplate DNA strands. The lid and a small portion of the rudder are disordered in the transcribing complex structure but are ordered in the free pol II structure. The lid and rudder may become ordered in the transcribing complex in conjunction with the second transition (*43*) and with the establishment of a stable, elongating complex.

Conclusions and prospects. The atomic structure of RNA polymerase II in the act of transcription reveals the protein-DNA and -RNA interactions underlying the process. The structure shows a right angle bend of the DNA path at the active center. This feature is understandable in retrospect. The bend orients the DNA-RNA hybrid optimally for transcription, which occurs along the direction of the hybrid axis. Nucleotides enter through the funnel and pore, add to the RNA at the end of the RNA-DNA hybrid, translocate through the hybrid-binding region, and exit beneath the rudder and lid.

Answers to many long-standing questions about the transcription mechanism may be found in the structure of the clamp. This mobile, multifunctional element does more than close over the nucleic acids in the active center to enhance the processivity of transcription. First, switch regions at the base of the clamp couple its closure to the presence of DNA-RNA hybrid in the active center. This coupling satisfies the dual requirement for retention of nucleic acids during transcript elongation and their release after termination. Second, through the rudder, lid, and zipper, the clamp plays a key role in the events of hybrid melting and template reannealing at the upstream end of the transcription bubble.

Experiments to test the proposed roles for these structural elements by site-directed mutagenesis are among the many that can now be designed on the basis of the structure. In addition, polymerase may be cocrystallized with synthetic transcription bubbles (*44*) and other forms of RNA and DNA.

References and Notes

1. P. Cramer *et al.*, *Science* **288**, 640 (2000).
2. G. Zhang *et al.*, *Cell* **98**, 811 (1999).
3. P. Cramer, D. A. Bushnell, R. D. Kornberg, *Science* **292**, 1882 (2001).
4. E. Nudler, *J. Mol. Biol.* **288**, 1 (1999).
5. N. Korzheva *et al.*, *Science* **289**, 619 (2000).
6. N. Naryshkin, A. Revyakin, Y. Kim, V. Mekler, R. H. Ebright, *Cell* **101**, 601 (2000).
7. C. I. Wooddell, R. R. Burgess, *Biochemistry* **39**, 13405 (2000).
8. M. S. Bartlett, M. Thomm, E. P. Geiduschek, *Nature Struct. Biol.* **7**, 782 (2000).
9. M. H. Sayre, H. Tschochner, R. D. Kornberg, *J. Biol. Chem.* **267**, 23376 (1992).
10. R. Conaway, J. Conaway, *Prog. Nucleic Acid Res. Mol. Biol.* **56**, 327 (1997).
11. A. M. Edwards, C. M. Kane, R. A. Young, R. D. Kornberg, *J. Biol. Chem.* **266**, 71 (1991).
12. T. R. Kadesch, M. J. Chamberlin, *J. Biol. Chem.* **257**, 5286 (1982).
13. A. Gnatt, J. Fu, R. D. Kornberg, *J. Biol. Chem.* **272**, 30799 (1997).
14. Plate-like monoclinic crystals of space group C2 with unit cell dimensions a = 157.3 Å, b = 220.7 Å, c = 191.3 Å, and β = 97.5° were grown by the sitting drop vapor diffusion method under the conditions previously developed for free pol II [J. Fu *et al.*, *Cell* **98**, 799 (1999)]. Crystals were transferred slowly to freezing buffer (*1*) and flash frozen in liquid nitrogen. Diffraction data were collected at a wavelength of 0.998 Å at beamline 9.2 at the Stanford Synchrotron Radiation Laboratory. Although diffraction to 3.1 Å resolution could be observed in two directions, anisotropy limited the useable data to 3.3 Å resolution.
15. Data processing with DENZO and SCALEPACK [Z. Otwinowski, W. Minor, *Methods Enzymol.* **276**, 307 (1996)] showed that the data collected at 0.998 Å were 100% complete in the resolution range 40 to 3.3 Å. A total of 96,867 unique reflections were measured. At a redundancy of 4.4, the R_{sym} was 11.1% (31.7% at 3.4 to 3.3 Å). The structure was solved by molecular replacement with AMORE [J. Navaza, *Acta Crystallogr. A* **50**, 157 (1994)]. A modified atomic pol II structure lacking the mobile clamp was used as search model. A single strong peak was obtained after rotation and translation searches (correlation coefficient = 59, *R* factor = 43%, 15 to 6.0 Å resolution).
16. Diffraction data were recollected at the zinc anomalous peak wavelength (1.283 Å) from the crystal used in structure determination. Initial phases were calculated from the pol II search model after rigid body refinement in CNS (*45*).
17. Model building was carried out with the program O [T. A. Jones, J. Y. Zou, S. W. Cowan, M. Kjeldgaard, *Acta Crystallogr. A* **47**, 110 (1991)] and refinement was carried out with CNS (*45*). For cross validation, 10% of the data were excluded from refinement. The four mobile modules defined for free pol II (*3*) were used for rigid body refinement, followed by bulk solvent correction and anisotropic scaling. After positional and restrained *B*-factor refinement, a free R-factor of 35% was obtained with all data. The resulting sigma-weighted electron density maps allowed building of switch 3 and rebuilding of the other switch regions. Loops that were present in free pol II but disordered in the transcribing complex were removed. The final protein electron density was generally of good quality and most side chains were visible. Some flexible regions, including the jaws, parts of Rpb8, and the upper portions of the wall and clamp, showed only main chain density. In these regions, the refined pol II structure was not rebuilt. A few rounds of model building and refinement of the protein lowered the free *R* factor to 31.0%. At this stage, difference density with a helical shape was observed for the nucleic acids in the hybrid region and phosphates and bases were revealed. The density originating at the active site metal was assigned to the RNA strand, and the opposite continuous density was assigned to the DNA template strand. A total of 22 nucleotides were placed individually, resulting in a 0.7% drop in the free *R* factor after refinement.
18. The 3.3 Å electron density map did not allow distinction of purine from pyrimidine bases. Placement of the particular sequences thus assumed complete RNA synthesis until the pause site and no back-tracking. Modeling resulted in a length of the downstream DNA that agrees with end-to-end packing of DNAs from neighboring complexes. The ambiguity in the assignment of nucleic acid sequences does not affect our conclusions because there are no base-specific protein contacts. The density map included a few weak, disconnected peaks in pore 1 that may arise from back-tracked RNA in a subpopulation of complexes or from incoming nucleoside triphosphates.
19. At the standard contour level of 1.0σ, only a few disconnected peaks are observed for the downstream DNA. At a contour level of 0.8σ, extended density features are observed, which identify the approximate helix axis and major groove of the downstream DNA, with only a few disconnected noise peaks in the surrounding solvent region. Inclusion of the DNA duplex placed in this way in the refinement led to an increase in the free *R* factor.
20. C. L. Poglitsch *et al.*, *Cell* **98**, 791 (1999).
21. J. R. Kiefer, C. Mao, J. C. Braman, L. S. Beese, *Nature* **391**, 304 (1998).
22. A. C. Rodriguez, H. W. Park, C. Mao, L. S. Beese, *J. Mol. Biol.* **299**, 447 (2000).
23. Rpb2 residues 468 to 476 and 503 to 508 are disordered in fork loop 1 and 2, respectively.
24. E. Nudler, A. Mustaev, E. Lukhtanov, A. Goldfarb, *Cell* **89**, 38 (1997).
25. M. L. Kireeva, N. Komissarova, D. S. Waugh, M. J. Kashlev, *J. Biol. Chem.* **275**, 6530 (2000).
26. The nucleic acid model was obtained by placing nucleotides manually into unbiased electron density peaks. At 3.3 Å resolution, the location of phosphate groups and the approximate axes through base pairs were revealed. After refinement, the positions of the nucleotides changed only slightly, showing that the final nucleic acid model reflects the experimental data and that the model is not primarily a result of the geometrical constraints applied during refinement. Although the available data define the overall hybrid conformation, stereochemical details are not revealed and the parameters of the hybrid helix must be viewed as approximate. The hybrid shows an average rise per residue of 3.2 Å {program CURVES [R. Lavery, H. Sklenar, *J. Biomol. Struct. Dyn.* **6**, 63 (1988)]}, compared with 2.8 and 3.4 Å for A- and B-DNA, respectively. The average minor groove width is 10.4 Å (CURVES), compared with 11 and 7.4 Å for A- and B-DNA, respectively. The root-mean-square (rms) deviation in phosphorus atom positions between the hybrid and canonical A- and B-DNA is 3.1 and 5.5 Å, respectively. The helical twist is 12.6 residues/turn {program NEWHELIX [K. Grzeskowiak, D. S. Goodsell, M. Kaczor-Grzeskowiak, D. Cascio, R. E. Dickerson, *Biochemistry* **32**, 8923 (1993)]}. The phosphorus atom positions show an rms deviation of 2.7 Å from the structure of a free hybrid (*27*).
27. N. C. Horton, B. C. Finzel, *J. Mol. Biol.* **264**, 521 (1996).
28. I. Sidorenkov, N. Komissarova, M. Kashlev, *Mol. Cell* **2**, 55 (1998).
29. The average atomic *B* factor is 97 A^2 for the hybrid, as compared with 63 $Å^2$ for the entire structure. The bases and backbone groups show similar *B* factors. This likely indicates mobility because static disorder, arising from the presence of complexes at different register, would be expected to result in low *B* factors for the backbone and higher *B* factors for the bases. We think that refinement of atomic *B* factors is justified at the given resolution and that the resulting *B* factors are meaningful, because refinement of all protein atoms, starting from a constant value of 30 $Å^2$, results in an overall *B* factor that is very close to that obtained for the free pol II structure at 2.8 Å resolution. Moreover, the general distribution of *B* factors is similar to that for the structure of free pol II.
30. These residues include arginines 320, 326, 839, and 840 and lysines 317, 323, 330, 343, and 830 of Rpb1 and arginines 476, 497, 766, 1020, 1096, and 1124 and lysines 210, 458, 507, 775, 865, 965, and 1102 of Rpb2.
31. L. S. Beese, T. A. Steitz, *EMBO J.* **10**, 25 (1991).
32. S. Doublie, S. Tabor, A. M. Long, C. C. Richardson, T. Ellenberger, *Nature* **391**, 251 (1998).
33. M. R. Sawaya, R. Prasad, S. H. Wilson, J. Kraut, H. Pelletier, *Biochemistry* **36**, 11205 (1997).
34. T. A. Steitz, *Curr. Opin. Struct. Biol.* **3**, 31 (1993).
35. ______, *Nature* **391**, 231 (1998).

36. Although the 3′-most residue of the RNA is in the position of a nucleotide just added to the chain, it must have undergone translocation and then returned to this position before crystallization. Translocation is necessary to create a site for the next nucleotide, whose absence from the reaction results in a paused complex.
37. G. M. Cheetham, T. A. Steitz, *Science* **286**, 2305 (1999).
38. Y. Tintut, J. T. Wang, J. D. Gralla, *J. Biol. Chem.* **270**, 24392 (1995).
39. L. Kinsella, C. Y. Hsu, W. Schulz, D. Dennis, *Biochemistry* **21**, 2719 (1982).
40. D. J. Jin, C. L. Turnbough, *J. Mol. Biol.* **236**, 72 (1994).
41. J. R. Levin, B. Krummel, M. J. Chamberlin, *J. Mol. Biol.* **196**, 85 (1987); B. Krummel, M. J. Chamberlin, *Biochemistry* **28**, 7829 (1989); F. C. P. Holstege, U. Fiedler, H. T. M. Timmers, *EMBO J.* **16**, 7468 (1997).
42. W. Gu, M. Wind, D. Reines, *Proc. Natl. Acad. Sci. U.S.A.* **93**, 6935 (1996).
43. Ordering of the rudder and lid may not be observed because of structural heterogeneity of the transcribing complexes in this region. Heterogeneity might be expected as a consequence of inefficient displacement of RNA from DNA-RNA hybrid during transcription of tailed templates.
44. S. S. Daube, P. H. von Hippel, *Science* **258**, 1320 (1992).
45. A. T. Brünger *et al.*, *Acta Crystallogr. D* **54**, 905 (1998).
46. For assistance at beamlines 1-5, 7-1, 9-1, and 9-2 of the Stanford Synchrotron Radiation Laboratory (SSRL), we thank H. Bellamy, A. Cohen, P. Ellis, P. Kuhn, T. McPhillips, M. Soltis, and the other members of the SSRL user support staff. This research is based in part on work done at SSRL, which is funded by the U.S. Department of Energy (DOE) Office of Basic Energy Sciences. The structural biology program is supported by the NIH National Center for Research Resources Biomedical Technology Program and the DOE Office of Biological and Environmental Research. We thank COMPAQ for providing a Unix workstation. We thank N. Thompson and R. Burgess for generously providing antibody for protein purification. We thank J. Puglisi and members of the Kornberg laboratory for comments on the manuscript. The contribution of A.L.G. was sponsored by USAMRMC Breast Cancer Initiative, DAMD17-97-7099, and does not necessarily reflect the policy of the government. P.C. was supported by a postdoctoral fellowship of the Deutsche Forschungsgemeinschaft (DFG). D.A.B. was supported by postdoctoral fellowship PF-00-014-01-GMC from the American Cancer Society. This research was supported by NIH grant GM49985 to R.D.K. Coordinates have been deposited at the Protein Data Bank (accession code 1I6H).

1 February 2001; accepted 28 March 2001
Published online 19 April 2001;
10.1126/science.1059495
Include this information when citing this paper.

Biophysical Chemistry 29 (1988) 17–29
Elsevier

17

BPC 01210

Approaching the molecular structure of ribosomes

A. Yonath [a,b] and H.G. Wittmann [c]

[a] *Department of Structural Chemistry, Weizmann Institute, Rehovot, Israel,* [b] *Max Planck Research Unit for Structural Molecular Biology, Hamburg, F.R.G. and* [c] *Max Planck Institute for Molecular Genetics, Berlin (West), Germany*

Accepted 15 October 1987

Ribosome; X-ray crystallography; Three-dimensional crystal; Image reconstruction; Subunit tunnel; Heavy-atom cluster

Fifteen forms of three-dimensional crystals and three forms of two-dimensional sheets from ribosomal particles have been grown. In all cases only biologically active particles could be crystallized, the crystalline material retaining its integrity and biological activity for months. Crystallographic data have been collected from crystals of 50 S ribosomal subunits, using synchrotron radiation, at temperatures between 19 and −180°C. Although at around 0°C in the synchrotron X-ray beam the crystals rapidly lose their high-resolution reflections, at cryo-temperature hardly any radiation damage occurs over long periods, and a complete set of diffraction data to about 6 Å resolution could be collected from a single crystal. Heavy-atom clusters were used for soaking as well as for specific binding to the surface of the ribosomal subunits prior to crystallization. The 50 S ribosomal subunits from a mutant of *Bacillus stearothermophilus* which lacks the ribosomal protein BL11 crystallize isomorphously with the native form. Models of the entire 70 S ribosome and of the 50 S subunit have been reconstructed from two-dimensional sheets at 47 and 30 Å, respectively. These models demonstrate the overall shape of the particles, the contact areas between large and small subunits, the space where protein biosynthesis may take place and a tunnel through the 50 S subunit which could provide a path for the nascent polypeptide chain.

1. Introduction

It had long been assumed that ribosomes, being flexible, unstable, asymmetric and of enormous size, could not be crytallized. On the other hand, it was found that under stressful conditions such as a cold shock or unsuitable diet, eukaryotic ribosomes form ordered arrays in vivo. The hypothesis that the temporary periodic organization of ribosomes occurs in living organisms when external conditions demand prolonged storage of potentially active ribosomes has led us to initiate crystallographic studies on intact ribosomal particles.

Ribosomes, the cell organelles on which protein biosynthesis takes place, consist of two different subunits which associate upon initiation of the biosynthetic process. Each ribosomal subunit is a defined assembly of proteins and ribonucleic acid chains. A typical bacterial ribosome has a molecular mass of 2.3 MDa. Its large (50 S) subunit is of 1.6 MDa and consists of 32–35 different proteins and two RNA chains. The small subunit (30 S) is of 700 kDa and contains about 21 proteins and one RNA chain.

As a result of an intensive systematic exploration of crystallization conditions and development of an innovative experimental technique for fine control of the volume of the crystallization mixture, procedures for in vitro growth of crystals of intact ribosomal particles have been developed [1–3]. These procedures, combined with sophisticated seeding, proved to be suitable for the reproducible production of ordered three-dimensional crystals and two-dimensional sheets of whole ribosomes as well as of small and large ribosomal subunits. Inactive ribosomal particles could not be

Correspondence address: A. Yonath, Department of Structural Chemistry, Weizmann Institute, Rehovot, Israel.

0301-4622/88/$03.50

crystallized. Moreover, in all cases, the crystalline material retains its integrity and biological activity for long periods in spite of the natural tendency of ribosomes to distintegrate and in contrast to the short lifetime of isolated ribosomes in solution.

So far we have obtained 15 different forms of three-dimensional crystals. These include two forms of 70 S ribosomes (from *Escherichia coli* and *Thermus thermophilus* [3–5]); two of the small (30 S) subunits of *Halobacterium marismortui* and *T. thermophilus* [3,5] as well as 11 crystal forms of large (50 S) ribosomal subunits; eight from *Bacillus stearothermophilus* [3,6–12], including particles to which a nascent polypeptide chain and a tRNA molecule are attached [13] and of 50 S subunits of a mutant which lacks one ribosomal protein [14]; and three forms of 50 S subunits from *H. marismortui* [15–17].

Crystals of 50 S subunits from *B. stearothermophilus* and *H. marismortui* are suitable for crystallographic studies. These bacteria are representatives of the eubacteria and archaebacteria, respectively. Thus, we expect that the results of our studies will not only shed light on the structure of ribosomes but also provide information about evolution.

Along with the growth of three-dimensional crystals we produced three forms of ordered two-dimensional sheets: two of native 50 S subunits from *B. stearothermophilus* [18–20], including particles to which a nascent protein and a tRNA molecule are attached [13] and one of 70 S ribosomes from the same bacterial species [21,22]. Three-dimensional image-reconstruction studies have been carried out on sheets of 50 S and 70 S particles [19,22], the results of which are summarized below.

Although ordered three-dimensional crystals and two-dimensional sheets can now be grown from virtually all preparations of active particles, the exact growth conditions for well-ordered crystals must still be varied slightly for each ribosomal preparation, due to the intricate nature of the particles. Moreover, for each crystal form the quality of the crystals depends, in a manner not yet fully characterized, on the procedure used for preparation of the ribosomal subunits and on the strain of a given bacterial species.

In this paper, we describe our crystallographic studies, present results obtained by three-dimensional image reconstruction from two-dimensional sheets and discuss their biological relevance.

2. Experimental

Growth of the different bacterial species, production of their ribosomes and biological characterization are described in refs. 6 and 15. The procedures used for crystal growth and the crystallographic studies are discussed in detail elsewhere [1,2,6,8–12,14–17]. The exact growth conditions are given in the legends to the various figures. Nonspecific derivatization was achieved by soaking crystals in solutions of 2–5 mM heavy-atom cluster for 6–8 h, followed by washing in transfer solutions [2] for 4–6 h. Two-dimensional sheets were grown according to refs. 18–22. The three-dimensional image-reconstruction studies have been described in refs. 19 and 22. For crystallographic studies we used synchrotron radiation at 19, 4, −10, −20 and −180°C

Radioactive *N*-ethylmaleimide was reacted with ribosomal particles to determine the accessibility and location of the sulfhydryl groups in ribosomal proteins of the 50 S subunits from *B. stearothermophilus* and *H. marismortui*. An 11-gold cluster and its radioactive derivative were prepared basically following the procedure described in ref. 23. Functional groups were attached to this cluster for covalent binding to ribosomal particles through accessible sulfhydryl groups (S. Weinstein and W. Jahn, manuscript in preparation). The extent of binding of the gold cluster was determined by measuring the radioactivity associated with the 50 S particles as well as by neutron activation (performed at Soreq Nuclear Research Laboratories, Israel). The proteins which bind the cluster and/or *N*-ethylmaleimide were identified by locating the radioactivity on two-dimensional gel electropherograms (S. Weinstein and H.S. Gewitz, manuscript in preparation).

3. Results and discussion

3.1. Crystallographic studies

3.1.1. Large (50 S) ribosomal particles from Bacillus stearothermophilus

Crystals of large ribosomal subunits from *B. stearothermophilus* grow at 4° C either from low-molecular-mass alcohols (fig. 1 and refs. 1, 2, 6 and 8–14), or from low concentrations of polyethylene glycol (fig. 2 and ref. 5). For the latter, the growth solution contained cations that are essential for maintaining the integrity and biological activity of ribosomes at somewhat higher concentrations than those commonly used for storage of the particles or for their crystallization with alcohols [6].

Growing crystals from volatile organic solvents imposes many technical difficulties on manipulation, data collection and heavy-atom derivatization. In fact, any handling of the crystals, such as their removal or reorientation, or replacing the growth medium with a different solution, is detrimental to the crystals. Thus, crystals from the 50 S subunits of wild-type and mutated *B. stearothermophilus* were grown directly in X-ray capillaries. They can reach a length of 2.0 mm and cross-section of 0.4 mm (fig. 1). Since most of the crystals grew with one of their faces adhering to the walls of the capillaries, it was possible to irradiate them without removing the original growth solution. Although the crystals tend to grow with their long axes parallel to the capillary axis, a fair number grow in different directions. Consequently, unit-cell constants have been determined, and diffraction data were obtained without manipulating the crystals. The crystals are loosely packed in an orthorhombic unit cell of 360 × 680 × 920 Å. Fresh crystals diffract to 10–13 Å resolution. However, the higher-resolution reflections decay within 5–10 min. This decay is typical of crystals of ribosomal particles which are irradiated above cryo-temperatures.

Oriented arcs and distinct spots, extending to 3.5 Å resolution, with spacings similar to those measured from diffraction patterns of ribosome gels and extracted rRNA [24–26], have been detected on a few diffraction patterns of single

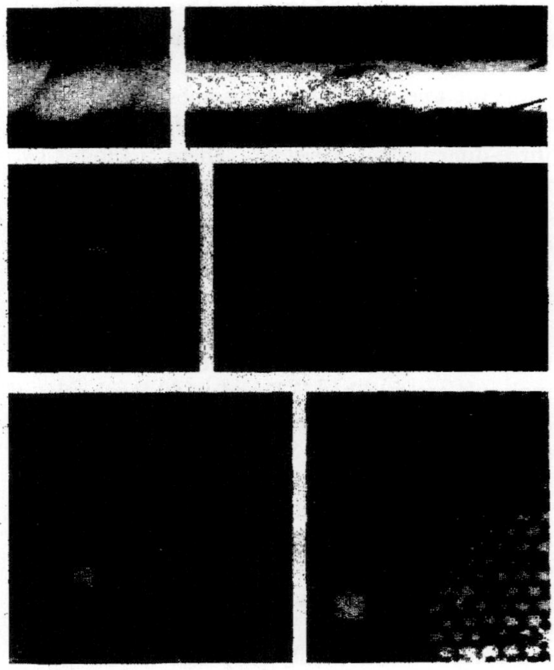

Fig. 1. (a) Crystals of 50 S ribosomal subunits from *B. stearothermophilus* grown in 0.5 mm X-ray capillaries by vapor diffusion at 4° C. Crystallization mixture, 20 μl of 50 S ribosomal subunits (10–20 mg/ml) in H-I buffer [9], 0.01 M spermine, 1% methanol, 10 mM Hepes or glycine buffer (pH 8.4), was equilibrated with a reservoir of 12% methanol, 12% ethylene glycol, 0.5 M NaCl (pH 8.4) (bar = 0.4 mm). (b) X-ray diffraction patterns from crystals similar to those shown in panel a, obtained at −4° C with synchrotron radiation (A1 station at Chess/Cornell University, operating at 5 GeV, current 30–40 mA) with a 0.3 mm collimated X-ray beam of wavelength 1.55 Å, on a Huber precession camera equipped with an He path. Exposure time, 3 min; crystal-to-film distance, 200 mm. (Left) 1° rotation photograph of $0kl$ zone, 680 × 920 Å. (Right) 0.4° rotation photograph of $hk0$ zone, 360 × 680 Å. (c) Electron micrographs of positively stained (2% uranyl acetate) thin sections of crystals similar to those shown in panel a that have been fixed in 0.2% glutaraldehyde and embedded in epon resin (ERL 4206). Optical diffraction patterns are inserted. (Left) Section approximately perpendicular to that shown on the right. Repeat distances measured from optical diffraction: 330 × 1050 Å. This corresponds to the $h0l$ zone (360 × 920 Å) in the X-ray patterns (bar = 1000 Å). (Right) Micrograph showing the characteristic open packing of this crystal form. The orthogonal choice of axes corresponds to the 680 × 920 Å zone observed in the X-ray diffraction patterns. Lattice spacing calculated from optical diffraction: 670 × 850 Å.

Fig. 2. Crystals of 50 S ribosomal particles from *B. stearothermophilus* grown from polyethylene glycol at 4°C in hanging drops (bar = 0.2 mm). (a) Crystallization solution, at pH 6.4, of 2.5% polyethylene glycol, 0.02 M $(NH_4)_2SO_4$, 0.02 M $MgCl_2$ and 50 S subunits in H-I buffer [9] was equilibrated with a reservoir of 0.15 M KCl together with all other materials in the droplet. (b) Crystallization solution, at pH 6.6, of 2.5% polyethylene glycol, 0.2 M KCl, 0.18 M $(NH_4)_2SO_4$, 0.03 M $MgCl_2$ and 50 S particles in H-I buffer [9] was equilibrated with a reservoir of 0.5 M KCl and all materials of the droplet.

crystals and on those of samples containing large numbers of microcrystals. For aligned crystals the average arc length was 60°. These patterns are similar to those observed for nucleosomes [27] and may arise from partial orientation of the nucleic acid component within the particle.

Because crystals of 50 S particles from *B. stearothermophilus* have been grown from polyethylene glycol (fig. 2) very recently, only preliminary crystallographic data are available. These crystals are compactly packed, diffract to about 16 Å resolution, and their X-ray diffraction patterns show periodic spacings of about 260, 320 and 700 Å. Since they have been grown using nonvolatile precipitants, they can be handled. Thus, they may be more suitable for subsequent crystallographic studies. Moreover, our ability to grow crystals of ribosomal particles of eubacteria using precipitants other than alcohols indicates that there is a fair chance that other ribosomal particles from eubacteria can be crystallized in a similar manner. Therefore, efficient crystallographic studies on

21

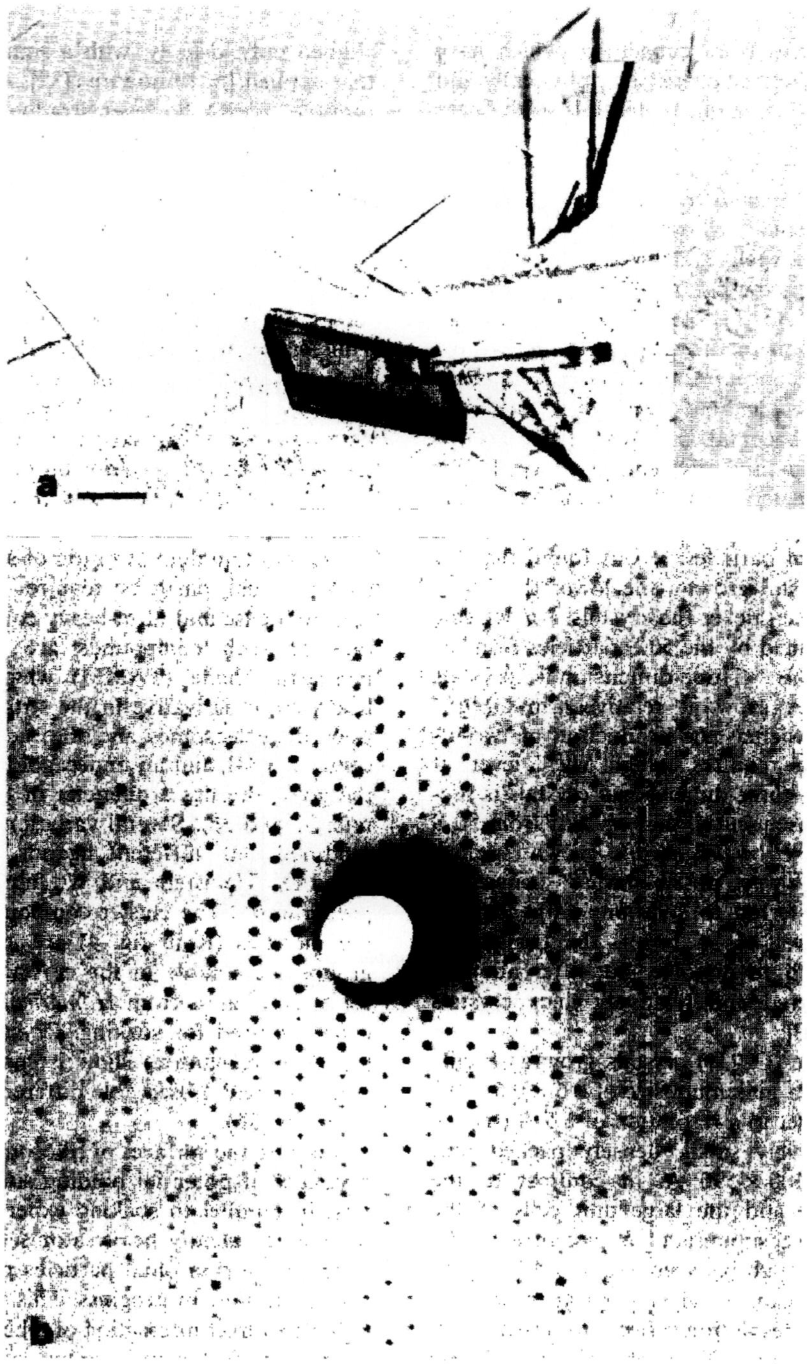

Fig. 3. (a) Crystals obtained from seeds which were grown within 3–5 days from droplets containing 4–5% polyethylene glycol, 1.2 M KCl, 0.5 M NH_4Cl, 0.05–0.10 M $MgCl_2$ and 10 mM spermidine (pH 5.0–5.6) in a crystallization drop containing 5% polyethylene glycol, 1.2 M KCl, 0.5 NH_4Cl, 0.03 M $MgCl_2$, at pH 5.6. This was equilibrated with 7% polyethylene glycol, 1.7 M KCl, 0.5 M NH_4Cl and 0.03 M $MgCl_2$ (pH 5.6) (bar = 0.2 mm). (b) A 1° rotation pattern of a crystal similar to the inserted one. The pattern was obtained at 85 K with synchrotron radiation (7.1 station at SSRL/Stanford University). Wavelength, 1.54 Å; exposure time, 2 min; crystal-to-film distance, 135 mm.

ribosomal particles from eubacteria, which have been well-characterized chemically, physically and biologically [28–31], might be feasible in the near future.

3.1.2. Ribosomal subunits from H. marismortui

Large and small ribosomal subunits from halobacteria have been crystallized in growth solutions which mimic the natural environment within these bacteria [5,15–17]. Thus, KCl, NH_4Cl and $MgCl_2$ are present in the crystallization solution at the minimum concentrations needed for storage of the ribosomes without loss of activity [15]. Advantage has been taken of the delicate equilibrium of mono- and divalent ions needed for growth of halobacteria as well as of the major role played by the Mg^{2+} concentration in crystallization of ribosomal particles. It was found that for spontaneous crystal growth, the lower the Mg^{2+} concentration, the thicker the crystals. An extreme example is provided by the 50 S particles from *B. stearothermophilus*. Three-dimensional crystals from this source grow at relatively low Mg^{2+} concentration, whereas the production of two-dimensional sheets requires a high Mg^{2+} level, at which growth of three-dimensional crystals is prevented [18]. Consequently, thin crystals from 50 S subunits from *H. marismortui*, grown spontaneously at the lowest possible Mg^{2+} concentration, were transferred to solutions with an even lower Mg^{2+} level. As a result, the transferred crystals appeared to dissolve, but after about 2 weeks well-ordered and relatively thick crystals were formed [2,3].

Orthorhombic ($C222_1$) crystals grow as fragile thin plates with a maximum size of $0.6 \times 0.6 \times 0.2$ mm. They diffract to a resolution of 5.5 Å (fig. 3), and have relatively small, densely packed unit cells of $214 \times 300 \times 590$ Å, in contrast to the 'open' structure and the large unit cells of the crystals from 50 S subunits of *B. stearothermophilus* (fig. 1). Although between -2 and 4°C up to 15 photographs can be taken from an individual crystal, the high-resolution reflections appear only on the first two to three X-ray photographs. Hence, over 260 crystals had to be irradiated in order to obtain a complete data set, and X-ray diffraction data were obtained from crystals which had been aligned only visually, with a similar algorithm to that applied for rhino virus [32]. At cryo-temperature (i.e., 85 K), however, irradiated crystals show scarcely any radiation damage for several days. Thus, for the first time a full data set could be collected from a single crystal.

Moreover, at cryo-temperature the lifetime of the crystals is sufficiently long to allow one to perform preliminary X-ray diffraction experiments using rotating anodes as X-ray generators. Thus, basic parameters such as resolution, unit-cell constants and isomorphism may be determined in the X-ray laboratory, and the progress of structure determination is expected to be less dependent on the availability of synchrotron radiation.

3.1.3. Phase determination

Heavy-atom derivatization of an object as large as ribosomal particles requires the use of extremely dense and ultra-heavy compounds. Examples of such compounds are tetrakis(acetoxymercuri)methane (TAMM) which was the key heavy-atom derivative in the structure determination of nucleosomes and the membrane reaction center [33,34], and an undecagold cluster in which the gold core has a diameter of 8.2 Å (fig. 4 and refs. 35 and 36). Several variations of this cluster, modified with different ligands, have been prepared (S. Weinstein and W. Jahn, manuscript in preparation). The cluster compounds in which all the moieties (R in fig. 4) are amine or alcohol groups are soluble in the crystallization solution for 50 S subunits from *H. marismortui*. Thus, they could be used for soaking. Crystallographic data (to 18 Å resolution) show isomorphous unit-cell constants with observable differences in the intensities (fig. 3b).

Because the surfaces of ribosomal particles have a variety of potential binding sites for such clusters, in parallel to soaking experiments, attempts to bind covalently heavy atoms to a few specific sites on the ribosomal particles prior to crystallization are now in progress. This may be achieved either by direct interaction of a heavy-atom cluster with chemically active groups such as -SH or the ends of rRNA [37] on the intact particles or by the covalent attachment of a cluster to natural or tailor-made carriers which bind specifically to ribosomes.

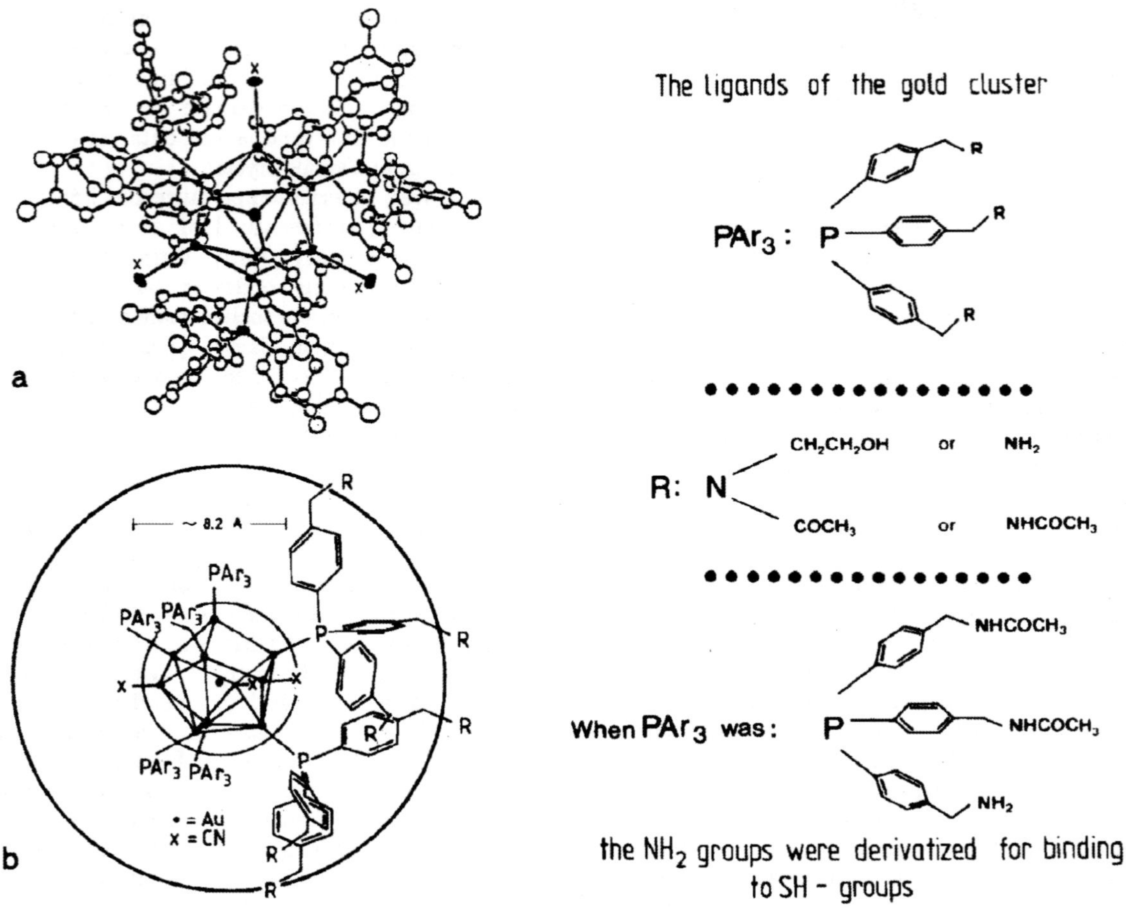

Fig. 4. (a) Postulated molecular structure of the gold cluster used by us based on the crystal structure of a similar cluster. (b) Semi-schematic representation of the gold cluster, depicting the gold core of 8.2 Å diameter and the arrangement of the ligands around it.

For direct binding to the surfaces of ribosomes, the following approaches were taken: Firstly, free sulfhydryls on the surface of the 50 S subunit were located by reaction with radioactive *N*-ethylmaleimide. The labeled proteins were identified by locating the radioactivity in two-dimensional gels. It was found that in the case of 50 S subunits from *B. stearothermophilus* there are mainly two proteins (BL11 and BL13) which bind *N*-ethylmaleimide. For *H. marismortui* a significant portion of the radioactivity was associated with a single protein. Secondly, the gold cluster described above was prepared in such a way that it could be bound to accessible -SH groups. Since this cluster is rather bulky, its accessibility was increased by the insertion of spacers, differing in length, into the cluster as well as into the free -SH groups on the ribosomal particles. Radioactive labeling of this cluster as well as neutron activation analysis of the gold element enabled us to determine the extent of binding of the cluster to the particles. The results of both analytical methods show that a spacer with a minimum length of about 10 Å between the -SH group of a ribosomal protein and the N atom on the cluster is needed for significant binding. Preliminary experiments indicate that the products of the derivatization reaction with 50 S particles can be crystallized.

As mentioned above, such clusters may also be bound to biochemical carriers. Examples of such carriers are antibiotics [38], DNA oligomers complementary to exposed single-stranded rRNA regions [39] and Fab fragments of antibodies specific to ribosomal proteins. Since most of the interactions of these compounds have been characterized biochemically, the crystallographic location of the heavy-atom compounds will not only be used for phase determination but also reveal the specific functional sites on the ribosome.

Alternatively, such clusters may be attached to selected sites on isolated ribosomal components which will subsequently be incorporated into particles lacking these particular components. Thus, a mutant of *B. stearothermophilus* which lacks protein BL11 was obtained by growing cells in the presence of thiostrepton at 60 °C (J. Schnier, H.S. Gewitz and B. Leighton, manuscript in preparation). The 50 S mutated ribosomal subunits yield sheets and crystals isomorphous with those of 50 S particles from the wild type [14]. This indicates that BL11, the missing protein, is not involved in crystal forces in the native crystals. Furthermore, protein BL11 has only one sulfhydryl group, and binding of *N*-ethylmaleimide to it does not reduce the biological activity and crystallizability of the particles.

Since protein BL11 is nearly globular [40], its location may be determined in a Patterson map with coefficients of $[F(\text{wild}) - F(\text{mutant})]$ and may serve, by itself, as a giant heavy-atom derivative. At preliminary stages of structure determination this approach may provide phase information and reveal the location of the missing protein.

Along these lines we have adopted a procedure for depleting the ribosomal subunits from several selected proteins [5]. The split proteins were, in turn, incorporated into the depleted core particles and the activity and crystallizability of the reconstituted particles were checked. Preliminary studies reconfirmed the results obtained with the mutant lacking protein BL11. Furthermore, there is a good correlation between the recovery of activity and the ability to crystallize. Thus, particles lacking protein BL12, which are inactive biologically, could not be crystallized, whereas the reconstructed particles produce crystals isomorphous with the native form.

3.2. Three-dimensional image reconstruction

The large size of ribosomal particles, which presents an obstacle to crystallographic studies, permits direct investigation by electron microscopy. Electron microscopy examination of crystalline ribosomal particles may show the location and orientation of particles within the crystals. A model obtained by three-dimensional image reconstruction of two-dimensional sheets may be used for gradual phasing of low-resolution crystallographic data. To this end, we have initiated three-dimensional image reconstruction studies.

Two-dimensional sheets from prokaryotic ribosomal subunits have firstly been grown in vitro from low-molecular-mass alcohols by vapor diffusion in hanging drops [18,42]. Recently, mixtures of salts and alcohols have been used for the rapid growth of better-ordered two-dimensional sheets in depression slides or on electron microscopy grids [19–22].

Well-ordered two-dimensional sheets from 70 S

Fig. 5. (a) Image of a two-dimensional sheet (×28000) of 70 S particles from *B. stearothermophilus* stained with gold-thioglucose, and an optical diffraction pattern from an area containing about 20×15 unit cells. (b) Computer graphic display of the outline of the reconstructed model of the 70 S ribosome at 47 Å resolution stained with gold-thioglucose. L and S indicate the 50 S and 30 S subunits, respectively. The arrows point to the interface between the two subunits (bar = 20 Å). (c) Computer graphic display of the outline of the reconstructed model of the 70 S ribosome at 42 Å resolution stained with uranyl acetate, in a similar orientation to that shown in panel b. UA denotes the regions to which uranyl acetate binds. L and S indicate the 50 S and 30 S subunits, respectively. The arrows point to the interface between the two subunits. Bar = 20 Å. (d) Outline of a 20 Å thick section in the middle of the reconstructed model of the 70 S ribosome. T indicates part of the tunnel. (e) Reconstructed model of the 50 S subunit at 30 Å resolution, obtained as described in fig. 6 and viewed in a projection which resembles models derived from electron microscopy studies of single particles.

ribosomes and from 50 S ribosomal subunits from *B. stearothermophilus* have been subjected to three-dimensional image reconstruction studies at 47 and 30 Å resolution, respectively [19,20]. In both cases the reconstructed particles have average dimensions similar to those determined using other physical methods [30]. On the basis of the known molecular mass of these particles and of the volume obtained from three-dimensional image reconstruction, the calculated densities are in good agreement with values tabulated [43] and calculated for crystals of the 50 S subunits from *H. marismortui* [17] and other nucleoproteins [44,45].

3.2.1. Two-dimensional sheets of 70 S ribosomes

Two-dimensional sheets of 70 S particles from *B. stearothermophilus* consist of dimers and are packed in relatively small unit cells (about 192 × 420 Å). Analysis of the models obtained by three-dimensional reconstruction of sheets stained with gold-thioglucose demonstrates the two ribosomal subunits. These are arranged around an empty space which is large enough to accommodate the components of protein biosynthesis, e.g., tRNAs and elongation factors. There is a similarity between the model of the small subunit obtained by visualization of single particles [30] and that revealed by our studies. However, isolated 30 S particles appear somewhat wider than those reconstructed within the 70 S particles. This may be a consequence of contact of the isolated particles with the flat electron microscope grid. In contrast, particles within the crystalline sheets are held together by crystalline forces. These form a network which may stabilize the conformation of the particle and diminish or even eliminate the influence of the flatness of the grids. The portion of the reconstructed 70 S particle which we assigned to the large subunit may be correlated to the image that we reconstructed from two-dimensional sheets of these subunits (fig. 5 and ref. 19).

Reconstruction from sheets of 70 S ribosomes stained with uranyl acetate led to a model which showed the features described above as well as regions where uranyl acetate, acting as a positive stain, was incorporated into the particle. This may indicate that in these regions the RNA which is the natural candidate for interacting with uranyl acetate is concentrated and exposed to the stain. Such regions are located on the body of the small subunit [46] and at the interface of the small and large subunits, in good agreement with other studies [47].

3.2.2. Two-dimensional sheets of 50 S ribosomal subunits

The two-dimensional sheets of the 50 S subunits consist of small unit cells, (around 140 × 330 Å). Optical diffraction patterns of electron micrographs of negatively stained specimens with gold-thioglucose extend to about 30 Å. Three-dimensional image reconstruction of these sheets yielded a model showing several projecting arms, two of which are longer than the rest. The arms are arranged radially around one edge near the presumed interface with the 30 S subunit. A narrow elongated cleft is formed between the projecting arms and turns into a tunnel of diameter up to 25 Å and 100–120 Å length (fig. 6). A similar feature was also detected on ribosomes from chick embryos by three-dimensional image reconstruction [47].

The functional significance of the tunnel remains to be determined. However, this tunnel, originating at the presumed site of actual protein biosynthesis and terminating on the other end of the particle, and being of a diameter large enough to accommodate a peptide of about 40 amino acids, appears to provide the path taken by the nascent polypeptide chain. Protection by the ribosome of newly formed peptides of about 40 amino acids from proteolytic enzymes was previously observed using biochemical methods [48–50]. It remains to be seen whether the tunnel terminates at a location compatible with that assigned by immunoelectron microscopy as the exit site for the growing polypeptide chain [51].

The tunnel is present in all reconstructions that we have performed, irrespective of the staining material (gold-thioglucose or uranyl acetate). In every reconstructed particle there is a region of low density which branches off the tunnel to form a Y (or V) shape and terminates on the opposite side of the particle (fig. 6). As yet, the exact nature of this region cannot be determined. It may be a loosely packed protein region, but in some recon-

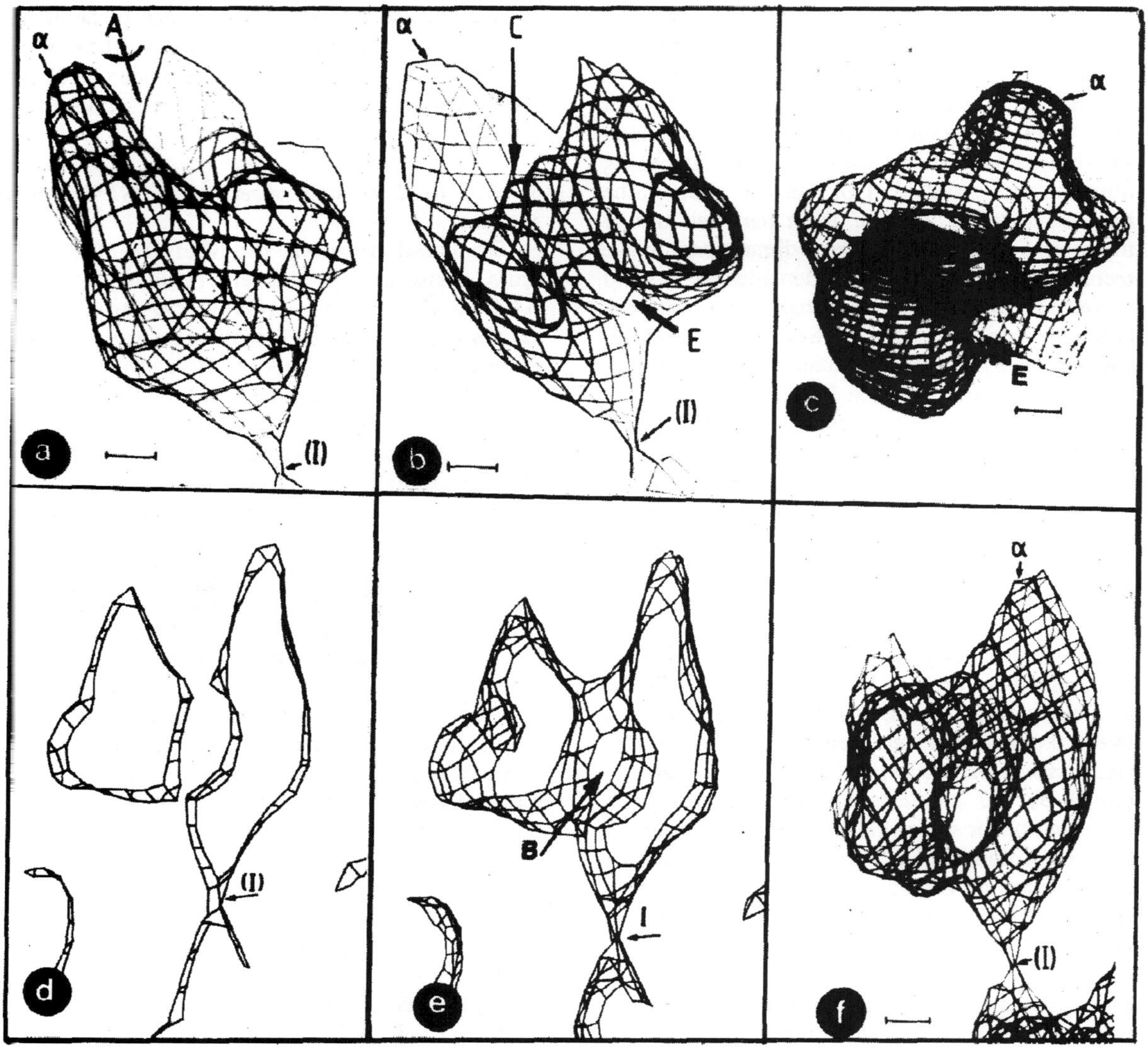

Fig. 6. Computer graphic display of the reconstructed model of the 50 S ribosomal subunit at 30 Å resolution. α denotes the longest arm; E indicates the exit site. (a) Side view of the model. The entire particle and part of a second one are shown. The arrow (I) points to the crystal contact between the two particles. A marks the approximate axis around which the model was turned to obtain the view shown in panel b (bar = 20 Å). (b) The model shown in panel a rotated about the axis A. C indicates the cleft between the projecting arms, at the site where it turns into the tunnel. (c) View into the tunnel from the cleft. (d) Outline of a 20 Å thick section in the middle of the reconstructed model, showing that the tunnel spans the entire particle. (e) Outline of a 40 Å thick section in the middle of the reconstructed model. The branching of the tunnel is seen (B). (f) Model viewed into the branch of the tunnel from the exit point.

structed models the density of this region is so low that it appears as a branch of the main tunnel.

Several models for the 50 S ribosomal subunits from *E. coli* have previously been suggested based on electron microscopical visualization, reconstruction or averaging of single particles [30]. Our model can be positioned in such a manner that its projected view resembles the usual image seen when single particles are investigated by electron microscopy (fig. 5). However, there are some discrepancies between our model and the others, i.e., ours does not have any flat surface. These, as in the case of 70 S particles, may stem from the basic differences between the subjective visualization of isolated particles in projection on the one hand, and the inherently more objective character of structure analysis by diffraction methods on the other.

In the reconstructed 70 S ribosomes from *B. stearothermophilus* there are some indications for the existence of a tunnel. Thus, portions of the tunnel could be detected in 20–40 Å thick sections through the particle. An example is shown in fig. 5. Since the 70 S particles used for these studies were harvested at the time of being active, it is feasible that nascent protein chains were attached to them. This may be the reason for the fact that the tunnel is only partially resolved.

4. Concluding remarks

We have demonstrated here that crystallographic studies may be carried out on intact ribosomal particles. We expect that these studies supported by electron microscopy and combined with available biophysical, biochemical and genetic knowledge will yield a reliable model for the ribosome and contribute to an understanding of the mechanism of protein biosynthesis at the molecular level.

Acknowledgements

We would like to thank Dr. H. Hope for introducing cryo-temperature crystallography, Drs M.A. Saper, K.S. Bartels, C. Kratky, G. Weber, Ms. P. Webster and Mr. K. von Buhlen for their efforts in data collection, Dr. K.R. Leonard and Mr. T. Arad for their contribution to the image reconstruction studies, Drs. S. Weinstein, W. Jahn and K. Amornajurasi for their work on the heavy-atom derivatization, Dr. F.L. Hirshfeld for his critical comments, Drs J. Sussman and B. Shaanan for assisting us with computing and display problems, Drs K. Wilson, H.D. Bartunik, J. Helliwell, M. Papiz, K. Moffat, W. Schildcamp, P. Pizackerley and E. Merrit for providing synchrotron radiation facilities, and H.S. Gewitz, J. Piefke, Y. Halfon, B. Romberg, G. Idan and H. Danz for technical assistance. This work was supported by BMFT (05 180 MP B0), NIH (GM 34360) and Minerva research grants.

References

1 A. Yonath, J. Müssig and H.G. Wittmann, J. Cell Biochem. 19 (1982) 145.
2 A. Yonath and H.G. Wittman, Methods Enzymol. (1988) in the press.
3 A. Yonath, F. Frolow, M. Shoham, J. Müssig, I. Makowski, C. Glotz, W. Jahn, S. Weinstein and H.G. Wittmann, J. Crystal Growth (1987) in the press.
4 H.G. Wittmann, J. Müssig, H.S. Gewitz, J. Piefke, H.J. Rheinberger and A. Yonath, FEBS Lett. 146 (1982) 217.
5 C. Glotz, J. Müssig, H.S. Gewitz, I. Makowski, T. Arad, A. Yonath and H.G. Wittmann, Biochem. Int. 15 (1987) 953.
6 A. Yonath, J. Müssig, B. Tesche, S. Lorenz, V.A. Erdmann and H.G. Wittmann, Biochem. Int. 1 (1980) 428.
7 A. Yonath, Trends Biochem. Sci. 9 (1984) 227.
8 A. Yonath, H.D. Bartunik, K.S. Bartels and H.G. Wittmann, J. Mol. Biol. 177 (1984) 201.
9 A. Yonath, M.A. Saper, I. Makowski, J. Müssig, J. Piefke, H.D. Bartunik, K.S. Bartels and H.G. Wittmann, J. Mol. Biol. 187 (1986) 633.
10 A. Yonath, M.A. Saper and H.G. Wittmann, in: Structure, function and genetics of ribosomes, eds. B. Hardesty and G. Kramer (Springer-Verlag, Heidelberg, 1986) p. 112.
11 A. Yonath, K.R. Leonard, S. Weinstein and H.G. Wittmann, Cold Spring Harbor Symp. (1987) in the press.
12 A. Yonath and H.G. Wittmann, Proceedings of the 5th Albany conversation, ed. M. Sarma (Adenine Press, Albany, NY, 1987) in the press.
13 H.S. Gewitz, C. Glotz, J. Müssig, J. Piefke, A. Yonath and H.G. Wittmann, Biochimie (1987) in the press.
14 A. Yonath, M.A. Saper, F. Frolow, I. Makowski and H.G. Wittmann, J. Mol. Biol. 192 (1986) 161.
15 A. Shevack, A., H.S. Gewitz, B. Hennemann, A. Yonath and H.G. Wittmann, FEBS Lett. 184 (1985) 68.
16 M. Shoham, J. Müssig, A. Shevack, T. Arad, H.G. Wittmann and A. Yonath, FEBS Lett. 208 (1986) 321.
17 I. Makowski, F. Frolow, M.A. Saper, H.G. Wittmann and A. Yonath, J. Mol. Biol. 193 (1987) 819.
18 T. Arad, K.R. Leonard, H.G. Wittmann and A. Yonath, EMBO J.3. (1984) 127.
19 A. Yonath, K.R. Leonard and H.G. Wittmann, Science 236 (1987) 813.
20 T. Arad, J. Piefke, H.S. Gewitz, B. Romberg, C. Glotz, J. Müssig, A. Yonath and H.G. Wittmann, Anal. Biochem. (1987) in the press.

21 J. Piefke, T. Arad, I. Makowski, H.S. Gewitz, B. Hennemann, A. Yonath and H.G. Wittmann, FEBS Lett. 209 (1986) 104.
22 T. Arad, J. Piefke, S. Weinstein, H.S. Gewitz, A. Yonath and H.G. Wittmann, Biochimie (1987) in the press.
23 P.A. Bartlett, B. Bauer and S.J. Singer, J. Am. Chem. Soc. 100 (1978) 5085.
24 A. Klug, K.C. Holmes and J.T. Finch, J. Mol. Biol. 3 (1961) 87.
25 G. Zubay and M.H.F. Wilkins, J. Mol. Biol. 2 (1960) 105.
26 R. Langridge and K.C. Holmes, J. Mol. Biol. 5 (1962) 611.
27 J.T. Finch, L.C. Lutter, D. Rhodes, R.S. Brown, B. Rushton, M. Levitt and A. Klug, Nature 269 (1977) 29.
28 G. Chambliss, G.R. Craven, J. Davies, L. Kahan and M. Nomura, Ribosomes: structure, function and genetics (University Park Press, Baltimore, 1980).
29 H.G. Wittmann, Annu. Rev. Biochem. 51 (1982) 155.
30 H.G. Wittmann, Annu. Rev. Biochem. 52 (1983) 35.
31 B. Hardesty and G. Kramer, Structure, function and genetics of ribosomes (Springer-Verlag, Heidelberg, 1986).
32 M.G. Rossmann and W. Erickson, J. Appl. Crystallogr. 16 (1983) 629.
33 J. Deisenhofer, O. Epp, K. Mikki, R. Huber and H. Michel, J. Mol. Biol. 180 (1984) 385.
34 T. Richmond, J.T. Finch, B. Rushton, D. Rhodes and A. Klug, Nature 311, (1984) 533.
35 P. Bellon, P.M. Manassero and M. Sansoni, J. Chem. Soc. Dalton. Trans. (1972) 1481.
36 J.S. Wall, J.F. Hainfeld, P.A. Barlett and S.J. Singer, Ultramicroscopy 8 (1982) 397.
37 O.W. Odom, Jr, D.R. Robbins, J. Lynch, D. Dottavio-Martin, G. Kramer and B. Hardesty, Biochemistry 19 (1980) 5941.
38 K.H. Nierhaus and H.G. Wittmann, Naturwissenschaften 67 (1980) 234.
39 W.E. Hill, B.E. Trappich and B. Tassanakajohn, in: Structure, function and genetics of ribosomes, eds. B. Hardesty and G. Kramer (Springer-Verlag, Heidelberg, 1986) p. 233.
40 L. Giri, W.E. Hill, H.G. Wittmann and B. Wittmann-Liebold, Adv. Protein Chem. 36 (1984) 1.
41 H.-S. Gewitz, C. Glotz, P. Goischke, B. Romberg, A. Yonath and H.G. Wittmann, Biochem. Int. 15 (1987) 1887.
42 M.W. Clark, K.R. Leonard and J.A. Lake, Science 216 (1982) 999.
43 B.W. Matthews, J. Mol. Biol. 33 (1968) 491.
44 J.M. Hogle, J. Mol. Biol. 160 (1982) 663.
45 L. Liljas, T. Unge, A. Jones, K. Fridborg, S. Lovgren, U. Skoglund and B. Strandberg, J. Mol. Biol. 159 (1982) 93.
46 R. Brimacombe, J. Atmadja, W. Stiege and D. Schüler, J. Mol. Biol. 199 (1988) 115.
47 R.A. Milligan and P.N.T. Unwin, Nature 319 (1986) 693.
48 L.I. Malkin and A. Rich, J. Mol. Biol. 26 (1967) 329.
49 G. Blobel and D.D. Sabatini, J. Cell Biol. 45 (1970) 130.
50 W.P. Smith, P.C. Tai and B.D. Davis, Proc. Natl. Acad. Sci. U.S.A. 75 (1978) 5922.
51 C. Bernabeau and J.A. Lake, Proc. Natl. Acad. Sci. U.S.A. 79 (1982) 3111.

Recognition of Cognate Transfer RNA by the 30*S* Ribosomal Subunit

James M. Ogle, Ditlev E. Brodersen, William M. Clemons Jr., Michael J. Tarry, Andrew P. Carter, V. Ramakrishnan*

Crystal structures of the 30*S* ribosomal subunit in complex with messenger RNA and cognate transfer RNA in the A site, both in the presence and absence of the antibiotic paromomycin, have been solved at between 3.1 and 3.3 angstroms resolution. Cognate transfer RNA (tRNA) binding induces global domain movements of the 30*S* subunit and changes in the conformation of the universally conserved and essential bases A1492, A1493, and G530 of 16*S* RNA. These bases interact intimately with the minor groove of the first two base pairs between the codon and anticodon, thus sensing Watson-Crick base-pairing geometry and discriminating against near-cognate tRNA. The third, or "wobble," position of the codon is free to accommodate certain noncanonical base pairs. By partially inducing these structural changes, paromomycin facilitates binding of near-cognate tRNAs.

During protein synthesis, the ribosome catalyzes the sequential addition of amino acids to a growing polypeptide chain, using mRNA as a template and aminoacylated tRNAs (aa-tRNAs) as substrates. Correct base pairing between the three bases of the codon on mRNA and those of the anticodon of the cognate aa-tRNA dictates the sequence of the polypeptide chain. Discrimination against noncognate tRNA, which generally has two or three mismatches in the base pairing, can be accounted for by the difference in the free energy of base pairing to the codon compared with cognate tRNA. For near-cognate tRNA, which usually involves a single mismatch, the free-energy difference in base pairing compared with cognate tRNA would predict an error rate that is one to two orders of magnitude higher than the actual error rate of protein synthesis (*1*), and it has long been recognized that the ribosome must improve the accuracy of protein synthesis by discriminating against near-cognate tRNAs (*2*). This discrimination involves the 30*S* subunit, which binds mRNA and the anticodon stem-loop (ASL) of tRNA.

At the beginning of the elongation cycle, which involves the addition of a new amino acid to a growing polypeptide chain, the aa-tRNA is presented to the ribosome as a ternary complex with elongation factor Tu (EF-Tu) and guanosine triphosphate (GTP). The selection of cognate tRNA is believed to occur in two stages—an initial recognition step and a proofreading step—that are separated by the irreversible hydrolysis of GTP by EF-Tu (*3–6*). In this scheme, the discrimination energy inherent in codon-anticodon base pairing is exploited twice to achieve the necessary accuracy. Recent experiments suggest that the binding of cognate rather than near-cognate tRNA results in higher rates of both GTP hydrolysis by EF-Tu, and accommodation, a process in which the acceptor arm of the aa-tRNA swings into the peptidyl transferase site after its release from EF-Tu (*7*, *8*). In both steps, the higher rate is proposed to be the result of structural changes in the ribosome induced by cognate tRNA. In the context of proofreading mechanisms alone, it is unclear whether additional structural discrimination by the ribosome, over that inherent in the energetics of codon-anticodon base pairing, is actually required in decoding. In an alternative view, it was proposed that the ribosome contributes to the specificity of decoding by recognizing the geometry of codon-anticodon base pairing (*9*, *10*), probably by direct recognition of the sugar phosphate backbone (*11*). This is consistent with evidence that the ribosome can discriminate between cognate and near-cognate tRNA with very high accuracy in a single step under near-equilibrium conditions (*12*), so that the recognition and proofreading steps may be needed not only for accuracy as proposed earlier, but for the combination of high speed and accuracy required in protein synthesis.

The heart of decoding involves recognition of Watson-Crick base pairing between the codon on mRNA and the anticodon found in positions 34 to 36 of the ASL of tRNA. This base pairing results in a short double helix that we call the codon-anticodon helix. Data from chemical protection (*13*), cross-linking (*14*), and genetic analysis (*15*) suggest that the decoding site is a region around the ribosomal A site that includes helix 44, the 530 loop, and helix 34 of 16*S* RNA, which are now known from the high-resolution structure of the 30*S* subunit to be close to one another (*16*, *17*).

Recent work on decoding has focused on an internal loop of helix 44 containing two universally conserved bases, A1492 and A1493. These bases are chemically protected at their N1 positions when RNA is bound in the A site of the 30*S* subunit (*13*), and were shown to be essential for viability in *Escherichia coli* and for A-site tRNA binding (*18*). This loop is also the binding site of the antibiotic paromomycin that reduces the fidelity of decoding (*19–21*). Nuclear magnetic resonance (NMR) structures of a fragment of RNA containing this internal loop showed that these bases were displaced into the minor groove of helix 44 in the presence of paromomycin (*20*, *22*). A direct interaction of the N1 of these bases with the 2′ OH of mRNA was inferred on the basis of biochemical experiments (*18*). Based on these data, it was suggested that the changes induced by the antibiotic mimic those induced by cognate tRNA, that the bases would interact via their minor-groove side with the minor groove of the codon-anticodon helix, and that they would sense the shape of the codon-anticodon helix during decoding (*18*, *20*, *22*). However, in the absence of a structure in the context of the whole 30*S* subunit, it was also possible to use existing data to propose an alternative and quite different model of the decoding site that involved interaction of the codon-anticodon helix with the major rather than minor groove of helix 44 (*23*).

A 7.8 Å resolution structure of the 70*S* ribosome in complex with mRNA and tRNA (*24*) showed that at that resolution, the conformation of the internal loop in helix 44 was similar to that in the paromomycin-bound NMR structure of the isolated fragment (*20*). However, A1492 and A1493 in this internal loop appeared too far from the codon-anticodon helix to interact with it directly (*24*), so their role and the general structural basis of decoding by the ribosome remained unclear.

More recently, the crystal structure of the 30*S* subunit complexed with paromomycin, streptomycin, and spectinomycin (*21*) showed that A1492 and A1493 are completely flipped out from their stacked position inside the internal loop rather than just displaced into the minor groove, as previously proposed (*20*). The displaced bases point into the A site in a position that would indeed allow direct monitoring of the minor groove of the codon-anticodon helix. On the basis of this structure, we suggested that during decoding, these bases would flip out even in the absence of paromomycin, in a way that would allow discrimination of cognate from near-cognate tRNA (*21*). However, this conformation would place the N1 of the adenines pointing toward tRNA, and hence unable to make the interactions with mRNA previously reported (*18*).

In this work, we directly examine the interactions of the 30*S* subunit with mRNA and an ASL of cognate tRNA bound in the A site by

MRC Laboratory of Molecular Biology, Hills Road, Cambridge CB2 2QH, UK.

*To whom correspondence should be addressed. E-mail: ramak@mrc-lmb.cam.ac.uk

x-ray crystallography.

Structure determination. Recently, we showed that the A site in our crystal form is accessible by diffusion to macromolecules as large as initiation factor IF1 (*25*). On the basis of this result, a 17-nucleotide fragment corresponding to the ASL of $tRNA^{Phe}$ with a GAA anticodon, along with a U_6 hexanucleotide that would place a codon for phenylalanine in the A site, were soaked into crystals of the 30*S* subunit (*26*). Diffraction data to 3.1 to 3.3 Å resolution were collected from such crystals both in the absence and presence of paromomycin, as well as from crystals of the 30*S* subunit soaked in paromomycin alone (Table 1). Initial refinement was done with the native 30*S* subunit structure (Protein Data Bank entry 1FJF) (*17*) as a starting model, with the occupancies of the atoms for parts of the structure in the vicinity of A-site tRNA initially set to zero to avoid possible bias (*26*). Clear density for the A-site codon and tRNA ASL, and elements of the 30*S* subunit that had changed conformation such as A1492, A1493, and G530 was visible in difference Fourier electron density maps (Fig. 1A) for the complexes both in the absence and presence of paromomycin. These elements were subsequently built into the maps for each of the individual structures, and a final refinement was done with all atoms visible in the electron density, including metal ions. Statistics for the refinement are shown in Table 1. A σ_A-weighted $2mF_o - DF_c$ map after refinement is shown in Fig. 1B.

Overview of mRNA and tRNA in the A site. The structure reveals details of mRNA and tRNA binding to the 30*S* subunit (Fig. 2). In our native 30*S* subunit structure, the 3′ end of 16*S* RNA is folded back into the message binding cleft of the 30*S* subunit, and thus mimics mRNA in the P and E sites (*21*). The P site is also occupied by a stem-loop that mimics tRNA and comes from helix 6 of a neighboring molecule in the crystal. These features are also seen in all the complexes described below. Additionally, in the structures with the ASL and U_6 hexanucleotide, we can clearly see the three nucleotides from the A-site codon, and slightly less well-defined density for a fourth nucleotide 3′ of the A site. The electron density for the mRNA between the P and A sites appears continuous despite the presence of a covalent break in the chain between the P- and A-site codons. As reported earlier at lower resolution (*24*), there is a distinct kink in the path of the mRNA between the A- and P-site codons. In addition, we can now see in detail how the ASL of cognate tRNA interacts with the A-site codon and the 30*S* subunit. The structure of the ASL, which does not contain any modified nucleosides, is very similar to that of isolated yeast phenylalanine tRNA (*27*, *28*) or of *Thermus* phenylalanine tRNA complexed with aminoacyl synthetase (*29*). The contacts of the ASL to the 530 loop, helix 34, and helix 44 of the ribosome arise solely from the anticodon loop.

Recognition of cognate tRNA in the ribosome by an induced fit. The binding of mRNA and cognate tRNA in the A site induces A1492 and A1493 to flip out of the internal loop of helix 44. This binding also causes the universally conserved base G530, which has been footprinted by A-site tRNA (*13*), to switch from the *syn* conformation present in the native structure to an *anti* conformation. In their new conformations, A1493 and A1492 interact, respectively, with the first and second base pairs of the codon-anticodon helix, whereas G530 interacts with both the second position of the anticodon and the third position of the codon. The result of these induced changes is that the first two base pairs of the codon-anticodon helix are closely monitored by the ribosome in a way that would be able to discriminate between Watson-Crick base pairing and mismatches, whereas the environment of the third, or "wobble," position appears to be suited for accommodating other base-pairing geometries (*30*, *31*).

A1493 packs tightly in the minor groove of the first base pair of the codon-anticodon helix, so that there are hydrogen bonds between the 2′ OH of the A1493 and the 2′ OH of the first nucleotide of the codon, as well as between N1 of A1493 and the 2′ OH of position 36 of the anticodon (Fig. 3a). Additionally, a hydrogen bond is evident from the 2′ OH of A1493 to the O2 of the uracil in the codon. Although this bond involves the base, it is not sequence specific because the O2 of pyrimidines and the N3 of purines are hydrogen bond acceptors that occupy equivalent positions in the minor groove of Watson-Crick base pairs (*32*). In the case of a GC or CG base pair at the first codon position, it would be possible to form an additional bond from the N3 of the A1493 to the N2 of the guanine.

A1492 and G530 together bind in and span the minor groove of the second base pair of the codon-anticodon helix (Fig. 3B), and their orientation relative to one another is stabilized in part by a hydrogen bond between their N1 positions. They make equivalent interactions with each strand of the codon-anticodon helix in that the 2′ OH of the codon hydrogen bonds to the N3 and the 2′ OH of A1492, whereas the 2′ OH of tRNA hydrogen bonds to the N3 and 2′ OH of G530. In addition, the N6 of A1492 is involved in hydrogen bonding to the 2′ OH of the universally conserved C518, as well as to the highly conserved Ser^{50} (*E. coli* Ser^{46}) of the nearby protein S12. The loop of S12 containing this serine (Pro^{45} to Ser^{50}; *E. coli* residues 41 to 46) is the site of mutations that affect the fidelity of translation (*33*). As with the first base pair, the close fit of van der Waals surfaces and various interactions in the minor groove would be sensitive to whether the second base pair had Watson-Crick geometry or not.

Unlike the first two base pairs, the geometry of the third base pair between the codon and anticodon, which in our structure is a GU base pair, is not as closely monitored by the ribosome (Fig. 3C). The 2′ OH of the codon makes a hydrogen bond with the O6 of G530, as well as an indirect metal-mediated interaction with the O2 of C518 and the main-chain carbonyl of Pro^{48} (*E. coli* Pro^{44}) of protein S12. In addition,

Table 1. Summary of crystallographic data.

Data set	U6/ASL	U6/ASL/ paromomycin	Paromomycin
	Data collection		
Beamline	SBC 19ID	SBC 19ID	SBC 19ID
Resolution limit (Å)	3.31	3.11	3.31
Unit cell			
a,b (Å)	401.57	400.80	401.16
c (Å)	176.11	175.88	176.36
No. of observations	968,646	1,170,885	1,079,968
No. of unique reflections	206,478	240,896	207,340
R_{sym}* (%)	12.2 (65.4)	11.7 (57.8)	16.0 (66.4)
Completeness* (%)	97.5 (93.6)	95.1 (92.7)	97.8 (98.0)
$\langle I \rangle / \langle \sigma_I \rangle$*	10.9 (2.6)	11.9 (2.8)	10.8 (2.7)
	Refinement		
Resolution range (Å)	316.23–3.31	316.23–3.11	316.23–3.31
Reflections excluded for cross-validation	5%	5%	5%
Number of nonhydrogen atoms	52,109	52,250	51,880
Proteins	19,178	19,178	19,237
RNA	32,809	32,904	32,486
Metals	122	126	115
R factor† (%)	23.1 (28.6)	23.2 (27.5)	22.9 (28.2)
σ_A coordinate error‡	0.56 (0.64)	0.48 (0.51)	0.53 (0.61)
Deviations from ideality			
rms deviations in bond lengths (Å)	0.007	0.007	0.007
rms deviations in bond angles (°)	1.23	1.21	1.23

*Values for the outermost resolution shell are given in parentheses. †Values in parentheses represent the free *R* factor. ‡Values in parentheses represent the cross-validated σ_A coordinate error.

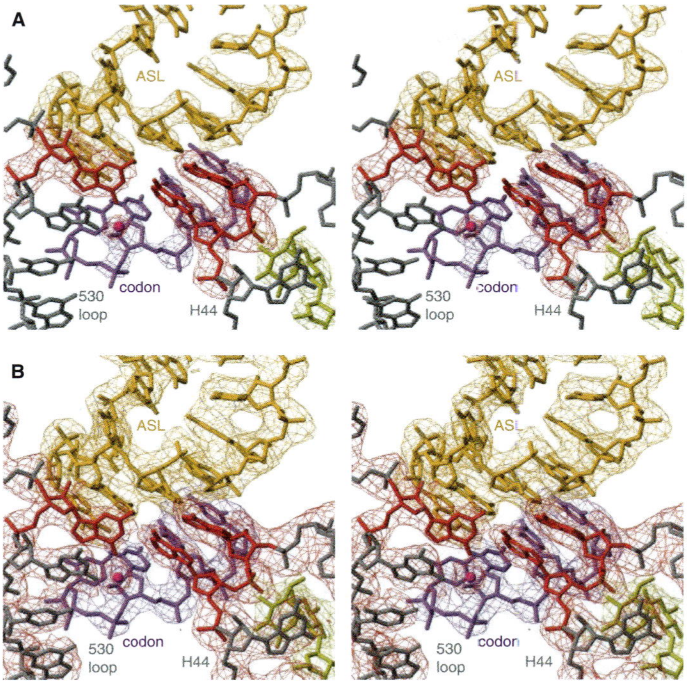

Fig. 1. Stereo views of the electron density for the 30*S* ribosomal subunit complex with mRNA and tRNA-ASL in the presence of paromomycin, around the site of the codon-anticodon interaction. Portions of the tRNA-ASL (gold), codon from a U_6 hexanucleotide (purple), and 530 loop and helix 44 (gray), along with the bases G530 (red, left), A1492 and A1493 (red, right), and a Mg^{2+} ion (magenta sphere), are shown. Part of paromomycin is in yellow in the lower right. (**A**) $mF_o - DF_c$ difference density based on a simulated annealing run, contoured at 3σ. The tRNA-ASL, U_6 hexanucleotide, paromomycin, and nucleotides 530 and 1492/3 were omitted from the refinement, and the density pertaining to each of these elements has been colored accordingly for clarity. (**B**) Refined $2mF_o - DF_c$ electron density contoured at 1.5σ. Figures 1, 2, and 4 were produced with O (*51*) and MOLMOL (*52*) and rendered with POV-Ray.

the pyrimidine O2 of the anticodon and the O6 of G530 are in close contact. The ribose of the anticodon packs against C1054 from helix 34 in the head of the ribosome. These interactions leave the minor groove of this base pair, especially on the anticodon side, relatively open as compared with the first two positions. Because noncanonical base pairs alter the shape and possible interactions in the minor groove, the structure shows why some of these are allowed in the wobble position. Also, the mRNA nucleotide in this position makes more interactions with the ribosome than does the tRNA, confirming predictions that its conformation should be the more restricted of the two (*31*).

The geometry of the first two base pairs of the codon-anticodon helix is very similar in the presence and absence of paromomycin, with a slightly less planar base pair in the first position in the presence of paromomycin. In the absence of paromomycin, the geometry for the third position deviates from that expected for a wobble pair (Fig. 3C), but the expected wobble geometry is seen in the higher resolution structure in the presence of paromomycin (Fig. 3D). Because we see occasional departures from ideality in other parts of the 30*S* subunit structure, higher resolution will be needed to make definitive statements about deviations from standard geometry.

The structure is consistent with most chemical protection (*13*) and genetic data (*15*) on A-site tRNA binding. The finding that A1492 and A1493 are required for viability (*18*) is rationalized by the observation that the smaller pyrimidines could not reach far enough to make interactions with tRNA or G530, and that guanine would not only have a steric clash at its N2 position, but would also not be able to hydrogen bond through its N1 in the same way. The

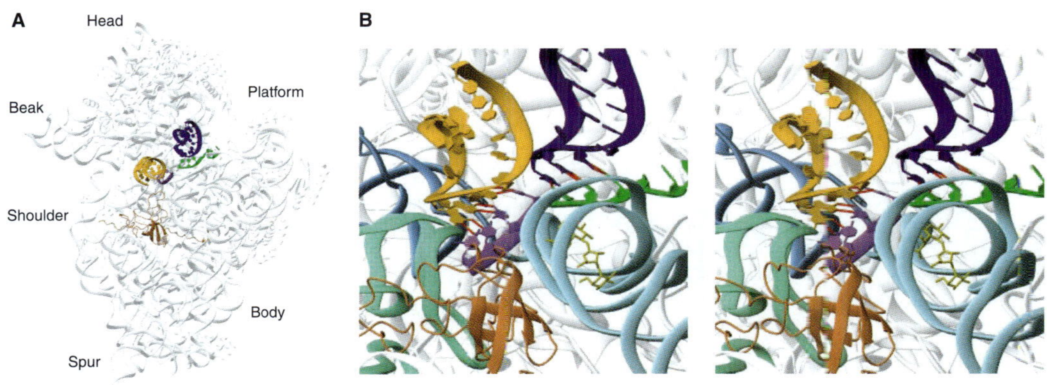

Fig. 2. Complex of the 30*S* subunit with mRNA from a U_6 hexanucleotide and a cognate tRNA-ASL. (**A**) Overview of the complex. The 50*S* interface side of the 30*S* subunit is facing the reader, and important elements have been given standard colors that are used throughout the figures, namely, ASL at the A site (gold), codon from the U_6 hexanucleotide at the A site (purple), 3′ end of 16*S* RNA that mimics mRNA at the P site (green), P site tRNA mimic introduced by helix 6 from a neighboring molecule (dark blue), and protein S12 (tan). (**B**) Stereo view showing details of the A and P sites, colored as in (A), with, in addition, helix 44 (cyan, right), helix 34 (blue, left), 530 loop (green, left), and paromomycin (yellow sticks, within helix 44). The hydrogen bonds responsible for the codon-anticodon interaction at both the A and P sites are shown as red lines.

requirement for guanine at position 530 (*34*) can be similarly rationalized, because its N1 is required to hydrogen bond with the N1 of A1492, and its O6 is required to hydrogen bond with the 2′ OH of the uridine at the third position of the codon. Although A1492 and A1493 flip out of helix 44 and monitor the minor groove of the codon-anticodon helix roughly as we had proposed, the details of the interactions are different from those in the two motifs found in the structure of 16*S* RNA that we had suggested as possible ways for adenines to recognize the codon-anticodon helix (*21*). Also, the N1 of these adenines interacts with the 2′ OH of tRNA or with the N1 of G530, rather than with the 2′ OH groups of mRNA, as proposed previously (*18*). Finally, the structure does not support a model in which the N6 of these adenines forms a hydrogen bond with the O2 of pyrimidines or N3 of purines on mRNA (*23*).

Both the structural interactions and the general strategy in decoding have similarities with other systems. The interactions that A1493 and A1492 make with the minor groove of the first two base pairs of the codon-anticodon helix are identical to tertiary interactions that occur in both the group I intron (*35*) and the structure of 23*S* RNA (*36*), and are classified as consecutive "Type I" and "Type II" A-minor motif interactions (*37*). These interactions always involve conserved adenines, and a major part of the stabilization energy arises from the complementarity of the adenines with the surface of the minor groove (*38*). Similar tertiary interactions involving tight packing of conserved adenines in minor grooves of adjacent helices have also been reported in the 16*S* RNA structure (*17*, *21*).

The strategy adopted by the ribosome to recognize correct base pairing has similarities and differences with other template-directed enzymes that have high fidelity such as DNA and RNA polymerases. These enzymes also make close interactions with the minor groove of the base pair to be formed. In DNA polymerases, conserved arginines or glutamines form hydrogen bonds to the O2 of pyrimidines or N3 of purines (*39*, *40*). In our structures, A1493 does make a similar hydrogen bond with the O2 of uracil in monitoring the first codon base. However, the ribosome mainly forms bonds with the ribose 2′ OH groups on both strands of the codon-anticodon helix, which are not available to DNA polymerases. Interestingly, T7 RNA polymerase also forms hydrogen bonds to 2′ OH moieties of the incoming ribonucleotide and the RNA transcript (*41*). It is clear that the ribosome, like polymerases, senses the shape and available interactions at the minor groove to ensure fidelity.

Effect of paromomycin. The effects of tRNA and paromomycin on the 30*S* subunit are shown schematically in Fig. 4. In the native structure (*17*), A1492 and A1493 are stacked in the interior of helix 44 (Fig. 4A), and G530 is stacked in the 530 loop in the *syn* conformation. The binding of paromomycin flips out A1492 and A1493 (Fig. 4B). The structure of the decoding site in the presence of paromomycin alone is essentially identical to that determined previously in the presence of paromomycin, streptomycin, and spectinomycin (*21*), showing that the changes induced in this region are solely due to the binding of paromomycin. However, in the absence of tRNA, paromomycin does not have an effect on G530, which remains in the *syn* conformation. The binding of A-site tRNA flips out G530 into the *anti* conformation (Fig. 4C); it also flips out A1492 and A1493, but they are not in exactly the same conformation as with paromomycin alone (compare Fig. 4, B and C). In the presence of cognate A-site tRNA, the additional presence of

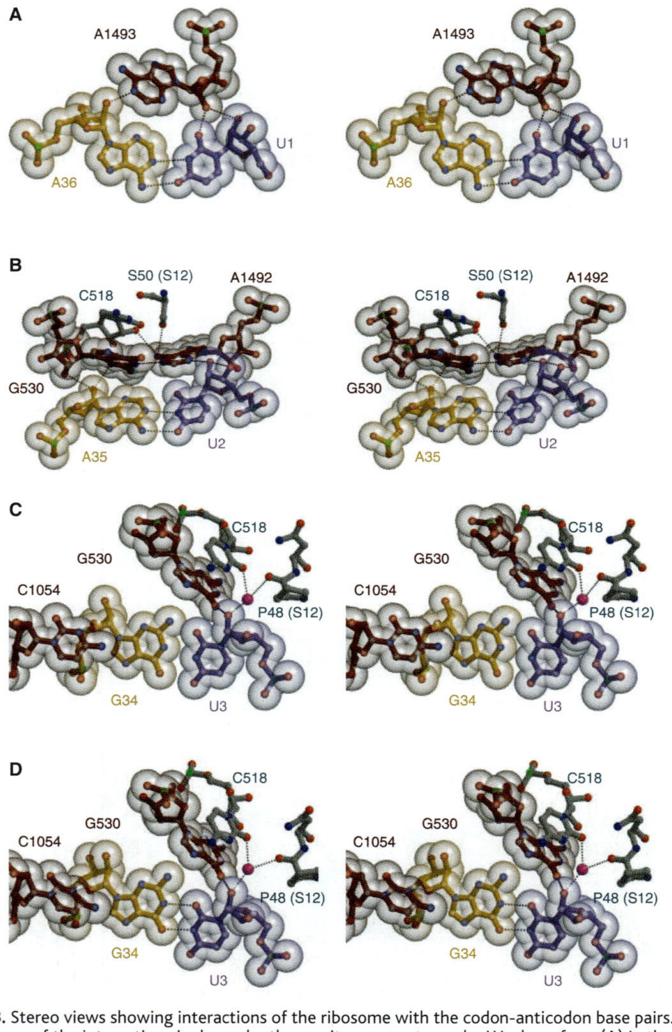

Fig. 3. Stereo views showing interactions of the ribosome with the codon-anticodon base pairs. The tightness of the interactions is shown by the semitransparent van der Waals surface. **(A)** In the first position, A1493 binds in the minor groove of the A36-U1 base pair. **(B)** In the second position, G530 and A1492 (both brown) act in concert to monitor the A35-U2 base pair. **(C)** The third (wobble) position, showing the G34-U3 base pair. C1054 stacks against G36 of the ASL. U3 interacts with G530, and indirectly through a Mg^{2+} ion (magenta) with C518 and residue Pro^{48} (*E. coli* Pro^{44}) from protein S12 (gray). The base pair seems closer to Watson-Crick geometry. **(D)** The third position in the presence of paromomycin, with the expected GU wobble pair. The interactions with the ribosome are similar to those in (C).

RESEARCH ARTICLES

paromomycin (Fig. 4D) does not result in further changes in the conformation of these bases (compare Fig. 4, C and D).

The structures provide firm evidence that the changes induced in the 30*S* subunit by cognate tRNA are energetically favorable because various compensating interactions can be made by the ribosome with the codon-anticodon helix, but only when the first two base pairs are canonical Watson-Crick base pairs. In the case of near-cognate tRNA, fewer of these interactions would be possible, so the induced changes would normally be less favorable. By paying part of the energetic cost of these induced changes, especially in displacing A1492 and A1493 from their internal loop in helix 44, paromomycin facilitates the binding of near-cognate tRNAs, along with the full changes induced as a result. This model is consistent with the finding that paromomycin accelerates precisely those steps (GTP hydrolysis and accommodation) for near-cognate tRNA that are normally slower as compared with cognate tRNA (*8*) and agrees with earlier proposals that paromomycin works by inducing structural changes in the decoding site (*21*, *22*), rather than by making direct interactions with mRNA or tRNA (*23*). The structure also explains the classical result that single-stranded DNA is not normally a good template for protein synthesis, but becomes so in the presence of the antibiotics neomycin and kanamycin, which are closely related to paromomycin (*42*). Because the 2′ OH groups are not present on DNA, the induced structural changes on cognate tRNA binding are not normally favorable because fewer interactions are possible. However, just as with normal translation, aminoglycosides can pay for part of the energetic cost of the induced changes and stabilize tRNA binding.

Other conformational changes during decoding. During accommodation, the A-site tRNA has to rotate significantly from its initial orientation as part of the ternary complex with EF-Tu to its final position in the A site, where its acceptor end is accessible to peptidyl transferase (*43*, *44*). From an examination of the A-site tRNA in 70*S* ribosome structures (*24*, *45*, *46*), it appears that the position of the ASL in our structure is close to the tRNA orientation after accommodation. Because most of the selectivity of cognate over near-cognate tRNA appears to come from the accommodation step (*7*), the structure we see is likely to be relevant for understanding the specificity of the ribosome for cognate tRNA. A high-resolution structure of the EF-Tu ternary complex bound to the ribosome will be required to understand details of the initial recognition step. However, it is likely that many of the changes induced by cognate tRNA are initiated during the binding of the ternary complex, and that the resulting interactions of the ribosome with the codon-anticodon helix persist during the subsequent movement of the tRNA.

A key question that remains to be answered is how the changes around the decoding site induced by cognate tRNA are transmitted to EF-Tu to trigger GTP hydrolysis. The binding of cognate tRNA results in a movement of the head and shoulder of the 30*S* subunit toward each other so that they close around the A-site tRNA. These changes are even greater when paromomycin is bound in addition to cognate tRNA, with as much as a 4 Å relative movement between the head and shoulder [see supplementary information (*47*)]. As with the binding of IF1 (*25*), it is likely that we do not see the full extent of conformational changes because of the constraints of the crystal lattice. It is possible that these changes reflect the types of changes in the 30*S* subunit that occur on the binding of the EF-Tu ternary complex, which may be transmitted to the 50*S* subunit and be part of the trigger for GTP hydrolysis by EF-Tu. Another possibility, suggested on the basis of experiments in which the acceptor arm and the ASL of cognate tRNA were present as two separate fragments, is that the signal to EF-Tu is transmitted through tRNA itself (*48*).

Conclusions. This work shows that the ribosome recognizes the geometry of codon-anticodon base pairing in a way that would discriminate against near-cognate tRNAs. The minor groove of the first and second base pairs between the codon and anticodon is closely monitored by a set of interactions that are induced by the binding of cognate tRNA. These interactions would be disrupted by mismatches, so that the induced structural changes would no longer be energetically favorable. The third or "wobble" position has less stringent constraints, and therefore can allow a broader range of base-pairing geometries, consistent with the requirements of the genetic code. The binding of paromomycin partially induces the changes that normally are induced by cognate tRNA. By

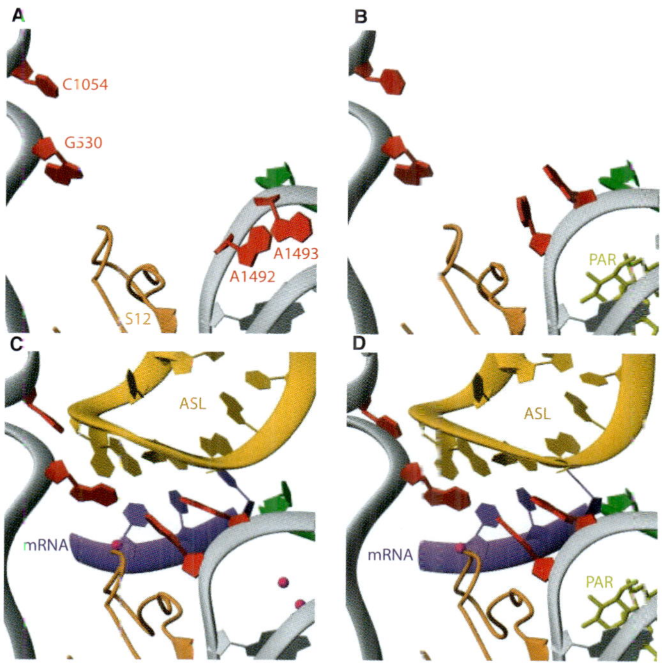

Fig. 4. Discrete states of the A site of the 30*S* subunit, as deduced from four different crystal structures. In all panels, the tRNA ASL (gold), A-site mRNA codon (purple), P-site mRNA (green), protein S12 (brown), and important bases (red) involved in conformational changes are shown. A few elements of the 16*S* RNA such as helix 44 (lower right), 530 loop (mid and lower left), and helix 34 (upper left) are also shown. **(A)** The native 30*S* subunit (*17*). A1492 and A1493 (red, right) are stacked in the interior of H44, and G530 is in the *syn* conformation. C1054 is shown in the upper left corner. **(B)** In the presence of paromomycin, the bases of A1492 and A1493 have been displaced into the A site by the binding of the antibiotic inside helix 44 (yellow, lower right). **(C)** When the codon and cognate tRNA-ASL bind in the A site, A1492 and A1493 flip out to monitor the codon-anticodon interaction, and G530 has switched to the *anti* conformation to interact with A1492, the anticodon in the second position, and the codon in the third. Two Mg^{2+} ions (magenta spheres) are visible near the region vacated by A1492 and A1493 in the interior of helix 44, and one is located close to the ribose of the codon in the wobble position (see Fig. 3, C and D). **(D)** When paromomycin is present in addition to tRNA-ASL and codon, the structure is very similar to that in (C).

paying part of the energetic cost of these induced changes, paromomycin facilitates the increased incorporation of amino acids from near-cognate tRNAs.

References and Notes

1. Reviewed in M. Ibba, D. Söll, *Science* **286**, 1893 (1999).
2. J. Davies, W. Gilbert, L. Gorini, *Proc. Natl. Acad. Sci. U.S.A.* **51**, 883 (1964).
3. J. J. Hopfield, *Proc. Natl. Acad. Sci. U.S.A.* **71**, 4135 (1974).
4. J. Ninio, *Biochimie* **57**, 587 (1975).
5. R. C. Thompson, P. J. Stone, *Proc. Natl. Acad. Sci. U.S.A.* **74**, 198 (1977).
6. T. Ruusala, M. Ehrenberg, C. G. Kurland, *EMBO J.* **1**, 741 (1982).
7. T. Pape, W. Wintermeyer, M. Rodnina, *EMBO J.* **18**, 3800 (1999).
8. ———, *Nature Struct. Biol.* **7**, 104 (2000).
9. M. Eigen, L. De Maeyer, *Naturwissenschaften* **53**, 50 (1966).
10. A. P. Potapov, *FEBS Lett.* **146**, 5 (1982).
11. ———, F. J. Triana-Alonso, K. H. Nierhaus, *J. Biol. Chem.* **270**, 17680 (1995).
12. R. C. Thompson, A. M. Karim, *Proc. Natl. Acad. Sci. U.S.A.* **79**, 4922 (1982).
13. D. Moazed, H. F. Noller, *J. Mol. Biol.* **211**, 135 (1990).
14. F. Mueller, H. Stark, M. van Heel, J. Rinke-Appel, R. Brimacombe, *J. Mol. Biol.* **271**, 566 (1997).
15. M. O'Connor *et al.*, *Biochem. Cell Biol.* **73**, 859. (1995).
16. F. Schluenzen *et al.*, *Cell* **102**, 615 (2000).
17. B. T. Wimberly *et al.*, *Nature* **407**, 327 (2000).
18. S. Yoshizawa, D. Fourmy, J. D. Puglisi, *Science* **285**, 1722 (1999).
19. D. Moazed, H. F. Noller, *Nature* **327**, 389 (1987).
20. D. Fourmy, M. I. Recht, S. C. Blanchard, J. D. Puglisi, *Science* **274**, 1367 (1996).
21. A. P. Carter *et al.*, *Nature* **407**, 340 (2000).
22. D. Fourmy, S. Yoshizawa, J. D. Puglisi, *J. Mol. Biol.* **277**, 333 (1998).
23. M. S. VanLoock, T. R. Easterwood, S. C. Harvey, *J. Mol. Biol.* **285**, 2069 (1999).
24. J. H. Cate, M. M. Yusupov, G. Z. Yusupova, T. N. Earnest, H. F. Noller, *Science* **285**, 2095 (1999).
25. A. P. Carter *et al.*, *Science* **291**, 498 (2001).
26. Crystals of the 30S subunit from *T. thermophilus* were grown as described (*17*). A tRNA^{Phe} ASL corresponding to the sequence GGGGAUUGAAAAUCCCC, and the U_6 hexanucleotide were synthesized chemically. The ASL and U_6 hexanucleotide were purified by gel electrophoresis. Crystals were transferred to a stabilizing buffer containing 26% 2-methyl-2,4-pentanediol (MPD) as described (*17*). Soaks of equimolar ASL and U_6 were done at a concentration of 100 or 300 μM in the same stabilizing buffer, while paromomycin, if present, was at a concentration of 80 μM. Initial screening of crystals was done at beamlines 14.1 and 14.2 at the Synchrotron Radiation Source at Daresbury Laboratory (Warrington, UK), where a 3.8 Å resolution data set on one of the soaks was also collected. All data to high resolution were collected at beamline SBC-19ID at the Advanced Photon Source at Argonne National Laboratory (Illinois), integrated and scaled with HKL-2000 (*49*). The atomic structure of the 30S subunit (*17*), omitting various parts near the A site, was used as a starting point for refinement against the three data sets with CNS (*50*). Initially, rigid-body refinement was done with the domains of 16S RNA and the individual proteins as separate rigid objects. This was followed by positional refinement, simulated annealing with torsion angle dynamics, and then two rounds of grouped B-factor refinement and refinement of the occupancies of metal ions. The ASL (15 and 11 residues in the presence and absence of paromomycin, respectively), mRNA, and any changes in the 30S subunit were built into the density in difference Fourier maps with the program O (*51*) before a final round of refinement. Coordinates have been deposited in the Protein Data Bank with accession numbers 1IBK (30S/paromomycin), 1IBL (30S/ASL/U_6/paromomycin), and 1IBM (30S/ASL/U_6). Coordinates of individual components will be made available on http://alf1.mrc-lmb.cam.ac.uk/~ramak/30S.
27. H. Shi, P. B. Moore, *RNA* **6**, 1091 (2000).
28. L. Jovine, S. Djordjevic, D. Rhodes, *J. Mol. Biol.* **301**, 401 (2000).
29. Y. Goldgur *et al.*, *Structure* **5**, 59 (1997).
30. F. H. C. Crick, *J. Mol. Biol.* **19**, 548 (1966).
31. S. Yokoyama, S. Nishimura, in *tRNA: Structure, Biosynthesis and Function*, D. Söll, U. RajBhandary, Eds. (American Society for Microbiology, Washington, DC, 1995), pp. 207–223.
32. N. C. Seeman, J. M. Rosenberg, A. Rich, *Proc. Natl. Acad. Sci. U.S.A.* **73**, 804 (1976).
33. Reviewed in C. G. Kurland, D. Hughes, M. Ehrenberg, in Escherichia coli *and* Salmonella typhimurium; *Cellular and Molecular Biology*, F. C. Neidhardt *et al.*, Eds. (American Society for Microbiology, Washington, DC, ed. 2, 1996), vol. 1, pp. 979–1004.
34. T. Powers, H. F. Noller, *Proc. Natl. Acad. Sci. U.S.A.* **87**, 1042 (1990).
35. J. Cate *et al.*, *Science* **273**, 1696 (1996).
36. N. Ban, P. Nissen, J. Hansen, P. B. Moore, T. A. Steitz, *Science* **289**, 905 (2000).
37. P. Nissen, J. Ippolito, N. Ban, P. B. Moore, T. A. Steitz, *Proc. Natl. Acad. Sci. U.S.A.* (online), 10 April 2001.
38. E. A. Doherty, R. T. Batey, B. Masquida, J. A. Doudna, *Nature Struct. Biol.* **8**, 339 (2001).
39. S. Doublie, S. Tabor, A. M. Long, C. C. Richardson, T. Ellenberger, *Nature* **391**, 251 (1998).
40. J. R. Kiefer, C. Mao, J. C. Braman, L. S. Beese, *Nature* **391**, 304 (1998).
41. G. M. Cheetham, T. A. Steitz, *Science* **286**, 2305 (1999).
42. B. J. McCarthy, J. J. Holland, *Proc. Natl. Acad. Sci. U.S.A.* **54**, 880 (1965).
43. H. Stark *et al.*, *Nature* **389**, 403 (1997).
44. R. K. Agrawal *et al.*, *J. Cell Biol.* **150**, 447 (2000).
45. R. K. Agrawal *et al.*, *Science* **271**, 1000 (1996).
46. H. Stark *et al.*, *Cell* **88**, 19 (1997).
47. Supplementary data are available on *Science* Online at www.sciencemag.org/cgi/content/full/292/5518/897/DC1.
48. O. Piepenburg *et al.*, *Biochemistry* **39**, 1734. (2000).
49. Z. Otwinowski, W. Minor, in *Methods in Enzymology*, C. W. J. Carter, R. M. Sweet, Eds. (Academic Press, New York, 1997), vol. 276, pp. 307–325.
50. A. T. Brünger *et al.*, *Acta Crystallogr. D Biol. Crystallogr.* **54**, 905 (1998).
51. T. A. Jones, M. Kjeldgaard, *Methods Enzymol.* **277B**, 173 (1997).
52. R. Koradi, M. Billeter, K. Wüthrich, *J. Mol. Graphics* **14**, 51 (1996).

53. We thank S. Ginell and A. Joachimiak (APS) and M. MacDonald and E. Duke (Daresbury) for their help and advice with synchrotron data collection; P. Nissen for advice on making Fig. 3; and B. T. Wimberly and K. Nagai for comments on the manuscript. Supported by the Medical Research Council (U.K.) and grant GM 44973 from the NIH to S. W. White and V.R.. D.E.B. was the recipient of a Human Frontier Science Program postdoctoral fellowship.

12 March 2001; accepted 3 April 2001

RESEARCH ARTICLES

The Complete Atomic Structure of the Large Ribosomal Subunit at 2.4 Å Resolution

Nenad Ban,[1]* Poul Nissen,[1]* Jeffrey Hansen,[1] Peter B. Moore,[1,2] Thomas A. Steitz[1,2,3]†

The large ribosomal subunit catalyzes peptide bond formation and binds initiation, termination, and elongation factors. We have determined the crystal structure of the large ribosomal subunit from *Haloarcula marismortui* at 2.4 angstrom resolution, and it includes 2833 of the subunit's 3045 nucleotides and 27 of its 31 proteins. The domains of its RNAs all have irregular shapes and fit together in the ribosome like the pieces of a three-dimensional jigsaw puzzle to form a large, monolithic structure. Proteins are abundant everywhere on its surface except in the active site where peptide bond formation occurs and where it contacts the small subunit. Most of the proteins stabilize the structure by interacting with several RNA domains, often using idiosyncratically folded extensions that reach into the subunit's interior.

In the last step of the gene expression pathway, genomic information encoded in messenger RNAs is translated into protein by a ribonucleoprotein called the ribosome (*1*). As in most other organisms, the prokaryotic ribosome (MW $\approx 2.6 \times 10^6$) is about two-thirds RNA and one-third protein and consists of two subunits, the larger of which is approximately twice the molecular weight of the smaller (*2*). The small subunit, which sediments at 30*S* in prokaryotes, mediates the interaction between mRNA codons and tRNA anticodons on which the fidelity of translation depends. The large subunit, which sediments at 50*S* in prokaryotes, includes the activity that catalyzes peptide bond formation—peptidyl transferase—and the binding site for the G-protein (GTP–binding protein) factors that assist in the initiation, elongation, and termination phases of protein synthesis.

Because the structures of several DNA and RNA polymerases have been determined at atomic resolution, the mechanisms of DNA and RNA synthesis are both well understood. Determination of the structure of the ribosome, however, has proven a daunting task. It is several times larger than the largest polymerase, and 100 times larger than lysozyme, the first enzyme to be understood at atomic resolution. Until now an atomic resolution structure for the ribosome has not been available, and as a result the mechanism of protein synthesis has remained a mystery.

Electron microscopy has contributed to our understanding of ribosome structure ever since the ribosome was discovered. In the last few years, three-dimensional (3D) electron microscopic images of the ribosome have been produced at resolutions sufficiently high to visualize many of the proteins and nucleic acids that assist in protein synthesis bound to the ribosome (*3*). Earlier this year, an approximate model of the RNA structure in the large subunit was constructed to fit a 7.5 Å resolution electron microscopic map of the 50*S* subunit from *Escherichia coli* as well as biochemical data (*4*).

Crystallization studies of the ribosome begun two decades ago by Yonath and Wittmann (*5*) and by the group at Pushchino (*6*) opened the possibility of using x-ray crystallography to determine the structure of the ribosome at atomic resolution. The first electron density map of the ribosome that showed features recognizable as duplex RNA was a 9 Å resolution x-ray crystallographic map of the large subunit from *Haloarcula marismortui* published 2 years ago (*7*). A year later, extension of the phasing of that map to 5 Å resolution made it possible to locate several proteins and nucleic acid sequences, the structures of which had been determined independently (*8*). At about the same time, with the use of similar crystallographic strategies, a 7.8 Å resolution map was generated of the entire *Thermus thermophilus* ribosome, showing the positions of tRNA molecules bound to its A, P, and E sites (*9*), and a 5.5 Å resolution map of the 30*S* subunit from *T. thermophilus* was obtained, which allowed the fitting of solved protein structures and the interpretation of some of its RNA features (*10*). Subsequently, an independently determined, 4.5 Å resolution map of the *T. thermophilus* 30*S* subunit was published, which was based, in part, on phases calculated from a model corresponding to 28% of the subunit mass that had been obtained with a 6 Å resolution experimental map (*11*). The interpretation of the subunit packing in the two 30*S* structures is not the same, even though the crystals used by the two groups appear to be identical.

Using a 2.4 Å resolution, experimentally phased, electron density map, we have produced an atomic structure of the *H. marismortui* 50*S* ribosomal. The model includes 2711 of the 2923 nucleotides of 23*S* ribosomal RNA (rRNA), all 122 nucleotides of its 5*S* rRNA, and structures for the 27 proteins that are well ordered in the subunit. Here, we describe the architecture of the subunit, the structure of its RNAs, and discuss the location, structures, and functions of its proteins.

The secondary structures of both 5*S* and 23*S* rRNA are remarkably close to those deduced for them by phylogenetic comparison. The secondary structure of the 23*S* rRNA divides it into six large domains, each of which has a highly asymmetric tertiary structure. The irregularities of their shapes notwithstanding, the domains fit together in an interlocking manner to yield a compact mass of RNA that is almost isometric. The proteins are dispersed throughout the structure and mostly concentrated on its surface, but they are largely absent from the regions of the subunit that are of primary functional significance to protein syntheses: the 30*S* subunit interface and the peptidyl transferase active site. The most surprising feature of many of these proteins is the extended, irregular structure of their loops and termini, which penetrate between RNA helices. The primary role of most proteins in the subunit appears to be stabilization of the 3D structure of its rRNA.

Structure determination. Several experimental approaches were used to extend the resolution of the electron density maps of the *H. marismortui* 50*S* ribosomal subunit from 5 to 2.4 Å. A back-extraction procedure was developed for reproducibly growing crystals that are much thicker than those available earlier and that diffract to at least 2.2 Å resolution. The twinning of crystals, which obstructed progress for many years (*8*), was eliminated by adjusting crystal stabilization conditions (*12*). All of the x-ray data used for high-resolution phasing were collected at the Brookhaven National Synchrotron Light Source except for two native data sets, which were collected at the Advanced Photon Source at Argonne (*13*) (Table 1). Osmium pentamine (132 sites) and iridium hexamine (84 sites) derivatives proved to be the most effective in producing isomorphous replacement and anomalous scattering phase information to 3.2 Å resolution (*14*). Intercrystal density averaging, which had contributed significantly at lower resolution, was not helpful beyond about 5 Å resolution. Electron density maps were dramatically improved, and

[1]Department of Molecular Biophysics & Biochemistry, and [2]Department of Chemistry, Yale University, and [3]Howard Hughes Medical Institute, New Haven, CT 06520–8114, USA.

*These two authors contributed equally to this work. †To whom correspondence should be addressed.

their resolutions were eventually extended to 2.4 Å with the solvent-flipping procedure in the CNS program (*15*, *16*).

Except for regions obscured by disorder, the experimentally phased 2.4 Å resolution electron density map was of sufficient quality that both protein and nucleic acid sequencing errors could be identified and corrected. Each nucleotide could be fitted individually, and the difference between A and G was usually clear without having to refer to the chemical sequence, as was the distinction between purines and pyrimidines (Fig. 1). Only a few of the many water molecules and metal ions evident in the electron density have been positioned so far.

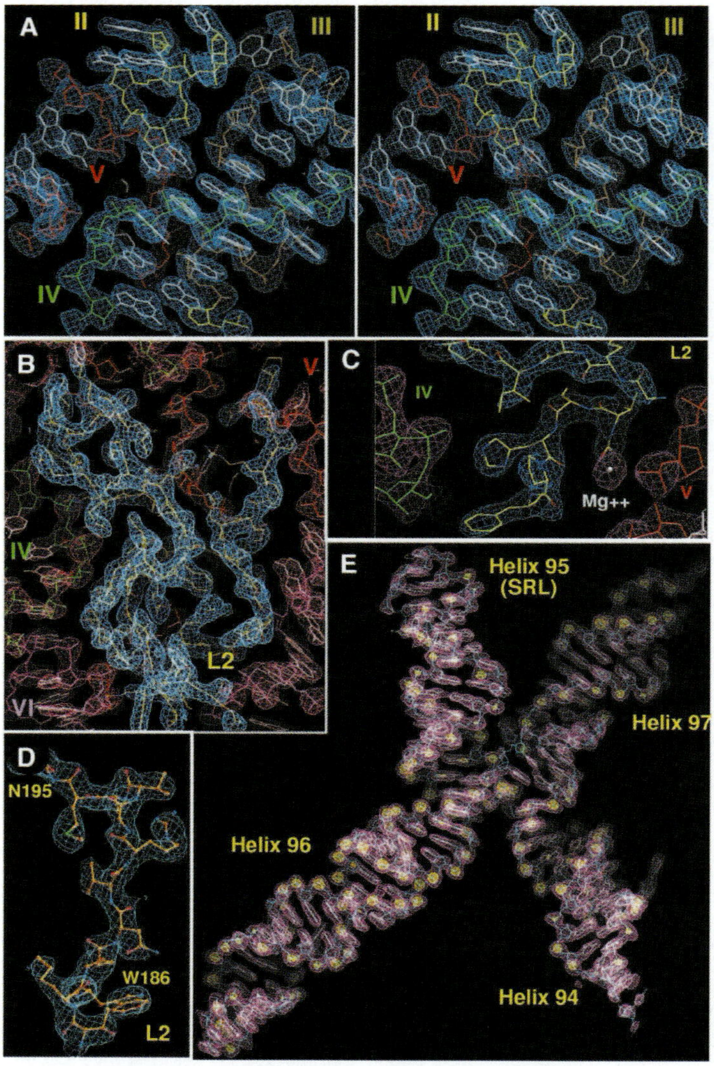

Fig. 1. Portions of the experimental 2.4 Å resolution electron density map. (**A**) A stereo view of a junction between 23*S* rRNA domains II, III, IV, and V having a complex structure that is clearly interpretable. The electron density is contoured at 2σ. The bases are white and the backbones are colored by domain as specified in Fig. 4. (**B**) The extended region of L3 interacting with its surrounding RNA, where the red RNA density is contoured at 2σ and the blue protein density is contoured at 1.5σ. (**C**) Detail in the L2 region showing a bound Mg^{2+} ion. (**D**) Detail from L2 showing amino acid side chains. (**E**) Helices 94 through 97 from domain VI. The red contour level is at 2σ, and the yellow contour at 6σ shows the positions of the higher electron density phosphate groups.

Subtraction of the atomic model from the experimental electron density map leaves no significant density except water and ions, showing that the model accounts for all the macromolecular density. Preliminary refinement of the model was achieved with experimental phase restraint in the program CNS (*16*). The model was further refined in real space against the 2.4 Å electron density map with the program TNT (*17*), which yielded a model with an *R* factor of 0.33. One additional round of mixed target refinement of both atomic positions and *B* factors with CNS led to the structure described here. The current free *R* factor is 0.26 (Table 1).

Sequence fitting and protein identification. Guided by the information available on the secondary structures of 23*S* rRNAs (*18*), the sequence of 23*S* rRNA was fit into the electron density map nucleotide by nucleotide starting from its sarcin/ricin loop sequence [A2691 to A2702 (*E. coli* numbers A2654 to A2665)] whose position had been determined at 5 Å resolution (*8*). The remaining RNA electron density neatly accommodated 5*S* rRNA. The interpretation of electron density corresponding to protein was more complicated because each protein region had to be identified chemically before the appropriate sequence could be fit into it; with the assistance of D. Klein, L. Min, S. Antolić, and M. Schmeing, ~4000 amino acid residues of 27 proteins were fit into electron density.

The *H. marismortui* 50*S* subunit appears to contain 31 proteins, and sequences for 28 of them exist in the Swiss-Prot data bank, including one called HMS6 or L7ae, which originally had been assigned to the small ribosomal subunit (*19*). The three remaining proteins were identified using the sequences of the ribosomal proteins from eukaryotes and other archaeal species as guides. No electron density was found for one of the *H. marismortui* large ribosomal subunit proteins in the sequence database, LX. Either the assignment of LX to the large subunit is in error, or LX is associated with a disordered region of the subunit. It is also possible that LX is absent from the subunits examined altogether.

The 2.4 Å resolution electron density map lacks clear electron density for proteins L1, L10, L11, and L12, the positions of which are known from earlier low-resolution x-ray and/or electron microscopic studies. These proteins are components of the two lateral protuberances of the subunit, which are both poorly ordered in these crystals. L1 is the sole protein component of one of them (*20*) and is visible in 9 Å resolution density maps of the subunit (*7*), but not at higher resolutions. L10, L11, and L12 are components of the other protuberance, which is often referred to as the L7/L12 stalk (*20*). L11 and the RNA to which it binds were located in the 5 Å resolution electron density map of the *H. marismortui* large subunit (*8*) using the independently determined crystal structures of that complex (*21, 22*). A protein fragment (~100 residues) associated with the RNA stalk that supports the L11 complex can be seen in the 2.4 Å resolution map. On the basis of its location, the fragment must be part of L10. No electron density corresponding to L12 was seen at any resolution, but the L12 tetramer is known to be attached to the ribosome through L10, and the L10/L12 assembly is known to be flexible under some circumstances (*23*), which may explain its invisibility here.

The structures of eubacterial homologs of proteins L2, L4, L6, L14, and L22 have previously been determined in whole or in part (Table 2). L2, L6, and L14 were initially located in the 5 Å resolution map (*8*). L4 and L22 have now been identified and positioned the same way. Electron density corresponding to most of the remaining proteins was assigned by comparing chain lengths and sequence motifs deduced from the electron density map with known sequence lengths. Occasionally, these comparisons were assisted by the information available on relative protein positions (*24*) and protein interactions with 23*S* rRNA and 5*S* rRNA (*25*). Each of the protein electron density regions so identified is well accounted for by the amino acid sequence assigned to it.

The most interesting of the proteins identified by sequence similarity was L7ae, which first appeared to be L30e. The L30e identification seemed plausible because the structure of yeast L30e superimposes neatly on the electron density of L7ae, and the structure of the RNA to which L7ae binds resembles that of the mRNA element to which yeast L30e binds (*26*). Nevertheless, the sequence of HMS6, which by sequence similarity is a member of the L7ae protein family, better fits the electron density. Four of the other proteins identified by sequence similarity, L24e, L37e, L37ae, and L44e, contain zinc finger motifs. The rat homologs of L37e and L37ae were predicted to be zinc finger proteins on the basis of their sequences (*27*), and this prediction helped identify their homologs in *H. marismortui*. Even though no *H. marismortui* sequences were available for the proteins L10e, L15e, and L37ae, they could be identified using the alignments of other available archaeal sequences.

General appearance of the subunit. In its rotated crown view (Fig. 2), the large ribosomal subunit, which is about 250 Å across, presents its surface that interacts with the small subunit to the viewer with the three projections that radiate from that surface pointed up. Although the protuberance that includes L1 is not visible in the 2.4 Å resolution electron density map, the structure of L1, which has been determined independently (*28*), has been positioned approximately in lower resolution maps (*7*) and is included here to orient the reader. It is evident that, except for its two lateral protuberances, the large ribosomal subunit is monolithic. There is no hint of a division of its structure into topologically separate domains. In addition, partly because it lacks obvious domain substructure but also because it is so large, it is impossible to comprehend looking at it as a whole. To convey a sense of how it is put together, the subunit must be dissected into its chemical components.

RNA secondary structure. All the base pairs in *H. marismortui* 23*S* rRNA stabilized by at least two hydrogen bonds were identified with a computer program that searched the structure for hydrogen bond donors and acceptors separated by less than 3.2 Å. Bases linked by at least two such bonds were considered paired if the angle between their normals was less than 45° and if the angle between bonds and base normals was also less than 45°. On the basis of the results of this analysis, R. Gutell and colleagues prepared a secondary structure diagram (Fig. 3) in the format standard for 23*S*/28*S* rRNAs. The secondary structure predicted for this molecule by phylogenetic comparison was remarkably accurate, but it did not find all of the tertiary pairings and failed to identify interactions involving conserved bases. In addition to base pairs of nearly every type, the RNA contains numerous examples of well-known secondary structure motifs such as base triples, tetraloops, and cross-strand purine stacks, but no dramatically new secondary structure motifs have been identified so far.

The secondary structure of this 23*S* rRNA consists of a central loop that is closed by a terminal stem, from which 11 more or less complicated stem-loops radiate. It is customary to describe the molecule as consisting of six domains and to number its helical stems sequentially starting from the 5′ end (Fig. 4) (*29*). The division of the molecule into domains as shown in Fig. 4 deviates from standard practice with respect to helix 25, which is usually considered part of domain I. Here, it is placed in domain II because it interacts more strongly with domain II than the other elements of domain I.

There are five sequences longer than 10 nucleotides in 23*S* rRNA whose structures cannot be determined from the 2.4 Å resolution map because of disorder. Together they account for 207 out of the 232 nucleotides missing from the final model. The disordered regions are: all of helix 1, the distal end of helix 38, helix 43/44 to which ribosomal protein L11 binds, the loop end of stem-loop 69, and helix 76/77/78, which is the RNA structure to which L1 binds. For completeness, these regions are included in Fig. 3 (in gray) with the secondary structures determined for them phylogenetically.

Overall architecture of rRNA. The six domains of 23*S* rRNA and 5*S* rRNA all have complicated convoluted shapes that fit together to produce a compact, monolithic RNA mass (Fig. 4, A and B). Thus, despite the organization of its RNAs at the secondary structure level, in three dimensions the large subunit is a single, gigantic domain. In this respect, it is quite different from the small subunit. Even in low-resolution electron micrographs the small subunit consists of three structural domains, each of which contains

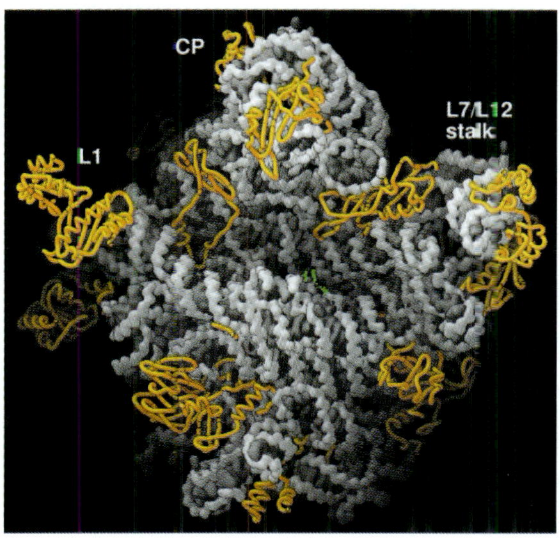

Fig. 2. The *H. marismortui* large ribosomal subunit in the rotated crown view. The L7/L12 stalk is to the right, the L1 stalk is to the left, and the central protuberance (CP) is at the top. In this view, the surface of the subunit that interacts with the small subunit faces the reader. RNA is shown in gray in a pseudo-space-filling rendering. The backbones of the proteins visible are rendered in gold. The Yarus inhibitor bound to the peptidyl transferase site of the subunit is indicated in green (*64*). The particle is approximately 250 Å across.

Secondary Structure: large subunit ribosomal RNA - 5′ half

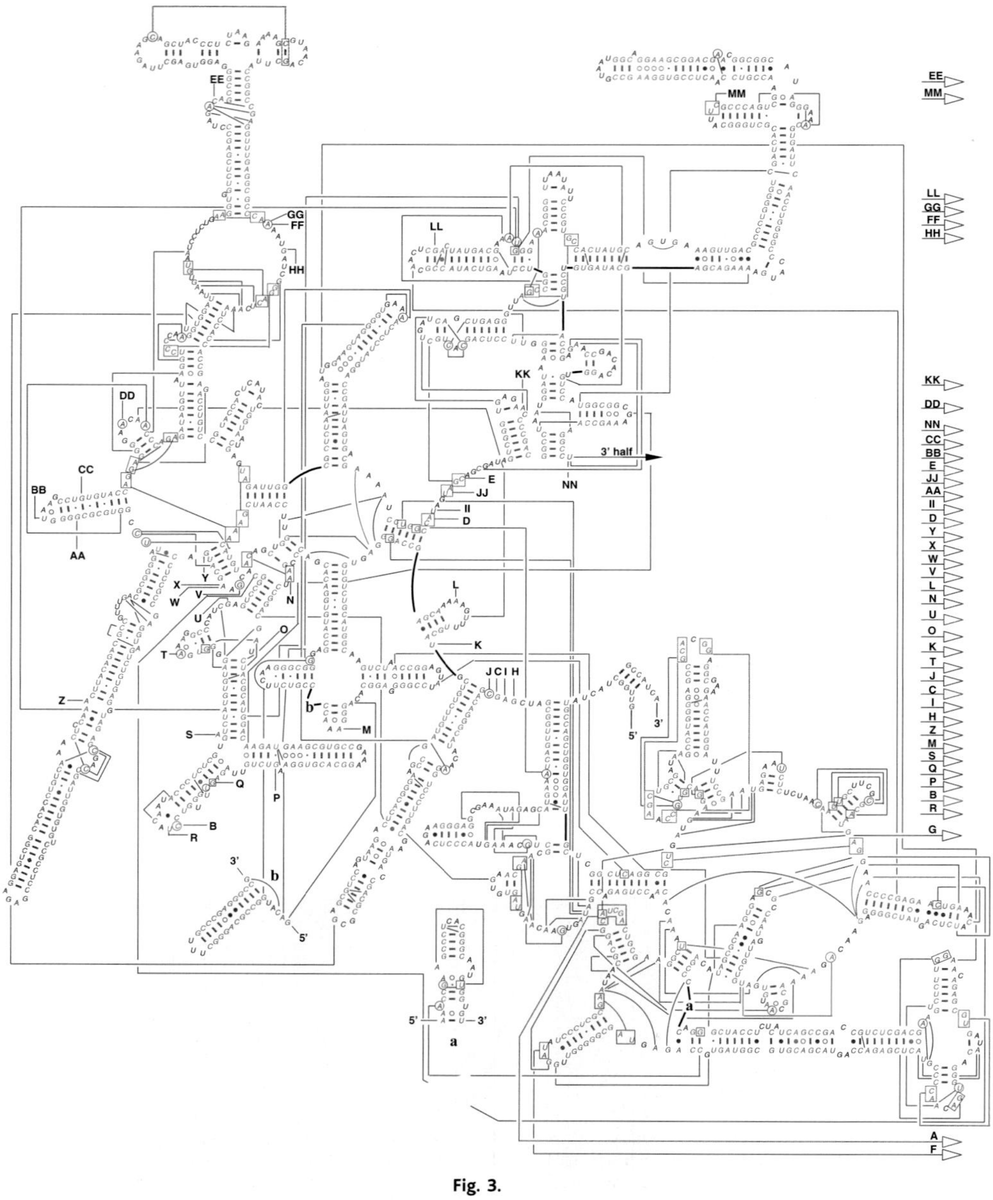

Fig. 3.

one of the three secondary structure domains of its RNA (*30*). This qualitative difference between the two subunits may reflect a requirement for conformational flexibility that is greater for the small subunit.

Domain I, which looks like a mushroom (Fig. 4E), lies in the back of the particle, behind and below the L1 region. The thin part of the domain starts in the vicinity of domain VI, which is the location of its first and last residues. Helices 1 and 25 span the particle in the back and then the domain expands into a larger, more globular structure below and behind the L1 region.

Domain II is the largest of the six 23*S* rRNA domains, accounting for most of the back of the particle. It has three protrusions that reach toward the subunit interface side of the particle (Fig. 4F). One of them (helix 42 to 44) is the RNA portion of the L7/L12 stalk, which is known to interact with elongation factors, is not well ordered in these crystals. The second domain II protrusion is helix 38, which is the longest, unbranched stem in the particle. It starts in the back of the particle, bends by about 90° and pro-

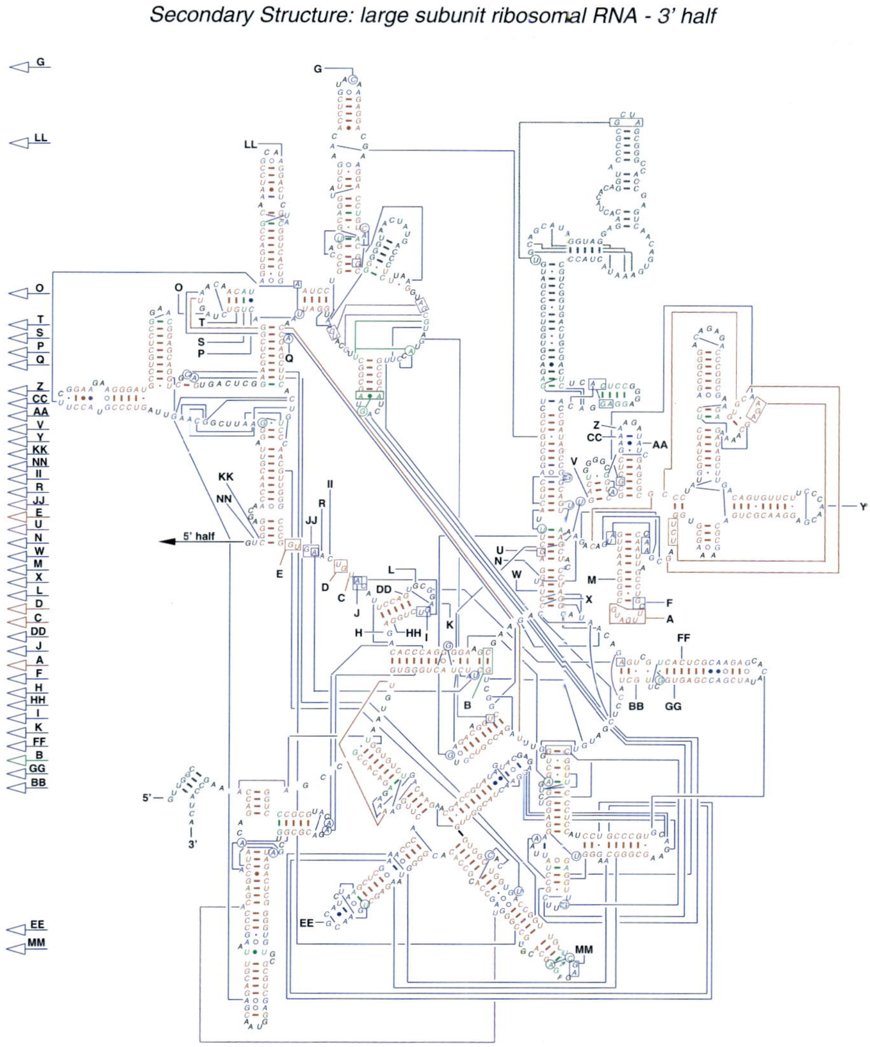

Fig. 3. The secondary structure of the 23*S* rRNA from *H. marismortui* is shown in a format made standard by R. Gutell and colleagues (*65*). It was prepared by Dr. Gutell to show all the base pairings seen in the crystal structure of the large subunit that are stabilized by at least two hydrogen bonds. Pairings shown in red were predicted and were observed. Those shown in green were predicted, but were not observed. Interactions shown in blue were observed, but were not predicted. Bases shown in black were not involved in pairing interactions. Sequences that cannot be visualized in the 2.4 Å resolution electron density map are depicted in gray with the secondary structures predicted for them.

trudes toward the small subunit between domains V and 5*S* rRNA. The third region (helix 32 to 35.1) points directly toward the small subunit and its terminus, the loop of stem-loop 34, interacts directly with the small ribosomal subunit (*31*). This loop emerges at the subunit interface between domains III and IV.

Domain III is a compact globular domain that occupies the bottom left region of the subunit in the crown view (Fig. 4G). It looks like a four-pointed star with the origin of the domain (stem-loop 48) and stem-loops 52, 57, and 58 forming the points. The most extensive contacts of domain III are with domain II, but it also interacts with domains I, IV, and VI. Unlike all the other domains, domain III hardly interacts with domain V at all; the sole contact is a van der Waals interaction involving a single base from each domain.

Domain IV accounts for most of the interface surface of the 50*S* subunit that contacts the 30*S* subunit (Fig. 4H). It forms a large diagonal patch of flat surface on that side of the subunit and connects to domains III and V in the back of the particle. Helices 67 through 71 constitute the most prominent feature of domain IV and form the front rim of the active site cleft, which is clearly visible at low resolution (Fig. 2). This is one of the few regions of the 23*S* rRNA that is not extensively stabilized by ribosomal proteins. Helix 69 in the middle of this ridge interacts with the long penultimate stem of 16*S* rRNA in the small ribosomal subunit (*9*).

Domain V, which is sandwiched between domains IV and II in the middle of the subunit is known to be intimately involved in the peptidyl transferase activity of the ribosome (*32*). Structurally, this domain can be divided into three regions (Fig. 4, I and J). The first starts with helix 75 and ultimately forms the binding site for protein L1. The second, which consists of helices 80 through 88, forms the bulk of the central protuberance region and is supported in the back by the 5*S* rRNA and domain II. The third region, which includes helices 89 through 93, extends toward domain VI and helps stabilize the elongation factor-binding region of the ribosome.

The smallest domain in 23*S* rRNA, domain VI, which forms a large part of the surface of the subunit immediately below the L7/L12 stalk, resembles a letter X with a horizontal bar at the bottom (Fig. 4K). The most interesting region of this domain is the sarcin-ricin loop (SRL) (stem-loop 95), the structure of which has been extensively studied in isolation (*33*, *34*). The SRL is essential for factor binding, and ribosomes can be inactivated by the cleavage of single covalent bonds in this loop (*35*). As suggested by nucleotide protection data, the major groove of this loop is exposed to solvent (*36*), and its conformation is stabilized by proteins and through interaction with domain V.

Table 1. Statistics for data collection, phase determination, and model construction. HA, heavy-atom concentration; ST, soaking time; Res, resolution; λ, wavelength; Obs, observations; Redun, redundancy; Compl, completeness; (*) last-resolution shell. R_{iso}: $\Sigma|F_{PH} - F_P|/F_{PH}$, where F_{PH} and F_P are the derivative and the native structure factor amplitudes, respectively. R_{sym}: $\Sigma\Sigma_i|I_{(h)i} - I_{(h)i}|/\Sigma\Sigma$: $I_{(h)i}$, where $I_{(h)}$ is the mean intensity after reflections. Phasing power: rms isomorphous difference divided by the rms residual lack of closure. R_{cullis}: $\Sigma(||F_{PH} - F_P| - |F_{H(calc)}||)/\Sigma|F_{PH} - F_P|$, where F_{PH} is the structure factor of the derivative and F_P is that of the native data. The summation is valid only for centric reflection. FOM (figure of merit): mean value of the cosine of the error in phase angles. Abbreviations: MIRAS, multiple isomorphous replacement, anomalous scattering; SAD, single wavelength anomalous diffraction.

Data statistics

	MIRAS1			MIRAS2			
	Native1	$Os(NH_3)_5^{2+}$	$UO_2F_5^{3-}$	Native2	$Ir(NH_3)_6^{3+}$	$Os(NH_3)_6^{3+}$	$Ta_6Br_{12}^{2+}$
HA (mM)	–	30.0	0.5	–	20.0	4.5	3.0
ST (hours)	–	1.5	4	–	24 hours	24 hours	24 hours
Sites no.	–	132	20	–	84	38	9
Res (Å)	90–2.4	40–3.5	40–3.8	30–2.9	30–3.2	30–3.5	30–3.8
(*)	(2.5–2.4)	(3.6–3.5)	(3.9–3.8)	(3.32–3.22)	(3.27–3.20)	(3.6–3.5)	(3.97–3.80)
λ (Å)	1.00	1.14	1.30	1.00	1.075	1.14	1.255
Obs	6,089,802	1,308,703	596,166	2,832,360	1,823,861	1,646,468	1,288,524
Unique	665,928	429,761	313,863	390,770	541,488	488,275	346,745
Redun (*)	9.1 (6.5)	3.0 (2.5)	1.9 (1.6)	7.2	3.4	4.3 (4.2)	3.7
Compl (*)	95.6 (71.0)	99.4 (96.8)	92.0 (54.1)	97.1	93.8	98.1 (99.0)	99.5
$I/\sigma I$ (*)	25.5 (1.9)	13.5 (3.3)	8.9 (1.6)	18.0 (6.4)	12.0 (2.6)	10.6 (2.7)	10.8 (3.2)
R_{merge} (*)	8.6 (69.1)	7.2 (32.0)	9.1 (37.9)	11.2 (36.9)	8.5 (29.5)	12.1 (46.0)	12.1 (40.5)
χ^2 (ano) (*)	–	2.8 (1.0)	1.5 (1.0)	–	2.63 (1.48)	1.8 (1.0)	2.42 (1.18)
R_{merge} (ano)	–	6.2	8.0	–	6.7	6.9	
R_{iso} (*)	–	14.1 (22.7)	26.4 (47.0)	–	12.9 (28.1)	19.5 (39.4)	

Phasing statistics

Resolution shells (Å): ~73,200 reflections per bin

	30.0–5.1	5.1–4.0	4.0–3.5	3.5–3.2	Total
MIRAS1 (FOM)	0.52	0.31	0.14	–	0.32
$Os(NH_3)_5^{2+}$					
Phasing power	0.87	0.72	0.66	–	0.75
Phasing power (SAD)	1.40	0.58	0.26	–	0.75
R_{cullis} (centric)	0.62	0.65	0.67	–	0.65
$UO_2F_5^{3-}$					
Phasing power	0.47	0.33	0.28	–	0.36
Phasing power (SAD)	0.46	0.25	–	–	0.36
R_{cullis} (centric)	0.72	0.77	0.75	–	0.75
MIRAS2 (FOM)	0.48	0.40	0.28	0.12	0.33
$Ir(NH_3)_6^{3+}$					
Phasing power	1.02	0.92	0.78	0.66	0.89
Phasing power (SAD)	2.02	1.60	1.22	0.83	1.47
R_{cullis} (centric)	0.58	0.63	0.70	0.74	0.63
$Os(NH_3)_6^{3+}$					
Phasing power	0.62	0.57	0.58	0.58	0.59
Phasing power (SAD)	0.47	0.39	–	–	0.42
R_{cullis} (centric)	0.78	0.78	0.78	0.76	0.78
$Ta_6Br_{12}^{2+}$ (used for SAD phasing only)					
Phasing power (SAD)	2.77	0.35	0.13	–	1.19
$FOM_{(MIRAS1+MIRAS2+SAD)}$	0.76	0.51	0.31	0.14	0.37

Model statistics

Resolution range (Å)	90.0–2.4	rms deviations:		Average *B* factors (Å²)	
Reflections	577,304	Bonds (Å)	0.0064	All atoms	37.4
R_{cryst} (%)	25.2	Angles (°)	1.19	23*S* rRNA	32.3
R_{free} (%)	26.1	Dihedrals (°)	28.8	5*S* rRNA	43.2
		Impropers (°)	1.68	Minimum/Max *B* factors (Å²)	7.0/107.9

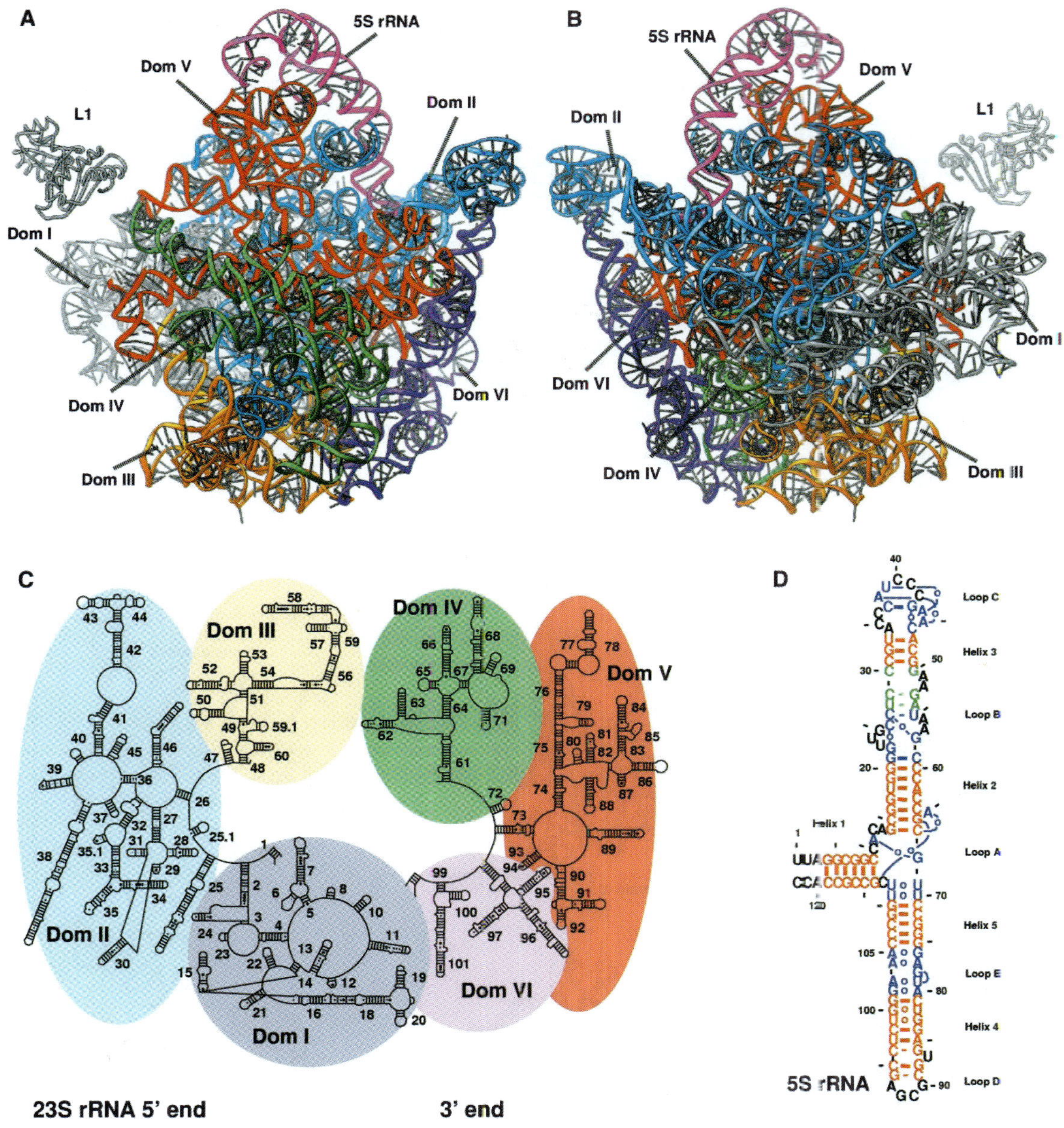

Fig. 4. The tertiary and secondary structures of the RNA in the *H. marismortui* large ribosomal subunit and its domains. (**A** and **B**) The RNA structure of the entire subunit. Domains are color-coded as shown in the schematic (**C**). (A) The subunit particle in its crown view. (B) The crown rotated by 180° about a vertical axis in the plane of the image. (C) Schematic secondary structure diagram of 23*S* rRNA with the domain coloring used throughout the figures and the helices numbered according to Leffers *et al.* (*29*). (**D**) The secondary structure of 5*S* rRNA from *H. marismortui*. Bases joined by thick lines represent Watson-Crick pairing, and those joined by a lower case "o" indicate non–Watson-Crick pairing. Bases joined by thin lines interact via a single hydrogen bond, whereas those in black are unpaired. Base pairings shown in red are phylogenetically predicted pairings that are now confirmed (*66*). Pairs shown in blue were observed but were not predicted, and pairs shown in green were predicted but were not observed. (**E** through **L**) Stereo views of the RNA domains of the 23*S* rRNA and of 5*S* rRNA. Each domain is color-coded from its 5′ end to its 3′ end to help the viewer follow its trajectory in three dimensions. The backbones are shown as ribbons and the bases as sticks. The surfaces where the most important interdomain interactions occur are shown in mono to the right: (E), Domain I; (F), domain II; (G), domain III; (H), domain IV; (I), domain V, crown view; (J), domain V, back view; (K), domain VI; and (L), 5*S* rRNA.

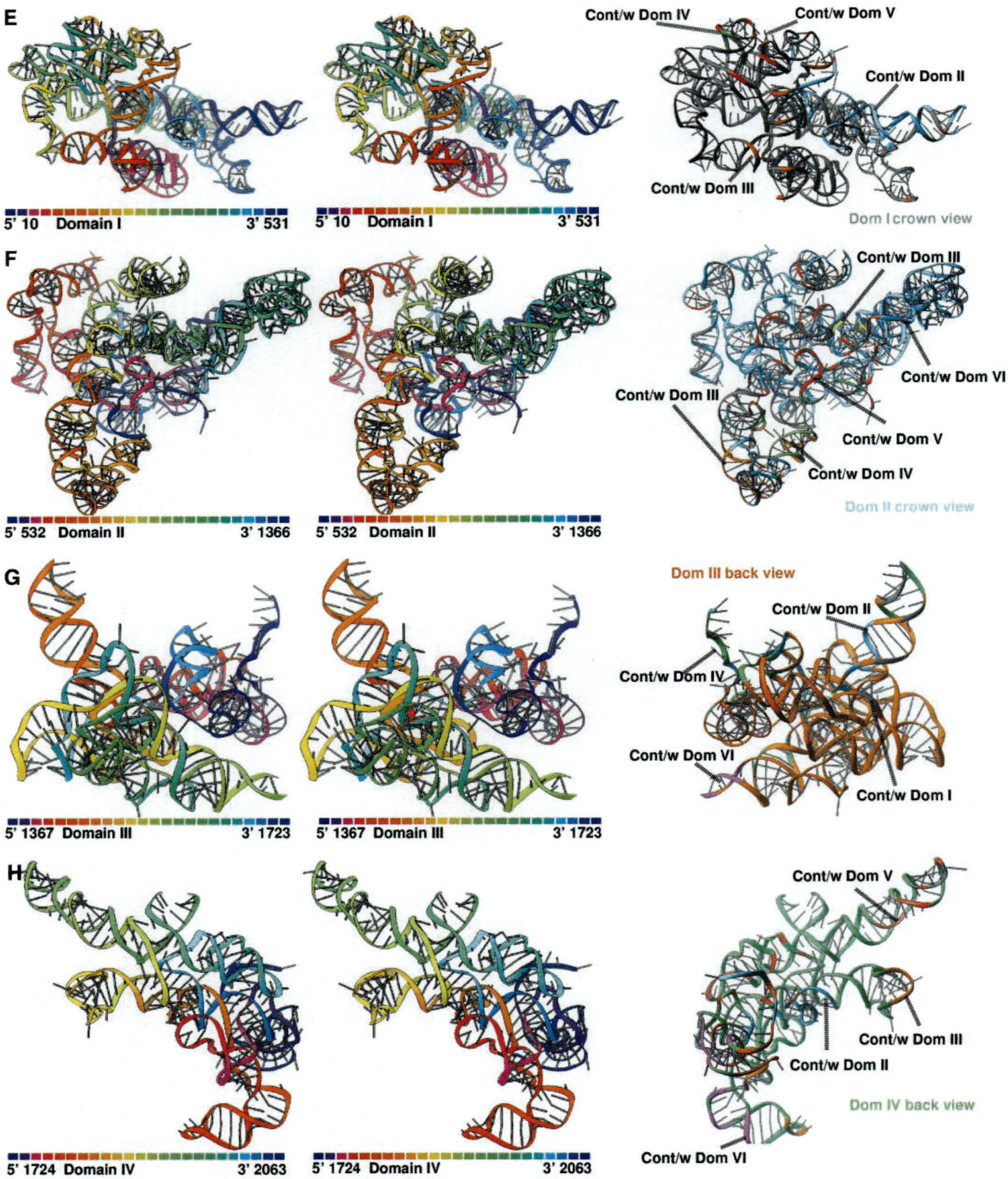

Fig. 4. (continued)

5*S* ribosomal RNA, which is effectively the seventh RNA domain in the subunit, consists of three stems radiating out from a common junction called loop A (Fig. 4D). In contrast to what is seen in the crystal structure of fragment 1 from *E. coli* 5*S* rRNA (*37*), the helix 2/3 arm of the molecule stacks on its helix 4/5 arm, not helix 1 (Fig. 4L). This arrangement results from a contorted

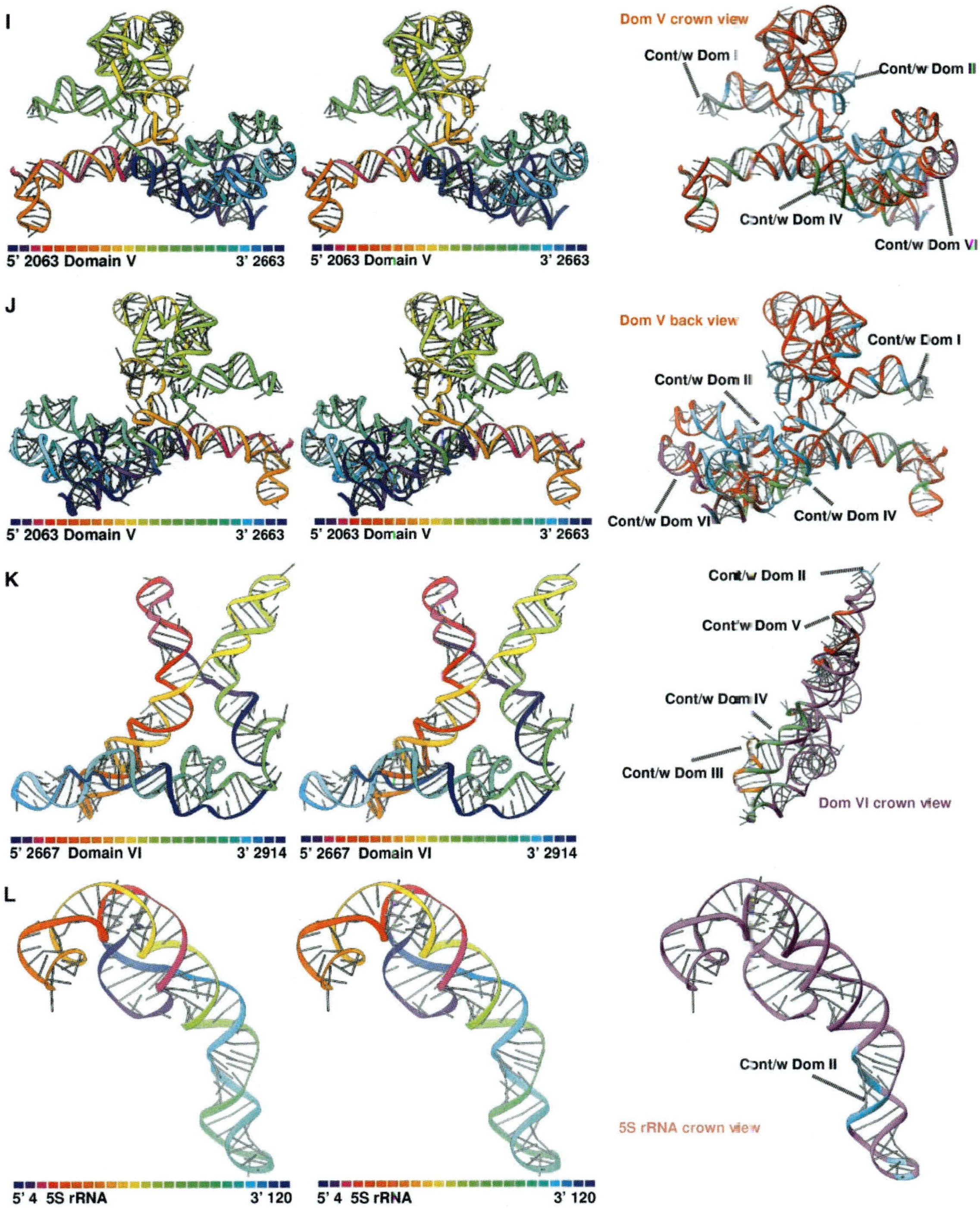

Fig. 4. (continued)

conformation of loop A residues that involves two stacked base triples. Indeed, from the secondary structure point of view, the loop A–helix 2/3 arm of 5*S* rRNA is remarkable, with a high concentration of unusual pairings leading to a convoluted RNA secondary structure.

Sequence conservation and interactions in 23*S* rRNA. Although 23*S*/28*S* rRNAs contain many conserved sequences, they also vary substantially in chain length. Shorter 23*S*/28*S* rRNAs are distinguished from their longer homologs by the truncation of, or even the elimination of, entire stem-loops, and by comparing sequences, one can identify a minimal structure that is shared by all (*38*). The expansion sequences in the 23*S* rRNA of *H. marismortui*, i.e., the sequences it contains that are larger than the minimum, are shown in Fig. 5 in green. They are largely absent from the subunit interface surface of the particle, but they are abundant on its back surface, far from its active sites. This is consistent with low-resolution electron microscopic observations, suggesting that the region of the large subunit whose structure is most conserved is the surface that interacts with the small subunit (*39*).

There are two classes of conserved sequences in 23*S* rRNA. One contains residues concentrated in the active site regions of the large subunit. The second class consists of much shorter sequences scattered throughout the particle (Fig. 5, red sequences). The SRL sequence in domain VI and the cluster of conserved residues belonging to domain V located at the bottom of the peptidyl transferase cleft are members of the first class. They are conserved because they are essential for substrate binding, factor binding, and catalytic activity. Most of the residues in the second class of conserved residues are involved in the inter- and intradomain interactions that stabilize the tertiary structure of 23*S* rRNA. Adenosines are disproportionately represented in this class. The predominance of adenosines among the conserved residues in rRNAs has been pointed out previously (*40*). Throughout the particle, adenosines are observed to participate in tertiary interactions by exploiting the smooth N1-C2-N3 face of the adenine base, which allows for very close packing and additional backbone-backbone interactions. In particular, a reoccurring pattern of two or more stacked adenosines that dock into the minor grooves of receptor helices seems to reveal a very basic principle in tertiary RNA structure formation and could be regarded as an equivalent of a hydrophobic core formation in glob-

Table 2. Large-subunit proteins from *Haloarcula marismortui*. The top block of proteins include all those known to have eubacterial homologs of the same name. The second block lists proteins found in the *H. marismortui* large ribosomal subunit that have only eukaryotic homologs (*19*). Their names are all followed by the letter "e" to distinguish them from eubacterial proteins that would otherwise have the same name. The third block are large-subunit proteins for which no *H. marismortui* sequence yet exists. They are identified by sequence homology with standard L names. [1]The structures of all or part of homologs of the following proteins were previously determined: L1 (*28*), L2 (*43*), L4 (*44*), L6 (*58*), L11 (*21, 22, 59*), L12 (*60*), L14 (*61*), L22 (*62*), and L30 (*63*). All other structures, except L10, have been newly determined in this study. [2]Rat homolog. Rat equivalents to *H. marismortui* protein are from (*26*). [3]Sequence chain length. [4]Conformation: glb, globular; ext, extension. [5]The protein interactions with the six domains of 23*S* rRNA, 5*S* rRNA, and other proteins are specified. (+) Implies that the interaction is substantial; (±) implies a weak, tangential interaction. Protein names in parentheses implies that the interactions are weak; otherwise, the interaction is substantial.

Name[1]	Hmlg[2]	Lgth[3]	Conf[4]	Interactions[5]							Proteins
				I	II	III	IV	V	VI	5*S*	
L1*	?	211	glb.					+			None
L2†	RL8	239	glb+ext		+	+	+	+			(L37ae)
L3	RL3	337	glb+ext		±		+	+	+		L14, L24e, (L13)
L4†	RL4	246	glb+ext	+	+			±			(L18e), (L24), (L37e)
L5	RL11	176	glb					+		+	L18
L6	RL9	177	glb		±			±	+		(L13)
L10*	RP0	348	glb?		+						L12
L11*	RL12	161	glb		+						None
L12*	RP1/2	115	glb								L10
L13	RL13a	145	glb		+			±	±		(L3), (L6)
L14	RL23	132	glb				+	+	+		L3, L24e
L15	RL27a	164	glb+ext	+	+			+			(L18e), (L32e)
L18	RL5	186	glb+ext		±			+		+	L5, L21e
L19	RL19	148	glb+ext		+	+	+		±		None
L22	RL17	154	glb+ext	+	±	+	+	+	+		None
L23	RL23a	84	glb	±		+					L29, (L39e)
L24	RL26	119	glb+ext	+							(L4)
L29	RL35	70	glb	+							L23
L30	RL7	154	glb		+					+	None
L18e	RL18	115	glb		+						(L4), (L15)
L21e	RL21	95	glb		+			+		±	L18
L24e	RL24	66	glb				±		+		L3, L14
L31e	RL31	91	glb			+	+		+		None
L32e	RL32	240	glb	±	+						(L15)
L37e	RL37	56	glb+ext	+	+	+	±				(L4)
L39e	RL39	49	ext	+		+					(L23)
L44e	RL36a	92	glb+ext	+			±	+			(L15e)
L7ae	RL7a	110	glb	±							L15e
L10e	RL10	163	glb		+			+		±	None
L15e	RL15	184	glb+ext	+	±	±	±	+			(L44e), L7ae
L37ae	RL37a	72	glb+ext		+	+	+				L2

*All entries so designated describe proteins that are not fully represented in the electron density maps described here. The summary information provided is derived from literature sources and is included here for completeness only. †The structure available for this protein in isolation does not include the extension(s) reported here.

ular protein domains. Common RNA structural motifs, such as the ribose zipper and the tetraloop-tetraloop receptor interaction, depend on this principle of adenosine packing. A manuscript in preparation describes these A-dependent interactions at greater length.

In addition to its reliance on A-dependent motifs, the tertiary structure of the domains of 23*S* rRNA and their relative positions are stabilized by familiar tertiary structure elements like pseudoknots and tetraloop-tetraloop receptor motifs (*41*, *42*). Thus, in many places, base pairs and triples stabilize the interactions of sequences belonging to different components of the secondary structure of 23*S* rRNA.

5*S* rRNA and 23*S* rRNA do not interact extensively with each other. The few RNA/RNA interactions that do occur involve the backbones of the helix 4/5 arm of 5*S* rRNA and of helix 38 of 23*S* rRNA. Most of the free energy and specificity of 5*S* rRNA binding to the large ribosomal subunit appears to depend on its extensive interactions with proteins that act as modeling clay, sticking it to the rest of ribosome.

Proteins. We have determined the structures of 27 proteins found in the large ribosomal subunit of *H. marismortui* (Table 2). Twenty-one of these protein structures have not been previously established for any homologs, and the structures of the six that do have homologs of known structure have been rebuilt into the electron density map with their *H. marismortui* sequences. In addition, there are structures available for homologs of *H. marismortui* L1, L11, and L12, which cannot be visualized in the 2.4 Å resolution electron density map. Only the structure of L10 is still unknown among the 31 proteins of this subunit.

Almost all of these structures are complete. Yet, an entire domain of L5 is missing from the electron density, presumably because of disorder. Further, L32e is also noteworthy. Its NH_2-terminal 97 residues are not seen in the electron density map, and the electron density map suggests that its COOH-terminal residue may be covalently bonded to the most NH_2-terminal of its visible residues.

Of the 30 large subunit ribosomal proteins whose structures are known, 17 are globular proteins, similar in character to thousands whose structures are in the Protein Data Bank (Table 2). The remaining 13 proteins either have globular bodies with extensions protruding from them ("glb+ext") or are entirely extended ("ext"). Their extensions often lack obvious tertiary structure and in many regions are devoid of significant secondary structure as well (Fig. 6). These extensions may explain why many ribosomal proteins have resisted crystallization in isolation. The exceptions that prove the rule are L2 and L4, both of which are proteins belonging to the "glb+ext" class. Protein L2 was crystallized and its structure solved only after its extensions had been removed (*43*), and the large loop of L4 that is extended in the ribosome is disordered in the crystal structure of intact L4 (*44*).

Except for proteins L1, L7, L10, and L11, which form the tips of the two lateral protuberances, the proteins of the 50*S* subunit do not extend significantly beyond the envelope defined by the RNA (Fig. 7). Their globular domains are found largely on the particle's exterior, often nestled in the gaps and crevices formed by the folding of the RNA. Thus, unlike the proteins in spherical viruses, the proteins of the large ribosomal subunit do not form a shell around the nucleic acid with which they associate, and unlike the proteins in nucleosomes, they do not become surrounded by nucleic acid, either. Instead, the proteins act like mortar filling the gaps and cracks between "RNA bricks."

The distribution of proteins on the subunit surface is nearly uniform, except for the active site cleft and the flat surface that interacts with the 30*S* subunit. In the crown view, the proteins lie around at the periphery of the subunit (Fig. 7A), but when viewed from the side opposite the 30*S* subunit binding site (the "back side"), they appear to form an almost uniform lattice over its entire surface (Fig. 7B). Similarly, the bottom surface of the subunit, which includes the exit of polypeptide tunnel, is studded with proteins (Fig. 7C). Indeed, the six proteins that surround the tunnel exit may play a role in protein secretion because they are part of the surface that faces the membrane and the translocon when membrane and secreted proteins are being synthesized (*45*).

Although Fig. 7 shows protein chains disappearing into the ribosome interior, the degree to which proteins penetrate the body of the particle can be fully appreciated only when the RNA is stripped away. The interior of the particle is not protein-free, but it is protein-poor compared with the surface of the particle. Extended tentacles of polypeptide, many of which emanate from globular domains on the surface, penetrate into the interior, filling the gaps between neighboring elements of RNA secondary structure (Fig. 8E). The bizarre structures of these extensions are explained by their interactions with RNA. A detailed analysis of these proteins and their interactions with RNA will be presented elsewhere.

Although extended, nonglobular structures are rare in the protein database, they are not unknown. Extended protein termini often form interprotein contacts, e.g., in viral capsids, presumably adopting fixed structures only upon capsid formation (*46*). The basic

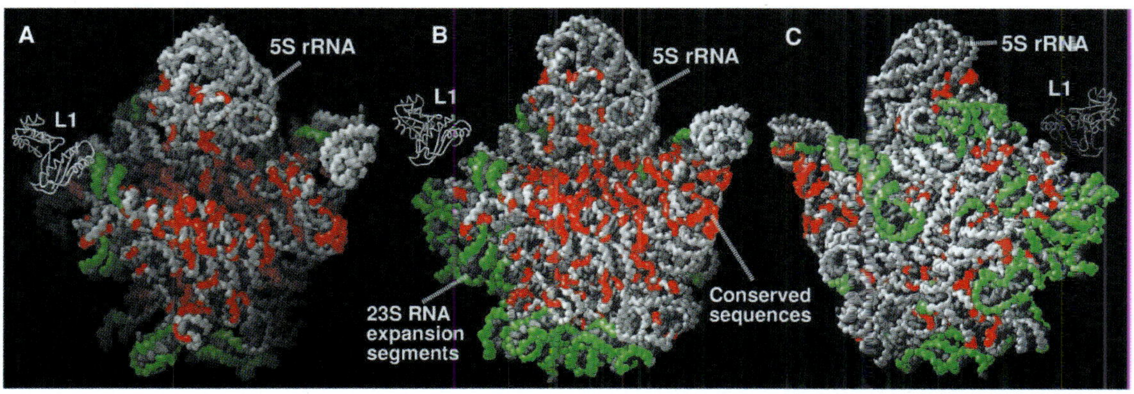

Fig. 5. Conserved residues and expansion sequences in the 23*S* rRNA of *H. marismortui*. The general, nonconserved RNA in these images is gray. Sequences that are found to be >95% conserved across the three phylogenetic kingdoms are shown in red. Sequences where expansion in the basic 23*S* structure is permitted are shown in green (*65*). **(A)** The particle rotated with respect to the crown view so that its active site cleft can be seen. **(B)** The crown view. **(C)** The back view of the particle, i.e., the crown view rotated 180° about its vertical axis.

"tails" of histones may behave the same way when nucleosomes form (*47*). The NH_2-terminal sequences of capsid proteins are often positively charged, and in virus crystal structures, the electron density for these sequences often disappears into the interior of the virus where they presumably interact with asymmetrically arranged nucleic acid. The interactions observed in the ribosome could be useful models for these viral interactions.

The interactions between extended polypeptides and RNA in the large subunit, which stabilize its massive nucleic acid structure, result in an intertwining of RNA and protein in the center of the subunit (Fig. 8, A and B). It is hard to imagine such an object assembling from its components efficiently in anything other than a highly ordered manner. Chaperones may well be required to prevent the aggregation of the extended regions of these proteins, which are likely to be disordered outside the context provided by rRNA, and to manage the folding of rRNA.

Mutations in some ribosomal proteins render bacteria resistant to certain antibiotics. One such example is a deletion of three amino acids in the β hairpin loop of protein L22 that renders bacteria resistant to erythromycin (*48*). Because this β hairpin is forming part of the surface of the tunnel wall, the mutation changes the surface properties of the polypeptide exit tunnel and may prevent the antibiotic from binding; alternatively, the mutation could be acting indirectly through RNA.

Protein and RNA interactions. Because protein permeates the large subunit extensively, there are only a few segments of the 23*S* rRNA that do not interact with protein at all. Of the 2923 nucleotides in 23*S* rRNA, 1157 make at least van der Waals contact with protein (Fig. 8D), and there are only 10 sequences longer than 20 nucleotides in which no nucleotide contacts protein. The longest such sequence contains 47 nucleotides, and is the part of domain IV that forms the ridge of the active site cleft.

The extent of the interactions between RNA and protein that occur when the large subunit assembles can be estimated quantitatively. Using the Richards algorithm (*49*) and a 1.7 Å radius probe to compute accessible surface areas, it can be shown that 180,000 $Å^2$ of surface become buried when the subunit forms from its isolated, but fully structured components. This is about half their total surface area. The average is about 6000 $Å^2$ per protein. Although this is an enormous amount compared with the surface buried when most protein oligomers form, it should be recognized that ribosome assembly must be accompanied by a large loss in conformational entropy that does not occur when most proteins oligomerize. The extended protein termini and loops of the ribosomal proteins are almost certainly flexible in isolation, and in the absence of protein, the RNA is probably quite flexible as well. Thus, the burial of a large amount of surface area may be required to provide the free energy required to immobilize the structures of these molecules.

All of the proteins in the particle except L12 interact directly with RNA, and all but 7 of the remaining 30 proteins interact with two rRNA domains or more (Table 2). The "champion" in this regard is L22, which is the only protein that interacts with RNA sequences belonging to all six domains of the 23*S* rRNA (Fig. 8C). The protein-mediated interactions between 5*S* rRNA and 23*S* rRNA are particularly extensive. Protein L18 attaches helix 1 and helix 2/3 of 5*S* rRNA to helix 87 of 23*S* rRNA. Protein L21e mediates an interaction between the same part of 5*S* rRNA and domains II and V. Protein L30 binds helix 4/5 region of 5*S* RNA to domain II. Loop C is linked to domain V by protein L5, and loop D is attached to domains II and V by protein L10e. Whatever else they may do, it is evident that an important function of these proteins is stabilization of the relative orientations of adjacent RNA domains. Several also help secure the tertiary structures of the domains with which they interact.

Because most ribosomal proteins interact with many RNA sequences and the number of proteins greatly exceeds the number of RNA domains, it can hardly come as a surprise that every rRNA domain interacts with multiple proteins (Table 2). Domain V, for example, interacts with 15 proteins, some intimately and a few in passing.

It is clear that the oligonucleotide binding experiments long relied on for information about the RNA binding properties of ribosomal proteins have underestimated their potential for interacting with RNA. The high-affinity RNA binding site identified on a protein by such an experiment may indeed be

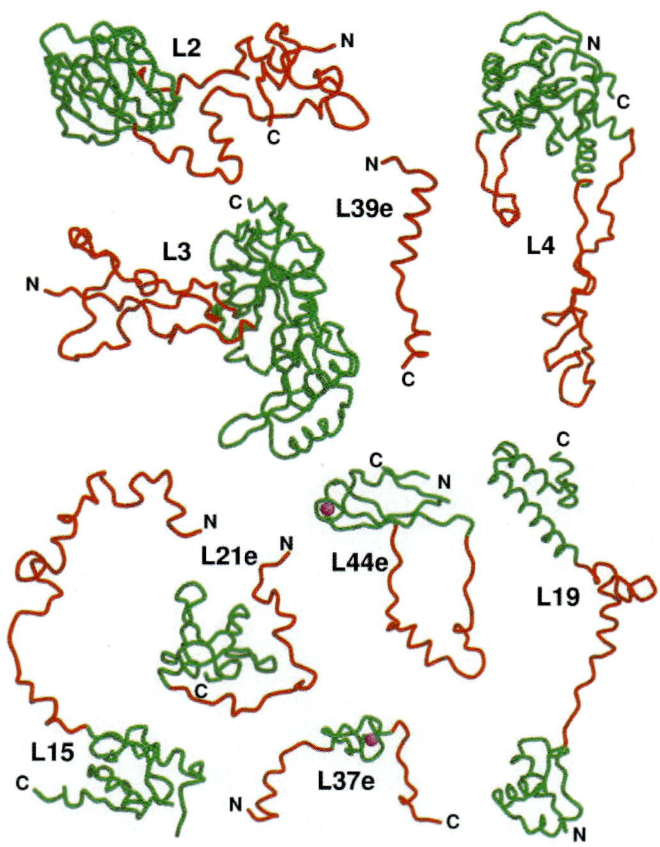

Fig. 6. The backbone structures of some large subunit ribosomal proteins that have nonglobular extensions. The globular domains of these proteins are shown in green, and their nonglobular extensions are depicted in red. The positions of the zinc ions in L44e and L37e are indicated by large dots in red.

important for ribosome assembly, but its many, weaker interactions with other sequences are likely to be missed, and they too may be vital for ribosome structure. Most ribosomal proteins crosslink RNA, and crosslinking is impossible without multiple interactions. Similar considerations may apply to proteins that are components of other ribonucleoproteins, such as the spliceosome.

Of the seven proteins that interact with only one domain, three (L1, L10, and L11) participate directly in the protein synthesis process. Rather than being included in the ribosome to ensure that the RNA adopts the proper conformation, it seems more appropriate to view the RNA as being structured to ensure the correct placement of these proteins. Another three (L24, L29, and L18e) interact with several secondary structure elements within the domains to which they bind, and presumably they function to stabilize the tertiary structures of their domains. The last of the single RNA domain proteins, L7ae, is puzzling. It cannot function as an RNA stabilizing protein because it interacts with only a single sequence in domain I, but it is far from the peptidyl transferase and factor binding sites. It is quite close to L1, however, which appears to be important for E-site function (5[illegible]), and maybe it is involved in that activity. It could also be involved in the 70*S* assembly, because L7ae was originally assigned as a small subunit protein (HMS6).

While many ribosomal proteins interact primarily with RNA, a few interact significantly with other proteins. The most striking

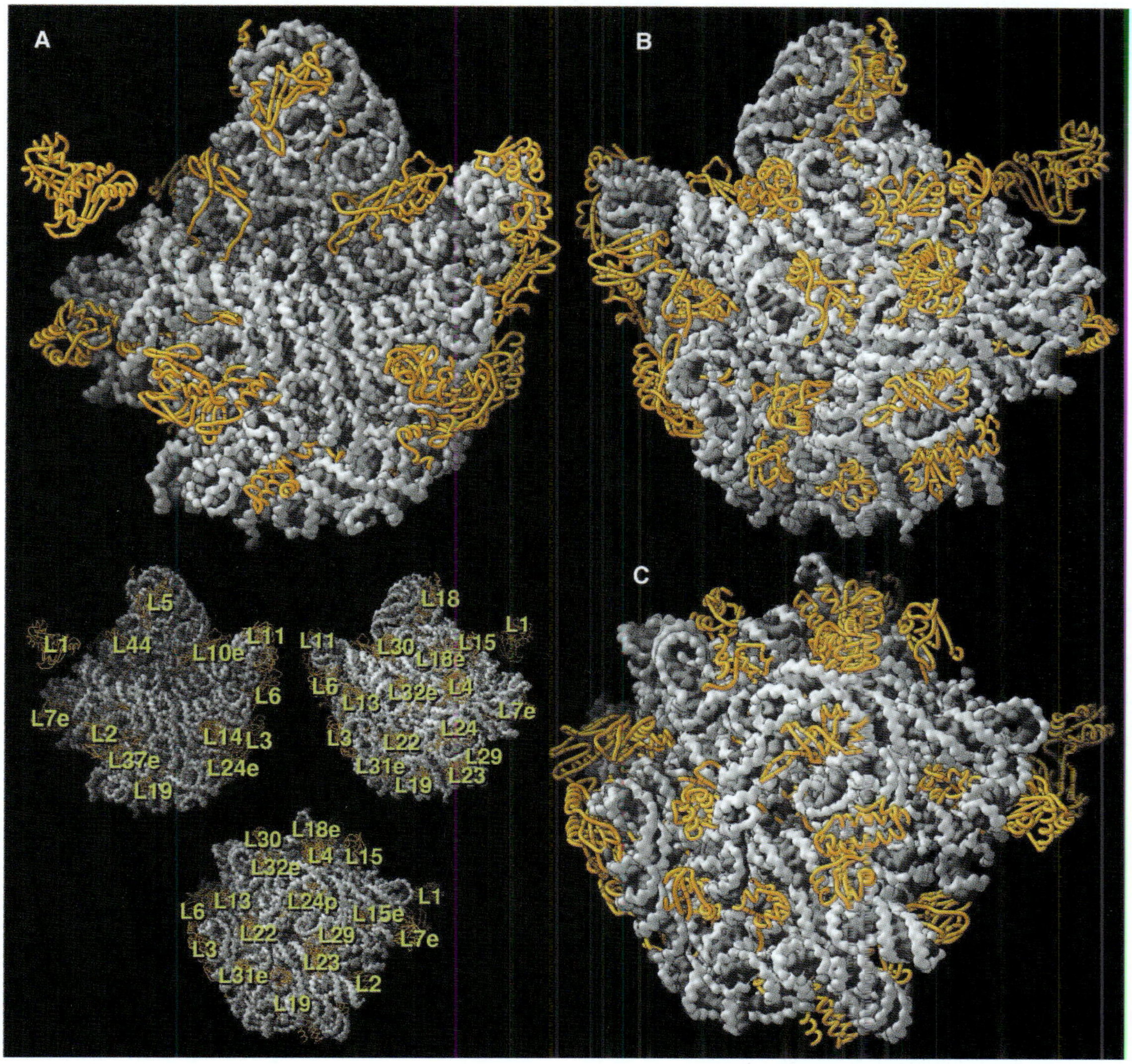

Fig. 7. Proteins that appear on the surface of the large ribosomal subunit. The RNA of the subunit is shown in gray and protein backbones are shown in gold. (**A**) The crown view of the subunit. (**B**) The back side of the subunit in the 180° rotated crown view orientation. (**C**) A view from the bottom of the subunit down the polypeptide tunnel exit which lies in the center. The proteins visible in each image are identified in the small images at the lower left of the figure. Figures were generated using RIBBONS (67).

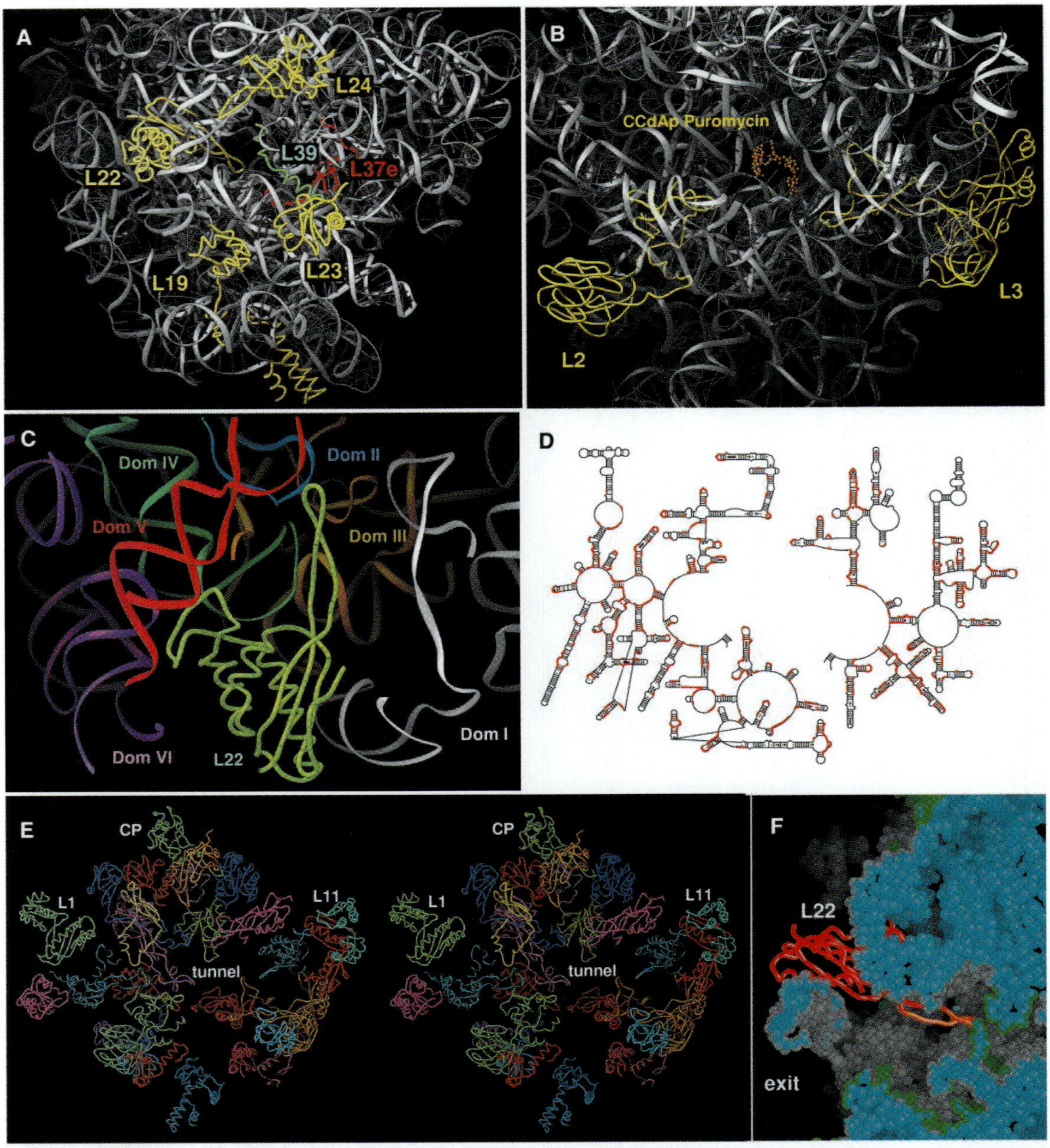

Fig. 8. The protein extensions into the RNA and multiple domain interactions. (**A**) Some of the proteins in the neighborhood of the polypeptide tunnel exit, showing the unusual extended structure of L39e (green) that enters the tunnel and L37e (red) that interpenetrates the RNA. L29, which is on top of L37e, has been removed. Protein L22 extends a long β hairpin extension inside the 23*S* rRNA. L24 has a similar extension but the entire protein is on the surface of the particle. L39 is the only protein in the subunit that lacks tertiary structure, whereas L37e has both NH_2- and COOH-terminal extensions. L19 is unique in having two globular domains on the surface of the subunit connected by an extended sequence that weaves through the RNA, shown as gray ribbons. (**B**) The nonglobular extensions of L2 and L3 reaching through the mass of 23*S* rRNA toward the peptidyl transferase site, which is marked by a CCdAp-puromycin molecule, the Yarus inhibitor (*64*). (**C**) L22 interacting with portions of all six of the domains of 23*S* rRNA. (**D**) Schematic of the 23*S* rRNA secondary structure showing the locations sequences (red) that make contact with protein. (**E**) Stereo view of the proteins of the large ribosomal subunit without the RNA. Proteins are colored as an aid to visualization only. (**F**) A cross section of the subunit in the area of the tunnel exit. Protein L22 is shown as ribbons in red, and the β hairpin loop where mutations confer erythromycin resistance is in orange. Atoms on the surface are gray, protein atoms are green, and atoms at the slice interface are blue.

structure generated by protein-protein interactions is the protein cluster composed of L3, L6, L13, L14, and L24e that is found close to the factor binding site. The surface of these proteins provides important interactions with factors. It may prove to be more generally the case that ribosomal proteins interacting primarily with RNA are principally stabilizing RNA structure, whereas some of those showing extensive protein-protein interactions may have additional binding functions.

The structure presented above illuminates both the strengths and weaknesses of approaches to complex assemblies that depend on determining the structures of components in isolation. The structures of the globular domains of homologs of the proteins in the large ribosomal subunit from *H. marismortui* are largely the same as those of the corresponding domains in the intact subunit, though adjustments in domain positions are sometimes required. Consequently, these structures were very useful for locating proteins and interpreting lower resolution electron density maps. However, for obvious reasons, the structures of the extended tails and loops of ribosomal proteins cannot be determined in the absence of the RNAs that give them structure, and the feasibility of strategies that depend on producing low–molecular weight RNA-protein complexes that have all the RNA contacts required to fix the structures of such proteins seems remote. The structures of RNA fragments also depend on their context. Whereas the sarcin/ricin loop has much the same structure in isolation (*33, 34*) as it does in the ribosome, the structure of 5*S* rRNA in isolation (*37*) differs in some respects from what is seen in the ribosome, and the structure of the isolated P loop (*51*) shows no resemblance to the structure of the P loop in the ribosome. Clearly, a "structural genomics" approach to the ribosome, which would have entailed determining the structures of all of the proteins and all possible rRNA fragments, neither would have provided the relevant structures of all of the pieces nor would it have shown their relative positions. Indeed, the structure of the large ribosomal subunit highlights the importance of structural studies of entire assemblies that show biological activity.

The analysis of the 50*S* ribosomal subunit structure presented here describes the overall architectural principles of RNA folding and its interaction with proteins, but many exciting details remain to be explored. The principles of protein-RNA interaction that should emerge from the 27 protein complexes with RNA have yet to be developed. On average, each of the 27 proteins has 3000 Å^2 of surface area in contact with RNA, which is comparable to the 2700 Å^2 of glutaminyl-tRNA synthetase that contact tRNA^{Gln} (*52*), so the number of interactions between RNA and protein to be analyzed in the large subunit structure is 30 times the number in this synthetase complex. Further, because the RNA structure of the large subunit will increase the RNA structural database by a factor of 4 to 5, most of the important RNA secondary and tertiary structural motifs to be found in nature may be represented. It will be interesting to see whether a complete analysis of this RNA structural database will enable the prediction of structures for other RNA sequences. Unknown at this time is the ease with which it will be possible to model by sequence homology the 50*S* ribosomal subunit rRNA from other species and kingdoms. However, the extensive sequence conservation in the 23*S* rRNA that forms the core active site and peptide tunnel regions suggests that reasonably accurate homology modeling based on this *H. marismortui* subunit structure may be feasible. Finally, enormous numbers of monovalent and divalent metal ions as well as water molecules are visible in this map. Analysis of their interactions with RNA should elucidate their roles in the formation and stabilization of RNA structure.

References and Notes

1. R. A. Garrett *et al.*, Eds., *The Ribosome: Structure, Function, Antibiotics and Cellular Interactions* (American Society for Microbiology, Washington, DC, 2000).
2. B. Wittmann-Liebold, in *Structure, Function, and Genetics of Ribosomes*, B. Hardesty and G. Kramer, Eds. (Springer-Verlag, New York, 1986), pp. 326–361.
3. R. K. Agrawal, P. Penczek, R. A. Grassucci, J. Frank, *Proc. Natl. Acad. Sci. U.S.A.* **95**, 6134 (1998); H. Stark, M. V. Rodnina, H.-J. Wieden, M. van Hell, W. Wintemeyer, *Cell* **100**, 301 (2000).
4. F. Mueller *et al.*, *J. Mol. Biol.* **298**, 35 (2000).
5. A. Yonath *et al.*, *Biochem. Int.* **2**, 428 (1980).
6. S. D. Trakanov *et al.*, *FEBS Lett.* **220**, 319 (1987).
7. N. Ban *et al.*, *Cell* **93**, 1105 (1998).
8. N. Ban *et al.*, *Nature* **400**, 841 (1999).
9. J. H. Cate, M. M. Yusupov, G. Z. Yusupova, T. N. Earnest, H. F. Noller, *Science* **285**, 2095 (1999).
10. W. M. Clemons Jr. *et al.*, *Nature* **400**, 833 (1999).
11. A. Tocilj *et al.*, *Proc. Natl. Acad. Sci. U.S.A.* **96**, 14252 (1999).
12. *H. marismortui* (American Type Culture Collection 43049) was grown, and ribosomes were prepared from it as described previously (*7, 53*). The buffers and precipitants used for crystal growth were those described earlier (*7, 54*), but with the following modifications. A crystallization solution was obtained by back-extraction of precipitated subunit at saturation into the crystallization buffer [1.2 M KCl, 0.5 M NH_4Cl, 100 mM Kacetate, 30 mM $MgCl_2$, 7% polyethylene glycol (PEG) 6000, 15 mM tris, 15 mM MES, and 1 mM $CdCl_2$ (pH 7.1)]. The crystals that resulted had maximum dimensions of 0.5 mm by 0.5 mm by 0.2 mm and were harvested after ~2 weeks. Crystals were stabilized by gradual transfer into a solution containing 12% PEG 6000, 22% ethylene glycol, 1.7 M NaCl, 0.5 M NH_4Cl, 100 mM potassium acetate, 30 mM $MgCl_2$, and 1 mM $CdCl_2$ (pH 6.2) at 4°C. The crystals were flash-frozen in liquid propane.
13. All data, except the two native data sets, were collected at the National Synchrotron Light Source (Brookhaven National Laboratory, Upton, NY) from crystals frozen at 100 K using beamlines X12b and X25 and recorded using a 345-mm MAR imaging plate. For each heavy-atom derivative, anomalous diffraction data were collected at the wavelength corresponding to the peak anomalous scattering, except for uranium, where a low energy was used. The beam size was 100 μm by 100 μm for most data collections at beamline X25 and 200 μm by 200 μm at beamline X12b. The crystals were aligned along the long axis of the unit cell (~575 Å) so that 1.0° oscillations could be used to collect reflections out to a maximum of 2.7 Å resolution at the edge of the MAR detector. At beamline X12b, the crystal-to-detector distances were varied depending on wavelength, crystal quality, and beam divergence so that maximum resolution data could be collected while avoiding overlapping of spots. At beamline X25, the detector was positioned on a rigid platform at 480 mm, which allowed data collection to 3.2 Å for iridium and osmium derivatives with the wavelength set at the anomalous edge. Native data to 2.4 Å resolution were collected at the structural biology beamline 19ID of the Advanced Photon Source (Argonne National Laboratory, Argonne, IL) using a 3 by 3 charge-coupled device (CCD) detector, an 80 μm by 80 μm beam size, and 0.4° oscillations. Data sets were processed with DENZO and SCALEPACK (*55*).
14. Heavy-atom–based phasing was extended to 3.2 Å resolution by combining MIR (multiple isomorphous replacement) phases calculated for two different isomorphous groups of data (MIR1 and MIR2) (Table 1) with single-wavelength anomalous dispersion (SAD) phases. Weights on the individual phase sets were adjusted on the Hendrickson-Lattman coefficients to prevent phase bias due to nonisomorphism. The best two derivatives were osmium pentamine and iridium hexamine, each of which contained a large number of binding sites (Table 1). Other derivatives with smaller numbers of sites further improved map quality. All phasing was done by maximum likelihood methods implemented in CNS with the exception of the Ta_6Br_{12} derivative, which was refined in SHARP (*56*), represented as spherically averaged electron density (Table 1). Phases were improved and extended from 3.3 to 2.4 Å by solvent flipping (*15*), with a scheme that progressively refined the solvent mask from 12 to 2.6 Å detail. RNA structure was built by positioning of the individual residues, and their rotomers were picked from a library derived from high-resolution structures (M. Kjeldgaard, unpublished data). Models were built with O (*57*).
15. J. P. Abrahams and A. G. W. Leslie, *Acta Crystallogr.* **D52**, 30 (1996).
16. A. T. Brunger *et al.*, *Acta Crystallogr.* **D54**, 905 (1998).
17. D. E. Tronrud, *Methods Enzymol.* **277** 306 (1997).
18. R. R. Gutell, in *Ribosomal RNA: Structure, Evolution, Processing and Function in Protein Biosynthesis*, A. Dahlberg and R. Zimmerman, Eds. (CRC Press, Boca Raton, FL, 1996), pp. 111–128.
19. B. Wittmann-Liebold *et al.*, in *The Ribosome: Structure, Function, and Evolution*, W. E. Hill *et al.*, Eds. (American Society for Microbiology, Washington, DC, 1990), pp. 598–616.
20. M. Oakes, E. Henderson, A. Scheinman, M. Clark, J. A. Lake, in *Structure, Function and Genetics of Ribosomes*, B. Hardesty and G. Kramer, Eds. (Springer-Verlag, New York, 1986), pp. 47–67; G. Stoeffler and M. Stoeffler-Meilicke, in *Structure, Function, and Genetics of Ribosomes*, B. Hardesty and G. Kramer, Eds. (Springer-Verlag, New York, 1986), pp. 28–46.
21. G. L. Conn, D. E. Draper, E. E. Lattman, A. G. Gittis, *Science* **284**, 1171 (1999).
22. B. T. Wimberly, R. Guymon, J. P. McCutcheon, S. W. White, V. R. Ramakrishnan, *Cell* **97**, 491 (1999).
23. W. Moller and J. A. Maassen, in *Structure, Function and Genetics of Ribosomes*, B. Hardesty and G. Kramer, Eds. (Springer-Verlag, New York, 1986), pp. 309–325.
24. J. Walleczek, D. Schuler, M. Stoeffler-Meilicke, R. Brimacombe, G. Stoeffler, *EMBO J.* **7**, 3571 (1988).
25. P. Ostergaard *et al.*, *J. Mol. Biol.* **284**, 227 (1998).
26. H. Mao, S. A. White, J. R. Williamson, *Nature Struct. Biol.* **6**, 1139 (1999).
27. I. G. Wool, Y.-L. Chan, A. Gluck, *Biochem. Cell Biol.* **73**, 933 (1995).
28. N. Nevskaya *et al.*, *Struct. Fold. Des.* **8**, 363 (2000).
29. H. Leffers, J. Kjems, L. Ostergaard, N. Larsen, R. A. Garrett, *J. Mol. Biol.* **195**, 43 (1987).
30. H. F. Noller *et al.*, in *The Ribosome: Structure, Function, and Evolution*, W. E. Hill *et al.*, Eds. (American Society for Microbiology, Washington, DC, 1990), pp. 73–92.

31. G. M. Culver, J. H. Cate, G. Zh. Yusupova, M. M. Yusupov, H. F. Noller, *Science* **285**, 2133 (1999).
32. R. A. Garrett and C. Rodriguez-Fonseca, in *Ribosomal RNA: Structure, Evolution, Processing and Function in Protein Biosynthesis*, R. A. Zimmerman and A. E. Dahlberg, Eds. (CRC Press, Boca Raton, FL, 1996), pp. 327–355.
33. A. A. Szewczak and P. B. Moore, *J. Mol. Biol.* **247**, 81 (1995).
34. C. C. Correll *et al.*, *Proc. Natl. Acad. Sci. U.S.A.* **95**, 13436 (1998).
35. I. G. Wool, A. Gluck, Y. Endo, *Trends Biochem. Sci.* **17**, 266 (1992).
36. D. Moazed, J. M. Roberston, H. F. Noller, *Nature* **334**, 362 (1988).
37. C. C. Correll, B. Freeborn, P. B. Moore, T. A. Steitz, *Cell* **91**, 705 (1997).
38. S. Gerbi, in *Ribosomal RNA: Structure, Evolution, Processing and Function in Protein Biosynthesis*, R. A. Zimmermann and A. E. Dahlberg, Eds. (CRC Press, Boca Raton, FL, 1996), pp. 71–88.
39. P. Dube *et al.*, *Structure* **6**, 389 (1998).
40. V. C. Ware *et al.*, *Nucleic Acids Res.* **22**, 7795 (1983).
41. P. B. Moore, *Annu. Rev. Biochem.* **68**, 287 (1999)
42. E. Westhof and V. Fritsch, *Structure* **8**, R55 (2000).
43. A. Nakagawa *et al.*, *EMBO J.* **18**, 1459 (1999).
44. M. Wahl, R. Huber, M. C. Wahl, *EMBO J.* **19**, 807 (2000).
45. R. Beckmann *et al.*, *Science* **278**, 2123 (1997).
46. M. G. Rossmann and J. E. Johnson, *Annu. Rev. Biochem.* **58**, 533 (1989); A. Liljas, *Int. Rev. Cytol.* **124**, 103 (1991).
47. K. Lugor *et al.*, *Nature* **389**, 251 (1997).
48. H. S. Chittum and W. S. Champney, *J. Bacteriol.* **176**, 6192 (1994).
49. B. Lee and F. M. Richards, *J. Mol. Biol.* **55**, 379 (1971).
50. R. K. Agrawal *et al.*, *J. Biol. Chem.* **274**, 8723 (1999).
51. E. V. Puglisi, R. Green, H. F. Noller, J. D. Puglisi, *Nature Struct. Biol.* **4**, 775 (1997).
52. M. A. Rould, J. J. Perona, D. Söll, T. A. Steitz, *Science* **246**, 1135 (1989).
53. A. Shevack, H. S. Gewitz, B. Hennemann, A. Yonath, H. G. Wittmann, *FEBS Lett.* **184**, 68 (1985).
54. K. vanBohlen *et al.*, *J. Mol. Biol.* **222**, 11 (1991).
55. Z. Otwinowski, in *Data Collection and Processing*, L. Sawyer, N. Isaacs, D. Bailey, Eds. (SERC Daresbury Laboratory, Warrington, UK, 1993), pp. 52–62.
56. E. de La Fortelle and G. Bricogne, *Methods Enzymol.* **276**, 472 (1997).
57. T. A. Jones, S. Cowan, J.-Y. Zou, M. Kjeldgaard, *Acta Crystallogr.* **A46**, 110 (1991).
58. B. L. Golden, V. Ramakrishnan, S. W. White, *EMBO J.* **12**, 4901 (1993).
59. M. A. Markus, A. P. Hinch, S. Huang, D. E. Draper, D. E. Torchia, *Nature Struct. Biol.* **4**, 70 (1997).
60. M. Leijonmarck, S. Eriksson, A. Liljas, *Nature* **286**, 824 (1980).
61. C. Davies, S. W. White, V. Ramakrishnan, *Structure* **4**, 55 (1996).
62. J. Unge *et al.*, *Structure* **6**, 1577 (1998).
63. K. S. Wilson, K. Appelt, J. Badger, I. Tanaka, S. W. White, *Proc. Natl. Acad. Sci. U.S.A.* **83**, 7251 (1986).
64. P. Nissen, J. Hansen, N. Ban, P. B. Moore, T. A. Steitz, *Science* **289**, 920 (2000).
65. R. R. Gutell *et al.*, in preparation. (Data can be found at www.rna.icmb.utexas.edu.)
66. M. Symanski, T. Specht, M. C. Barciszewska, J. Barciszewski, V. A. Erdmann, *Nucleic Acids Res.* **26**, 156 (1998).
67. M. Carson, *Methods Enzymol.* **277**, 493 (1997).
68. We thank B. Freeborn for her skilled technical assistance in preparing 50S ribosomal subunit material and crystals. We are indebted to D. Klein, M. Lu, S. Antoć, and M. Schmeing for their help with the fitting of protein sequences into electron density. We thank M. Kjeldgaard for providing us with a prerelease version of O adapted for RNA model building, J. Cate for contributing the iridium hexamine, R. Gutell and J. Cannone for preparing the secondary structure diagram for 23*S* rRNA, M. Wahl for sending us coordinates for L4 before their release, J. Williamson for sending us coordinates of the L30e-RNA complex before their release, R. Sweet, L. Berman, and M. Capel for their assistance with data collection at the National Synchrotron Light Source, and A. Joachimiak and the staff of 19ID at the Advanced Photon Source. Supported by grants from NIH to T.A.S. (GM22778) and P.B.M. (GM54216) and by a grant from the Agouron Institute to T.A.S. and P.B.M. N.B. is supported by a Burroughs Welcome Fund Career Award. Complete coordinates for 23*S* and 5*S* rRNAs and α carbon coordinates for the 27 proteins discussed have been deposited in the Protein Data Bank. The accession number is 1FFK for the amplitudes, experimental phases, and coordinates.

29 June 2000; accepted 24 July 2000

VOLUME 53, NUMBER 20 PHYSICAL REVIEW LETTERS 12 NOVEMBER 1984

Metallic Phase with Long-Range Orientational Order and No Translational Symmetry

D. Shechtman and I. Blech
Department of Materials Engineering, Israel Institute of Technology–Technion, 3200 Haifa, Israel

and

D. Gratias
Centre d'Etudes de Chimie Métallurgique, Centre National de la Recherche Scientifique, F-94400 Vitry, France

and

J. W. Cahn
Center for Materials Science, National Bureau of Standards, Gaithersburg, Maryland 20760
(Received 9 October 1984)

We have observed a metallic solid (Al–14-at.%-Mn) with long-range orientational order, but with icosahedral point group symmetry, which is inconsistent with lattice translations. Its diffraction spots are as sharp as those of crystals but cannot be indexed to any Bravais lattice. The solid is metastable and forms from the melt by a first-order transition.

PACS numbers: 61.50.Em, 61.55.Hg, 64.70.Ew

We report herein the existence of a metallic solid which diffracts electrons like a single crystal but has point group symmetery $m\bar{3}\bar{5}$ (icosahedral) which is inconsistent with lattice translations. If the specimen is rotated through the angles of this point group (Fig. 1), selected-area electron diffraction patterns clearly display the six fivefold, ten threefold, and fifteen twofold axes characteristic[1] of icosahedral symmetry (Fig. 2). Grains up to 2 μm in size with this structure form in rapidly cooled alloys of Al with 10–14 at.% Mn, Fe, or Cr. We will refer to the phase as the icosahedral phase. Microdiffraction from many different volume elements of a grain and dark-field imaging from various diffraction spots confirm that entire grains have long-range orientational order. If the orientational order decays with distance, its correlation length is far greater than the grain size. We have thus a solid metallic phase with no translational order and with with long-range orientational order.

The remarkable sharpness of the diffraction spots (Fig. 2) indicates a high coherency in the spatial interference, comparable to the one usually encountered in crystals. The diffraction data are qualitatively well fitted by a model consisting of a random packing of nonoverlapping parallel icosahedra attached by edges.[2] The invariance of the local orientational symmetry from site to site and the finite number of possible translations between two adjacent icosahedra seem to be sufficient for insuring highly coherent interferences. Icosahedra are a common packing unit in intermetallic crystals with the smaller transition element at the center surrounded by twelve larger atoms arranged like the corners of an icosahedron.[3] The symmetries of the crystals dictate that the several icosahedra in a unit cell have different orientations and allow them to be distorted, leaving the overall crystal consistent with the well-known crystallographic point and space groups. Even though icosahedral symmetry is of great importance as an approximate site symmetry in crystals, it cannot survive the imposition of lattice translations: Crystals cannot and do not exhibit the icosahedral point group symmetry.

Elementary crystallography indicates that fivefold

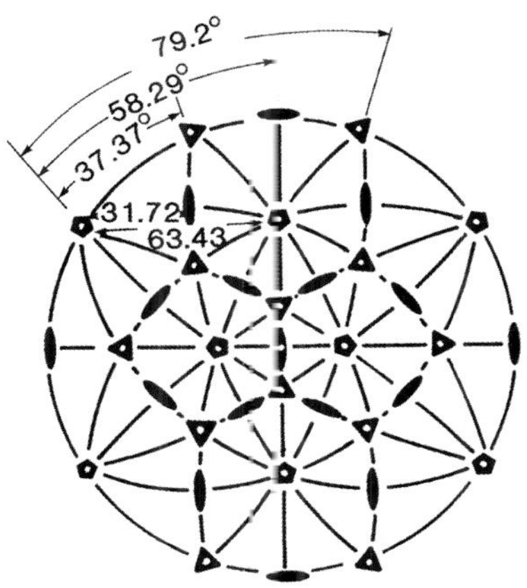

FIG. 1. Stereographic projection of the symmetry elements of the icosahedral group $m\bar{3}\bar{5}$.

VOLUME 53, NUMBER 20 PHYSICAL REVIEW LETTERS 12 NOVEMBER 1984

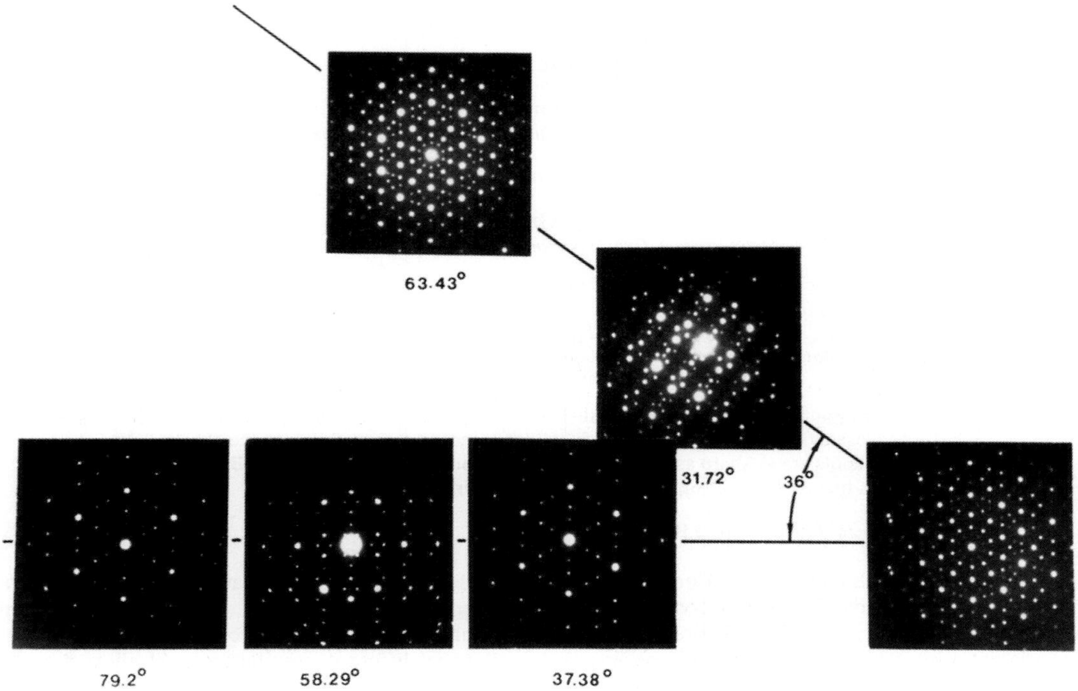

FIG. 2. Selected-area electron diffraction patterns taken from a single grain of the icosahedral phase. Rotations match those in Fig. 1.

axes are inconsistent with translational order.[4] Crystals with an apparent fivefold axis do occur as multiple twins.[5] Icosahedral symmetry could conceivably occur by multiple twinning, but at least five twins are required in order to obtain an icosahedral diffraction set from a conventional crystal. Experiments in the electron microscope and with x-ray diffraction contradict the twin hypothesis:

(1) A set of dark-field images taken from any reflection reveals no twins and in all cases the whole grain is illuminated.

(2) A convergent-beam diffraction pattern along the fivefold zone taken from an area 20 nm in diameter at any point of the grain displays all the reflections that appear in the selected-area diffraction pattern. This is also true when the thickness of the specimen is of the order of 10 nm.

(3) An x-ray diffraction pattern (Cu $K\alpha$ source) was taken from a single-phase sample of the material containing many grains of various orientations. Had the phase consisted of a multiply twinned crystalline structure, it should have been possible to index the powder pattern regardless of the twins. The pattern obtained from the icosahedral phase could not be indexed to any Bravais lattice.

On the basis of these experiments we conclude that the icosahedral phase does not consist of multiply twinned regular crystal structures.

The icosahedral phase forms during rapid cooling of the melt by a nucleation and growth mechanism. This mechanism is characteristic of a first-order transition because the two phases coexist along a moving interface. Each particle nucleates at a center and grows out from there. The atomic rearrangements that result in the orientational order of the icosahedral phase occur at this interface, and the two adjoining phases differ in entropy and, for some alloys, in composition. If the transition were higher order, ordering would occur everywhere in the liquid instead of being confined to interfaces. Our evidence for the first-order character is morphological. We examined alloys with 10–14 at.% Mn. Samples with 10% to 12% Mn showed many nodular grains separated from each other by crystalline films of fcc Al. The grains were approximately spherical in shape but were deeply indented with radial streaks composed of fcc Al crystals. The morphology is similar to a commonly observed one in rapid solidification in which crystals nucleate at many centers and grow until the remaining liquid solidifies, except that instead of isolated crystals we have isolated icosahedral grains embedded in solidified Al.

Another aspect of the first-order character of the transition is that segregation accompanies the growth of the icosahedral phase. In the 10%- and 12%-Mn samples the growing phase rejects Al into the liquid. The indentations in the nodules are characteristic of the cellular morphology caused by an instability in the diffusion layer surrounding the solid resulting from segregation during growth.[6] In the 14%-Mn alloys the icosahedral grains occupy almost all of the volume of the specimen. Only small amounts of fcc Al occur along some grain boundaries where the grains grew to impingement. Some radial streaking still points to the nucleation centers from which the grains grew. We conclude that little segregation occurs in this alloy and that the icosahedral phase contains 14% Mn.

The obvious cellular morphology indicates that the growth of the icosahedral phase is slow enough to permit diffusional segregation on the scale of 1 μm. With the diffusion coefficient in the liqud of order 10^{-8} m^2/s this indicates interface velocities of order 10^{-2} m/s and formation times of 10^{-4} s. This is 2 to 3 orders of magnitude slower than the fastest known crystallization velocities of a metallic liquid during rapid cooling, but it is a typical upper limit of velocity of crystallization with composition changes.[7] There is thus plenty of time for atomic rearrangements to occur at the interface between the melt and the icosahedral phase.

The icosahedral phase in rapidly solidified Al-Mn alloys is remarkably resistant to crystallization. Prolonged heating of 6 h at 300 °C and 1 h at 350 °C produced no detectable crystallization, but 1 h at 400 °C caused crystallization to the stable Al_6Mn phase. We conclude that the icosahedral phase is a truly metastable phase which nucleates and grows for a range of cooling rates which are slow enough to permit its formation but rapid enough to prevent crystallization, either from the melt or from the icosahedral phase after its formation.

The icosahedral phase has symmetries intermediate between those of a crystal and a liquid. It differs from other intermediate phases in that it is both solid, like a metallic glass, and that it has long-range orientational order. Many intermediate phases do have orientational order, but usually it is only local, and the transition to such phases is continuous.[8,9] The possibility of an icosahedral phase with long range order was inferred from a computer simulation,[10,11] and a first-order liquid-to-icosahedral phase transition has been predicted from a mean-field theory.[11,12]

We thank F. S. Biancaniello for spinning the alloys, C. R. Hubbard for x-ray experiments, and B. Burton for a critical review of the manuscript. This work was sponsored by the Defense Advanced Research Projects Agency and the National Science Foundation. This work was performed while one of us (D.S.) was a guest worker at the National Bureau of Standards and while another of us (D.G.) was at the Institute for Theoretical Physics, University of California at Santa Barbara.

[1] *International Tables for Crystallography* (Reidel, Higham, Mass., 1983), Vol. A, p. 179.

[2] D. Shechtman and I. Blech, to be published.

[3] F. C. Frank and J. S. Kasper, Acta Crystallogr. **11**, 184 (1958), and **12**, 483 (1959).

[4] M. J. Buerger, *Elementary Crystallography* (MIT Press, Cambridge, 1978).

[5] G. Friedel, *Lecon de Crystallographie* (Masson, Paris, 1921).

[6] W. W. Mullins and R. F. Sekerka, J. Appl. Phys. **35**, 444 (1963).

[7] W. J. Boettinger, F. S. Biancaniello, G. Kalonji, and J. W. Cahn, in *Rapid Solidification Processing II,* edited by R. Mehrabian *et al.* (Claitors, Baton Rouge, 1980), p. 50.

[8] D. R. Nelson, Phys. Rev. B **28**, 5515 (1983).

[9] J. Sadoc, J. Phys. (Paris), Colloq. **41**, C8-326 (1980).

[10] P. J. Steinhardt, D. R. Nelson and M. Rouchetti, Phys. Rev. Lett. **47**, 1297 (1981).

[11] P. J. Steinhardt, D. R. Nelson and M. Rouchetti, Phys. Rev. B **28**, 784 (1983).

[12] A. D. J. Haymet, Phys. Rev. B **27**, 1725 (1983).

ARTICLE

doi:10.1038/nature10361

Crystal structure of the β_2 adrenergic receptor–Gs protein complex

Søren G. F. Rasmussen[1,2*], Brian T. DeVree[3*], Yaozhong Zou[1], Andrew C. Kruse[1], Ka Young Chung[1], Tong Sun Kobilka[1], Foon Sun Thian[1], Pil Seok Chae[4], Els Pardon[5,6], Diane Calinski[3], Jesper M. Mathiesen[1], Syed T. A. Shah[7], Joseph A. Lyons[7], Martin Caffrey[7], Samuel H. Gellman[4], Jan Steyaert[5,6], Georgios Skiniotis[8], William I. Weis[1,9], Roger K. Sunahara[3] & Brian K. Kobilka[1]

G protein-coupled receptors (GPCRs) are responsible for the majority of cellular responses to hormones and neurotransmitters as well as the senses of sight, olfaction and taste. The paradigm of GPCR signalling is the activation of a heterotrimeric GTP binding protein (G protein) by an agonist-occupied receptor. The β_2 adrenergic receptor (β_2AR) activation of Gs, the stimulatory G protein for adenylyl cyclase, has long been a model system for GPCR signalling. Here we present the crystal structure of the active state ternary complex composed of agonist-occupied monomeric β_2AR and nucleotide-free Gs heterotrimer. The principal interactions between the β_2AR and Gs involve the amino- and carboxy-terminal α-helices of Gs, with conformational changes propagating to the nucleotide-binding pocket. The largest conformational changes in the β_2AR include a 14 Å outward movement at the cytoplasmic end of transmembrane segment 6 (TM6) and an α-helical extension of the cytoplasmic end of TM5. The most surprising observation is a major displacement of the α-helical domain of Gαs relative to the Ras-like GTPase domain. This crystal structure represents the first high-resolution view of transmembrane signalling by a GPCR.

Introduction

The β_2 adrenergic receptor (β_2AR) has been a model system for the large and diverse family of G protein-coupled receptors (GPCRs) for over 40 years. It was one of the first GPCRs to be characterized by radioligand binding, and it was the first neurotransmitter receptor to be cloned[1] and structurally determined by crystallography[2,3]. The β_2AR was initially identified based on its physiological and pharmacological properties, but it was not known if receptors and G proteins were separate entities, or parts of the same protein[4]. Subsequent biochemical studies led to the isolation and purification of functional β_2AR and Gs, the stimulatory G protein that activates adenylyl cyclase, and the reconstitution of this signalling complex in phospholipid vesicles[5,6]. The cooperative interactions of β_2AR and Gs observed in ligand binding assays formed the foundation of the ternary complex model of GPCR activation[7,8]. In the ternary complex consisting of agonist, receptor and G protein, the affinity of the receptor for agonist is enhanced and the specificity of the G protein for guanine nucleotides changes in favour of GTP over GDP. The GPCR field has evolved markedly since these initial studies. Isolation of the genes and cDNAs for the β_2AR and other GPCRs using protein sequencing and expression cloning led to the expansion of the family by homology cloning. More recently, sequencing of the human genome led to the identification of over 800 GPCR genes[9]. Experimental tools for identifying protein–protein interactions and for expression and silencing of genes have revealed a complex network of cellular signalling and regulatory pathways including G protein-independent activation of cytosolic kinases[10,11]. Nevertheless, the β_2AR continues to be a relevant model for most aspects of GPCR pharmacology, signalling and regulation.

Notwithstanding the remarkable advances in this field, we still know relatively little about the structural basis for transmembrane signalling by GPCRs. Figure 1 shows the G protein cycle for the β_2AR–Gs complex. Agonist binding to the β_2AR promotes interactions with GDP-bound Gs$\alpha\beta\gamma$ heterotrimer, leading to the exchange of GDP for GTP, and the functional dissociation of Gs into Gα-GTP and G$\beta\gamma$ subunits. The separate Gα-GTP and G$\beta\gamma$ subunits can modulate the activity of different cellular effectors (channels, kinases or other enzymes). The intrinsic GTPase activity of Gαs leads to hydrolysis of GTP to GDP and the reassociation of Gα-GDP and G$\beta\gamma$ subunits, and the termination of signalling. The active state of a GPCR can be defined as that conformation that couples to and stabilizes a nucleotide-free G protein. In this agonist-β_2AR–Gs ternary complex, Gs has a higher affinity for GTP than GDP, and the β_2AR has an approximately 100-fold higher affinity for agonists than does β_2AR alone. In an effort to understand the structural basis for GPCR signalling, we crystallized the β_2AR–Gs complex.

Crystallization of the β_2AR–Gs complex

The first challenge for crystallogenesis was to prepare a stable β_2AR–Gs complex in detergent solution. The β_2AR and Gs couple efficiently in lipid bilayers, but not in detergents used to solubilize and purify these proteins. We found that a relatively stable β_2AR–Gs complex could be prepared by mixing purified GDP-Gs (approximately 100 µM final concentration) with a molar excess of purified β_2AR bound to a high affinity agonist (BI-167107, Boehringer Ingelheim)[12] in dodecylmaltoside solution. Apyrase, a non-selective purine pyrophosphatase, was added to hydrolyse GDP released from Gs on forming a complex with the β_2AR. Removal of GDP was essential because both GDP and GTP

[1]Department of Molecular and Cellular Physiology, Stanford University School of Medicine, Stanford, California 94305, USA. [2]Department of Neuroscience and Pharmacology, The Panum Institute, University of Copenhagen, 2200 Copenhagen N, Denmark. [3]Department of Pharmacology, University of Michigan Medical School, Ann Arbor, Michigan 48109, USA. [4]Department of Chemistry, University of Wisconsin, Madison, Wisconsin 53706, USA. [5]Department of Molecular and Cellular Interactions, Vlaams Instituut voor Biotechnologie (VIB), Vrije Universiteit Brussel, B-1050 Brussel, Belgium. [6]Structural Biology Brussels, Vrije Universiteit Brussel, B-1050 Brussels, Belgium. [7]Membrane Structural and Functional Biology Group, Schools of Medicine and Biochemistry & Immunology, Trinity College, Dublin 2, Ireland. [8]Life Sciences Institute and Department of Biological Chemistry, University of Michigan, Ann Arbor, Michigan 48109, USA. [9]Department of Structural Biology, Stanford University School of Medicine, Stanford, California 94305, USA.
*These authors contributed equally to this work.

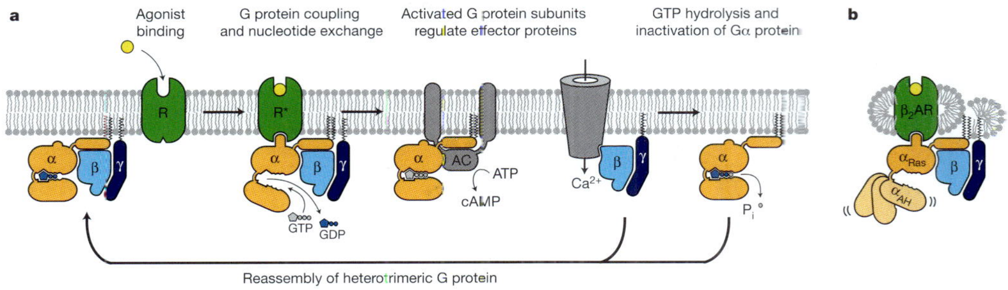

Figure 1 | G protein cycle for the β_2AR–Gs complex. a, Extracellular agonist binding to the β_2AR leads to conformational rearrangements of the cytoplasmic ends of transmembrane segments that enable the Gs heterotrimer (α, β, and γ) to bind the receptor. GDP is released from the α subunit upon formation of β_2AR–Gs complex. The GTP binds to the nucleotide-free α subunit resulting in dissociation of the α and $\beta\gamma$ subunits from the receptor. The subunits regulate their respective effector proteins adenylyl cyclase (AC) and Ca^{2+} channels. The Gs heterotrimer reassembles from α and $\beta\gamma$ subunits following hydrolysis of GTP to GDP in the α subunit. **b**, The purified nucleotide-free β_2AR–Gs protein complex maintained in detergent micelles. The Gαs subunit consists of two domains, the Ras domain (αRas) and the α-helical domain (αAH). Both are involved in nucleotide binding. In the nucleotide-free state, the αAH domain has a variable position relative the αRas domain.

can disrupt the high-affinity interaction between β_2AR and Gs (Supplementary Fig. 1a). The complex was subsequently purified by sequential antibody affinity chromatography and size-exclusion chromatography. The stability of the complex was enhanced by exchanging it into a recently developed maltose neopentyl glycol detergent MNG-3 (NG310, Anatrace)[13]. The complex could be incubated at room temperature for 24 h without any noticeable degradation; however, initial efforts to crystallize the complex using sparse matrix screens in detergent micelles, bicelles and lipidic cubic phase (LCP) failed.

To further assess the quality of the complex, we analysed the protein by single particle electron microscopy[34]. The results confirmed that the complex was monodisperse, but revealed two potential problems for obtaining diffraction of quality crystals. First, the detergent used to stabilize the complex formed a large micelle, leaving little polar surface on the extracellular side of the β_2AR–Gs complex for the formation of crystal lattice contacts. Our initial approach to this problem, which was to generate antibodies to the extracellular surface, was not successful. As an alternative approach, we replaced the unstructured amino terminus of the β_2AR with T4 lysozyme (T4L). We previously used T4L to facilitate crystallogenesis of the inactive β_2AR by inserting T4L between the cytoplasmic ends of TM5 and TM6 (ref. 3). Several different amino-terminal fusion proteins were prepared and single particle electron microscopy was used to identify a fusion with a relatively fixed orientation of T4L in relation to the β_2AR.

The second problem revealed by single particle electron microscopy analysis was increased variability in the positioning of the α-helical component of the Gαs subunit. Gαs consists of two domains, the Ras-like GTPase domain (GαsRas), which interacts with the β_2AR and the Gβ subunit, and the α-helical domain (GαsAH)[14]. The interface of the two Gαs subdomains forms the nucleotide-binding pocket (Fig. 1), and electron microscopy two-dimensional (2D) averages and three-dimensional (3D) reconstructions show that in the absence of guanine nucleotide, GαsAH has a variable position relative to the complex of T4L–β_2AR–GαsRas–G$\beta\gamma$ (Fig. 1b)[34].

We attributed the variable position of GαsAH to the empty nucleotide-binding pocket. However, as noted above both GDP and non-hydrolysable GTP analogues disrupt the β_2AR–Gs complex (Supplementary Fig. 1). The addition of the pyrophosphate analogue phosphonoformate (foscarnet) led to a significant increase in stabilization of GαsAH as determined by electron microscopy analysis of the detergent-solubilized complex[34]. Crystallization trials were carried out in LCP using a modified monolein (7.7 MAG, see Methods) designed to accommodate the large hydrophilic component of the T4L–β_2AR–Gs complex[15]. Although we were able to obtain small crystals that diffracted to 7 Å, we were unable to improve their quality through the use of additives and other modifications.

In an effort to generate an antibody that would further stabilize the complex and facilitate crystallogenesis, we crosslinked β_2AR and the Gs heterotrimer with a small, homobifunctional amine-reactive cross-linker and used this stabilized complex to immunize llamas. Llamas and other camelids produce antibodies devoid of light chains. The single domain antigen binding fragments of these heavy-chain-only antibodies, known as nanobodies, are small (15 kDa), rigid, and are easily cloned and expressed in *Escherichia coli* (Methods)[16]. We obtained a nanobody (Nb35) that binds to the complex and prevents dissociation of the complex by GTPγS (Supplementary Fig. 1). The T4L–β_2AR–Gs–Nb35 complex was used to obtain crystals that grew to 250 µm (Supplementary Fig. 2) in LCP (7.7 MAG) and diffracted to 2.9 Å. A 3.2 Å data set was obtained from 20 crystals and the structure was determined by molecular replacement (Methods).

The β_2AR–Gs complex crystallized in primitive monoclinic space group $P2_1$, with a single complex in each asymmetric unit. Figure 2a shows the crystallographic packing interactions. Complexes are arrayed in alternating aqueous and lipidic layers with lattice contacts formed almost exclusively between soluble components of the complex, leaving receptor molecules suspended between G protein layers and widely separated from one another in the plane of the membrane. Extensive lattice contacts are formed among all the soluble proteins, probably accounting for the strong overall diffraction and remarkably clear electron density for the G protein. Nb35 and T4L facilitated crystal formation. Nb35 packs at the interface of the Gβ and Gα subunits, with the complementarity determining region (CDR) 1 interacting primarily with Gβ and a long CDR3 loop interacting with both Gβ and Gα subunits. The framework regions of Nb35 from one complex also interact with Gα subunits from two adjacent complexes. T4L is linked to the β_2AR only through amino-terminal fusion, but packs against the amino terminus of the Gβ subunit of one complex, the carboxy terminus of the Gγ subunit of another complex, and the Gα subunit of yet another complex. Figure 2b shows the structure of the complete complex including T4L and Nb35, and Fig. 2c shows the β_2AR–Gs complex alone.

Structure of the active-state β_2AR

The β_2AR–Gs structure provides the first high-resolution insight into the mechanism of signal transduction across the plasma membrane by a GPCR, and the structural basis for the functional properties of the ternary complex. Figure 3a compares the structures of the agonist-bound receptor in the β_2AR–Gs complex and the inactive

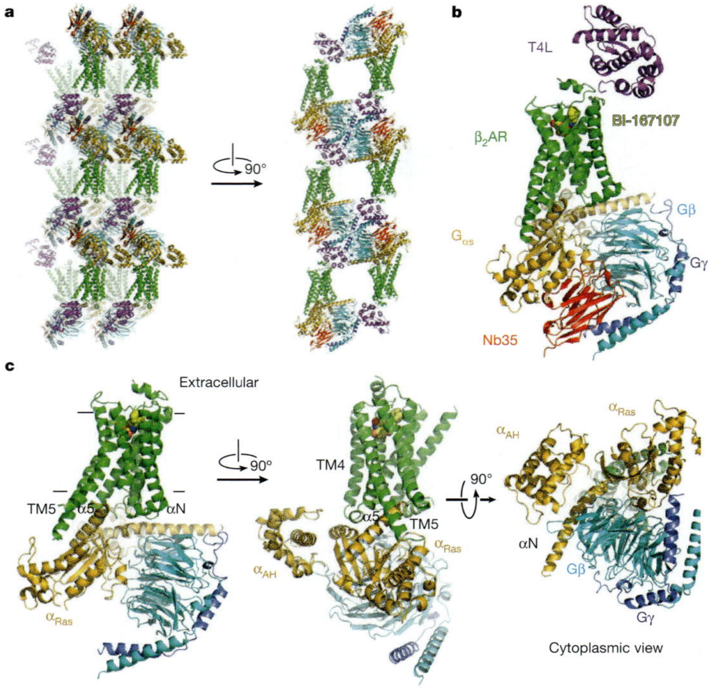

Figure 2 | Overall structure of the β_2AR–Gs complex. **a**, Lattice packing of the complex shows alternating layers of receptor and G protein within the crystal. Abundant contacts are formed among proteins within the aqueous layers. **b**, The overall structure of the asymmetric unit contents shows the β_2AR (green) bound to an agonist (yellow spheres) and engaged in extensive interactions with Gαs (orange). Gαs together with Gβ (cyan) and Gγ (purple) constitute the heterotrimeric G protein Gs. A Gs-binding nanobody (red) binds the G protein between the α and β subunits. The nanobody (Nb35) facilitates crystallization, as does T4 lysozyme (magenta) fused to the amino terminus of the β_2AR. **c**, The biological complex omitting crystallization aids, showing its location and orientation within a cell membrane.

carazolol-bound β_2AR. The largest difference between the inactive and active structures is a 14 Å outward movement of TM6 when measured at the Cα carbon of E268. There is a smaller outward movement and extension of the cytoplasmic end of the TM5 helix by 7 residues. A stretch of 26 amino acids in the third intracellular loop (ICL3) is disordered. Another notable difference between inactive and active structures is the second intracellular loop (ICL2), which forms an extended loop in the inactive β_2AR structure and an α-helix in the β_2AR–Gs complex. This helix is also observed in the β_2AR–Nb80 structure (Fig. 3b); however, it may not be a feature that is unique to the active state, because it is also observed in the inactive structure of the highly homologous avian β_1AR (ref. 17).

The quality of the electron density maps for the β_2AR is highest at the β_2AR–GαsRas interface, and much weaker for the extracellular half. The extracellular half of the receptor is not stabilized by any packing interactions either laterally with adjacent receptors in the membrane or through the extracellular surface. Instead, the extracellular region is indirectly tethered to the well-packed soluble components by the amino-terminal fusion to T4 lysozyme (Fig. 2a). Given the flexible and dynamic nature of GPCRs, the absence of stabilizing packing interactions may lead to structural heterogeneity in the extracellular half of the receptor and, consequently, to the limited quality of the electron density maps. However, the overall structure of the β_2AR in the T4L–β_2AR–Gs complex is very similar to our recent active-state structure of β_2AR stabilized by a G protein mimetic nanobody (Nb80)[12]. In the β_2AR–Nb80 crystal, each receptor molecule has extensive packing interactions with adjacent receptors and the quality of the electron density maps for the agonist-bound β_2AR in this complex is remarkably good for a 3.5 Å structure. Therefore, the β_2AR–Nb80 structure allows us to confidently model BI-167107 here, and provide a more reliable view of the conformational rearrangements of amino acids around the ligand-binding pocket and between the ligand-binding pocket and the Gs-coupling interface[12].

The overall root mean square deviation between the β_2AR components in the β_2AR–Gs and β_2AR–Nb80 structures is approximately 0.6 Å, and they differ most at the cytoplasmic ends of transmembrane helices 5 and 6 where they interact with the different proteins (Fig. 3b–d). The largest divergence is a 3 Å outward movement at the end of helix 6 in the β_2AR–Gs complex. However, the differences between these two structures are very small at the level of the most highly conserved amino acids (E/DRY and NPxxY), which are located at the cytoplasmic ends of the transmembrane segments (Fig. 3c, d). These conserved sequences have been proposed to be important for activation or for maintaining the receptor in the inactive state[18]. Of these residues, only Arg 131 differs significantly between these two structures. In β_2AR–Nb80 Arg 131 interacts with Nb80, whereas in the β_2AR–Gs structure Arg 131 packs against Tyr 391 of Gαs (Supplementary Fig. 3). The high structural similarity is in agreement with the functional similarity of these two proteins. The β_2AR–Nb80 complex shows the same high affinity for the agonist isoproterenol as does the β_2AR–Gs complex[12], consistent with high structural homology around the ligand binding pocket.

The active state of the β_2AR is stabilized by extensive interactions with GαsRas (Fig. 4). There are no direct interactions with Gβ or Gγ subunits. The total buried surface of the β_2AR–GαsRas interface is 2,576 Å^2 (1,300 Å^2 for GαsRas and 1,276 Å^2 for the β_2AR). This interface is formed by ICL2, TM5 and TM6 of the β_2AR, and by α5-helix, the αN–β1 junction, the top of the β3-strand, and the α4-helix of GαsRas (see Supplementary Table 1 for specific interactions). Some of the β_2AR sequences involved in this interaction have been shown to have a role in G protein coupling; however, there is no clear consensus sequence for Gs-coupling specificity when these segments are aligned with other GPCRs. Perhaps this is not surprising considering that the β_2AR also couples to Gi and that many GPCRs couple to more than one G protein isoform. Of the 21 amino acids of Gs that are within 4 Å of the β_2AR, only five are identical between Gs and Gi, and all of these

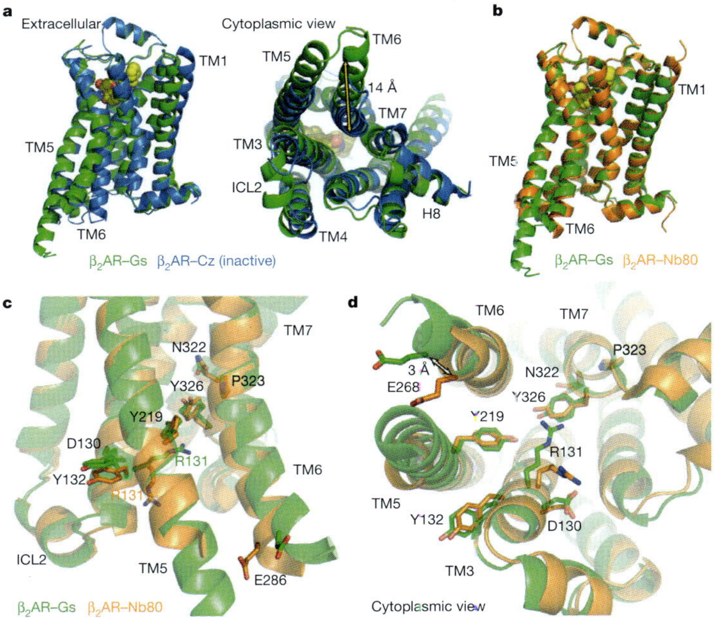

Figure 3 | Comparison of active and inactive β_2AR structures. **a**, Side and cytoplasmic views of the β_2AR–Gs structure (green) compared to the inactive carazolol-bound β_2AR structure[3] (blue). Significant structural changes are seen for the intracellular domains of TM5 and TM6. TM5 is extended by two helical turns whereas TM6 is moved outward by 14 Å as measured at the α-carbons of Glu 268 (yellow arrow) in the two structures. **b**, β_2AR–Gs compared with the nanobody-stabilized active state β_2AR–Nb80 structure[12] (orange). **c**, The positions of residues in the E/DRY and NPxxY motifs and other key residues of the β_2AR–Gs and β_2AR–Nb80 structures. All residues occupy very similar positions except Arg 131 which in the β_2AR–Nb80 structure interacts with the nanobody. **d**, View from the cytoplasmic side of residues shown in **c**.

are in the carboxy-terminal α helix. The structural basis for G protein coupling specificity must therefore involve more subtle features of the secondary and tertiary structure. Nevertheless, a noteworthy interaction involves Phe 139, which is located at the beginning of the ICL2 helix and sits in a hydrophobic pocket formed by Gαs His 41 at the beginning of the β1-strand, Val 217 at the start of the β3-strand and Phe 376, Cys 379, Arg 380 and Ile 383 in the α5-helix (Fig. 4c). The β_2AR mutant F139A has severely impaired coupling to Gs[19]. The residue corresponding to Phe 139 is a Phe or Leu on almost all Gs coupled receptors, but is more variable in GPCRs known to couple to other G proteins. Of interest, the ICL2 helix is stabilized by an interaction between Asp 130 of the conserved DRY sequence and Tyr 141 in the middle of the ICL2 helix (Fig. 4c). Tyr 141 has been shown to be a substrate for the insulin receptor tyrosine kinase[20]; however the functional significance of this phosphorylation is currently unknown.

The lack of direct interactions between the β_2AR and Gβγ is somewhat unexpected given that a heterotrimer is required for efficient coupling to a GPCR. Whereas Gβ does not interact directly with the β_2AR, it has an indirect but important role in coupling by stabilizing the amino-terminal α helix of Gαs (Fig. 2c). Several models involving GPCR dimers propose that one of the protomers interacts predominantly with Gα while the other interacts with Gβγ[21–23]. Consistent with these models, biochemical and biophysical evidence suggests that Gαi2 forms a stable complex with a LTB4 receptor dimer[24].

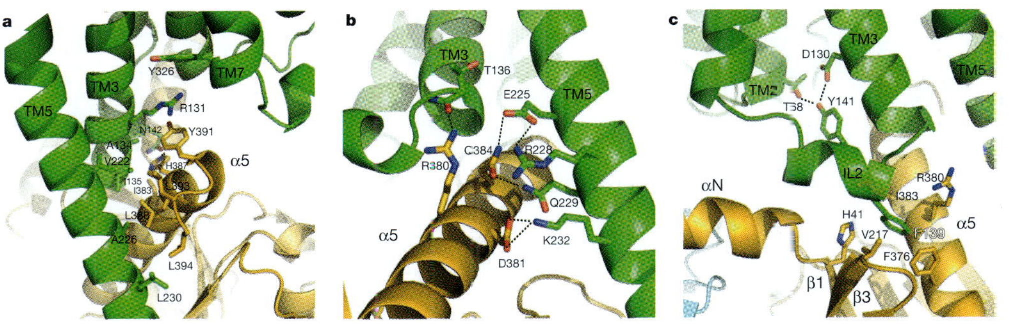

Figure 4 | Receptor-G protein interactions. **a**, **b**, The α5-helix of Gαs docks into a cavity formed on the intracellular side of the receptor by the opening of transmembrane helices 5 and 6. **a**, Within the transmembrane core, the interactions are primarily non-polar. An exception involves packing of Tyr 391 of the α5-helix against Arg 131 of the conserved DRY sequence in TM3 (see also Supplementary Fig. 3). Arg 131 also packs against Tyr 326 of the conserved NPxxY sequence in TM7. **b**, As α5-helix exits the receptor it forms a network of polar interactions with TM5 and TM3. **c**, Receptor residues Thr 68 and Asp 130 interact with the ICL2 helix of the β_2AR via Tyr 141, positioning the helix so that Phe 139 of the receptor docks into a hydrophobic pocket on the G protein surface, thereby structurally linking receptor–G protein interactions with the highly conserved DRY motif of the β_2AR.

Whereas the β_2AR efficiently activates Gs as a monomer, extensive biochemical and biophysical evidence supports the existence of β_2AR dimers or oligomers in living cells[25]. Therefore, we cannot exclude the possibility that in cell membranes one protomer of a β_2AR dimer may interact with the Gβγ subunit.

Structure of activated Gs

The most surprising observation in the β_2AR–Gs complex is the large displacement of the GαsAH relative to GαsRas (an approximately 127° rotation about the junction between the domains) (Fig. 5a). GαsAH moves as a rigid body as shown by the alignment of β_2AR–Gs with GαsAH from the crystal structure of Gαs–GTPγS[26] (Supplementary Fig. 4). In the structure of Gαs–GTPγS, the nucleotide-binding pocket is formed by the interface between GαsRas and GαsAH. Guanine nucleotide binding stabilizes the interaction between these two domains. The loss of this stabilizing effect of guanine nucleotide binding is consistent with the high flexibility observed for GαsAH in single particle electron microscopy analysis of the detergent-solubilized complex[34]. It is also in agreement with the increase in deuterium exchange at the interface between these two domains upon formation of the complex[35]. Recently double electron-electron resonance (DEER) spectroscopy was used to document large (up to 20 Å) changes in distance between nitroxide probes positioned on the Ras and α-helical domains of Gi upon formation of a complex with light-activated rhodopsin[27]. Finally, it has been shown that GαsRas and GαsAH can form a functional GTPase when expressed together as separate proteins[28]. Therefore, it is perhaps not surprising that GαsAH is displaced relative to GαsRas; however, its location in this crystal structure most probably reflects only one of an ensemble of conformations that it can adopt under physiological conditions, but has been stabilized by crystal packing interactions (Supplementary Fig. 5).

A potential concern is that Nb35, which was used to facilitate crystallogenesis, may be responsible for the displacement of the GαsAH. However, Nb35, which binds at the interface between Gβ and GαsRas (Fig. 2b and Supplementary Fig. 6), would not be expected to interact with the GαsAH or interfere with its interactions with GαsRas. None of the Nb35 contacts on the Ras domain are involved in interactions with GαsAH on the basis of the crystal structure of Gαs–GTPγS (1AZT). Moreover, if we superimpose the structures of the Ras domains of Gαs–GTPγS (1AZT) and β_2AR–Gs, there is no overlap between Nb35 and the α-helical domain of Gαs–GTPγS (Supplementary Fig. 6). Similarly, if we align the Gβ subunits of the Gi-GDP heterotrimer (1GP2) with that of β_2AR–Gs, there is no overlap between Nb35 and the α-helical domain of Gi (Supplementary Fig. 6). This analysis is in agreement with single particle electron microscopy studies which provide further evidence that Nb35 does not disrupt interactions between GαsAH and GαsRas[34].

The conformational links between the β_2AR and the nucleotide-binding pocket primarily involve the amino- and carboxy-terminal helices of Gαs (Fig. 4). Figure 5b focuses on the region of GαsRas that undergoes the largest conformational change when comparing the structure of GαsRas from the β_2AR–Gs complex with that from the Gαs–GTPγS complex[26]. The largest difference is observed for the α5-helix, which is displaced 6 Å towards the receptor and rotated as the carboxy-terminal end projects into transmembrane core of the β_2AR. Previous studies using a variety of approaches have demonstrated the important role of the α5-helix in GPCR–G protein interactions[29,30]. Associated with movement of the α5-helix, the β6-α5 loop, which interacts with the guanine ring in the Gαs–GTPγS structure, is displaced outward, away from the nucleotide-binding pocket (Fig. 5b–d). The movement of the α5-helix is also associated with changes in interactions between this helix and the β6-strand, the αN-β1 loop, and the α1-helix. The β1-strand forms another link between the β_2AR and the nucleotide-binding pocket. The carboxy-terminal end of this strand changes conformation around Gly 47, and there are further changes in the β1-α1 loop (P-loop) that coordinates the β-phosphate in the GDP and GTP-bound forms (Fig. 5b–d). The observations in the crystal structure are in agreement with deuterium exchange experiments where there is enhanced deuterium exchange in the β1-strand and the amino-terminal end of the α5-helix upon formation of the nucleotide-free β_2AR–Gs complex[35]. The deuterium exchange-mass spectrometry (DXMS) studies provide additional insights into the dynamic nature of these conformational changes in Gs upon complex formation[35].

The structure of a GDP-bound Gs heterotrimer has not been determined, so it is not possible to directly compare the Gαs–Gβγ interface before and after formation of the β_2AR–Gs complex. On the basis of the structure of the GDP-bound Gi heterotrimer[31], we do not observe large changes in interactions between GαsRas and Gβγ upon formation of the complex with β_2AR. This is also consistent with deuterium exchange studies[35]. As discussed above, Nb35 binds at the interface between GαsRas and Gβ (Fig. 2b); therefore, we cannot exclude the possibility that Nb35 may influence the relative orientation of the GαsRas–Gβγ interface in the crystal structure.

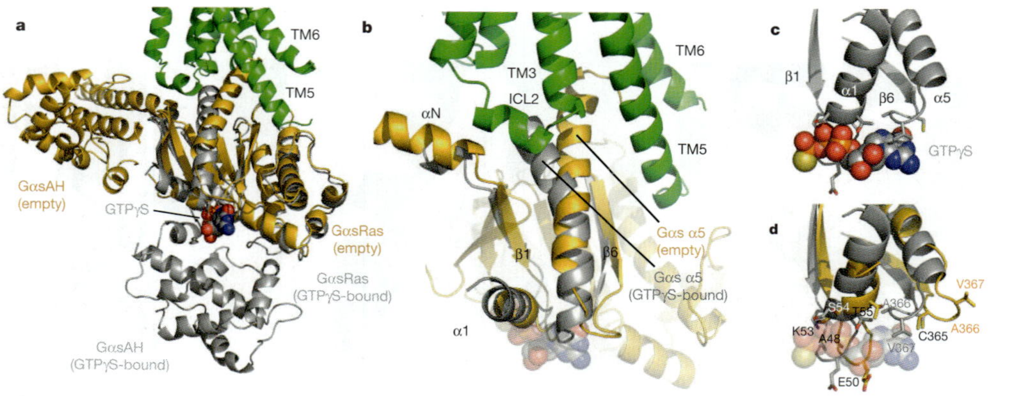

Figure 5 | Conformational changes in Gαs. **a**, A comparison of Gαs in the β_2AR–Gs complex (orange) with the GTPγS-bound Gαs (grey)[26] (PDB ID: 1AZT). GTPγS is shown as spheres. The helical domain of Gαs (GαsAH) shows a marked displacement relative to its position in the GTPγS-bound state. **b**, The α5-helix of Gαs is rotated and displaced towards the β_2AR, perturbing the β6-α5 loop, which otherwise forms part of the GTPγS-binding pocket. **c**, The β1-α1 loop (P-loop) and β6-α5 loop of Gαs interact with the phosphates and purine ring, respectively, of GTPγS in the Gαs–GTPγS structure. **d**, The β1-α1 and β6-α5 loops are rearranged in the nucleotide-free β_2AR–Gs structure.

Assembly of the β_2AR–Gs complex

Clues to the initial stages of complex formation may come from the recent active state structures of rhodopsin[32,33]. Figure 6a, b compares the active-state structure of β_2AR in the β_2AR–Gs complex with the recent structure of metarhodopsin II bound to a peptide representing the carboxy terminus of transducin[32]. The conformational changes in TM5 and TM6 are smaller in metarhodopsin II, and the position of the carboxy-terminal α helix of transducin is tilted by approximately 30° relative to the position of the homologous region of Gs. These may represent fundamental differences in the receptor–G protein interactions between these two proteins, but given the conservation of the G-protein binding pocket, the changes more probably reflect the more extensive contacts formed with the intact G protein. The position of the transducin peptide in metarhodopsin II may represent the initial interaction between a GDP-bound G protein and a GPCR. We have attempted to reproduce a similar complex between the β_2AR and a synthetic peptide representing the carboxy-terminal 20 amino acids of Gs, but did not observe any effect of this peptide on receptor function, possibly due to the solubility and behaviour of the peptide in solution. However, when the carboxy-terminal 20 amino acids of Gs are fused to the carboxy terminus of the β_2AR (Fig. 6c), we observe a 27-fold increase in agonist affinity (Fig. 6d). This effect is only 3.5-fold smaller than the effect we observe on agonist binding affinity in the β_2AR–Gs complex, and demonstrates that there is a functional interaction between the peptide and receptor that may represent an initial stage in β_2AR–Gs complex formation. Figure 6e and f presents a possible sequence of interactions of β_2AR and Gs when forming the nucleotide-free complex. The first interaction of the β_2AR with the Gs heterotrimer would require a movement of the carboxy terminus of the α5-helix away from the β6-strand to permit interactions with the β_2AR similar to those observed in metarhodopsin II (Fig. 6e). The availability of the carboxy terminus of the α5-helix for interactions with the β_2AR is supported by deuterium exchange studies[35] showing that this segment is more dynamic in the Gs-GDP heterotrimer than would be expected from the crystal structure of Gαs[26]. The subsequent formation of more extensive interactions between the β_2AR ICL2 and the amino terminus of Gαs requires a rotation of GαsRas relative to the receptor and would be associated with further conformational changes in both β_2AR and GαsRas (Fig. 6f). We cannot say when GDP is released during the formation of the complex; however, we speculate that uncoupling of the GαsAH from GαsRas is a consequence of nucleotide release or at least a coincident event. This binding model is in agreement with deuterium exchange experiments[35].

The β_2AR–Gs complex crystal structure provides the first high-resolution view of transmembrane signalling for a GPCR. We now have a framework to design experiments to investigate the mechanism of complex formation, GTP binding and complex dissociation. Of particular interest will be studies designed to determine the functional significance of the large movement of GαsAH relative to GαsRas that is observed in the β_2AR–Gs complex. A better understanding of the structural basis for G protein activation may provide new approaches for drug discovery. The high degree of structural homology within the ligand-binding pocket has posed challenges for developing highly selective drugs for specific GPCR targets. In contrast, there is relatively low homology at the interface between the β_2AR and Gαs, so identifying sequence and structural features that define specificity for particular G proteins may enable the development of selective inhibitors of specific GPCR–G protein interactions.

METHODS SUMMARY

The β_2AR–Gs complex was crystallized from β_2AR and Gs protein expressed in insect cells. Crystallogenesis was aided by fusing T4 lysozyme to the amino terminus of the β_2AR and the addition of a nanobody (Nb35) that binds at the interface between the Gα and Gβ subunits. Crystals were grown in a LCP using 7.7 MAG, a lipid that accommodates membrane proteins with larger hydrophilic surfaces[15]. Diffraction data were measured at beamline 23ID-B of the Advanced Photon Source and the structure was solved by molecular replacement. For more experimental details see Methods.

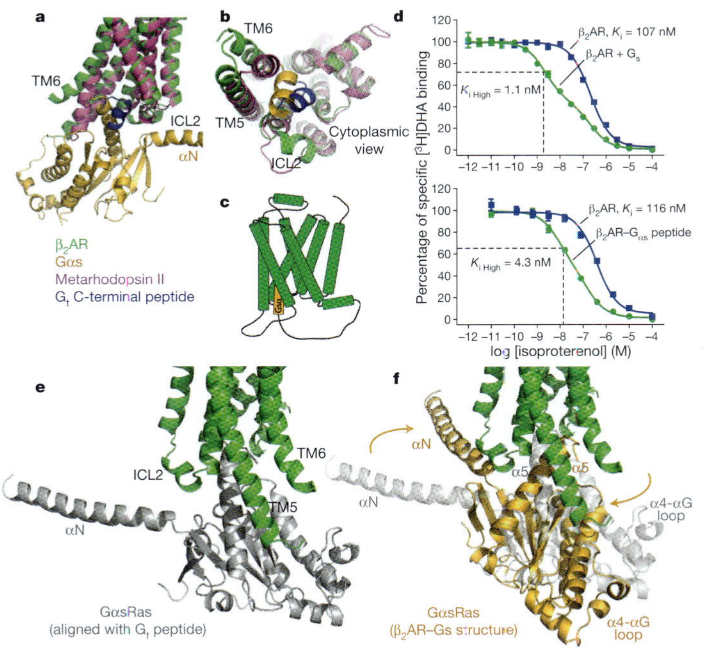

Figure 6 | Possible sequence of β_2AR–Gs complex formation. a, b, Comparison of the β_2AR–Gs structure (green and orange) with metarhodopsin II[32] (PDB ID: 3PQR) (purple) bound with the carboxy-terminal peptide of transducin (G_t) (blue). TM7 has been omitted in panel **a** to better visualize the G proteins. **c,** Cartoon of the β_2AR–Gαs peptide fusion construct used in the binding experiments (**d**). **d,** Competition binding experiments between [^{3}H]dihydroalprenolol ([^{3}H]DHA) and full agonist isoproterenol. Top panel shows binding data (reproduced from ref. 12) on β_2AR reconstituted in HDL particles with and without Gs heterotrimer. The fraction of β_2AR in the $K_{i\,High}$ state for the β_2AR with Gs is 0.55. Bottom panel shows binding to β_2AR and a β_2AR–Gαs peptide fusion expressed in *Sf9* cell membranes. The fraction of β_2AR in the $K_{i\,High}$ state for the β_2AR–Gαs peptide fusion is 0.68. **e,** The initial interaction of agonist-bound β_2AR and GαsRas may involve an orientation of the carboxy terminus of GαsRas similar to that of the carboxy-terminal peptide of transducin in the structure of metarhodopsin II. **f,** The final position of GαsRas on the β_2AR as observed in the β_2AR–Gs complex.

Full Methods and any associated references are available in the online version of the paper at www.nature.com/nature.

Received 8 June; accepted 11 July 2011.
Published online 19 July 2011.

1. Dixon, R. A. *et al.* Cloning of the gene and cDNA for mammalian β-adrenergic receptor and homology with rhodopsin. *Nature* **321,** 75–79 (1986).
2. Rasmussen, S. G. *et al.* Crystal structure of the human β_2 adrenergic G-protein-coupled receptor. *Nature* **450,** 383–387 (2007).
3. Rosenbaum, D. M. *et al.* GPCR engineering yields high-resolution structural insights into β_2-adrenergic receptor function. *Science* **318,** 1266–1273 (2007).
4. Lefkowitz, R. J. Seven transmembrane receptors: something old, something new. *Acta Physiol. (Oxf.)* **190,** 9–19 (2007).
5. Brandt, D. R., Asano, T., Pedersen, S. E. & Ross, E. M. Reconstitution of catecholamine-stimulated guanosinetriphosphatase activity. *Biochemistry* **22,** 4357–4362 (1983).
6. Cerione, R. A. *et al.* The mammalian β_2-adrenergic receptor: reconstitution of functional interactions between pure receptor and pure stimulatory nucleotide binding protein of the adenylate cyclase system. *Biochemistry* **23,** 4519–4525 (1984).
7. Ross, E. M., Maguire, M. E., Sturgill, T. W., Biltonen, R. L. & Gilman, A. G. Relationship between the β-adrenergic receptor and adenylate cyclase. Studies of ligand binding and enzyme activity in purified membranes of S49 lymphoma cells. *J. Biol. Chem.* **252,** 5761–5775 (1977).
8. De Lean, A., Stadel, J. M. & Lefkowitz, R. J. A ternary complex model explains the agonist-specific binding properties of the adenylate cyclase-coupled β-adrenergic receptor. *J. Biol. Chem.* **255,** 7108–7117 (1980).
9. Fredriksson, R., Lagerstrom, M. C., Lundin, L. G. & Schioth, H. B. The G-protein-coupled receptors in the human genome form five main families. Phylogenetic analysis, paralogon groups, and fingerprints. *Mol. Pharmacol.* **63,** 1256–1272 (2003).
10. Azzi, M. *et al.* β-arrestin-mediated activation of MAPK by inverse agonists reveals distinct active conformations for G protein-coupled receptors. *Proc. Natl Acad. Sci. USA* **100,** 11406–11411 (2003).
11. Lefkowitz, R. J. & Shenoy, S. K. Transduction of receptor signals by β-arrestins. *Science* **308,** 512–517 (2005).
12. Rasmussen, S. G. *et al.* Structure of a nanobody-stabilized active state of the β_2 adrenoceptor. *Nature* **469,** 175–180 (2011).
13. Chae, P. S. *et al.* Maltose-neopentyl glycol (MNG) amphiphiles for solubilization, stabilization and crystallization of membrane proteins. *Nature Methods* **7,** 1003–1008 (2010).
14. Sprang, S. R. G protein mechanisms: insights from structural analysis. *Annu. Rev. Biochem.* **66,** 639–678 (1997).
15. Misquitta, L. V. *et al.* Membrane protein crystallization in lipidic mesophases with tailored bilayers. *Structure* **12,** 2113–2124 (2004).
16. Domanska, K. *et al.* Atomic structure of a nanobody-trapped domain-swapped dimer of an amyloidogenic β2-microglobulin variant. *Proc. Natl Acad. Sci. USA* **108,** 1314–1319 (2011).
17. Warne, T. *et al.* Structure of a β_1-adrenergic G-protein-coupled receptor. *Nature* **454,** 486–491 (2008).
18. Hofmann, K. P. *et al.* A G protein-coupled receptor at work: the rhodopsin model. *Trends Biochem. Sci.* **34,** 540–552 (2009).
19. Moro, O., Lameh, J., Hogger, P. & Sadee, W. Hydrophobic amino acid in the i2 loop plays a key role in receptor-G protein coupling. *J. Biol. Chem.* **268,** 22273–22276 (1993).
20. Baltensperger, K. *et al.* The β-adrenergic receptor is a substrate for the insulin receptor tyrosine kinase. *J. Biol. Chem.* **271,** 1061–1064 (1996).
21. Jastrzebska, B., Tsybovsky, Y. & Palczewski, K. Complexes between photoactivated rhodopsin and transducin: progress and questions. *Biochem. J.* **428,** 1–10 (2010).
22. Johnston, C. A. & Siderovski, D. P. Receptor-mediated activation of heterotrimeric G-proteins: current structural insights. *Mol. Pharmacol.* **72,** 219–230 (2007).
23. Breitwieser, G. E. G protein-coupled receptor oligomerization: implications for G protein activation and cell signaling. *Circ. Res.* **94,** 17–27 (2004).
24. Banères, J. L. & Parello, J. Structure-based analysis of GPCR function: evidence for a novel pentameric assembly between the dimeric leukotriene B4 receptor BLT1 and the G-protein. *J. Mol. Biol.* **329,** 815–829 (2003).
25. Angers, S., Salahpour, A. & Bouvier, M. Dimerization: an emerging concept for G protein-coupled receptor ontogeny and function. *Annu. Rev. Pharmacol. Toxicol.* **42,** 409–435 (2002).
26. Sunahara, R. K., Tesmer, J. J., Gilman, A. G. & Sprang, S. R. Crystal structure of the adenylyl cyclase activator $G_{s\alpha}$. *Science* **278,** 1943–1947 (1997).
27. Van Eps, N. *et al.* Interaction of a G protein with an activated receptor opens the interdomain interface in the α subunit. *Proc. Natl Acad. Sci. USA* **108,** 9420–9424 (2011).
28. Markby, D. W., Onrust, R. & Bourne, H. R. Separate GTP binding and GTPase activating domains of a Gα subunit. *Science* **262,** 1895–1901 (1993).
29. Conklin, B. R. & Bourne, H. R. Structural elements of Gα subunits that interact with Gβγ, receptors, and effectors. *Cell* **73,** 631–641 (1993).
30. Oldham, W. M. & Hamm, H. E. Heterotrimeric G protein activation by G-protein-coupled receptors. *Nature Rev. Mol. Cell Biol.* **9,** 60–71 (2008).
31. Wall, M. A. *et al.* The structure of the G protein heterotrimer $G_{i\alpha1}\beta_1\gamma_2$. *Cell* **83,** 1047–1058 (1995).
32. Choe, H. W. *et al.* Crystal structure of metarhodopsin II. *Nature* **471,** 651–655 (2011).
33. Standfuss, J. *et al.* The structural basis of agonist-induced activation in constitutively active rhodopsin. *Nature* **471,** 656–660 (2011).
34. Westfield, G. *et al.* Structurall flexibility of the Gαs α-helical domain in the β2-adrenoceptor Gs complex. *Proc. Nati Acad. Sci. USA* doi:10.1073/pnas.1113645108 (in the press).
35. Chung, K. Y. *et al.* β2 adrenergic receptor-induced conformational changes in the heterotrimeric G protein Gs. *Nature* doi:10.1038/nature10488 (this issue).

Supplementary Information is linked to the online version of the paper at www.nature.com/nature.

Acknowledgements We acknowledge support from National Institutes of Health Grants NS028471 (B.K.K.) and GM083118 (B.K.K. and R.K.S.), GM56169 (W.I.W.), P01 GM75913 (S.H.G), T32-GM008270 and P60DK-20572 (R.K.S.), GM75915, P50GM073210 and U54GM094599 (M.C.), Science Foundation Ireland (07/IN.1/B1836) and FP7 COST Action CM0902 (M.C.),the Mathers Foundation (B.K.K. and W.I.W.), the Lundbeck Foundation (Junior Group Leader Fellowship, S.G.F.R.), the University of Michigan Biomedical Sciences Scholars Program (R.K.S.), the Fund for Scientific Research of Flanders (FWO-Vlaanderen) and the Institute for the encouragement of Scientific Research and Innovation of Brussels (ISRIB) (E.P. and J.S.), The Danish Council for Independent Research, Medical Sciences (J.M.M.). We thank D. Grayson and A. Coughlan for help with lipid synthesis.

Author Contributions S.G.F.R. performed the final stages of β_2AR purification; assisted with β_2AR and Gs protein virus production and expression in insect cell cultures; worked out conditions to form and stabilize the β_2AR–Gs complex following screening, identification and characterization of the BI-167107 agonist and MNG-3 detergent; developed the β_2AR–Gs complex purification strategy with B.K.K. and characterized the stability of the complex under a variety of conditions; purified and analysed all preparations of the β_2AR–Gs complex used for crystallography, DXMS and electron microscopy studies, immunization, and nanobody selection; expressed and purified nanobodies and characterized β_2AR–Gs–Nb binding by size exclusion chromatography; set up crystallization trials in detergent solution, bicelles and lipidic cubic phase; crystallized the T4L–β_2AR–Gs, T4L–β_2AR–Gs–Nb37 and T4L–β_2AR–Gs–Nb35 complexes; optimized crystallization conditions and grew crystals for data collection; assisted with data collection and manuscript preparation. B.T.D. managed Gs heterotrimer subunit virus production and titration; expressed and purified Gs protein; with R.K.S. he identified the use of apyrase in forming the β_2AR–Gs complex and foscarnet/pyrophosphate during crystallogenesis; reconstituted the β_2AR–Gs complex and receptor alone in high density lipoprotein particles which were used for the initial nanobody selection. He assisted with data collection. Y.Z. designed, generated and optimized the T4L–β_2AR fusion protein, characterized its expression and functional properties, and prepared baculovirus for large scale expression. A.C.K. collected crystals, collected and processed diffraction data, solved and refined the structure, and assisted with manuscript preparation. K.Y.C. developed the cross-linking conditions for the purified β_2AR–Gs complex used for immunization of llamas. E.P. performed llama immunization, cloned and expressed nanobodies and performed initial selections. J.S. supervised nanobody production. D.C. assisted with Gs heterotrimer expression and purification. J.M.M. generated the β_2AR–Gsα peptide fusion construct, expressed it in insect cell membranes and performed competition binding experiments. F.S.T. expressed β_2AR in insect cell cultures and with T.S.K. performed the initial stage of β_2AR purification. S.T.A.S., J.A.L. and M.C. provided the 7.7 MAG lipid and helpful suggestions for lipidic mesophase crystallization using this lipid. P.S.C. and S.H.G. provided MNG-3 detergent for stabilization of the β_2AR–Gs complex. G.S. provided the essential feedback from the electron microscopy studies that helped guide the crystallization effort. W.I.W. oversaw data processing, structure determination and refinement. R.K.S. supervised Gs protein production, provided valuable ideas and insights into Gs structure and function, and assisted with data collection and manuscript preparation. B.K.K. was responsible for overall project strategy and management, harvested crystals and assisted with collection of diffraction data, and wrote the manuscript.

Author Information Coordinates and structure factors for the β_2AR–Gs complex are deposited in the Protein Data Bank (accession code 3SN6). Reprints and permissions information is available at www.nature.com/reprints. The authors declare no competing financial interests. Readers are welcome to comment on the online version of this article at www.nature.com/nature. Correspondence and requests for materials should be addressed to B.K.K. (kobilka@stanford.edu) or R.K.S. (sunahara@umich.edu).

RESEARCH ARTICLE

METHODS

Expression and purification of β_2AR, Gs heterotrimer and nanobody-35. An amino-terminally fused T4 lysozyme–β_2AR construct with β_2AR truncated in position 365 (T4L–β_2AR, described in detail below) was expressed in *Sf9* insect cell cultures infected with recombinant baculovirus (BestBac, Expression Systems), and solubilized in *n*-dodecyl-β-D-maltoside (DDM) according to methods described previously[36] (see Supplementary Fig. 7 for purification overview). A β_2AR construct truncated after residue 365 (β_2AR-365) was used for the majority of the analytical experiments. M1 Flag affinity chromatography (Sigma) served as the initial purification step followed by alprenolol-Sepharose chromatography for selection of functional receptor. A subsequent M1 Flag affinity chromatography step was used to exchange receptor-bound alprenolol for high-affinity agonist BI-167107. The agonist-bound receptor was eluted, dialysed against buffer (20 mM HEPES, pH 7.5, 100 mM NaCl, 0.1% DDM and 10 μM BI-167107), treated with lambda phosphatase (New England Biolabs), and concentrated to approximately 50 mg ml^{-1} with a 50 kDa molecular weight cut off (MWCO) Millipore concentrator. Prior to spin concentration, the β_2AR-365 construct, but not T4L–β_2AR, was treated with PNGaseF (New England Biolabs) to remove amino-terminal amino-linked glycosylation. The purified receptor was routinely analysed by SDS–PAGE/Coomassie brilliant blue staining (see Supplementary Fig. 8a).

Bovine Gαs short, His$_6$-rat Gβ$_1$ and bovine Gγ$_2$ were expressed in HighFive insect cells (Invitrogen) grown in Insect Xpress serum-free media (Lonza). Cultures were grown to a density of 1.5 million cells per ml and then infected with three separate *Autographa californica* nuclear polyhedrosis viruses each containing the gene for one of the G protein subunits at a 1:1 multiplicity of infection (the viruses were a gift from A. Gilman). After 40–48 h of incubation the infected cells were harvested by centrifugation and resuspended in 75 ml lysis buffer (50 mM HEPES, pH 8.0, 65 mM NaCl, 1.1 mM $MgCl_2$, 1 mM EDTA, 1× PTT (35 μg ml^{-1} phenylmethanesulphonyl fluoride, 32 μg ml^{-1} tosyl phenylalanyl chloromethyl ketone, 32 μg ml^{-1} tosyl lysyl chloromethyl ketone), 1× LS (3.2 μg ml^{-1} leupeptin and 3.2 μg ml^{-1} soybean trypsin inhibitor), 5 mM β-mercaptoethanol (β-ME), and 10 μM GDP) per litre of culture volume. The suspension was pressurized with 600 p.s.i. N_2 for 40 min in a nitrogen cavitation bomb (Parr Instrument Company). After depressurization, the lysate was centrifuged to remove nuclei and unlysed cells, and then ultracentrifuged at 180,000*g* for 40 min. The pelleted membranes were resuspended in 30 ml wash buffer (50 mM HEPES, pH 8.0, 50 mM NaCl, 100 μM $MgCl_2$, 1× PTT, 1× LS, 5 mM β-ME, 10 μM GDP) per litre culture volume using a Dounce homogenizer and centrifuged again at 180,000*g* for 40 min. The washed pellet was resuspended in a minimal volume of wash buffer and flash-frozen with liquid nitrogen.

The frozen membranes were thawed and diluted to a total protein concentration of 5 mg ml^{-1} with fresh wash buffer. Sodium cholate detergent was added to the suspension at a final concentration of 1.0%, $MgCl_2$ was added to a final concentration of 5 mM, and 0.05 mg of purified protein phosphatase 5 (prepared in house) was added per litre of culture volume. The sample was stirred on ice for 40 min, and then centrifuged at 180,000*g* for 40 min to remove insoluble debris. The supernatant was diluted fivefold with Ni-NTA load buffer (20 mM HEPES, pH 8.0, 363 mM NaCl, 1.25 mM $MgCl_2$, 6.25 mM imidazole, 0.2% Anzergent 3-12, 1× PTT, 1× LS, 5 mM β-ME, 10 μM GDP), taking care to add the buffer slowly to avoid dropping the cholate concentration below its critical micelle concentration too quickly. Ni-NTA resin (3 ml; Qiagen) pre-equilibrated in Ni-NTA wash buffer 1 (20 mM HEPES, pH 8.0, 300 mM NaCl, 2 mM $MgCl_2$, 5 mM imidazole, 0.2% cholate, 0.15% Anzergent 3-12, 1× PTT, 1× LS, 5 mM β-ME, 10 μM GDP) per litre culture volume was added and the sample was stirred on ice for 20 min. The resin was collected into a gravity column and washed with 4× column volumes of Ni-NTA wash buffer 1, Ni-NTA wash buffer 2 (20 mM HEPES, pH 8.0, 50 mM NaCl, 1 mM $MgCl_2$, 10 mM imidazole, 0.15% Anzergent 3-12, 0.1% DDM, 1× PTT, 1× LS, 5 mM β-ME, 10 μM GDP), and Ni-NTA wash buffer 3 (20 mM HEPES, pH 8.0, 50 mM NaCl, 1 mM $MgCl_2$, 5 mM imidazole, 0.1% DDM, 1× PTT, 1× LS, 5 mM β-ME, 10 μM GDP). The protein was eluted with Ni-NTA elution buffer (20 mM HEPES, pH 8.0, 40 mM NaCl, 1 mM $MgCl_2$, 200 mM imidazole, 0.1% DDM, 1× PTT, 1× LS, 5 mM β-ME, 10 μM GDP). Protein-containing fractions were pooled and $MnCl_2$ was added to a final concentration of 100 μM. Purified lambda protein phosphatase (50 μg; prepared in house) was added per litre of culture volume and the eluate was incubated on ice with stirring for 30 min. The eluate was passed through a 0.22-μm filter and loaded directly onto a MonoQ HR 16/10 column (GE Healthcare) equilibrated in MonoQ buffer A (20 mM HEPES, pH 8.0, 50 mM NaCl, 100 μM $MgCl_2$, 0.1% DDM, 5 mM β-ME, 1× PTT). The column was washed with 150 ml buffer A at 5 ml min^{-1} and bound proteins were eluted over 350 ml with a linear gradient up to 28% MonoQ buffer B (same as buffer A except with 1 M NaCl). Fractions were collected in tubes spotted with enough GDP to make a final concentration of 10 μM. The Gs-containing fractions were concentrated to 2 ml using a stirred ultrafiltration cell (Amicon) with a 10 kDa nominal molecular weight limit (NMWL) regenerated cellulose membrane (Millipore). The concentrated sample was run on a Superdex 200 prep grade XK 16/70 column (GE Healthcare) equilibrated in S200 buffer (20 mM HEPES, pH 8.0, 100 mM NaCl, 1.1 mM $MgCl_2$, 1 mM EDTA, 0.012% DDM, 100 μM TCEP, 2 μM GDP). The fractions containing pure Gs were pooled, glycerol was added to 10% final concentration, and then the protein was concentrated to at least 10 mg ml^{-1} using a 30 kDa MWCO centrifugal ultrafiltration device (Millipore). The concentrated sample was then aliquoted, flash frozen, and stored at −80 °C. A typical yield of final, purified Gs heterotrimer from 8 l of cell culture volume was 6 mg.

Nanobody-35 (Nb35) was expressed in the periplasm of *E. coli* strain WK6, extracted, and purified by nickel affinity chromatography according to previously described methods[12] followed by ion-exchange chromatography (Supplementary Fig. 9a) using a Mono S 10/100 GL column (GE Healthcare). Selected Nb35 fractions were dialysed against buffer (10 mM HEPES, pH 7.5, 100 mM NaCl) and concentrated to approximately 65 mg ml^{-1} with a 10 kDa MWCO Millipore concentrator.

Complex formation, stabilization and purification. Formation of a stable complex (see Supplementary Fig. 10) was accomplished by mixing Gs heterotrimer at approximately 100 μM concentration with BI-167107-bound T4L–β_2AR (or β_2AR-365) in molar excess (approximately 130 μM) in 2 ml buffer (10 mM HEPES, pH 7.5, 100 mM NaCl, 0.1% DDM, 1 mM EDTA, 3 mM $MgCl_2$, 10 μM BI-167107) and incubating for 3 h at room temperature. BI-167107, which was identified from screening and characterizing approximately 50 different β_2AR agonists (data not shown), has a dissociation half-time of approximately 30 h, providing a higher degree of stabilization to the active G protein-bound receptor than other full agonists such as isoproterenol[12]. To maintain the high-affinity nucleotide-free state of the complex, apyrase (25 mU ml^{-1}, NEB) was added after 90 min to hydrolyse residual GDP released from Gαs upon binding to the receptor. GMP resulting from hydrolysis of GDP by apyrase has very poor affinity for the G protein in the complex. Rebinding of GDP can cause dissociation of the β_2AR–Gs complex (Supplementary Fig. 11a).

The β_2AR–Gs complex in DDM shows significant dissociation after 48 h at 4 °C (Supplementary Fig. 11a). We screened and characterized over 50 amphiphiles (data not shown) and identified MNG-3 (refs 12, 13; NG-310, Affymetrix-Anatrace) and its closely related analogues as detergents that substantially stabilize the complex (Supplementary Fig. 11a). The complex was exchanged into MNG-3 by adding the β_2AR–Gs mixture (2 ml) to 8 ml buffer (20 mM HEPES, pH 7.5, 100 mM NaCl, 10 μM BI-167107) containing 1% MNG-3 for 1 h at room temperature.

At this stage the mixture contains the β_2AR–Gs complex, non-functional Gs, and an excess of β_2AR. To separate functional β_2AR–Gs complex from non-functional Gs, and to complete the detergent exchange, the β_2AR–Gs complex was immobilized on M1 Flag resin and washed in buffer (20 mM HEPES, pH 7.5, 100 mM NaCl, 10 μM BI-167107, and 3 mM $CaCl_2$) containing 0.2% MNG-3. To prevent cysteine bridge-mediated aggregation of β_2AR–Gs complexes, 100 μM TCEP was added to the eluted protein before concentrating it with a 50 kDa MWCO Millipore concentrator. Of note, it was discovered later that crystal growth improved at even higher TCEP concentrations (above 1 mM) compared to 100 μM TCEP, and that the integrity of the β_2AR–Gs complex in MNG-3 was stable to 10 mM TCEP as measured by gel filtration analysis (Supplementary Fig. 12). In contrast, DDM-solubilized β_2AR loses its ability to bind the high-affinity antagonist [^{3}H]dihydroalprenolol ([^{3}H]DHA) in 10 mM TCEP (data not shown), probably due to disruption of extracellular disulphide bonds. Iodoacetamide could not be used to block reactive cysteines on Gs α and β subunits as it caused dissociation of the β_2AR–Gs complex (Supplementary Fig. 12b). The final size exclusion chromatography procedure to separate excess free receptor from the β_2AR–Gs complex (Supplementary Fig. 8b) was performed on a Superdex 200 10/300 GL column (GE Healthcare) equilibrated with buffer containing 0.02% MNG-3, 10 mM HEPES, pH 7.5, 100 mM NaCl, 10 μM BI-167107 and 100 μM TCEP. Peak fractions were pooled (Supplementary Fig. 8b) and concentrated to approximately 90 mg ml^{-1} with a 100 kDa MWCO Viva-spin concentrator and analysed by SDS–PAGE/Coomassie brilliant blue staining (Supplementary Fig. 8a) and gel filtration (Supplementary Fig. 8c). To confirm a pure, homogeneous and dephosphorylated preparation, the β_2AR–Gs complex was routinely analysed by ion exchange chromatography (Supplementary Fig. 8d).

Protein engineering. To increase the probability of obtaining crystals of the β_2AR–Gs complex we set out to increase the polar surface area on the extracellular side of the receptor using two strategies. The first approach, to generate extracellular binding antibodies, was not successful. The second approach was to replace the flexible and presumably unstructured amino terminus with the globular protein T4 lysozyme (T4L) used previously to crystallize and solve the carazolol-bound receptor[3] The construct used here (T4L–β_2AR) contained the cleavable signal sequence followed by the M1 Flag epitope (DYKDDDDA), the Tobacco etch virus (TEV) protease recognition sequence (ENLYFQG), bacteriophage T4 lysozyme from N2 through Y161 including C54T and C97A mutations, and a two residue alanine linker fused to the human β_2AR sequence D29 through G365. The PNGaseF-inaccessible glycosylation site of the β_2AR at N187 was mutated to Glu. M96 and M98 in the first extracellular loop were each replaced by Thr to increase the otherwise low expression level of T4L–β_2AR. The threonine mutations did not affect ligand binding affinity for [^{3}H]DHA, but caused a small, approximately twofold decrease in affinity for isoproterenol (data not shown).

The β_2AR–Gs peptide fusion construct used for [^{3}H]DHA competition binding with isoproterenol was constructed from the receptor truncated at position 365 and fused to the last 21 amino acids of the Gαs subunit (amino acids 374–394, except for C379A). A Gly-Ser is inserted between the receptor and the peptide. Also an extended TEV protease site (**SENLYFQGS**) was introduced in the β_2AR between G360 and G361.

Stabilization of Gs with nanobodies. From negative stain electron microscopy imaging[34], we observed that the α-helical domain of Gαs was flexible and therefore possibly responsible for poor crystal quality. Targeted stabilization of this domain was addressed by immunizing two llamas (*Lama glama*) with the bis(sulphosuccinimidyl)glutarate (BS2G, Pierce) cross-linked β_2AR-Gs-BI-167107 ternary complex. Peripheral blood lymphocytes were isolated from the immunized animals to extract total RNA, prepare cDNA and construct a Nanobody phage display library according to published methods[16]. Nb35 and Nb37 were enriched by two rounds of biopanning on the β_2AR-Gs-BI-167107 ternary complex embedded in biotinylated high-density lipoprotein particles[37]. Nb35 and Nb37 were selected for further characterization because they bind the β_2AR-Gs-BI-167107 ternary complex but not the free receptor in an ELISA assay. Nanobody binding to the β_2AR–Gs complex was confirmed by size exclusion chromatography (Supplementary Fig. 1d), and it was noted that both nanobodies protected the complex from dissociation by GTPγS, suggestive of a stabilizing Gs–Nb interaction (Supplementary Fig. 1d).

Crystallization. BI-167107 bound T4L–β_2AR–Gs complex and Nb35 were mixed in 1:1.2 molar ratio. The small molar excess of Nb35 was verified by analytical gel filtration (see Supplementary Fig. 9b). The mixture incubated for 1 h at room temperature before mixing with 7.7 MAG (provided by M. Caffrey) containing 10% cholesterol (C8667, Sigma) in 1:1 protein solution to lipid ratio (w/w) using the twin-syringe mixing method reported previously[38]. The concentration of T4L–β_2AR–Gs–Nb35 complex in 7.7 MAG was approximately 25 mg ml^{-1}. We believe the detergent MNG-3 stabilizes the T4L–β_2AR–Gs complex during its incorporation into the lipid cubic phase. This may be due to the high affinity of MNG-3 for the receptor. The β_2AR in MNG-3 maintains its structural integrity even when diluted below the CMC of the detergent, in contrast to β_2AR in DDM, which rapidly loses binding activity (Supplementary Fig. 11b). Moreover, MNG-3 improved crystal size and quality, as previously reported[12,13,39]. The protein:lipid mixture was delivered through an LCP dispensing robot (Gryphon, Art Robbins Instruments) in 40 nl drops to either 24-well or 96-well glass sandwich plates and overlaid en-bloc with 0.8 μl precipitant solution. Multiple crystallization leads were initially identified using in-house screens partly based on reagents from the StockOptions Salt kit (Hampton Research). Crystals for data collection were grown in 18 to 22% PEG 400, 100 mM MES, pH 6.5 (Supplementary Fig. 1c), 350 to 450 mM potassium nitrate, 10 mM foscarnet (Supplementary Fig. 1b), 1 mM TCEP (Supplementary Fig. 12c), and 10 μM BI-167107. Crystals reached full size within 3–4 days at 20 °C and were picked from a sponge-like mesophase and flash-frozen in liquid nitrogen without additional cryoprotectant.

Microcrystallography data collection and processing. Diffraction data were measured at the Advanced Photon Source beamline 23 ID-B. Hundreds of crystals were screened, and a final data set was compiled using diffraction wedges of typically 10 degrees from 20 strongly diffracting crystals (Supplementary Table 2). All data reduction was performed using HKL2000 (ref. 40). Although in many cases diffraction to beyond 3 Å was seen in initial frames, radiation damage and anisotropic diffraction resulted in low completeness in higher resolution shells. Analysis of the final data set by the UCLA diffraction anisotropy server[41] indicated that diffraction along the a^* axis was superior to that in other directions. On the basis of an $F/\sigma(F)$ cutoff of 3 along each reciprocal space axis, reflections were subjected to an anisotropic truncation with resolution limits of 2.9, 3.2 and 3.2 Å along a^*, b^* and c^* before use in refinement. We report this structure to an overall resolution of 3.2 Å. Despite the low completeness in the highest resolution shells (Supplementary Table 3) inclusion of these reflections gave substantial improvements in map quality and lower R_{free} during refinement.

Structure solution and refinement. The structure was solved by molecular replacement using Phaser[42,43]. The order of the molecular replacement search was found to be critical in solving the structure. In order, the search models used were: the β and γ subunits from a Gi heterotrimer (PDB ID: 1GP2), the Gs α Ras-like domain (PDB ID: 1AZT), the active-state β_2AR (PDB ID: 3P0G), a β_2AR-binding nanobody (PDB ID: 3P0G), T4 lysozyme (PDB ID: 2RH1), and the Gs α-helical domain (PDB ID: 1AZT). Following the determination of the initial structure by molecular replacement, rigid body refinement and simulated annealing were performed in Phenix[44] and BUSTER[45], followed by restrained refinement and manual rebuilding in Coot[46]. After iterative refinement and manual adjustments, the structure was refined in CNS using the DEN method[47]. Although the resolution of this structure exceeds that for which DEN is typically most useful, the presence of several poorly resolved regions indicated that the incorporation of additional information to guide refinement could provide better results. The DEN reference models used were those used for molecular replacement, with the exception of NB35, which was well ordered and for which no higher resolution structure is available. Side chains were omitted from 53 residues for which there was no electron density past Cβ below a low contour level of 0.7σ in a $2Fo - Fc$ map. Figures were prepared using PyMOL (The PyMOL Molecular Graphics System, Version 1.3, Schrödinger, LLC.). MolProbity was used to determine Ramachandran statistics[48].

Competition binding. Membranes expressing the β_2AR or the β_2AR–Gs peptide fusion were prepared from baculovirus-infected *Sf9* cells and [^{3}H]DHA-binding performed as previously described[49]. For competition binding, membranes were incubated with [^{3}H]DHA (1.1 nM final) and increasing concentrations of (−)-isoproterenol for 1 h before harvesting onto GF/B filters. Competition data were fitted to a two-site binding model and isoproterenol high and low K_i values and fractions calculated using GraphPad prism.

36. Kobilka, B. K. Amino and carboxyl terminal modifications to facilitate the production and purification of a G protein-coupled receptor. *Anal. Biochem.* **231,** 269–271 (1995).
37. Whorton, M. R. *et al.* A monomeric G protein-coupled receptor isolated in a high-density lipoprotein particle efficiently activates its G protein. *Proc. Natl Acad. Sci. USA* **104,** 7682–7687 (2007).
38. Caffrey, M. & Cherezov, V. Crystallizing membrane proteins using lipidic mesophases. *Nature Protocols* **4,** 706–731 (2009).
39. Rosenbaum, D. M. *et al.* Structure and function of an irreversible agonist-β_2 adrenoceptor complex. *Nature* **469,** 236–240 (2011).
40. Otwinowski, Z. & Minor, W. Processing of X-ray diffraction data collected in oscillation mode. *Methods Enzymol.* **276,** 307–326 (1997).
41. Strong, M. *et al.* Toward the structural genomics of complexes: crystal structure of a PE/PPE protein complex from *Mycobacterium tuberculosis*. *Proc. Natl Acad. Sci. USA* **103,** 8060–8065 (2006).
42. McCoy, A. J. Solving structures of protein complexes by molecular replacement with *Phaser*. *Acta Crystallogr. D* **63,** 32–41 (2007).
43. McCoy, A. J. *et al.* *Phaser* crystallographic software. *J. Appl. Cryst.* **40,** 658–674 (2007).
44. Afonine, P. V., Grosse-Kunstleve, R. W. & Adams, P. D. A robust bulk-solvent correction and anisotropic scaling procedure. *Acta Crystallogr. D* **61,** 850–855 (2005).
45. Blanc, E. *et al.* Refinement of severely incomplete structures with maximum likelihood in *BUSTER-TNT*. *Acta Crystallogr. D* **60,** 2210–2221 (2004).
46. Emsley, P. & Cowtan, K. *Coot*: model-building tools for molecular graphics. *Acta Crystallogr. D* **60,** 2126–2132 (2004).
47. Schröder, G. F., Levitt, M. & Brunger, A. T. Super-resolution biomolecular crystallography with low-resolution data. *Nature* **464,** 1218–1222 (2010).
48. Chen, V. B. *et al.* *MolProbity*: all-atom structure validation for macromolecular crystallography. *Acta Crystallogr. D* **66,** 12–21 (2010).
49. Swaminath, G., Steenhuis, J., Kobilka, B. & Lee, T. W. Allosteric modulation of β_2-adrenergic receptor by Zn^{2+}. *Mol. Pharmacol.* **61,** 65–72 (2002).

Index